REFERENCE PAGES

Algebra

Arithmetic Operations

$$a(b+c) = ab + ac \qquad \frac{a}{b} + \frac{c}{d} = \frac{ad+bc}{bd}$$

$$\frac{a+c}{b} = \frac{a}{b} + \frac{c}{b} \qquad \frac{\frac{a}{b}}{\frac{c}{d}} = \frac{a}{b} \times \frac{d}{c} = \frac{ad}{bc}$$

Exponents and Radicals

$$x^m x^n = x^{m+n} \qquad \frac{x^m}{x^n} = x^{m-n}$$

$$(x^m)^n = x^{mn} \qquad x^{-n} = \frac{1}{x^n}$$

$$(xy)^n = x^n y^n \qquad \left(\frac{x}{y}\right)^n = \frac{x^n}{y^n}$$

$$x^{1/n} = \sqrt[n]{x} \qquad x^{m/n} = \sqrt[n]{x^m} = (\sqrt[n]{x})^m$$

$$\sqrt[n]{xy} = \sqrt[n]{x}\sqrt[n]{y} \qquad \sqrt[n]{\frac{x}{y}} = \frac{\sqrt[n]{x}}{\sqrt[n]{y}}$$

Factoring Special Polynomials

$$x^2 - y^2 = (x+y)(x-y)$$
$$x^3 + y^3 = (x+y)(x^2 - xy + y^2)$$
$$x^3 - y^3 = (x-y)(x^2 + xy + y^2)$$

Binomial Theorem

$$(x+y)^2 = x^2 + 2xy + y^2 \qquad (x-y)^2 = x^2 - 2xy + y^2$$
$$(x+y)^3 = x^3 + 3x^2 y + 3xy^2 + y^3$$
$$(x-y)^3 = x^3 - 3x^2 y + 3xy^2 - y^3$$
$$(x+y)^n = x^n + nx^{n-1}y + \frac{n(n-1)}{2}x^{n-2}y^2$$
$$+ \cdots + \binom{n}{k} x^{n-k}y^k + \cdots + nxy^{n-1} + y^n$$

where $\binom{n}{k} = \dfrac{n(n-1)\cdots(n-k+1)}{1 \cdot 2 \cdot 3 \cdot \cdots \cdot k}$

Quadratic Formula

If $ax^2 + bx + c = 0$, then $x = \dfrac{-b \pm \sqrt{b^2 - 4ac}}{2a}$.

Inequalities and Absolute Value

If $a < b$ and $b < c$, then $a < c$.

If $a < b$, then $a + c < b + c$.

If $a < b$ and $c > 0$, then $ca < cb$.

If $a < b$ and $c < 0$, then $ca > cb$.

If $a > 0$, then

$|x| = a$ means $x = a$ or $x = -a$

$|x| < a$ means $-a < x < a$

$|x| > a$ means $x > a$ or $x < -a$

Geometry

Geometric Formulas

Formulas for area A, circumference C, and volume V:

Triangle
$A = \frac{1}{2}bh$
$= \frac{1}{2}ab \sin \theta$

Circle
$A = \pi r^2$
$C = 2\pi r$

Sector of Circle
$A = \frac{1}{2}r^2 \theta$
$s = r\theta$ (θ in radians)

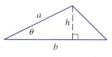

Sphere
$V = \frac{4}{3}\pi r^3$
$A = 4\pi r^2$

Cylinder
$V = \pi r^2 h$

Cone
$V = \frac{1}{3}\pi r^2 h$
$A = \pi r \sqrt{r^2 + h^2}$

Distance and Midpoint Formulas

Distance between $P_1(x_1, y_1)$ and $P_2(x_2, y_2)$:

$$d = \sqrt{(x_2 - x_1)^2 + (y_2 - y_1)^2}$$

Midpoint of $\overline{P_1 P_2}$: $\left(\dfrac{x_1 + x_2}{2}, \dfrac{y_1 + y_2}{2}\right)$

Lines

Slope of line through $P_1(x_1, y_1)$ and $P_2(x_2, y_2)$:

$$m = \frac{y_2 - y_1}{x_2 - x_1}$$

Point-slope equation of line through $P_1(x_1, y_1)$ with slope m:

$$y - y_1 = m(x - x_1)$$

Slope-intercept equation of line with slope m and y-intercept b:

$$y = mx + b$$

Circles

Equation of the circle with center (h, k) and radius r:

$$(x - h)^2 + (y - k)^2 = r^2$$

REFERENCE PAGES

Trigonometry

Angle Measurement

π radians = $180°$

$1° = \dfrac{\pi}{180}$ rad 1 rad $= \dfrac{180°}{\pi}$

$s = r\theta$

(θ in radians)

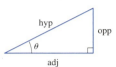

Right Angle Trigonometry

$\sin\theta = \dfrac{\text{opp}}{\text{hyp}}$ $\csc\theta = \dfrac{\text{hyp}}{\text{opp}}$

$\cos\theta = \dfrac{\text{adj}}{\text{hyp}}$ $\sec\theta = \dfrac{\text{hyp}}{\text{adj}}$

$\tan\theta = \dfrac{\text{opp}}{\text{adj}}$ $\cot\theta = \dfrac{\text{adj}}{\text{opp}}$

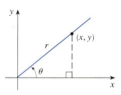

Trigonometric Functions

$\sin\theta = \dfrac{y}{r}$ $\csc\theta = \dfrac{r}{y}$

$\cos\theta = \dfrac{x}{r}$ $\sec\theta = \dfrac{r}{x}$

$\tan\theta = \dfrac{y}{x}$ $\cot\theta = \dfrac{x}{y}$

Graphs of Trigonometric Functions

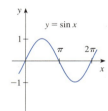
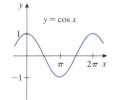
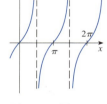
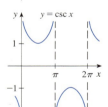
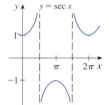
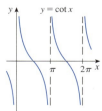

Trigonometric Functions of Important Angles

θ	radians	$\sin\theta$	$\cos\theta$	$\tan\theta$
0°	0	0	1	0
30°	$\pi/6$	$1/2$	$\sqrt{3}/2$	$\sqrt{3}/3$
45°	$\pi/4$	$\sqrt{2}/2$	$\sqrt{2}/2$	1
60°	$\pi/3$	$\sqrt{3}/2$	$1/2$	$\sqrt{3}$
90°	$\pi/2$	1	0	—

Fundamental Identities

$\csc\theta = \dfrac{1}{\sin\theta}$ $\sec\theta = \dfrac{1}{\cos\theta}$

$\tan\theta = \dfrac{\sin\theta}{\cos\theta}$ $\cot\theta = \dfrac{\cos\theta}{\sin\theta}$

$\cot\theta = \dfrac{1}{\tan\theta}$ $\sin^2\theta + \cos^2\theta = 1$

$1 + \tan^2\theta = \sec^2\theta$ $1 + \cot^2\theta = \csc^2\theta$

$\sin(-\theta) = -\sin\theta$ $\cos(-\theta) = \cos\theta$

$\tan(-\theta) = -\tan\theta$ $\sin\left(\dfrac{\pi}{2} - \theta\right) = \cos\theta$

$\cos\left(\dfrac{\pi}{2} - \theta\right) = \sin\theta$ $\tan\left(\dfrac{\pi}{2} - \theta\right) = \cot\theta$

The Law of Sines

$\dfrac{\sin A}{a} = \dfrac{\sin B}{b} = \dfrac{\sin C}{c}$

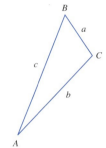

The Law of Cosines

$a^2 = b^2 + c^2 - 2bc\cos A$

$b^2 = a^2 + c^2 - 2ac\cos B$

$c^2 = a^2 + b^2 - 2ab\cos C$

Addition and Subtraction Formulas

$\sin(x + y) = \sin x \cos y + \cos x \sin y$

$\sin(x - y) = \sin x \cos y - \cos x \sin y$

$\cos(x + y) = \cos x \cos y - \sin x \sin y$

$\cos(x - y) = \cos x \cos y + \sin x \sin y$

$\tan(x + y) = \dfrac{\tan x + \tan y}{1 - \tan x \tan y}$

$\tan(x - y) = \dfrac{\tan x - \tan y}{1 + \tan x \tan y}$

Double-Angle Formulas

$\sin 2x = 2\sin x \cos x$

$\cos 2x = \cos^2 x - \sin^2 x = 2\cos^2 x - 1 = 1 - 2\sin^2 x$

$\tan 2x = \dfrac{2\tan x}{1 - \tan^2 x}$

Half-Angle Formulas

$\sin^2 x = \dfrac{1 - \cos 2x}{2}$ $\cos^2 x = \dfrac{1 + \cos 2x}{2}$

single variable
CALCULUS

Volume 2: Chapters 5—12

ABOUT THE COVER

The art on the cover was created by Bill Ralph, a mathematician who uses modern mathematics to produce visual representations of "dynamical systems." Examples of dynamical systems in nature include the weather, blood pressure, the motions of the planets, and other phenomena that involve continual change. Such systems, which tend to be unpredictable and even chaotic at times, are modeled mathematically using the concepts of composition and iteration of functions. The process of creating the cover art starts with a photograph of a violin. The color values at each point on the photograph are then converted into numbers and a particular function is evaluated at each of those numbers giving a new number at each point of the photograph. The same function is then evaluated at each of these new numbers. Repeating this process produces a sequence of numbers called *iterates* of the function. The original photograph is then "repainted" using colors determined by certain properties of this sequence of iterates and the mathematical concept of "dimension." The final image is the result of mingling photographic reality with the complex behavior of a dynamical system.

single variable CALCULUS
fifth edition

Volume 2: Chapters 5–12

JAMES STEWART
McMaster University

Australia • Canada • Mexico • Singapore • Spain
United Kingdom • United States

THOMSON

BROOKS/COLE

Mathematics Editor ◦ *Bob Pirtle*	Production Service ◦ *TECH·arts of Colorado, Inc.*
Assistant Editor ◦ *Stacy Green*	Interior Design ◦ *TECH·arts of Colorado, Inc.*
Editorial Assistant ◦ *Jessica Zimmerman*	Copy Editor ◦ *Kathi Townes*
Technology Project Manager ◦ *Earl Perry*	Interior Illustration ◦ *Brian Betsill*
Marketing Manager ◦ *Karin Sandberg*	Cover Designer ◦ *Denise Davidson*
Marketing Assistant ◦ *Stephanie Taylor*	Cover Image ◦ *Bill Ralph*
Advertising Project Manager ◦ *Bryan Vann*	Cover Printer ◦ *Lehigh Press*
Project Manager, Editorial Production ◦ *Kirk Bomont*	Compositor ◦ *Stephanie Kuhns*
Print/Media Buyer ◦ *Jessica Reed*	Printer ◦ *Quebecor World–Versailles*

COPYRIGHT © 2005 Brooks/Cole, a division of Thomson Learning, Inc. Thomson Learning™ is a trademark used herein under license.

ALL RIGHTS RESERVED. No part of this work covered by the copyright hereon may be reproduced or used in any form or by any means—graphic, electronic, or mechanical, including but not limited to photocopying, recording, taping, Web distribution, information networks, or information storage and retrieval systems—without the written permission of the publisher.

Printed in the United States of America

1 2 3 4 5 6 7 06 05 04

> For more information about our products, contact us at:
> **Thomson Learning Academic Resource Center**
> **1-800-423-0563**
>
> For permission to use material from this text or product, submit a request online at <http://www.thomsonrights.com>. Any additional questions about permissions can be submitted by email to thomsonrights@thomson.com.

COPYRIGHT © 2003 Thomson Learning, Inc. All Rights Reserved. Thomson Learning *WebTutor*™ is a trademark of Thomson Learning, Inc.

Trademarks
Derive is a registered trademark of Soft Warehouse, Inc.
Journey Through is a trademark used herein under license.
Maple is a registered trademark of Waterloo Maple, Inc.
Mathematica is a registered trademark of Wolfram Research, Inc.
Tools for Enriching is a trademark used herein under license.

Credits continue on page A81.

ISBN 0-534-49677-6

Thomson Brooks/Cole
10 Davis Drive
Belmont, CA 94002
USA

Asia
Thomson Learning
5 Shenton Way #01-01
UIC Building
Singapore 068808

Australia / New Zealand
Thomson Learning
102 Dodds Street
Southbank, Victoria 3006
Australia

Canada
Nelson
1120 Birchmount Road
Toronto, Ontario M1K 5G4
Canada

Europe/Middle East/Africa
Thomson Learning
High Holborn House
50/51 Bedford Row
London WC1R 4LR
United Kingdom

Latin America
Thomson Learning
Seneca, 53
Colonia Polanco
11560 Mexico D.F.
Mexico

Spain / Portugal
Paraninfo
Calle/Magallanes, 25
28015 Madrid, Spain

To my students, past and present

Contents

Preface   xii

To the Student   xxiv

5 Integrals   314

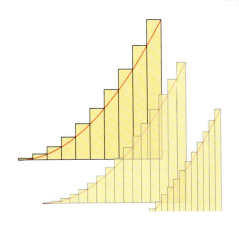

5.1   Areas and Distances   315

5.2   The Definite Integral   326

      Discovery Project ∘ Area Functions   339

5.3   The Fundamental Theorem of Calculus   340

5.4   Indefinite Integrals and the Net Change Theorem   350

      Writing Project ∘ Newton, Leibniz, and the Invention of Calculus   359

5.5   The Substitution Rule   360

     Review   368

Problems Plus   372

6 Applications of Integration   374

6.1   Areas between Curves   375

6.2   Volumes   382

6.3   Volumes by Cylindrical Shells   393

6.4   Work   398

6.5   Average Value of a Function   402

     Review   406

Problems Plus   408

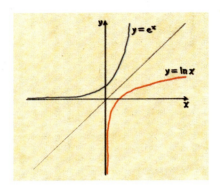

7 Inverse Functions:
Exponential, Logarithmic, and Inverse Trigonometric Functions 412

7.1 Inverse Functions 413

Instructors may cover either Sections 7.2–7.4 or Sections 7.2*–7.4*. See the Preface.

7.2	Exponential Functions and Their Derivatives 421	7.2*	The Natural Logarithmic Function 451
7.3	Logarithmic Functions 434	7.3*	The Natural Exponential Function 460
7.4	Derivatives of Logarithmic Functions 441	7.4*	General Logarithmic and Exponential Functions 467

7.5 Inverse Trigonometric Functions 477
    Applied Project ▫ Where to Sit at the Movies 485

7.6 Hyperbolic Functions 486

7.7 Indeterminate Forms and L'Hospital's Rule 493
    Writing Project ▫ The Origins of L'Hospital's Rule 504

Review 504

Problems Plus 508

8 Techniques of Integration 510

8.1 Integration by Parts 511

8.2 Trigonometric Integrals 518

8.3 Trigonometric Substitution 525

8.4 Integration of Rational Functions by Partial Fractions 532

8.5 Strategy for Integration 541

8.6 Integration Using Tables and Computer Algebra Systems 547
    Discovery Project ▫ Patterns in Integrals 553

8.7 Approximate Integration 554

8.8 Improper Integrals 566

Review 576

Problems Plus 579

CONTENTS ix

9 Further Applications of Integration   582

9.1 Arc Length   583
       Discovery Project ▫ Arc Length Contest   590

9.2 Area of a Surface of Revolution   590
       Discovery Project ▫ Rotating on a Slant   596

9.3 Applications to Physics and Engineering   597

9.4 Applications to Economics and Biology   607

9.5 Probability   611

Review   618

Problems Plus   620

10 Differential Equations   622

10.1 Modeling with Differential Equations   623

10.2 Direction Fields and Euler's Method   628

10.3 Separable Equations   637
       Applied Project ▫ How Fast Does a Tank Drain?   645
       Applied Project ▫ Which Is Faster, Going Up or Coming Down?   646

10.4 Exponential Growth and Decay   647
       Applied Project ▫ Calculus and Baseball   658

10.5 The Logistic Equation   659

10.6 Linear Equations   668

10.7 Predator-Prey Systems   674

Review   680

Problems Plus   684

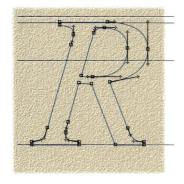

11 Parametric Equations and Polar Coordinates   686

11.1 Curves Defined by Parametric Equations   687
       Laboratory Project ▫ Running Circles around Circles   695

11.2 Calculus with Parametric Curves   696
       Laboratory Project ▫ Bézier Curves   705

11.3 Polar Coordinates 705

11.4 Areas and Lengths in Polar Coordinates 715

11.5 Conic Sections 720

11.6 Conic Sections in Polar Coordinates 728

Review 732

Problems Plus 735

12 Infinite Sequences and Series 736

12.1 Sequences 737

    Laboratory Project ▫ Logistic Sequences 749

12.2 Series 749

12.3 The Integral Test and Estimates of Sums 759

12.4 The Comparison Tests 766

12.5 Alternating Series 771

12.6 Absolute Convergence and the Ratio and Root Tests 776

12.7 Strategy for Testing Series 783

12.8 Power Series 785

12.9 Representations of Functions as Power Series 790

12.10 Taylor and Maclaurin Series 796

    Laboratory Project ▫ An Elusive Limit 808

12.11 The Binomial Series 808

    Writing Project ▫ How Newton Discovered the Binomial Series 812

12.12 Applications of Taylor Polynomials 812

    Applied Project ▫ Radiation from the Stars 821

Review 822

Problems Plus 825

Appendixes A1

A	Numbers, Inequalities, and Absolute Values	A2
B	Coordinate Geometry and Lines	A10
C	Graphs of Second-Degree Equations	A16
D	Trigonometry	A24
E	Sigma Notation	A34
F	Proofs of Theorems	A39
G	Complex Numbers	A41
H	Answers to Odd-Numbered Exercises	A49

Index A83

Preface

> A great discovery solves a great problem but there is a grain of discovery in the solution of any problem. Your problem may be modest; but if it challenges your curiosity and brings into play your inventive faculties, and if you solve it by your own means, you may experience the tension and enjoy the triumph of discovery.
>
> GEORGE POLYA

This volume of Single Variable Calculus, Fifth Edition, contains Chapters 5–12. Volume 1 contains Chapters 1–6.

The art of teaching, Mark Van Doren said, is the art of assisting discovery. I have tried to write a book that assists students in discovering calculus—both for its practical power and its surprising beauty. In this edition, as in the first four editions, I aim to convey to the student a sense of the utility of calculus and develop technical competence, but I also strive to give some appreciation for the intrinsic beauty of the subject. Newton undoubtedly experienced a sense of triumph when he made his great discoveries. I want students to share some of that excitement.

The emphasis is on understanding concepts. I think that nearly everybody agrees that this should be the primary goal of calculus instruction. In fact, the impetus for the current calculus reform movement came from the Tulane Conference in 1986, which formulated as their first recommendation:

Focus on conceptual understanding.

I have tried to implement this goal through the *Rule of Three*: "Topics should be presented geometrically, numerically, and algebraically." Visualization, numerical and graphical experimentation, and other approaches have changed how we teach conceptual reasoning in fundamental ways. More recently, the Rule of Three has been expanded to become the *Rule of Four* by emphasizing the verbal, or descriptive, point of view as well.

In writing the fifth edition my premise has been that it is possible to achieve conceptual understanding and still retain the best traditions of traditional calculus. The book contains elements of reform, but within the context of a traditional curriculum. (Instructors who prefer a more streamlined curriculum should look at my book *Calculus: Concepts and Contexts, Second Edition*.)

What's New in the Fifth Edition

By way of preparing to write the fifth edition of this text, I spent a year teaching calculus from the fourth edition at the University of Toronto. I listened carefully to my students' questions and my colleagues' suggestions. And as I prepared each lecture I sometimes realized that an additional example was needed, or a sentence could be clarified, or a section could use a few more exercises of a certain type. In addition, I paid attention to the suggestions sent to me by many users and to the comments of the reviewers.

An unusual source of new problems was a phone call I received from a friend of mine, Richard Armstrong. Richard is a partner in an engineering consulting firm and advises clients who build hospitals and hotels. He told me that, in certain parts of the world, sprinkler systems for large buildings are supplied with water by tanks located on the roofs of the buildings. Of course he knew that the water pressure decreases as the water level decreases, but he needed to be able to quantify this effect so his clients could guarantee a certain pressure for a certain period of time. I told him how he could solve his problem by solving a separable differential equation, but it occurred to me that his problem could be developed into a rather nice project when combined with other ideas. (See the project on page 645).

The structure of *Single Variable Calculus, Fifth Edition,* remains largely unchanged, but there are hundreds of improvements, small and large:

- Two sections in Chapter 11 have been combined.

- New phrases and margin notes have been added to clarify the exposition.

- A number of pieces of art have been redrawn.

- The data in examples and exercises have been updated to be more timely.

- Examples have been added. For instance, I added the new Example 1 in Section 5.3 (pages 340–341) because students have a tough time grasping the idea of a function defined by an integral with a variable limit of integration. I think it helps to look at Example 1 before considering the Fundamental Theorem of Calculus.

- Extra steps have been provided in some of the existing examples.

- More than 25% of the exercises in each chapter are new. Here are a few of my favorites:

Exercise	Page	Exercise	Page	Exercise	Page
3.1.34	133	5.4.52	357	7.6.67	493
7.7.84	503	8.7.36	565	10.1.11–12	628
11.3.47–48	714	12.9.40	796	12.12.35	820

    I've also added new problems to the Problems Plus sections. See, for instance, Problem 20 on page 221, Problem 13 on page 509, Problems 9 and 10 on page 621, and Problems 20 and 22 on page 827.

- Four new projects have been added. The one on page 198 asks students to design a roller coaster so the track is smooth at transition points. The project on page 590, the idea for which I thank Larry Riddle, is actually a contest in which the winning curve has the smallest arc length (within a certain class of curves).

- A CD called *Tools for Enriching Calculus* (*TEC*) is available for use with the fifth edition. See the description on page xv.

- Conscious of the need to control the size of the book, I've put additional topics (with exercises) on the revamped web site www.stewartcalculus.com (see the description on page xvi) rather than in the text itself. These include the new topics Fourier Series and Formulas for the Remainder Term in Taylor Series, as well as topics that appeared in previous editions: Review of Basic Algebra, Rotation of Axes, and Lies My Calculator and Computer Told Me.

Features

Conceptual Exercises The most important way to foster conceptual understanding is through the problems that we assign. To that end I have devised various types of new problems. Some exercise sets begin with requests to explain the meanings of the basic concepts of the section. (See, for instance, the first few exercises in Sections 2.2, 2.5, 2.6, and 12.2.) Similarly, all the review sections begin with a *Concept Check* and a *True-False Quiz*. Other exercises test conceptual understanding through graphs or tables (see Exercises 3.1.1–3, 3.2.33–36, 3.8.1–4, 10.1.11–12, 11.1.24–27, and 12.10.2).

Another type of exercise uses verbal description to test conceptual understanding (see Exercises 2.5.8, 3.2.46, 4.3.47–48, and 8.8.67). I particularly value problems that combine and compare graphical, numerical, and algebraic approaches (see Exercises 4.4.33–34, 3.4.23, and 10.5.2).

Graded Exercise Sets Each exercise set is carefully graded, progressing from basic conceptual exercises and skill-development problems to more challenging problems involving applications and proofs.

Real-World Data My assistants and I spent a great deal of time looking in libraries, contacting companies and government agencies, and searching the Internet for interesting real-world data to introduce, motivate, and illustrate the concepts of calculus. As a result, many of the examples and exercises deal with functions defined by such numerical data or graphs. See, for instance, Figures 1, 11, and 12 in Section 1.1 (seismograms from the Northridge earthquake), Exercise 3.2.34 (percentage of the population under age 18), Exercise 5.1.14 (velocity of the space shuttle *Endeavour*), and Figure 4 in Section 5.4 (San Francisco power consumption).

Projects One way of involving students and making them active learners is to have them work (perhaps in groups) on extended projects that give a feeling of substantial accomplishment when completed. I have included four kinds of projects: *Applied Projects* involve applications that are designed to appeal to the imagination of students. The project after Section 10.3 asks whether a ball thrown upward takes longer to reach its maximum height or to fall back to its original height. (The answer might surprise you.) *Laboratory Projects* involve technology; the one following Section 11.2 shows how to use Bézier curves to design shapes that represent letters for a laser printer. *Writing Projects* ask students to compare present-day methods with those of the founders of calculus—Fermat's method for finding tangents, for instance. Suggested references are supplied. *Discovery Projects* anticipate results to be discussed later or encourage discovery through pattern recognition (see the one following Section 8.6). Additional projects can be found in the *Instructor's Guide* (see, for instance, Group Exercise 5.1: Position from Samples) and also in the *CalcLabs* supplements.

Problem Solving Students usually have difficulties with problems for which there is no single well-defined procedure for obtaining the answer. I think nobody has improved very much on George Polya's four-stage problem-solving strategy and, accordingly, I have included a version of his problem-solving principles following Chapter 1. They are applied, both explicitly and implicitly, throughout the book. After the other chapters I have placed sections called *Problems Plus,* which feature examples of how to tackle challenging calculus problems. In selecting the varied problems for these sections I kept in mind the following advice from David Hilbert: "A mathematical problem should be difficult in order to entice us, yet not inaccessible lest it mock our efforts." When I put these challenging problems on assignments and tests I grade them in a different way. Here I reward a student significantly for ideas toward a solution and for recognizing which problem-solving principles are relevant.

Dual Treatment of Exponential and Logarithmic Functions

There are two possible ways of treating the exponential and logarithmic functions and each method has its passionate advocates. Because one often finds advocates of both approaches teaching the same course, I include full treatments of both methods. In Sections 7.2, 7.3, and 7.4 the exponential function is defined first, followed by the logarithmic function as its inverse. (Students have seen these functions introduced this way since high school.) In the alternative approach, presented in Sections 7.2*, 7.3*, and 7.4*, the logarithm is defined as an integral and the exponential function is its inverse. This latter method is, of course, less intuitive but more elegant. You can use whichever treatment you prefer.

If the first approach is taken, then much of Chapter 7 can be covered before Chapters 5 and 6, if desired. To accommodate this choice of presentation there are specially identified problems involving integrals of exponential and logarithmic functions at the end of the appropriate sections of Chapters 5 and 6. This order of presentation allows a faster-paced course to teach the transcendental functions and the definite integral in the first semester of the course.

For instructors who would like to go even further in this direction I have prepared an alternate edition of this book, called *Single Variable Calculus, Early Transcendentals, Fifth Edition,* in which the exponential and logarithmic functions are introduced in the first chapter. Their limits and derivatives are found in the second and third chapters at the same time as polynomials and the other elementary functions.

Technology

The availability of technology makes it not less important but more important to clearly understand the concepts that underlie the images on the screen. But, when properly used, graphing calculators and computers are powerful tools for discovering and understanding those concepts. This textbook can be used either with or without technology and I use two special symbols to indicate clearly when a particular type of machine is required. The icon ⌐⌐ indicates an exercise that definitely requires the use of such technology, but that is not to say that it can't be used on the other exercises as well. The symbol [CAS] is reserved for problems in which the full resources of a computer algebra system (like Derive, Maple, Mathematica, or the TI-89/92) are required. But technology doesn't make pencil and paper obsolete. Hand calculation and sketches are often preferable to technology for illustrating and reinforcing some concepts. Both instructors and students need to develop the ability to decide where the hand or the machine is appropriate.

Tools for Enriching™ Calculus

The CD-ROM called *TEC* is a companion to the text and is intended to enrich and complement its contents. Developed by Harvey Keynes at the University of Minnesota and Dan Clegg at Palomar College, *TEC* uses a discovery and exploratory approach. In sections of the book where technology is particularly appropriate, marginal icons direct students to *TEC* modules that provide a laboratory environment in which they can explore the topic in different ways and at different levels. Instructors can choose to become involved at several different levels, ranging from simply encouraging students to use the modules for independent exploration, to assigning specific exercises from those included with each module, or to creating additional exercises, labs, and projects that make use of the modules.

TEC also includes *homework hints* for representative exercises (usually odd-numbered) in every section of the text, indicated by printing the exercise number in red. These hints are usually presented in the form of questions and try to imitate an effective teaching assistant by functioning as a silent tutor. They are constructed so as not to reveal any more of the actual solution than is minimally necessary to make further progress.

xvi ▌ PREFACE

Web Site: www.stewartcalculus.com   This site has been renovated and now includes the following.

- Algebra Review, with tutorial
- Additional Topics (complete with exercise sets): Fourier Series, Formulas for the Remainder Term in Taylor Series, Rotation of Axes, Lies My Calculator and Computer Told Me
- Drill exercises that appeared in previous editions, together with their solutions
- Problems Plus from prior editions
- Links, for particular topics, to outside web resources
- History of Mathematics, with links to the better historical web sites
- Downloadable versions of *CalcLabs* for Derive and TI graphing calculators

Content

A Preview of Calculus   The book begins with an overview of the subject and includes a list of questions to motivate the study of calculus.

1 ▫ Functions and Models   From the beginning, multiple representations of functions are stressed: verbal, numerical, visual, and algebraic. A discussion of mathematical models leads to a review of the standard functions from these four points of view.

2 ▫ Limits and Rates of Change   The material on limits is motivated by a prior discussion of the tangent and velocity problems. Limits are treated from descriptive, graphical, numerical, and algebraic points of view. Section 2.4, on the precise ε-δ definition of a limit, is an optional section.

3 ▫ Derivatives   The material on derivatives is covered in two sections in order to give students more time to get used to the idea of a derivative as a function. The examples and exercises explore the meanings of derivatives in various contexts.

4 ▫ Applications of Differentiation   The basic facts concerning extreme values and shapes of curves are deduced from the Mean Value Theorem. Graphing with technology emphasizes the interaction between calculus and calculators and the analysis of families of curves. Some substantial optimization problems are provided, including an explanation of why you need to raise your head 42° to see the top of a rainbow.

5 ▫ Integrals   The area problem and the distance problem serve to motivate the definite integral, with sigma notation introduced as needed. (Full coverage of sigma notation is provided in Appendix E.) Emphasis is placed on explaining the meanings of integrals in various contexts and on estimating their values from graphs and tables.

6 ▫ Applications of Integration   Here I present the applications of integration—area, volume, work, average value—that can reasonably be done without specialized techniques of integration. General methods are emphasized. The goal is for students to be able to divide a quantity into small pieces, estimate with Riemann sums, and recognize the limit as an integral.

7 ▫ Inverse Functions:
Exponential, Logarithmic, and Inverse Trigonometric Functions   As discussed more fully on page xv, only one of the two treatments of these functions need be covered—either with exponential functions first or with the logarithm defined as a definite integral. If Chapter 10 is not going to be taught, then Section 10.4 on exponential growth and decay can be covered after Section 7.4.

8 ▫ Techniques of Integration   All the standard methods are covered but, of course, the real challenge is to be able to recognize which technique is best used in a given situation. Accordingly, in Section 8.5, I present a strategy for integration. The use of computer algebra systems is discussed in Section 8.6.

9 ▫ Further Applications of Integration	Here are the applications of integration—arc length and surface area—for which it is useful to have available all the techniques of integration, as well as applications to biology, economics, and physics (hydrostatic force and centers of mass). I have also included a section on probability. There are more applications here than can realistically be covered in a given course. Instructors should select applications suitable for their students and for which they themselves have enthusiasm.
10 ▫ Differential Equations	Modeling is the theme that unifies this introductory treatment of differential equations. Direction fields and Euler's method are studied before separable and linear equations are solved explicitly, so that qualitative, numerical, and analytic approaches are given equal consideration. These methods are applied to the exponential, logistic, and other models for population growth. The first five or six sections of this chapter serve as a good introduction to first-order differential equations. An optional final section uses predator-prey models to illustrate systems of differential equations.
11 ▫ Parametric Equations and Polar Coordinates	The sections on areas and tangents for parametric curves and arc length and surface area have been streamlined and combined as *Calculus with Parametric Curves*. Such curves are well suited to laboratory projects; the two presented here involve families of curves and Bézier curves. A brief treatment of conic sections in polar coordinates prepares the way for Kepler's Laws in Chapter 14.
12 ▫ Infinite Sequences and Series	The convergence tests have intuitive justifications (see page 759) as well as formal proofs. Numerical estimates of sums of series are based on which test was used to prove convergence. The emphasis is on Taylor series and polynomials and their applications to physics. Error estimates include those from graphing devices.

Ancillaries

Single Variable Calculus, Fifth Edition, is supported by a complete set of ancillaries developed under my direction. Each piece has been designed to enhance student understanding and to facilitate creative instruction. The tables on pages xxii–xxiii describe each of these ancillaries.

Acknowledgments

The preparation of this and previous editions has involved much time spent reading the reasoned (but sometimes contradictory) advice from a large number of astute reviewers. I greatly appreciate the time they spent to understand my motivation for the approach taken. I have learned something from each of them.

Fifth Edition Reviewers

Martina Bode,
　Northwestern University
Philip L. Bowers,
　Florida State University
Scott Chapman,
　Trinity University
Charles N. Curtis,
　Missouri Southern State College
Elias Deeba,
　University of Houston–Downtown

Greg Dresden,
　Washington and Lee University
Martin Erickson,
　Truman State University
Laurene V. Fausett,
　Georgia Southern University
Norman Feldman,
　Sonoma State University
José D. Flores,
　The University of South Dakota

Howard B. Hamilton,
   California State University, Sacramento
Gary W. Harrison,
   College of Charleston
Randall R. Holmes,
   Auburn University
James F. Hurley,
   University of Connecticut
Matthew A. Isom,
   Arizona State University
Zsuzsanna M. Kadas,
   St. Michael's College
Frederick W. Keene,
   Pasadena City College
Robert L. Kelley,
   University of Miami
John C. Lawlor,
   University of Vermont
Christopher C. Leary,
   State University of New York at Geneseo
Gerald Y. Matsumoto,
   American River College
Michael Montaño,
   Riverside Community College

Hussain S. Nur,
   California State University, Fresno
Mike Penna,
   Indiana University–Purdue University Indianapolis
John Ringland,
   State University of New York at Buffalo
E. Arthur Robinson, Jr.,
   The George Washington University
Rob Root
   Lafayette College
Teresa Morgan Smith,
   Blinn College
Donald W. Solomon,
   University of Wisconsin–Milwaukee
Kristin Stoley,
   Blinn College
Paul Xavier Uhlig,
   St. Mary's University, San Antonio
Dennis H. Wortman,
   University of Massachusetts, Boston
Xian Wu,
   University of South Carolina

Previous Edition Reviewers

B. D. Aggarwala,
   University of Calgary
John Alberghini,
   Manchester Community College
Michael Albert,
   Carnegie-Mellon University
Daniel Anderson,
   University of Iowa
Donna J. Bailey,
   Northeast Missouri State University
Wayne Barber,
   Chemeketa Community College
Neil Berger,
   University of Illinois, Chicago
David Berman,
   University of New Orleans
Richard Biggs,
   University of Western Ontario
Robert Blumenthal,
   Oglethorpe University
Barbara Bohannon,
   Hofstra University

Jay Bourland,
   Colorado State University
Stephen W. Brady,
   Wichita State University
Michael Breen,
   Tennessee Technological University
Stephen Brown
Robert N. Bryan,
   University of Western Ontario
David Buchthal,
   University of Akron
Jorge Cassio,
   Miami-Dade Community College
Jack Ceder,
   University of California, Santa Barbara
James Choike,
   Oklahoma State University
Barbara Cortzen,
   DePaul University
Carl Cowen,
   Purdue University

Philip S. Crooke,
Vanderbilt University
Daniel Cyphert,
Armstrong State College
Robert Dahlin
Gregory J. Davis,
University of Wisconsin–Green Bay
Daniel DiMaria,
Suffolk Community College
Seymour Ditor,
University of Western Ontario
Daniel Drucker,
Wayne State University
Kenn Dunn,
Dalhousie University
Dennis Dunninger,
Michigan State University
Bruce Edwards,
University of Florida
David Ellis,
San Francisco State University
John Ellison,
Grove City College
Garret Etgen,
University of Houston
Theodore G. Faticoni,
Fordham University
Newman Fisher,
San Francisco State University
William Francis,
Michigan Technological University
James T. Franklin,
Valencia Community College, East
Stanley Friedlander,
Bronx Community College
Patrick Gallagher,
Columbia University–New York
Paul Garrett,
*University of Minnesota–
Minneapolis*
Frederick Gass,
Miami University of Ohio
Bruce Gilligan,
University of Regina
Matthias K. Gobbert,
*University of Maryland,
Baltimore County*
Gerald Goff,
Oklahoma State University
Stuart Goldenberg,
California Polytechnic State University
John A. Graham,
Buckingham Browne & Nichols School
Richard Grassl,
University of New Mexico
Michael Gregory,
University of North Dakota
Charles Groetsch,
University of Cincinnati
Salim M. Haïdar,
Grand Valley State University
D. W. Hall,
Michigan State University
Robert L. Hall,
University of Wisconsin–Milwaukee
Darel Hardy,
Colorado State University
Melvin Hausner,
New York University/Courant Institute
Curtis Herink,
Mercer University
Russell Herman,
*University of North Carolina
at Wilmington*
Allen Hesse,
Rochester Community College
Gerald Janusz,
*University of Illinois at
Urbana-Champaign*
John H. Jenkins,
*Embry-Riddle Aeronautical
University, Prescott Campus*
Clement Jeske,
University of Wisconsin, Platteville
Carl Jockusch,
*University of Illinois at
Urbana-Champaign*
Jan E. H. Johansson,
University of Vermont
Jerry Johnson,
Oklahoma State University
Matt Kaufman
Matthias Kawski,
Arizona State University
Virgil Kowalik,
Texas A&I University
Kevin Kreider,
University of Akron
Leonard Krop,
DePaul University
Mark Krusemeyer,
Carleton College

David Leeming,
  University of Victoria
Sam Lesseig,
  Northeast Missouri State University
Phil Locke,
  University of Maine
Joan McCarter,
  Arizona State University
Phil McCartney,
  Northern Kentucky University
Igor Malyshev,
  San Jose State University
Larry Mansfield,
  Queens College
Mary Martin,
  Colgate University
Nathaniel F. G. Martin,
  University of Virginia
Tom Metzger,
  University of Pittsburgh
Teri Jo Murphy,
  University of Oklahoma
Richard Nowakowski,
  Dalhousie University
Wayne N. Palmer,
  Utica College
Vincent Panico,
  University of the Pacific
F. J. Papp,
  University of Michigan–Dearborn
Mark Pinsky,
  Northwestern University
Lothar Redlin,
  The Pennsylvania State University
Tom Rishel,
  Cornell University
Joel W. Robbin,
  University of Wisconsin–Madison
Richard Rockwell,
  Pacific Union College
Richard Ruedemann,
  Arizona State University
David Ryeburn,
  Simon Fraser University
Richard St. Andre,
  Central Michigan University
Ricardo Salinas,
  San Antonio College

Robert Schmidt,
  South Dakota State University
Eric Schreiner,
  Western Michigan University
Mihr J. Shah,
  Kent State University–Trumbull
Theodore Shifrin,
  University of Georgia
Wayne Skrapek,
  University of Saskatchewan
Larry Small,
  Los Angeles Pierce College
William Smith,
  University of North Carolina
Edward Spitznagel,
  Washington University
Joseph Stampfli,
  Indiana University
M. B. Tavakoli,
  Chaffey College
Stan Ver Nooy,
  University of Oregon
Andrei Verona,
  *California State University–
  Los Angeles*
Russell C. Walker,
  Carnegie Mellon University
William L. Walton,
  McCallie School
Jack Weiner,
  University of Guelph
Alan Weinstein,
  *University of California,
  Berkeley*
Theodore W. Wilcox,
  Rochester Institute of Technology
Steven Willard,
  University of Alberta
Robert Wilson,
  University of Wisconsin–Madison
Jerome Wolbert,
  *University of Michigan–
  Ann Arbor*
Mary Wright,
  *Southern Illinois University–
  Carbondale*
Paul M. Wright,
  Austin Community College

In addition, I would like to thank George Bergman, Stuart Goldenberg, Emile LeBlanc, Gerald Leibowitz, Charles Pugh, Marina Ratner, Peter Rosenthal, and Alan Weinstein for their suggestions; Dan Clegg for his research in libraries and on the Internet; Arnold Good for his treatment of optimization problems with implicit differentiation; Al Shenk and Dennis Zill for permission to use exercises from their calculus texts; COMAP for permission to use project material; George Bergman, David Bleecker, Dan Clegg, Victor Kaftal, Anthony Lam, Jamie Lawson, Ira Rosenholtz, Lowell Smylie, and Larry Wallen for ideas for exercises; Dan Drucker for the roller derby project; Tom Farmer, Fred Gass, John Ramsay, Larry Riddle, and Philip Straffin for ideas for projects; Dan Anderson and Dan Drucker for solving the new exercises; and Jeff Cole and Dan Clegg for their careful preparation and proofreading of the answer manuscript. I'm grateful to Jeff Cole for suggesting ways to improve the exercises. Dan Clegg acted as my assistant throughout; he proofread, made suggestions, and contributed many of the new exercises.

In addition, I thank those who have contributed to past editions: Ed Barbeau, Fred Brauer, Andy Bulman-Fleming, Bob Burton, Tom DiCiccio, Garret Etgen, Chris Fisher, Gene Hecht, Harvey Keynes, Kevin Kreider, E. L. Koh, Zdislav Kovarik, David Leep, Lothar Redlin, Carl Riehm, Doug Shaw, and Saleem Watson.

I also thank Stephanie Kuhns, Kathi Townes, and Brian Betsill of TECHarts for their production services and the following Brooks/Cole staff: Kirk Bomont, editorial production project manager; Karin Sandberg, Stephanie Taylor, and Bryan Vann, marketing team; Stacy Green, assistant editor, and Jessica Zimmerman, editorial assistant; Earl Perry, technology project manager, and Jessica Reed, print/media buyer. They have all done an outstanding job.

I have been very fortunate to have worked with some of the best mathematics editors in the business over the past two decades: Ron Munro, Harry Campbell, Craig Barth, Jeremy Hayhurst, Gary Ostedt, and now Bob Pirtle. Bob continues in that tradition of editors who, while offering sound advice and ample assistance, trust my instincts and allow me to write the books that I want to write.

JAMES STEWART

Ancillaries for Instructors

Instructor's Resource CD-ROM
ISBN 0-534-39340-3

Contains Electronic Instructor's Guide, Electronic Solutions, BCA Testing, Instructions for BCA Homework, and electronic transparencies (CalcLink).
- http://bca.brookscole.com

Tools for Enriching Calculus CD-ROM
by Harvey B. Keynes, James Stewart, and Dan Clegg
ISBN 0-534-39731-X

TEC provides a laboratory environment in which students can explore selected topics. TEC also includes homework hints for representative exercises.

Instructor's Guide.
by Douglas Shaw, Harvey B. Keynes, and James Stewart
ISBN 0-534-39363-2

Each section of the main text is discussed from several viewpoints and contains suggested time to allot, points to stress, text discussion topics, core materials for lecture, workshop/discussion suggestions, group work exercises in a form suitable for handout, and suggested homework problems. An electronic version is available on the Instructor's Resource CD-ROM.

Instructor's Guide for AP® Calculus
by Douglas Shaw and Robert Gerver, contributing author
ISBN 0-534-39341-1

Taking the perspective of optimizing preparation for the AP exam, each section of the main text is discussed from several viewpoints and contains suggested time to allot, points to stress, daily quizzes, core materials for lecture, workshop/discussion suggestions, group work exercises in a form suitable for handout, tips for the AP exam, and suggested homework problems.

Complete Solutions Manual
Single Variable
by Daniel Anderson, Jeffery A. Cole, and Daniel Drucker
ISBN 0-534-39368-3

Includes worked-out solutions to all exercises in the text.

Printed Test Items
ISBN 0-534-39365-9

Contains multiple-choice and short-answer test items that key directly to the text.

Brooks/Cole Assessment (BCA) Testing
ISBN 0-534-39364-0

Available online or on CD-ROM, browser-based BCA is fully integrated text-specific testing, tutorial, and course management software. With no need for plug-ins or downloads, BCA offers algorithmically generated problem values and machine-graded free response mathematics.
- http://bca.brookscole.com

Text-Specific Videos
ISBN 0-534-39392-6

Text-specific videotape sets, available at no charge to adopters, consisting of one tape per text chapter. Each tape features a 10- to 20-minute problem-solving lesson for each section of the chapter.

Transparencies by James Stewart
Single Variable ISBN 0-534-39337-3

Full-color, large-scale sheets of reproductions of material from the text.

Ancillaries for Instructors and Students

Stewart Specialty Web Site
www.stewartcalculus.com

Contents: *Algebra Review with tutorial* ▫ *Additional Topics* ▫ *Drill exercises* ▫ *Problems Plus* ▫ *Web Links* ▫ *History of Mathematics* ▫ *Downloadable versions of* CalcLabs *for Derive and TI graphing calculators*

BCA Homework
ISBN 0-534-39381-0

BCA Homework allows instructors to assign machine-gradable homework problems that help students identify where they need additional help. That assistance is available through worked-out solutions that guide students through the steps of problem solving, or via live online tutoring at vMentor. The tutors at this online service will skillfully guide students through a problem, using unique two-way audio and whiteboard features.
- http://bca.brookscole.com

The Brooks/Cole Mathematics Resource Center Web Site
http://mathematics.brookscole.com

When you adopt a Thomson–Brooks/Cole mathematics text, you and your students will have access to a variety of teaching and learning resources. This Web site features everything from

|||| Electronic items |||| Printed items

xxii

book-specific resources to newsgroups. It's a great way to make teaching and learning an interactive and intriguing experience.

WebTutor™ on WebCT
ISBN 0-534-39390-X

Lecture notes, discussion threads, and quizzes on WebCT.

Journey Through Calculus
by Bill Ralph in conjunction with James Stewart
Student Version ISBN 0-534-26220-1
Instructor's Version ISBN 0-534-36823-9

A Calculus CD-ROM that brings together activities, tutorials, testing, a computer algebra system, and calculus content into one unified environment.

Student Resources

Tools for Enriching™ Calculus CD-ROM
by Harvey B. Keynes, James Stewart, and Dan Clegg
ISBN 0-534-39731-X

TEC provides a laboratory environment in which students can explore selected topics. TEC also includes homework hints for representative exercises.

Interactive Video SkillBuilder CD-ROM
ISBN 0-534-39347-0

Think of it as portable office hours! The Interactive Video Skillbuilder *CD-ROM contains more than eight hours of video instruction. The problems worked during each video lesson are shown next to the viewing screen so that students can try working them before watching the solution. To help students evaluate their progress, each section contains a ten-question Web quiz (the results of which can be emailed to the instructor) and each chapter contains a chapter test, with answers to each problem. Also included is* MathCue Tutorial *software. This dual-platform software presents and scores problems and tutors students by displaying annotated, step-by-step solutions. Problem sets may be customized.*

Study Guide
Single Variable
by Richard St. Andre
ISBN 0-534-39367-5

Contains a short list of key concepts, a short list of skills to master, a brief introduction to the ideas of the section, an elaboration of the concepts and skills, including extra worked-out

examples, and links in the margin to earlier and later material in the text and Study Guide.

Student Solutions Manual
Single Variable
by Daniel Anderson, Jeffery A. Cole, and Daniel Drucker
ISBN 0-534-39369-1

Provides completely worked-out solutions to all odd-numbered exercises within the text, giving students a way to check their answers and ensure that they took the correct steps to arrive at an answer.

CalcLabs with Maple
Single Variable
by Philip Yasskin, Albert Boggess, David Barrow, Maurice Rahe, Jeffery Morgan, Michael Stecher, Art Belmonte, and Kirby Smith
ISBN 0-534-39370-5

CalcLabs with Mathematica
Single Variable
by Selwyn Hollis
ISBN 0-534-39371-3

Each of these comprehensive lab manuals will help students learn to effectively use the technology tools available to them. Each lab contains clearly explained exercises and a variety of labs and projects to accompany the text.

A Companion to Calculus
by Dennis Ebersole, Doris Schattschneider, Alicia Sevilla, and Kay Somers
ISBN 0-534-26592-8

Written to improve algebra and problem-solving skills of students taking a calculus course, every chapter in this companion is keyed to a calculus topic, providing conceptual background and specific algebra techniques needed to understand and solve calculus problems related to that topic. It is designed for calculus courses that integrate the review of precalculus concepts or for individual use.

Linear Algebra for Calculus
by Konrad J. Heuvers, William P. Francis, John H. Kuisti, Deborah F. Lockhart, Daniel S. Moak, and Gene M. Ortner
ISBN 0-534-25248-6

This comprehensive book, designed to supplement the calculus course, provides an introduction to and review of the basic ideas of linear algebra.

|||| Electronic items |||| Printed items

To the Student

Reading a calculus textbook is different from reading a newspaper or a novel, or even a physics book. Don't be discouraged if you have to read a passage more than once in order to understand it. You should have pencil and paper and calculator at hand to sketch a diagram or make a calculation.

Some students start by trying their homework problems and read the text only if they get stuck on an exercise. I suggest that a far better plan is to read and understand a section of the text before attempting the exercises. In particular, you should look at the definitions to see the exact meanings of the terms. And before you read each example, I suggest that you cover up the solution and try solving the problem yourself. You'll get a lot more from looking at the solution if you do so.

Part of the aim of this course is to train you to think logically. Learn to write the solutions of the exercises in a connected, step-by-step fashion with explanatory sentences—not just a string of disconnected equations or formulas.

The answers to the odd-numbered exercises appear at the back of the book, in Appendix H. Some exercises ask for a verbal explanation or interpretation or description. In such cases there is no single correct way of expressing the answer, so don't worry that you haven't found the definitive answer. In addition, there are often several different forms in which to express a numerical or algebraic answer, so if your answer differs from mine, don't immediately assume you're wrong. For example, if the answer given in the back of the book is $\sqrt{2} - 1$ and you obtain $1/(1 + \sqrt{2})$, then you're right and rationalizing the denominator will show that the answers are equivalent.

The icon ⊞ indicates an exercise that definitely requires the use of either a graphing calculator or a computer with graphing software. (Section 1.4 discusses the use of these graphing devices and some of the pitfalls that you may encounter.) But that doesn't mean that graphing devices can't be used to check your work on the other exercises as well. The symbol [CAS] is reserved for problems in which the full resources of a computer algebra system (like Derive, Maple, Mathematica, or the TI-89/92) are required. You will also encounter the symbol ⊘, which warns you against committing an error. I have placed this symbol in the margin in situations where I have observed that a large proportion of my students tend to make the same mistake.

The icon ⓙ indicates a reference to the CD-ROM *Journey Through Calculus*. The symbols in the margin refer you to the location in *Journey* where a concept is introduced through an interactive exploration or animation.

The CD-ROM *Tools for Enriching Calculus* (see inside front cover for availability) is referred to by means of the symbol ⓣ. It directs you to modules in which you can explore aspects of calculus for which the computer is particularly useful. TEC also provides *Homework Hints* for representative exercises that are indicated by printing the exercise number in red: **23.** These homework hints ask you questions that allow you to make progress toward a solution without actually giving you the answer. You need to pursue each hint in an active manner with pencil and paper to work out the details. If a particular hint doesn't enable you to solve the problem, you can click to reveal the next hint.

The CD-ROM *Interactive Video Skillbuilder* (see inside front cover for availability) contains videos of instructors explaining two or three of the examples in every section of the text. Also on the CD is a video in which I offer advice on how to succeed in your calculus course.

I recommend that you keep this book for reference purposes after you finish the course. Because you will likely forget some of the specific details of calculus, the book will serve as a useful reminder when you need to use calculus in subsequent courses. And, because this book contains more material than can be covered in any one course, it can also serve as a valuable resource for a working scientist or engineer.

Calculus is an exciting subject, justly considered to be one of the greatest achievements of the human intellect. I hope you will discover that it is not only useful but also intrinsically beautiful.

JAMES STEWART

single variable
CALCULUS

Volume 2: Chapters 5–12

Chapter 5

To compute an area we approximate a region by rectangles and let the number of rectangles become large. The precise area is the limit of these sums of areas of rectangles.

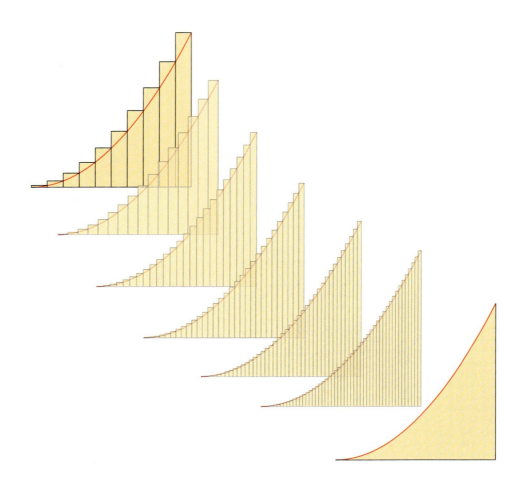

Integrals

IIII Now is a good time to read (or reread) *A Preview of Calculus* (see page 2). It discusses the unifying ideas of calculus and helps put in perspective where we have been and where we are going.

In Chapter 2 we used the tangent and velocity problems to introduce the derivative, which is the central idea in differential calculus. In much the same way, this chapter starts with the area and distance problems and uses them to formulate the idea of a definite integral, which is the basic concept of integral calculus. We will see in Chapters 6 and 9 how to use the integral to solve problems concerning volumes, lengths of curves, population predictions, cardiac output, forces on a dam, work, consumer surplus, and baseball, among many others.

There is a connection between integral calculus and differential calculus. The Fundamental Theorem of Calculus relates the integral to the derivative, and we will see in this chapter that it greatly simplifies the solution of many problems.

5.1 Areas and Distances

In this section we discover that in trying to find the area under a curve or the distance traveled by a car, we end up with the same special type of limit.

The Area Problem

We begin by attempting to solve the *area problem:* Find the area of the region S that lies under the curve $y = f(x)$ from a to b. This means that S, illustrated in Figure 1, is bounded by the graph of a continuous function f [where $f(x) \geq 0$], the vertical lines $x = a$ and $x = b$, and the x-axis.

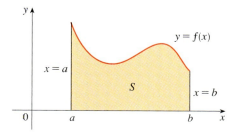

FIGURE 1
$S = \{(x, y) \mid a \leq x \leq b, 0 \leq y \leq f(x)\}$

In trying to solve the area problem we have to ask ourselves: What is the meaning of the word *area*? This question is easy to answer for regions with straight sides. For a rectangle, the area is defined as the product of the length and the width. The area of a triangle is half the base times the height. The area of a polygon is found by dividing it into triangles (as in Figure 2) and adding the areas of the triangles.

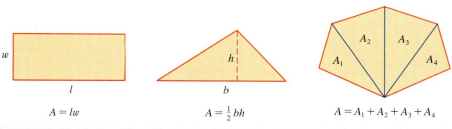

FIGURE 2
$A = lw$
$A = \tfrac{1}{2} bh$
$A = A_1 + A_2 + A_3 + A_4$

However, it isn't so easy to find the area of a region with curved sides. We all have an intuitive idea of what the area of a region is. But part of the area problem is to make this intuitive idea precise by giving an exact definition of area.

Recall that in defining a tangent we first approximated the slope of the tangent line by slopes of secant lines and then we took the limit of these approximations. We pursue a similar idea for areas. We first approximate the region S by rectangles and then we take the limit of the areas of these rectangles as we increase the number of rectangles. The following example illustrates the procedure.

EXAMPLE 1 Use rectangles to estimate the area under the parabola $y = x^2$ from 0 to 1 (the parabolic region S illustrated in Figure 3).

Try placing rectangles to estimate the area.
Resources / Module 6
/ What Is Area?
/ Estimating Area under a Parabola

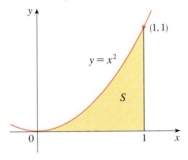

FIGURE 3

SOLUTION We first notice that the area of S must be somewhere between 0 and 1 because S is contained in a square with side length 1, but we can certainly do better than that. Suppose we divide S into four strips S_1, S_2, S_3, and S_4 by drawing the vertical lines $x = \frac{1}{4}$, $x = \frac{1}{2}$, and $x = \frac{3}{4}$ as in Figure 4(a).

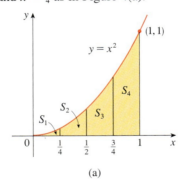 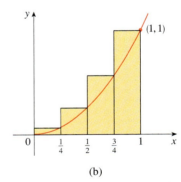

FIGURE 4
(a)   (b)

We can approximate each strip by a rectangle whose base is the same as the strip and whose height is the same as the right edge of the strip [see Figure 4(b)]. In other words, the heights of these rectangles are the values of the function $f(x) = x^2$ at the right end points of the subintervals $\left[0, \frac{1}{4}\right]$, $\left[\frac{1}{4}, \frac{1}{2}\right]$, $\left[\frac{1}{2}, \frac{3}{4}\right]$, and $\left[\frac{3}{4}, 1\right]$.

Each rectangle has width $\frac{1}{4}$ and the heights are $\left(\frac{1}{4}\right)^2$, $\left(\frac{1}{2}\right)^2$, $\left(\frac{3}{4}\right)^2$, and 1^2. If we let R_4 be the sum of the areas of these approximating rectangles, we get

$$R_4 = \tfrac{1}{4} \cdot \left(\tfrac{1}{4}\right)^2 + \tfrac{1}{4} \cdot \left(\tfrac{1}{2}\right)^2 + \tfrac{1}{4} \cdot \left(\tfrac{3}{4}\right)^2 + \tfrac{1}{4} \cdot 1^2 = \tfrac{15}{32} = 0.46875$$

From Figure 4(b) we see that the area A of S is less than R_4, so

$$A < 0.46875$$

Instead of using the rectangles in Figure 4(b) we could use the smaller rectangles in Figure 5 whose heights are the values of f at the left endpoints of the subintervals. (The leftmost rectangle has collapsed because its height is 0.) The sum of the areas of these

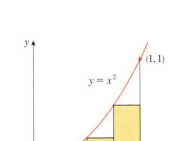

FIGURE 5

approximating rectangles is

$$L_4 = \tfrac{1}{4} \cdot 0^2 + \tfrac{1}{4} \cdot \left(\tfrac{1}{4}\right)^2 + \tfrac{1}{4} \cdot \left(\tfrac{1}{2}\right)^2 + \tfrac{1}{4} \cdot \left(\tfrac{3}{4}\right)^2 = \tfrac{7}{32} = 0.21875$$

We see that the area of S is larger than L_4, so we have lower and upper estimates for A:

$$0.21875 < A < 0.46875$$

We can repeat this procedure with a larger number of strips. Figure 6 shows what happens when we divide the region S into eight strips of equal width.

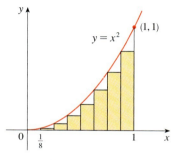
(a) Using left endpoints

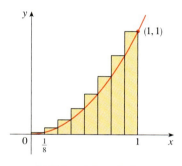
(b) Using right endpoints

FIGURE 6
Approximating S with eight rectangles

By computing the sum of the areas of the smaller rectangles (L_8) and the sum of the areas of the larger rectangles (R_8), we obtain better lower and upper estimates for A:

$$0.2734375 < A < 0.3984375$$

So one possible answer to the question is to say that the true area of S lies somewhere between 0.2734375 and 0.3984375.

We could obtain better estimates by increasing the number of strips. The table at the left shows the results of similar calculations (with a computer) using n rectangles whose heights are found with left endpoints (L_n) or right endpoints (R_n). In particular, we see by using 50 strips that the area lies between 0.3234 and 0.3434. With 1000 strips we narrow it down even more: A lies between 0.3328335 and 0.3338335. A good estimate is obtained by averaging these numbers: $A \approx 0.3333335$.

n	L_n	R_n
10	0.2850000	0.3850000
20	0.3087500	0.3587500
30	0.3168519	0.3501852
50	0.3234000	0.3434000
100	0.3283500	0.3383500
1000	0.3328335	0.3338335

From the values in the table in Example 1, it looks as if R_n is approaching $\tfrac{1}{3}$ as n increases. We confirm this in the next example.

EXAMPLE 2 For the region S in Example 1, show that the sum of the areas of the upper approximating rectangles approaches $\tfrac{1}{3}$, that is,

$$\lim_{n \to \infty} R_n = \tfrac{1}{3}$$

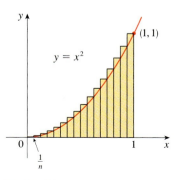

FIGURE 7

SOLUTION R_n is the sum of the areas of the n rectangles in Figure 7. Each rectangle has width $1/n$ and the heights are the values of the function $f(x) = x^2$ at the points $1/n, 2/n, 3/n, \ldots, n/n$; that is, the heights are $(1/n)^2, (2/n)^2, (3/n)^2, \ldots, (n/n)^2$.

Thus

$$R_n = \frac{1}{n}\left(\frac{1}{n}\right)^2 + \frac{1}{n}\left(\frac{2}{n}\right)^2 + \frac{1}{n}\left(\frac{3}{n}\right)^2 + \cdots + \frac{1}{n}\left(\frac{n}{n}\right)^2$$

$$= \frac{1}{n} \cdot \frac{1}{n^2}(1^2 + 2^2 + 3^2 + \cdots + n^2)$$

$$= \frac{1}{n^3}(1^2 + 2^2 + 3^2 + \cdots + n^2)$$

The ideas in Examples 1 and 2 are explored in Module 5.1/5.2/8.7 for a variety of functions.

Here we need the formula for the sum of the squares of the first n positive integers:

$$\boxed{1} \qquad 1^2 + 2^2 + 3^2 + \cdots + n^2 = \frac{n(n+1)(2n+1)}{6}$$

Perhaps you have seen this formula before. It is proved in Example 5 in Appendix E. Putting Formula 1 into our expression for R_n, we get

$$R_n = \frac{1}{n^3} \cdot \frac{n(n+1)(2n+1)}{6} = \frac{(n+1)(2n+1)}{6n^2}$$

Thus, we have

|||| Here we are computing the limit of the sequence $\{R_n\}$. Sequences were discussed in *A Preview of Calculus* and will be studied in detail in Chapter 12. Their limits are calculated in the same way as limits at infinity (Section 4.4). In particular, we know that

$$\lim_{n \to \infty} \frac{1}{n} = 0$$

$$\lim_{n \to \infty} R_n = \lim_{n \to \infty} \frac{(n+1)(2n+1)}{6n^2}$$

$$= \lim_{n \to \infty} \frac{1}{6}\left(\frac{n+1}{n}\right)\left(\frac{2n+1}{n}\right)$$

$$= \lim_{n \to \infty} \frac{1}{6}\left(1 + \frac{1}{n}\right)\left(2 + \frac{1}{n}\right)$$

$$= \frac{1}{6} \cdot 1 \cdot 2 = \frac{1}{3}$$

It can be shown that the lower approximating sums also approach $\frac{1}{3}$, that is,

$$\lim_{n \to \infty} L_n = \frac{1}{3}$$

From Figures 8 and 9 it appears that, as n increases, both L_n and R_n become better and better approximations to the area of S. Therefore, we *define* the area A to be the limit of the

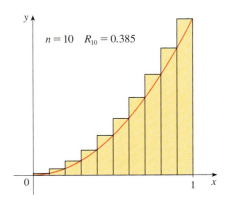
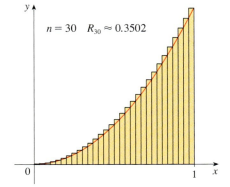
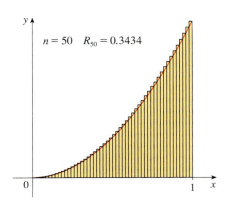

FIGURE 8

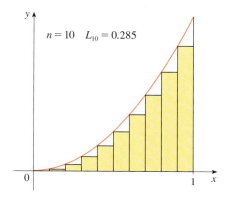
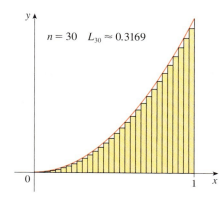
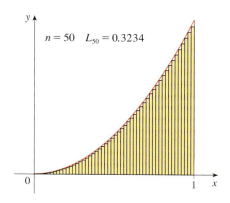

FIGURE 9
The area is the number that is smaller than all upper sums and larger than all lower sums

sums of the areas of the approximating rectangles, that is,

$$A = \lim_{n \to \infty} R_n = \lim_{n \to \infty} L_n = \tfrac{1}{3}$$

Let's apply the idea of Examples 1 and 2 to the more general region S of Figure 1. We start by subdividing S into n strips $S_1, S_2, \ldots, S_n$ of equal width as in Figure 10. The width of the interval $[a, b]$ is $b - a$, so the width of each of the n strips is

$$\Delta x = \frac{b - a}{n}$$

These strips divide the interval $[a, b]$ into n subintervals

$$[x_0, x_1], \quad [x_1, x_2], \quad [x_2, x_3], \quad \ldots, \quad [x_{n-1}, x_n]$$

where $x_0 = a$ and $x_n = b$. The right endpoints of the subintervals are

$$x_1 = a + \Delta x, \quad x_2 = a + 2\Delta x, \quad x_3 = a + 3\Delta x, \quad \ldots$$

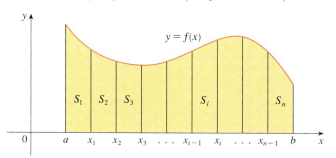

FIGURE 10

Let's approximate the ith strip S_i by a rectangle with width Δx and height $f(x_i)$, which is the value of f at the right endpoint (see Figure 11). Then the area of the ith rectangle

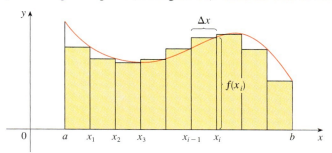

FIGURE 11

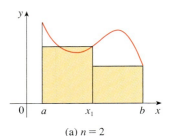
(a) $n = 2$

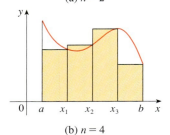
(b) $n = 4$

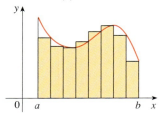
(c) $n = 8$

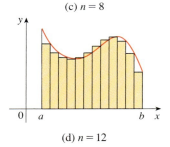
(d) $n = 12$

FIGURE 12

is $f(x_i)\,\Delta x$. What we think of intuitively as the area of S is approximated by the sum of the areas of these rectangles, which is

$$R_n = f(x_1)\,\Delta x + f(x_2)\,\Delta x + \cdots + f(x_n)\,\Delta x$$

Figure 12 shows this approximation for $n = 2, 4, 8$, and 12. Notice that this approximation appears to become better and better as the number of strips increases, that is, as $n \to \infty$. Therefore, we define the area A of the region S in the following way.

2 Definition The **area** A of the region S that lies under the graph of the continuous function f is the limit of the sum of the areas of approximating rectangles:

$$A = \lim_{n \to \infty} R_n = \lim_{n \to \infty} [f(x_1)\,\Delta x + f(x_2)\,\Delta x + \cdots + f(x_n)\,\Delta x]$$

It can be proved that the limit in Definition 2 always exists, since we are assuming that f is continuous. It can also be shown that we get the same value if we use left endpoints:

$$\boxed{3} \qquad A = \lim_{n \to \infty} L_n = \lim_{n \to \infty} [f(x_0)\,\Delta x + f(x_1)\,\Delta x + \cdots + f(x_{n-1})\,\Delta x]$$

In fact, instead of using left endpoints or right endpoints, we could take the height of the ith rectangle to be the value of f at *any* number x_i^* in the ith subinterval $[x_{i-1}, x_i]$. We call the numbers $x_1^*, x_2^*, \ldots, x_n^*$ the **sample points**. Figure 13 shows approximating rectangles when the sample points are not chosen to be endpoints. So a more general expression for the area of S is

$$\boxed{4} \qquad A = \lim_{n \to \infty} [f(x_1^*)\,\Delta x + f(x_2^*)\,\Delta x + \cdots + f(x_n^*)\,\Delta x]$$

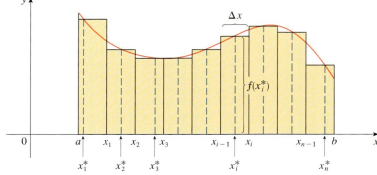

FIGURE 13

We often use **sigma notation** to write sums with many terms more compactly. For instance,

$$\sum_{i=1}^{n} f(x_i)\,\Delta x = f(x_1)\,\Delta x + f(x_2)\,\Delta x + \cdots + f(x_n)\,\Delta x$$

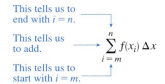

|||| If you need practice with sigma notation, look at the examples and try some of the exercises in Appendix E.

So the expressions for area in Equations 2, 3, and 4 can be written as follows:

$$A = \lim_{n \to \infty} \sum_{i=1}^{n} f(x_i) \, \Delta x$$

$$A = \lim_{n \to \infty} \sum_{i=1}^{n} f(x_{i-1}) \, \Delta x$$

$$A = \lim_{n \to \infty} \sum_{i=1}^{n} f(x_i^*) \, \Delta x$$

We could also rewrite Formula 1 in the following way:

$$\sum_{i=1}^{n} i^2 = \frac{n(n+1)(2n+1)}{6}$$

EXAMPLE 3 Let A be the area of the region that lies under the graph of $f(x) = \cos x$ between $x = 0$ and $x = b$, where $0 \leq b \leq \pi/2$.
(a) Using right endpoints, find an expression for A as a limit. Do not evaluate the limit.
(b) Estimate the area for the case $b = \pi/2$ by taking the sample points to be midpoints and using four subintervals.

SOLUTION
(a) Since $a = 0$, the width of a subinterval is

$$\Delta x = \frac{b - 0}{n} = \frac{b}{n}$$

So $x_1 = b/n$, $x_2 = 2b/n$, $x_3 = 3b/n$, $x_i = ib/n$, and $x_n = nb/n$. The sum of the areas of the approximating rectangles is

$$R_n = f(x_1) \, \Delta x + f(x_2) \, \Delta x + \cdots + f(x_n) \, \Delta x$$
$$= (\cos x_1) \, \Delta x + (\cos x_2) \, \Delta x + \cdots + (\cos x_n) \, \Delta x$$
$$= \left(\cos \frac{b}{n}\right) \frac{b}{n} + \left(\cos \frac{2b}{n}\right) \frac{b}{n} + \cdots + \left(\cos \frac{nb}{n}\right) \frac{b}{n}$$

According to Definition 2, the area is

$$A = \lim_{n \to \infty} R_n = \lim_{n \to \infty} \frac{b}{n} \left(\cos \frac{b}{n} + \cos \frac{2b}{n} + \cos \frac{3b}{n} + \cdots + \cos \frac{nb}{n} \right)$$

Using sigma notation we could write

$$A = \lim_{n \to \infty} \frac{b}{n} \sum_{i=1}^{n} \cos \frac{ib}{n}$$

It is very difficult to evaluate this limit directly by hand, but with the aid of a computer algebra system it isn't hard (see Exercise 25). In Section 5.3 we will be able to find A more easily using a different method.

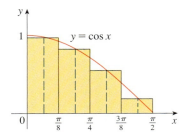

FIGURE 14

(b) With $n = 4$ and $b = \pi/2$ we have $\Delta x = (\pi/2)/4 = \pi/8$, so the subintervals are $[0, \pi/8]$, $[\pi/8, \pi/4]$, $[\pi/4, 3\pi/8]$, and $[3\pi/8, \pi/2]$. The midpoints of these subintervals are

$$x_1^* = \frac{\pi}{16} \qquad x_2^* = \frac{3\pi}{16} \qquad x_3^* = \frac{5\pi}{16} \qquad x_4^* = \frac{7\pi}{16}$$

and the sum of the areas of the four approximating rectangles (see Figure 14) is

$$M_4 = \sum_{i=1}^{4} f(x_i^*) \Delta x$$

$$= f(\pi/16) \Delta x + f(3\pi/16) \Delta x + f(5\pi/16) \Delta x + f(7\pi/16) \Delta x$$

$$= \left(\cos \frac{\pi}{16}\right) \frac{\pi}{8} + \left(\cos \frac{3\pi}{16}\right) \frac{\pi}{8} + \left(\cos \frac{5\pi}{16}\right) \frac{\pi}{8} + \left(\cos \frac{7\pi}{16}\right) \frac{\pi}{8}$$

$$= \frac{\pi}{8}\left(\cos \frac{\pi}{16} + \cos \frac{3\pi}{16} + \cos \frac{5\pi}{16} + \cos \frac{7\pi}{16}\right) \approx 1.006$$

So an estimate for the area is

$$A \approx 1.006$$

The Distance Problem

Now let's consider the *distance problem:* Find the distance traveled by an object during a certain time period if the velocity of the object is known at all times. (In a sense this is the inverse problem of the velocity problem that we discussed in Section 2.1.) If the velocity remains constant, then the distance problem is easy to solve by means of the formula

$$\text{distance} = \text{velocity} \times \text{time}$$

But if the velocity varies, it's not so easy to find the distance traveled. We investigate the problem in the following example.

EXAMPLE 4 Suppose the odometer on our car is broken and we want to estimate the distance driven over a 30-second time interval. We take speedometer readings every five seconds and record them in the following table:

Time (s)	0	5	10	15	20	25	30
Velocity (mi/h)	17	21	24	29	32	31	28

In order to have the time and the velocity in consistent units, let's convert the velocity readings to feet per second (1 mi/h = 5280/3600 ft/s):

Time (s)	0	5	10	15	20	25	30
Velocity (ft/s)	25	31	35	43	47	46	41

During the first five seconds the velocity doesn't change very much, so we can estimate the distance traveled during that time by assuming that the velocity is constant. If we take the velocity during that time interval to be the initial velocity (25 ft/s), then we obtain the approximate distance traveled during the first five seconds:

$$25 \text{ ft/s} \times 5 \text{ s} = 125 \text{ ft}$$

Similarly, during the second time interval the velocity is approximately constant and we take it to be the velocity when $t = 5$ s. So our estimate for the distance traveled from $t = 5$ s to $t = 10$ s is

$$31 \text{ ft/s} \times 5 \text{ s} = 155 \text{ ft}$$

If we add similar estimates for the other time intervals, we obtain an estimate for the total distance traveled:

$$25 \times 5 + 31 \times 5 + 35 \times 5 + 43 \times 5 + 47 \times 5 + 46 \times 5 = 1135 \text{ ft}$$

We could just as well have used the velocity at the *end* of each time period instead of the velocity at the beginning as our assumed constant velocity. Then our estimate becomes

$$31 \times 5 + 35 \times 5 + 43 \times 5 + 47 \times 5 + 46 \times 5 + 41 \times 5 = 1215 \text{ ft}$$

If we had wanted a more accurate estimate, we could have taken velocity readings every two seconds, or even every second.

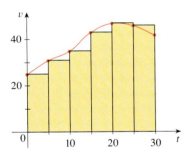

FIGURE 15

Perhaps the calculations in Example 4 remind you of the sums we used earlier to estimate areas. The similarity is explained when we sketch a graph of the velocity function of the car in Figure 15 and draw rectangles whose heights are the initial velocities for each time interval. The area of the first rectangle is $25 \times 5 = 125$, which is also our estimate for the distance traveled in the first five seconds. In fact, the area of each rectangle can be interpreted as a distance because the height represents velocity and the width represents time. The sum of the areas of the rectangles in Figure 15 is $L_6 = 1135$, which is our initial estimate for the total distance traveled.

In general, suppose an object moves with velocity $v = f(t)$, where $a \leq t \leq b$ and $f(t) \geq 0$ (so the object always moves in the positive direction). We take velocity readings at times $t_0 \,(= a), t_1, t_2, \ldots, t_n \,(= b)$ so that the velocity is approximately constant on each subinterval. If these times are equally spaced, then the time between consecutive readings is $\Delta t = (b - a)/n$. During the first time interval the velocity is approximately $f(t_0)$ and so the distance traveled is approximately $f(t_0) \Delta t$. Similarly, the distance traveled during the second time interval is about $f(t_1) \Delta t$ and the total distance traveled during the time interval $[a, b]$ is approximately

$$f(t_0) \Delta t + f(t_1) \Delta t + \cdots + f(t_{n-1}) \Delta t = \sum_{i=1}^{n} f(t_{i-1}) \Delta t$$

If we use the velocity at right endpoints instead of left endpoints, our estimate for the total distance becomes

$$f(t_1) \Delta t + f(t_2) \Delta t + \cdots + f(t_n) \Delta t = \sum_{i=1}^{n} f(t_i) \Delta t$$

The more frequently we measure the velocity, the more accurate our estimates become, so it seems plausible that the *exact* distance d traveled is the *limit* of such expressions:

$$\boxed{5} \qquad d = \lim_{n \to \infty} \sum_{i=1}^{n} f(t_{i-1}) \Delta t = \lim_{n \to \infty} \sum_{i=1}^{n} f(t_i) \Delta t$$

We will see in Section 5.4 that this is indeed true.

5.1 Exercises

1. (a) By reading values from the given graph of f, use five rectangles to find a lower estimate and an upper estimate for the area under the given graph of f from $x = 0$ to $x = 10$. In each case sketch the rectangles that you use.
    (b) Find new estimates using 10 rectangles in each case.

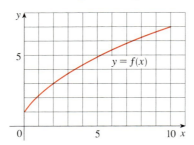

2. (a) Use six rectangles to find estimates of each type for the area under the given graph of f from $x = 0$ to $x = 12$.
    (i) L_6 (sample points are left endpoints)
    (ii) R_6 (sample points are right endpoints)
    (iii) M_6 (sample points are midpoints)
    (b) Is L_6 an underestimate or overestimate of the true area?
    (c) Is R_6 an underestimate or overestimate of the true area?
    (d) Which of the numbers L_6, R_6, or M_6 gives the best estimate? Explain.

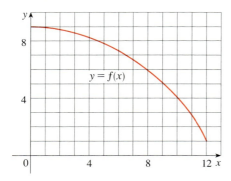

3. (a) Estimate the area under the graph of $f(x) = 1/x$ from $x = 1$ to $x = 5$ using four approximating rectangles and right endpoints. Sketch the graph and the rectangles. Is your estimate an underestimate or an overestimate?
    (b) Repeat part (a) using left endpoints.

4. (a) Estimate the area under the graph of $f(x) = 25 - x^2$ from $x = 0$ to $x = 5$ using five approximating rectangles and right endpoints. Sketch the graph and the rectangles. Is your estimate an underestimate or an overestimate?
    (b) Repeat part (a) using left endpoints.

5. (a) Estimate the area under the graph of $f(x) = 1 + x^2$ from $x = -1$ to $x = 2$ using three rectangles and right endpoints. Then improve your estimate by using six rectangles. Sketch the curve and the approximating rectangles.
    (b) Repeat part (a) using left endpoints.
    (c) Repeat part (a) using midpoints.
    (d) From your sketches in parts (a)–(c), which appears to be the best estimate?

6. (a) Graph the function $f(x) = 1/(1 + x^2)$, $-2 \le x \le 2$.
    (b) Estimate the area under the graph of f using four approximating rectangles and taking the sample points to be
    (i) right endpoints   (ii) midpoints
    In each case sketch the curve and the rectangles.
    (c) Improve your estimates in part (b) by using eight rectangles.

7–8 With a programmable calculator (or a computer), it is possible to evaluate the expressions for the sums of areas of approximating rectangles, even for large values of n, using looping. (On a TI use the Is> command or a For-EndFor loop, on a Casio use Isz, on an HP or in BASIC use a FOR-NEXT loop.) Compute the sum of the areas of approximating rectangles using equal subintervals and right endpoints for $n = 10, 30,$ and 50. Then guess the value of the exact area.

7. The region under $y = \sin x$ from 0 to π

8. The region under $y = 1/x^2$ from 1 to 2

9. Some computer algebra systems have commands that will draw approximating rectangles and evaluate the sums of their areas, at least if x_i^* is a left or right endpoint. (For instance, in Maple use leftbox, rightbox, leftsum, and rightsum.)
    (a) If $f(x) = \sqrt{x}$, $1 \le x \le 4$, find the left and right sums for $n = 10, 30,$ and 50.
    (b) Illustrate by graphing the rectangles in part (a).
    (c) Show that the exact area under f lies between 4.6 and 4.7.

10. (a) If $f(x) = \sin(\sin x)$, $0 \le x \le \pi/2$, use the commands discussed in Exercise 9 to find the left and right sums for $n = 10, 30,$ and 50.

(b) Illustrate by graphing the rectangles in part (a).

(c) Show that the exact area under f lies between 0.87 and 0.91.

11. The speed of a runner increased steadily during the first three seconds of a race. Her speed at half-second intervals is given in the table. Find lower and upper estimates for the distance that she traveled during these three seconds.

t (s)	0	0.5	1.0	1.5	2.0	2.5	3.0
v (ft/s)	0	6.2	10.8	14.9	18.1	19.4	20.2

12. Speedometer readings for a motorcycle at 12-second intervals are given in the table.

(a) Estimate the distance traveled by the motorcycle during this time period using the velocities at the beginning of the time intervals.

(b) Give another estimate using the velocities at the end of the time periods.

(c) Are your estimates in parts (a) and (b) upper and lower estimates? Explain.

t (s)	0	12	24	36	48	60
v (ft/s)	30	28	25	22	24	27

13. Oil leaked from a tank at a rate of $r(t)$ liters per hour. The rate decreased as time passed and values of the rate at 2-hour time intervals are shown in the table. Find lower and upper estimates for the total amount of oil that leaked out.

t (h)	0	2	4	6	8	10
$r(t)$ (L/h)	8.7	7.6	6.8	6.2	5.7	5.3

14. When we estimate distances from velocity data, it is sometimes necessary to use times $t_0, t_1, t_2, t_3, \ldots$ that are not equally spaced. We can still estimate distances using the time periods $\Delta t_i = t_i - t_{i-1}$. For example, on May 7, 1992, the space shuttle *Endeavour* was launched on mission STS-49, the purpose of which was to install a new perigee kick motor in an Intelsat communications satellite. The table, provided by NASA, gives the velocity data for the shuttle between liftoff and the jettisoning of the solid rocket boosters.

Event	Time (s)	Velocity (ft/s)
Launch	0	0
Begin roll maneuver	10	185
End roll maneuver	15	319
Throttle to 89%	20	447
Throttle to 67%	32	742
Throttle to 104%	59	1325
Maximum dynamic pressure	62	1445
Solid rocket booster separation	125	4151

Use these data to estimate the height above the Earth's surface of the space shuttle *Endeavour*, 62 seconds after liftoff.

15. The velocity graph of a braking car is shown. Use it to estimate the distance traveled by the car while the brakes are applied.

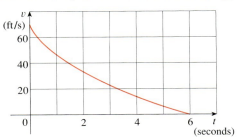

16. The velocity graph of a car accelerating from rest to a speed of 120 km/h over a period of 30 seconds is shown. Estimate the distance traveled during this period.

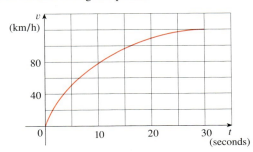

17–19 Use Definition 2 to find an expression for the area under the graph of f as a limit. Do not evaluate the limit.

17. $f(x) = \sqrt[4]{x}, \quad 1 \le x \le 16$

18. $f(x) = 1 + x^4, \quad 2 \le x \le 5$

19. $f(x) = x \cos x, \quad 0 \le x \le \pi/2$

20–21 Determine a region whose area is equal to the given limit. Do not evaluate the limit.

20. $\lim_{n \to \infty} \sum_{i=1}^{n} \frac{2}{n} \left(5 + \frac{2i}{n} \right)^{10}$

21. $\lim_{n \to \infty} \sum_{i=1}^{n} \frac{\pi}{4n} \tan \frac{i\pi}{4n}$

22. (a) Use Definition 2 to find an expression for the area under the curve $y = x^3$ from 0 to 1 as a limit.

(b) The following formula for the sum of the cubes of the first n integers is proved in Appendix E. Use it to evaluate the limit in part (a).

$$1^3 + 2^3 + 3^3 + \cdots + n^3 = \left[\frac{n(n+1)}{2} \right]^2$$

CAS **23.** (a) Express the area under the curve $y = x^5$ from 0 to 2 as a limit.
(b) Use a computer algebra system to find the sum in your expression from part (a).
(c) Evaluate the limit in part (a).

CAS **24.** (a) Express the area under the curve $y = x^4 + 5x^2 + x$ from 2 to 7 as a limit.
(b) Use a computer algebra system to evaluate the sum in part (a).
(c) Use a computer algebra system to find the exact area by evaluating the limit of the expression in part (b).

CAS **25.** Find the exact area under the cosine curve $y = \cos x$ from $x = 0$ to $x = b$, where $0 \leq b \leq \pi/2$. (Use a computer algebra system both to evaluate the sum and compute the limit in Example 3.) In particular, what is the area if $b = \pi/2$? Compare your answer with the estimate obtained in Example 3(b).

26. (a) Let A_n be the area of a polygon with n equal sides inscribed in a circle with radius r. By dividing the polygon into n congruent triangles with central angle $2\pi/n$, show that $A_n = \frac{1}{2}nr^2 \sin(2\pi/n)$.
(b) Show that $\lim_{n\to\infty} A_n = \pi r^2$. [*Hint:* Use Equation 3.5.2.]

5.2 The Definite Integral

We saw in Section 5.1 that a limit of the form

$$\boxed{1} \quad \lim_{n\to\infty} \sum_{i=1}^{n} f(x_i^*)\,\Delta x = \lim_{n\to\infty} [f(x_1^*)\,\Delta x + f(x_2^*)\,\Delta x + \cdots + f(x_n^*)\,\Delta x]$$

arises when we compute an area. We also saw that it arises when we try to find the distance traveled by an object. It turns out that this same type of limit occurs in a wide variety of situations even when f is not necessarily a positive function. In Chapters 6 and 9 we will see that limits of the form (1) also arise in finding lengths of curves, volumes of solids, centers of mass, force due to water pressure, and work, as well as other quantities. We therefore give this type of limit a special name and notation.

> **2 Definition of a Definite Integral** If f is a continuous function defined for $a \leq x \leq b$, we divide the interval $[a, b]$ into n subintervals of equal width $\Delta x = (b - a)/n$. We let $x_0 \, (= a), x_1, x_2, \ldots, x_n \, (= b)$ be the endpoints of these subintervals and we let $x_1^*, x_2^*, \ldots, x_n^*$ be any **sample points** in these subintervals, so x_i^* lies in the ith subinterval $[x_{i-1}, x_i]$. Then the **definite integral of f from a to b** is
>
> $$\int_a^b f(x)\,dx = \lim_{n\to\infty} \sum_{i=1}^{n} f(x_i^*)\,\Delta x$$

Because we have assumed that f is continuous, it can be proved that the limit in Definition 2 always exists and gives the same value no matter how we choose the sample points x_i^*. (See Note 4 for a precise definition of this type of limit.) If we take the sample points to be right endpoints, then $x_i^* = x_i$ and the definition of an integral becomes

$$\boxed{3} \quad \int_a^b f(x)\,dx = \lim_{n\to\infty} \sum_{i=1}^{n} f(x_i)\,\Delta x$$

If we choose the sample points to be left endpoints, then $x_i^* = x_{i-1}$ and the definition becomes

$$\int_a^b f(x)\,dx = \lim_{n \to \infty} \sum_{i=1}^n f(x_{i-1})\,\Delta x$$

Alternatively, we could choose x_i^* to be the midpoint of the subinterval or any other number between x_{i-1} and x_i.

Although most of the functions that we encounter are continuous, the limit in Definition 2 also exists if f has a finite number of removable or jump discontinuities (but not infinite discontinuities). (See Section 2.5.) So we can also define the definite integral for such functions.

NOTE 1 ▫ The symbol $\int$ was introduced by Leibniz and is called an **integral sign**. It is an elongated S and was chosen because an integral is a limit of sums. In the notation $\int_a^b f(x)\,dx$, $f(x)$ is called the **integrand** and a and b are called the **limits of integration**; a is the **lower limit** and b is the **upper limit**. The symbol dx has no official meaning by itself; $\int_a^b f(x)\,dx$ is all one symbol. The procedure of calculating an integral is called **integration**.

NOTE 2 ▫ The definite integral $\int_a^b f(x)\,dx$ is a number; it does not depend on x. In fact, we could use any letter in place of x without changing the value of the integral:

$$\int_a^b f(x)\,dx = \int_a^b f(t)\,dt = \int_a^b f(r)\,dr$$

NOTE 3 ▫ The sum

$$\sum_{i=1}^n f(x_i^*)\,\Delta x$$

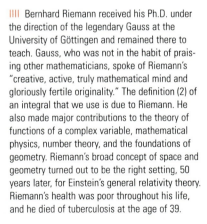

Bernhard Riemann received his Ph.D. under the direction of the legendary Gauss at the University of Göttingen and remained there to teach. Gauss, who was not in the habit of praising other mathematicians, spoke of Riemann's "creative, active, truly mathematical mind and gloriously fertile originality." The definition (2) of an integral that we use is due to Riemann. He also made major contributions to the theory of functions of a complex variable, mathematical physics, number theory, and the foundations of geometry. Riemann's broad concept of space and geometry turned out to be the right setting, 50 years later, for Einstein's general relativity theory. Riemann's health was poor throughout his life, and he died of tuberculosis at the age of 39.

that occurs in Definition 2 is called a **Riemann sum** after the German mathematician Bernhard Riemann (1826–1866). We know that if f happens to be positive, then the Riemann sum can be interpreted as a sum of areas of approximating rectangles (see Figure 1). By comparing Definition 2 with the definition of area in Section 5.1, we see that the definite integral $\int_a^b f(x)\,dx$ can be interpreted as the area under the curve $y = f(x)$ from a to b. (See Figure 2.)

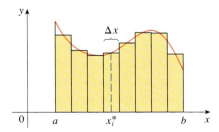

FIGURE 1
If $f(x) \geq 0$, the Riemann sum $\sum f(x_i^*)\,\Delta x$ is the sum of areas of rectangles.

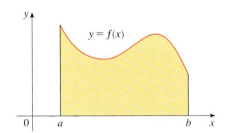

FIGURE 2
If $f(x) \geq 0$, the integral $\int_a^b f(x)\,dx$ is the area under the curve $y = f(x)$ from a to b.

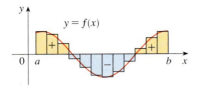

FIGURE 3
$\Sigma f(x_i^*) \Delta x$ is an approximation to the net area

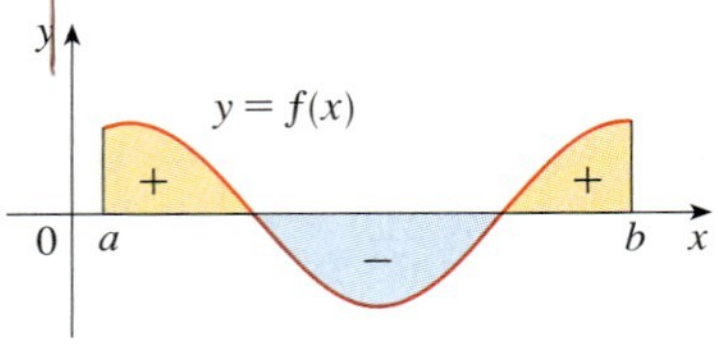

FIGURE 4
$\int_a^b f(x)\, dx$ is the net area

If f takes on both positive and negative values, as in Figure 3, then the Riemann sum is the sum of the areas of the rectangles that lie above the x-axis and the *negatives* of the areas of the rectangles that lie below the x-axis (the areas of the gold rectangles *minus* the areas of the blue rectangles). When we take the limit of such Riemann sums, we get the situation illustrated in Figure 4. A definite integral can be interpreted as a **net area**, that is, a difference of areas:

$$\int_a^b f(x)\, dx = A_1 - A_2$$

where A_1 is the area of the region above the x-axis and below the graph of f, and A_2 is the area of the region below the x-axis and above the graph of f.

NOTE 4 ◦ In the spirit of the precise definition of the limit of a function in Section 2.4, we can write the precise meaning of the limit that defines the integral in Definition 2 as follows:

For every number $\varepsilon > 0$ there is an integer N such that

$$\left| \int_a^b f(x)\, dx - \sum_{i=1}^n f(x_i^*)\, \Delta x \right| < \varepsilon$$

for every integer $n > N$ and for every choice of x_i^* in $[x_{i-1}, x_i]$.

This means that a definite integral can be approximated to within any desired degree of accuracy by a Riemann sum.

NOTE 5 ◦ Although we have defined $\int_a^b f(x)\, dx$ by dividing $[a, b]$ into subintervals of equal width, there are situations in which it is advantageous to work with subintervals of unequal width. For instance, in Exercise 14 in Section 5.1 NASA provided velocity data at times that were not equally spaced, but we were still able to estimate the distance traveled. And there are methods for numerical integration that take advantage of unequal subintervals.

If the subinterval widths are $\Delta x_1, \Delta x_2, \ldots, \Delta x_n$, we have to ensure that all these widths approach 0 in the limiting process. This happens if the largest width, max Δx_i, approaches 0. So in this case the definition of a definite integral becomes

$$\int_a^b f(x)\, dx = \lim_{\max \Delta x_i \to 0} \sum_{i=1}^n f(x_i^*)\, \Delta x_i$$

EXAMPLE 1 Express

$$\lim_{n \to \infty} \sum_{i=1}^n (x_i^3 + x_i \sin x_i)\, \Delta x$$

as an integral on the interval $[0, \pi]$.

SOLUTION Comparing the given limit with the limit in Definition 2, we see that they will be identical if we choose

$$f(x) = x^3 + x \sin x \quad \text{and} \quad x_i^* = x_i$$

(So the sample points are right endpoints and the given limit is of the form of Equation 3.) We are given that $a = 0$ and $b = \pi$. Therefore, by Definition 2 or Equation 3, we have

$$\lim_{n \to \infty} \sum_{i=1}^n (x_i^3 + x_i \sin x_i)\, \Delta x = \int_0^\pi (x^3 + x \sin x)\, dx$$

Later, when we apply the definite integral to physical situations, it will be important to recognize limits of sums as integrals, as we did in Example 1. When Leibniz chose the notation for an integral, he chose the ingredients as reminders of the limiting process. In general, when we write

$$\lim_{n \to \infty} \sum_{i=1}^{n} f(x_i^*) \, \Delta x = \int_a^b f(x) \, dx$$

we replace $\lim \Sigma$ by $\int$, x_i^* by x, and Δx by dx.

Evaluating Integrals

When we use the definition to evaluate a definite integral, we need to know how to work with sums. The following three equations give formulas for sums of powers of positive integers. Equation 4 may be familiar to you from a course in algebra. Equations 5 and 6 were discussed in Section 5.1 and are proved in Appendix E.

$$\boxed{4} \qquad \sum_{i=1}^{n} i = \frac{n(n+1)}{2}$$

$$\boxed{5} \qquad \sum_{i=1}^{n} i^2 = \frac{n(n+1)(2n+1)}{6}$$

$$\boxed{6} \qquad \sum_{i=1}^{n} i^3 = \left[\frac{n(n+1)}{2} \right]^2$$

The remaining formulas are simple rules for working with sigma notation:

$$\boxed{7} \qquad \sum_{i=1}^{n} c = nc$$

$$\boxed{8} \qquad \sum_{i=1}^{n} ca_i = c \sum_{i=1}^{n} a_i$$

$$\boxed{9} \qquad \sum_{i=1}^{n} (a_i + b_i) = \sum_{i=1}^{n} a_i + \sum_{i=1}^{n} b_i$$

$$\boxed{10} \qquad \sum_{i=1}^{n} (a_i - b_i) = \sum_{i=1}^{n} a_i - \sum_{i=1}^{n} b_i$$

|||| Formulas 7–10 are proved by writing out each side in expanded form. The left side of Equation 8 is
$$ca_1 + ca_2 + \cdots + ca_n$$
The right side is
$$c(a_1 + a_2 + \cdots + a_n)$$
These are equal by the distributive property. The other formulas are discussed in Appendix E.

EXAMPLE 2
(a) Evaluate the Riemann sum for $f(x) = x^3 - 6x$ taking the sample points to be right endpoints and $a = 0$, $b = 3$, and $n = 6$.

(b) Evaluate $\int_0^3 (x^3 - 6x) \, dx$.

Try more problems like this one.
Resources / Module 6
/ What Is Area?
/ Problems and Tests

SOLUTION
(a) With $n = 6$ the interval width is

$$\Delta x = \frac{b-a}{n} = \frac{3-0}{6} = \frac{1}{2}$$

330 ‖ CHAPTER 5 INTEGRALS

and the right endpoints are $x_1 = 0.5$, $x_2 = 1.0$, $x_3 = 1.5$, $x_4 = 2.0$, $x_5 = 2.5$, and $x_6 = 3.0$. So the Riemann sum is

$$R_6 = \sum_{i=1}^{6} f(x_i)\,\Delta x$$
$$= f(0.5)\,\Delta x + f(1.0)\,\Delta x + f(1.5)\,\Delta x + f(2.0)\,\Delta x + f(2.5)\,\Delta x + f(3.0)\,\Delta x$$
$$= \tfrac{1}{2}(-2.875 - 5 - 5.625 - 4 + 0.625 + 9)$$
$$= -3.9375$$

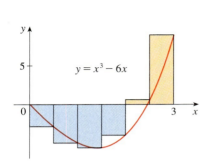

FIGURE 5

Notice that f is not a positive function and so the Riemann sum does not represent a sum of areas of rectangles. But it does represent the sum of the areas of the gold rectangles (above the x-axis) minus the sum of the areas of the blue rectangles (below the x-axis) in Figure 5.

(b) With n subintervals we have

$$\Delta x = \frac{b-a}{n} = \frac{3}{n}$$

Thus $x_0 = 0$, $x_1 = 3/n$, $x_2 = 6/n$, $x_3 = 9/n$, and, in general, $x_i = 3i/n$. Since we are using right endpoints, we can use Equation 3:

$$\int_0^3 (x^3 - 6x)\,dx = \lim_{n \to \infty} \sum_{i=1}^{n} f(x_i)\,\Delta x = \lim_{n \to \infty} \sum_{i=1}^{n} f\!\left(\frac{3i}{n}\right)\frac{3}{n}$$
$$= \lim_{n \to \infty} \frac{3}{n} \sum_{i=1}^{n} \left[\left(\frac{3i}{n}\right)^3 - 6\!\left(\frac{3i}{n}\right)\right] \qquad \text{(Equation 8 with } c = 3/n\text{)}$$
$$= \lim_{n \to \infty} \frac{3}{n} \sum_{i=1}^{n} \left[\frac{27}{n^3} i^3 - \frac{18}{n} i\right]$$
$$= \lim_{n \to \infty} \left[\frac{81}{n^4} \sum_{i=1}^{n} i^3 - \frac{54}{n^2} \sum_{i=1}^{n} i\right] \qquad \text{(Equations 10 and 8)}$$
$$= \lim_{n \to \infty} \left\{\frac{81}{n^4}\left[\frac{n(n+1)}{2}\right]^2 - \frac{54}{n^2}\frac{n(n+1)}{2}\right\}$$
$$= \lim_{n \to \infty} \left[\frac{81}{4}\left(1 + \frac{1}{n}\right)^2 - 27\!\left(1 + \frac{1}{n}\right)\right]$$
$$= \frac{81}{4} - 27 = -\frac{27}{4} = -6.75$$

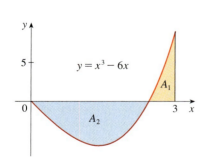

FIGURE 6

$\int_0^3 (x^3 - 6x)\,dx = A_1 - A_2 = -6.75$

This integral can't be interpreted as an area because f takes on both positive and negative values. But it can be interpreted as the difference of areas $A_1 - A_2$, where A_1 and A_2 are shown in Figure 6.

Figure 7 illustrates the calculation by showing the positive and negative terms in the right Riemann sum R_n for $n = 40$. The values in the table show the Riemann sums approaching the exact value of the integral, -6.75, as $n \to \infty$.

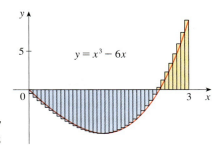

FIGURE 7
$R_{40} \approx -6.3998$

n	R_n
40	-6.3998
100	-6.6130
500	-6.7229
1000	-6.7365
5000	-6.7473

A much simpler method for evaluating the integral in Example 2 will be given in Section 5.3.

|||| Because $f(x) = x^4$ is positive, the integral in Example 3 represents the area shown in Figure 8.

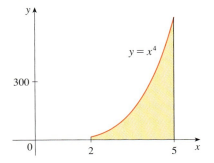

FIGURE 8

EXAMPLE 3
(a) Set up an expression for $\int_2^5 x^4\, dx$ as a limit of sums.
(b) Use a computer algebra system to evaluate the expression.

SOLUTION
(a) Here we have $f(x) = x^4$, $a = 2$, $b = 5$, and

$$\Delta x = \frac{b - a}{n} = \frac{3}{n}$$

So $x_0 = 2$, $x_1 = 2 + 3/n$, $x_2 = 2 + 6/n$, $x_3 = 2 + 9/n$, and

$$x_i = 2 + \frac{3i}{n}$$

From Equation 3, we get

$$\int_2^5 x^4\, dx = \lim_{n \to \infty} \sum_{i=1}^n f(x_i)\, \Delta x = \lim_{n \to \infty} \sum_{i=1}^n f\left(2 + \frac{3i}{n}\right) \frac{3}{n}$$

$$= \lim_{n \to \infty} \frac{3}{n} \sum_{i=1}^n \left(2 + \frac{3i}{n}\right)^4$$

(b) If we ask a computer algebra system to evaluate the sum and simplify, we obtain

$$\sum_{i=1}^n \left(2 + \frac{3i}{n}\right)^4 = \frac{2062n^4 + 3045n^3 + 1170n^2 - 27}{10n^3}$$

Now we ask the CAS to evaluate the limit:

$$\int_2^5 x^4\, dx = \lim_{n \to \infty} \frac{3}{n} \sum_{i=1}^n \left(2 + \frac{3i}{n}\right)^4 = \lim_{n \to \infty} \frac{3(2062n^4 + 3045n^3 + 1170n^2 - 27)}{10n^4}$$

$$= \frac{3(2062)}{10} = \frac{3093}{5} = 618.6$$

We will learn a much easier method for the evaluation of integrals in the next section.

EXAMPLE 4 Evaluate the following integrals by interpreting each in terms of areas.

(a) $\int_0^1 \sqrt{1-x^2}\, dx$

(b) $\int_0^3 (x-1)\, dx$

SOLUTION
(a) Since $f(x) = \sqrt{1-x^2} \geq 0$, we can interpret this integral as the area under the curve $y = \sqrt{1-x^2}$ from 0 to 1. But, since $y^2 = 1 - x^2$, we get $x^2 + y^2 = 1$, which shows that the graph of f is the quarter-circle with radius 1 in Figure 9. Therefore

$$\int_0^1 \sqrt{1-x^2}\, dx = \tfrac{1}{4}\pi(1)^2 = \frac{\pi}{4}$$

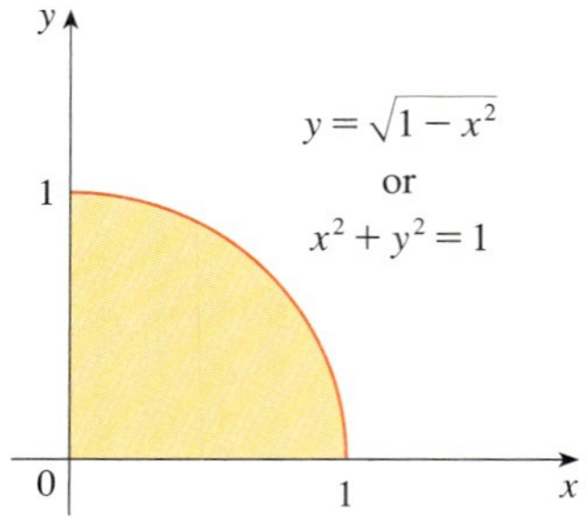

(In Section 8.3 we will be able to *prove* that the area of a circle of radius r is πr^2.)

(b) The graph of $y = x - 1$ is the line with slope 1 shown in Figure 10. We compute the integral as the difference of the areas of the two triangles:

$$\int_0^3 (x-1)\, dx = A_1 - A_2 = \tfrac{1}{2}(2 \cdot 2) - \tfrac{1}{2}(1 \cdot 1) = 1.5$$

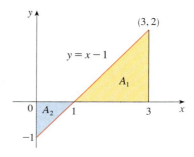

FIGURE 9

FIGURE 10

The Midpoint Rule

We often choose the sample point x_i^* to be the right endpoint of the ith subinterval because it is convenient for computing the limit. But if the purpose is to find an *approximation* to an integral, it is usually better to choose x_i^* to be the midpoint of the interval, which we denote by $\bar{x}_i$. Any Riemann sum is an approximation to an integral, but if we use midpoints we get the following approximation.

Module 5.1/5.2/8.7 shows how the Midpoint Rule estimates improve as n increases.

Midpoint Rule

$$\int_a^b f(x)\, dx \approx \sum_{i=1}^n f(\bar{x}_i)\, \Delta x = \Delta x\, [f(\bar{x}_1) + \cdots + f(\bar{x}_n)]$$

where $\quad \Delta x = \dfrac{b-a}{n}$

and $\quad \bar{x}_i = \tfrac{1}{2}(x_{i-1} + x_i) =$ midpoint of $[x_{i-1}, x_i]$

EXAMPLE 5 Use the Midpoint Rule with $n = 5$ to approximate $\int_1^2 \dfrac{1}{x}\, dx$.

SOLUTION The endpoints of the five subintervals are 1, 1.2, 1.4, 1.6, 1.8, and 2.0, so the midpoints are 1.1, 1.3, 1.5, 1.7, and 1.9. The width of the subintervals is $\Delta x = (2-1)/5 = \frac{1}{5}$, so the Midpoint Rule gives

$$\int_1^2 \frac{1}{x}\,dx \approx \Delta x [f(1.1) + f(1.3) + f(1.5) + f(1.7) + f(1.9)]$$

$$= \frac{1}{5}\left(\frac{1}{1.1} + \frac{1}{1.3} + \frac{1}{1.5} + \frac{1}{1.7} + \frac{1}{1.9}\right)$$

$$\approx 0.691908$$

Since $f(x) = 1/x > 0$ for $1 \le x \le 2$, the integral represents an area, and the approximation given by the Midpoint Rule is the sum of the areas of the rectangles shown in Figure 11.

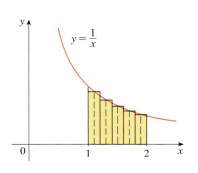

FIGURE 11

At the moment we don't know how accurate the approximation in Example 5 is, but in Section 8.7 we will learn a method for estimating the error involved in using the Midpoint Rule. At that time we will discuss other methods for approximating definite integrals.

If we apply the Midpoint Rule to the integral in Example 2, we get the picture in Figure 12. The approximation $M_{40} \approx -6.7563$ is much closer to the true value -6.75 than the right endpoint approximation, $R_{40} \approx -6.3998$, shown in Figure 7.

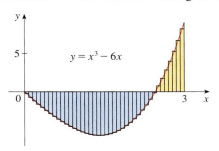

FIGURE 12
$M_{40} \approx -6.7563$

Properties of the Definite Integral

When we defined the definite integral $\int_a^b f(x)\,dx$, we implicitly assumed that $a < b$. But the definition as a limit of Riemann sums makes sense even if $a > b$. Notice that if we reverse a and b, then Δx changes from $(b-a)/n$ to $(a-b)/n$. Therefore

$$\int_b^a f(x)\,dx = -\int_a^b f(x)\,dx$$

If $a = b$, then $\Delta x = 0$ and so

$$\int_a^a f(x)\,dx = 0$$

We now develop some basic properties of integrals that will help us to evaluate integrals in a simple manner. We assume that f and g are continuous functions.

334 CHAPTER 5 INTEGRALS

Properties of the Integral

1. $\int_a^b c\,dx = c(b-a)$, where c is any constant

2. $\int_a^b [f(x) + g(x)]\,dx = \int_a^b f(x)\,dx + \int_a^b g(x)\,dx$

3. $\int_a^b cf(x)\,dx = c\int_a^b f(x)\,dx$, where c is any constant

4. $\int_a^b [f(x) - g(x)]\,dx = \int_a^b f(x)\,dx - \int_a^b g(x)\,dx$

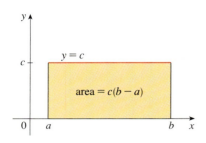

FIGURE 13
$\int_a^b c\,dx = c(b-a)$

Property 1 says that the integral of a constant function $f(x) = c$ is the constant times the length of the interval. If $c > 0$ and $a < b$, this is to be expected because $c(b - a)$ is the area of the shaded rectangle in Figure 13.

Property 2 says that the integral of a sum is the sum of the integrals. For positive functions it says that the area under $f + g$ is the area under f plus the area under g. Figure 14 helps us understand why this is true: In view of how graphical addition works, the corresponding vertical line segments have equal height.

In general, Property 2 follows from Equation 3 and the fact that the limit of a sum is the sum of the limits:

$$\int_a^b [f(x) + g(x)]\,dx = \lim_{n\to\infty} \sum_{i=1}^n [f(x_i) + g(x_i)]\,\Delta x$$

$$= \lim_{n\to\infty} \left[\sum_{i=1}^n f(x_i)\,\Delta x + \sum_{i=1}^n g(x_i)\,\Delta x\right]$$

$$= \lim_{n\to\infty} \sum_{i=1}^n f(x_i)\,\Delta x + \lim_{n\to\infty} \sum_{i=1}^n g(x_i)\,\Delta x$$

$$= \int_a^b f(x)\,dx + \int_a^b g(x)\,dx$$

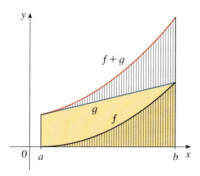

FIGURE 14
$\int_a^b [f(x) + g(x)]\,dx =$
$\int_a^b f(x)\,dx + \int_a^b g(x)\,dx$

|||| Property 3 seems intuitively reasonable because we know that multiplying a function by a positive number c stretches or shrinks its graph vertically by a factor of c. So it stretches or shrinks each approximating rectangle by a factor c and therefore it has the effect of multiplying the area by c.

Property 3 can be proved in a similar manner and says that the integral of a constant times a function is the constant times the integral of the function. In other words, a constant (but *only* a constant) can be taken in front of an integral sign. Property 4 is proved by writing $f - g = f + (-g)$ and using Properties 2 and 3 with $c = -1$.

EXAMPLE 6 Use the properties of integrals to evaluate $\int_0^1 (4 + 3x^2)\,dx$.

SOLUTION Using Properties 2 and 3 of integrals, we have

$$\int_0^1 (4 + 3x^2)\,dx = \int_0^1 4\,dx + \int_0^1 3x^2\,dx = \int_0^1 4\,dx + 3\int_0^1 x^2\,dx$$

We know from Property 1 that

$$\int_0^1 4\,dx = 4(1-0) = 4$$

and we found in Example 2 in Section 5.1 that $\int_0^1 x^2 \, dx = \frac{1}{3}$. So

$$\int_0^1 (4 + 3x^2) \, dx = \int_0^1 4 \, dx + 3 \int_0^1 x^2 \, dx$$

$$= 4 + 3 \cdot \tfrac{1}{3} = 5$$

The next property tells us how to combine integrals of the same function over adjacent intervals:

5. $\quad \int_a^c f(x) \, dx + \int_c^b f(x) \, dx = \int_a^b f(x) \, dx$

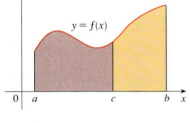

FIGURE 15

This is not easy to prove in general, but for the case where $f(x) \geq 0$ and $a < c < b$ Property 5 can be seen from the geometric interpretation in Figure 15: The area under $y = f(x)$ from a to c plus the area from c to b is equal to the total area from a to b.

EXAMPLE 7 If it is known that $\int_0^{10} f(x) \, dx = 17$ and $\int_0^8 f(x) \, dx = 12$, find $\int_8^{10} f(x) \, dx$.

SOLUTION By Property 5, we have

$$\int_0^8 f(x) \, dx + \int_8^{10} f(x) \, dx = \int_0^{10} f(x) \, dx$$

so

$$\int_8^{10} f(x) \, dx = \int_0^{10} f(x) \, dx - \int_0^8 f(x) \, dx = 17 - 12 = 5$$

Notice that Properties 1–5 are true whether $a < b$, $a = b$, or $a > b$. The following properties, in which we compare sizes of functions and sizes of integrals, are true only if $a \leq b$.

Comparison Properties of the Integral

6. If $f(x) \geq 0$ for $a \leq x \leq b$, then $\int_a^b f(x) \, dx \geq 0$.

7. If $f(x) \geq g(x)$ for $a \leq x \leq b$, then $\int_a^b f(x) \, dx \geq \int_a^b g(x) \, dx$.

8. If $m \leq f(x) \leq M$ for $a \leq x \leq b$, then

$$m(b - a) \leq \int_a^b f(x) \, dx \leq M(b - a)$$

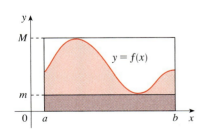

FIGURE 16

If $f(x) \geq 0$, then $\int_a^b f(x) \, dx$ represents the area under the graph of f, so the geometric interpretation of Property 6 is simply that areas are positive. But the property can be proved from the definition of an integral (Exercise 64). Property 7 says that a bigger function has a bigger integral. It follows from Properties 6 and 4 because $f - g \geq 0$.

Property 8 is illustrated by Figure 16 for the case where $f(x) \geq 0$. If f is continuous we could take m and M to be the absolute minimum and maximum values of f on the interval $[a, b]$. In this case Property 8 says that the area under the graph of f is greater than the area of the rectangle with height m and less than the area of the rectangle with height M.

Proof of Property 8 Since $m \leq f(x) \leq M$, Property 7 gives

$$\int_a^b m \, dx \leq \int_a^b f(x) \, dx \leq \int_a^b M \, dx$$

Using Property 1 to evaluate the integrals on the left and right sides, we obtain

$$m(b - a) \leq \int_a^b f(x) \, dx \leq M(b - a)$$

Property 8 is useful when all we want is a rough estimate of the size of an integral without going to the bother of using the Midpoint Rule.

EXAMPLE 8 Use Property 8 to estimate $\int_1^4 \sqrt{x} \, dx$.

SOLUTION Since $f(x) = \sqrt{x}$ is an increasing function, its absolute minimum on $[1, 4]$ is $m = f(1) = 1$ and its absolute maximum on $[1, 4]$ is $M = f(4) = \sqrt{4} = 2$. Thus, Property 8 gives

$$1(4 - 1) \leq \int_1^4 \sqrt{x} \, dx \leq 2(4 - 1)$$

or

$$3 \leq \int_1^4 \sqrt{x} \, dx \leq 6$$

The result of Example 8 is illustrated in Figure 17. The area under $y = \sqrt{x}$ from 1 to 4 is greater than the area of the lower rectangle and less than the area of the large rectangle.

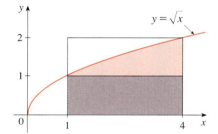

FIGURE 17

5.2 Exercises

1. Evaluate the Riemann sum for $f(x) = 2 - x^2$, $0 \leq x \leq 2$, with four subintervals, taking the sample points to be right endpoints. Explain, with the aid of a diagram, what the Riemann sum represents.

2. If $f(x) = 3x - 7$, $0 \leq x \leq 3$, evaluate the Riemann sum with $n = 6$, taking the sample points to be left endpoints. What does the Riemann sum represent? Illustrate with a diagram.

3. If $f(x) = \sqrt{x} - 2$, $1 \leq x \leq 6$, find the Riemann sum with $n = 5$ correct to six decimal places, taking the sample points to be midpoints. What does the Riemann sum represent? Illustrate with a diagram.

4. (a) Find the Riemann sum for $f(x) = x - 2 \sin 2x$, $0 \leq x \leq 3$, with six terms, taking the sample points to be right endpoints. (Give your answer correct to six decimal places.) Explain what the Riemann sum represents with the aid of a sketch.
   (b) Repeat part (a) with midpoints as the sample points.

5. The graph of a function f is given. Estimate $\int_0^8 f(x) \, dx$ using four subintervals with (a) right endpoints, (b) left endpoints, and (c) midpoints.

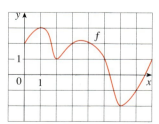

6. The graph of g is shown. Estimate $\int_{-3}^{3} g(x)\, dx$ with six subintervals using (a) right endpoints, (b) left endpoints, and (c) midpoints.

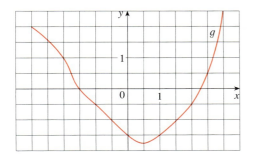

7. A table of values of an increasing function f is shown. Use the table to find lower and upper estimates for $\int_{0}^{25} f(x)\, dx$.

x	0	5	10	15	20	25
$f(x)$	-42	-37	-25	-6	15	36

8. The table gives the values of a function obtained from an experiment. Use them to estimate $\int_{0}^{6} f(x)\, dx$ using three equal subintervals with (a) right endpoints, (b) left endpoints, and (c) midpoints. If the function is known to be a decreasing function, can you say whether your estimates are less than or greater than the exact value of the integral?

x	0	1	2	3	4	5	6
$f(x)$	9.3	9.0	8.3	6.5	2.3	-7.6	-10.5

9–12 ▪ Use the Midpoint Rule with the given value of n to approximate the integral. Round the answer to four decimal places.

9. $\int_{2}^{10} \sqrt{x^3 + 1}\, dx, \quad n = 4$

10. $\int_{0}^{\pi} \sec(x/3)\, dx, \quad n = 6$

11. $\int_{0}^{1} \sin(x^2)\, dx, \quad n = 5$

12. $\int_{1}^{5} \dfrac{x - 1}{x + 1}\, dx, \quad n = 4$

CAS 13. If you have a CAS that evaluates midpoint approximations and graphs the corresponding rectangles (use `middlesum` and `middlebox` commands in Maple), check the answer to Exercise 11 and illustrate with a graph. Then repeat with $n = 10$ and $n = 20$.

14. With a programmable calculator or computer (see the instructions for Exercise 7 in Section 5.1), compute the left and right Riemann sums for the function $f(x) = \sin(x^2)$ on the interval $[0, 1]$ with $n = 100$. Explain why these estimates show that

$$0.306 < \int_{0}^{1} \sin(x^2)\, dx < 0.315$$

Deduce that the approximation using the Midpoint Rule with $n = 5$ in Exercise 11 is accurate to two decimal places.

15. Use a calculator or computer to make a table of values of right Riemann sums R_n for the integral $\int_{0}^{\pi} \sin x\, dx$ with $n = 5, 10, 50$, and 100. What value do these numbers appear to be approaching?

16. Use a calculator or computer to make a table of values of left and right Riemann sums L_n and R_n for the integral $\int_{0}^{2} \sqrt{1 + x^4}\, dx$ with $n = 5, 10, 50$, and 100. Between what two numbers must the value of the integral lie? Can you make a similar statement for the integral $\int_{-1}^{2} \sqrt{1 + x^4}\, dx$? Explain.

17–20 ▪ Express the limit as a definite integral on the given interval.

17. $\lim\limits_{n \to \infty} \sum\limits_{i=1}^{n} x_i \sin x_i\, \Delta x, \quad [0, \pi]$

18. $\lim\limits_{n \to \infty} \sum\limits_{i=1}^{n} \dfrac{x_i}{1 + x_i}\, \Delta x, \quad [1, 5]$

19. $\lim\limits_{n \to \infty} \sum\limits_{i=1}^{n} \sqrt{2x_i^* + (x_i^*)^2}\, \Delta x, \quad [1, 8]$

20. $\lim\limits_{n \to \infty} \sum\limits_{i=1}^{n} [4 - 3(x_i^*)^2 + 6(x_i^*)^5]\, \Delta x, \quad [0, 2]$

21–25 ▪ Use the form of the definition of the integral given in Equation 3 to evaluate the integral.

21. $\int_{-1}^{5} (1 + 3x)\, dx$

22. $\int_{1}^{4} (x^2 + 2x - 5)\, dx$

23. $\int_{0}^{2} (2 - x^2)\, dx$

24. $\int_{0}^{5} (1 + 2x^3)\, dx$

25. $\int_{1}^{2} x^3\, dx$

26. (a) Find an approximation to the integral $\int_{0}^{4} (x^2 - 3x)\, dx$ using a Riemann sum with right endpoints and $n = 8$.
(b) Draw a diagram like Figure 3 to illustrate the approximation in part (a).
(c) Use Equation 3 to evaluate $\int_{0}^{4} (x^2 - 3x)\, dx$.
(d) Interpret the integral in part (c) as a difference of areas and illustrate with a diagram like Figure 4.

27. Prove that $\int_{a}^{b} x\, dx = \dfrac{b^2 - a^2}{2}$.

28. Prove that $\int_{a}^{b} x^2\, dx = \dfrac{b^3 - a^3}{3}$.

29–30 ▪ Express the integral as a limit of Riemann sums. Do not evaluate the limit.

29. $\int_{2}^{6} \dfrac{x}{1 + x^5}\, dx$

30. $\int_{0}^{2\pi} x^2 \sin x\, dx$

CAS 31–32 Express the integral as a limit of sums. Then evaluate, using a computer algebra system to find both the sum and the limit.

31. $\int_0^\pi \sin 5x \, dx$

32. $\int_2^{10} x^6 \, dx$

33. The graph of f is shown. Evaluate each integral by interpreting it in terms of areas.

(a) $\int_0^2 f(x) \, dx$ (b) $\int_0^5 f(x) \, dx$

(c) $\int_5^7 f(x) \, dx$ (d) $\int_0^9 f(x) \, dx$

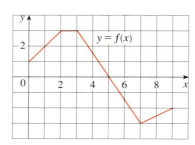

34. The graph of g consists of two straight lines and a semicircle. Use it to evaluate each integral.

(a) $\int_0^2 g(x) \, dx$ (b) $\int_2^6 g(x) \, dx$ (c) $\int_0^7 g(x) \, dx$

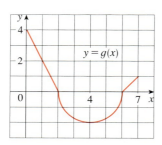

35–40 Evaluate the integral by interpreting it in terms of areas.

35. $\int_0^3 (\tfrac{1}{2}x - 1) \, dx$

36. $\int_{-2}^2 \sqrt{4 - x^2} \, dx$

37. $\int_{-3}^0 (1 + \sqrt{9 - x^2}) \, dx$

38. $\int_{-1}^3 (3 - 2x) \, dx$

39. $\int_{-1}^2 |x| \, dx$

40. $\int_0^{10} |x - 5| \, dx$

41. Given that $\int_4^9 \sqrt{x} \, dx = \tfrac{38}{3}$, what is $\int_9^4 \sqrt{t} \, dt$?

42. Evaluate $\int_1^1 x^2 \cos x \, dx$.

43. In Example 2 in Section 5.1 we showed that $\int_0^1 x^2 \, dx = \tfrac{1}{3}$. Use this fact and the properties of integrals to evaluate $\int_0^1 (5 - 6x^2) \, dx$.

44. Use the properties of integrals and the result of Example 3 to evaluate $\int_2^5 (1 + 3x^4) \, dx$.

45. Use the results of Exercises 27 and 28 and the properties of integrals to evaluate $\int_1^4 (2x^2 - 3x + 1) \, dx$.

46. Use the result of Exercise 27 and the fact that $\int_0^{\pi/2} \cos x \, dx = 1$ (from Exercise 25 in Section 5.1), together with the properties of integrals, to evaluate $\int_0^{\pi/2} (2 \cos x - 5x) \, dx$.

47. Write as a single integral in the form $\int_a^b f(x) \, dx$:

$$\int_{-2}^2 f(x) \, dx + \int_2^5 f(x) \, dx - \int_{-2}^{-1} f(x) \, dx$$

48. If $\int_1^5 f(x) \, dx = 12$ and $\int_4^5 f(x) \, dx = 3.6$, find $\int_1^4 f(x) \, dx$.

49. If $\int_0^9 f(x) \, dx = 37$ and $\int_0^9 g(x) \, dx = 16$, find $\int_0^9 [2f(x) + 3g(x)] \, dx$.

50. Find $\int_0^5 f(x) \, dx$ if

$$f(x) = \begin{cases} 3 & \text{for } x < 3 \\ x & \text{for } x \geq 3 \end{cases}$$

51–54 Use the properties of integrals to verify the inequality without evaluating the integrals.

51. $\int_0^{\pi/4} \sin^3 x \, dx \leq \int_0^{\pi/4} \sin^2 x \, dx$

52. $\int_1^2 \sqrt{5 - x} \, dx \geq \int_1^2 \sqrt{x + 1} \, dx$

53. $2 \leq \int_{-1}^1 \sqrt{1 + x^2} \, dx \leq 2\sqrt{2}$

54. $\dfrac{\pi}{6} \leq \int_{\pi/6}^{\pi/2} \sin x \, dx \leq \dfrac{\pi}{3}$

55–60 Use Property 8 to estimate the value of the integral.

55. $\int_1^2 \dfrac{1}{x} \, dx$

56. $\int_0^2 \sqrt{x^3 + 1} \, dx$

57. $\int_{\pi/4}^{\pi/3} \tan x \, dx$

58. $\int_0^2 (x^3 - 3x + 3) \, dx$

59. $\int_{-1}^1 \sqrt{1 + x^4} \, dx$

60. $\int_{\pi/4}^{3\pi/4} \sin^2 x \, dx$

61–62 Use properties of integrals, together with Exercises 27 and 28, to prove the inequality.

61. $\int_1^3 \sqrt{x^4 + 1} \, dx \geq \dfrac{26}{3}$

62. $\int_0^{\pi/2} x \sin x \, dx \leq \dfrac{\pi^2}{8}$

63. Prove Property 3 of integrals.

64. Prove Property 6 of integrals.

65. If f is continuous on $[a, b]$, show that

$$\left| \int_a^b f(x) \, dx \right| \leq \int_a^b |f(x)| \, dx$$

[*Hint:* $-|f(x)| \leq f(x) \leq |f(x)|$.]

66. Use the result of Exercise 65 to show that

$$\left| \int_0^{2\pi} f(x) \sin 2x \, dx \right| \leq \int_0^{2\pi} |f(x)| \, dx$$

67–68 ▮▮▮▮ Express the limit as a definite integral.

67. $\lim\limits_{n \to \infty} \sum\limits_{i=1}^{n} \dfrac{i^4}{n^5}$ [*Hint:* Consider $f(x) = x^4$.]

68. $\lim\limits_{n \to \infty} \dfrac{1}{n} \sum\limits_{i=1}^{n} \dfrac{1}{1 + (i/n)^2}$

69. Find $\int_1^2 x^{-2} \, dx$. *Hint:* Choose x_i^* to be the geometric mean of x_{i-1} and x_i (that is, $x_i^* = \sqrt{x_{i-1} x_i}$) and use the identity

$$\frac{1}{m(m+1)} = \frac{1}{m} - \frac{1}{m+1}$$

DISCOVERY PROJECT

Area Functions

1. (a) Draw the line $y = 2t + 1$ and use geometry to find the area under this line, above the t-axis, and between the vertical lines $t = 1$ and $t = 3$.
   (b) If $x > 1$, let $A(x)$ be the area of the region that lies under the line $y = 2t + 1$ between $t = 1$ and $t = x$. Sketch this region and use geometry to find an expression for $A(x)$.
   (c) Differentiate the area function $A(x)$. What do you notice?

2. (a) If $x \geq -1$, let

$$A(x) = \int_{-1}^{x} (1 + t^2) \, dt$$

$A(x)$ represents the area of a region. Sketch that region.
   (b) Use the result of Exercise 28 in Section 5.2 to find an expression for $A(x)$.
   (c) Find $A'(x)$. What do you notice?
   (d) If $x \geq -1$ and h is a small positive number, then $A(x + h) - A(x)$ represents the area of a region. Describe and sketch the region.
   (e) Draw a rectangle that approximates the region in part (d). By comparing the areas of these two regions, show that

$$\frac{A(x + h) - A(x)}{h} \approx 1 + x^2$$

   (f) Use part (e) to give an intuitive explanation for the result of part (c).

3. (a) Draw the graph of the function $f(x) = \cos(x^2)$ in the viewing rectangle $[0, 2]$ by $[-1.25, 1.25]$.
   (b) If we define a new function g by

$$g(x) = \int_0^x \cos(t^2) \, dt$$

then $g(x)$ is the area under the graph of f from 0 to x [until $f(x)$ becomes negative, at which point $g(x)$ becomes a difference of areas]. Use part (a) to determine the value of x at which $g(x)$ starts to decrease. [Unlike the integral in Problem 2, it is impossible to evaluate the integral defining g to obtain an explicit expression for $g(x)$.]
   (c) Use the integration command on your calculator or computer to estimate $g(0.2)$, $g(0.4)$, $g(0.6), \ldots, g(1.8), g(2)$. Then use these values to sketch a graph of g.
   (d) Use your graph of g from part (c) to sketch the graph of g' using the interpretation of $g'(x)$ as the slope of a tangent line. How does the graph of g' compare with the graph of f?

4. Suppose f is a continuous function on the interval $[a, b]$ and we define a new function g by the equation

$$g(x) = \int_a^x f(t)\, dt$$

Based on your results in Problems 1–3, conjecture an expression for $g'(x)$.

5.3 The Fundamental Theorem of Calculus

The Fundamental Theorem of Calculus is appropriately named because it establishes a connection between the two branches of calculus: differential calculus and integral calculus. Differential calculus arose from the tangent problem, whereas integral calculus arose from a seemingly unrelated problem, the area problem. Newton's teacher at Cambridge, Isaac Barrow (1630–1677), discovered that these two problems are actually closely related. In fact, he realized that differentiation and integration are inverse processes. The Fundamental Theorem of Calculus gives the precise inverse relationship between the derivative and the integral. It was Newton and Leibniz who exploited this relationship and used it to develop calculus into a systematic mathematical method. In particular, they saw that the Fundamental Theorem enabled them to compute areas and integrals very easily without having to compute them as limits of sums as we did in Sections 5.1 and 5.2.

Investigate the area function interactively.
Resources / Module 6
/ Areas and Derivatives
/ Area as a Function

The first part of the Fundamental Theorem deals with functions defined by an equation of the form

$$\boxed{1} \qquad g(x) = \int_a^x f(t)\, dt$$

where f is a continuous function on $[a, b]$ and x varies between a and b. Observe that g depends only on x, which appears as the variable upper limit in the integral. If x is a fixed number, then the integral $\int_a^x f(t)\, dt$ is a definite number. If we then let x vary, the number $\int_a^x f(t)\, dt$ also varies and defines a function of x denoted by $g(x)$.

If f happens to be a positive function, then $g(x)$ can be interpreted as the area under the graph of f from a to x, where x can vary from a to b. (Think of g as the "area so far" function; see Figure 1.)

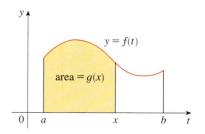

FIGURE 1

EXAMPLE 1 If f is the function whose graph is shown in Figure 2 and $g(x) = \int_0^x f(t)\, dt$, find the values of $g(0)$, $g(1)$, $g(2)$, $g(3)$, $g(4)$, and $g(5)$. Then sketch a rough graph of g.

SOLUTION First we notice that $g(0) = \int_0^0 f(t)\, dt = 0$. From Figure 3 we see that $g(1)$ is the area of a triangle:

$$g(1) = \int_0^1 f(t)\, dt = \tfrac{1}{2}(1 \cdot 2) = 1$$

To find $g(2)$ we add to $g(1)$ the area of a rectangle:

$$g(2) = \int_0^2 f(t)\, dt = \int_0^1 f(t)\, dt + \int_1^2 f(t)\, dt = 1 + (1 \cdot 2) = 3$$

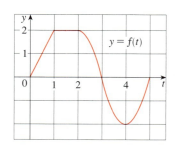

FIGURE 2

We estimate that the area under f from 2 to 3 is about 1.3, so

$$g(3) = g(2) + \int_2^3 f(t)\, dt \approx 3 + 1.3 = 4.3$$

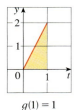
$g(1) = 1$

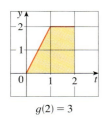
$g(2) = 3$

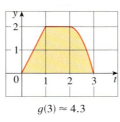
$g(3) \approx 4.3$

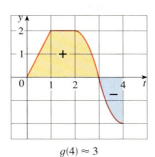
$g(4) \approx 3$

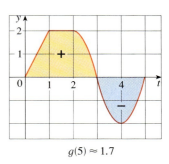
$g(5) \approx 1.7$

FIGURE 3

For $t > 3$, $f(t)$ is negative and so we start subtracting areas:

$$g(4) = g(3) + \int_3^4 f(t)\, dt \approx 4.3 + (-1.3) = 3.0$$

$$g(5) = g(4) + \int_4^5 f(t)\, dt \approx 3 + (-1.3) = 1.7$$

We use these values to sketch the graph of g in Figure 4. Notice that, because $f(t)$ is positive for $t < 3$, we keep adding area for $t < 3$ and so g is increasing up to $x = 3$, where it attains a maximum value. For $x > 3$, g decreases because $f(t)$ is negative.

If we take $f(t) = t$ and $a = 0$, then, using Exercise 27 in Section 5.2, we have

$$g(x) = \int_0^x t\, dt = \frac{x^2}{2}$$

Notice that $g'(x) = x$, that is, $g' = f$. In other words, if g is defined as the integral of f by Equation 1, then g turns out to be an antiderivative of f, at least in this case. And if we sketch the derivative of the function g shown in Figure 4 by estimating slopes of tangents, we get a graph like that of f in Figure 2. So we suspect that $g' = f$ in Example 1 too.

To see why this might be generally true we consider any continuous function f with $f(x) \geq 0$. Then $g(x) = \int_a^x f(t)\, dt$ can be interpreted as the area under the graph of f from a to x, as in Figure 1.

In order to compute $g'(x)$ from the definition of derivative we first observe that, for $h > 0$, $g(x + h) - g(x)$ is obtained by subtracting areas, so it is the area under the graph of f from x to $x + h$ (the gold area in Figure 5). For small h you can see from the figure that this area is approximately equal to the area of the rectangle with height $f(x)$ and width h:

$$g(x + h) - g(x) \approx h f(x)$$

so

$$\frac{g(x + h) - g(x)}{h} \approx f(x)$$

FIGURE 4

$g(x) = \int_a^x f(t)\, dt$

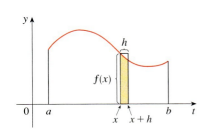

FIGURE 5

Intuitively, we therefore expect that

$$g'(x) = \lim_{h \to 0} \frac{g(x+h) - g(x)}{h} = f(x)$$

The fact that this is true, even when f is not necessarily positive, is the first part of the Fundamental Theorem of Calculus.

|||| We abbreviate the name of this theorem as FTC1. In words, it says that the derivative of a definite integral with respect to its upper limit is the integrand evaluated at the upper limit.

The Fundamental Theorem of Calculus, Part 1 If f is continuous on $[a, b]$, then the function g defined by

$$g(x) = \int_a^x f(t)\, dt \qquad a \le x \le b$$

is continuous on $[a, b]$ and differentiable on (a, b), and $g'(x) = f(x)$.

Proof If x and $x + h$ are in (a, b), then

$$g(x+h) - g(x) = \int_a^{x+h} f(t)\, dt - \int_a^x f(t)\, dt$$

$$= \left(\int_a^x f(t)\, dt + \int_x^{x+h} f(t)\, dt \right) - \int_a^x f(t)\, dt \qquad \text{(by Property 5)}$$

$$= \int_x^{x+h} f(t)\, dt$$

and so, for $h \ne 0$,

$$\boxed{2} \qquad \frac{g(x+h) - g(x)}{h} = \frac{1}{h} \int_x^{x+h} f(t)\, dt$$

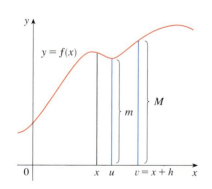

FIGURE 6

For now let us assume that $h > 0$. Since f is continuous on $[x, x + h]$, the Extreme Value Theorem says that there are numbers u and v in $[x, x + h]$ such that $f(u) = m$ and $f(v) = M$, where m and M are the absolute minimum and maximum values of f on $[x, x + h]$. (See Figure 6.)

By Property 8 of integrals, we have

$$mh \le \int_x^{x+h} f(t)\, dt \le Mh$$

that is,

$$f(u)h \le \int_x^{x+h} f(t)\, dt \le f(v)h$$

Since $h > 0$, we can divide this inequality by h:

$$f(u) \le \frac{1}{h} \int_x^{x+h} f(t)\, dt \le f(v)$$

Now we use Equation 2 to replace the middle part of this inequality:

$$\boxed{3} \qquad f(u) \le \frac{g(x+h) - g(x)}{h} \le f(v)$$

Inequality 3 can be proved in a similar manner for the case $h < 0$. (See Exercise 55.)

Module 5.3 provides visual evidence for FTC1.

Now we let $h \to 0$. Then $u \to x$ and $v \to x$, since u and v lie between x and $x + h$. Therefore

$$\lim_{h \to 0} f(u) = \lim_{u \to x} f(u) = f(x)$$

and

$$\lim_{h \to 0} f(v) = \lim_{v \to x} f(v) = f(x)$$

because f is continuous at x. We conclude, from (3) and the Squeeze Theorem, that

$$\boxed{4} \qquad g'(x) = \lim_{h \to 0} \frac{g(x + h) - g(x)}{h} = f(x)$$

If $x = a$ or b, then Equation 4 can be interpreted as a one-sided limit. Then Theorem 3.2.4 (modified for one-sided limits) shows that g is continuous on $[a, b]$.

Using Leibniz notation for derivatives, we can write FTC1 as

$$\boxed{5} \qquad \frac{d}{dx} \int_a^x f(t) \, dt = f(x)$$

when f is continuous. Roughly speaking, Equation 5 says that if we first integrate f and then differentiate the result, we get back to the original function f.

EXAMPLE 2 Find the derivative of the function $g(x) = \int_0^x \sqrt{1 + t^2} \, dt$.

SOLUTION Since $f(t) = \sqrt{1 + t^2}$ is continuous, Part 1 of the Fundamental Theorem of Calculus gives

$$g'(x) = \sqrt{1 + x^2}$$

EXAMPLE 3 Although a formula of the form $g(x) = \int_a^x f(t) \, dt$ may seem like a strange way of defining a function, books on physics, chemistry, and statistics are full of such functions. For instance, the **Fresnel function**

$$S(x) = \int_0^x \sin(\pi t^2/2) \, dt$$

is named after the French physicist Augustin Fresnel (1788–1827), who is famous for his works in optics. This function first appeared in Fresnel's theory of the diffraction of light waves, but more recently it has been applied to the design of highways.

Part 1 of the Fundamental Theorem tells us how to differentiate the Fresnel function:

$$S'(x) = \sin(\pi x^2/2)$$

This means that we can apply all the methods of differential calculus to analyze S (see Exercise 49).

Figure 7 shows the graphs of $f(x) = \sin(\pi x^2/2)$ and the Fresnel function $S(x) = \int_0^x f(t)\, dt$. A computer was used to graph S by computing the value of this integral for many values of x. It does indeed look as if $S(x)$ is the area under the graph of f from 0 to x [until $x \approx 1.4$ when $S(x)$ becomes a difference of areas]. Figure 8 shows a larger part of the graph of S.

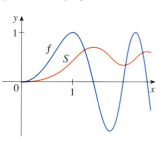

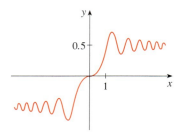

FIGURE 7
$f(x) = \sin(\pi x^2/2)$
$S(x) = \int_0^x \sin(\pi t^2/2)\, dt$

FIGURE 8
The Fresnel function $S(x) = \int_0^x \sin(\pi t^2/2)\, dt$

If we now start with the graph of S in Figure 7 and think about what its derivative should look like, it seems reasonable that $S'(x) = f(x)$. [For instance, S is increasing when $f(x) > 0$ and decreasing when $f(x) < 0$.] So this gives a visual confirmation of Part 1 of the Fundamental Theorem of Calculus.

EXAMPLE 4 Find $\dfrac{d}{dx} \int_1^{x^4} \sec t\, dt$.

SOLUTION Here we have to be careful to use the Chain Rule in conjunction with FTC1. Let $u = x^4$. Then

$$\frac{d}{dx} \int_1^{x^4} \sec t\, dt = \frac{d}{dx} \int_1^u \sec t\, dt$$

$$= \frac{d}{du}\left(\int_1^u \sec t\, dt\right) \frac{du}{dx} \quad \text{(by the Chain Rule)}$$

$$= \sec u\, \frac{du}{dx} \quad \text{(by FTC1)}$$

$$= \sec(x^4) \cdot 4x^3$$

In Section 5.2 we computed integrals from the definition as a limit of Riemann sums and we saw that this procedure is sometimes long and difficult. The second part of the Fundamental Theorem of Calculus, which follows easily from the first part, provides us with a much simpler method for the evaluation of integrals.

The Fundamental Theorem of Calculus, Part 2 If f is continuous on $[a, b]$, then

$$\int_a^b f(x)\, dx = F(b) - F(a)$$

where F is any antiderivative of f, that is, a function such that $F' = f$.

|||| We abbreviate this theorem as FTC2.

Proof   Let $g(x) = \int_a^x f(t) \, dt$. We know from Part 1 that $g'(x) = f(x)$; that is, g is an antiderivative of f. If F is any other antiderivative of f on $[a, b]$, then we know from Corollary 4.2.7 that F and g differ by a constant:

6 $$F(x) = g(x) + C$$

for $a < x < b$. But both F and g are continuous on $[a, b]$ and so, by taking limits of both sides of Equation 6 (as $x \to a^+$ and $x \to b^-$), we see that it also holds when $x = a$ and $x = b$.

If we put $x = a$ in the formula for $g(x)$, we get

$$g(a) = \int_a^a f(t) \, dt = 0$$

So, using Equation 6 with $x = b$ and $x = a$, we have

$$F(b) - F(a) = [g(b) + C] - [g(a) + C]$$
$$= g(b) - g(a) = g(b) = \int_a^b f(t) \, dt$$

Part 2 of the Fundamental Theorem states that if we know an antiderivative F of f, then we can evaluate $\int_a^b f(x) \, dx$ simply by subtracting the values of F at the endpoints of the interval $[a, b]$. It's very surprising that $\int_a^b f(x) \, dx$, which was defined by a complicated procedure involving all of the values of $f(x)$ for $a \le x \le b$, can be found by knowing the values of $F(x)$ at only two points, a and b.

Although the theorem may be surprising at first glance, it becomes plausible if we interpret it in physical terms. If $v(t)$ is the velocity of an object and $s(t)$ is its position at time t, then $v(t) = s'(t)$, so s is an antiderivative of v. In Section 5.1 we considered an object that always moves in the positive direction and made the guess that the area under the velocity curve is equal to the distance traveled. In symbols:

$$\int_a^b v(t) \, dt = s(b) - s(a)$$

That is exactly what FTC2 says in this context.

EXAMPLE 5   Evaluate the integral $\int_{-2}^{1} x^3 \, dx$.

SOLUTION   The function $f(x) = x^3$ is continuous on $[-2, 1]$ and we know from Section 4.10 that an antiderivative is $F(x) = \frac{1}{4}x^4$, so Part 2 of the Fundamental Theorem gives

$$\int_{-2}^{1} x^3 \, dx = F(1) - F(-2) = \tfrac{1}{4}(1)^4 - \tfrac{1}{4}(-2)^4 = -\tfrac{15}{4}$$

Notice that FTC2 says we can use *any* antiderivative F of f. So we may as well use the simplest one, namely $F(x) = \frac{1}{4}x^4$, instead of $\frac{1}{4}x^4 + 7$ or $\frac{1}{4}x^4 + C$.

We often use the notation

$$F(x) \Big]_a^b = F(b) - F(a)$$

So the equation of FTC2 can be written as

$$\int_a^b f(x)\,dx = F(x)\Big]_a^b \qquad \text{where} \qquad F' = f$$

Other common notations are $F(x)\big|_a^b$ and $[F(x)]_a^b$.

EXAMPLE 6 Find the area under the parabola $y = x^2$ from 0 to 1.

SOLUTION An antiderivative of $f(x) = x^2$ is $F(x) = \tfrac{1}{3}x^3$. The required area A is found using Part 2 of the Fundamental Theorem:

$$A = \int_0^1 x^2\,dx = \frac{x^3}{3}\bigg]_0^1 = \frac{1^3}{3} - \frac{0^3}{3} = \frac{1}{3}$$

|||| In applying the Fundamental Theorem we use a particular antiderivative F of f. It is not necessary to use the most general antiderivative.

If you compare the calculation in Example 6 with the one in Example 2 in Section 5.1, you will see that the Fundamental Theorem gives a *much* shorter method.

EXAMPLE 7 Find the area under the cosine curve from 0 to b, where $0 \leq b \leq \pi/2$.

SOLUTION Since an antiderivative of $f(x) = \cos x$ is $F(x) = \sin x$, we have

$$A = \int_0^b \cos x\,dx = \sin x\big]_0^b = \sin b - \sin 0 = \sin b$$

In particular, taking $b = \pi/2$, we have proved that the area under the cosine curve from 0 to $\pi/2$ is $\sin(\pi/2) = 1$. (See Figure 9.)

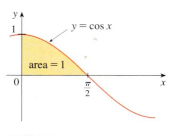

FIGURE 9

When the French mathematician Gilles de Roberval first found the area under the sine and cosine curves in 1635, this was a very challenging problem that required a great deal of ingenuity. If we didn't have the benefit of the Fundamental Theorem, we would have to compute a difficult limit of sums using obscure trigonometric identities (or a computer algebra system as in Exercise 25 in Section 5.1). It was even more difficult for Roberval because the apparatus of limits had not been invented in 1635. But in the 1660s and 1670s, when the Fundamental Theorem was discovered by Barrow and exploited by Newton and Leibniz, such problems became very easy, as you can see from Example 7.

EXAMPLE 8 What is wrong with the following calculation?

$$\int_{-1}^{3} \frac{1}{x^2}\,dx = \frac{x^{-1}}{-1}\bigg]_{-1}^{3} = -\frac{1}{3} - 1 = -\frac{4}{3}$$

SOLUTION To start, we notice that this calculation must be wrong because the answer is negative but $f(x) = 1/x^2 \geq 0$ and Property 6 of integrals says that $\int_a^b f(x)\,dx \geq 0$ when $f \geq 0$. The Fundamental Theorem of Calculus applies to continuous functions. It can't be applied here because $f(x) = 1/x^2$ is not continuous on $[-1, 3]$. In fact, f has an infinite discontinuity at $x = 0$, so

$$\int_{-1}^{3} \frac{1}{x^2}\,dx \qquad \text{does not exist}$$

Differentiation and Integration as Inverse Processes

We end this section by bringing together the two parts of the Fundamental Theorem.

> **The Fundamental Theorem of Calculus**   Suppose f is continuous on $[a, b]$.
>
> 1. If $g(x) = \int_a^x f(t)\, dt$, then $g'(x) = f(x)$.
>
> 2. $\int_a^b f(x)\, dx = F(b) - F(a)$, where F is any antiderivative of f, that is, $F' = f$.

We noted that Part 1 can be rewritten as

$$\frac{d}{dx}\int_a^x f(t)\, dt = f(x)$$

which says that if f is integrated and then the result is differentiated, we arrive back at the original function f. Since $F'(x) = f(x)$, Part 2 can be rewritten as

$$\int_a^b F'(x)\, dx = F(b) - F(a)$$

This version says that if we take a function F, first differentiate it, and then integrate the result, we arrive back at the original function F, but in the form $F(b) - F(a)$. Taken together, the two parts of the Fundamental Theorem of Calculus say that differentiation and integration are inverse processes. Each undoes what the other does.

The Fundamental Theorem of Calculus is unquestionably the most important theorem in calculus and, indeed, it ranks as one of the great accomplishments of the human mind. Before it was discovered, from the time of Eudoxus and Archimedes to the time of Galileo and Fermat, problems of finding areas, volumes, and lengths of curves were so difficult that only a genius could meet the challenge. But now, armed with the systematic method that Newton and Leibniz fashioned out of the Fundamental Theorem, we will see in the chapters to come that these challenging problems are accessible to all of us.

5.3 Exercises

1. Explain exactly what is meant by the statement that "differentiation and integration are inverse processes."

2. Let $g(x) = \int_0^x f(t)\, dt$, where f is the function whose graph is shown.
    (a) Evaluate $g(x)$ for $x = 0, 1, 2, 3, 4, 5$, and 6.
    (b) Estimate $g(7)$.
    (c) Where does g have a maximum value? Where does it have a minimum value?
    (d) Sketch a rough graph of g.

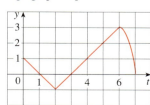

3. Let $g(x) = \int_0^x f(t)\, dt$, where f is the function whose graph is shown.
   (a) Evaluate $g(0)$, $g(1)$, $g(2)$, $g(3)$, and $g(6)$.
   (b) On what interval is g increasing?
   (c) Where does g have a maximum value?
   (d) Sketch a rough graph of g.

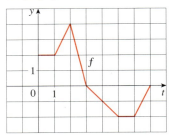

4. Let $g(x) = \int_{-3}^x f(t)\, dt$, where f is the function whose graph is shown.
   (a) Evaluate $g(-3)$ and $g(3)$.
   (b) Estimate $g(-2)$, $g(-1)$, and $g(0)$.
   (c) On what interval is g increasing?
   (d) Where does g have a maximum value?
   (e) Sketch a rough graph of g.
   (f) Use the graph in part (e) to sketch the graph of $g'(x)$. Compare with the graph of f.

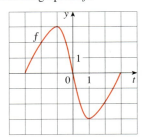

5–6 Sketch the area represented by $g(x)$. Then find $g'(x)$ in two ways: (a) by using Part 1 of the Fundamental Theorem and (b) by evaluating the integral using Part 2 and then differentiating.

5. $g(x) = \int_1^x t^2\, dt$

6. $g(x) = \int_0^x (1 + \sqrt{t})\, dt$

7–18 Use Part 1 of the Fundamental Theorem of Calculus to find the derivative of the function.

7. $g(x) = \int_0^x \sqrt{1 + 2t}\, dt$

8. $g(x) = \int_1^x (2 + t^4)^5\, dt$

9. $g(y) = \int_2^y t^2 \sin t\, dt$

10. $g(u) = \int_3^u \dfrac{1}{x + x^2}\, dx$

11. $F(x) = \int_x^2 \cos(t^2)\, dt$

$\left[\text{Hint: } \int_x^2 \cos(t^2)\, dt = -\int_2^x \cos(t^2)\, dt \right]$

12. $F(x) = \int_x^{10} \tan\theta\, d\theta$

13. $h(x) = \int_2^{1/x} \sin^4 t\, dt$

14. $h(x) = \int_0^{x^2} \sqrt{1 + r^3}\, dr$

15. $y = \int_3^{\sqrt{x}} \dfrac{\cos t}{t}\, dt$

16. $y = \int_1^{\cos x} (t + \sin t)\, dt$

17. $y = \int_{1-3x}^1 \dfrac{u^3}{1 + u^2}\, du$

18. $y = \int_{1/x^2}^0 \sin^3 t\, dt$

19–36 Use Part 2 of the Fundamental Theorem of Calculus to evaluate the integral, or explain why it does not exist.

19. $\int_{-1}^3 x^5\, dx$

20. $\int_{-2}^5 6\, dx$

21. $\int_2^8 (4x + 3)\, dx$

22. $\int_0^4 (1 + 3y - y^2)\, dy$

23. $\int_0^1 x^{4/5}\, dx$

24. $\int_1^8 \sqrt[3]{x}\, dx$

25. $\int_1^2 \dfrac{3}{t^4}\, dt$

26. $\int_{-2}^3 x^{-5}\, dx$

27. $\int_{-5}^5 \dfrac{2}{x^3}\, dx$

28. $\int_\pi^{2\pi} \cos\theta\, d\theta$

29. $\int_0^2 x(2 + x^5)\, dx$

30. $\int_1^4 \dfrac{1}{\sqrt{x}}\, dx$

31. $\int_0^{\pi/4} \sec^2 t\, dt$

32. $\int_0^1 (3 + x\sqrt{x})\, dx$

33. $\int_\pi^{2\pi} \csc^2\theta\, d\theta$

34. $\int_0^{\pi/6} \csc\theta \cot\theta\, d\theta$

35. $\int_0^2 f(x)\, dx$ where $f(x) = \begin{cases} x^4 & \text{if } 0 \leq x < 1 \\ x^5 & \text{if } 1 \leq x \leq 2 \end{cases}$

36. $\int_{-\pi}^\pi f(x)\, dx$ where $f(x) = \begin{cases} x & \text{if } -\pi \leq x \leq 0 \\ \sin x & \text{if } 0 < x \leq \pi \end{cases}$

37–40 Use a graph to give a rough estimate of the area of the region that lies beneath the given curve. Then find the exact area.

37. $y = \sqrt[3]{x}$, $0 \leq x \leq 27$

38. $y = x^{-4}$, $1 \leq x \leq 6$

39. $y = \sin x$, $0 \leq x \leq \pi$

40. $y = \sec^2 x$, $0 \leq x \leq \pi/3$

41–42 ▪ Evaluate the integral and interpret it as a difference of areas. Illustrate with a sketch.

41. $\int_{-1}^{2} x^3\, dx$

42. $\int_{\pi/4}^{5\pi/2} \sin x\, dx$

43–46 ▪ Find the derivative of the function.

43. $g(x) = \int_{2x}^{3x} \dfrac{u^2 - 1}{u^2 + 1}\, du$

$\left[\text{Hint: } \int_{2x}^{3x} f(u)\, du = \int_{2x}^{0} f(u)\, du + \int_{0}^{3x} f(u)\, du \right]$

44. $g(x) = \int_{\tan x}^{x^2} \dfrac{1}{\sqrt{2 + t^4}}\, dt$

45. $y = \int_{\sqrt{x}}^{x^3} \sqrt{t}\,\sin t\, dt$

46. $y = \int_{\cos x}^{5x} \cos(u^2)\, du$

47. If $F(x) = \int_{1}^{x} f(t)\, dt$, where $f(t) = \int_{1}^{t^2} \dfrac{\sqrt{1+u^4}}{u}\, du$, find $F''(2)$.

48. Find the interval on which the curve
$$y = \int_0^x \dfrac{1}{1 + t + t^2}\, dt$$
is concave upward.

49. The Fresnel function S was defined in Example 3 and graphed in Figures 7 and 8.
  (a) At what values of x does this function have local maximum values?
  (b) On what intervals is the function concave upward?
  (c) Use a graph to solve the following equation correct to two decimal places:
  $$\int_0^x \sin(\pi t^2 / 2)\, dt = 0.2$$

CAS

CAS **50.** The **sine integral function**
$$\mathrm{Si}(x) = \int_0^x \dfrac{\sin t}{t}\, dt$$
is important in electrical engineering. [The integrand $f(t) = (\sin t)/t$ is not defined when $t = 0$, but we know that its limit is 1 when $t \to 0$. So we define $f(0) = 1$ and this makes f a continuous function everywhere.]
  (a) Draw the graph of Si.
  (b) At what values of x does this function have local maximum values?
  (c) Find the coordinates of the first inflection point to the right of the origin.

  (d) Does this function have horizontal asymptotes?
  (e) Solve the following equation correct to one decimal place:
  $$\int_0^x \dfrac{\sin t}{t}\, dt = 1$$

51–52 ▪ Let $g(x) = \int_0^x f(t)\, dt$, where f is the function whose graph is shown.
  (a) At what values of x do the local maximum and minimum values of g occur?
  (b) Where does g attain its absolute maximum value?
  (c) On what intervals is g concave downward?
  (d) Sketch the graph of g.

51.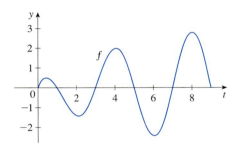

52.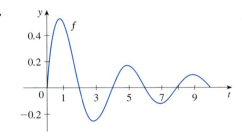

53–54 ▪ Evaluate the limit by first recognizing the sum as a Riemann sum for a function defined on $[0, 1]$.

53. $\displaystyle\lim_{n \to \infty} \sum_{i=1}^{n} \dfrac{i^3}{n^4}$

54. $\displaystyle\lim_{n \to \infty} \dfrac{1}{n}\left(\sqrt{\dfrac{1}{n}} + \sqrt{\dfrac{2}{n}} + \sqrt{\dfrac{3}{n}} + \cdots + \sqrt{\dfrac{n}{n}} \right)$

55. Justify (3) for the case $h < 0$.

56. If f is continuous and g and h are differentiable functions, find a formula for
$$\dfrac{d}{dx} \int_{g(x)}^{h(x)} f(t)\, dt$$

57. (a) Show that $1 \leq \sqrt{1 + x^3} \leq 1 + x^3$ for $x \geq 0$.
  (b) Show that $1 \leq \int_0^1 \sqrt{1 + x^3}\, dx \leq 1.25$.

58. Let

$$f(x) = \begin{cases} 0 & \text{if } x < 0 \\ x & \text{if } 0 \leq x \leq 1 \\ 2 - x & \text{if } 1 < x \leq 2 \\ 0 & \text{if } x > 2 \end{cases}$$

and

$$g(x) = \int_0^x f(t)\, dt$$

(a) Find an expression for $g(x)$ similar to the one for $f(x)$.
(b) Sketch the graphs of f and g.
(c) Where is f differentiable? Where is g differentiable?

59. Find a function f and a number a such that

$$6 + \int_a^x \frac{f(t)}{t^2}\, dt = 2\sqrt{x}$$

for all $x > 0$.

60. Suppose h is a function such that $h(1) = -2$, $h'(1) = 2$, $h''(1) = 3$, $h(2) = 6$, $h'(2) = 5$, $h''(2) = 13$, and h'' is continuous everywhere. Evaluate $\int_1^2 h''(u)\, du$.

61. A manufacturing company owns a major piece of equipment that depreciates at the (continuous) rate $f = f(t)$, where t is the time measured in months since its last overhaul. Because a fixed cost A is incurred each time the machine is overhauled, the company wants to determine the optimal time T (in months) between overhauls.
(a) Explain why $\int_0^t f(s)\, ds$ represents the loss in value of the machine over the period of time t since the last overhaul.
(b) Let $C = C(t)$ be given by

$$C(t) = \frac{1}{t}\left[A + \int_0^t f(s)\, ds\right]$$

What does C represent and why would the company want to minimize C?
(c) Show that C has a minimum value at the numbers $t = T$ where $C(T) = f(T)$.

62. A high-tech company purchases a new computing system whose initial value is V. The system will depreciate at the rate $f = f(t)$ and will accumulate maintenance costs at the rate $g = g(t)$, where t is the time measured in months. The company wants to determine the optimal time to replace the system.
(a) Let

$$C(t) = \frac{1}{t}\int_0^t [f(s) + g(s)]\, ds$$

Show that the critical numbers of C occur at the numbers t where $C(t) = f(t) + g(t)$.
(b) Suppose that

$$f(t) = \begin{cases} \dfrac{V}{15} - \dfrac{V}{450}t & \text{if } 0 < t \leq 30 \\ 0 & \text{if } t > 30 \end{cases}$$

and

$$g(t) = \frac{Vt^2}{12{,}900} \qquad t > 0$$

Determine the length of time T for the total depreciation $D(t) = \int_0^t f(s)\, ds$ to equal the initial value V.
(c) Determine the absolute minimum of C on $(0, T]$.
(d) Sketch the graphs of C and $f + g$ in the same coordinate system, and verify the result in part (a) in this case.

The following exercises are intended only for those who have already covered Chapter 7.

63–68 ▪ Evaluate the integral.

63. $\displaystyle\int_1^9 \frac{1}{2x}\, dx$

64. $\displaystyle\int_0^1 10^x\, dx$

65. $\displaystyle\int_{1/2}^{\sqrt{3}/2} \frac{6}{\sqrt{1 - t^2}}\, dt$

66. $\displaystyle\int_0^1 \frac{4}{t^2 + 1}\, dt$

67. $\displaystyle\int_{-1}^{1} e^{u+1}\, du$

68. $\displaystyle\int_1^2 \frac{4 + u^2}{u^3}\, du$

5.4 Indefinite Integrals and the Net Change Theorem

We saw in Section 5.3 that the second part of the Fundamental Theorem of Calculus provides a very powerful method for evaluating the definite integral of a function, assuming that we can find an antiderivative of the function. In this section we introduce a notation for antiderivatives, review the formulas for antiderivatives, and use them to evaluate definite integrals. We also reformulate FTC2 in a way that makes it easier to apply to science and engineering problems.

Indefinite Integrals

Both parts of the Fundamental Theorem establish connections between antiderivatives and definite integrals. Part 1 says that if f is continuous, then $\int_a^x f(t)\, dt$ is an antiderivative of f. Part 2 says that $\int_a^b f(x)\, dx$ can be found by evaluating $F(b) - F(a)$, where F is an antiderivative of f.

We need a convenient notation for antiderivatives that makes them easy to work with. Because of the relation given by the Fundamental Theorem between antiderivatives and integrals, the notation $\int f(x)\, dx$ is traditionally used for an antiderivative of f and is called an **indefinite integral**. Thus

$$\int f(x)\, dx = F(x) \qquad \text{means} \qquad F'(x) = f(x)$$

For example, we can write

$$\int x^2\, dx = \frac{x^3}{3} + C \qquad \text{because} \qquad \frac{d}{dx}\left(\frac{x^3}{3} + C\right) = x^2$$

So we can regard an indefinite integral as representing an entire *family* of functions (one antiderivative for each value of the constant C).

⊘ You should distinguish carefully between definite and indefinite integrals. A definite integral $\int_a^b f(x)\, dx$ is a number, whereas an indefinite integral $\int f(x)\, dx$ is a function (or family of functions). The connection between them is given by Part 2 of the Fundamental Theorem. If f is continuous on $[a, b]$, then

$$\int_a^b f(x)\, dx = \int f(x)\, dx \Big]_a^b$$

The effectiveness of the Fundamental Theorem depends on having a supply of antiderivatives of functions. We therefore restate the Table of Antidifferentiation Formulas from Section 4.10, together with a few others, in the notation of indefinite integrals. Any formula can be verified by differentiating the function on the right side and obtaining the integrand. For instance

$$\int \sec^2 x\, dx = \tan x + C \qquad \text{because} \qquad \frac{d}{dx}(\tan x + C) = \sec^2 x$$

1 Table of Indefinite Integrals

$$\int cf(x)\, dx = c\int f(x)\, dx \qquad\qquad \int [f(x) + g(x)]\, dx = \int f(x)\, dx + \int g(x)\, dx$$

$$\int k\, dx = kx + C \qquad\qquad \int x^n\, dx = \frac{x^{n+1}}{n+1} + C \quad (n \neq -1)$$

$$\int \sin x\, dx = -\cos x + C \qquad\qquad \int \cos x\, dx = \sin x + C$$

$$\int \sec^2 x\, dx = \tan x + C \qquad\qquad \int \csc^2 x\, dx = -\cot x + C$$

$$\int \sec x \tan x\, dx = \sec x + C \qquad\qquad \int \csc x \cot x\, dx = -\csc x + C$$

Recall from Theorem 4.10.1 that the most general antiderivative *on a given interval* is obtained by adding a constant to a particular antiderivative. **We adopt the convention that when a formula for a general indefinite integral is given, it is valid only on an interval.** Thus, we write

$$\int \frac{1}{x^2}\, dx = -\frac{1}{x} + C$$

with the understanding that it is valid on the interval $(0, \infty)$ or on the interval $(-\infty, 0)$. This is true despite the fact that the general antiderivative of the function $f(x) = 1/x^2$, $x \ne 0$, is

$$F(x) = \begin{cases} -\dfrac{1}{x} + C_1 & \text{if } x < 0 \\ -\dfrac{1}{x} + C_2 & \text{if } x > 0 \end{cases}$$

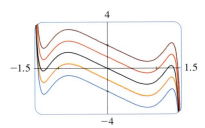

|||| The indefinite integral in Example 1 is graphed in Figure 1 for several values of C. The value of C is the y-intercept.

FIGURE 1

EXAMPLE 1 Find the general indefinite integral

$$\int (10x^4 - 2\sec^2 x)\, dx$$

SOLUTION Using our convention and Table 1, we have

$$\int (10x^4 - 2\sec^2 x)\, dx = 10\int x^4\, dx - 2\int \sec^2 x\, dx$$

$$= 10\,\frac{x^5}{5} - 2\tan x + C$$

$$= 2x^5 - 2\tan x + C$$

You should check this answer by differentiating it.

EXAMPLE 2 Evaluate $\displaystyle\int \frac{\cos\theta}{\sin^2\theta}\, d\theta$.

SOLUTION This indefinite integral isn't immediately apparent in Table 1, so we use trigonometric identities to rewrite the function before integrating:

$$\int \frac{\cos\theta}{\sin^2\theta}\, d\theta = \int \left(\frac{1}{\sin\theta}\right)\left(\frac{\cos\theta}{\sin\theta}\right) d\theta$$

$$= \int \csc\theta \cot\theta\, d\theta = -\csc\theta + C$$

EXAMPLE 3 Evaluate $\displaystyle\int_0^3 (x^3 - 6x)\, dx$.

SOLUTION Using FTC2 and Table 1, we have

$$\int_0^3 (x^3 - 6x)\, dx = \frac{x^4}{4} - 6\,\frac{x^2}{2}\bigg]_0^3$$

$$= \left(\tfrac{1}{4}\cdot 3^4 - 3\cdot 3^2\right) - \left(\tfrac{1}{4}\cdot 0^4 - 3\cdot 0^2\right)$$

$$= \tfrac{81}{4} - 27 - 0 + 0 = -6.75$$

Compare this calculation with Example 2(b) in Section 5.2.

EXAMPLE 4 Find $\int_0^{12} (x - 12 \sin x) \, dx$.

SOLUTION The Fundamental Theorem gives

$$\int_0^{12} (x - 12 \sin x) \, dx = \frac{x^2}{2} - 12(-\cos x) \Big]_0^{12}$$

$$= \tfrac{1}{2}(12)^2 + 12(\cos 12 - \cos 0)$$

$$= 72 + 12 \cos 12 - 12$$

$$= 60 + 12 \cos 12$$

This is the exact value of the integral. If a decimal approximation is desired, we can use a calculator to approximate cos 12. Doing so, we get

$$\int_0^{12} (x - 12 \sin x) \, dx \approx 70.1262$$

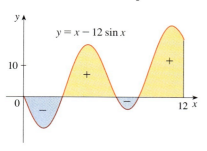

|||| Figure 2 shows the graph of the integrand in Example 4. We know from Section 5.2 that the value of the integral can be interpreted as the sum of the areas labeled with a plus sign minus the areas labeled with a minus sign.

FIGURE 2

EXAMPLE 5 Evaluate $\int_1^9 \dfrac{2t^2 + t^2 \sqrt{t} - 1}{t^2} \, dt$.

SOLUTION First we need to write the integrand in a simpler form by carrying out the division:

$$\int_1^9 \frac{2t^2 + t^2 \sqrt{t} - 1}{t^2} \, dt = \int_1^9 (2 + t^{1/2} - t^{-2}) \, dt$$

$$= 2t + \frac{t^{3/2}}{\frac{3}{2}} - \frac{t^{-1}}{-1} \bigg]_1^9 = 2t + \tfrac{2}{3} t^{3/2} + \frac{1}{t} \bigg]_1^9$$

$$= \left[2 \cdot 9 + \tfrac{2}{3}(9)^{3/2} + \tfrac{1}{9} \right] - \left(2 \cdot 1 + \tfrac{2}{3} \cdot 1^{3/2} + \tfrac{1}{1} \right)$$

$$= 18 + 18 + \tfrac{1}{9} - 2 - \tfrac{2}{3} - 1 = 32 \tfrac{4}{9}$$

|||| **Applications**

Part 2 of the Fundamental Theorem says that if f is continuous on $[a, b]$, then

$$\int_a^b f(x) \, dx = F(b) - F(a)$$

where F is any antiderivative of f. This means that $F' = f$, so the equation can be rewritten as

$$\int_a^b F'(x) \, dx = F(b) - F(a)$$

We know that $F'(x)$ represents the rate of change of $y = F(x)$ with respect to x and $F(b) - F(a)$ is the change in y when x changes from a to b. [Note that y could, for instance, increase, then decrease, then increase again. Although y might change in both directions, $F(b) - F(a)$ represents the *net* change in y.] So we can reformulate FTC2 in words as follows.

> **The Net Change Theorem** The integral of a rate of change is the net change:
> $$\int_a^b F'(x)\,dx = F(b) - F(a)$$

This principle can be applied to all of the rates of change in the natural and social sciences that we discussed in Section 3.4. Here are a few instances of this idea:

- If $V(t)$ is the volume of water in a reservoir at time t, then its derivative $V'(t)$ is the rate at which water flows into the reservoir at time t. So
$$\int_{t_1}^{t_2} V'(t)\,dt = V(t_2) - V(t_1)$$
is the change in the amount of water in the reservoir between time t_1 and time t_2.

- If $[C](t)$ is the concentration of the product of a chemical reaction at time t, then the rate of reaction is the derivative $d[C]/dt$. So
$$\int_{t_1}^{t_2} \frac{d[C]}{dt}\,dt = [C](t_2) - [C](t_1)$$
is the change in the concentration of C from time t_1 to time t_2.

- If the mass of a rod measured from the left end to a point x is $m(x)$, then the linear density is $\rho(x) = m'(x)$. So
$$\int_a^b \rho(x)\,dx = m(b) - m(a)$$
is the mass of the segment of the rod that lies between $x = a$ and $x = b$.

- If the rate of growth of a population is dn/dt, then
$$\int_{t_1}^{t_2} \frac{dn}{dt}\,dt = n(t_2) - n(t_1)$$
is the net change in population during the time period from t_1 to t_2. (The population increases when births happen and decreases when deaths occur. The net change takes into account both births and deaths.)

- If $C(x)$ is the cost of producing x units of a commodity, then the marginal cost is the derivative $C'(x)$. So
$$\int_{x_1}^{x_2} C'(x)\,dx = C(x_2) - C(x_1)$$
is the increase in cost when production is increased from x_1 units to x_2 units.

- If an object moves along a straight line with position function $s(t)$, then its velocity is $v(t) = s'(t)$, so

2
$$\int_{t_1}^{t_2} v(t)\,dt = s(t_2) - s(t_1)$$

is the net change of position, or *displacement*, of the particle during the time period from t_1 to t_2. In Section 5.1 we guessed that this was true for the case where the object moves in the positive direction, but now we have proved that it is always true.

- If we want to calculate the distance traveled during the time interval, we have to consider the intervals when $v(t) \geq 0$ (the particle moves to the right) and also the intervals when $v(t) \leq 0$ (the particle moves to the left). In both cases the distance is computed by integrating $|v(t)|$, the speed. Therefore

$$\boxed{3} \qquad \int_{t_1}^{t_2} |v(t)|\, dt = \text{total distance traveled}$$

Figure 3 shows how both displacement and distance traveled can be interpreted in terms of areas under a velocity curve.

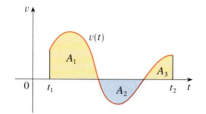

$$\text{displacement} = \int_{t_1}^{t_2} v(t)\, dt = A_1 - A_2 + A_3$$

$$\text{distance} = \int_{t_1}^{t_2} |v(t)|\, dt = A_1 + A_2 + A_3$$

FIGURE 3

- The acceleration of the object is $a(t) = v'(t)$, so

$$\int_{t_1}^{t_2} a(t)\, dt = v(t_2) - v(t_1)$$

is the change in velocity from time t_1 to time t_2.

Resources / Module 7
/ Physics and Engineering
/ Start of Spy Tracks

EXAMPLE 6 A particle moves along a line so that its velocity at time t is $v(t) = t^2 - t - 6$ (measured in meters per second).
(a) Find the displacement of the particle during the time period $1 \leq t \leq 4$.
(b) Find the distance traveled during this time period.

SOLUTION
(a) By Equation 2, the displacement is

$$s(4) - s(1) = \int_1^4 v(t)\, dt = \int_1^4 (t^2 - t - 6)\, dt$$

$$= \left[\frac{t^3}{3} - \frac{t^2}{2} - 6t \right]_1^4 = -\frac{9}{2}$$

This means that the particle moved 4.5 m toward the left.

(b) Note that $v(t) = t^2 - t - 6 = (t-3)(t+2)$ and so $v(t) \leq 0$ on the interval $[1, 3]$ and $v(t) \geq 0$ on $[3, 4]$. Thus, from Equation 3, the distance traveled is

|||| To integrate the absolute value of $v(t)$, we use Property 5 of integrals from Section 5.2 to split the integral into two parts, one where $v(t) \leq 0$ and one where $v(t) \geq 0$.

$$\int_1^4 |v(t)|\, dt = \int_1^3 [-v(t)]\, dt + \int_3^4 v(t)\, dt$$

$$= \int_1^3 (-t^2 + t + 6)\, dt + \int_3^4 (t^2 - t - 6)\, dt$$

$$= \left[-\frac{t^3}{3} + \frac{t^2}{2} + 6t \right]_1^3 + \left[\frac{t^3}{3} - \frac{t^2}{2} - 6t \right]_3^4$$

$$= \frac{61}{6} \approx 10.17 \text{ m}$$

EXAMPLE 7 Figure 4 shows the power consumption in the city of San Francisco for a day in September (P is measured in megawatts; t is measured in hours starting at midnight). Estimate the energy used on that day.

FIGURE 4

Pacific Gas & Electric

SOLUTION Power is the rate of change of energy: $P(t) = E'(t)$. So, by the Net Change Theorem,

$$\int_0^{24} P(t)\, dt = \int_0^{24} E'(t)\, dt = E(24) - E(0)$$

is the total amount of energy used that day. We approximate the value of the integral using the Midpoint Rule with 12 subintervals and $\Delta t = 2$:

$$\int_0^{24} P(t)\, dt \approx [P(1) + P(3) + P(5) + \cdots + P(21) + P(23)]\, \Delta t$$

$$\approx (440 + 400 + 420 + 620 + 790 + 840 + 850$$

$$+ 840 + 810 + 690 + 670 + 550)(2)$$

$$= 15{,}840$$

The energy used was approximately 15,840 megawatt-hours.

|||| A note on units

How did we know what units to use for energy in Example 7? The integral $\int_0^{24} P(t)\, dt$ is defined as the limit of sums of terms of the form $P(t_i^*)\, \Delta t$. Now $P(t_i^*)$ is measured in megawatts and Δt is measured in hours, so their product is measured in megawatt-hours. The same is true of the limit. In general, the unit of measurement for $\int_a^b f(x)\, dx$ is the product of the unit for $f(x)$ and the unit for x.

5.4 Exercises

1–4 |||| Verify by differentiation that the formula is correct.

1. $\displaystyle \int \frac{x}{\sqrt{x^2+1}}\, dx = \sqrt{x^2+1} + C$

2. $\displaystyle \int x \cos x\, dx = x \sin x + \cos x + C$

3. $\displaystyle \int \frac{1}{\sqrt{(a^2-x^2)^3}}\, dx = \frac{x}{a^2\sqrt{a^2-x^2}} + C$

4. $\displaystyle \int \frac{1}{x^2\sqrt{x^2+a^2}}\, dx = -\frac{\sqrt{x^2+a^2}}{a^2 x} + C$

5–14 |||| Find the general indefinite integral.

5. $\displaystyle \int x^{-3/4}\, dx$

6. $\displaystyle \int \sqrt[3]{x}\, dx$

7. $\displaystyle \int (x^3 + 6x + 1)\, dx$

8. $\displaystyle \int x(1 + 2x^4)\, dx$

9. $\int (1-t)(2+t^2)\, dt$

10. $\int \left(u^2 + 1 + \dfrac{1}{u^2}\right) du$

11. $\int (2 - \sqrt{x})^2\, dx$

12. $\int (\sin\theta + 3\cos\theta)\, d\theta$

13. $\int \dfrac{\sin x}{1 - \sin^2 x}\, dx$

14. $\int \dfrac{\sin 2x}{\sin x}\, dx$

15–16 Find the general indefinite integral. Illustrate by graphing several members of the family on the same screen.

15. $\int x\sqrt{x}\, dx$

16. $\int (\cos x - 2\sin x)\, dx$

17–40 Evaluate the integral.

17. $\int_0^2 (6x^2 - 4x + 5)\, dx$

18. $\int_1^3 (1 + 2x - 4x^3)\, dx$

19. $\int_{-3}^0 (5y^4 - 6y^2 + 14)\, dy$

20. $\int_{-2}^0 (u^5 - u^3 + u^2)\, du$

21. $\int_{-2}^2 (3u + 1)^2\, du$

22. $\int_0^4 (2v + 5)(3v - 1)\, dv$

23. $\int_1^4 \sqrt{t}(1 + t)\, dt$

24. $\int_0^9 \sqrt{2t}\, dt$

25. $\int_{-2}^{-1} \left(4y^3 + \dfrac{2}{y^3}\right) dy$

26. $\int_1^2 \dfrac{y + 5y^7}{y^3}\, dy$

27. $\int_0^1 x(\sqrt[3]{x} + \sqrt[4]{x})\, dx$

28. $\int_1^2 \left(x + \dfrac{1}{x}\right)^2 dx$

29. $\int_1^4 \sqrt{\dfrac{5}{x}}\, dx$

30. $\int_1^9 \dfrac{3x - 2}{\sqrt{x}}\, dx$

31. $\int_0^\pi (4\sin\theta - 3\cos\theta)\, d\theta$

32. $\int_{\pi/4}^{\pi/3} \sec\theta \tan\theta\, d\theta$

33. $\int_0^{\pi/4} \dfrac{1 + \cos^2\theta}{\cos^2\theta}\, d\theta$

34. $\int_0^{\pi/3} \dfrac{\sin\theta + \sin\theta \tan^2\theta}{\sec^2\theta}\, d\theta$

35. $\int_1^{64} \dfrac{1 + \sqrt[3]{x}}{\sqrt{x}}\, dx$

36. $\int_0^1 (1 + x^2)^3\, dx$

37. $\int_0^1 (\sqrt[4]{x^5} + \sqrt[5]{x^4})\, dx$

38. $\int_1^8 \dfrac{x - 1}{\sqrt[3]{x^2}}\, dx$

39. $\int_{-1}^2 (x - 2|x|)\, dx$

40. $\int_0^{3\pi/2} |\sin x|\, dx$

41. Use a graph to estimate the x-intercepts of the curve $y = x + x^2 - x^4$. Then use this information to estimate the area of the region that lies under the curve and above the x-axis.

42. Repeat Exercise 41 for the curve $y = 2x + 3x^4 - 2x^6$.

SECTION 5.4 INDEFINITE INTEGRALS AND THE NET CHANGE THEOREM ▮ 357

43. The area of the region that lies to the right of the y-axis and to the left of the parabola $x = 2y - y^2$ (the shaded region in the figure) is given by the integral $\int_0^2 (2y - y^2)\, dy$. (Turn your head clockwise and think of the region as lying below the curve $x = 2y - y^2$ from $y = 0$ to $y = 2$.) Find the area of the region.

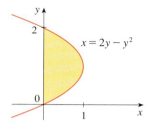

44. The boundaries of the shaded region are the y-axis, the line $y = 1$, and the curve $y = \sqrt[4]{x}$. Find the area of this region by writing x as a function of y and integrating with respect to y (as in Exercise 43).

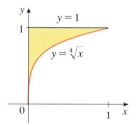

45. If $w'(t)$ is the rate of growth of a child in pounds per year, what does $\int_5^{10} w'(t)\, dt$ represent?

46. The current in a wire is defined as the derivative of the charge: $I(t) = Q'(t)$. (See Example 3 in Section 3.4.) What does $\int_a^b I(t)\, dt$ represent?

47. If oil leaks from a tank at a rate of $r(t)$ gallons per minute at time t, what does $\int_0^{120} r(t)\, dt$ represent?

48. A honeybee population starts with 100 bees and increases at a rate of $n'(t)$ bees per week. What does $100 + \int_0^{15} n'(t)\, dt$ represent?

49. In Section 4.8 we defined the marginal revenue function $R'(x)$ as the derivative of the revenue function $R(x)$, where x is the number of units sold. What does $\int_{1000}^{5000} R'(x)\, dx$ represent?

50. If $f(x)$ is the slope of a trail at a distance of x miles from the start of the trail, what does $\int_3^5 f(x)\, dx$ represent?

51. If x is measured in meters and $f(x)$ is measured in newtons, what are the units for $\int_0^{100} f(x)\, dx$?

52. If the units for x are feet and the units for $a(x)$ are pounds per foot, what are the units for da/dx? What units does $\int_2^8 a(x)\, dx$ have?

53–54 The velocity function (in meters per second) is given for a particle moving along a line. Find (a) the displacement and (b) the distance traveled by the particle during the given time interval.

53. $v(t) = 3t - 5, \quad 0 \leq t \leq 3$

54. $v(t) = t^2 - 2t - 8, \quad 1 \leq t \leq 6$

55–56 The acceleration function (in m/s²) and the initial velocity are given for a particle moving along a line. Find (a) the velocity at time t and (b) the distance traveled during the given time interval.

55. $a(t) = t + 4, \quad v(0) = 5, \quad 0 \leq t \leq 10$

56. $a(t) = 2t + 3, \quad v(0) = -4, \quad 0 \leq t \leq 3$

57. The linear density of a rod of length 4 m is given by $\rho(x) = 9 + 2\sqrt{x}$ measured in kilograms per meter, where x is measured in meters from one end of the rod. Find the total mass of the rod.

58. Water flows from the bottom of a storage tank at a rate of $r(t) = 200 - 4t$ liters per minute, where $0 \leq t \leq 50$. Find the amount of water that flows from the tank during the first 10 minutes.

59. The velocity of a car was read from its speedometer at 10-second intervals and recorded in the table. Use the Midpoint Rule to estimate the distance traveled by the car.

t (s)	v (mi/h)	t (s)	v (mi/h)
0	0	60	56
10	38	70	53
20	52	80	50
30	58	90	47
40	55	100	45
50	51		

60. Suppose that a volcano is erupting and readings of the rate $r(t)$ at which solid materials are spewed into the atmosphere are given in the table. The time t is measured in seconds and the units for $r(t)$ are tonnes (metric tons) per second.

t	0	1	2	3	4	5	6
$r(t)$	2	10	24	36	46	54	60

(a) Give upper and lower estimates for the quantity $Q(6)$ of erupted materials after 6 seconds.
(b) Use the Midpoint Rule to estimate $Q(6)$.

61. The marginal cost of manufacturing x yards of a certain fabric is $C'(x) = 3 - 0.01x + 0.000006x^2$ (in dollars per yard). Find the increase in cost if the production level is raised from 2000 yards to 4000 yards.

62. Water flows in and out of a storage tank. A graph of the rate of change $r(t)$ of the volume of water in the tank, in liters per day, is shown. If the amount of water in the tank at time $t = 0$ is 25,000 L, use the Midpoint Rule to estimate the amount of water four days later.

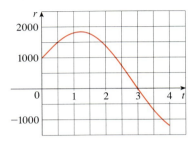

63. Economists use a cumulative distribution called a *Lorenz curve* to describe the distribution of income between households in a given country. Typically, a Lorenz curve is defined on [0, 1] with endpoints (0, 0) and (1, 1), and is continuous, increasing, and concave upward. The points on this curve are determined by ranking all households by income and then computing the percentage of households whose income is less than or equal to a given percentage of the total income of the country. For example, the point $(a/100, b/100)$ is on the Lorenz curve if the bottom $a\%$ of the households receive less than or equal to $b\%$ of the total income. *Absolute equality* of income distribution would occur if the bottom $a\%$ of the households receive $a\%$ of the income, in which case the Lorenz curve would be the line $y = x$. The area between the Lorenz curve and the line $y = x$ measures how much the income distribution differs from absolute equality. The *coefficient of inequality* is the ratio of the area between the Lorenz curve and the line $y = x$ to the area under $y = x$.

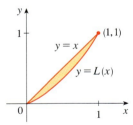

(a) Show that the coefficient of inequality is twice the area between the Lorenz curve and the line $y = x$, that is, show that

$$\text{coefficient of inequality} = 2 \int_0^1 [x - L(x)]\, dx$$

(b) The income distribution for a certain country is represented by the Lorenz curve defined by the equation

$$L(x) = \tfrac{5}{12}x^2 + \tfrac{7}{12}x$$

What is the percentage of total income received by the bottom 50% of the households? Find the coefficient of inequality.

64. On May 7, 1992, the space shuttle *Endeavour* was launched on mission STS-49, the purpose of which was to install a new perigee kick motor in an Intelsat communications satellite. The table gives the velocity data for the shuttle between liftoff and the jettisoning of the solid rocket boosters.

Event	Time (s)	Velocity (ft/s)
Launch	0	0
Begin roll maneuver	10	185
End roll maneuver	15	319
Throttle to 89%	20	447
Throttle to 67%	32	742
Throttle to 104%	59	1325
Maximum dynamic pressure	62	1445
Solid rocket booster separation	125	4151

(a) Use a graphing calculator or computer to model these data by a third-degree polynomial.
(b) Use the model in part (a) to estimate the height reached by the *Endeavour*, 125 seconds after liftoff.

The following exercises are intended only for those who have already covered Chapter 7.

65–67 Evaluate the integral.

65. $\int_1^e \dfrac{x^2 + x + 1}{x}\, dx$

66. $\int_4^9 \left(\sqrt{x} + \dfrac{1}{\sqrt{x}}\right)^2 dx$

67. $\int \left(x^2 + 1 + \dfrac{1}{x^2 + 1}\right) dx$

68. The area labeled B is three times the area labeled A. Express b in terms of a.

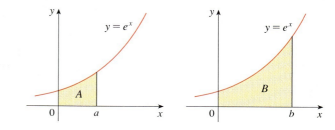

WRITING PROJECT

Newton, Leibniz, and the Invention of Calculus

We sometimes read that the inventors of calculus were Sir Isaac Newton (1642–1727) and Gottfried Wilhelm Leibniz (1646–1716). But we know that the basic ideas behind integration were investigated 2500 years ago by ancient Greeks such as Eudoxus and Archimedes, and methods for finding tangents were pioneered by Pierre Fermat (1601–1665), Isaac Barrow (1630–1677), and others. Barrow, Newton's teacher at Cambridge, was the first to understand the inverse relationship between differentiation and integration. What Newton and Leibniz did was to use this relationship, in the form of the Fundamental Theorem of Calculus, in order to develop calculus into a systematic mathematical discipline. It is in this sense that Newton and Leibniz are credited with the invention of calculus.

Read about the contributions of these men in one or more of the given references and write a report on one of the following three topics. You can include biographical details, but the main thrust of your report should be a description, in some detail, of their methods and notations. In particular, you should consult one of the sourcebooks, which give excerpts from the original publications of Newton and Leibniz, translated from Latin to English.

- The Role of Newton in the Development of Calculus
- The Role of Leibniz in the Development of Calculus
- The Controversy between the Followers of Newton and Leibniz over Priority in the Invention of Calculus

References

1. Carl Boyer and Uta Merzbach, *A History of Mathematics* (New York: Wiley, 1987), Chapter 19.

2. Carl Boyer, *The History of the Calculus and Its Conceptual Development* (New York: Dover, 1959), Chapter V.
3. C. H. Edwards, *The Historical Development of the Calculus* (New York: Springer-Verlag, 1979), Chapters 8 and 9.
4. Howard Eves, *An Introduction to the History of Mathematics,* 6th ed. (New York: Saunders, 1990), Chapter 11.
5. C. C. Gillispie, ed., *Dictionary of Scientific Biography* (New York: Scribner's, 1974). See the article on Leibniz by Joseph Hofmann in Volume VIII and the article on Newton by I. B. Cohen in Volume X.
6. Victor Katz, *A History of Mathematics: An Introduction* (New York: HarperCollins, 1993), Chapter 12.
7. Morris Kline, *Mathematical Thought from Ancient to Modern Times* (New York: Oxford University Press, 1972), Chapter 17.

Sourcebooks

1. John Fauvel and Jeremy Gray, eds., *The History of Mathematics: A Reader* (London: MacMillan Press, 1987), Chapters 12 and 13.
2. D. E. Smith, ed., *A Sourcebook in Mathematics* (New York: Dover, 1959), Chapter V.
3. D. J. Struik, ed., *A Sourcebook in Mathematics, 1200–1800* (Princeton, N.J.: Princeton University Press, 1969), Chapter V.

5.5 The Substitution Rule

Because of the Fundamental Theorem, it's important to be able to find antiderivatives. But our antidifferentiation formulas don't tell us how to evaluate integrals such as

$$\boxed{1} \qquad \int 2x\sqrt{1 + x^2}\, dx$$

To find this integral we use the problem-solving strategy of *introducing something extra*. Here the "something extra" is a new variable; we change from the variable x to a new variable u. Suppose that we let u be the quantity under the root sign in (1), $u = 1 + x^2$. Then the differential of u is $du = 2x\, dx$. Notice that if the dx in the notation for an integral were to be interpreted as a differential, then the differential $2x\, dx$ would occur in (1) and, so, formally, without justifying our calculation, we could write

▌ Differentials were defined in Section 3.10. If $u = f(x)$, then
$$du = f'(x)\, dx$$

$$\boxed{2} \qquad \int 2x\sqrt{1 + x^2}\, dx = \int \sqrt{1 + x^2}\, 2x\, dx = \int \sqrt{u}\, du$$
$$= \tfrac{2}{3} u^{3/2} + C = \tfrac{2}{3}(x^2 + 1)^{3/2} + C$$

But now we can check that we have the correct answer by using the Chain Rule to differentiate the final function of Equation 2:

$$\frac{d}{dx}\left[\tfrac{2}{3}(x^2 + 1)^{3/2} + C\right] = \tfrac{2}{3} \cdot \tfrac{3}{2}(x^2 + 1)^{1/2} \cdot 2x = 2x\sqrt{x^2 + 1}$$

In general, this method works whenever we have an integral that we can write in the form $\int f(g(x))g'(x)\, dx$. Observe that if $F' = f$, then

$$\boxed{3} \qquad \int F'(g(x))g'(x)\, dx = F(g(x)) + C$$

because, by the Chain Rule,

$$\frac{d}{dx}[F(g(x))] = F'(g(x))g'(x)$$

If we make the "change of variable" or "substitution" $u = g(x)$, then from Equation 3 we have

$$\int F'(g(x))g'(x)\, dx = F(g(x)) + C = F(u) + C = \int F'(u)\, du$$

or, writing $F' = f$, we get

$$\int f(g(x))g'(x)\, dx = \int f(u)\, du$$

Thus, we have proved the following rule.

> **4 The Substitution Rule** If $u = g(x)$ is a differentiable function whose range is an interval I and f is continuous on I, then
>
> $$\int f(g(x))g'(x)\, dx = \int f(u)\, du$$

Notice that the Substitution Rule for integration was proved using the Chain Rule for differentiation. Notice also that if $u = g(x)$, then $du = g'(x)\, dx$, so a way to remember the Substitution Rule is to think of dx and du in (4) as differentials.

Thus, the Substitution Rule says: **It is permissible to operate with dx and du after integral signs as if they were differentials.**

EXAMPLE 1 Find $\int x^3 \cos(x^4 + 2)\, dx$.

SOLUTION We make the substitution $u = x^4 + 2$ because its differential is $du = 4x^3\, dx$, which, apart from the constant factor 4, occurs in the integral. Thus, using $x^3\, dx = du/4$ and the Substitution Rule, we have

$$\int x^3 \cos(x^4 + 2)\, dx = \int \cos u \cdot \tfrac{1}{4}\, du = \tfrac{1}{4} \int \cos u\, du$$

$$= \tfrac{1}{4} \sin u + C$$

$$= \tfrac{1}{4} \sin(x^4 + 2) + C$$

|||| Check the answer by differentiating it.

Notice that at the final stage we had to return to the original variable x.

The idea behind the Substitution Rule is to replace a relatively complicated integral by a simpler integral. This is accomplished by changing from the original variable x to a new variable u that is a function of x. Thus, in Example 1 we replaced the integral $\int x^3 \cos(x^4 + 2)\, dx$ by the simpler integral $\tfrac{1}{4} \int \cos u\, du$.

The main challenge in using the Substitution Rule is to think of an appropriate substitution. You should try to choose u to be some function in the integrand whose differential also occurs (except for a constant factor). This was the case in Example 1. If that is not possible, try choosing u to be some complicated part of the integrand. Finding the right substitution is a bit of an art. It's not unusual to guess wrong; if your first guess doesn't work, try another substitution.

EXAMPLE 2 Evaluate $\int \sqrt{2x+1}\, dx$.

SOLUTION 1 Let $u = 2x + 1$. Then $du = 2\, dx$, so $dx = du/2$. Thus, the Substitution Rule gives

$$\int \sqrt{2x+1}\, dx = \int \sqrt{u}\, \frac{du}{2} = \tfrac{1}{2} \int u^{1/2}\, du$$

$$= \frac{1}{2} \cdot \frac{u^{3/2}}{3/2} + C = \tfrac{1}{3} u^{3/2} + C$$

$$= \tfrac{1}{3}(2x+1)^{3/2} + C$$

SOLUTION 2 Another possible substitution is $u = \sqrt{2x+1}$. Then

$$du = \frac{dx}{\sqrt{2x+1}} \qquad \text{so} \qquad dx = \sqrt{2x+1}\, du = u\, du$$

(Or observe that $u^2 = 2x + 1$, so $2u\, du = 2\, dx$.) Therefore

$$\int \sqrt{2x+1}\, dx = \int u \cdot u\, du = \int u^2\, du$$

$$= \frac{u^3}{3} + C = \tfrac{1}{3}(2x+1)^{3/2} + C$$

EXAMPLE 3 Find $\int \dfrac{x}{\sqrt{1-4x^2}}\, dx$.

SOLUTION Let $u = 1 - 4x^2$. Then $du = -8x\, dx$, so $x\, dx = -\tfrac{1}{8} du$ and

$$\int \frac{x}{\sqrt{1-4x^2}}\, dx = -\tfrac{1}{8} \int \frac{du}{\sqrt{u}} = -\tfrac{1}{8} \int u^{-1/2}\, du$$

$$= -\tfrac{1}{8}(2\sqrt{u}) + C = -\tfrac{1}{4}\sqrt{1-4x^2} + C$$

The answer to Example 3 could be checked by differentiation, but instead let's check it with a graph. In Figure 1 we have used a computer to graph both the integrand $f(x) = x/\sqrt{1-4x^2}$ and its indefinite integral $g(x) = -\tfrac{1}{4}\sqrt{1-4x^2}$ (we take the case $C = 0$). Notice that $g(x)$ decreases when $f(x)$ is negative, increases when $f(x)$ is positive, and has its minimum value when $f(x) = 0$. So it seems reasonable, from the graphical evidence, that g is an antiderivative of f.

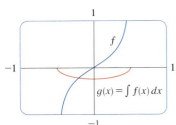

FIGURE 1

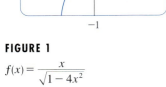

EXAMPLE 4 Calculate $\int \cos 5x\, dx$.

SOLUTION If we let $u = 5x$, then $du = 5\, dx$, so $dx = \tfrac{1}{5} du$. Therefore

$$\int \cos 5x\, dx = \tfrac{1}{5} \int \cos u\, du = \tfrac{1}{5} \sin u + C = \tfrac{1}{5} \sin 5x + C$$

EXAMPLE 5 Find $\int \sqrt{1 + x^2}\, x^5\, dx$.

SOLUTION An appropriate substitution becomes more obvious if we factor x^5 as $x^4 \cdot x$. Let $u = 1 + x^2$. Then $du = 2x\, dx$, so $x\, dx = du/2$. Also $x^2 = u - 1$, so $x^4 = (u - 1)^2$:

$$\int \sqrt{1 + x^2}\, x^5\, dx = \int \sqrt{1 + x^2}\, x^4 \cdot x\, dx$$

$$= \int \sqrt{u}\, (u - 1)^2\, \frac{du}{2} = \tfrac{1}{2} \int \sqrt{u}\, (u^2 - 2u + 1)\, du$$

$$= \tfrac{1}{2} \int (u^{5/2} - 2u^{3/2} + u^{1/2})\, du$$

$$= \tfrac{1}{2} \left(\tfrac{2}{7} u^{7/2} - 2 \cdot \tfrac{2}{5} u^{5/2} + \tfrac{2}{3} u^{3/2} \right) + C$$

$$= \tfrac{1}{7}(1 + x^2)^{7/2} - \tfrac{2}{5}(1 + x^2)^{5/2} + \tfrac{1}{3}(1 + x^2)^{3/2} + C$$

Definite Integrals

When evaluating a *definite* integral by substitution, two methods are possible. One method is to evaluate the indefinite integral first and then use the Fundamental Theorem. For instance, using the result of Example 2, we have

$$\int_0^4 \sqrt{2x + 1}\, dx = \int \sqrt{2x + 1}\, dx \bigg]_0^4 = \tfrac{1}{3}(2x + 1)^{3/2} \bigg]_0^4$$

$$= \tfrac{1}{3}(9)^{3/2} - \tfrac{1}{3}(1)^{3/2} = \tfrac{1}{3}(27 - 1) = \tfrac{26}{3}$$

Another method, which is usually preferable, is to change the limits of integration when the variable is changed.

|||| This rule says that when using a substitution in a definite integral, we must put everything in terms of the new variable u, not only x and dx but also the limits of integration. The new limits of integration are the values of u that correspond to $x = a$ and $x = b$.

5 The Substitution Rule for Definite Integrals If g' is continuous on $[a, b]$ and f is continuous on the range of $u = g(x)$, then

$$\int_a^b f(g(x))g'(x)\, dx = \int_{g(a)}^{g(b)} f(u)\, du$$

Proof Let F be an antiderivative of f. Then, by (3), $F(g(x))$ is an antiderivative of $f(g(x))g'(x)$, so by Part 2 of the Fundamental Theorem, we have

$$\int_a^b f(g(x))g'(x)\, dx = F(g(x)) \bigg]_a^b = F(g(b)) - F(g(a))$$

But, applying FTC2 a second time, we also have

$$\int_{g(a)}^{g(b)} f(u)\, du = F(u) \bigg]_{g(a)}^{g(b)} = F(g(b)) - F(g(a))$$

EXAMPLE 6 Evaluate $\int_0^4 \sqrt{2x + 1}\, dx$ using (5).

SOLUTION Using the substitution from Solution 1 of Example 2, we have $u = 2x + 1$ and $dx = du/2$. To find the new limits of integration we note that

$$\text{when } x = 0,\ u = 1 \quad \text{and} \quad \text{when } x = 4,\ u = 9$$

Therefore
$$\int_0^4 \sqrt{2x+1}\,dx = \int_1^9 \tfrac{1}{2}\sqrt{u}\,du$$

|||| The geometric interpretation of Example 6 is shown in Figure 2. The substitution $u = 2x + 1$ stretches the interval $[0, 4]$ by a factor of 2 and translates it to the right by 1 unit. The Substitution Rule shows that the two areas are equal.

$$= \tfrac{1}{2} \cdot \tfrac{2}{3} u^{3/2} \Big]_1^9$$

$$= \tfrac{1}{3}(9^{3/2} - 1^{3/2}) = \tfrac{26}{3}$$

Observe that when using (5) we do not return to the variable x after integrating. We simply evaluate the expression in u between the appropriate values of u.

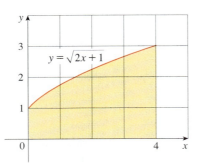
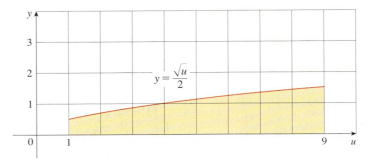

FIGURE 2

EXAMPLE 7 Evaluate $\displaystyle\int_1^2 \frac{dx}{(3-5x)^2}$.

|||| The integral given in Example 7 is an abbreviation for
$$\int_1^2 \frac{1}{(3-5x)^2}\,dx$$

SOLUTION Let $u = 3 - 5x$. Then $du = -5\,dx$, so $dx = -du/5$. When $x = 1$, $u = -2$ and when $x = 2$, $u = -7$. Thus

$$\int_1^2 \frac{dx}{(3-5x)^2} = -\frac{1}{5}\int_{-2}^{-7}\frac{du}{u^2}$$

$$= -\frac{1}{5}\left[-\frac{1}{u}\right]_{-2}^{-7} = \frac{1}{5u}\bigg]_{-2}^{-7}$$

$$= \frac{1}{5}\left(-\frac{1}{7} + \frac{1}{2}\right) = \frac{1}{14}$$

|||| **Symmetry**

The next theorem uses the Substitution Rule for Definite Integrals (5) to simplify the calculation of integrals of functions that possess symmetry properties.

> **6 Integrals of Symmetric Functions** Suppose f is continuous on $[-a, a]$.
> (a) If f is even $[f(-x) = f(x)]$, then $\int_{-a}^a f(x)\,dx = 2\int_0^a f(x)\,dx$.
> (b) If f is odd $[f(-x) = -f(x)]$, then $\int_{-a}^a f(x)\,dx = 0$.

Proof We split the integral in two:

$$\boxed{7} \quad \int_{-a}^a f(x)\,dx = \int_{-a}^0 f(x)\,dx + \int_0^a f(x)\,dx = -\int_0^{-a} f(x)\,dx + \int_0^a f(x)\,dx$$

In the first integral on the far right side we make the substitution $u = -x$. Then

$du = -dx$ and when $x = -a$, $u = a$. Therefore

$$-\int_0^{-a} f(x)\,dx = -\int_0^a f(-u)(-du) = \int_0^a f(-u)\,du$$

and so Equation 7 becomes

$$\boxed{8} \qquad \int_{-a}^a f(x)\,dx = \int_0^a f(-u)\,du + \int_0^a f(x)\,dx$$

(a) If f is even, then $f(-u) = f(u)$ so Equation 8 gives

$$\int_{-a}^a f(x)\,dx = \int_0^a f(u)\,du + \int_0^a f(x)\,dx = 2\int_0^a f(x)\,dx$$

(b) If f is odd, then $f(-u) = -f(u)$ and so Equation 8 gives

$$\int_{-a}^a f(x)\,dx = -\int_0^a f(u)\,du + \int_0^a f(x)\,dx = 0$$

Theorem 6 is illustrated by Figure 3. For the case where f is positive and even, part (a) says that the area under $y = f(x)$ from $-a$ to a is twice the area from 0 to a because of symmetry. Recall that an integral $\int_a^b f(x)\,dx$ can be expressed as the area above the x-axis and below $y = f(x)$ minus the area below the axis and above the curve. Thus, part (b) says the integral is 0 because the areas cancel.

EXAMPLE 8 Since $f(x) = x^6 + 1$ satisfies $f(-x) = f(x)$, it is even and so

$$\int_{-2}^{2} (x^6 + 1)\,dx = 2\int_0^2 (x^6 + 1)\,dx$$
$$= 2\left[\tfrac{1}{7}x^7 + x\right]_0^2 = 2\left(\tfrac{128}{7} + 2\right) = \tfrac{284}{7}$$

EXAMPLE 9 Since $f(x) = (\tan x)/(1 + x^2 + x^4)$ satisfies $f(-x) = -f(x)$, it is odd and so

$$\int_{-1}^1 \frac{\tan x}{1 + x^2 + x^4}\,dx = 0$$

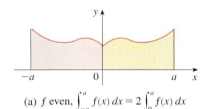

(a) f even, $\int_{-a}^a f(x)\,dx = 2\int_0^a f(x)\,dx$

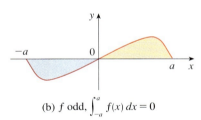

(b) f odd, $\int_{-a}^a f(x)\,dx = 0$

FIGURE 3

5.5 Exercises

1–6 Evaluate the integral by making the given substitution.

1. $\int \cos 3x\,dx$, $u = 3x$

2. $\int x(4 + x^2)^{10}\,dx$, $u = 4 + x^2$

3. $\int x^2\sqrt{x^3 + 1}\,dx$, $u = x^3 + 1$

4. $\int \dfrac{\sin\sqrt{x}}{\sqrt{x}}\,dx$, $u = \sqrt{x}$

5. $\int \dfrac{4}{(1 + 2x)^3}\,dx$, $u = 1 + 2x$

6. $\int \cos^4\theta \sin\theta\,d\theta$, $u = \cos\theta$

7–32 ▌▌▌▌ Evaluate the indefinite integral.

7. $\int 2x(x^2 + 3)^4 \, dx$

8. $\int x^2(x^3 + 5)^9 \, dx$

9. $\int (3x - 2)^{20} \, dx$

10. $\int (2 - x)^6 \, dx$

11. $\int \dfrac{1 + 4x}{\sqrt{1 + x + 2x^2}} \, dx$

12. $\int \dfrac{x}{(x^2 + 1)^2} \, dx$

13. $\int \dfrac{3}{(2y + 1)^5} \, dy$

14. $\int \dfrac{1}{(5t + 4)^{2.7}} \, dt$

15. $\int \sqrt{4 - t} \, dt$

16. $\int y^3 \sqrt{2y^4 - 1} \, dy$

17. $\int \sin \pi t \, dt$

18. $\int \sec 2\theta \tan 2\theta \, d\theta$

19. $\int \dfrac{\cos \sqrt{t}}{\sqrt{t}} \, dt$

20. $\int \sqrt{x} \sin(1 + x^{3/2}) \, dx$

21. $\int \cos \theta \sin^6 \theta \, d\theta$

22. $\int (1 + \tan \theta)^5 \sec^2 \theta \, d\theta$

23. $\int \dfrac{z^2}{\sqrt[3]{1 + z^3}} \, dz$

24. $\int \dfrac{ax + b}{\sqrt{ax^2 + 2bx + c}} \, dx$

25. $\int \sqrt{\cot x} \, \csc^2 x \, dx$

26. $\int \dfrac{\cos(\pi/x)}{x^2} \, dx$

27. $\int \sec^3 x \tan x \, dx$

28. $\int \sqrt[3]{x^3 + 1} \, x^5 \, dx$

29. $\int x^a \sqrt{b + cx^{a+1}} \, dx \quad (c \neq 0, \, a \neq -1)$

30. $\int \sin t \, \sec^2(\cos t) \, dt$

31. $\int \dfrac{x}{\sqrt[4]{x + 2}} \, dx$

32. $\int \dfrac{x^2}{\sqrt{1 - x}} \, dx$

33–36 ▌▌▌▌ Evaluate the indefinite integral. Illustrate and check that your answer is reasonable by graphing both the function and its antiderivative (take $C = 0$).

33. $\int \dfrac{3x - 1}{(3x^2 - 2x + 1)^4} \, dx$

34. $\int \dfrac{x}{\sqrt{x^2 + 1}} \, dx$

35. $\int \sin^3 x \cos x \, dx$

36. $\int \tan^2 \theta \sec^2 \theta \, d\theta$

37–54 ▌▌▌▌ Evaluate the definite integral, if it exists.

37. $\int_0^2 (x - 1)^{25} \, dx$

38. $\int_0^7 \sqrt{4 + 3x} \, dx$

39. $\int_0^1 x^2(1 + 2x^3)^5 \, dx$

40. $\int_0^{\sqrt{\pi}} x \cos(x^2) \, dx$

41. $\int_0^\pi \sec^2(t/4) \, dt$

42. $\int_{1/6}^{1/2} \csc \pi t \cot \pi t \, dt$

43. $\int_{-\pi/6}^{\pi/6} \tan^3 \theta \, d\theta$

44. $\int_0^2 \dfrac{dx}{(2x - 3)^2}$

45. $\int_0^{\pi/3} \dfrac{\sin \theta}{\cos^2 \theta} \, d\theta$

46. $\int_{-\pi/2}^{\pi/2} \dfrac{x^2 \sin x}{1 + x^6} \, dx$

47. $\int_0^{13} \dfrac{dx}{\sqrt[3]{(1 + 2x)^2}}$

48. $\int_0^{\pi/2} \cos x \sin(\sin x) \, dx$

49. $\int_1^2 x\sqrt{x - 1} \, dx$

50. $\int_0^4 \dfrac{x}{\sqrt{1 + 2x}} \, dx$

51. $\int_0^4 \dfrac{dx}{(x - 2)^3}$

52. $\int_0^a x\sqrt{a^2 - x^2} \, dx$

53. $\int_0^a x\sqrt{x^2 + a^2} \, dx \quad (a > 0)$

54. $\int_{-a}^a x\sqrt{x^2 + a^2} \, dx$

55–56 ▌▌▌▌ Use a graph to give a rough estimate of the area of the region that lies under the given curve. Then find the exact area.

55. $y = \sqrt{2x + 1}, \quad 0 \leq x \leq 1$

56. $y = 2 \sin x - \sin 2x, \quad 0 \leq x \leq \pi$

57. Evaluate $\int_{-2}^2 (x + 3)\sqrt{4 - x^2} \, dx$ by writing it as a sum of two integrals and interpreting one of those integrals in terms of an area.

58. Evaluate $\int_0^1 x\sqrt{1 - x^4} \, dx$ by making a substitution and interpreting the resulting integral in terms of an area.

59. Breathing is cyclic and a full respiratory cycle from the beginning of inhalation to the end of exhalation takes about 5 s. The maximum rate of air flow into the lungs is about 0.5 L/s. This explains, in part, why the function $f(t) = \tfrac{1}{2}\sin(2\pi t/5)$ has often been used to model the rate of air flow into the lungs. Use this model to find the volume of inhaled air in the lungs at time t.

60. Alabama Instruments Company has set up a production line to manufacture a new calculator. The rate of production of these calculators after t weeks is

$$\dfrac{dx}{dt} = 5000\left(1 - \dfrac{100}{(t + 10)^2}\right) \text{ calculators/week}$$

(Notice that production approaches 5000 per week as time goes on, but the initial production is lower because of the workers' unfamiliarity with the new techniques.) Find the number of calculators produced from the beginning of the third week to the end of the fourth week.

61. If f is continuous and $\int_0^4 f(x)\,dx = 10$, find $\int_0^2 f(2x)\,dx$.

62. If f is continuous and $\int_0^9 f(x)\,dx = 4$, find $\int_0^3 xf(x^2)\,dx$.

63. Suppose f is continuous on $\mathbb{R}$.
(a) Prove that
$$\int_a^b f(-x)\,dx = \int_{-b}^{-a} f(x)\,dx$$
For the case where $f(x) \geq 0$ and $0 < a < b$, draw a diagram to interpret this equation geometrically as an equality of areas.
(b) Prove that
$$\int_a^b f(x+c)\,dx = \int_{a+c}^{b+c} f(x)\,dx$$
For the case where $f(x) \geq 0$, draw a diagram to interpret this equation geometrically as an equality of areas.

64. Show that the area under the graph of $y = \sin\sqrt{x}$ from 0 to 4 is the same as the area under the graph of $y = 2x \sin x$ from 0 to 2.

65. If a and b are positive numbers, show that
$$\int_0^1 x^a(1-x)^b\,dx = \int_0^1 x^b(1-x)^a\,dx$$

66. Use the substitution $u = \pi - x$ to show that
$$\int_0^\pi x f(\sin x)\,dx = \frac{\pi}{2}\int_0^\pi f(\sin x)\,dx$$

The following exercises are intended only for those who have already covered Chapter 7.

67–82 ▪ Evaluate the integral.

67. $\displaystyle\int \frac{dx}{5 - 3x}$

68. $\displaystyle\int \frac{x}{x^2 + 1}\,dx$

69. $\displaystyle\int \frac{(\ln x)^2}{x}\,dx$

70. $\displaystyle\int \frac{\tan^{-1}x}{1 + x^2}\,dx$

71. $\displaystyle\int e^x\sqrt{1 + e^x}\,dx$

72. $\displaystyle\int e^{\cos t}\sin t\,dt$

73. $\displaystyle\int \frac{dx}{x \ln x}$

74. $\displaystyle\int \frac{e^x}{e^x + 1}\,dx$

75. $\displaystyle\int \cot x\,dx$

76. $\displaystyle\int \frac{\sin x}{1 + \cos^2 x}\,dx$

77. $\displaystyle\int \frac{1 + x}{1 + x^2}\,dx$

78. $\displaystyle\int \frac{x}{1 + x^4}\,dx$

79. $\displaystyle\int_1^2 \frac{e^{1/x}}{x^2}\,dx$

80. $\displaystyle\int_0^1 xe^{-x^2}\,dx$

81. $\displaystyle\int_e^{e^4} \frac{dx}{x\sqrt{\ln x}}$

82. $\displaystyle\int_0^{1/2} \frac{\sin^{-1}x}{\sqrt{1-x^2}}\,dx$

83. Use Exercise 66 to evaluate the integral
$$\int_0^\pi \frac{x \sin x}{1 + \cos^2 x}\,dx$$

5 Review

CONCEPT CHECK

1. (a) Write an expression for a Riemann sum of a function f. Explain the meaning of the notation that you use.
   (b) If $f(x) \geq 0$, what is the geometric interpretation of a Riemann sum? Illustrate with a diagram.
   (c) If $f(x)$ takes on both positive and negative values, what is the geometric interpretation of a Riemann sum? Illustrate with a diagram.

2. (a) Write the definition of the definite integral of a continuous function from a to b.
   (b) What is the geometric interpretation of $\int_a^b f(x)\, dx$ if $f(x) \geq 0$?
   (c) What is the geometric interpretation of $\int_a^b f(x)\, dx$ if $f(x)$ takes on both positive and negative values? Illustrate with a diagram.

3. State both parts of the Fundamental Theorem of Calculus.

4. (a) State the Net Change Theorem.
   (b) If $r(t)$ is the rate at which water flows into a reservoir, what does $\int_{t_1}^{t_2} r(t)\, dt$ represent?

5. Suppose a particle moves back and forth along a straight line with velocity $v(t)$, measured in feet per second, and acceleration $a(t)$.
   (a) What is the meaning of $\int_{60}^{120} v(t)\, dt$?
   (b) What is the meaning of $\int_{60}^{120} |v(t)|\, dt$?
   (c) What is the meaning of $\int_{60}^{120} a(t)\, dt$?

6. (a) Explain the meaning of the indefinite integral $\int f(x)\, dx$.
   (b) What is the connection between the definite integral $\int_a^b f(x)\, dx$ and the indefinite integral $\int f(x)\, dx$?

7. Explain exactly what is meant by the statement that "differentiation and integration are inverse processes."

8. State the Substitution Rule. In practice, how do you use it?

TRUE-FALSE QUIZ

Determine whether the statement is true or false. If it is true, explain why. If it is false, explain why or give an example that disproves the statement.

1. If f and g are continuous on $[a, b]$, then
$$\int_a^b [f(x) + g(x)]\, dx = \int_a^b f(x)\, dx + \int_a^b g(x)\, dx$$

2. If f and g are continuous on $[a, b]$, then
$$\int_a^b [f(x)g(x)]\, dx = \left(\int_a^b f(x)\, dx\right)\left(\int_a^b g(x)\, dx\right)$$

3. If f is continuous on $[a, b]$, then
$$\int_a^b 5f(x)\, dx = 5\int_a^b f(x)\, dx$$

4. If f is continuous on $[a, b]$, then
$$\int_a^b xf(x)\, dx = x\int_a^b f(x)\, dx$$

5. If f is continuous on $[a, b]$ and $f(x) \geq 0$, then
$$\int_a^b \sqrt{f(x)}\, dx = \sqrt{\int_a^b f(x)\, dx}$$

6. If f' is continuous on $[1, 3]$, then $\int_1^3 f'(v)\, dv = f(3) - f(1)$.

7. If f and g are continuous and $f(x) \geq g(x)$ for $a \leq x \leq b$, then
$$\int_a^b f(x)\, dx \geq \int_a^b g(x)\, dx$$

8. If f and g are differentiable and $f(x) \geq g(x)$ for $a < x < b$, then $f'(x) \geq g'(x)$ for $a < x < b$.

9. $\int_{-1}^{1} \left(x^5 - 6x^9 + \dfrac{\sin x}{(1 + x^4)^2}\right) dx = 0$

10. $\int_{-5}^{5} (ax^2 + bx + c)\, dx = 2\int_0^5 (ax^2 + c)\, dx$

11. $\int_{-2}^{1} \dfrac{1}{x^4}\, dx = -\dfrac{3}{8}$

12. $\int_0^2 (x - x^3)\, dx$ represents the area under the curve $y = x - x^3$ from 0 to 2.

13. All continuous functions have derivatives.

14. All continuous functions have antiderivatives.

EXERCISES

1. Use the given graph of f to find the Riemann sum with six subintervals. Take the sample points to be (a) left endpoints and (b) midpoints. In each case draw a diagram and explain what the Riemann sum represents.

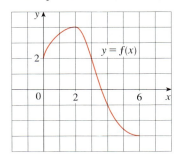

2. (a) Evaluate the Riemann sum for
$$f(x) = x^2 - x \qquad 0 \le x \le 2$$
with four subintervals, taking the sample points to be right endpoints. Explain, with the aid of a diagram, what the Riemann sum represents.
(b) Use the definition of a definite integral (with right endpoints) to calculate the value of the integral
$$\int_0^2 (x^2 - x)\, dx$$
(c) Use the Fundamental Theorem to check your answer to part (b).
(d) Draw a diagram to explain the geometric meaning of the integral in part (b).

3. Evaluate
$$\int_0^1 \left(x + \sqrt{1 - x^2}\right) dx$$
by interpreting it in terms of areas.

4. Express
$$\lim_{n \to \infty} \sum_{i=1}^n \sin x_i \, \Delta x$$
as a definite integral on the interval $[0, \pi]$ and then evaluate the integral.

5. If $\int_0^6 f(x)\, dx = 10$ and $\int_0^4 f(x)\, dx = 7$, find $\int_4^6 f(x)\, dx$.

[CAS] **6.** (a) Write $\int_1^5 (x + 2x^5)\, dx$ as a limit of Riemann sums, taking the sample points to be right endpoints. Use a computer algebra system to evaluate the sum and to compute the limit.
(b) Use the Fundamental Theorem to check your answer to part (a).

7. The figure shows the graphs of f, f', and $\int_0^x f(t)\, dt$. Identify each graph, and explain your choices.

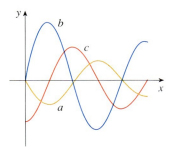

8. Evaluate:
(a) $\int_0^{\pi/2} \dfrac{d}{dx} \left(\sin \dfrac{x}{2} \cos \dfrac{x}{3} \right) dx$
(b) $\dfrac{d}{dx} \int_0^{\pi/2} \sin \dfrac{x}{2} \cos \dfrac{x}{3}\, dx$
(c) $\dfrac{d}{dx} \int_x^{\pi/2} \sin \dfrac{t}{2} \cos \dfrac{t}{3}\, dt$

9–28 ▮▮▮▮ Evaluate the integral, if it exists.

9. $\int_1^2 (8x^3 + 3x^2)\, dx$

10. $\int_0^T (x^4 - 8x + 7)\, dx$

11. $\int_0^1 (1 - x^9)\, dx$

12. $\int_0^1 (1 - x)^9\, dx$

13. $\int_1^9 \dfrac{\sqrt{u} - 2u^2}{u}\, du$

14. $\int_0^1 (\sqrt[4]{u} + 1)^2\, du$

15. $\int_0^1 y(y^2 + 1)^5\, dy$

16. $\int_0^2 y^2 \sqrt{1 + y^3}\, dy$

17. $\int_1^5 \dfrac{dt}{(t - 4)^2}$

18. $\int_0^1 \sin(3\pi t)\, dt$

19. $\int_0^1 v^2 \cos(v^3)\, dv$

20. $\int_{-1}^1 \dfrac{\sin x}{1 + x^2}\, dx$

21. $\int \dfrac{x + 2}{\sqrt{x^2 + 4x}}\, dx$

22. $\int \csc^2 3t\, dt$

23. $\int \sin \pi t \cos \pi t\, dt$

24. $\int \sin x \cos(\cos x)\, dx$

25. $\int_0^{\pi/8} \sec 2\theta \tan 2\theta\, d\theta$

26. $\int_0^{\pi/4} (1 + \tan t)^3 \sec^2 t\, dt$

27. $\int_0^3 |x^2 - 4|\, dx$

28. $\int_0^4 |\sqrt{x} - 1|\, dx$

29–30 ▪ Evaluate the indefinite integral. Illustrate and check that your answer is reasonable by graphing both the function and its antiderivative (take $C = 0$).

29. $\displaystyle\int \frac{\cos x}{\sqrt{1 + \sin x}}\, dx$

30. $\displaystyle\int \frac{x^3}{\sqrt{x^2 + 1}}\, dx$

31. Use a graph to give a rough estimate of the area of the region that lies under the curve $y = x\sqrt{x}$, $0 \leq x \leq 4$. Then find the exact area.

32. Graph the function $f(x) = \cos^2 x \, \sin^3 x$ and use the graph to guess the value of the integral $\int_0^{2\pi} f(x)\, dx$. Then evaluate the integral to confirm your guess.

33–38 ▪ Find the derivative of the function.

33. $\displaystyle F(x) = \int_1^x \sqrt{1 + t^4}\, dt$

34. $\displaystyle F(x) = \int_\pi^x \tan(s^2)\, ds$

35. $\displaystyle g(x) = \int_0^{x^3} \frac{t}{\sqrt{1 + t^3}}\, dt$

36. $\displaystyle g(x) = \int_1^{\cos x} \sqrt[3]{1 - t^2}\, dt$

37. $\displaystyle y = \int_{\sqrt{x}}^x \frac{\cos \theta}{\theta}\, d\theta$

38. $\displaystyle y = \int_{2x}^{3x+1} \sin(t^4)\, dt$

39–40 ▪ Use Property 8 of integrals to estimate the value of the integral.

39. $\displaystyle\int_1^3 \sqrt{x^2 + 3}\, dx$

40. $\displaystyle\int_3^5 \frac{1}{x + 1}\, dx$

41–42 ▪ Use the properties of integrals to verify the inequality.

41. $\displaystyle\int_0^1 x^2 \cos x\, dx \leq \frac{1}{3}$

42. $\displaystyle\int_{\pi/4}^{\pi/2} \frac{\sin x}{x}\, dx \leq \frac{\sqrt{2}}{2}$

43. Use the Midpoint Rule with $n = 5$ to approximate $\int_0^1 \sqrt{1 + x^3}\, dx$.

44. A particle moves along a line with velocity function $v(t) = t^2 - t$, where v is measured in meters per second. Find (a) the displacement and (b) the distance traveled by the particle during the time interval $[0, 5]$.

45. Let $r(t)$ be the rate at which the world's oil is consumed, where t is measured in years starting at $t = 0$ on January 1, 2000, and $r(t)$ is measured in barrels per year. What does $\int_0^3 r(t)\, dt$ represent?

46. A radar gun was used to record the speed of a runner at the times given in the table. Use the Midpoint Rule to estimate the distance the runner covered during those 5 seconds.

t (s)	v (m/s)	t (s)	v (m/s)
0	0	3.0	10.51
0.5	4.67	3.5	10.67
1.0	7.34	4.0	10.76
1.5	8.86	4.5	10.81
2.0	9.73	5.0	10.81
2.5	10.22		

47. A population of honeybees increased at a rate of $r(t)$ bees per week, where the graph of r is as shown. Use the Midpoint Rule with six subintervals to estimate the increase in the bee population during the first 24 weeks.

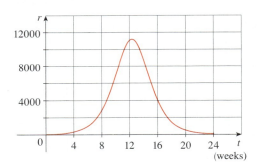

48. Let

$$f(x) = \begin{cases} -x - 1 & \text{if } -3 \leq x \leq 0 \\ -\sqrt{1 - x^2} & \text{if } 0 \leq x \leq 1 \end{cases}$$

Evaluate $\int_{-3}^1 f(x)\, dx$ by interpreting the integral as a difference of areas.

49. Find an antiderivative F of $f(x) = x^2 \sin(x^2)$ such that $F(1) = 0$.

50. The Fresnel function $S(x) = \int_0^x \sin(\frac{1}{2}\pi t^2)\, dt$ was introduced in Section 5.3. Fresnel also used the function

$$C(x) = \int_0^x \cos(\tfrac{1}{2}\pi t^2)\, dt$$

in his theory of the diffraction of light waves.
(a) On what intervals is C increasing?
(b) On what intervals is C concave upward?
[CAS] (c) Use a graph to solve the following equation correct to two decimal places:

$$\int_0^x \cos(\tfrac{1}{2}\pi t^2)\, dt = 0.7$$

[CAS] (d) Plot the graphs of C and S on the same screen. How are these graphs related?

51. If f is a continuous function such that

$$\int_0^x f(t)\, dt = x \sin x + \int_0^x \frac{f(t)}{1 + t^2}\, dt$$

for all x, find an explicit formula for $f(x)$.

52. Find a function f and a value of the constant a such that

$$2 \int_a^x f(t)\, dt = 2 \sin x - 1$$

53. If f' is continuous on $[a, b]$, show that

$$2 \int_a^b f(x) f'(x)\, dx = [f(b)]^2 - [f(a)]^2$$

54. Find $\displaystyle\lim_{h \to 0} \frac{1}{h} \int_2^{2+h} \sqrt{1 + t^3}\, dt$.

55. If f is continuous on $[0, 1]$, prove that

$$\int_0^1 f(x)\, dx = \int_0^1 f(1 - x)\, dx$$

56. Evaluate

$$\lim_{n \to \infty} \frac{1}{n}\left[\left(\frac{1}{n}\right)^9 + \left(\frac{2}{n}\right)^9 + \left(\frac{3}{n}\right)^9 + \cdots + \left(\frac{n}{n}\right)^9\right]$$

PROBLEMS PLUS

Before you look at the solution of the following example, cover it up and first try to solve the problem yourself.

EXAMPLE 1 Evaluate $\lim\limits_{x \to 3} \left(\dfrac{x}{x-3} \int_3^x \dfrac{\sin t}{t}\, dt \right)$.

SOLUTION Let's start by having a preliminary look at the ingredients of the function. What happens to the first factor, $x/(x-3)$, when x approaches 3? The numerator approaches 3 and the denominator approaches 0, so we have

$$\dfrac{x}{x-3} \to \infty \quad \text{as} \quad x \to 3^+ \quad \text{and} \quad \dfrac{x}{x-3} \to -\infty \quad \text{as} \quad x \to 3^-$$

The second factor approaches $\int_3^3 (\sin t)/t\, dt$, which is 0. It's not clear what happens to the function as a whole. (One factor is becoming large while the other is becoming small.) So how do we proceed?

■ The principles of problem solving are discussed on page 58.

One of the principles of problem solving is *recognizing something familiar*. Is there a part of the function that reminds us of something we've seen before? Well, the integral

$$\int_3^x \dfrac{\sin t}{t}\, dt$$

has x as its upper limit of integration and that type of integral occurs in Part 1 of the Fundamental Theorem of Calculus:

$$\dfrac{d}{dx} \int_a^x f(t)\, dt = f(x)$$

This suggests that differentiation might be involved.

Once we start thinking about differentiation, the denominator $(x-3)$ reminds us of something else that should be familiar: One of the forms of the definition of the derivative in Chapter 3 is

$$F'(a) = \lim_{x \to a} \dfrac{F(x) - F(a)}{x - a}$$

and with $a = 3$ this becomes

$$F'(3) = \lim_{x \to 3} \dfrac{F(x) - F(3)}{x - 3}$$

So what is the function F in our situation? Notice that if we define

$$F(x) = \int_3^x \dfrac{\sin t}{t}\, dt$$

then $F(3) = 0$. What about the factor x in the numerator? That's just a red herring, so let's factor it out and put together the calculation:

$$\lim_{x \to 3} \left(\dfrac{x}{x-3} \int_3^x \dfrac{\sin t}{t}\, dt \right) = \lim_{x \to 3} x \cdot \lim_{x \to 3} \dfrac{\int_3^x \dfrac{\sin t}{t}\, dt}{x-3}$$

■ Another approach is to use l'Hospital's Rule.

$$= 3 \lim_{x \to 3} \dfrac{F(x) - F(3)}{x - 3}$$

$$= 3F'(3)$$

$$= 3 \dfrac{\sin 3}{3} \quad \text{(FTC1)}$$

$$= \sin 3$$

PROBLEMS

1. If $x \sin \pi x = \int_0^{x^2} f(t)\, dt$, where f is a continuous function, find $f(4)$.

2. In this problem we approximate the sine function on the interval $[0, \pi]$ by three quadratic functions, each of which has the same zeros as the sine function on this interval.
   (a) Find a quadratic function f such that $f(0) = f(\pi) = 0$ and which has the same maximum value as sin on $[0, \pi]$.
   (b) Find a quadratic function g such that $g(0) = g(\pi) = 0$ and which has the same rate of change as the sine function at 0 and π.
   (c) Find a quadratic function h such that $h(0) = h(\pi) = 0$ and the area under h from 0 to π is the same as for the sine function.
   (d) Illustrate by graphing f, g, h, and the sine function in the same viewing rectangle $[0, \pi]$ by $[0, 1]$. Identify which graph belongs to each function.

3. Show that $\dfrac{1}{17} \leq \int_1^2 \dfrac{1}{1+x^4}\, dx \leq \dfrac{7}{24}$.

4. Suppose the curve $y = f(x)$ passes through the origin and the point $(1, 1)$. Find the value of the integral $\int_0^1 f'(x)\, dx$.

5. Find a function f such that $f(1) = -1$, $f(4) = 7$, and $f'(x) > 3$ for all x, or prove that such a function cannot exist.

6. (a) Graph several members of the family of functions $f(x) = (2cx - x^2)/c^3$ for $c > 0$ and look at the regions enclosed by these curves and the x-axis. Make a conjecture about how the areas of these regions are related.
   (b) Prove your conjecture in part (a).
   (c) Take another look at the graphs in part (a) and use them to sketch the curve traced out by the vertices (highest points) of the family of functions. Can you guess what kind of curve this is?
   (d) Find the equation of the curve you sketched in part (c).

7. If $f(x) = \int_0^{g(x)} \dfrac{1}{\sqrt{1+t^3}}\, dt$, where $g(x) = \int_0^{\cos x}[1 + \sin(t^2)]\, dt$, find $f'(\pi/2)$.

8. If $f(x) = \int_0^x x^2 \sin(t^2)\, dt$, find $f'(x)$.

9. Find the interval $[a, b]$ for which the value of the integral $\int_a^b (2 + x - x^2)\, dx$ is a maximum.

10. Use an integral to estimate the sum $\sum_{i=1}^{10000} \sqrt{i}$.

11. (a) Evaluate $\int_0^n [\![x]\!]\, dx$, where n is a positive integer.
    (b) Evaluate $\int_a^b [\![x]\!]\, dx$, where a and b are real numbers with $0 \leq a < b$.

12. Find $\dfrac{d^2}{dx^2} \int_0^x \left(\int_1^{\sin t} \sqrt{1 + u^4}\, du \right) dt$.

13. If f is a differentiable function such that $\int_0^x f(t)\, dt = [f(x)]^2$ for all x, find f.

14. A circular disk of radius r is used in an evaporator and is rotated in a vertical plane. If it is to be partially submerged in the liquid so as to maximize the exposed wetted area of the disk, show that the center of the disk should be positioned at a height $r/\sqrt{1 + \pi^2}$ above the surface of the liquid.

15. Prove that if f is continuous, then $\int_0^x f(u)(x - u)\, du = \int_0^x \left(\int_0^u f(t)\, dt \right) du$.

16. The figure shows a region consisting of all points inside a square that are closer to the center than to the sides of the square. Find the area of the region.

17. Evaluate $\displaystyle\lim_{n\to\infty} \left(\dfrac{1}{\sqrt{n}\sqrt{n+1}} + \dfrac{1}{\sqrt{n}\sqrt{n+2}} + \cdots + \dfrac{1}{\sqrt{n}\sqrt{n+n}} \right)$.

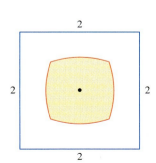

FIGURE FOR PROBLEM 16

Chapter 6

The volume of a sphere is the limit of sums of volumes of approximating cylinders.

Applications of Integration

In this chapter we explore some of the applications of the definite integral by using it to compute areas between curves, volumes of solids, and the work done by a varying force. The common theme is the following general method, which is similar to the one we used to find areas under curves: We break up a quantity Q into a large number of small parts. We next approximate each small part by a quantity of the form $f(x_i^*)\,\Delta x$ and thus approximate Q by a Riemann sum. Then we take the limit and express Q as an integral. Finally we evaluate the integral using the Fundamental Theorem of Calculus or the Midpoint Rule.

6.1 Areas between Curves

In Chapter 5 we defined and calculated areas of regions that lie under the graphs of functions. Here we use integrals to find areas of regions that lie between the graphs of two functions.

Consider the region S that lies between two curves $y = f(x)$ and $y = g(x)$ and between the vertical lines $x = a$ and $x = b$, where f and g are continuous functions and $f(x) \geq g(x)$ for all x in $[a, b]$. (See Figure 1.)

Just as we did for areas under curves in Section 5.1, we divide S into n strips of equal width and then we approximate the ith strip by a rectangle with base Δx and height $f(x_i^*) - g(x_i^*)$. (See Figure 2. If we like, we could take all of the sample points to be right endpoints, in which case $x_i^* = x_i$.) The Riemann sum

$$\sum_{i=1}^{n} [f(x_i^*) - g(x_i^*)]\,\Delta x$$

is therefore an approximation to what we intuitively think of as the area of S.

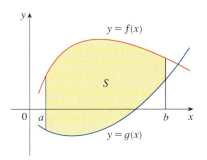

FIGURE 1
$S = \{(x, y) \mid a \leq x \leq b, g(x) \leq y \leq f(x)\}$

Guess the area of an island.
Resources / Module 7
/ Areas
/ Start of Areas

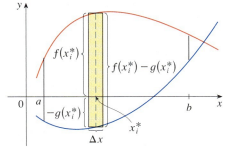 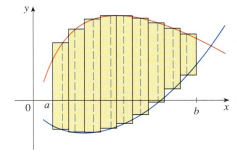

FIGURE 2 (a) Typical rectangle   (b) Approximating rectangles

This approximation appears to become better and better as $n \to \infty$. Therefore, we define the **area** A of S as the limiting value of the sum of the areas of these approximating rectangles.

$$\boxed{1 \qquad A = \lim_{n \to \infty} \sum_{i=1}^{n} [f(x_i^*) - g(x_i^*)]\,\Delta x}$$

We recognize the limit in (1) as the definite integral of $f - g$. Therefore, we have the following formula for area.

> **2** The area A of the region bounded by the curves $y = f(x)$, $y = g(x)$, and the lines $x = a$, $x = b$, where f and g are continuous and $f(x) \geq g(x)$ for all x in $[a, b]$, is
> $$A = \int_a^b [f(x) - g(x)]\, dx$$

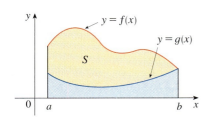

FIGURE 3
$A = \int_a^b f(x)\, dx - \int_a^b g(x)\, dx$

Notice that in the special case where $g(x) = 0$, S is the region under the graph of f and our general definition of area (1) reduces to our previous definition (Definition 2 in Section 5.1).

In the case where both f and g are positive, you can see from Figure 3 why (2) is true:

$$A = [\text{area under } y = f(x)] - [\text{area under } y = g(x)]$$
$$= \int_a^b f(x)\, dx - \int_a^b g(x)\, dx = \int_a^b [f(x) - g(x)]\, dx$$

EXAMPLE 1 Find the area of the region bounded above by $y = x^2 + 1$, bounded below by $y = x$, and bounded on the sides by $x = 0$ and $x = 1$.

SOLUTION The region is shown in Figure 4. The upper boundary curve is $y = x^2 + 1$ and the lower boundary curve is $y = x$. So we use the area formula (2) with $f(x) = x^2 + 1$, $g(x) = x$, $a = 0$, and $b = 1$:

$$A = \int_0^1 [(x^2 + 1) - x]\, dx = \int_0^1 (x^2 - x + 1)\, dx$$
$$= \frac{x^3}{3} - \frac{x^2}{2} + x \bigg]_0^1 = \frac{1}{3} - \frac{1}{2} + 1 = \frac{5}{6}$$

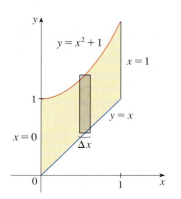

FIGURE 4

In Figure 4 we drew a typical approximating rectangle with width Δx as a reminder of the procedure by which the area is defined in (1). In general, when we set up an integral for an area, it's helpful to sketch the region to identify the top curve y_T, the bottom curve y_B, and a typical approximating rectangle as in Figure 5. Then the area of a typical rectangle is $(y_T - y_B)\, \Delta x$ and the equation

$$A = \lim_{n \to \infty} \sum_{i=1}^n (y_T - y_B)\, \Delta x = \int_a^b (y_T - y_B)\, dx$$

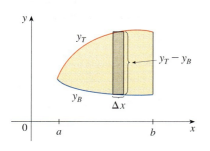

FIGURE 5

summarizes the procedure of adding (in a limiting sense) the areas of all the typical rectangles.

Notice that in Figure 5 the left-hand boundary reduces to a point, whereas in Figure 3 the right-hand boundary reduces to a point. In the next example both of the side boundaries reduce to a point, so the first step is to find a and b.

EXAMPLE 2 Find the area of the region enclosed by the parabolas $y = x^2$ and $y = 2x - x^2$.

SOLUTION We first find the points of intersection of the parabolas by solving their equations simultaneously. This gives $x^2 = 2x - x^2$, or $2x^2 - 2x = 0$. Thus $2x(x - 1) = 0$, so $x = 0$ or 1. The points of intersection are $(0, 0)$ and $(1, 1)$.

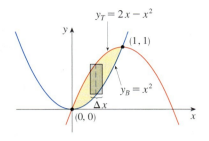

FIGURE 6

We see from Figure 6 that the top and bottom boundaries are

$$y_T = 2x - x^2 \quad \text{and} \quad y_B = x^2$$

The area of a typical rectangle is

$$(y_T - y_B)\,\Delta x = (2x - x^2 - x^2)\,\Delta x$$

and the region lies between $x = 0$ and $x = 1$. So the total area is

$$A = \int_0^1 (2x - 2x^2)\,dx = 2\int_0^1 (x - x^2)\,dx$$

$$= 2\left[\frac{x^2}{2} - \frac{x^3}{3}\right]_0^1$$

$$= 2\left(\frac{1}{2} - \frac{1}{3}\right) = \frac{1}{3}$$

Sometimes it's difficult, or even impossible, to find the points of intersection of two curves exactly. As shown in the following example, we can use a graphing calculator or computer to find approximate values for the intersection points and then proceed as before.

EXAMPLE 3 Find the approximate area of the region bounded by the curves $y = x/\sqrt{x^2 + 1}$ and $y = x^4 - x$.

SOLUTION If we were to try to find the exact intersection points, we would have to solve the equation

$$\frac{x}{\sqrt{x^2 + 1}} = x^4 - x$$

This looks like a very difficult equation to solve exactly (in fact, it's impossible), so instead we use a graphing device to draw the graphs of the two curves in Figure 7. One intersection point is the origin. We zoom in toward the other point of intersection and find that $x \approx 1.18$. (If greater accuracy is required, we could use Newton's method or a rootfinder, if available on our graphing device.) Thus, an approximation to the area between the curves is

$$A \approx \int_0^{1.18} \left[\frac{x}{\sqrt{x^2 + 1}} - (x^4 - x)\right] dx$$

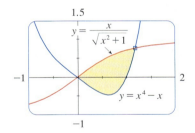

FIGURE 7

To integrate the first term we use the substitution $u = x^2 + 1$. Then $du = 2x\,dx$, and when $x = 1.18$, we have $u \approx 2.39$. So

$$A \approx \frac{1}{2}\int_1^{2.39} \frac{du}{\sqrt{u}} - \int_0^{1.18} (x^4 - x)\,dx$$

$$= \sqrt{u}\,\Big]_1^{2.39} - \left[\frac{x^5}{5} - \frac{x^2}{2}\right]_0^{1.18}$$

$$= \sqrt{2.39} - 1 - \frac{(1.18)^5}{5} + \frac{(1.18)^2}{2}$$

$$\approx 0.785$$

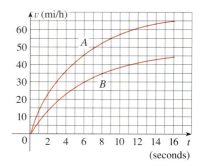

FIGURE 8

EXAMPLE 4 Figure 8 shows velocity curves for two cars, A and B, that start side by side and move along the same road. What does the area between the curves represent? Use the Midpoint Rule to estimate it.

SOLUTION We know from Section 5.4 that the area under the velocity curve A represents the distance traveled by car A during the first 16 seconds. Similarly, the area under curve B is the distance traveled by car B during that time period. So the area between these curves, which is the difference of the areas under the curves, is the distance between the cars after 16 seconds. We read the velocities from the graph and convert them to feet per second (1 mi/h $= \frac{5280}{3600}$ ft/s).

t	0	2	4	6	8	10	12	14	16
v_A	0	34	54	67	76	84	89	92	95
v_B	0	21	34	44	51	56	60	63	65
$v_A - v_B$	0	13	20	23	25	28	29	29	30

We use the Midpoint Rule with $n = 4$ intervals, so that $\Delta t = 4$. The midpoints of the intervals are $\bar{t}_1 = 2$, $\bar{t}_2 = 6$, $\bar{t}_3 = 10$, and $\bar{t}_4 = 14$. We estimate the distance between the cars after 16 seconds as follows:

$$\int_0^{16} (v_A - v_B)\, dt \approx \Delta t\, [13 + 23 + 28 + 29]$$

$$= 4(93) = 372 \text{ ft}$$

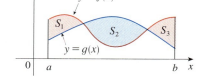

FIGURE 9

If we are asked to find the area between the curves $y = f(x)$ and $y = g(x)$ where $f(x) \geq g(x)$ for some values of x but $g(x) \geq f(x)$ for other values of x, then we split the given region S into several regions $S_1, S_2, \ldots$ with areas $A_1, A_2, \ldots$ as shown in Figure 9. We then define the area of the region S to be the sum of the areas of the smaller regions $S_1, S_2, \ldots$, that is, $A = A_1 + A_2 + \cdots$. Since

$$|f(x) - g(x)| = \begin{cases} f(x) - g(x) & \text{when } f(x) \geq g(x) \\ g(x) - f(x) & \text{when } g(x) \geq f(x) \end{cases}$$

we have the following expression for A.

3 The area between the curves $y = f(x)$ and $y = g(x)$ and between $x = a$ and $x = b$ is

$$A = \int_a^b |f(x) - g(x)|\, dx$$

When evaluating the integral in (3), however, we must still split it into integrals corresponding to $A_1, A_2, \ldots$.

EXAMPLE 5 Find the area of the region bounded by the curves $y = \sin x$, $y = \cos x$, $x = 0$, and $x = \pi/2$.

SOLUTION The points of intersection occur when $\sin x = \cos x$, that is, when $x = \pi/4$ (since $0 \leq x \leq \pi/2$). The region is sketched in Figure 10. Observe that $\cos x \geq \sin x$

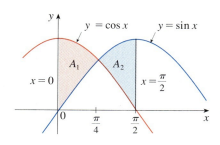

FIGURE 10

when $0 \le x \le \pi/4$ but $\sin x \ge \cos x$ when $\pi/4 \le x \le \pi/2$. Therefore, the required area is

$$A = \int_0^{\pi/2} |\cos x - \sin x|\, dx = A_1 + A_2$$

$$= \int_0^{\pi/4} (\cos x - \sin x)\, dx + \int_{\pi/4}^{\pi/2} (\sin x - \cos x)\, dx$$

$$= [\sin x + \cos x]_0^{\pi/4} + [-\cos x - \sin x]_{\pi/4}^{\pi/2}$$

$$= \left(\frac{1}{\sqrt{2}} + \frac{1}{\sqrt{2}} - 0 - 1\right) + \left(-0 - 1 + \frac{1}{\sqrt{2}} + \frac{1}{\sqrt{2}}\right)$$

$$= 2\sqrt{2} - 2$$

In this particular example we could have saved some work by noticing that the region is symmetric about $x = \pi/4$ and so

$$A = 2A_1 = 2\int_0^{\pi/4} (\cos x - \sin x)\, dx$$

Some regions are best treated by regarding x as a function of y. If a region is bounded by curves with equations $x = f(y)$, $x = g(y)$, $y = c$, and $y = d$, where f and g are continuous and $f(y) \ge g(y)$ for $c \le y \le d$ (see Figure 11), then its area is

$$A = \int_c^d [f(y) - g(y)]\, dy$$

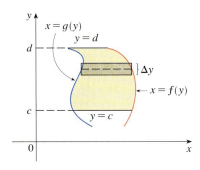

FIGURE 11

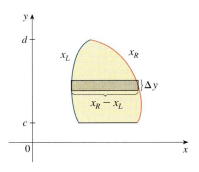

FIGURE 12

If we write x_R for the right boundary and x_L for the left boundary, then, as Figure 12 illustrates, we have

$$A = \int_c^d (x_R - x_L)\, dy$$

Here a typical approximating rectangle has dimensions $x_R - x_L$ and Δy.

EXAMPLE 6 Find the area enclosed by the line $y = x - 1$ and the parabola $y^2 = 2x + 6$.

SOLUTION By solving the two equations we find that the points of intersection are $(-1, -2)$ and $(5, 4)$. We solve the equation of the parabola for x and notice from Figure 13 that the left and right boundary curves are

$$x_L = \tfrac{1}{2}y^2 - 3 \qquad x_R = y + 1$$

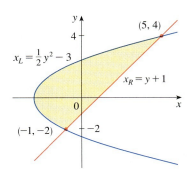

FIGURE 13

We must integrate between the appropriate y-values, $y = -2$ and $y = 4$. Thus

$$A = \int_{-2}^{4} (x_R - x_L)\,dy = \int_{-2}^{4} \left[(y + 1) - \left(\tfrac{1}{2}y^2 - 3\right)\right] dy$$

$$= \int_{-2}^{4} \left(-\tfrac{1}{2}y^2 + y + 4\right) dy = -\frac{1}{2}\left(\frac{y^3}{3}\right) + \frac{y^2}{2} + 4y \Big]_{-2}^{4}$$

$$= -\tfrac{1}{6}(64) + 8 + 16 - \left(\tfrac{4}{3} + 2 - 8\right) = 18$$

We could have found the area in Example 6 by integrating with respect to x instead of y, but the calculation is much more involved. It would have meant splitting the region in two and computing the areas labeled A_1 and A_2 in Figure 14. The method we used in Example 6 is *much* easier.

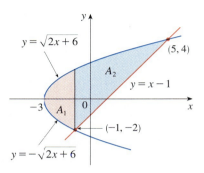

FIGURE 14

6.1 Exercises

1–4 ▪ Find the area of the shaded region.

1.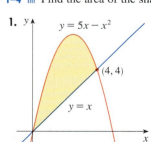

2.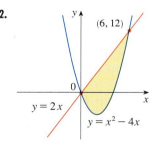

3.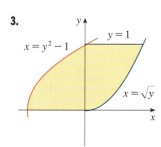

4.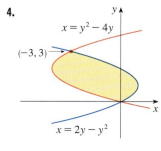

5–26 ▪ Sketch the region enclosed by the given curves. Decide whether to integrate with respect to x or y. Draw a typical approximating rectangle and label its height and width. Then find the area of the region.

5. $y = x + 1$, $\quad y = 9 - x^2$, $\quad x = -1$, $\quad x = 2$

6. $y = \sin x$, $\quad y = x$, $\quad x = \pi/2$, $\quad x = \pi$

7. $y = x$, $\quad y = x^2$

8. $y = x^2$, $\quad y = x^4$

9. $y = \sqrt{x + 3}$, $\quad y = (x + 3)/2$

10. $y = 1 + \sqrt{x}$, $\quad y = (3 + x)/3$

11. $y = x^2$, $\quad y^2 = x$

12. $y = x$, $\quad y = \sqrt[3]{x}$

13. $y = 12 - x^2$, $\quad y = x^2 - 6$

14. $y = x^3 - x$, $\quad y = 3x$

15. $y = \sqrt{x}$, $\quad y = \tfrac{1}{2}x$, $\quad x = 9$

16. $y = 8 - x^2$, $\quad y = x^2$, $\quad x = -3$, $\quad x = 3$

17. $x = 2y^2$, $\quad x + y = 1$

18. $4x + y^2 = 12$, $x = y$

19. $x = 1 - y^2$, $x = y^2 - 1$

20. $y = \sin(\pi x/2)$, $y = x$

21. $y = \cos x$, $y = \sin 2x$, $x = 0$, $x = \pi/2$

22. $y = \sin x$, $y = \sin 2x$, $x = 0$, $x = \pi/2$

23. $y = \cos x$, $y = 1 - 2x/\pi$

24. $y = |x|$, $y = x^2 - 2$

25. $y = x$, $x + 2y = 0$, $2x + y = 3$

26. $y = \sin \pi x$, $y = x^2 - x$, $x = 2$

27–28 ▮▮▮▮ Use calculus to find the area of the triangle with the given vertices.

27. $(0, 0)$, $(2, 1)$, $(-1, 6)$

28. $(0, 5)$, $(2, -2)$, $(5, 1)$

29–30 ▮▮▮▮ Evaluate the integral and interpret it as the area of a region. Sketch the region.

29. $\int_{-1}^{1} |x^3 - x| \, dx$

30. $\int_{0}^{4} |\sqrt{x + 2} - x| \, dx$

31–32 ▮▮▮▮ Use the Midpoint Rule with $n = 4$ to approximate the area of the region bounded by the given curves.

31. $y = \sin^2(\pi x/4)$, $y = \cos^2(\pi x/4)$, $0 \leq x \leq 1$

32. $y = \sqrt[3]{16 - x^3}$, $y = x$, $x = 0$

33–36 ▮▮▮▮ Use a graph to find approximate x-coordinates of the points of intersection of the given curves. Then find (approximately) the area of the region bounded by the curves.

33. $y = x^2$, $y = 2 \cos x$

34. $y = x^4$, $y = 3x - x^3$

35. $y = \sqrt{x + 1}$, $y = x^2$

36. $y = x^4 - 1$, $y = x \sin(x^2)$

CAS 37. Use a computer algebra system to find the exact area enclosed by the curves $y = x^5 - 6x^3 + 4x$ and $y = x$.

38. Sketch the region in the xy-plane defined by the inequalities $x - 2y^2 \geq 0$, $1 - x - |y| \geq 0$ and find its area.

39. Racing cars driven by Chris and Kelly are side by side at the start of a race. The table shows the velocities of each car (in miles per hour) during the first ten seconds of the race. Use the Midpoint Rule to estimate how much farther Kelly travels than Chris does during the first ten seconds.

t	v_C	v_K	t	v_C	v_K
0	0	0	6	69	80
1	20	22	7	75	86
2	32	37	8	81	93
3	46	52	9	86	98
4	54	61	10	90	102
5	62	71			

40. The widths (in meters) of a kidney-shaped swimming pool were measured at 2-meter intervals as indicated in the figure. Use the Midpoint Rule to estimate the area of the pool.

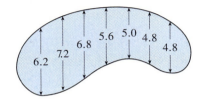

41. Two cars, A and B, start side by side and accelerate from rest. The figure shows the graphs of their velocity functions.
(a) Which car is ahead after one minute? Explain.
(b) What is the meaning of the area of the shaded region?
(c) Which car is ahead after two minutes? Explain.
(d) Estimate the time at which the cars are again side by side.

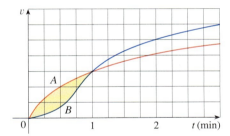

42. The figure shows graphs of the marginal revenue function R' and the marginal cost function C' for a manufacturer. [Recall from Section 4.8 that $R(x)$ and $C(x)$ represent the revenue and cost when x units are manufactured. Assume that R and C are measured in thousands of dollars.] What is the meaning of the area of the shaded region? Use the Midpoint Rule to estimate the value of this quantity.

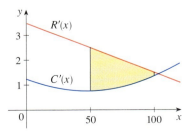

43. The curve with equation $y^2 = x^2(x+3)$ is called **Tschirnhausen's cubic**. If you graph this curve you will see that part of the curve forms a loop. Find the area enclosed by the loop.

44. Find the area of the region bounded by the parabola $y = x^2$, the tangent line to this parabola at $(1, 1)$, and the x-axis.

45. Find the number b such that the line $y = b$ divides the region bounded by the curves $y = x^2$ and $y = 4$ into two regions with equal area.

46. (a) Find the number a such that the line $x = a$ bisects the area under the curve $y = 1/x^2$, $1 \leq x \leq 4$.
(b) Find the number b such that the line $y = b$ bisects the area in part (a).

47. Find the values of c such that the area of the region enclosed by the parabolas $y = x^2 - c^2$ and $y = c^2 - x^2$ is 576.

48. Suppose that $0 < c < \pi/2$. For what value of c is the area of the region enclosed by the curves $y = \cos x$, $y = \cos(x - c)$, and $x = 0$ equal to the area of the region enclosed by the curves $y = \cos(x - c)$, $x = \pi$, and $y = 0$?

The following exercises are intended only for those who have already covered Chapter 7.

49–51 ⅲ Sketch the region bounded by the given curves and find the area of the region.

49. $y = 1/x$, $y = 1/x^2$, $x = 2$

50. $y = \sin x$, $y = e^x$, $x = 0$, $x = \pi/2$

51. $y = x^2$, $y = 2/(x^2 + 1)$

52. For what values of m do the line $y = mx$ and the curve $y = x/(x^2 + 1)$ enclose a region? Find the area of the region.

6.2 Volumes

In trying to find the volume of a solid we face the same type of problem as in finding areas. We have an intuitive idea of what volume means, but we must make this idea precise by using calculus to give an exact definition of volume.

We start with a simple type of solid called a **cylinder** (or, more precisely, a *right cylinder*). As illustrated in Figure 1(a), a cylinder is bounded by a plane region B_1, called the **base**, and a congruent region B_2 in a parallel plane. The cylinder consists of all points on line segments that are perpendicular to the base and join B_1 to B_2. If the area of the base is A and the height of the cylinder (the distance from B_1 to B_2) is h, then the volume V of the cylinder is defined as

$$V = Ah$$

In particular, if the base is a circle with radius r, then the cylinder is a circular cylinder with volume $V = \pi r^2 h$ [see Figure 1(b)], and if the base is a rectangle with length l and width w, then the cylinder is a rectangular box (also called a *rectangular parallelepiped*) with volume $V = lwh$ [see Figure 1(c)].

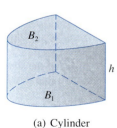

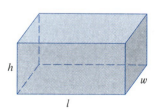

FIGURE 1 (a) Cylinder $V = Ah$ (b) Circular cylinder $V = \pi r^2 h$ (c) Rectangular box $V = lwh$

For a solid S that isn't a cylinder we first "cut" S into pieces and approximate each piece by a cylinder. We estimate the volume of S by adding the volumes of the cylinders. We arrive at the exact volume of S through a limiting process in which the number of pieces becomes large.

We start by intersecting S with a plane and obtaining a plane region that is called a **cross-section** of S. Let $A(x)$ be the area of the cross-section of S in a plane P_x perpendicular to the x-axis and passing through the point x, where $a \leq x \leq b$. (See Figure 2. Think of slicing S with a knife through x and computing the area of this slice.) The cross-sectional area $A(x)$ will vary as x increases from a to b.

Watch an animation of Figure 2.
    Resources / Module 7
        / Volumes
            / Volumes by Cross-Section

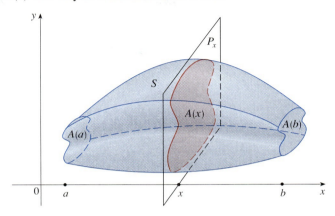

FIGURE 2

Let's divide S into n "slabs" of equal width Δx by using the planes $P_{x_1}, P_{x_2}, \ldots$ to slice the solid. (Think of slicing a loaf of bread.) If we choose sample points x_i^* in $[x_{i-1}, x_i]$, we can approximate the ith slab S_i (the part of S that lies between the planes $P_{x_{i-1}}$ and P_{x_i}) by a cylinder with base area $A(x_i^*)$ and "height" Δx. (See Figure 3.)

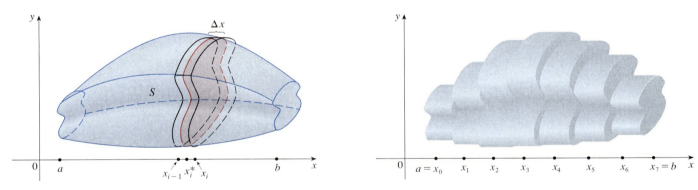

FIGURE 3

The volume of this cylinder is $A(x_i^*) \Delta x$, so an approximation to our intuitive conception of the volume of the ith slab S_i is

$$V(S_i) \approx A(x_i^*) \Delta x$$

Adding the volumes of these slabs, we get an approximation to the total volume (that is, what we think of intuitively as the volume):

$$V \approx \sum_{i=1}^{n} A(x_i^*) \Delta x$$

This approximation appears to become better and better as $n \to \infty$. (Think of the slices as becoming thinner and thinner.) Therefore, we *define* the volume as the limit of these sums as $n \to \infty$. But we recognize the limit of Riemann sums as a definite integral and so we have the following definition.

384 |||| CHAPTER 6 APPLICATIONS OF INTEGRATION

|||| It can be proved that this definition is independent of how S is situated with respect to the x-axis. In other words, no matter how we slice S with parallel planes, we always get the same answer for V.

Definition of Volume Let S be a solid that lies between $x = a$ and $x = b$. If the cross-sectional area of S in the plane P_x, through x and perpendicular to the x-axis, is $A(x)$, where A is a continuous function, then the **volume** of S is

$$V = \lim_{n \to \infty} \sum_{i=1}^{n} A(x_i^*) \, \Delta x = \int_a^b A(x) \, dx$$

When we use the volume formula $V = \int_a^b A(x) \, dx$ it is important to remember that $A(x)$ is the area of a moving cross-section obtained by slicing through x perpendicular to the x-axis.

Notice that, for a cylinder, the cross-sectional area is constant: $A(x) = A$ for all x. So our definition of volume gives $V = \int_a^b A \, dx = A(b - a)$; this agrees with the formula $V = Ah$.

EXAMPLE 1 Show that the volume of a sphere of radius r is

$$V = \tfrac{4}{3}\pi r^3$$

SOLUTION If we place the sphere so that its center is at the origin (see Figure 4), then the plane P_x intersects the sphere in a circle whose radius (from the Pythagorean Theorem) is $y = \sqrt{r^2 - x^2}$. So the cross-sectional area is

$$A(x) = \pi y^2 = \pi(r^2 - x^2)$$

Using the definition of volume with $a = -r$ and $b = r$, we have

$$V = \int_{-r}^{r} A(x) \, dx = \int_{-r}^{r} \pi(r^2 - x^2) \, dx$$

$$= 2\pi \int_0^r (r^2 - x^2) \, dx \qquad \text{(The integrand is even.)}$$

$$= 2\pi \left[r^2 x - \frac{x^3}{3} \right]_0^r = 2\pi \left(r^3 - \frac{r^3}{3} \right)$$

$$= \tfrac{4}{3}\pi r^3$$

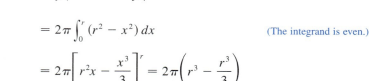

FIGURE 4

Watch an animation of Figure 5.
Resources / Module 7
/ Volumes
/ Volumes

Figure 5 illustrates the definition of volume when the solid is a sphere with radius $r = 1$. From the result of Example 1, we know that the volume of the sphere is $\tfrac{4}{3}\pi \approx 4.18879$. Here the slabs are circular cylinders, or *disks*, and the three parts of Figure 5

(a) Using 5 disks, $V \approx 4.2726$

(b) Using 10 disks, $V \approx 4.2097$

(c) Using 20 disks, $V \approx 4.1940$

FIGURE 5 Approximating the volume of a sphere with radius 1

show the geometric interpretations of the Riemann sums

$$\sum_{i=1}^{n} A(\bar{x}_i)\,\Delta x = \sum_{i=1}^{n} \pi(1^2 - \bar{x}_i^2)\,\Delta x$$

when $n = 5$, 10, and 20 if we choose the sample points x_i^* to be the midpoints $\bar{x}_i$. Notice that as we increase the number of approximating cylinders, the corresponding Riemann sums become closer to the true volume.

EXAMPLE 2 Find the volume of the solid obtained by rotating about the x-axis the region under the curve $y = \sqrt{x}$ from 0 to 1. Illustrate the definition of volume by sketching a typical approximating cylinder.

SOLUTION The region is shown in Figure 6(a). If we rotate about the x-axis, we get the solid shown in Figure 6(b). When we slice through the point x, we get a disk with radius $\sqrt{x}$. The area of this cross-section is

$$A(x) = \pi(\sqrt{x})^2 = \pi x$$

and the volume of the approximating cylinder (a disk with thickness Δx) is

$$A(x)\,\Delta x = \pi x\,\Delta x$$

The solid lies between $x = 0$ and $x = 1$, so its volume is

$$V = \int_0^1 A(x)\,dx = \int_0^1 \pi x\,dx = \pi \frac{x^2}{2}\bigg]_0^1 = \frac{\pi}{2}$$

|||| Did we get a reasonable answer in Example 2? As a check on our work, let's replace the given region by a square with base $[0, 1]$ and height 1. If we rotate this square, we get a cylinder with radius 1, height 1, and volume $\pi \cdot 1^2 \cdot 1 = \pi$. We computed that the given solid has half this volume. That seems about right.

See a volume of revolution being formed.
Resources / Module 7
/ Volumes
/ Volumes of Revolution

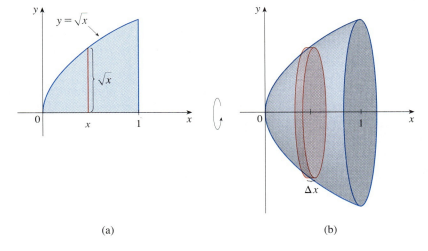

FIGURE 6   (a)   (b)

EXAMPLE 3 Find the volume of the solid obtained by rotating the region bounded by $y = x^3$, $y = 8$, and $x = 0$ about the y-axis.

SOLUTION The region is shown in Figure 7(a) and the resulting solid is shown in Figure 7(b). Because the region is rotated about the y-axis, it makes sense to slice the solid perpendicular to the y-axis and therefore to integrate with respect to y. If we slice at height y, we get a circular disk with radius x, where $x = \sqrt[3]{y}$. So the area of a cross-section through y is

$$A(y) = \pi x^2 = \pi(\sqrt[3]{y})^2 = \pi y^{2/3}$$

and the volume of the approximating cylinder pictured in Figure 7(b) is

$$A(y)\,\Delta y = \pi y^{2/3}\,\Delta y$$

Since the solid lies between $y = 0$ and $y = 8$, its volume is

$$V = \int_0^8 A(y)\,dy = \int_0^8 \pi y^{2/3}\,dy$$

$$= \pi \left[\tfrac{3}{5} y^{5/3}\right]_0^8 = \frac{96\pi}{5}$$

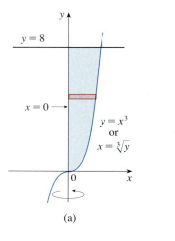
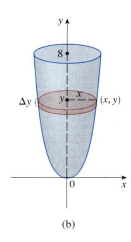

FIGURE 7 (a) (b)

EXAMPLE 4 The region $\mathcal{R}$ enclosed by the curves $y = x$ and $y = x^2$ is rotated about the x-axis. Find the volume of the resulting solid.

SOLUTION The curves $y = x$ and $y = x^2$ intersect at the points $(0, 0)$ and $(1, 1)$. The region between them, the solid of rotation, and a cross-section perpendicular to the x-axis are shown in Figure 8. A cross-section in the plane P_x has the shape of a *washer* (an annular ring) with inner radius x^2 and outer radius x, so we find the cross-sectional area by subtracting the area of the inner circle from the area of the outer circle:

$$A(x) = \pi x^2 - \pi (x^2)^2 = \pi (x^2 - x^4)$$

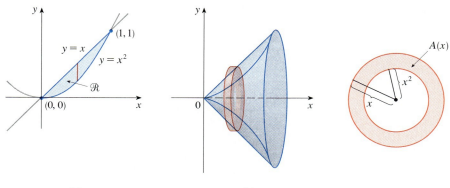

FIGURE 8 (a) (b) (c)

Therefore, we have

$$V = \int_0^1 A(x)\, dx = \int_0^1 \pi(x^2 - x^4)\, dx$$

$$= \pi \left[\frac{x^3}{3} - \frac{x^5}{5} \right]_0^1 = \frac{2\pi}{15}$$

Module 6.2/6.3 illustrates the formation and computation of volumes using disks and washers.

EXAMPLE 5 Find the volume of the solid obtained by rotating the region in Example 4 about the line $y = 2$.

SOLUTION The solid and a cross-section are shown in Figure 9. Again the cross-section is a washer, but this time the inner radius is $2 - x$ and the outer radius is $2 - x^2$. The cross-sectional area is

$$A(x) = \pi(2 - x^2)^2 - \pi(2 - x)^2$$

and so the volume of S is

$$V = \int_0^1 A(x)\, dx = \pi \int_0^1 \left[(2 - x^2)^2 - (2 - x)^2 \right] dx$$

$$= \pi \int_0^1 (x^4 - 5x^2 + 4x)\, dx = \pi \left[\frac{x^5}{5} - 5\frac{x^3}{3} + 4\frac{x^2}{2} \right]_0^1 = \frac{8\pi}{15}$$

The solids in Examples 1–5 are all called **solids of revolution** because they are obtained by revolving a region about a line. In general, we calculate the volume of a solid of revolution by using the basic defining formula

$$V = \int_a^b A(x)\, dx \quad \text{or} \quad V = \int_c^d A(y)\, dy$$

and we find the cross-sectional area $A(x)$ or $A(y)$ in one of the following ways:

- If the cross-section is a disk (as in Examples 1–3), we find the radius of the disk (in terms of x or y) and use

$$A = \pi(\text{radius})^2$$

- If the cross-section is a washer (as in Examples 4 and 5), we find the inner radius r_{in} and outer radius r_{out} from a sketch (as in Figures 9 and 10) and compute the area of the washer by subtracting the area of the inner disk from the area of the outer disk:

$$A = \pi(\text{outer radius})^2 - \pi(\text{inner radius})^2$$

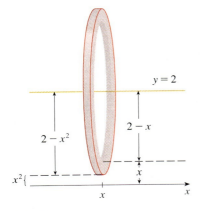

FIGURE 9

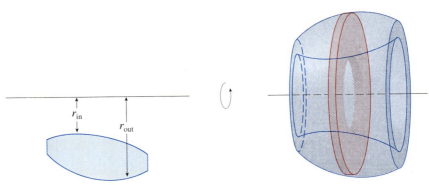

FIGURE 10

The next example gives a further illustration of the procedure.

EXAMPLE 6 Find the volume of the solid obtained by rotating the region in Example 4 about the line $x = -1$.

SOLUTION Figure 11 shows a horizontal cross-section. It is a washer with inner radius $1 + y$ and outer radius $1 + \sqrt{y}$, so the cross-sectional area is

$$A(y) = \pi(\text{outer radius})^2 - \pi(\text{inner radius})^2$$
$$= \pi(1 + \sqrt{y})^2 - \pi(1 + y)^2$$

The volume is

$$V = \int_0^1 A(y)\, dy$$
$$= \pi \int_0^1 \left[(1 + \sqrt{y})^2 - (1 + y)^2\right] dy$$
$$= \pi \int_0^1 \left(2\sqrt{y} - y - y^2\right) dy$$
$$= \pi \left[\frac{4y^{3/2}}{3} - \frac{y^2}{2} - \frac{y^3}{3}\right]_0^1 = \frac{\pi}{2}$$

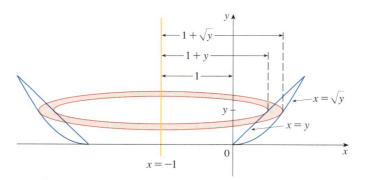

FIGURE 11

We now find the volumes of three solids that are *not* solids of revolution.

EXAMPLE 7 Figure 12 shows a solid with a circular base of radius 1. Parallel cross-sections perpendicular to the base are equilateral triangles. Find the volume of the solid.

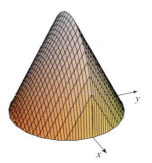

FIGURE 12
Computer-generated picture of the solid in Example 7

SOLUTION Let's take the circle to be $x^2 + y^2 = 1$. The solid, its base, and a typical cross-section at a distance x from the origin are shown in Figure 13.

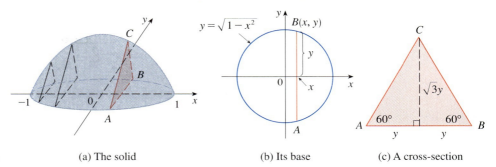

FIGURE 13   (a) The solid   (b) Its base   (c) A cross-section

Since B lies on the circle, we have $y = \sqrt{1 - x^2}$ and so the base of the triangle ABC is $|AB| = 2\sqrt{1 - x^2}$. Since the triangle is equilateral, we see from Figure 13(c) that its height is $\sqrt{3}\, y = \sqrt{3}\sqrt{1 - x^2}$. The cross-sectional area is therefore

$$A(x) = \tfrac{1}{2} \cdot 2\sqrt{1 - x^2} \cdot \sqrt{3}\sqrt{1 - x^2} = \sqrt{3}\,(1 - x^2)$$

and the volume of the solid is

$$V = \int_{-1}^{1} A(x)\,dx = \int_{-1}^{1} \sqrt{3}\,(1 - x^2)\,dx$$

$$= 2\int_{0}^{1} \sqrt{3}\,(1 - x^2)\,dx = 2\sqrt{3}\left[x - \frac{x^3}{3}\right]_{0}^{1} = \frac{4\sqrt{3}}{3}$$

Resources / Module 7
  / Volumes
    / Start of Mystery of the Topless Pyramid

EXAMPLE 8 Find the volume of a pyramid whose base is a square with side L and whose height is h.

SOLUTION We place the origin O at the vertex of the pyramid and the x-axis along its central axis as in Figure 14. Any plane P_x that passes through x and is perpendicular to the x-axis intersects the pyramid in a square with side of length s, say. We can express s in terms of x by observing from the similar triangles in Figure 15 that

$$\frac{x}{h} = \frac{s/2}{L/2} = \frac{s}{L}$$

and so $s = Lx/h$. [Another method is to observe that the line OP has slope $L/(2h)$ and so its equation is $y = Lx/(2h)$.] Thus, the cross-sectional area is

$$A(x) = s^2 = \frac{L^2}{h^2} x^2$$

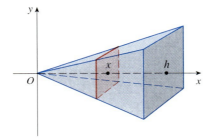

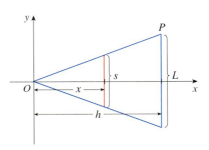

FIGURE 14   **FIGURE 15**

390 ■ CHAPTER 6 APPLICATIONS OF INTEGRATION

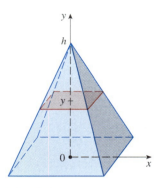

FIGURE 16

The pyramid lies between $x = 0$ and $x = h$, so its volume is

$$V = \int_0^h A(x)\,dx = \int_0^h \frac{L^2}{h^2} x^2\,dx$$

$$= \frac{L^2}{h^2} \frac{x^3}{3} \bigg]_0^h = \frac{L^2 h}{3}$$

NOTE ▫ We didn't need to place the vertex of the pyramid at the origin in Example 8. We did so merely to make the equations simple. If, instead, we had placed the center of the base at the origin and the vertex on the positive y-axis, as in Figure 16, you can verify that we would have obtained the integral

$$V = \int_0^h \frac{L^2}{h^2} (h - y)^2\,dy = \frac{L^2 h}{3}$$

EXAMPLE 9 A wedge is cut out of a circular cylinder of radius 4 by two planes. One plane is perpendicular to the axis of the cylinder. The other intersects the first at an angle of 30° along a diameter of the cylinder. Find the volume of the wedge.

SOLUTION If we place the x-axis along the diameter where the planes meet, then the base of the solid is a semicircle with equation $y = \sqrt{16 - x^2}$, $-4 \leq x \leq 4$. A cross-section perpendicular to the x-axis at a distance x from the origin is a triangle ABC, as shown in Figure 17, whose base is $y = \sqrt{16 - x^2}$ and whose height is $|BC| = y \tan 30° = \sqrt{16 - x^2}/\sqrt{3}$. Thus, the cross-sectional area is

$$A(x) = \tfrac{1}{2}\sqrt{16 - x^2} \cdot \frac{1}{\sqrt{3}} \sqrt{16 - x^2} = \frac{16 - x^2}{2\sqrt{3}}$$

and the volume is

$$V = \int_{-4}^{4} A(x)\,dx = \int_{-4}^{4} \frac{16 - x^2}{2\sqrt{3}}\,dx$$

$$= \frac{1}{\sqrt{3}} \int_0^4 (16 - x^2)\,dx = \frac{1}{\sqrt{3}} \left[16x - \frac{x^3}{3} \right]_0^4$$

$$= \frac{128}{3\sqrt{3}}$$

For another method see Exercise 62.

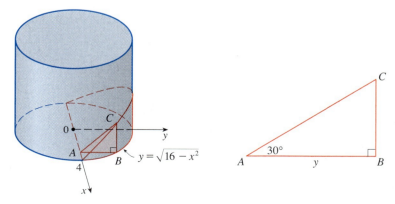

FIGURE 17

6.2 Exercises

1–18 Find the volume of the solid obtained by rotating the region bounded by the given curves about the specified line. Sketch the region, the solid, and a typical disk or washer.

1. $y = x^2$, $x = 1$, $y = 0$; about the x-axis
2. $x + 2y = 2$, $x = 0$ $y = 0$; about the x-axis
3. $y = 1/x$, $x = 1$, $x = 2$, $y = 0$; about the x-axis
4. $y = \sqrt{x - 1}$, $x = 2$, $x = 5$, $y = 0$; about the x-axis
5. $y = x^2$, $0 \leq x \leq 2$, $y = 4$, $x = 0$; about the y-axis
6. $x = y - y^2$, $x = 0$; about the y-axis
7. $y = x^2$, $y^2 = x$; about the x-axis
8. $y = \sec x$, $y = 1$, $x = -1$, $x = 1$; about the x-axis
9. $y^2 = x$, $x = 2y$; about the y-axis
10. $y = x^{2/3}$, $x = 1$, $y = 0$; about the y-axis
11. $y = x$, $y = \sqrt{x}$; about $y = 1$
12. $y = x^2$, $y = 4$; about $y = 4$
13. $y = x^4$, $y = 1$; about $y = 2$
14. $y = 1/x^2$, $y = 0$, $x = 1$, $x = 3$; about $y = -1$
15. $x = y^2$, $x = 1$; about $x = 1$
16. $y = x$, $y = \sqrt{x}$; about $x = 2$
17. $y = x^2$, $x = y^2$; about $x = -1$
18. $y = x$, $y = 0$, $x = 2$, $x = 4$; about $x = 1$

19–30 Refer to the figure and find the volume generated by rotating the given region about the specified line.

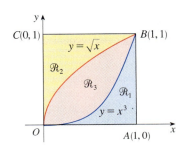

19. $\mathcal{R}_1$ about OA
20. $\mathcal{R}_1$ about OC
21. $\mathcal{R}_1$ about AB
22. $\mathcal{R}_1$ about BC
23. $\mathcal{R}_2$ about OA
24. $\mathcal{R}_2$ about OC
25. $\mathcal{R}_2$ about AB
26. $\mathcal{R}_2$ about BC
27. $\mathcal{R}_3$ about OA
28. $\mathcal{R}_3$ about OC
29. $\mathcal{R}_3$ about AB
30. $\mathcal{R}_3$ about BC

31–36 Set up, but do not evaluate, an integral for the volume of the solid obtained by rotating the region bounded by the given curves about the specified line.

31. $y = \tan^3 x$, $y = 1$, $x = 0$; about $y = 1$
32. $y = (x - 2)^4$, $8x - y = 16$; about $x = 10$
33. $y = 0$, $y = \sin x$, $0 \leq x \leq \pi$; about $y = 1$
34. $y = 0$, $y = \sin x$, $0 \leq x \leq \pi$; about $y = -2$
35. $x^2 - y^2 = 1$, $x = 3$; about $x = -2$
36. $2x + 3y = 6$, $(y - 1)^2 = 4 - x$; about $x = -5$

37–38 Use a graph to find approximate x-coordinates of the points of intersection of the given curves. Then find (approximately) the volume of the solid obtained by rotating about the x-axis the region bounded by these curves.

37. $y = x^2$, $y = \sqrt{x + 1}$
38. $y = x^4$, $y = 3x - x^3$

CAS **39–40** Use a computer algebra system to find the exact volume of the solid obtained by rotating the region bounded by the given curves about the specified line.

39. $y = \sin^2 x$, $y = 0$, $0 \leq x \leq \pi$; about $y = -1$
40. $y = x^2 - 2x$, $y = x \cos(\pi x/4)$; about $y = 2$

41–44 Each integral represents the volume of a solid. Describe the solid.

41. $\pi \int_0^{\pi/2} \cos^2 x \, dx$
42. $\pi \int_2^5 y \, dy$
43. $\pi \int_0^1 (y^4 - y^8) \, dy$
44. $\pi \int_0^{\pi/2} [(1 + \cos x)^2 - 1^2] \, dx$

45. A CAT scan produces equally spaced cross-sectional views of a human organ that provide information about the organ otherwise obtained only by surgery. Suppose that a CAT scan of a human liver shows cross-sections spaced 1.5 cm apart. The liver is 15 cm long and the cross-sectional areas, in square centimeters, are 0, 18, 58, 79, 94, 106, 117, 128, 63, 39, and 0. Use the Midpoint Rule to estimate the volume of the liver.

46. A log 10 m long is cut at 1-meter intervals and its cross-sectional areas A (at a distance x from the end of the log) are

listed in the table. Use the Midpoint Rule with $n = 5$ to estimate the volume of the log.

x (m)	A (m^2)	x (m)	A (m^2)
0	0.68	6	0.53
1	0.65	7	0.55
2	0.64	8	0.52
3	0.61	9	0.50
4	0.58	10	0.48
5	0.59		

47–59 ▮▮▮▮ Find the volume of the described solid S.

47. A right circular cone with height h and base radius r

48. A frustum of a right circular cone with height h, lower base radius R, and top radius r

49. A cap of a sphere with radius r and height h

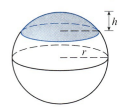

50. A frustum of a pyramid with square base of side b, square top of side a, and height h

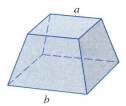

What happens if $a = b$? What happens if $a = 0$?

51. A pyramid with height h and rectangular base with dimensions b and $2b$

52. A pyramid with height h and base an equilateral triangle with side a (a tetrahedron)

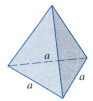

53. A tetrahedron with three mutually perpendicular faces and three mutually perpendicular edges with lengths 3 cm, 4 cm, and 5 cm

54. The base of S is a circular disk with radius r. Parallel cross-sections perpendicular to the base are squares.

55. The base of S is an elliptical region with boundary curve $9x^2 + 4y^2 = 36$. Cross-sections perpendicular to the x-axis are isosceles right triangles with hypotenuse in the base.

56. The base of S is the parabolic region $\{(x, y) \mid x^2 \leq y \leq 1\}$. Cross-sections perpendicular to the y-axis are equilateral triangles.

57. S has the same base as in Exercise 56, but cross-sections perpendicular to the y-axis are squares.

58. The base of S is the triangular region with vertices $(0, 0)$, $(3, 0)$, and $(0, 2)$. Cross-sections perpendicular to the y-axis are semicircles.

59. S has the same base as in Exercise 58, but cross-sections perpendicular to the y-axis are isosceles triangles with height equal to the base.

60. The base of S is a circular disk with radius r. Parallel cross-sections perpendicular to the base are isosceles triangles with height h and unequal side in the base.
(a) Set up an integral for the volume of S.
(b) By interpreting the integral as an area, find the volume of S.

61. (a) Set up an integral for the volume of a solid *torus* (the donut-shaped solid shown in the figure) with radii r and R.
(b) By interpreting the integral as an area, find the volume of the torus.

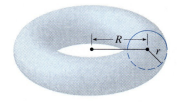

62. Solve Example 9 taking cross-sections to be parallel to the line of intersection of the two planes.

63. (a) Cavalieri's Principle states that if a family of parallel planes gives equal cross-sectional areas for two solids S_1 and S_2, then the volumes of S_1 and S_2 are equal. Prove this principle.
(b) Use Cavalieri's Principle to find the volume of the oblique cylinder shown in the figure.

64. Find the volume common to two circular cylinders, each with radius r, if the axes of the cylinders intersect at right angles.

65. Find the volume common to two spheres, each with radius r, if the center of each sphere lies on the surface of the other sphere.

66. A bowl is shaped like a hemisphere with diameter 30 cm. A ball with diameter 10 cm is placed in the bowl and water is poured into the bowl to a depth of h centimeters. Find the volume of water in the bowl.

67. A hole of radius r is bored through a cylinder of radius $R > r$ at right angles to the axis of the cylinder. Set up, but do not evaluate, an integral for the volume cut out.

68. A hole of radius r is bored through the center of a sphere of radius $R > r$. Find the volume of the remaining portion of the sphere.

69. Some of the pioneers of calculus, such as Kepler and Newton, were inspired by the problem of finding the volumes of wine barrels. (In fact Kepler published a book *Stereometria doliorum* in 1715 devoted to methods for finding the volumes of barrels.) They often approximated the shape of the sides by parabolas.
(a) A barrel with height h and maximum radius R is constructed by rotating about the x-axis the parabola $y = R - cx^2$, $-h/2 \le x \le h/2$, where c is a positive constant. Show that the radius of each end of the barrel is $r = R - d$, where $d = ch^2/4$.
(b) Show that the volume enclosed by the barrel is
$$V = \tfrac{1}{3}\pi h\left(2R^2 + r^2 - \tfrac{2}{5}d^2\right)$$

70. Suppose that a region $\mathcal{R}$ has area A and lies above the x-axis. When $\mathcal{R}$ is rotated about the x-axis, it sweeps out a solid with volume V_1. When $\mathcal{R}$ is rotated about the line $y = -k$ (where k is a positive number), it sweeps out a solid with volume V_2. Express V_2 in terms of V_1, k, and A.

6.3 Volumes by Cylindrical Shells

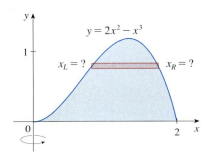

FIGURE 1

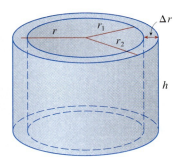

FIGURE 2

Some volume problems are very difficult to handle by the methods of the preceding section. For instance, let's consider the problem of finding the volume of the solid obtained by rotating about the y-axis the region bounded by $y = 2x^2 - x^3$ and $y = 0$. (See Figure 1.) If we slice perpendicular to the y-axis, we get a washer. But to compute the inner radius and the outer radius of the washer, we would have to solve the cubic equation $y = 2x^2 - x^3$ for x in terms of y; that's not easy.

Fortunately, there is a method, called the **method of cylindrical shells**, that is easier to use in such a case. Figure 2 shows a cylindrical shell with inner radius r_1, outer radius r_2, and height h. Its volume V is calculated by subtracting the volume V_1 of the inner cylinder from the volume V_2 of the outer cylinder:

$$\begin{aligned} V &= V_2 - V_1 \\ &= \pi r_2^2 h - \pi r_1^2 h = \pi(r_2^2 - r_1^2)h \\ &= \pi(r_2 + r_1)(r_2 - r_1)h \\ &= 2\pi \frac{r_2 + r_1}{2} h(r_2 - r_1) \end{aligned}$$

If we let $\Delta r = r_2 - r_1$ (the thickness of the shell) and $r = \tfrac{1}{2}(r_2 + r_1)$ (the average radius of the shell), then this formula for the volume of a cylindrical shell becomes

$$\boxed{1 \qquad V = 2\pi r h\, \Delta r}$$

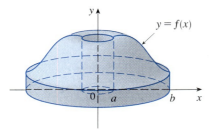

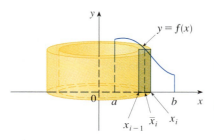

FIGURE 3

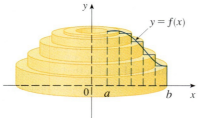

FIGURE 4

and it can be remembered as

$$V = [\text{circumference}][\text{height}][\text{thickness}]$$

Now let S be the solid obtained by rotating about the y-axis the region bounded by $y = f(x)$ [where $f(x) \geq 0$], $y = 0$, $x = a$, and $x = b$, where $b > a \geq 0$. (See Figure 3.) We divide the interval $[a, b]$ into n subintervals $[x_{i-1}, x_i]$ of equal width Δx and let $\bar{x}_i$ be the midpoint of the ith subinterval. If the rectangle with base $[x_{i-1}, x_i]$ and height $f(\bar{x}_i)$ is rotated about the y-axis, then the result is a cylindrical shell with average radius $\bar{x}_i$, height $f(\bar{x}_i)$, and thickness Δx (see Figure 4), so by Formula 1 its volume is

$$V_i = (2\pi \bar{x}_i)[f(\bar{x}_i)]\,\Delta x$$

Therefore, an approximation to the volume V of S is given by the sum of the volumes of these shells:

$$V \approx \sum_{i=1}^{n} V_i = \sum_{i=1}^{n} 2\pi \bar{x}_i f(\bar{x}_i)\,\Delta x$$

This approximation appears to become better as $n \to \infty$. But, from the definition of an integral, we know that

$$\lim_{n \to \infty} \sum_{i=1}^{n} 2\pi \bar{x}_i f(\bar{x}_i)\,\Delta x = \int_a^b 2\pi x f(x)\,dx$$

Thus, the following appears plausible:

---

2 The volume of the solid in Figure 3, obtained by rotating about the y-axis the region under the curve $y = f(x)$ from a to b, is

$$V = \int_a^b 2\pi x f(x)\,dx \qquad \text{where } 0 \leq a < b$$

The argument using cylindrical shells makes Formula 2 seem reasonable, but later we will be able to prove it (see Exercise 65 in Section 8.1).

The best way to remember Formula 2 is to think of a typical shell, cut and flattened as in Figure 5, with radius x, circumference $2\pi x$, height $f(x)$, and thickness Δx or dx:

$$\int_a^b \underbrace{(2\pi x)}_{\text{circumference}} \underbrace{[f(x)]}_{\text{height}}\,dx$$

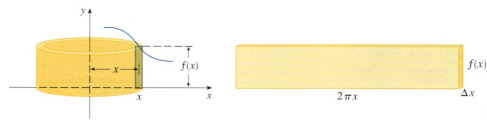

FIGURE 5

This type of reasoning will be helpful in other situations, such as when we rotate about lines other than the y-axis.

EXAMPLE 1 Find the volume of the solid obtained by rotating about the y-axis the region bounded by $y = 2x^2 - x^3$ and $y = 0$.

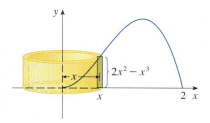

FIGURE 6

SOLUTION From the sketch in Figure 6 we see that a typical shell has radius x, circumference $2\pi x$, and height $f(x) = 2x^2 - x^3$. So, by the shell method, the volume is

$$V = \int_0^2 (2\pi x)(2x^2 - x^3)\, dx = 2\pi \int_0^2 (2x^3 - x^4)\, dx$$

$$= 2\pi \left[\tfrac{1}{2}x^4 - \tfrac{1}{5}x^5\right]_0^2 = 2\pi(8 - \tfrac{32}{5}) = \tfrac{16}{5}\pi$$

It can be verified that the shell method gives the same answer as slicing.

|||| Figure 7 shows a computer-generated picture of the solid whose volume we computed in Example 1.

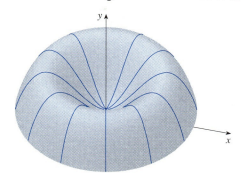

FIGURE 7

NOTE ○ Comparing the solution of Example 1 with the remarks at the beginning of this section, we see that the method of cylindrical shells is much easier than the washer method for this problem. We did not have to find the coordinates of the local maximum and we did not have to solve the equation of the curve for x in terms of y. However, in other examples the methods of the preceding section may be easier.

EXAMPLE 2 Find the volume of the solid obtained by rotating about the y-axis the region between $y = x$ and $y = x^2$.

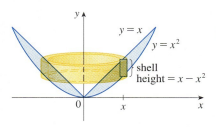

FIGURE 8

SOLUTION The region and a typical shell are shown in Figure 8. We see that the shell has radius x, circumference $2\pi x$, and height $x - x^2$. So the volume is

$$V = \int_0^1 (2\pi x)(x - x^2)\, dx = 2\pi \int_0^1 (x^2 - x^3)\, dx$$

$$= 2\pi \left[\frac{x^3}{3} - \frac{x^4}{4}\right]_0^1 = \frac{\pi}{6}$$

Module 6.2/6.3 compares shells with disks and washers.

As the following example shows, the shell method works just as well if we rotate about the x-axis. We simply have to draw a diagram to identify the radius and height of a shell.

EXAMPLE 3 Use cylindrical shells to find the volume of the solid obtained by rotating about the x-axis the region under the curve $y = \sqrt{x}$ from 0 to 1.

SOLUTION This problem was solved using disks in Example 2 in Section 6.2. To use shells we relabel the curve $y = \sqrt{x}$ (in the figure in that example) as $x = y^2$ in Figure 9. For rotation about the x-axis we see that a typical shell has radius y, circumference $2\pi y$, and height $1 - y^2$. So the volume is

$$V = \int_0^1 (2\pi y)(1 - y^2)\, dy = 2\pi \int_0^1 (y - y^3)\, dy$$

$$= 2\pi \left[\frac{y^2}{2} - \frac{y^4}{4}\right]_0^1 = \frac{\pi}{2}$$

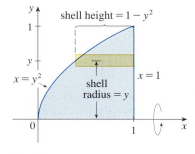

FIGURE 9

In this problem the disk method was simpler.

EXAMPLE 4 Find the volume of the solid obtained by rotating the region bounded by $y = x - x^2$ and $y = 0$ about the line $x = 2$.

SOLUTION Figure 10 shows the region and a cylindrical shell formed by rotation about the line $x = 2$. It has radius $2 - x$, circumference $2\pi(2 - x)$, and height $x - x^2$.

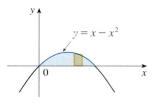
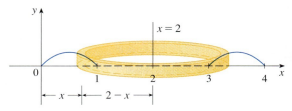

FIGURE 10

The volume of the given solid is

$$V = \int_0^1 2\pi(2 - x)(x - x^2)\, dx = 2\pi \int_0^1 (x^3 - 3x^2 + 2x)\, dx$$

$$= 2\pi \left[\frac{x^4}{4} - x^3 + x^2 \right]_0^1 = \frac{\pi}{2}$$

6.3 Exercises

1. Let S be the solid obtained by rotating the region shown in the figure about the y-axis. Explain why it is awkward to use slicing to find the volume V of S. Sketch a typical approximating shell. What are its circumference and height? Use shells to find V.

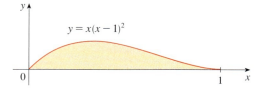

2. Let S be the solid obtained by rotating the region shown in the figure about the y-axis. Sketch a typical cylindrical shell and find its circumference and height. Use shells to find the volume of S. Do you think this method is preferable to slicing? Explain.

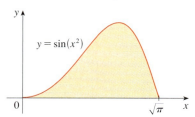

3–7 ▪▪▪▪ Use the method of cylindrical shells to find the volume generated by rotating the region bounded by the given curves about the y-axis. Sketch the region and a typical shell.

3. $y = 1/x$, $y = 0$, $x = 1$, $x = 2$

4. $y = x^2$, $y = 0$, $x = 1$

5. $y = x^2$, $0 \le x \le 2$, $y = 4$, $x = 0$

6. $y = 3 + 2x - x^2$, $x + y = 3$

7. $y = 4(x - 2)^2$, $y = x^2 - 4x + 7$

8. Let V be the volume of the solid obtained by rotating about the y-axis the region bounded by $y = \sqrt{x}$ and $y = x^2$. Find V both by slicing and by cylindrical shells. In both cases draw a diagram to explain your method.

9–14 ▪▪▪▪ Use the method of cylindrical shells to find the volume of the solid obtained by rotating the region bounded by the given curves about the x-axis. Sketch the region and a typical shell.

9. $x = 1 + y^2$, $x = 0$, $y = 1$, $y = 2$

10. $x = \sqrt{y}$, $x = 0$, $y = 1$

11. $y = x^3$, $y = 8$, $x = 0$

12. $x = 4y^2 - y^3$, $x = 0$

13. $y = 4x^2$, $2x + y = 6$

14. $x + y = 3$, $x = 4 - (y - 1)^2$

15–20 ▪▪▪▪ Use the method of cylindrical shells to find the volume generated by rotating the region bounded by the given curves about the specified axis. Sketch the region and a typical shell.

15. $y = x^2$, $y = 0$, $x = 1$, $x = 2$; about $x = 1$

16. $y = x^2$, $y = 0$, $x = -2$, $x = -1$; about the y-axis

17. $y = x^2$, $y = 0$, $x = 1$, $x = 2$; about $x = 4$

18. $y = 4x - x^2$, $y = 8x - 2x^2$; about $x = -2$

19. $y = \sqrt{x - 1}$, $y = 0$, $x = 5$; about $y = 3$

20. $y = x^2$, $x = y^2$; about $y = -1$

21–26 ▮▮▮▮ Set up, but do not evaluate, an integral for the volume of the solid obtained by rotating the region bounded by the given curves about the specified axis.

21. $y = \sin x$, $y = 0$, $x = 2\pi$, $x = 3\pi$; about the y-axis

22. $y = x$, $y = 4x - x^2$; about $x = 7$

23. $y = x^4$, $y = \sin(\pi x/2)$; about $x = -1$

24. $y = 1/(1 + x^2)$, $y = 0$, $x = 0$, $x = 2$; about $x = 2$

25. $x = \sqrt{\sin y}$, $0 \le y \le \pi$, $x = 0$; about $y = 4$

26. $x^2 - y^2 = 7$, $x = 4$; about $y = 5$

27. Use the Midpoint Rule with $n = 4$ to estimate the volume obtained by rotating about the y-axis the region under the curve $y = \tan x$, $0 \le x \le \pi/4$.

28. If the region shown in the figure is rotated about the y-axis to form a solid, use the Midpoint Rule with $n = 5$ to estimate the volume of the solid.

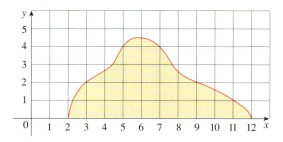

29–32 ▮▮▮▮ Each integral represents the volume of a solid. Describe the solid.

29. $\int_0^3 2\pi x^5 \, dx$

30. $2\pi \int_0^2 \dfrac{y}{1 + y^2} \, dy$

31. $\int_0^1 2\pi(3 - y)(1 - y^2) \, dy$

32. $\int_0^{\pi/4} 2\pi(\pi - x)(\cos x - \sin x) \, dx$

33–34 ▮▮▮▮ Use a graph to estimate the x-coordinates of the points of intersection of the given curves. Then use this information to estimate the volume of the solid obtained by rotating about the y-axis the region enclosed by these curves.

33. $y = 0$, $y = x + x^2 - x^4$

34. $y = x^4$, $y = 3x - x^3$

CAS 35–36 ▮▮▮▮ Use a computer algebra system to find the exact volume of the solid obtained by rotating the region bounded by the given curves about the specified line.

35. $y = \sin^2 x$, $y = \sin^4 x$, $0 \le x \le \pi$; about $x = \pi/2$

36. $y = x^3 \sin x$, $y = 0$, $0 \le x \le \pi$; about $x = -1$

37–42 ▮▮▮▮ The region bounded by the given curves is rotated about the specified axis. Find the volume of the resulting solid by any method.

37. $y = x^2 + x - 2$, $y = 0$; about the x-axis

38. $y = x^2 - 3x + 2$, $y = 0$; about the y-axis

39. $y = 5$, $y = x^2 - 5x + 9$; about $x = -1$

40. $x = 1 - y^4$, $x = 0$; about $x = 2$

41. $x^2 + (y - 1)^2 = 1$; about the y-axis

42. $x^2 + (y - 1)^2 = 1$; about the x-axis

43–45 ▮▮▮▮ Use cylindrical shells to find the volume of the solid.

43. A sphere of radius r

44. The solid torus of Exercise 61 in Section 6.2

45. A right circular cone with height h and base radius r

46. Suppose you make napkin rings by drilling holes with different diameters through two wooden balls (which also have different diameters). You discover that both napkin rings have the same height h, as shown in the figure.
(a) Guess which ring has more wood in it.
(b) Check your guess: Use cylindrical shells to compute the volume of a napkin ring created by drilling a hole with radius r through the center of a sphere of radius R and express the answer in terms of h.

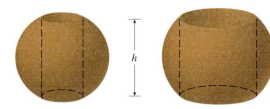

6.4 Work

The term *work* is used in everyday language to mean the total amount of effort required to perform a task. In physics it has a technical meaning that depends on the idea of a *force*. Intuitively, you can think of a force as describing a push or pull on an object—for example, a horizontal push of a book across a table or the downward pull of Earth's gravity on a ball. In general, if an object moves along a straight line with position function $s(t)$, then the **force** F on the object (in the same direction) is defined by Newton's Second Law of Motion as the product of its mass m and its acceleration:

$$\boxed{1} \qquad F = m\frac{d^2s}{dt^2}$$

In the SI metric system, the mass is measured in kilograms (kg), the displacement in meters (m), the time in seconds (s), and the force in newtons (N = kg·m/s²). Thus, a force of 1 N acting on a mass of 1 kg produces an acceleration of 1 m/s². In the U.S. Customary system the fundamental unit is chosen to be the unit of force, which is the pound.

In the case of constant acceleration, the force F is also constant and the work done is defined to be the product of the force F and the distance d that the object moves:

$$\boxed{2} \qquad W = Fd \qquad \text{work} = \text{force} \times \text{distance}$$

If F is measured in newtons and d in meters, then the unit for W is a newton-meter, which is called a joule (J). If F is measured in pounds and d in feet, then the unit for W is a foot-pound (ft-lb), which is about 1.36 J.

EXAMPLE 1
(a) How much work is done in lifting a 1.2-kg book off the floor to put it on a desk that is 0.7 m high? Use the fact that the acceleration due to gravity is $g = 9.8$ m/s².
(b) How much work is done in lifting a 20-lb weight 6 ft off the ground?

SOLUTION
(a) The force exerted is equal and opposite to that exerted by gravity, so Equation 1 gives

$$F = mg = (1.2)(9.8) = 11.76 \text{ N}$$

and then Equation 2 gives the work done as

$$W = Fd = (11.76)(0.7) \approx 8.2 \text{ J}$$

(b) Here the force is given as $F = 20$ lb, so the work done is

$$W = Fd = 20 \cdot 6 = 120 \text{ ft-lb}$$

Notice that in part (b), unlike part (a), we did not have to multiply by g because we were given the *weight* (which is a force) and not the mass of the object.

Equation 2 defines work as long as the force is constant, but what happens if the force is variable? Let's suppose that the object moves along the x-axis in the positive direction, from $x = a$ to $x = b$, and at each point x between a and b a force $f(x)$ acts on the object, where f is a continuous function. We divide the interval $[a, b]$ into n subintervals with endpoints $x_0, x_1, \ldots, x_n$ and equal width Δx. We choose a sample point x_i^* in the ith subinterval $[x_{i-1}, x_i]$. Then the force at that point is $f(x_i^*)$. If n is large, then Δx is small, and since f is continuous, the values of f don't change very much over the interval $[x_{i-1}, x_i]$. In other words, f is almost constant on the interval and so the work W_i that is done in moving the particle from x_{i-1} to x_i is approximately given by Equation 2:

$$W_i \approx f(x_i^*) \, \Delta x$$

Thus, we can approximate the total work by

$$\boxed{3} \qquad W \approx \sum_{i=1}^{n} f(x_i^*) \, \Delta x$$

It seems that this approximation becomes better as we make n larger. Therefore, we define the **work done in moving the object from a to b** as the limit of this quantity as $n \to \infty$. Since the right side of (3) is a Riemann sum, we recognize its limit as being a definite integral and so

$$\boxed{4} \qquad W = \lim_{n \to \infty} \sum_{i=1}^{n} f(x_i^*) \, \Delta x = \int_a^b f(x) \, dx$$

EXAMPLE 2 When a particle is located a distance x feet from the origin, a force of $x^2 + 2x$ pounds acts on it. How much work is done in moving it from $x = 1$ to $x = 3$?

SOLUTION
$$W = \int_1^3 (x^2 + 2x) \, dx = \frac{x^3}{3} + x^2 \Big]_1^3 = \frac{50}{3}$$

The work done is $16\frac{2}{3}$ ft-lb.

In the next example we use a law from physics: **Hooke's Law** states that the force required to maintain a spring stretched x units beyond its natural length is proportional to x:

$$f(x) = kx$$

where k is a positive constant (called the **spring constant**). Hooke's Law holds provided that x is not too large (see Figure 1).

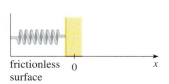

(a) Natural position of spring

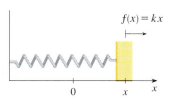

(b) Stretched position of spring

FIGURE 1
Hooke's Law

EXAMPLE 3 A force of 40 N is required to hold a spring that has been stretched from its natural length of 10 cm to a length of 15 cm. How much work is done in stretching the spring from 15 cm to 18 cm?

SOLUTION According to Hooke's Law, the force required to hold the spring stretched x meters beyond its natural length is $f(x) = kx$. When the spring is stretched from 10 cm to 15 cm, the amount stretched is 5 cm $=$ 0.05 m. This means that $f(0.05) = 40$, so

$$0.05k = 40 \qquad k = \frac{40}{0.05} = 800$$

Thus, $f(x) = 800x$ and the work done in stretching the spring from 15 cm to 18 cm is

$$W = \int_{0.05}^{0.08} 800x \, dx = 800 \frac{x^2}{2} \Big]_{0.05}^{0.08}$$

$$= 400[(0.08)^2 - (0.05)^2] = 1.56 \text{ J}$$

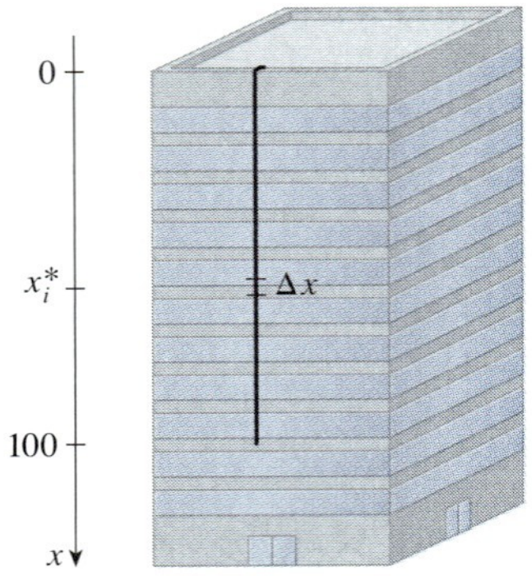

400 ❘❘❘❘ CHAPTER 6 APPLICATIONS OF INTEGRATION

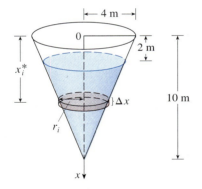

FIGURE 2

❘❘❘❘ If we had placed the origin at the bottom of the cable and the x-axis upward, we would have gotten

$$W = \int_0^{100} 2(100 - x)\, dx$$

which gives the same answer.

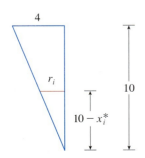

FIGURE 3

FIGURE 4

EXAMPLE 4 A 200-lb cable is 100 ft long and hangs vertically from the top of a tall building. How much work is required to lift the cable to the top of the building?

SOLUTION Here we don't have a formula for the force function, but we can use an argument similar to the one that led to Definition 4.

Let's place the origin at the top of the building and the x-axis pointing downward as in Figure 2. We divide the cable into small parts with length Δx. If x_i^* is a point in the ith such interval, then all points in the interval are lifted by approximately the same amount, namely x_i^*. The cable weighs 2 pounds per foot, so the weight of the ith part is $2\Delta x$. Thus, the work done on the ith part, in foot-pounds, is

$$\underbrace{(2\Delta x)}_{\text{force}}\ \ \underbrace{x_i^*}_{\text{distance}} = 2x_i^*\, \Delta x$$

We get the total work done by adding all these approximations and letting the number of parts become large (so $\Delta x \to 0$):

$$W = \lim_{n \to \infty} \sum_{i=1}^{n} 2x_i^*\, \Delta x = \int_0^{100} 2x\, dx$$

$$= x^2 \big]_0^{100} = 10{,}000 \text{ ft-lb}$$

EXAMPLE 5 A tank has the shape of an inverted circular cone with height 10 m and base radius 4 m. It is filled with water to a height of 8 m. Find the work required to empty the tank by pumping all of the water to the top of the tank. (The density of water is 1000 kg/m³.)

SOLUTION Let's measure depths from the top of the tank by introducing a vertical coordinate line as in Figure 3. The water extends from a depth of 2 m to a depth of 10 m and so we divide the interval $[2, 10]$ into n subintervals with endpoints $x_0, x_1, \ldots, x_n$ and choose x_i^* in the ith subinterval. This divides the water into n layers. The ith layer is approximated by a circular cylinder with radius r_i and height Δx. We can compute r_i from similar triangles, using Figure 4, as follows:

$$\frac{r_i}{10 - x_i^*} = \frac{4}{10} \qquad r_i = \tfrac{2}{5}(10 - x_i^*)$$

Thus, an approximation to the volume of the ith layer of water is

$$V_i \approx \pi r_i^2\, \Delta x = \frac{4\pi}{25}(10 - x_i^*)^2\, \Delta x$$

and so its mass is

$$m_i = \text{density} \times \text{volume}$$

$$\approx 1000 \cdot \frac{4\pi}{25}(10 - x_i^*)^2\, \Delta x = 160\pi(10 - x_i^*)^2\, \Delta x$$

The force required to raise this layer must overcome the force of gravity and so

$$F_i = m_i g \approx (9.8)160\pi(10 - x_i^*)^2\, \Delta x$$

$$\approx 1570\pi(10 - x_i^*)^2\, \Delta x$$

Each particle in the layer must travel a distance of approximately x_i^*. The work W_i done to raise this layer to the top is approximately the product of the force F_i and the distance x_i^*:

$$W_i \approx F_i x_i^* \approx 1570\pi x_i^*(10 - x_i^*)^2\, \Delta x$$

To find the total work done in emptying the entire tank, we add the contributions of each of the n layers and then take the limit as $n \to \infty$:

$$W = \lim_{n \to \infty} \sum_{i=1}^{n} 1570\pi x_i^*(10 - x_i^*)^2 \Delta x = \int_2^{10} 1570\pi x(10 - x)^2 \, dx$$

$$= 1570\pi \int_2^{10} (100x - 20x^2 + x^3) \, dx = 1570\pi \left[50x^2 - \frac{20x^3}{3} + \frac{x^4}{4} \right]_2^{10}$$

$$= 1570\pi \left(\tfrac{2048}{3} \right) \approx 3.4 \times 10^6 \text{ J}$$

6.4 Exercises

1. Find the work done in pushing a car a distance of 8 m while exerting a constant force of 900 N.

2. How much work is done by a weightlifter in raising a 60-kg barbell from the floor to a height of 2 m?

3. A particle is moved along the x-axis by a force that measures $10/(1 + x)^2$ pounds at a point x feet from the origin. Find the work done in moving the particle from the origin to a distance of 9 ft.

4. When a particle is located a distance x meters from the origin, a force of $\cos(\pi x/3)$ newtons acts on it. How much work is done in moving the particle from $x = 1$ to $x = 2$? Interpret your answer by considering the work done from $x = 1$ to $x = 1.5$ and from $x = 1.5$ to $x = 2$.

5. Shown is the graph of a force function (in newtons) that increases to its maximum value and then remains constant. How much work is done by the force in moving an object a distance of 8 m?

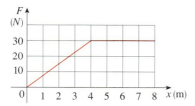

6. The table shows values of a force function $f(x)$, where x is measured in meters and $f(x)$ in newtons. Use the Midpoint Rule to estimate the work done by the force in moving an object from $x = 4$ to $x = 20$.

x	4	6	8	10	12	14	16	18	20
$f(x)$	5	5.8	7.0	8.8	9.6	8.2	6.7	5.2	4.1

7. A force of 10 lb is required to hold a spring stretched 4 in. beyond its natural length. How much work is done in stretching it from its natural length to 6 in. beyond its natural length?

8. A spring has a natural length of 20 cm. If a 25-N force is required to keep it stretched to a length of 30 cm, how much work is required to stretch it from 20 cm to 25 cm?

9. Suppose that 2 J of work is needed to stretch a spring from its natural length of 30 cm to a length of 42 cm. How much work is needed to stretch it from 35 cm to 40 cm?

10. If the work required to stretch a spring 1 ft beyond its natural length is 12 ft-lb, how much work is needed to stretch it 9 in. beyond its natural length?

11. How far beyond its natural length will a force of 30 N keep the spring in Exercise 9 stretched?

12. If 6 J of work is needed to stretch a spring from 10 cm to 12 cm and another 10 J is needed to stretch it from 12 cm to 14 cm, what is the natural length of the spring?

13–20 Show how to approximate the required work by a Riemann sum. Then express the work as an integral and evaluate it.

13. A heavy rope, 50 ft long, weighs 0.5 lb/ft and hangs over the edge of a building 120 ft high.
    (a) How much work is done in pulling the rope to the top of the building?
    (b) How much work is done in pulling half the rope to the top of the building?

14. A chain lying on the ground is 10 m long and its mass is 80 kg. How much work is required to raise one end of the chain to a height of 6 m?

15. A cable that weighs 2 lb/ft is used to lift 800 lb of coal up a mineshaft 500 ft deep. Find the work done.

16. A bucket that weighs 4 lb and a rope of negligible weight are used to draw water from a well that is 80 ft deep. The bucket is filled with 40 lb of water and is pulled up at a rate of 2 ft/s, but water leaks out of a hole in the bucket at a rate of 0.2 lb/s. Find the work done in pulling the bucket to the top of the well.

17. A leaky 10-kg bucket is lifted from the ground to a height of 12 m at a constant speed with a rope that weighs 0.8 kg/m. Initially the bucket contains 36 kg of water, but the water leaks at a constant rate and finishes draining just as the bucket reaches the 12 m level. How much work is done?

18. A 10-ft chain weighs 25 lb and hangs from a ceiling. Find the work done in lifting the lower end of the chain to the ceiling so that it's level with the upper end.

19. An aquarium 2 m long, 1 m wide, and 1 m deep is full of water. Find the work needed to pump half of the water out of the aquarium. (Use the fact that the density of water is 1000 kg/m³.)

20. A circular swimming pool has a diameter of 24 ft, the sides are 5 ft high, and the depth of the water is 4 ft. How much work is required to pump all of the water out over the side? (Use the fact that water weighs 62.5 lb/ft³.)

21–24 A tank is full of water. Find the work required to pump the water out of the outlet. In Exercises 23 and 24 use the fact that water weighs 62.5 lb/ft³.

21.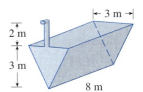

22.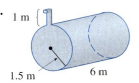

23. 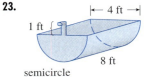 semicircle

24.  hemisphere

25. Suppose that for the tank in Exercise 21 the pump breaks down after 4.7×10^5 J of work has been done. What is the depth of the water remaining in the tank?

26. Solve Exercise 22 if the tank is half full of oil that has a density of 920 kg/m³.

27. When gas expands in a cylinder with radius r, the pressure at any given time is a function of the volume: $P = P(V)$. The force exerted by the gas on the piston (see the figure) is the product of the pressure and the area: $F = \pi r^2 P$. Show that the work done by the gas when the volume expands from volume V_1 to volume V_2 is

$$W = \int_{V_1}^{V_2} P \, dV$$

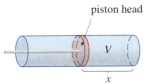

28. In a steam engine the pressure P and volume V of steam satisfy the equation $PV^{1.4} = k$, where k is a constant. (This is true for adiabatic expansion, that is, expansion in which there is no heat transfer between the cylinder and its surroundings.) Use Exercise 27 to calculate the work done by the engine during a cycle when the steam starts at a pressure of 160 lb/in² and a volume of 100 in³ and expands to a volume of 800 in³.

29. Newton's Law of Gravitation states that two bodies with masses m_1 and m_2 attract each other with a force

$$F = G \frac{m_1 m_2}{r^2}$$

where r is the distance between the bodies and G is the gravitational constant. If one of the bodies is fixed, find the work needed to move the other from $r = a$ to $r = b$.

30. Use Newton's Law of Gravitation to compute the work required to launch a 1000-kg satellite vertically to an orbit 1000 km high. You may assume that Earth's mass is 5.98×10^{24} kg and is concentrated at its center. Take the radius of Earth to be 6.37×10^6 m and $G = 6.67 \times 10^{-11}$ N·m²/kg².

6.5 Average Value of a Function

FIGURE 1

It is easy to calculate the average value of finitely many numbers $y_1, y_2, \ldots, y_n$:

$$y_{\text{ave}} = \frac{y_1 + y_2 + \cdots + y_n}{n}$$

But how do we compute the average temperature during a day if infinitely many temperature readings are possible? Figure 1 shows the graph of a temperature function $T(t)$, where t is measured in hours and T in °C, and a guess at the average temperature, T_{ave}.

In general, let's try to compute the average value of a function $y = f(x)$, $a \le x \le b$. We start by dividing the interval $[a, b]$ into n equal subintervals, each with length $\Delta x = (b - a)/n$. Then we choose points $x_1^*, \ldots, x_n^*$ in successive subintervals and calculate the average of the numbers $f(x_1^*), \ldots, f(x_n^*)$:

$$\frac{f(x_1^*) + \cdots + f(x_n^*)}{n}$$

(For example, if f represents a temperature function and $n = 24$, this means that we take temperature readings every hour and then average them.) Since $\Delta x = (b-a)/n$, we can write $n = (b-a)/\Delta x$ and the average value becomes

$$\frac{f(x_1^*) + \cdots + f(x_n^*)}{\dfrac{b-a}{\Delta x}} = \frac{1}{b-a}[f(x_1^*)\,\Delta x + \cdots + f(x_n^*)\,\Delta x]$$

$$= \frac{1}{b-a}\sum_{i=1}^{n} f(x_i^*)\,\Delta x$$

If we let n increase, we would be computing the average value of a large number of closely spaced values. (For example, we would be averaging temperature readings taken every minute or even every second.) The limiting value is

$$\lim_{n \to \infty} \frac{1}{b-a} \sum_{i=1}^{n} f(x_i^*)\,\Delta x = \frac{1}{b-a}\int_a^b f(x)\,dx$$

by the definition of a definite integral.

Therefore, we define the **average value of f** on the interval $[a, b]$ as

$$f_{\text{ave}} = \frac{1}{b-a}\int_a^b f(x)\,dx$$

EXAMPLE 1 Find the average value of the function $f(x) = 1 + x^2$ on the interval $[-1, 2]$.

SOLUTION With $a = -1$ and $b = 2$ we have

$$f_{\text{ave}} = \frac{1}{b-a}\int_a^b f(x)\,dx = \frac{1}{2-(-1)}\int_{-1}^{2}(1+x^2)\,dx$$

$$= \frac{1}{3}\left[x + \frac{x^3}{3}\right]_{-1}^{2} = 2$$

The question arises: Is there a number c at which the value of f is exactly equal to the average value of the function, that is, $f(c) = f_{\text{ave}}$? The following theorem says that this is true for continuous functions.

The Mean Value Theorem for Integrals If f is continuous on $[a, b]$, then there exists a number c in $[a, b]$ such that

$$\int_a^b f(x)\,dx = f(c)(b-a)$$

The Mean Value Theorem for Integrals is a consequence of the Mean Value Theorem for derivatives and the Fundamental Theorem of Calculus. The proof is outlined in Exercise 23.

The geometric interpretation of the Mean Value Theorem for Integrals is that, for *positive* functions f, there is a number c such that the rectangle with base $[a, b]$ and height $f(c)$ has the same area as the region under the graph of f from a to b. (See Figure 2 and the more picturesque interpretation in the margin note.)

|||| You can always chop off the top of a (two-dimensional) mountain at a certain height and use it to fill in the valleys so that the mountain becomes completely flat.

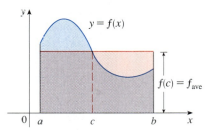

FIGURE 2

EXAMPLE 2 Since $f(x) = 1 + x^2$ is continuous on the interval $[-1, 2]$, the Mean Value Theorem for Integrals says there is a number c in $[-1, 2]$ such that

$$\int_{-1}^{2} (1 + x^2)\, dx = f(c)[2 - (-1)]$$

In this particular case we can find c explicitly. From Example 1 we know that $f_{\text{ave}} = 2$, so the value of c satisfies

$$f(c) = f_{\text{ave}} = 2$$

Therefore $\qquad 1 + c^2 = 2 \qquad$ so $\qquad c^2 = 1$

Thus, in this case there happen to be two numbers $c = \pm 1$ in the interval $[-1, 2]$ that work in the Mean Value Theorem for Integrals.

Examples 1 and 2 are illustrated by Figure 3.

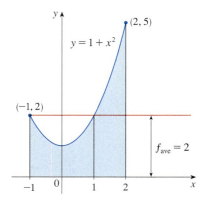

FIGURE 3

EXAMPLE 3 Show that the average velocity of a car over a time interval $[t_1, t_2]$ is the same as the average of its velocities during the trip.

SOLUTION If $s(t)$ is the displacement of the car at time t, then, by definition, the average velocity of the car over the interval is

$$\frac{\Delta s}{\Delta t} = \frac{s(t_2) - s(t_1)}{t_2 - t_1}$$

On the other hand, the average value of the velocity function on the interval is

$$v_{\text{ave}} = \frac{1}{t_2 - t_1} \int_{t_1}^{t_2} v(t)\, dt = \frac{1}{t_2 - t_1} \int_{t_1}^{t_2} s'(t)\, dt$$

$$= \frac{1}{t_2 - t_1} [s(t_2) - s(t_1)] \qquad \text{(by the Net Change Theorem)}$$

$$= \frac{s(t_2) - s(t_1)}{t_2 - t_1} = \text{average velocity}$$

6.5 Exercises

1–8 Find the average value of the function on the given interval.

1. $f(x) = x^2$, $[-1, 1]$
2. $f(x) = x - x^2$, $[0, 2]$
3. $g(x) = \cos x$, $[0, \pi/2]$
4. $g(x) = x^2\sqrt{1 + x^3}$, $[0, 2]$
5. $f(t) = t\sqrt{1 + t^2}$, $[0, 5]$
6. $f(\theta) = \sec \theta \tan \theta$, $[0, \pi/4]$
7. $h(x) = \cos^4 x \sin x$, $[0, \pi]$
8. $h(r) = 3/(1 + r)^2$, $[1, 6]$

9–12
(a) Find the average value of f on the given interval.
(b) Find c such that $f_{\text{ave}} = f(c)$.
(c) Sketch the graph of f and a rectangle whose area is the same as the area under the graph of f.

9. $f(x) = (x - 3)^2$, $[2, 5]$
10. $f(x) = \sqrt{x}$, $[0, 4]$
11. $f(x) = 2 \sin x - \sin 2x$, $[0, \pi]$
12. $f(x) = 2x/(1 + x^2)^2$, $[0, 2]$

13. If f is continuous and $\int_1^3 f(x)\, dx = 8$, show that f takes on the value 4 at least once on the interval $[1, 3]$.

14. Find the numbers b such that the average value of $f(x) = 2 + 6x - 3x^2$ on the interval $[0, b]$ is equal to 3.

15. The table gives values of a continuous function. Use the Midpoint Rule to estimate the average value of f on $[20, 50]$.

x	20	25	30	35	40	45	50
$f(x)$	42	38	31	29	35	48	60

16. The velocity graph of an accelerating car is shown.
(a) Estimate the average velocity of the car during the first 12 seconds.
(b) At what time was the instantaneous velocity equal to the average velocity?

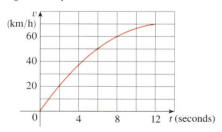

17. In a certain city the temperature (in °F) t hours after 9 A.M. was modeled by the function

$$T(t) = 50 + 14 \sin \frac{\pi t}{12}$$

Find the average temperature during the period from 9 A.M. to 9 P.M.

18. The temperature of a metal rod, 5 m long, is $4x$ (in °C) at a distance x meters from one end of the rod. What is the average temperature of the rod?

19. The linear density in a rod 8 m long is $12/\sqrt{x + 1}$ kg/m, where x is measured in meters from one end of the rod. Find the average density of the rod.

20. If a freely falling body starts from rest, then its displacement is given by $s = \frac{1}{2}gt^2$. Let the velocity after a time T be v_T. Show that if we compute the average of the velocities with respect to t we get $v_{\text{ave}} = \frac{1}{2}v_T$, but if we compute the average of the velocities with respect to s we get $v_{\text{ave}} = \frac{2}{3}v_T$.

21. Use the result of Exercise 59 in Section 5.5 to compute the average volume of inhaled air in the lungs in one respiratory cycle.

22. The velocity v of blood that flows in a blood vessel with radius R and length l at a distance r from the central axis is

$$v(r) = \frac{P}{4\eta l}(R^2 - r^2)$$

where P is the pressure difference between the ends of the vessel and η is the viscosity of the blood (see Example 7 in Section 3.4). Find the average velocity (with respect to r) over the interval $0 \le r \le R$. Compare the average velocity with the maximum velocity.

23. Prove the Mean Value Theorem for Integrals by applying the Mean Value Theorem for derivatives (see Section 4.2) to the function $F(x) = \int_a^x f(t)\, dt$.

24. If $f_{\text{ave}}[a, b]$ denotes the average value of f on the interval $[a, b]$ and $a < c < b$, show that

$$f_{\text{ave}}[a, b] = \frac{c - a}{b - a} f_{\text{ave}}[a, c] + \frac{b - c}{b - a} f_{\text{ave}}[c, b]$$

6 Review

CONCEPT CHECK

1. (a) Draw two typical curves $y = f(x)$ and $y = g(x)$, where $f(x) \geq g(x)$ for $a \leq x \leq b$. Show how to approximate the area between these curves by a Riemann sum and sketch the corresponding approximating rectangles. Then write an expression for the exact area.
   (b) Explain how the situation changes if the curves have equations $x = f(y)$ and $x = g(y)$, where $f(y) \geq g(y)$ for $c \leq y \leq d$.

2. Suppose that Sue runs faster than Kathy throughout a 1500-meter race. What is the physical meaning of the area between their velocity curves for the first minute of the race?

3. (a) Suppose S is a solid with known cross-sectional areas. Explain how to approximate the volume of S by a Riemann sum. Then write an expression for the exact volume.
   (b) If S is a solid of revolution, how do you find the cross-sectional areas?

4. (a) What is the volume of a cylindrical shell?
   (b) Explain how to use cylindrical shells to find the volume of a solid of revolution.
   (c) Why might you want to use the shell method instead of slicing?

5. Suppose that you push a book across a 6-meter-long table by exerting a force $f(x)$ at each point from $x = 0$ to $x = 6$. What does $\int_0^6 f(x)\,dx$ represent? If $f(x)$ is measured in newtons, what are the units for the integral?

6. (a) What is the average value of a function f on an interval $[a, b]$?
   (b) What does the Mean Value Theorem for Integrals say? What is its geometric interpretation?

EXERCISES

1–6 Find the area of the region bounded by the given curves.

1. $y = x^2 - x - 6$, $y = 0$
2. $y = 20 - x^2$, $y = x^2 - 12$
3. $y = 1 - x^2$, $y = 1 - \sqrt{x}$
4. $x + y = 0$, $x = y^2 + 3y$
5. $y = \sin(\pi x/2)$, $y = x^2 - 2x$
6. $y = \sqrt{x}$, $y = x^2$, $x = 2$

7–11 Find the volume of the solid obtained by rotating the region bounded by the given curves about the specified axis.

7. $y = 2x$, $y = x^2$; about the x-axis
8. $x = 1 + y^2$, $y = x - 3$; about the y-axis
9. $x = 0$, $x = 9 - y^2$; about $x = -1$
10. $y = x^2 + 1$, $y = 9 - x^2$; about $y = -1$
11. $x^2 - y^2 = a^2$, $x = a + h$ (where $a > 0$, $h > 0$); about the y-axis

12–14 Set up, but do not evaluate, an integral for the volume of the solid obtained by rotating the region bounded by the given curves about the specified axis.

12. $y = \cos x$, $y = 0$, $x = 3\pi/2$, $x = 5\pi/2$; about the y-axis
13. $y = x^3$, $y = x^2$; about $y = 1$
14. $y = x^3$, $y = 8$, $x = 0$; about $x = 2$

15. Find the volumes of the solids obtained by rotating the region bounded by the curves $y = x$ and $y = x^2$ about the following lines:
    (a) The x-axis   (b) The y-axis   (c) $y = 2$

16. Let $\mathcal{R}$ be the region in the first quadrant bounded by the curves $y = x^3$ and $y = 2x - x^2$. Calculate the following quantities.
    (a) The area of $\mathcal{R}$
    (b) The volume obtained by rotating $\mathcal{R}$ about the x-axis
    (c) The volume obtained by rotating $\mathcal{R}$ about the y-axis

17. Let $\mathcal{R}$ be the region bounded by the curves $y = \tan(x^2)$, $x = 1$, and $y = 0$. Use the Midpoint Rule with $n = 4$ to estimate the following.
    (a) The area of $\mathcal{R}$
    (b) The volume obtained by rotating $\mathcal{R}$ about the x-axis

18. Let $\mathcal{R}$ be the region bounded by the curves $y = 1 - x^2$ and $y = x^6 - x + 1$. Estimate the following quantities.
    (a) The x-coordinates of the points of intersection of the curves
    (b) The area of $\mathcal{R}$
    (c) The volume generated when $\mathcal{R}$ is rotated about the x-axis
    (d) The volume generated when $\mathcal{R}$ is rotated about the y-axis

19–22 Each integral represents the volume of a solid. Describe the solid.

19. $\displaystyle\int_0^{\pi/2} 2\pi x \cos x\,dx$

20. $\displaystyle\int_0^{\pi/2} 2\pi \cos^2 x\,dx$

21. $\int_0^2 2\pi y(4 - y^2)\, dy$

22. $\int_0^1 \pi[(2 - x^2)^2 - (2 - \sqrt{x})^2]\, dx$

23. The base of a solid is a circular disk with radius 3. Find the volume of the solid if parallel cross-sections perpendicular to the base are isosceles right triangles with hypotenuse lying along the base.

24. The base of a solid is the region bounded by the parabolas $y = x^2$ and $y = 2 - x^2$. Find the volume of the solid if the cross-sections perpendicular to the x-axis are squares with one side lying along the base.

25. The height of a monument is 20 m. A horizontal cross-section at a distance x meters from the top is an equilateral triangle with side $x/4$ meters. Find the volume of the monument.

26. (a) The base of a solid is a square with vertices located at $(1, 0), (0, 1), (-1, 0)$, and $(0, -1)$. Each cross-section perpendicular to the x-axis is a semicircle. Find the volume of the solid.
(b) Show that by cutting the solid of part (a), we can rearrange it to form a cone. Thus compute its volume more simply.

27. A force of 30 N is required to maintain a spring stretched from its natural length of 12 cm to a length of 15 cm. How much work is done in stretching the spring from 12 cm to 20 cm?

28. A 1600-lb elevator is suspended by a 200-ft cable that weighs 10 lb/ft. How much work is required to raise the elevator from the basement to the third floor, a distance of 30 ft?

29. A tank full of water has the shape of a paraboloid of revolution as shown in the figure; that is, its shape is obtained by rotating a parabola about a vertical axis.
(a) If its height is 4 ft and the radius at the top is 4 ft, find the work required to pump the water out of the tank.
(b) After 4000 ft-lb of work has been done, what is the depth of the water remaining in the tank?

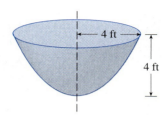

30. Find the average value of the function $f(t) = t \sin(t^2)$ on the interval $[0, 10]$.

31. If f is a continuous function, what is the limit as $h \to 0$ of the average value of f on the interval $[x, x + h]$?

32. Let $\mathcal{R}_1$ be the region bounded by $y = x^2$, $y = 0$, and $x = b$, where $b > 0$. Let $\mathcal{R}_2$ be the region bounded by $y = x^2$, $x = 0$, and $y = b^2$.
(a) Is there a value of b such that $\mathcal{R}_1$ and $\mathcal{R}_2$ have the same area?
(b) Is there a value of b such that $\mathcal{R}_1$ sweeps out the same volume when rotated about the x-axis and the y-axis?
(c) Is there a value of b such that $\mathcal{R}_1$ and $\mathcal{R}_2$ sweep out the same volume when rotated about the x-axis?
(d) Is there a value of b such that $\mathcal{R}_1$ and $\mathcal{R}_2$ sweep out the same volume when rotated about the y-axis?

PROBLEMS PLUS

1. (a) Find a positive continuous function f such that the area under the graph of f from 0 to t is $A(t) = t^3$ for all $t > 0$.
    (b) A solid is generated by rotating about the x-axis the region under the curve $y = f(x)$, where f is a positive function and $x \geq 0$. The volume generated by the part of the curve from $x = 0$ to $x = b$ is b^2 for all $b > 0$. Find the function f.

2. There is a line through the origin that divides the region bounded by the parabola $y = x - x^2$ and the x-axis into two regions with equal area. What is the slope of that line?

3. The figure shows a horizontal line $y = c$ intersecting the curve $y = 8x - 27x^3$. Find the number c such that the areas of the shaded regions are equal.

4. A cylindrical glass of radius r and height L is filled with water and then tilted until the water remaining in the glass exactly covers its base.
    (a) Determine a way to "slice" the water into parallel rectangular cross-sections and then set up a definite integral for the volume of the water in the glass.
    (b) Determine a way to "slice" the water into parallel cross-sections that are trapezoids and then set up a definite integral for the volume of the water.
    (c) Find the volume of water in the glass by evaluating one of the integrals in part (a) or part (b).
    (d) Find the volume of the water in the glass from purely geometric considerations.
    (e) Suppose the glass is tilted until the water exactly covers half the base. In what direction can you "slice" the water into triangular cross-sections? Rectangular cross-sections? Cross-sections that are segments of circles? Find the volume of water in the glass.

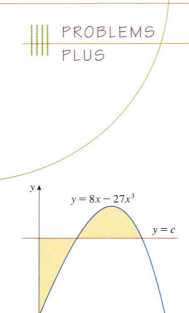

FIGURE FOR PROBLEM 3

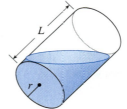 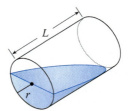

5. (a) Show that the volume of a segment of height h of a sphere of radius r is
    $$V = \tfrac{1}{3}\pi h^2 (3r - h)$$
    (b) Show that if a sphere of radius 1 is sliced by a plane at a distance x from the center in such a way that the volume of one segment is twice the volume of the other, then x is a solution of the equation
    $$3x^3 - 9x + 2 = 0$$
    where $0 < x < 1$. Use Newton's method to find x accurate to four decimal places.
    (c) Using the formula for the volume of a segment of a sphere, it can be shown that the depth x to which a floating sphere of radius r sinks in water is a root of the equation
    $$x^3 - 3rx^2 + 4r^3s = 0$$
    where s is the specific gravity of the sphere. Suppose a wooden sphere of radius 0.5 m has specific gravity 0.75. Calculate, to four-decimal-place accuracy, the depth to which the sphere will sink.
    (d) A hemispherical bowl has radius 5 inches and water is running into the bowl at the rate of 0.2 in³/s.
      (i) How fast is the water level in the bowl rising at the instant the water is 3 inches deep?
      (ii) At a certain instant, the water is 4 inches deep. How long will it take to fill the bowl?

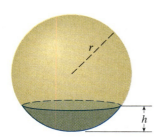

FIGURE FOR PROBLEM 5

6. Archimedes' Principle states that the buoyant force on an object partially or fully submerged in a fluid is equal to the weight of the fluid that the object displaces. Thus, for an object of density ρ_0 floating partly submerged in a fluid of density ρ_f, the buoyant force is given by $F = \rho_f g \int_{-h}^{0} A(y)\,dy$, where g is the acceleration due to gravity and $A(y)$ is the area of a typical cross-section of the object. The weight of the object is given by

$$W = \rho_0 g \int_{-h}^{L-h} A(y)\,dy$$

(a) Show that the percentage of the volume of the object above the surface of the liquid is

$$100\,\frac{\rho_f - \rho_0}{\rho_f}$$

(b) The density of ice is 917 kg/m³ and the density of seawater is 1030 kg/m³. What percentage of the volume of an iceberg is above water?

(c) An ice cube floats in a glass filled to the brim with water. Does the water overflow when the ice melts?

(d) A sphere of radius 0.4 m and having negligible weight is floating in a large freshwater lake. How much work is required to completely submerge the sphere? The density of the water is 1000 kg/m³.

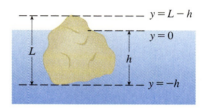

7. Water in an open bowl evaporates at a rate proportional to the area of the surface of the water. (This means that the rate of decrease of the volume is proportional to the area of the surface.) Show that the depth of the water decreases at a constant rate, regardless of the shape of the bowl.

8. A sphere of radius 1 overlaps a smaller sphere of radius r in such a way that their intersection is a circle of radius r. (In other words, they intersect in a great circle of the small sphere.) Find r so that the volume inside the small sphere and outside the large sphere is as large as possible.

9. The figure shows a curve C with the property that, for every point P on the middle curve $y = 2x^2$, the areas A and B are equal. Find an equation for C.

10. A paper drinking cup filled with water has the shape of a cone with height h and semivertical angle θ (see the figure). A ball is placed carefully in the cup, thereby displacing some of the water and making it overflow. What is the radius of the ball that causes the greatest volume of water to spill out of the cup?

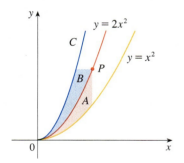

FIGURE FOR PROBLEM 9

11. A *clepsydra*, or water clock, is a glass container with a small hole in the bottom through which water can flow. The "clock" is calibrated for measuring time by placing markings on the container corresponding to water levels at equally spaced times. Let $x = f(y)$ be continuous on the interval $[0, b]$ and assume that the container is formed by rotating the graph of f about the y-axis. Let V denote the volume of water and h the height of the water level at time t.
   (a) Determine V as a function of h.
   (b) Show that
   $$\frac{dV}{dt} = \pi[f(h)]^2 \frac{dh}{dt}$$
   (c) Suppose that A is the area of the hole in the bottom of the container. It follows from Torricelli's Law that the rate of change of the volume of the water is given by
   $$\frac{dV}{dt} = kA\sqrt{h}$$
   where k is a negative constant. Determine a formula for the function f such that dh/dt is a constant C. What is the advantage in having $dh/dt = C$?

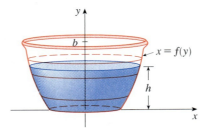

12. A cylindrical container of radius r and height L is partially filled with a liquid whose volume is V. If the container is rotated about its axis of symmetry with constant angular speed ω, then the container will induce a rotational motion in the liquid around the same axis. Eventually, the liquid will be rotating at the same angular speed as the container. The surface of the liquid will be convex, as indicated in the figure, because the centrifugal force on the liquid particles increases with the distance from the axis of the container. It can be shown that the surface of the liquid is a paraboloid of revolution generated by rotating the parabola
$$y = h + \frac{\omega^2 x^2}{2g}$$
about the y-axis, where g is the acceleration due to gravity.
(a) Determine h as a function of ω.
(b) At what angular speed will the surface of the liquid touch the bottom? At what speed will it spill over the top?
(c) Suppose the radius of the container is 2 ft, the height is 7 ft, and the container and liquid are rotating at the same constant angular speed. The surface of the liquid is 5 ft below the top of the tank at the central axis and 4 ft below the top of the tank 1 ft out from the central axis.
    (i) Determine the angular speed of the container and the volume of the fluid.
    (ii) How far below the top of the tank is the liquid at the wall of the container?

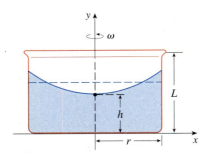

FIGURE FOR PROBLEM 12

13. If the tangent at a point P on the curve $y = x^3$ intersects the curve again at Q, let A be the area of the region bounded by the curve and the line segment PQ. Let B be the area of the region defined in the same way starting with Q instead of P. What is the relationship between A and B?

14. Suppose we are planning to make a taco from a round tortilla with diameter 8 inches by bending the tortilla so that it is shaped as if it is partially wrapped around a circular cylinder. We will fill the tortilla to the edge (but no more) with meat, cheese, and other ingredients. Our problem is to decide how to curve the tortilla in order to maximize the volume of food it can hold.

(a) We start by placing a circular cylinder of radius r along a diameter of the tortilla and folding the tortilla around the cylinder. Let x represent the distance from the center of the tortilla to a point P on the diameter (see the figure). Show that the cross-sectional area of the filled taco in the plane through P perpendicular to the axis of the cylinder is

$$A(x) = r\sqrt{16 - x^2} - \tfrac{1}{2}r^2 \sin\left(\frac{2}{r}\sqrt{16 - x^2}\right)$$

and write an expression for the volume of the filled taco.

(b) Determine (approximately) the value of r that maximizes the volume of the taco. (Use a graphical approach with your CAS.)

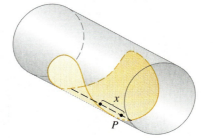

Chapter 7

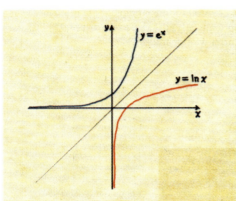

The graphs of the functions of this chapter appear as reflections of each other—the natural exponential and logarithmic functions, the restricted tangent and inverse tangent functions, and the hyperbolic sine function and its inverse.

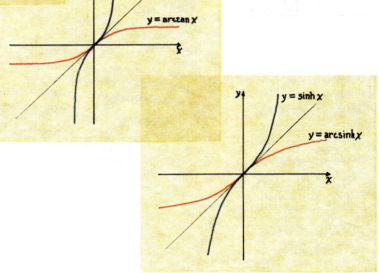

Inverse Functions:
Exponential, Logarithmic, and Inverse Trigonometric Functions

The common theme that links the functions of this chapter is that they occur as pairs of inverse functions. In particular, two of the most important functions that occur in mathematics and its applications are the exponential function $f(x) = a^x$ and its inverse function, the logarithmic function $g(x) = \log_a x$. In this chapter we investigate their properties, compute their derivatives, and use them to describe exponential growth and decay in biology, physics, chemistry, and other sciences. We also study the inverses of trigonometric and hyperbolic functions. Finally, we look at a method (l'Hospital's Rule) for computing difficult limits and apply it to sketching curves.

There are two possible ways of defining the exponential and logarithmic functions and developing their properties and derivatives. One is to start with the exponential function (defined as in algebra or precalculus courses) and then define the logarithm as its inverse. That is the approach taken in Sections 7.2, 7.3, and 7.4 and is probably the most intuitive method. The other way is to start by defining the logarithm as an integral and then define the exponential function as its inverse. This approach is followed in Sections 7.2*, 7.3*, and 7.4* and, although it is less intuitive, many instructors prefer it because it is more rigorous and the properties follow more easily. You need only read one of these two approaches (whichever your instructor recommends).

7.1 Inverse Functions

Table 1 gives data from an experiment in which a bacteria culture started with 100 bacteria in a limited nutrient medium; the size of the bacteria population was recorded at hourly intervals. The number of bacteria N is a function of the time t: $N = f(t)$.

Suppose, however, that the biologist changes her point of view and becomes interested in the time required for the population to reach various levels. In other words, she is thinking of t as a function of N. This function is called the *inverse function* of f, denoted by f^{-1}, and read "f inverse." Thus, $t = f^{-1}(N)$ is the time required for the population level to reach N. The values of f^{-1} can be found by reading Table 1 from right to left or by consulting Table 2. For instance, $f^{-1}(550) = 6$ because $f(6) = 550$.

TABLE 1 N as a function of t

t (hours)	$N = f(t)$ = population at time t
0	100
1	168
2	259
3	358
4	445
5	509
6	550
7	573
8	586

TABLE 2 t as a function of N

N	$t = f^{-1}(N)$ = time to reach N bacteria
100	0
168	1
259	2
358	3
445	4
509	5
550	6
573	7
586	8

Not all functions possess inverses. Let's compare the functions f and g whose arrow diagrams are shown in Figure 1.

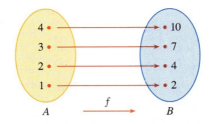
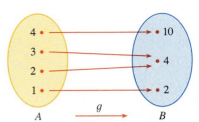

FIGURE 1

Note that f never takes on the same value twice (any two inputs in A have different outputs), whereas g does take on the same value twice (both 2 and 3 have the same output, 4). In symbols,

$$g(2) = g(3)$$

but $\qquad f(x_1) \neq f(x_2) \qquad$ whenever $x_1 \neq x_2$

Functions that have this property are called *one-to-one functions*.

|1| **Definition** A function f is called a **one-to-one function** if it never takes on the same value twice; that is,

$$f(x_1) \neq f(x_2) \qquad \text{whenever } x_1 \neq x_2$$

|||| In the language of inputs and outputs, this definition says that f is one-to-one if each output corresponds to only one input.

If a horizontal line intersects the graph of f in more than one point, then we see from Figure 2 that there are numbers x_1 and x_2 such that $f(x_1) = f(x_2)$. This means that f is not one-to-one. Therefore, we have the following geometric method for determining whether a function is one-to-one.

Horizontal Line Test A function is one-to-one if and only if no horizontal line intersects its graph more than once.

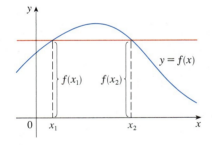

FIGURE 2
This function is not one-to-one because $f(x_1) = f(x_2)$.

EXAMPLE 1 Is the function $f(x) = x^3$ one-to-one?

SOLUTION 1 If $x_1 \neq x_2$, then $x_1^3 \neq x_2^3$ (two different numbers can't have the same cube). Therefore, by Definition 1, $f(x) = x^3$ is one-to-one.

SOLUTION 2 From Figure 3 we see that no horizontal line intersects the graph of $f(x) = x^3$ more than once. Therefore, by the Horizontal Line Test, f is one-to-one.

EXAMPLE 2 Is the function $g(x) = x^2$ one-to-one?

SOLUTION 1 This function is not one-to-one because, for instance,

$$g(1) = 1 = g(-1)$$

and so 1 and -1 have the same output.

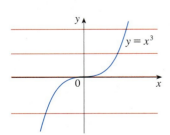

FIGURE 3
$f(x) = x^3$ is one-to-one.

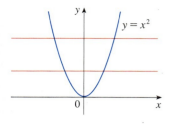

FIGURE 4
$g(x) = x^2$ is not one-to-one.

SOLUTION 2 From Figure 4 we see that there are horizontal lines that intersect the graph of g more than once. Therefore, by the Horizontal Line Test, g is not one-to-one.

One-to-one functions are important because they are precisely the functions that possess inverse functions according to the following definition.

[2] Definition Let f be a one-to-one function with domain A and range B. Then its **inverse function** f^{-1} has domain B and range A and is defined by

$$f^{-1}(y) = x \iff f(x) = y$$

for any y in B.

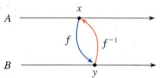

FIGURE 5

This definition says that if f maps x into y, then f^{-1} maps y back into x. (If f were not one-to-one, then f^{-1} would not be uniquely defined.) The arrow diagram in Figure 5 indicates that f^{-1} reverses the effect of f. Note that

$$\text{domain of } f^{-1} = \text{range of } f$$
$$\text{range of } f^{-1} = \text{domain of } f$$

For example, the inverse function of $f(x) = x^3$ is $f^{-1}(x) = x^{1/3}$ because if $y = x^3$, then

$$f^{-1}(y) = f^{-1}(x^3) = (x^3)^{1/3} = x$$

⊘ **CAUTION** ▫ Do not mistake the -1 in f^{-1} for an exponent. Thus

$$f^{-1}(x) \quad \text{does } not \text{ mean} \quad \frac{1}{f(x)}$$

The reciprocal $1/f(x)$ could, however, be written as $[f(x)]^{-1}$.

EXAMPLE 3 If $f(1) = 5$, $f(3) = 7$, and $f(8) = -10$, find $f^{-1}(7)$, $f^{-1}(5)$, and $f^{-1}(-10)$.

SOLUTION From the definition of f^{-1} we have

$$f^{-1}(7) = 3 \quad \text{because} \quad f(3) = 7$$
$$f^{-1}(5) = 1 \quad \text{because} \quad f(1) = 5$$
$$f^{-1}(-10) = 8 \quad \text{because} \quad f(8) = -10$$

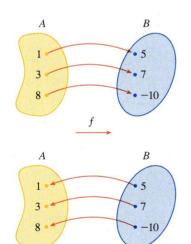

FIGURE 6
The inverse function reverses inputs and outputs.

The diagram in Figure 6 makes it clear how f^{-1} reverses the effect of f in this case.

The letter x is traditionally used as the independent variable, so when we concentrate on f^{-1} rather than on f, we usually reverse the roles of x and y in Definition 2 and write

$$\boxed{3} \qquad f^{-1}(x) = y \iff f(y) = x$$

By substituting for y in Definition 2 and substituting for x in (3), we get the following **cancellation equations**:

$$\boxed{4} \qquad \begin{aligned} f^{-1}(f(x)) &= x \quad \text{for every } x \text{ in } A \\ f(f^{-1}(x)) &= x \quad \text{for every } x \text{ in } B \end{aligned}$$

The first cancellation equation says that if we start with x, apply f, and then apply f^{-1}, we arrive back at x, where we started (see the machine diagram in Figure 7). Thus, f^{-1} undoes what f does. The second equation says that f undoes what f^{-1} does.

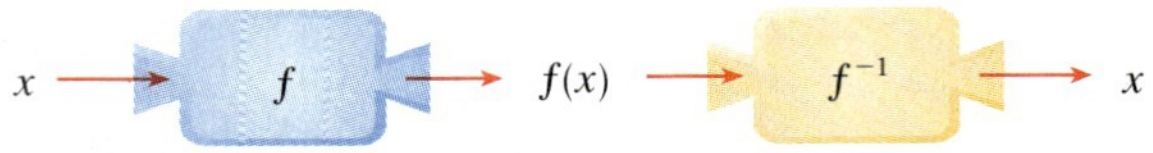

FIGURE 7

$$x \longrightarrow \boxed{f} \longrightarrow f(x) \longrightarrow \boxed{f^{-1}} \longrightarrow x$$

For example, if $f(x) = x^3$, then $f^{-1}(x) = x^{1/3}$ and the cancellation equations become

$$f^{-1}(f(x)) = (x^3)^{1/3} = x$$
$$f(f^{-1}(x)) = (x^{1/3})^3 = x$$

These equations simply say that the cube function and the cube root function cancel each other when applied in succession.

Now let's see how to compute inverse functions. If we have a function $y = f(x)$ and are able to solve this equation for x in terms of y, then according to Definition 2 we must have $x = f^{-1}(y)$. If we want to call the independent variable x, we then interchange x and y and arrive at the equation $y = f^{-1}(x)$.

$\boxed{5}$ **How to Find the Inverse Function of a One-To-One Function f**

STEP 1 Write $y = f(x)$.

STEP 2 Solve this equation for x in terms of y (if possible).

STEP 3 To express f^{-1} as a function of x, interchange x and y. The resulting equation is $y = f^{-1}(x)$.

EXAMPLE 4 Find the inverse function of $f(x) = x^3 + 2$.

SOLUTION According to (5) we first write

$$y = x^3 + 2$$

Then we solve this equation for x:

$$x^3 = y - 2$$
$$x = \sqrt[3]{y - 2}$$

|||| In Example 4, notice how f^{-1} reverses the effect of f. The function f is the rule "Cube, then add 2"; f^{-1} is the rule "Subtract 2, then take the cube root."

Finally, we interchange x and y:

$$y = \sqrt[3]{x - 2}$$

Therefore, the inverse function is $f^{-1}(x) = \sqrt[3]{x - 2}$.

The principle of interchanging x and y to find the inverse function also gives us the method for obtaining the graph of f^{-1} from the graph of f. Since $f(a) = b$ if and only if $f^{-1}(b) = a$, the point (a, b) is on the graph of f if and only if the point (b, a) is on the graph of f^{-1}. But we get the point (b, a) from (a, b) by reflecting about the line $y = x$. (See Figure 8.)

Therefore, as illustrated by Figure 9:

> The graph of f^{-1} is obtained by reflecting the graph of f about the line $y = x$.

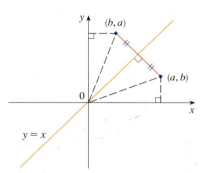

FIGURE 8

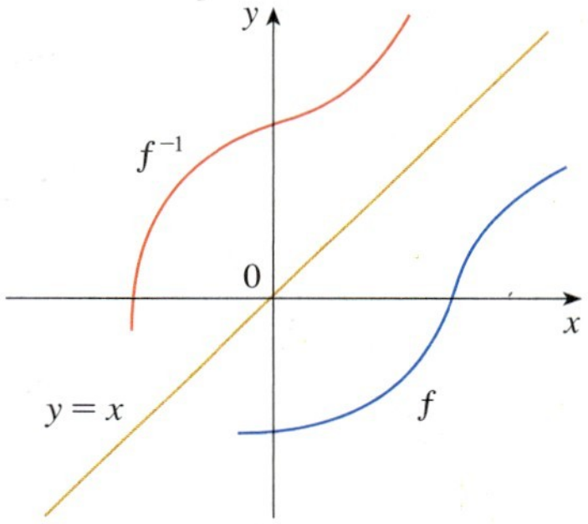

FIGURE 9

EXAMPLE 5 Sketch the graphs of $f(x) = \sqrt{-1 - x}$ and its inverse function using the same coordinate axes.

SOLUTION First we sketch the curve $y = \sqrt{-1 - x}$ (the top half of the parabola $y^2 = -1 - x$, or $x = -y^2 - 1$) and then we reflect about the line $y = x$ to get the graph of f^{-1}. (See Figure 10.) As a check on our graph, notice that the expression for f^{-1} is $f^{-1}(x) = -x^2 - 1$, $x \geq 0$. So the graph of f^{-1} is the right half of the parabola $y = -x^2 - 1$ and this seems reasonable from Figure 10.

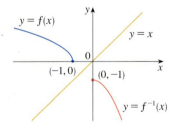

FIGURE 10

|||| **The Calculus of Inverse Functions**

Now let's look at inverse functions from the point of view of calculus. Suppose that f is both one-to-one and continuous. We think of a continuous function as one whose graph has no break in it. (It consists of just one piece.) Since the graph of f^{-1} is obtained from the graph of f by reflecting about the line $y = x$, the graph of f^{-1} has no break in it either (see Figure 9). Thus, we might expect that f^{-1} is also a continuous function.

This geometrical argument does not prove the following theorem but at least it makes the theorem plausible. A proof can be found in Appendix F.

> **6 Theorem** If f is a one-to-one continuous function defined on an interval, then its inverse function f^{-1} is also continuous.

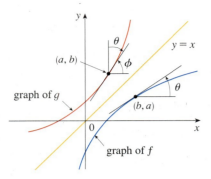

FIGURE 11

Now suppose that f is a one-to-one differentiable function. Geometrically we can think of a differentiable function as one whose graph has no corner or kink in it. We get the graph of f^{-1} by reflecting the graph of f about the line $y = x$, so the graph of f^{-1} has no corner or kink in it either. We therefore expect that f^{-1} is also differentiable (except where its tangents are vertical). In fact, we can predict the value of the derivative of f^{-1} at a given point by a geometric argument. In Figure 11 the graphs of f and its inverse $g = f^{-1}$ are shown. If $f(b) = a$, then $g(a) = f^{-1}(a) = b$ and $g'(a)$ is the slope of the tangent to the graph of g at (a, b), which is $\tan \phi$. Likewise, $f'(b) = \tan \theta$. From Figure 11 we see that $\theta + \phi = \pi/2$, so

$$g'(a) = \tan \phi = \tan\left(\frac{\pi}{2} - \theta\right) = \frac{1}{\tan \theta} = \frac{1}{f'(b)}$$

that is,

$$g'(a) = \frac{1}{f'(g(a))}$$

7 Theorem If f is a one-to-one differentiable function with inverse function $g = f^{-1}$ and $f'(g(a)) \neq 0$, then the inverse function is differentiable at a and

$$g'(a) = \frac{1}{f'(g(a))}$$

Proof Write the definition of derivative as in Equation 3.1.3:

$$g'(a) = \lim_{x \to a} \frac{g(x) - g(a)}{x - a}$$

By (3) we have

$$g(x) = y \iff f(y) = x$$

and

$$g(a) = b \iff f(b) = a$$

Since f is differentiable, it is continuous, so $g = f^{-1}$ is continuous by Theorem 6. Thus, if $x \to a$, then $g(x) \to g(a)$, that is, $y \to b$. Therefore

$$g'(a) = \lim_{x \to a} \frac{g(x) - g(a)}{x - a} = \lim_{y \to b} \frac{y - b}{f(y) - f(b)}$$

$$= \lim_{y \to b} \frac{1}{\frac{f(y) - f(b)}{y - b}} = \frac{1}{\lim_{y \to b} \frac{f(y) - f(b)}{y - b}}$$

$$= \frac{1}{f'(b)} = \frac{1}{f'(g(a))}$$

NOTE 1 ▫ Replacing a by the general number x in the formula of Theorem 7, we get

$$\boxed{8} \qquad g'(x) = \frac{1}{f'(g(x))}$$

If we write $y = g(x)$, then $f(y) = x$, so Equation 8, when expressed in Leibniz notation, becomes

$$\frac{dy}{dx} = \frac{1}{\dfrac{dx}{dy}}$$

NOTE 2 ▫ If it is known in advance that f^{-1} is differentiable, then its derivative can be computed more easily than in the proof of Theorem 7 by using implicit differentiation. If $y = f^{-1}(x)$, then $f(y) = x$. Differentiating the equation $f(y) = x$ implicitly with respect to x, remembering that y is a function of x, and using the Chain Rule, we get

$$f'(y) \frac{dy}{dx} = 1$$

Therefore
$$\frac{dy}{dx} = \frac{1}{f'(y)} = \frac{1}{\dfrac{dx}{dy}}$$

EXAMPLE 6 Although the function $y = x^2$, $x \in \mathbb{R}$, is not one-to-one and therefore does not have an inverse function, we can turn it into a one-to-one function by restricting its domain. For instance, the function $f(x) = x^2$, $0 \leq x \leq 2$, is one-to-one (by the Horizontal Line Test) and has domain $[0, 2]$ and range $[0, 4]$. (See Figure 12.) Thus, f has an inverse function $g = f^{-1}$ with domain $[0, 4]$ and range $[0, 2]$.

Without computing a formula for g' we can still calculate $g'(1)$. Since $f(1) = 1$, we have $g(1) = 1$. Also $f'(x) = 2x$. So by Theorem 7 we have

$$g'(1) = \frac{1}{f'(g(1))} = \frac{1}{f'(1)} = \frac{1}{2}$$

In this case it is easy to find g explicitly. In fact, $g(x) = \sqrt{x}$, $0 \leq x \leq 4$. [In general, we could use the method given by (5).] Then $g'(x) = 1/(2\sqrt{x})$, so $g'(1) = \frac{1}{2}$, which agrees with the preceding computation. The functions f and g are graphed in Figure 13.

EXAMPLE 7 If $f(x) = 2x + \cos x$, find $(f^{-1})'(1)$.

SOLUTION Notice that f is one-to-one because

$$f'(x) = 2 - \sin x > 0$$

and so f is increasing. To use Theorem 7 we need to know $f^{-1}(1)$ and we can find it by inspection:

$$f(0) = 1 \quad \Rightarrow \quad f^{-1}(1) = 0$$

Therefore $\qquad (f^{-1})'(1) = \dfrac{1}{f'(f^{-1}(1))} = \dfrac{1}{f'(0)} = \dfrac{1}{2 - \sin 0} = \dfrac{1}{2}$

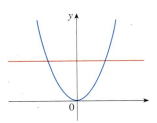

(a) $y = x^2$, $x \in \mathbb{R}$

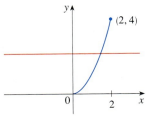

(b) $f(x) = x^2$, $0 \leq x \leq 2$

FIGURE 12

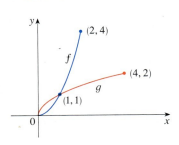

FIGURE 13

7.1 Exercises

1. (a) What is a one-to-one function?
   (b) How can you tell from the graph of a function whether it is one-to-one?

2. (a) Suppose f is a one-to-one function with domain A and range B. How is the inverse function f^{-1} defined? What is the domain of f^{-1}? What is the range of f^{-1}?
   (b) If you are given a formula for f, how do you find a formula for f^{-1}?
   (c) If you are given the graph of f, how do you find the graph of f^{-1}?

3–16 ■ A function is given by a table of values, a graph, a formula, or a verbal description. Determine whether it is one-to-one.

3.
x	1	2	3	4	5	6
$f(x)$	1.5	2.0	3.6	5.3	2.8	2.0

4.
x	1	2	3	4	5	6
$f(x)$	1	2	4	8	16	32

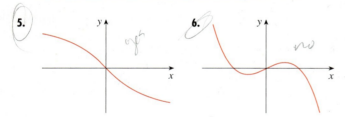

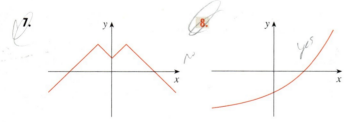

9. $f(x) = \frac{1}{2}(x + 5)$
10. $f(x) = 1 + 4x - x^2$
11. $g(x) = \sqrt{x}$
12. $g(x) = |x|$
13. $h(x) = x^4 + 5$
14. $h(x) = x^4 + 5$, $0 \le x \le 2$
15. $f(t)$ is the height of a football t seconds after kickoff.
16. $f(t)$ is your height at age t.

17–18 ■ Use a graph to decide whether f is one-to-one.

17. $f(x) = x^3 - x$
18. $f(x) = x^3 + x$

19. If f is a one-to-one function such that $f(2) = 9$, what is $f^{-1}(9)$?

20. If $f(x) = x + \cos x$, find $f^{-1}(1)$.

21. If $h(x) = x + \sqrt{x}$, find $h^{-1}(6)$.

22. The graph of f is given.
   (a) Why is f one-to-one?
   (b) State the domain and range of f^{-1}.
   (c) Estimate the value of $f^{-1}(1)$.

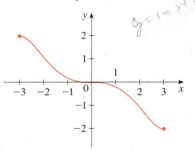

23. The formula $C = \frac{5}{9}(F - 32)$, where $F \ge -459.67$, expresses the Celsius temperature C as a function of the Fahrenheit temperature F. Find a formula for the inverse function and interpret it. What is the domain of the inverse function?

24. In the theory of relativity, the mass of a particle with speed v is
$$m = f(v) = \frac{m_0}{\sqrt{1 - v^2/c^2}}$$
where m_0 is the rest mass of the particle and c is the speed of light in a vacuum. Find the inverse function of f and explain its meaning.

25–30 ■ Find a formula for the inverse of the function.

25. $f(x) = 3 - 2x$
26. $f(x) = \dfrac{4x - 1}{2x + 3}$
27. $f(x) = \sqrt{10 - 3x}$
28. $y = 2x^3 + 3$
29. $y = \dfrac{1 - \sqrt{x}}{1 + \sqrt{x}}$
30. $f(x) = 2x^2 - 8x$, $x \ge 2$

31–32 ■ Find an explicit formula for f^{-1} and use it to graph f^{-1}, f, and the line $y = x$ on the same screen. To check your work, see whether the graphs of f and f^{-1} are reflections about the line.

31. $f(x) = 1 - 2/x^2$, $x > 0$
32. $f(x) = \sqrt{x^2 + 2x}$, $x > 0$

33. Use the given graph of f to sketch the graph of f^{-1}.

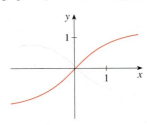

34. Use the given graph of f to sketch the graphs of f^{-1} and $1/f$.

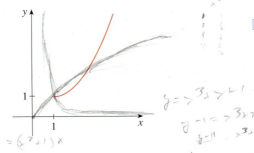

35–38
(a) Show that f is one-to-one.
(b) Use Theorem 7 to find $g'(a)$, where $g = f^{-1}$.
(c) Calculate $g(x)$ and state the domain and range of g.
(d) Calculate $g'(a)$ from the formula in part (c) and check that it agrees with the result of part (b).
(e) Sketch the graphs of f and g on the same axes.

35. $f(x) = x^3$, $a = 8$
36. $f(x) = \sqrt{x-2}$, $a = 2$
37. $f(x) = 9 - x^2$, $0 \leq x \leq 3$, $a = 8$
38. $f(x) = 1/(x-1)$, $x > 1$, $a = 2$

39–42 Find $(f^{-1})'(a)$.

39. $f(x) = x^3 + x + 1$, $a = 1$
40. $f(x) = x^5 - x^3 + 2x$, $a = 2$
41. $f(x) = 3 + x^2 + \tan(\pi x/2)$, $-1 < x < 1$, $a = 3$
42. $f(x) = \sqrt{x^3 + x^2 + x + 1}$, $a = 2$

43. Suppose g is the inverse function of f and $f(4) = 5$, $f'(4) = \frac{2}{3}$. Find $g'(5)$.

44. Suppose g is the inverse function of a differentiable function f and let $G(x) = 1/g(x)$. If $f(3) = 2$ and $f'(3) = \frac{1}{9}$, find $G'(2)$.

45. Use a computer algebra system to find an explicit expression for the inverse of the function $f(x) = \sqrt{x^3 + x^2 + x + 1}$. (Your CAS will produce three possible expressions. Explain why two of them are irrelevant in this context.)

46. Show that $h(x) = \sin x$, $x \in \mathbb{R}$, is not one-to-one, but its restriction $f(x) = \sin x$, $-\pi/2 \leq x \leq \pi/2$, is one-to-one. Compute the derivative of $f^{-1} = \sin^{-1}$ by the method of Note 2.

47. (a) If we shift a curve to the left, what happens to its reflection about the line $y = x$? In view of this geometric principle, find an expression for the inverse of $g(x) = f(x + c)$, where f is a one-to-one function.
(b) Find an expression for the inverse of $h(x) = f(cx)$, where $c \neq 0$.

48. (a) If f is a one-to-one, twice differentiable function with inverse function g, show that

$$g''(x) = -\frac{f''(g(x))}{[f'(g(x))]^3}$$

(b) Deduce that if f is increasing and concave upward, then its inverse function is concave downward.

7.2 Exponential Functions and Their Derivatives

| If your instructor has assigned Sections 7.2*, 7.3*, and 7.4*, you don't need to read Sections 7.2–7.4 (pp. 421–450).

The function $f(x) = 2^x$ is called an *exponential function* because the variable, x, is the exponent. It should not be confused with the power function $g(x) = x^2$, in which the variable is the base.

In general, an **exponential function** is a function of the form

$$f(x) = a^x$$

where a is a positive constant. Let's recall what this means. If $x = n$, a positive integer, then

$$a^n = \underbrace{a \cdot a \cdot \cdots \cdot a}_{n \text{ factors}}$$

If $x = 0$, then $a^0 = 1$, and if $x = -n$, where n is a positive integer, then

$$a^{-n} = \frac{1}{a^n}$$

If x is a rational number, $x = p/q$, where p and q are integers and $q > 0$, then

$$a^x = a^{p/q} = \sqrt[q]{a^p} = \left(\sqrt[q]{a}\right)^p$$

But what is the meaning of a^x if x is an irrational number? For instance, what is meant by $2^{\sqrt{3}}$ or 5^π?

To help us answer this question we first look at the graph of the function $y = 2^x$, where x is rational. A representation of this graph is shown in Figure 1. We want to enlarge the domain of $y = 2^x$ to include both rational and irrational numbers.

There are holes in the graph in Figure 1 corresponding to irrational values of x. We want to fill in the holes by defining $f(x) = 2^x$, where $x \in \mathbb{R}$, so that f is an increasing function. In particular, since the irrational number $\sqrt{3}$ satisfies

$$1.7 < \sqrt{3} < 1.8$$

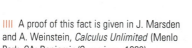

FIGURE 1
Representation of $y = 2^x$, x rational

we must have

$$2^{1.7} < 2^{\sqrt{3}} < 2^{1.8}$$

and we know what $2^{1.7}$ and $2^{1.8}$ mean because 1.7 and 1.8 are rational numbers. Similarly, if we use better approximations for $\sqrt{3}$, we obtain better approximations for $2^{\sqrt{3}}$:

$$1.73 < \sqrt{3} < 1.74 \quad \Rightarrow \quad 2^{1.73} < 2^{\sqrt{3}} < 2^{1.74}$$
$$1.732 < \sqrt{3} < 1.733 \quad \Rightarrow \quad 2^{1.732} < 2^{\sqrt{3}} < 2^{1.733}$$
$$1.7320 < \sqrt{3} < 1.7321 \quad \Rightarrow \quad 2^{1.7320} < 2^{\sqrt{3}} < 2^{1.7321}$$
$$1.73205 < \sqrt{3} < 1.73206 \quad \Rightarrow \quad 2^{1.73205} < 2^{\sqrt{3}} < 2^{1.73206}$$

$$\vdots \qquad \vdots \qquad \vdots \qquad \vdots$$

|||| A proof of this fact is given in J. Marsden and A. Weinstein, *Calculus Unlimited* (Menlo Park, CA: Benjamin/Cummings, 1980).

It can be shown that there is exactly one number that is greater than all of the numbers

$$2^{1.7}, \quad 2^{1.73}, \quad 2^{1.732}, \quad 2^{1.7320}, \quad 2^{1.73205}, \quad \ldots$$

and less than all of the numbers

$$2^{1.8}, \quad 2^{1.74}, \quad 2^{1.733}, \quad 2^{1.7321}, \quad 2^{1.73206}, \quad \ldots$$

We define $2^{\sqrt{3}}$ to be this number. Using the preceding approximation process we can compute it correct to six decimal places:

$$2^{\sqrt{3}} \approx 3.321997$$

Similarly, we can define 2^x where x is any irrational number. Figure 2 shows how all the holes in Figure 1 have been filled to complete the graph of the function $f(x) = 2^x$, $x \in \mathbb{R}$.

In general, if a is any positive number, we define

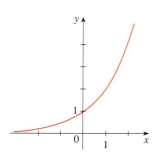

FIGURE 2
$y = 2^x$, x real

$$\boxed{1} \qquad a^x = \lim_{r \to x} a^r \qquad r \text{ rational}$$

This definition makes sense because any irrational number can be approximated as closely as we like by a rational number. For instance, because $\sqrt{3}$ has the decimal representation $\sqrt{3} = 1.7320508\ldots$, Definition 1 says that $2^{\sqrt{3}}$ is the limit of the sequence of numbers

$$2^{1.7}, \quad 2^{1.73}, \quad 2^{1.732}, \quad 2^{1.7320}, \quad 2^{1.73205}, \quad 2^{1.732050}, \quad 2^{1.7320508}, \quad \ldots$$

Similarly, 5^{π} is the limit of the sequence of numbers

$$5^{3.1}, \quad 5^{3.14}, \quad 5^{3.141}, \quad 5^{3.1415}, \quad 5^{3.14159}, \quad 5^{3.141592}, \quad 5^{3.1415926}, \quad \ldots$$

It can be shown that Definition 1 uniquely specifies a^x and makes the function $f(x) = a^x$ continuous.

The graphs of members of the family of functions $y = a^x$ are shown in Figure 3 for various values of the base a. Notice that all of these graphs pass through the same point $(0, 1)$ because $a^0 = 1$ for $a \neq 0$. Notice also that as the base a gets larger, the exponential function grows more rapidly (for $x > 0$).

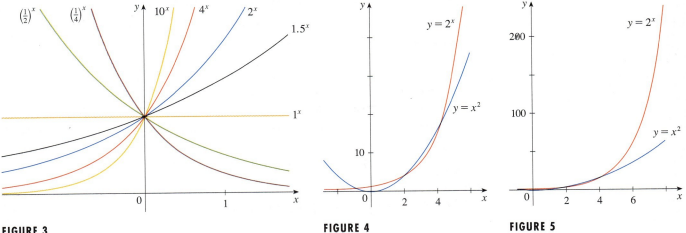

FIGURE 3
Members of the family of exponential functions

FIGURE 4

FIGURE 5

Figure 4 shows how the exponential function $y = 2^x$ compares with the power function $y = x^2$. The graphs intersect three times, but ultimately the exponential curve $y = 2^x$ grows far more rapidly than the parabola $y = x^2$. (See also Figure 5.)

You can see from Figure 3 that there are basically three kinds of exponential functions $y = a^x$. If $0 < a < 1$, the exponential function decreases; if $a = 1$, it is a constant; and if $a > 1$, it increases. These three cases are illustrated in Figure 6. Since $(1/a)^x = 1/a^x = a^{-x}$, the graph of $y = (1/a)^x$ is just the reflection of the graph of $y = a^x$ about the y-axis.

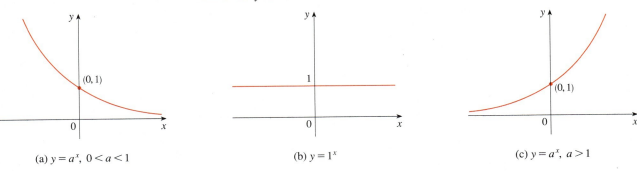

(a) $y = a^x$, $0 < a < 1$

(b) $y = 1^x$

(c) $y = a^x$, $a > 1$

FIGURE 6

The properties of the exponential function are summarized in the following theorem.

> **[2] Theorem** If $a > 0$ and $a \neq 1$, then $f(x) = a^x$ is a continuous function with domain $\mathbb{R}$ and range $(0, \infty)$. In particular, $a^x > 0$ for all x. If $0 < a < 1$, $f(x) = a^x$ is a decreasing function; if $a > 1$, f is an increasing function. If $a, b > 0$ and $x, y \in \mathbb{R}$, then
>
> **1.** $a^{x+y} = a^x a^y$ **2.** $a^{x-y} = \dfrac{a^x}{a^y}$ **3.** $(a^x)^y = a^{xy}$ **4.** $(ab)^x = a^x b^x$

The reason for the importance of the exponential function lies in properties 1–4, which are called the **Laws of Exponents**. If x and y are rational numbers, then these laws are well known from elementary algebra. For arbitrary real numbers x and y these laws can be deduced from the special case where the exponents are rational by using Equation 1.

The following limits can be read from the graphs shown in Figure 6 or proved from the definition of a limit at infinity. (See Exercise 73 in Section 7.3.)

> **[3]**   If $a > 1$, then $\lim\limits_{x \to \infty} a^x = \infty$ and $\lim\limits_{x \to -\infty} a^x = 0$
>
>   If $0 < a < 1$, then $\lim\limits_{x \to \infty} a^x = 0$ and $\lim\limits_{x \to -\infty} a^x = \infty$

In particular, if $a \neq 1$, then the x-axis is a horizontal asymptote of the graph of the exponential function $y = a^x$.

EXAMPLE 1
(a) Find $\lim_{x \to \infty} (2^{-x} - 1)$.
(b) Sketch the graph of the function $y = 2^{-x} - 1$.

SOLUTION

(a) $$\lim_{x \to \infty} (2^{-x} - 1) = \lim_{x \to \infty} \left[\left(\tfrac{1}{2}\right)^x - 1\right]$$
$$= 0 - 1 \qquad \text{[by (3) with } a = \tfrac{1}{2} < 1\text{]}$$
$$= -1$$

(b) We write $y = \left(\tfrac{1}{2}\right)^x - 1$ as in part (a). The graph of $y = \left(\tfrac{1}{2}\right)^x$ is shown in Figure 3, so we shift it down one unit to obtain the graph of $y = \left(\tfrac{1}{2}\right)^x - 1$ shown in Figure 7. (For a review of shifting graphs, see Section 1.3.) Part (a) shows that the line $y = -1$ is a horizontal asymptote.

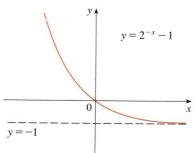

FIGURE 7

Applications of Exponential Functions

The exponential function occurs very frequently in mathematical models of nature and society. Here we indicate briefly how it arises in the description of population growth and radioactive decay. In Chapter 10 we will pursue these and other applications in greater detail.

In Section 3.4 we considered a bacteria population that doubles every hour and saw that if the initial population is n_0, then the population after t hours is given by the function $f(t) = n_0 2^t$. This population function is a constant multiple of the exponential function $y = 2^t$, so it exhibits the rapid growth that we observed in Figures 2 and 5. Under ideal conditions (unlimited space and nutrition and freedom from disease) this exponential growth is typical of what actually occurs in nature.

What about the human population? Table 1 shows data for the population of the world in the 20th century and Figure 8 shows the corresponding scatter plot.

TABLE 1

Year	Population (millions)
1900	1650
1910	1750
1920	1860
1930	2070
1940	2300
1950	2560
1960	3040
1970	3710
1980	4450
1990	5280
2000	6080

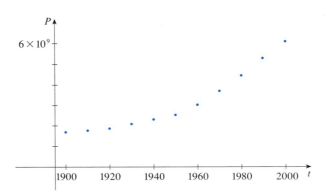

FIGURE 8 Scatter plot for world population growth

The pattern of the data points in Figure 8 suggests exponential growth, so we use a graphing calculator with exponential regression capability to apply the method of least squares and obtain the exponential model

$$P = (0.008079266) \cdot (1.013731)^t$$

Figure 9 shows the graph of this exponential function together with the original data points. We see that the exponential curve fits the data reasonably well. The period of relatively slow population growth is explained by the two world wars and the Great Depression of the 1930s.

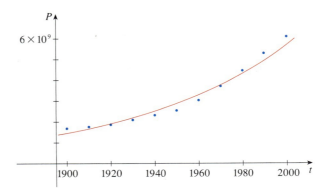

FIGURE 9
Exponential model for population growth

Exponential functions also occur in the study of the decay of radioactive substances. For instance, when physicists say the *half-life* of strontium-90, ^{90}Sr, is 25 years they mean that half of any given quantity of ^{90}Sr will disintegrate in 25 years. So if the initial mass of a sample of ^{90}Sr is 24 mg and the mass that remains after t years is $m(t)$, then

$$m(25) = \frac{1}{2}(24) \qquad m(50) = \frac{1}{2^2}(24)$$

$$m(75) = \frac{1}{2^3}(24) \qquad m(100) = \frac{1}{2^4}(24)$$

From this pattern, we see that the mass remaining after t years is

$$\boxed{4} \qquad m(t) = \frac{1}{2^{t/25}}(24) = 24 \cdot 2^{-t/25}$$

This is an exponential function with base $a = 2^{-1/25}$. (See Exercise 55 and Section 10.4.)

Derivatives of Exponential Functions

Let's try to compute the derivative of the exponential function $f(x) = a^x$ using the definition of a derivative:

$$f'(x) = \lim_{h \to 0} \frac{f(x+h) - f(x)}{h} = \lim_{h \to 0} \frac{a^{x+h} - a^x}{h}$$

$$= \lim_{h \to 0} \frac{a^x a^h - a^x}{h} = \lim_{h \to 0} \frac{a^x(a^h - 1)}{h}$$

The factor a^x doesn't depend on h, so we can take it in front of the limit:

$$f'(x) = a^x \lim_{h \to 0} \frac{a^h - 1}{h}$$

Notice that the limit is the value of the derivative of f at 0, that is,

$$\lim_{h \to 0} \frac{a^h - 1}{h} = f'(0)$$

Therefore, we have shown that if the exponential function $f(x) = a^x$ is differentiable at 0, then it is differentiable everywhere and

$$\boxed{5} \qquad f'(x) = f'(0) a^x$$

This equation says that *the rate of change of any exponential function is proportional to the function itself.* (The slope is proportional to the height.)

Numerical evidence for the existence of $f'(0)$ is given in the table at the left for the cases $a = 2$ and $a = 3$. (Values are stated correct to four decimal places. For the case $a = 2$, see also Example 3 in Section 3.1.) It appears that the limits exist and

h	$\frac{2^h - 1}{h}$	$\frac{3^h - 1}{h}$
0.1	0.7177	1.1612
0.01	0.6956	1.1047
0.001	0.6934	1.0992
0.0001	0.6932	1.0987

$$\text{for } a = 2, \quad f'(0) = \lim_{h \to 0} \frac{2^h - 1}{h} \approx 0.69$$

$$\text{for } a = 3, \quad f'(0) = \lim_{h \to 0} \frac{3^h - 1}{h} \approx 1.10$$

In fact, it can be proved that these limits exist and, correct to six decimal places, the values are

$$\boxed{6} \qquad \frac{d}{dx}(2^x)\bigg|_{x=0} \approx 0.693147 \qquad \frac{d}{dx}(3^x)\bigg|_{x=0} \approx 1.098612$$

Thus, from Equation 5 we have

$$\boxed{7} \qquad \frac{d}{dx}(2^x) \approx (0.69)2^x \qquad \frac{d}{dx}(3^x) \approx (1.10)3^x$$

Of all possible choices for the base a in Equation 5, the simplest differentiation formula occurs when $f'(0) = 1$. In view of the estimates of $f'(0)$ for $a = 2$ and $a = 3$, it seems reasonable that there is a number a between 2 and 3 for which $f'(0) = 1$. It is traditional to denote this value by the letter e. Thus, we have the following definition.

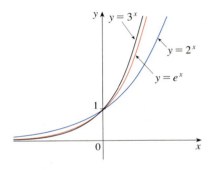

FIGURE 10

> **8 Definition of the Number e**
>
> e is the number such that $\displaystyle\lim_{h \to 0} \frac{e^h - 1}{h} = 1$

Geometrically, this means that of all the possible exponential functions $y = a^x$, the function $f(x) = e^x$ is the one whose tangent line at $(0, 1)$ has a slope $f'(0)$ that is exactly 1 (see Figures 10 and 11).

If we put $a = e$ and, therefore, $f'(0) = 1$ in Equation 5, it becomes the following important differentiation formula.

> **9 Derivative of the Natural Exponential Function**
>
> $\dfrac{d}{dx}(e^x) = e^x$

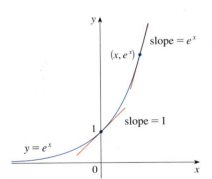

FIGURE 11

Thus, the exponential function $f(x) = e^x$ has the property that it is its own derivative. The geometrical significance of this fact is that the slope of a tangent line to the curve $y = e^x$ at any point is equal to the y-coordinate of the point (see Figure 11).

EXAMPLE 2 Differentiate the function $y = e^{\tan x}$.

SOLUTION To use the Chain Rule, we let $u = \tan x$. Then we have $y = e^u$, so

$$\frac{dy}{dx} = \frac{dy}{du}\frac{du}{dx} = e^u \frac{du}{dx} = e^{\tan x} \sec^2 x$$

In general if we combine Formula 9 with the Chain Rule, as in Example 2, we get

$$\boxed{10} \qquad \frac{d}{dx}(e^u) = e^u \frac{du}{dx}$$

EXAMPLE 3 Find y' if $y = e^{-4x} \sin 5x$.

SOLUTION Using Formula 10 and the Product Rule, we have

$$y' = e^{-4x}(\cos 5x)(5) + (\sin 5x)e^{-4x}(-4) = e^{-4x}(5 \cos 5x - 4 \sin 5x)$$

We have seen that e is a number that lies somewhere between 2 and 3, but we can use Equation 5 to estimate the numerical value of e more accurately. Let $e = 2^c$. Then $e^x = 2^{cx}$. If $f(x) = 2^x$, then from Equation 5 we have $f'(x) = k2^x$, where the value of k is $f'(0) \approx 0.693147$. Thus, by the Chain Rule,

$$e^x = \frac{d}{dx}(e^x) = \frac{d}{dx}(2^{cx}) = k2^{cx} \frac{d}{dx}(cx) = ck2^{cx}$$

Putting $x = 0$, we have $1 = ck$, so $c = 1/k$ and

$$e = 2^{1/k} \approx 2^{1/0.693147} \approx 2.71828$$

It can be shown that the approximate value to 20 decimal places is

$$e \approx 2.71828182845904523536$$

The decimal expansion of e is nonrepeating because e is an irrational number.

EXAMPLE 4 In Example 6 in Section 3.4 we considered a population of bacteria cells in a homogeneous nutrient medium. We showed that if the population doubles every hour, then the population after t hours is

$$n = n_0 2^t$$

where n_0 is the initial population. Now we can use (5) and (6) to compute the growth rate:

|||| The rate of growth is proportional to the size of the population.

$$\frac{dn}{dt} \approx n_0(0.693147)2^t$$

For instance, if the initial population is $n_0 = 1000$ cells, then the growth rate after two hours is

$$\frac{dn}{dt}\bigg|_{t=2} \approx (1000)(0.693147)2^t\big|_{t=2}$$

$$= (4000)(0.693147) \approx 2773 \text{ cells/h}$$

EXAMPLE 5 Find the absolute maximum value of the function $f(x) = xe^{-x}$.

SOLUTION We differentiate to find any critical numbers:

$$f'(x) = xe^{-x}(-1) + e^{-x}(1) = e^{-x}(1 - x)$$

Since exponential functions are always positive, we see that $f'(x) > 0$ when $1 - x > 0$, that is, when $x < 1$. Similarly, $f'(x) < 0$ when $x > 1$. By the First Derivative Test for Absolute Extreme Values, f has an absolute maximum value when $x = 1$ and the value is

$$f(1) = (1)e^{-1} = \frac{1}{e} \approx 0.37$$

Exponential Graphs

The exponential function $f(x) = e^x$ is one of the most frequently occurring functions in calculus and its applications, so it is important to be familiar with its graph (Figure 12) and properties. We summarize these properties as follows, using the fact that this function is just a special case of the exponential functions considered in Theorem 2 but with base $a = e > 1$.

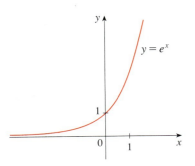

FIGURE 12
The natural exponential function

11 Properties of the Natural Exponential Function The exponential function $f(x) = e^x$ is an increasing continuous function with domain $\mathbb{R}$ and range $(0, \infty)$. Thus, $e^x > 0$ for all x. Also

$$\lim_{x \to -\infty} e^x = 0 \qquad \lim_{x \to \infty} e^x = \infty$$

So the x-axis is a horizontal asymptote of $f(x) = e^x$.

EXAMPLE 6 Find $\lim_{x \to \infty} \dfrac{e^{2x}}{e^{2x} + 1}$.

SOLUTION We divide numerator and denominator by e^{2x}:

$$\lim_{x \to \infty} \frac{e^{2x}}{e^{2x} + 1} = \lim_{x \to \infty} \frac{1}{1 + e^{-2x}} = \frac{1}{1 + \lim_{x \to \infty} e^{-2x}}$$

$$= \frac{1}{1 + 0} = 1$$

We have used the fact that $t = -2x \to -\infty$ as $x \to \infty$ and so

$$\lim_{x \to \infty} e^{-2x} = \lim_{t \to -\infty} e^{t} = 0$$

EXAMPLE 7 Use the first and second derivatives of $f(x) = e^{1/x}$, together with asymptotes, to sketch its graph.

SOLUTION Notice that the domain of f is $\{x \mid x \neq 0\}$, so we check for vertical asymptotes by computing the left and right limits as $x \to 0$. As $x \to 0^+$, we know that $t = 1/x \to \infty$, so

$$\lim_{x \to 0^+} e^{1/x} = \lim_{t \to \infty} e^{t} = \infty$$

and this shows that $x = 0$ is a vertical asymptote. As $x \to 0^-$, we have $t = 1/x \to -\infty$, so

$$\lim_{x \to 0^-} e^{1/x} = \lim_{t \to -\infty} e^{t} = 0$$

As $x \to \pm\infty$, we have $1/x \to 0$ and so

$$\lim_{x \to \pm\infty} e^{1/x} = e^0 = 1$$

This shows that $y = 1$ is a horizontal asymptote.
    Now let's compute the derivative. The Chain Rule gives

$$f'(x) = -\frac{e^{1/x}}{x^2}$$

Since $e^{1/x} > 0$ and $x^2 > 0$ for all $x \neq 0$, we have $f'(x) < 0$ for all $x \neq 0$. Thus, f is decreasing on $(-\infty, 0)$ and on $(0, \infty)$. There is no critical number, so the function has no maximum or minimum. The second derivative is

$$f''(x) = -\frac{x^2 e^{1/x}(-1/x^2) - e^{1/x}(2x)}{x^4} = \frac{e^{1/x}(2x + 1)}{x^4}$$

Since $e^{1/x} > 0$ and $x^4 > 0$, we have $f''(x) > 0$ when $x > -\frac{1}{2}$ $(x \neq 0)$ and $f''(x) < 0$ when $x < -\frac{1}{2}$. So the curve is concave downward on $\left(-\infty, -\frac{1}{2}\right)$ and concave upward on $\left(-\frac{1}{2}, 0\right)$ and on $(0, \infty)$. The inflection point is $\left(-\frac{1}{2}, e^{-2}\right)$.

To sketch the graph of f we first draw the horizontal asymptote $y = 1$ (as a dashed line), together with the parts of the curve near the asymptotes in a preliminary sketch [Figure 13(a)]. These parts reflect the information concerning limits and the fact that f is decreasing on both $(-\infty, 0)$ and $(0, \infty)$. Notice that we have indicated that $f(x) \to 0$ as $x \to 0^-$ even though $f(0)$ does not exist. In Figure 13(b) we finish the sketch by incorporating the information concerning concavity and the inflection point. In Figure 13(c) we check our work with a graphing device.

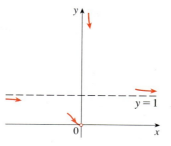 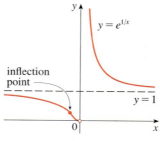 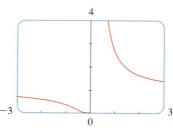

FIGURE 13 (a) Preliminary sketch   (b) Finished sketch   (c) Computer confirmation

Integration

Because the exponential function $y = e^x$ has a simple derivative, its integral is also simple:

$$\boxed{12} \quad \int e^x \, dx = e^x + C$$

EXAMPLE 8 Evaluate $\int x^2 e^{x^3} \, dx$.

SOLUTION We substitute $u = x^3$. Then $du = 3x^2 \, dx$, so $x^2 \, dx = \frac{1}{3} du$ and

$$\int x^2 e^{x^3} \, dx = \frac{1}{3} \int e^u \, du = \frac{1}{3} e^u + C = \frac{1}{3} e^{x^3} + C$$

EXAMPLE 9 Find the area under the curve $y = e^{-3x}$ from 0 to 1.

SOLUTION The area is

$$A = \int_0^1 e^{-3x} \, dx = -\frac{1}{3} e^{-3x} \Big]_0^1 = \frac{1}{3}(1 - e^{-3})$$

7.2 Exercises

1. (a) Write an equation that defines the exponential function with base $a > 0$.
   (b) What is the domain of this function?
   (c) If $a \neq 1$, what is the range of this function?
   (d) Sketch the general shape of the graph of the exponential function for each of the following cases.
   (i) $a > 1$ (ii) $a = 1$ (iii) $0 < a < 1$

2. (a) How is the number e defined?
   (b) What is an approximate value for e?
   (c) What is the natural exponential function?

3–6 Graph the given functions on a common screen. How are these graphs related?

3. $y = 2^x$, $y = e^x$, $y = 5^x$, $y = 20^x$

4. $y = e^x$, $y = e^{-x}$, $y = 8^x$, $y = 8^{-x}$

5. $y = 3^x$, $y = 10^x$, $y = \left(\frac{1}{3}\right)^x$, $y = \left(\frac{1}{10}\right)^x$

6. $y = 0.9^x$, $y = 0.6^x$, $y = 0.3^x$, $y = 0.1^x$

7–12 Make a rough sketch of the graph of the function. Do not use a calculator. Just use the graphs given in Figures 3 and 12 and, if necessary, the transformations of Section 1.3.

7. $y = 4^x - 3$
8. $y = 4^{x-3}$
9. $y = -2^{-x}$
10. $y = 1 + 2e^x$
11. $y = 3 - e^x$
12. $y = 2 + 5(1 - e^{-x})$

13. Starting with the graph of $y = e^x$, write the equation of the graph that results from
    (a) shifting 2 units downward
    (b) shifting 2 units to the right
    (c) reflecting about the x-axis
    (d) reflecting about the y-axis
    (e) reflecting about the x-axis and then about the y-axis

14. Starting with the graph of $y = e^x$, find the equation of the graph that results from
    (a) reflecting about the line $y = 4$
    (b) reflecting about the line $x = 2$

15–16 Find the domain of each function.

15. (a) $f(x) = \dfrac{1}{1 + e^x}$ (b) $f(x) = \dfrac{1}{1 - e^x}$

16. (a) $g(t) = \sin(e^{-t})$ (b) $g(t) = \sqrt{1 - 2^t}$

17–18 Find the exponential function $f(x) = Ca^x$ whose graph is given.

17.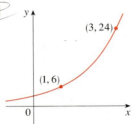

18.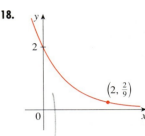

19. Suppose the graphs of $f(x) = x^2$ and $g(x) = 2^x$ are drawn on a coordinate grid where the unit of measurement is 1 inch. Show that, at a distance 2 ft to the right of the origin, the height of the graph of f is 48 ft but the height of the graph of g is about 265 mi.

20. Compare the rates of growth of the functions $f(x) = x^5$ and $g(x) = 5^x$ by graphing both functions in several viewing rectangles. Find all points of intersection of the graphs correct to one decimal place.

21. Compare the functions $f(x) = x^{10}$ and $g(x) = e^x$ by graphing both f and g in several viewing rectangles. When does the graph of g finally surpass the graph of f?

22. Use a graph to estimate the values of x such that $e^x > 1{,}000{,}000{,}000$.

23–28 Find the limit.

23. $\lim\limits_{x \to \infty} (1.001)^x$
24. $\lim\limits_{x \to \infty} e^{-x^2}$
25. $\lim\limits_{x \to \infty} \dfrac{e^{3x} - e^{-3x}}{e^{3x} + e^{-3x}}$
26. $\lim\limits_{x \to (\pi/2)^+} e^{\tan x}$
27. $\lim\limits_{x \to 2^+} e^{3/(2-x)}$
28. $\lim\limits_{x \to 2^-} e^{3/(2-x)}$

29–42 Differentiate the function.

29. $f(x) = x^2 e^x$
30. $y = \dfrac{e^x}{1 + x}$
31. $y = e^{ax^3}$
32. $y = e^u(\cos u + cu)$
33. $f(u) = e^{1/u}$
34. $g(x) = \sqrt{x}\, e^x$
35. $F(t) = e^{t \sin 2t}$
36. $y = e^{k \tan \sqrt{x}}$
37. $y = \sqrt{1 + 2e^{3x}}$
38. $y = \cos(e^{\pi x})$

39. $y = e^{e^x}$

40. $y = \sqrt{1 + xe^{-2x}}$

41. $y = \dfrac{ae^x + b}{ce^x + d}$

42. $y = \dfrac{e^x + e^{-x}}{e^x - e^{-x}}$

43–44 ▪ Find an equation of the tangent line to the curve at the given point.

43. $y = e^{2x} \cos \pi x$, $(0, 1)$

44. $y = e^x/x$, $(1, e)$

45. Find y' if $e^{x^2 y} = x + y$.

46. Find an equation of the tangent line to the curve $xe^y + ye^x = 1$ at the point $(0, 1)$.

47. Show that the function $y = e^x + e^{-x/2}$ satisfies the differential equation $2y'' - y' - y = 0$.

48. Show that the function $y = Ae^{-x} + Bxe^{-x}$ satisfies the differential equation $y'' + 2y' + y = 0$.

49. For what values of r does the function $y = e^{rx}$ satisfy the equation $y'' + 6y' + 8y = 0$?

50. Find the values of λ for which $y = e^{\lambda x}$ satisfies the equation $y + y' = y''$.

51. If $f(x) = e^{2x}$, find a formula for $f^{(n)}(x)$.

52. Find the thousandth derivative of $f(x) = xe^{-x}$.

53. (a) Use the Intermediate Value Theorem to show that there is a root of the equation $e^x + x = 0$.
(b) Use Newton's method to find the root of the equation in part (a) correct to six decimal places.

54. Use a graph to find an initial approximation (to one decimal place) to the root of the equation $e^{-x^2} = x^3 + x - 3$. Then use Newton's method to find the root correct to six decimal places.

55. If the initial mass of a sample of ^{90}Sr is 24 mg, then the mass after t years, from Equation 4, is
$$m(t) = 24 \cdot 2^{-t/25}$$
(a) Find the mass remaining after 40 years.
(b) Use Equation 7 and the Chain Rule to estimate the rate at which the mass decays after 40 years.
(c) Use a graphing device to estimate the time required for the mass to be reduced to 5 mg.

56. For the period from 1980 to 2000, the percentage of households in the United States with at least one VCR has been modeled by the function
$$V(t) = \dfrac{85}{1 + 53e^{-0.5t}}$$
where the time t is measured in years since midyear 1980, so $0 \leq t \leq 20$. Use a graph to estimate the time at which the number of VCRs was increasing most rapidly. Then use derivatives to give a more accurate estimate.

57. Under certain circumstances a rumor spreads according to the equation
$$p(t) = \dfrac{1}{1 + ae^{-kt}}$$
where $p(t)$ is the proportion of the population that knows the rumor at time t and a and k are positive constants. [In Section 10.5 we will see that this is a reasonable model for $p(t)$.]
(a) Find $\lim_{t \to \infty} p(t)$.
(b) Find the rate of spread of the rumor.
(c) Graph p for the case $a = 10$, $k = 0.5$ with t measured in hours. Use the graph to estimate how long it will take for 80% of the population to hear the rumor.

58. An object is attached to the end of a vibrating spring and its displacement from its equilibrium position is $y = 8e^{-t/2} \sin 4t$, where t is measured in seconds and y is measured in centimeters.
(a) Graph the displacement function together with the functions $y = 8e^{-t/2}$ and $y = -8e^{-t/2}$. How are these graphs related? Can you explain why?
(b) Use the graph to estimate the maximum value of the displacement. Does it occur when the graph touches the graph of $y = 8e^{-t/2}$?
(c) What is the velocity of the object when it first returns to its equilibrium position?
(d) Use the graph to estimate the time after which the displacement is no more than 2 cm from equilibrium.

59. The flash unit on a camera operates by storing charge on a capacitor and releasing it suddenly when the flash is set off. The following data describe the charge Q remaining on the capacitor (measured in microcoulombs, μC) at time t (measured in seconds).

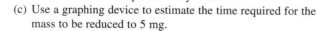

t	0.00	0.02	0.04	0.06	0.08	0.10
Q	100.00	81.87	67.03	54.88	44.93	36.76

(a) Use a graphing calculator or computer to find an exponential model for the charge. (See Section 1.2.)
(b) The derivative $Q'(t)$ represents the electric current (measured in microamperes, μA) flowing from the capacitor to the flash bulb. Use part (a) to estimate the current when $t = 0.04$ s. Compare with the result of Example 2 in Section 2.1.

60. The table gives the U.S. population from 1790 to 1860.

Year	Population	Year	Population
1790	3,929,000	1830	12,861,000
1800	5,308,000	1840	17,063,000
1810	7,240,000	1850	23,192,000
1820	9,639,000	1860	31,443,000

(a) Use a graphing calculator or computer to fit an exponential function to the data. Graph the data points and the exponential model. How good is the fit?

(b) Estimate the rates of population growth in 1800 and 1850 by averaging slopes of secant lines.
(c) Use the exponential model in part (a) to estimate the rates of growth in 1800 and 1850. Compare these estimates with the ones in part (b).
(d) Use the exponential model to predict the population in 1870. Compare with the actual population of 38,558,000. Can you explain the discrepancy?

61. Find the absolute maximum value of the function $f(x) = x - e^x$.

62. Find the absolute minimum value of the function $g(x) = e^x/x$, $x > 0$.

63–64 Find (a) the intervals of increase or decrease, (b) the intervals of concavity, and (c) the points of inflection.

63. $f(x) = xe^x$ **64.** $f(x) = x^2 e^x$

65–67 Discuss the curve using the guidelines of Section 4.5.

65. $y = e^{-1/(x+1)}$ **66.** $y = e^{2x} - e^x$

67. $y = e^{3x} + e^{-2x}$

68–69 Draw a graph of f that shows all the important aspects of the curve. Estimate the local maximum and minimum values and then use calculus to find these values exactly. Use a graph of f'' to estimate the inflection points.

68. $f(x) = e^{\cos x}$ **69.** $f(x) = e^{x^3-x}$

70. The family of bell-shaped curves

$$y = \frac{1}{\sigma\sqrt{2\pi}} e^{-(x-\mu)^2/(2\sigma^2)}$$

occurs in probability and statistics, where it is called the *normal density function*. The constant μ is called the *mean* and the positive constant σ is called the *standard deviation*. For simplicity, let's scale the function so as to remove the factor $1/(\sigma\sqrt{2\pi})$ and let's analyze the special case where $\mu = 0$. So we study the function

$$f(x) = e^{-x^2/(2\sigma^2)}$$

(a) Find the asymptote, maximum value, and inflection points of f.
(b) What role does σ play in the shape of the curve?
(c) Illustrate by graphing four members of this family on the same screen.

71–78 Evaluate the integral.

71. $\int_0^5 e^{-3x}\,dx$ **72.** $\int_0^1 xe^{-x^2}\,dx$

73. $\int e^x\sqrt{1+e^x}\,dx$ **74.** $\int \sec^2 x\, e^{\tan x}\,dx$

75. $\int \dfrac{e^x + 1}{e^x}\,dx$ **76.** $\int \dfrac{e^{1/x}}{x^2}\,dx$

77. $\int \dfrac{e^{\sqrt{x}}}{\sqrt{x}}\,dx$ **78.** $\int e^x \sin(e^x)\,dx$

79. Find, correct to three decimal places, the area of the region bounded by the curves $y = e^x$, $y = e^{3x}$, and $x = 1$.

80. Find $f(x)$ if $f''(x) = 3e^x + 5\sin x$, $f(0) = 1$, and $f'(0) = 2$.

81. Find the volume of the solid obtained by rotating about the x-axis the region bounded by the curves $y = e^x$, $y = 0$, $x = 0$, and $x = 1$.

82. Find the volume of the solid obtained by rotating about the y-axis the region bounded by the curves $y = e^{-x^2}$, $y = 0$, $x = 0$, and $x = 1$.

83. If $f(x) = 3 + x + e^x$, find $(f^{-1})'(4)$.

84. Evaluate $\displaystyle\lim_{x\to\pi}\dfrac{e^{\sin x} - 1}{x - \pi}$.

85. (a) Show that $e^x \geq 1 + x$ if $x \geq 0$.
[*Hint:* Show that $f(x) = e^x - (1 + x)$ is increasing for $x > 0$.]
(b) Deduce that $\frac{4}{3} \leq \int_0^1 e^{x^2}\,dx \leq e$.

86. (a) Use the inequality of Exercise 85(a) to show that, for $x \geq 0$,
$$e^x \geq 1 + x + \tfrac{1}{2}x^2$$
(b) Use part (a) to improve the estimate of $\int_0^1 e^{x^2}\,dx$ given in Exercise 85(b).

87. (a) Use mathematical induction to prove that for $x \geq 0$ and any positive integer n,
$$e^x \geq 1 + x + \frac{x^2}{2!} + \cdots + \frac{x^n}{n!}$$
(b) Use part (a) to show that $e > 2.7$.
(c) Use part (a) to show that
$$\lim_{x\to\infty}\frac{e^x}{x^k} = \infty$$
for any positive integer k.

7.3 Logarithmic Functions

If $a > 0$ and $a \neq 1$, the exponential function $f(x) = a^x$ is either increasing or decreasing and so it is one-to-one. It therefore has an inverse function f^{-1}, which is called the **logarithmic function with base a** and is denoted by $\log_a$. If we use the formulation of an inverse function given by (7.1.3),

$$f^{-1}(x) = y \iff f(y) = x$$

then we have

$$\boxed{1} \qquad \log_a x = y \iff a^y = x$$

Thus, if $x > 0$, $\log_a x$ is the exponent to which the base a must be raised to give x.

EXAMPLE 1 Evaluate (a) $\log_3 81$, (b) $\log_{25} 5$, and (c) $\log_{10} 0.001$.

SOLUTION
(a) $\log_3 81 = 4$ because $3^4 = 81$
(b) $\log_{25} 5 = \frac{1}{2}$ because $25^{1/2} = 5$
(c) $\log_{10} 0.001 = -3$ because $10^{-3} = 0.001$

The cancellation equations (7.1.4), when applied to $f(x) = a^x$ and $f^{-1}(x) = \log_a x$, become

$$\boxed{2} \qquad \log_a(a^x) = x \quad \text{for every } x \in \mathbb{R}$$
$$a^{\log_a x} = x \quad \text{for every } x > 0$$

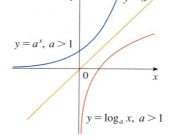

FIGURE 1

The logarithmic function $\log_a$ has domain $(0, \infty)$ and range $\mathbb{R}$ and is continuous since it is the inverse of a continuous function, namely, the exponential function. Its graph is the reflection of the graph of $y = a^x$ about the line $y = x$.

Figure 1 shows the case where $a > 1$. (The most important logarithmic functions have base $a > 1$.) The fact that $y = a^x$ is a very rapidly increasing function for $x > 0$ is reflected in the fact that $y = \log_a x$ is a very slowly increasing function for $x > 1$.

Figure 2 shows the graphs of $y = \log_a x$ with various values of the base a. Since $\log_a 1 = 0$, the graphs of all logarithmic functions pass through the point $(1, 0)$.

The following theorem summarizes the properties of logarithmic functions.

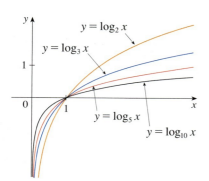

FIGURE 2

$\boxed{3}$ **Theorem** If $a > 1$, the function $f(x) = \log_a x$ is a one-to-one, continuous, increasing function with domain $(0, \infty)$ and range $\mathbb{R}$. If $x, y > 0$ and r is any real number, then

1. $\log_a(xy) = \log_a x + \log_a y$
2. $\log_a\left(\dfrac{x}{y}\right) = \log_a x - \log_a y$
3. $\log_a(x^r) = r \log_a x$

Properties 1, 2, and 3 follow from the corresponding properties of exponential functions given in Section 7.2.

EXAMPLE 2 Use the properties of logarithms in Theorem 3 to evaluate the following.
(a) $\log_4 2 + \log_4 32$ (b) $\log_2 80 - \log_2 5$

SOLUTION
(a) Using Property 1 in Theorem 3, we have
$$\log_4 2 + \log_4 32 = \log_4(2 \cdot 32) = \log_4 64 = 3$$
since $4^3 = 64$.

(b) Using Property 2 we have
$$\log_2 80 - \log_2 5 = \log_2\left(\tfrac{80}{5}\right) = \log_2 16 = 4$$
since $2^4 = 16$.

The limits of exponential functions given in Section 7.2 are reflected in the following limits of logarithmic functions. (Compare with Figure 1.)

4 If $a > 1$, then
$$\lim_{x \to \infty} \log_a x = \infty \qquad \text{and} \qquad \lim_{x \to 0^+} \log_a x = -\infty$$

In particular, the y-axis is a vertical asymptote of the curve $y = \log_a x$.

EXAMPLE 3 Find $\lim_{x \to 0} \log_{10}(\tan^2 x)$.

SOLUTION As $x \to 0$, we know that $t = \tan^2 x \to \tan^2 0 = 0$ and the values of t are positive. So by (4) with $a = 10 > 1$, we have
$$\lim_{x \to 0} \log_{10}(\tan^2 x) = \lim_{t \to 0^+} \log_{10} t = -\infty$$

Natural Logarithms

|||| NOTATION FOR LOGARITHMS
Most textbooks in calculus and the sciences, as well as calculators, use the notation $\ln x$ for the natural logarithm and $\log x$ for the "common logarithm," $\log_{10} x$. In the more advanced mathematical and scientific literature and in computer languages, however, the notation $\log x$ usually denotes the natural logarithm.

Of all possible bases a for logarithms, we will see in the next section that the most convenient choice of a base is the number e, which was defined in Section 7.2. The logarithm with base e is called the **natural logarithm** and has a special notation:

$$\log_e x = \ln x$$

If we put $a = e$ and replace $\log_e$ with "ln" in (1) and (2), then the defining properties of the natural logarithm function become

5 $$\ln x = y \iff e^y = x$$

6 $$\ln(e^x) = x \qquad x \in \mathbb{R}$$
$$e^{\ln x} = x \qquad x > 0$$

In particular, if we set $x = 1$, we get

$$\ln e = 1$$

EXAMPLE 4 Find x if $\ln x = 5$.

SOLUTION 1 From (5) we see that

$$\ln x = 5 \quad \text{means} \quad e^5 = x$$

Therefore, $x = e^5$.

(If you have trouble working with the "ln" notation, just replace it by $\log_e$. Then the equation becomes $\log_e x = 5$; so, by the definition of logarithm, $e^5 = x$.)

SOLUTION 2 Start with the equation

$$\ln x = 5$$

and apply the exponential function to both sides of the equation:

$$e^{\ln x} = e^5$$

But the second cancellation equation in (6) says that $e^{\ln x} = x$. Therefore, $x = e^5$.

EXAMPLE 5 Solve the equation $e^{5-3x} = 10$.

SOLUTION We take natural logarithms of both sides of the equation and use (6):

$$\ln(e^{5-3x}) = \ln 10$$

$$5 - 3x = \ln 10$$

$$3x = 5 - \ln 10$$

$$x = \tfrac{1}{3}(5 - \ln 10)$$

Since the natural logarithm is found on scientific calculators, we can approximate the solution to four decimal places: $x \approx 0.8991$.

EXAMPLE 6 Express $\ln a + \tfrac{1}{2}\ln b$ as a single logarithm.

SOLUTION Using Properties 3 and 1 of logarithms, we have

$$\ln a + \tfrac{1}{2}\ln b = \ln a + \ln b^{1/2}$$
$$= \ln a + \ln \sqrt{b}$$
$$= \ln(a\sqrt{b})$$

The following formula shows that logarithms with any base can be expressed in terms of the natural logarithm.

7 Change of Base Formula For any positive number a ($a \neq 1$), we have

$$\log_a x = \frac{\ln x}{\ln a}$$

Proof Let $y = \log_a x$. Then, from (1), we have $a^y = x$. Taking natural logarithms of both sides of this equation, we get $y \ln a = \ln x$. Therefore

$$y = \frac{\ln x}{\ln a}$$

Scientific calculators have a key for natural logarithms, so Formula 7 enables us to use a calculator to compute a logarithm with any base (as shown in the next example). Similarly, Formula 7 allows us to graph any logarithmic function on a graphing calculator or computer (see Exercises 20–22).

EXAMPLE 7 Evaluate $\log_8 5$ correct to six decimal places.

SOLUTION Formula 7 gives

$$\log_8 5 = \frac{\ln 5}{\ln 8} \approx 0.773976$$

EXAMPLE 8 In Section 7.2 we showed that the mass of ^{90}Sr that remains from a 24-mg sample after t years is $m = f(t) = 24 \cdot 2^{-t/25}$. Find the inverse of this function and interpret it.

SOLUTION We need to solve the equation $m = 24 \cdot 2^{-t/25}$ for t. We start by taking natural logarithms of both sides:

$$\ln m = \ln(24 \cdot 2^{-t/25}) = \ln 24 + \ln(2^{-t/25})$$

$$\ln m = \ln 24 - \frac{t}{25} \ln 2$$

$$\frac{t}{25} \ln 2 = \ln 24 - \ln m$$

$$t = \frac{25}{\ln 2} (\ln 24 - \ln m)$$

So the inverse function is

$$f^{-1}(m) = \frac{25}{\ln 2} (\ln 24 - \ln m)$$

This function gives the time required for the mass to decay to m milligrams. In particular, the time required for the mass to be reduced to 5 mg is

$$t = f^{-1}(5) = \frac{25}{\ln 2} (\ln 24 - \ln 5) \approx 56.58 \text{ years}$$

The graphs of the exponential function $y = e^x$ and its inverse function, the natural logarithm function, are shown in Figure 3. Because the curve $y = e^x$ crosses the y-axis with a slope of 1, it follows that the reflected curve $y = \ln x$ crosses the x-axis with a slope of 1.

In common with all other logarithmic functions with base greater than 1, the natural logarithm is a continuous, increasing function defined on $(0, \infty)$ and the y-axis is a vertical asymptote.

If we put $a = e$ in (4), then we have the following limits:

$$\boxed{8} \qquad \lim_{x \to \infty} \ln x = \infty \qquad \lim_{x \to 0^+} \ln x = -\infty$$

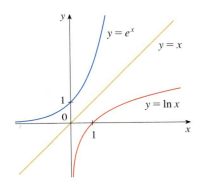

FIGURE 3

EXAMPLE 9 Sketch the graph of the function $y = \ln(x - 2) - 1$.

SOLUTION We start with the graph of $y = \ln x$ as given in Figure 3. Using the transformations of Section 1.3, we shift it 2 units to the right to get the graph of $y = \ln(x - 2)$ and then we shift it 1 unit downward to get the graph of $y = \ln(x - 2) - 1$. (See Figure 4.) Notice that the line $x = 2$ is a vertical asymptote since

$$\lim_{x \to 2^+} [\ln(x - 2) - 1] = -\infty$$

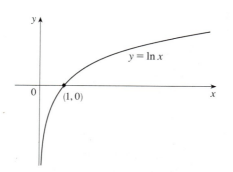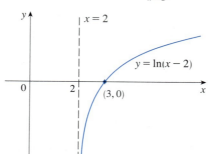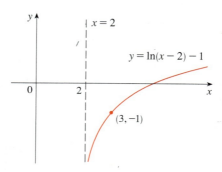

FIGURE 4

We have seen that $\ln x \to \infty$ as $x \to \infty$. But this happens *very* slowly. In fact, $\ln x$ grows more slowly than any positive power of x. To illustrate this fact, we compare approximate values of the functions $y = \ln x$ and $y = x^{1/2} = \sqrt{x}$ in the following table and we graph them in Figures 5 and 6.

x	1	2	5	10	50	100	500	1000	10,000	100,000
$\ln x$	0	0.69	1.61	2.30	3.91	4.6	6.2	6.9	9.2	11.5
$\sqrt{x}$	1	1.41	2.24	3.16	7.07	10.0	22.4	31.6	100	316
$\dfrac{\ln x}{\sqrt{x}}$	0	0.49	0.72	0.73	0.55	0.46	0.28	0.22	0.09	0.04

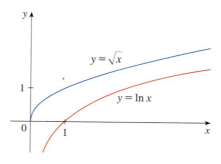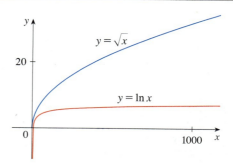

FIGURE 5   **FIGURE 6**

You can see that initially the graphs of $y = \sqrt{x}$ and $y = \ln x$ grow at comparable rates, but eventually the root function far surpasses the logarithm. In fact, we will be able to show in Section 7.7 that

$$\lim_{x \to \infty} \frac{\ln x}{x^p} = 0$$

for any positive power p. So for large x, the values of $\ln x$ are very small compared with x^p. (See Exercise 74.)

7.3 Exercises

1. (a) How is the logarithmic function $y = \log_a x$ defined?
   (b) What is the domain of this function?
   (c) What is the range of this function?
   (d) Sketch the general shape of the graph of the function $y = \log_a x$ if $a > 1$.

2. (a) What is the natural logarithm?
   (b) What is the common logarithm?
   (c) Sketch the graphs of the natural logarithm function and the natural exponential function with a common set of axes.

3–8 Find the exact value of each expression.

3. (a) $\log_{10} 1000$ (b) $\log_{16} 4$
4. (a) $\ln e^{-100}$ (b) $\log_3 81$
5. (a) $\log_5 \frac{1}{25}$ (b) $e^{\ln 15}$
6. (a) $\log_{10} 0.1$ (b) $\log_8 320 - \log_8 5$
7. (a) $\log_{12} 3 + \log_{12} 48$ (b) $\log_2 5 - \log_2 90 + 2\log_2 3$
8. (a) $2^{(\log_2 3 + \log_2 5)}$ (b) $e^{3\ln 2}$

9–12 Use the properties of logarithms to expand the quantity.

9. $\log_2 \left(\dfrac{x^3 y}{z^2} \right)$
10. $\ln \sqrt{a(b^2 + c^2)}$
11. $\ln(uv)^{10}$
12. $\ln \dfrac{3x^2}{(x+1)^5}$

13–18 Express the quantity as a single logarithm.

13. $\log_{10} a - \log_{10} b + \log_{10} c$
14. $\ln(x + y) + \ln(x - y) - 2\ln z$
15. $2\ln 4 - \ln 2$
16. $\ln 3 + \frac{1}{3}\ln 8$
17. $\frac{1}{2}\ln x - 5\ln(x^2 + 1)$
18. $\ln x + a\ln y - b\ln z$

19. Use Formula 7 to evaluate each logarithm correct to six decimal places.
    (a) $\log_{12} e$ (b) $\log_6 13.54$ (c) $\log_2 \pi$

20–22 Use Formula 7 to graph the given functions on a common screen. How are these graphs related?

20. $y = \log_2 x$, $y = \log_4 x$, $y = \log_6 x$, $y = \log_8 x$
21. $y = \log_{1.5} x$, $y = \ln x$, $y = \log_{10} x$, $y = \log_{50} x$
22. $y = \ln x$, $y = \log_{10} x$, $y = e^x$, $y = 10^x$

23–28 Make a rough sketch of the graph of each function. Do not use a calculator. Just use the graphs given in Figures 1, 2, and 3 and, if necessary, the transformations of Section 1.3.

23. $y = \log_{10}(x + 5)$
24. $y = \log_2(x - 3)$
25. $y = -\ln x$
26. $y = \ln(10x)$
27. $y = 5 + \ln(x - 2)$
28. $y = \ln|x|$

29–38 Solve each equation for x.

29. (a) $2\ln x = 1$ (b) $e^{-x} = 5$
30. (a) $e^{2x+3} - 7 = 0$ (b) $\ln(5 - 2x) = -3$
31. (a) $5^{x-3} = 10$ (b) $\log_{10}(x + 1) = 4$
32. (a) $e^{3x+1} = k$ (b) $\log_2(mx) = c$
33. $\ln(\ln x) = 1$
34. $e^{e^x} = 10$
35. $2\ln x = \ln 2 + \ln(3x - 4)$
36. $\ln(2x + 1) = 2 - \ln x$
37. $e^{ax} = Ce^{bx}$, where $a \neq b$
38. $7e^x - e^{2x} = 12$

39–42 Find the solution of the equation correct to four decimal places.

39. $e^{2+5x} = 100$
40. $\ln(1 + \sqrt{x}) = 2$
41. $\ln(e^x - 2) = 3$
42. $3^{1/(x-4)} = 7$

43–44 Solve each inequality for x.

43. (a) $e^x < 10$ (b) $\ln x > -1$
44. (a) $2 < \ln x < 9$ (b) $e^{2-3x} > 4$

45. Suppose that the graph of $y = \log_2 x$ is drawn on a coordinate grid where the unit of measurement is an inch. How many miles to the right of the origin do we have to move before the height of the curve reaches 3 ft?

46. The velocity of a particle that moves in a straight line under the influence of viscous forces is $v(t) = ce^{-kt}$, where c and k are positive constants.
    (a) Show that the acceleration is proportional to the velocity.
    (b) Explain the significance of the number c.
    (c) At what time is the velocity equal to half the initial velocity?

47. The geologist C. F. Richter defined the magnitude of an earthquake to be $\log_{10}(I/S)$, where I is the intensity of the quake (measured by the amplitude of a seismograph 100 km from the epicenter) and S is the intensity of a "standard" earthquake (where the amplitude is only 1 micron $= 10^{-4}$ cm). The 1989 Loma Prieta earthquake that shook San Francisco had a magnitude of 7.1 on the Richter scale. The 1906 San Francisco earthquake was 16 times as intense. What was its magnitude on the Richter scale?

48. A sound so faint that it can just be heard has intensity $I_0 = 10^{-12}$ watt/m^2 at a frequency of 1000 hertz (Hz). The loudness, in decibels (dB), of a sound with intensity I is then

defined to be $L = 10 \log_{10}(I/I_0)$. Amplified rock music is measured at 120 dB, whereas the noise from a motor-driven lawn mower is measured at 106 dB. Find the ratio of the intensity of the rock music to that of the mower.

49. If a bacteria population starts with 100 bacteria and doubles every three hours, then the number of bacteria after t hours is $n = f(t) = 100 \cdot 2^{t/3}$.
(a) Find the inverse of this function and explain its meaning.
(b) When will the population reach 50,000?

50. When a camera flash goes off, the batteries immediately begin to recharge the flash's capacitor, which stores electric charge given by
$$Q(t) = Q_0(1 - e^{-t/a})$$
(The maximum charge capacity is Q_0 and t is measured in seconds.)
(a) Find the inverse of this function and explain its meaning.
(b) How long does it take to recharge the capacitor to 90% of capacity if $a = 2$?

51–56 ▪ Find the limit.

51. $\lim_{x \to 2^-} \ln(2 - x)$

52. $\lim_{x \to 3^+} \log_{10}(x^2 - 5x + 6)$

53. $\lim_{x \to 0} \ln(\cos x)$

54. $\lim_{x \to 0^+} \ln(\sin x)$

55. $\lim_{x \to \infty} [\ln(1 + x^2) - \ln(1 + x)]$

56. $\lim_{x \to \infty} [\ln(2 + x) - \ln(1 + x)]$

57–58 ▪ Find the domain and range of the function.

57. $f(x) = \log_2(5x - 3)$

58. $G(t) = \ln(e^t - 2)$

59–60 ▪ Find (a) the domain of f and (b) f^{-1} and its domain.

59. $f(x) = \sqrt{3 - e^{2x}}$

60. $f(x) = \ln(2 + \ln x)$

61–66 ▪ Find the inverse function.

61. $y = \ln(x + 3)$

62. $y = 2^{10^x}$

63. $f(x) = e^{x^3}$

64. $y = (\ln x)^2$, $x \geq 1$

65. $y = \dfrac{10^x}{10^x + 1}$

66. $y = \dfrac{1 + e^x}{1 - e^x}$

67. On what interval is the function $f(x) = e^{3x} - e^x$ increasing?

68. On what interval is the curve $y = 2e^x - e^{-3x}$ concave downward?

69. (a) Show that the function $f(x) = \ln(x + \sqrt{x^2 + 1})$ is an odd function.
(b) Find the inverse function of f.

70. Find an equation of the tangent to the curve $y = e^{-x}$ that is perpendicular to the line $2x - y = 8$.

71. Show that the equation $x^{1/\ln x} = 2$ has no solution. What can you say about the function $f(x) = x^{1/\ln x}$?

72. Any function of the form $f(x) = [g(x)]^{h(x)}$, where $g(x) > 0$, can be analyzed as a power of e by writing $g(x) = e^{\ln g(x)}$ so that $f(x) = e^{h(x) \ln g(x)}$. Using this device, calculate each limit.
(a) $\lim_{x \to \infty} x^{\ln x}$
(b) $\lim_{x \to 0^+} x^{-\ln x}$
(c) $\lim_{x \to 0^+} x^{1/x}$
(d) $\lim_{x \to \infty} (\ln 2x)^{-\ln x}$

73. Let $a > 1$. Prove, using Definitions 4.4.6 and 4.4.7, that
(a) $\lim_{x \to -\infty} a^x = 0$
(b) $\lim_{x \to \infty} a^x = \infty$

74. (a) Compare the rates of growth of $f(x) = x^{0.1}$ and $g(x) = \ln x$ by graphing both f and g in several viewing rectangles. When does the graph of f finally surpass the graph of g?
(b) Graph the function $h(x) = (\ln x)/x^{0.1}$ in a viewing rectangle that displays the behavior of the function as $x \to \infty$.
(c) Find a number N such that
$$\frac{\ln x}{x^{0.1}} < 0.1 \quad \text{whenever} \quad x > N$$

75. Solve the inequality $\ln(x^2 - 2x - 2) \leq 0$.

76. A **prime number** is a positive integer that has no factors other than 1 and itself. The first few primes are 2, 3, 5, 7, 11, 13, 17, ... We denote by $\pi(n)$ the number of primes that are less than or equal to n. For instance, $\pi(15) = 6$ because there are six primes smaller than 15.
(a) Calculate the numbers $\pi(25)$ and $\pi(100)$.
[*Hint:* To find $\pi(100)$, first compile a list of the primes up to 100 using the *sieve of Eratosthenes:* Write the numbers from 2 to 100 and cross out all multiples of 2. Then cross out all multiples of 3. The next remaining number is 5, so cross out all remaining multiples of it, and so on.]
(b) By inspecting tables of prime numbers and tables of logarithms, the great mathematician K. F. Gauss made the guess in 1792 (when he was 15) that the number of primes up to n is approximately $n/\ln n$ when n is large. More precisely, he conjectured that
$$\lim_{n \to \infty} \frac{\pi(n)}{n/\ln n} = 1$$
This was finally proved, a hundred years later, by Jacques Hadamard and Charles de la Vallée Poussin and is called the **Prime Number Theorem**. Provide evidence for the truth of this theorem by computing the ratio of $\pi(n)$ to $n/\ln n$ for $n = 100, 1000, 10^4, 10^5, 10^6$, and 10^7. Use the following data: $\pi(1000) = 168$, $\pi(10^4) = 1229$, $\pi(10^5) = 9592$, $\pi(10^6) = 78{,}498$, $\pi(10^7) = 664{,}579$.
(c) Use the Prime Number Theorem to estimate the number of primes up to a billion.

7.4 Derivatives of Logarithmic Functions

In this section we find the derivatives of the logarithmic functions $y = \log_a x$ and the exponential functions $y = a^x$. We start with the natural logarithmic function $y = \ln x$. We know that it is differentiable because it is the inverse of the differentiable function $y = e^x$.

$$\boxed{1} \qquad \frac{d}{dx}(\ln x) = \frac{1}{x}$$

Proof Let $y = \ln x$. Then

$$e^y = x$$

Differentiating this equation implicitly with respect to x, we get

$$e^y \frac{dy}{dx} = 1$$

and so

$$\frac{dy}{dx} = \frac{1}{e^y} = \frac{1}{x}$$

EXAMPLE 1 Differentiate $y = \ln(x^3 + 1)$.

SOLUTION To use the Chain Rule we let $u = x^3 + 1$. Then $y = \ln u$, so

$$\frac{dy}{dx} = \frac{dy}{du}\frac{du}{dx} = \frac{1}{u}\frac{du}{dx} = \frac{1}{x^3 + 1}(3x^2) = \frac{3x^2}{x^3 + 1}$$

In general, if we combine Formula 1 with the Chain Rule as in Example 1, we get

$$\boxed{2} \qquad \frac{d}{dx}(\ln u) = \frac{1}{u}\frac{du}{dx} \qquad \text{or} \qquad \frac{d}{dx}[\ln g(x)] = \frac{g'(x)}{g(x)}$$

EXAMPLE 2 Find $\dfrac{d}{dx}\ln(\sin x)$.

SOLUTION Using (2), we have

$$\frac{d}{dx}\ln(\sin x) = \frac{1}{\sin x}\frac{d}{dx}(\sin x) = \frac{1}{\sin x}\cos x = \cot x$$

EXAMPLE 3 Differentiate $f(x) = \sqrt{\ln x}$.

SOLUTION This time the logarithm is the inner function, so the Chain Rule gives

$$f'(x) = \tfrac{1}{2}(\ln x)^{-1/2}\frac{d}{dx}(\ln x) = \frac{1}{2\sqrt{\ln x}} \cdot \frac{1}{x} = \frac{1}{2x\sqrt{\ln x}}$$

EXAMPLE 4 Find $\dfrac{d}{dx} \ln \dfrac{x+1}{\sqrt{x-2}}$.

SOLUTION 1

$$\dfrac{d}{dx} \ln \dfrac{x+1}{\sqrt{x-2}} = \dfrac{1}{\dfrac{x+1}{\sqrt{x-2}}} \dfrac{d}{dx} \dfrac{x+1}{\sqrt{x-2}}$$

$$= \dfrac{\sqrt{x-2}}{x+1} \dfrac{\sqrt{x-2} \cdot 1 - (x+1)(\tfrac{1}{2})(x-2)^{-1/2}}{x-2}$$

$$= \dfrac{x - 2 - \tfrac{1}{2}(x+1)}{(x+1)(x-2)} = \dfrac{x-5}{2(x+1)(x-2)}$$

SOLUTION 2 If we first simplify the given function using the laws of logarithms, then the differentiation becomes easier:

$$\dfrac{d}{dx} \ln \dfrac{x+1}{\sqrt{x-2}} = \dfrac{d}{dx} \left[\ln(x+1) - \tfrac{1}{2} \ln(x-2) \right]$$

$$= \dfrac{1}{x+1} - \dfrac{1}{2}\left(\dfrac{1}{x-2}\right)$$

(This answer can be left as written, but if we used a common denominator we would see that it gives the same answer as in Solution 1.)

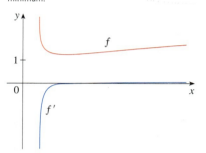

FIGURE 1

Figure 1 shows the graph of the function f of Example 4 together with the graph of its derivative. It gives a visual check on our calculation. Notice that $f'(x)$ is large negative when f is rapidly decreasing and $f'(x) = 0$ when f has a minimum.

EXAMPLE 5 Find the absolute minimum value of $f(x) = x^2 \ln x$.

SOLUTION The domain is $(0, \infty)$ and the Product Rule gives

$$f'(x) = x^2 \cdot \dfrac{1}{x} + 2x \ln x = x(1 + 2 \ln x)$$

Therefore, $f'(x) = 0$ when $2 \ln x = -1$, that is, $\ln x = -\tfrac{1}{2}$, or $x = e^{-1/2}$. Also, $f'(x) > 0$ when $x > e^{-1/2}$ and $f'(x) < 0$ for $0 < x < e^{-1/2}$. So by the First Derivative Test for Absolute Extreme Values, $f(1/\sqrt{e}) = -1/(2e)$ is the absolute minimum.

EXAMPLE 6 Discuss the curve $y = \ln(4 - x^2)$ using the guidelines of Section 4.5.

SOLUTION
A. The domain is

$$\{x \mid 4 - x^2 > 0\} = \{x \mid x^2 < 4\} = \{x \mid |x| < 2\} = (-2, 2)$$

B. The y-intercept is $f(0) = \ln 4$. To find the x-intercept we set

$$y = \ln(4 - x^2) = 0$$

We know that $\ln 1 = \log_e 1 = 0$ (since $e^0 = 1$), so we have $4 - x^2 = 1 \Rightarrow x^2 = 3$ and therefore the x-intercepts are $\pm\sqrt{3}$.
C. Since $f(-x) = f(x)$, f is even and the curve is symmetric about the y-axis.
D. We look for vertical asymptotes at the endpoints of the domain. Since $4 - x^2 \to 0^+$ as $x \to 2^-$ and also as $x \to -2^+$, we have

$$\lim_{x \to 2^-} \ln(4 - x^2) = -\infty \qquad \text{and} \qquad \lim_{x \to -2^+} \ln(4 - x^2) = -\infty$$

Thus, the lines $x = 2$ and $x = -2$ are vertical asymptotes.

E.
$$f'(x) = \frac{-2x}{4 - x^2}$$

Since $f'(x) > 0$ when $-2 < x < 0$ and $f'(x) < 0$ when $0 < x < 2$, f is increasing on $(-2, 0)$ and decreasing on $(0, 2)$.

F. The only critical number is $x = 0$. Since f' changes from positive to negative at 0, $f(0) = \ln 4$ is a local maximum by the First Derivative Test.

G.
$$f''(x) = \frac{(4 - x^2)(-2) + 2x(-2x)}{(4 - x^2)^2} = \frac{-8 - 2x^2}{(4 - x^2)^2}$$

Since $f''(x) < 0$ for all x, the curve is concave downward on $(-2, 2)$ and has no inflection point.

H. Using this information, we sketch the curve in Figure 2.

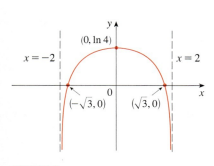

FIGURE 2
$y = \ln(4 - x^2)$

EXAMPLE 7 Find $f'(x)$ if $f(x) = \ln|x|$.

SOLUTION Since

$$f(x) = \begin{cases} \ln x & \text{if } x > 0 \\ \ln(-x) & \text{if } x < 0 \end{cases}$$

it follows that

$$f'(x) = \begin{cases} \dfrac{1}{x} & \text{if } x > 0 \\ \dfrac{1}{-x}(-1) = \dfrac{1}{x} & \text{if } x < 0 \end{cases}$$

Thus, $f'(x) = 1/x$ for all $x \neq 0$.

The result of Example 7 is worth remembering:

$$\boxed{3} \qquad \frac{d}{dx}(\ln|x|) = \frac{1}{x}$$

The corresponding integration formula is

$$\boxed{4} \qquad \int \frac{1}{x}\,dx = \ln|x| + C$$

Notice that this fills the gap in the rule for integrating power functions:

$$\int x^n\,dx = \frac{x^{n+1}}{n + 1} + C \qquad \text{if } n \neq -1$$

The missing case $(n = -1)$ is supplied by Formula 4.

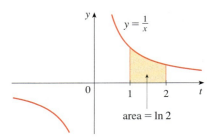

FIGURE 3

EXAMPLE 8 Find, correct to three decimal places, the area of the region under the hyperbola $xy = 1$ from $x = 1$ to $x = 2$.

SOLUTION The given region is shown in Figure 3. Using Formula 4 (without the absolute value sign, since $x > 0$), we see that the area is

$$A = \int_1^2 \frac{1}{x}\,dx = \ln x \Big]_1^2$$

$$= \ln 2 - \ln 1 = \ln 2 \approx 0.693$$

EXAMPLE 9 Evaluate $\displaystyle\int \frac{x}{x^2 + 1}\,dx$.

SOLUTION We make the substitution $u = x^2 + 1$ because the differential $du = 2x\,dx$ occurs (except for the constant factor 2). Thus, $x\,dx = \frac{1}{2}\,du$ and

$$\int \frac{x}{x^2+1}\,dx = \tfrac{1}{2}\int \frac{du}{u} = \tfrac{1}{2}\ln|u| + C$$

$$= \tfrac{1}{2}\ln|x^2 + 1| + C = \tfrac{1}{2}\ln(x^2 + 1) + C$$

Notice that we removed the absolute value signs because $x^2 + 1 > 0$ for all x. We could use the properties of logarithms to write the answer as

$$\ln\sqrt{x^2 + 1} + C$$

but this isn't necessary.

|||| Since the function $f(x) = (\ln x)/x$ in Example 10 is positive for $x > 1$, the integral represents the area of the shaded region in Figure 4.

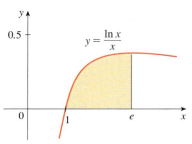

FIGURE 4

EXAMPLE 10 Calculate $\displaystyle\int_1^e \frac{\ln x}{x}\,dx$.

SOLUTION We let $u = \ln x$ because its differential $du = dx/x$ occurs in the integral. When $x = 1$, $u = \ln 1 = 0$; when $x = e$, $u = \ln e = 1$. Thus

$$\int_1^e \frac{\ln x}{x}\,dx = \int_0^1 u\,du = \frac{u^2}{2}\Big]_0^1 = \frac{1}{2}$$

EXAMPLE 11 Calculate $\displaystyle\int \tan x\,dx$.

SOLUTION First we write tangent in terms of sine and cosine:

$$\int \tan x\,dx = \int \frac{\sin x}{\cos x}\,dx$$

This suggests that we should substitute $u = \cos x$ since then $du = -\sin x\,dx$ and so $\sin x\,dx = -du$:

$$\int \tan x\,dx = \int \frac{\sin x}{\cos x}\,dx = -\int \frac{du}{u}$$

$$= -\ln|u| + C = -\ln|\cos x| + C$$

Since $-\ln|\cos x| = \ln(1/|\cos x|) = \ln|\sec x|$, the result of Example 11 can also be written as

5
$$\int \tan x \, dx = \ln|\sec x| + C$$

General Logarithmic and Exponential Functions

Formula 7 in Section 7.3 expresses a logarithmic function with base a in terms of the natural logarithmic function:

$$\log_a x = \frac{\ln x}{\ln a}$$

Since $\ln a$ is a constant, we can differentiate as follows:

$$\frac{d}{dx}(\log_a x) = \frac{d}{dx}\frac{\ln x}{\ln a} = \frac{1}{\ln a}\frac{d}{dx}(\ln x) = \frac{1}{x \ln a}$$

6
$$\frac{d}{dx}(\log_a x) = \frac{1}{x \ln a}$$

EXAMPLE 12

$$\frac{d}{dx}\log_{10}(2 + \sin x) = \frac{1}{(2 + \sin x)\ln 10}\frac{d}{dx}(2 + \sin x) = \frac{\cos x}{(2 + \sin x)\ln 10}$$

From Formula 6 we see one of the main reasons that natural logarithms (logarithms with base e) are used in calculus: The differentiation formula is simplest when $a = e$ because $\ln e = 1$.

Exponential Functions with Base a In Section 7.2 we showed that the derivative of the general exponential function $f(x) = a^x$, $a > 0$, is a constant multiple of itself:

$$f'(x) = f'(0)a^x \qquad \text{where} \qquad f'(0) = \lim_{h \to 0}\frac{a^h - 1}{h}$$

We are now in a position to show that the value of the constant is $f'(0) = \ln a$.

7
$$\frac{d}{dx}(a^x) = a^x \ln a$$

Proof We use the fact that $e^{\ln a} = a$:

$$\frac{d}{dx}(a^x) = \frac{d}{dx}(e^{\ln a})^x = \frac{d}{dx}e^{(\ln a)x} = e^{(\ln a)x}\frac{d}{dx}(\ln a)x$$

$$= (e^{\ln a})^x(\ln a) = a^x \ln a$$

In Example 6 in Section 3.4 we considered a population of bacteria cells that doubles every hour and saw that the population after t hours is $n = n_0 2^t$, where n_0 is the initial population. Formula 7 enables us to find the growth rate:

$$\frac{dn}{dt} = n_0 2^t \ln 2$$

EXAMPLE 13 Combining Formula 7 with the Chain Rule, we have

$$\frac{d}{dx}\left(10^{x^2}\right) = 10^{x^2}(\ln 10) \frac{d}{dx}(x^2) = (2 \ln 10) x 10^{x^2}$$

The integration formula that follows from Formula 7 is

$$\int a^x \, dx = \frac{a^x}{\ln a} + C \qquad a \neq 1$$

EXAMPLE 14
$$\int_0^5 2^x \, dx = \frac{2^x}{\ln 2} \Bigg]_0^5 = \frac{2^5}{\ln 2} - \frac{2^0}{\ln 2} = \frac{31}{\ln 2}$$

Logarithmic Differentiation

The calculation of derivatives of complicated functions involving products, quotients, or powers can often be simplified by taking logarithms. The method used in the following example is called **logarithmic differentiation**.

EXAMPLE 15 Differentiate $y = \dfrac{x^{3/4}\sqrt{x^2 + 1}}{(3x + 2)^5}$.

SOLUTION We take logarithms of both sides of the equation and use properties of logarithms to simplify:

$$\ln y = \tfrac{3}{4} \ln x + \tfrac{1}{2} \ln(x^2 + 1) - 5 \ln(3x + 2)$$

Differentiating implicitly with respect to x gives

$$\frac{1}{y}\frac{dy}{dx} = \frac{3}{4} \cdot \frac{1}{x} + \frac{1}{2} \cdot \frac{2x}{x^2 + 1} - 5 \cdot \frac{3}{3x + 2}$$

Solving for dy/dx, we get

$$\frac{dy}{dx} = y\left(\frac{3}{4x} + \frac{x}{x^2 + 1} - \frac{15}{3x + 2}\right)$$

Because we have an explicit expression for y, we can substitute and write

$$\frac{dy}{dx} = \frac{x^{3/4}\sqrt{x^2 + 1}}{(3x + 2)^5}\left(\frac{3}{4x} + \frac{x}{x^2 + 1} - \frac{15}{3x + 2}\right)$$

|||| If we hadn't used logarithmic differentiation in Example 15, we would have had to use both the Quotient Rule and the Product Rule. The resulting calculation would have been horrendous.

Steps in Logarithmic Differentiation
1. Take logarithms of both sides of an equation $y = f(x)$ and use properties of logarithms to simplify.
2. Differentiate implicitly with respect to x.
3. Solve the resulting equation for y'.

If $f(x) < 0$ for some values of x, then $\ln f(x)$ is not defined, but we can write $|y| = |f(x)|$ and use Equation 3. We illustrate this procedure by proving the general version of the Power Rule, as promised in Section 3.3.

The Power Rule If n is any real number and $f(x) = x^n$, then
$$f'(x) = nx^{n-1}$$

|||| If $x = 0$, we can show that $f'(0) = 0$ for $n > 1$ directly from the definition of a derivative.

Proof Let $y = x^n$ and use logarithmic differentiation:
$$\ln|y| = \ln|x|^n = n\ln|x| \qquad x \neq 0$$

Therefore
$$\frac{y'}{y} = \frac{n}{x}$$

Hence
$$y' = n\frac{y}{x} = n\frac{x^n}{x} = nx^{n-1}$$

   You should distinguish carefully between the Power Rule $[(d/dx)\,x^n = nx^{n-1}]$, where the base is variable and the exponent is constant, and the rule for differentiating exponential functions $[(d/dx)\,a^x = a^x \ln a]$, where the base is constant and the exponent is variable. In general there are four cases for exponents and bases:

1. $\dfrac{d}{dx}(a^b) = 0$ (a and b are constants)

2. $\dfrac{d}{dx}[f(x)]^b = b[f(x)]^{b-1}f'(x)$

3. $\dfrac{d}{dx}[a^{g(x)}] = a^{g(x)}(\ln a)g'(x)$

4. To find $(d/dx)[f(x)]^{g(x)}$, logarithmic differentiation can be used, as in the next example.

|||| Figure 5 illustrates Example 16 by showing the graphs of $f(x) = x^{\sqrt{x}}$ and its derivative.

EXAMPLE 16 Differentiate $y = x^{\sqrt{x}}$.

SOLUTION 1 Using logarithmic differentiation, we have
$$\ln y = \ln x^{\sqrt{x}} = \sqrt{x}\,\ln x$$

$$\frac{y'}{y} = \sqrt{x} \cdot \frac{1}{x} + (\ln x)\frac{1}{2\sqrt{x}}$$

$$y' = y\left(\frac{1}{\sqrt{x}} + \frac{\ln x}{2\sqrt{x}}\right) = x^{\sqrt{x}}\left(\frac{2 + \ln x}{2\sqrt{x}}\right)$$

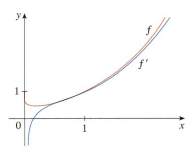

FIGURE 5

SOLUTION 2 Another method is to write $x^{\sqrt{x}} = (e^{\ln x})^{\sqrt{x}}$:

$$\frac{d}{dx}(x^{\sqrt{x}}) = \frac{d}{dx}(e^{\sqrt{x}\ln x}) = e^{\sqrt{x}\ln x}\frac{d}{dx}(\sqrt{x}\ln x)$$

$$= x^{\sqrt{x}}\left(\frac{2 + \ln x}{2\sqrt{x}}\right) \quad \text{(as in Solution 1)}$$

The Number e as a Limit

We have shown that if $f(x) = \ln x$, then $f'(x) = 1/x$. Thus, $f'(1) = 1$. We now use this fact to express the number e as a limit.

From the definition of a derivative as a limit, we have

$$f'(1) = \lim_{h \to 0}\frac{f(1+h) - f(1)}{h} = \lim_{x \to 0}\frac{f(1+x) - f(1)}{x}$$

$$= \lim_{x \to 0}\frac{\ln(1+x) - \ln 1}{x} = \lim_{x \to 0}\frac{1}{x}\ln(1+x)$$

$$= \lim_{x \to 0}\ln(1+x)^{1/x}$$

Because $f'(1) = 1$, we have

$$\lim_{x \to 0}\ln(1+x)^{1/x} = 1$$

Then, by Theorem 2.5.8 and the continuity of the exponential functions, we have

$$e = e^1 = e^{\lim_{x \to 0}\ln(1+x)^{1/x}} = \lim_{x \to 0}e^{\ln(1+x)^{1/x}} = \lim_{x \to 0}(1+x)^{1/x}$$

8
$$e = \lim_{x \to 0}(1+x)^{1/x}$$

Formula 8 is illustrated by the graph of the function $y = (1 + x)^{1/x}$ in Figure 6 and a table of values for small values of x.

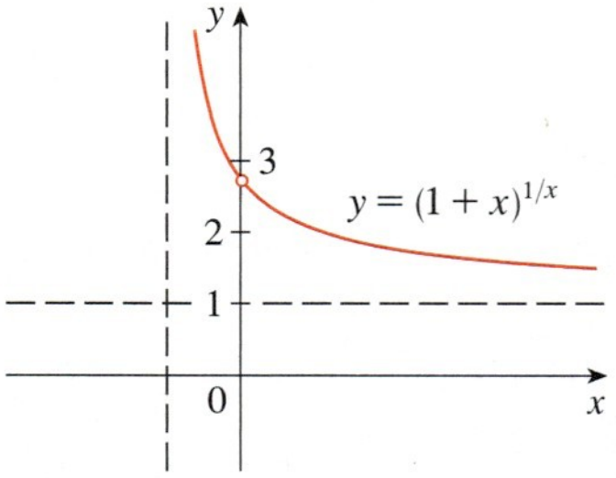

FIGURE 6

x	$(1 + x)^{1/x}$
0.1	2.59374246
0.01	2.70481383
0.001	2.71692393
0.0001	2.71814593
0.00001	2.71826824
0.000001	2.71828047
0.0000001	2.71828169
0.00000001	2.71828181

SECTION 7.4 DERIVATIVES OF LOGARITHMIC FUNCTIONS ■ 449

If we put $n = 1/x$ in Formula 8, then $n \to \infty$ as $x \to 0^+$ and so an alternative expression for e is

$$\boxed{9 \quad e = \lim_{n \to \infty}\left(1 + \frac{1}{n}\right)^n}$$

7.4 Exercises

1. Explain why the natural logarithmic function $y = \ln x$ is used much more frequently in calculus than the other logarithmic functions $y = \log_a x$.

2–24 ■ Differentiate the function.

2. $f(x) = \ln(x^2 + 10)$
3. $f(\theta) = \ln(\cos \theta)$
4. $f(x) = \cos(\ln x)$
5. $f(x) = \log_2(1 - 3x)$
6. $f(x) = \log_{10}\left(\dfrac{x}{x-1}\right)$
7. $f(x) = \sqrt[5]{\ln x}$
8. $f(x) = \ln \sqrt[5]{x}$
9. $f(x) = \sqrt{x} \ln x$
10. $f(t) = \dfrac{1 + \ln t}{1 - \ln t}$
11. $F(t) = \ln \dfrac{(2t+1)^3}{(3t-1)^4}$
12. $h(x) = \ln(x + \sqrt{x^2 - 1})$
13. $g(x) = \ln \dfrac{a-x}{a+x}$
14. $F(y) = y \ln(1 + e^y)$
15. $f(u) = \dfrac{\ln u}{1 + \ln(2u)}$
16. $y = \ln(x^4 \sin^2 x)$
17. $h(t) = t^3 - 3^t$
18. $y = 10^{\tan \theta}$
19. $y = \ln|2 - x - 5x^2|$
20. $G(u) = \ln \sqrt{\dfrac{3u+2}{3u-2}}$
21. $y = \ln(e^{-x} + xe^{-x})$
22. $y = [\ln(1 + e^x)]^2$
23. $y = 5^{-1/x}$
24. $y = 2^{3^{x^2}}$

25–28 ■ Find y' and y''.

25. $y = x \ln x$
26. $y = \dfrac{\ln x}{x^2}$
27. $y = \log_{10} x$
28. $y = \ln(\sec x + \tan x)$

29–32 ■ Differentiate f and find the domain of f.

29. $f(x) = \dfrac{x}{1 - \ln(x-1)}$
30. $f(x) = \dfrac{1}{1 + \ln x}$
31. $f(x) = x^2 \ln(1 - x^2)$
32. $f(x) = \ln \ln \ln x$

33. If $f(x) = \dfrac{x}{\ln x}$, find $f'(e)$.

34. If $f(x) = x^2 \ln x$, find $f'(1)$.

35–36 ■ Find an equation of the tangent line to the curve at the given point.

35. $y = \ln \ln x$, $(e, 0)$
36. $y = \ln(x^3 - 7)$, $(2, 0)$

37–38 ■ Find $f'(x)$. Check that your answer is reasonable by comparing the graphs of f and f'.

37. $f(x) = \sin x + \ln x$
38. $f(x) = x^{\cos x}$

39–50 ■ Use logarithmic differentiation to find the derivative of the function.

39. $y = (2x + 1)^5(x^4 - 3)^6$
40. $y = \sqrt{x}\, e^{x^2}(x^2 + 1)^{10}$

450 CHAPTER 7 INVERSE FUNCTIONS

41. $y = \dfrac{\sin^2 x \, \tan^4 x}{(x^2 + 1)^2}$

42. $y = \sqrt[4]{\dfrac{x^2 + 1}{x^2 - 1}}$

43. $y = x^x$

44. $y = x^{1/x}$

45. $y = x^{\sin x}$

46. $y = (\sin x)^x$

47. $y = (\ln x)^x$

48. $y = x^{\ln x}$

49. $y = x^{e^x}$

50. $y = (\ln x)^{\cos x}$

51. Find y' if $y = \ln(x^2 + y^2)$.

52. Find y' if $x^y = y^x$.

53. Find a formula for $f^{(n)}(x)$ if $f(x) = \ln(x - 1)$.

54. Find $\dfrac{d^9}{dx^9}(x^8 \ln x)$.

55–56 ▪ Use a graph to estimate the roots of the equation. Then use these estimates as the initial approximations in Newton's method to find the roots correct to six decimal places.

55. $(x - 4)^2 = \ln x$

56. $\ln(4 - x^2) = x$

57. Find the intervals of concavity and the inflection points of the function $f(x) = (\ln x)/\sqrt{x}$.

58. Find the absolute minimum value of the function $f(x) = x \ln x$.

59–62 ▪ Discuss the curve under the guidelines of Section 4.5.

59. $y = \ln(\sin x)$

60. $y = \ln(\tan^2 x)$

61. $y = \ln(1 + x^2)$

62. $y = \ln(x^2 - 3x + 2)$

CAS 63. If $f(x) = \ln(2x + x \sin x)$, use the graphs of f, f', and f'' to estimate the intervals of increase and the inflection points of f on the interval $(0, 15]$.

64. Investigate the family of curves $f(x) = \ln(x^2 + c)$. What happens to the inflection points and asymptotes as c changes? Graph several members of the family to illustrate what you discover.

65–76 ▪ Evaluate the integral.

65. $\displaystyle\int_2^4 \dfrac{3}{x}\, dx$

66. $\displaystyle\int_1^2 \dfrac{4 + u^2}{u^3}\, du$

67. $\displaystyle\int_1^2 \dfrac{dt}{8 - 3t}$

68. $\displaystyle\int_4^9 \left(\sqrt{x} + \dfrac{1}{\sqrt{x}}\right)^2 dx$

69. $\displaystyle\int_1^e \dfrac{x^2 + x + 1}{x}\, dx$

70. $\displaystyle\int_e^6 \dfrac{dx}{x \ln x}$

71. $\displaystyle\int \dfrac{2 - x^2}{6x - x^3}\, dx$

72. $\displaystyle\int \dfrac{\cos x}{2 + \sin x}\, dx$

73. $\displaystyle\int \dfrac{(\ln x)^2}{x}\, dx$

74. $\displaystyle\int \dfrac{e^x}{e^x + 1}\, dx$

75. $\displaystyle\int_1^2 10^t \, dt$

76. $\displaystyle\int x 2^{x^2} dx$

77. Show that $\int \cot x \, dx = \ln|\sin x| + C$ by (a) differentiating the right side of the equation and (b) using the method of Example 11.

78. Find, correct to three decimal places, the area of the region above the hyperbola $y = 2/(x - 2)$, below the x-axis, and between the lines $x = -4$ and $x = -1$.

79. Find the volume of the solid obtained by rotating the region under the curve
$$y = \dfrac{1}{\sqrt{x + 1}}$$
from 0 to 1 about the x-axis.

80. Find the volume of the solid obtained by rotating the region under the curve
$$y = \dfrac{1}{x^2 + 1}$$
from 0 to 3 about the y-axis.

81. The work done by a gas when it expands from volume V_1 to volume V_2 is $W = \int_{V_1}^{V_2} P\, dV$, where $P = P(V)$ is the pressure as a function of the volume V. (See Exercise 27 in Section 6.4.) Boyle's Law states that when a quantity of gas expands at constant temperature, $PV = C$, where C is a constant. If the initial volume is 600 cm³ and the initial pressure is 150 kPa, find the work done by the gas when it expands at constant temperature to 1000 cm³.

82. Find f if $f''(x) = x^{-2}$, $x > 0$, $f(1) = 0$, and $f(2) = 0$.

83. If g is the inverse function of $f(x) = 2x + \ln x$, find $g'(2)$.

84. If $f(x) = e^x + \ln x$ and $h(x) = f^{-1}(x)$, find $h'(e)$.

85. For what values of m do the line $y = mx$ and the curve $y = x/(x^2 + 1)$ enclose a region? Find the area of the region.

86. (a) Find the linear approximation to $f(x) = \ln x$ near 1.
(b) Illustrate part (a) by graphing f and its linearization.
(c) For what values of x is the linear approximation accurate to within 0.1?

87. Use the definition of derivative to prove that
$$\lim_{x \to 0} \dfrac{\ln(1 + x)}{x} = 1$$

88. Show that $\displaystyle\lim_{n \to \infty}\left(1 + \dfrac{x}{n}\right)^n = e^x$ for any $x > 0$.

7.2* The Natural Logarithmic Function

If your instructor has assigned Sections 7.2–7.4, you need not read Sections 7.2*, 7.3*, and 7.4* (pp. 451–476).

In this section we define the natural logarithm as an integral and then show that it obeys the usual laws of logarithms. The Fundamental Theorem makes it easy to differentiate this function.

> **1 Definition** The **natural logarithmic function** is the function defined by
> $$\ln x = \int_1^x \frac{1}{t}\, dt \qquad x > 0$$

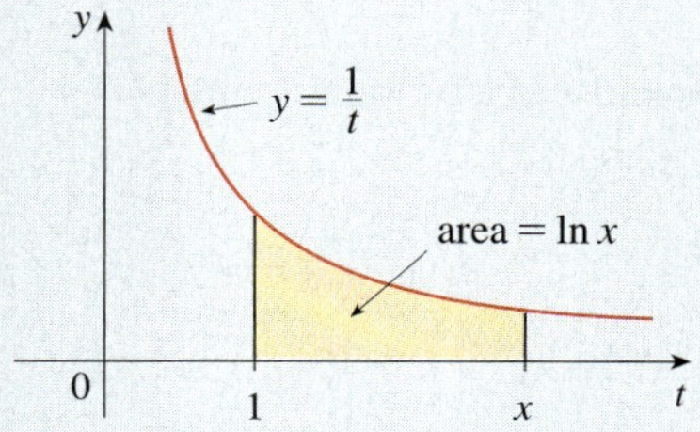

The existence of this function depends on the fact that the integral of a continuous function always exists. If $x > 1$, then $\ln x$ can be interpreted geometrically as the area under the hyperbola $y = 1/t$ from $t = 1$ to $t = x$. (See Figure 1.) For $x = 1$, we have

$$\ln 1 = \int_1^1 \frac{1}{t}\, dt = 0$$

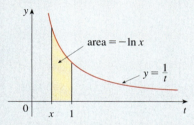

FIGURE 1

For $0 < x < 1$,

$$\ln x = \int_1^x \frac{1}{t}\, dt = -\int_x^1 \frac{1}{t}\, dt < 0$$

and so $\ln x$ is the negative of the area shaded in Figure 2.

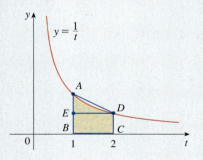

FIGURE 2

EXAMPLE 1
(a) By comparing areas, show that $\frac{1}{2} < \ln 2 < \frac{3}{4}$.
(b) Use the Midpoint Rule with $n = 10$ to estimate the value of $\ln 2$.

SOLUTION
(a) We can interpret $\ln 2$ as the area under the curve $y = 1/t$ from 1 to 2. From Figure 3 we see that this area is larger than the area of rectangle $BCDE$ and smaller than the area of trapezoid $ABCD$. Thus, we have

$$\tfrac{1}{2} \cdot 1 < \ln 2 < 1 \cdot \tfrac{1}{2}\left(1 + \tfrac{1}{2}\right)$$

$$\tfrac{1}{2} < \ln 2 < \tfrac{3}{4}$$

(b) If we use the Midpoint Rule with $f(t) = 1/t$, $n = 10$, and $\Delta t = 0.1$, we get

$$\ln 2 = \int_1^2 \frac{1}{t}\, dt \approx (0.1)[f(1.05) + f(1.15) + \cdots + f(1.95)]$$

$$= (0.1)\left(\frac{1}{1.05} + \frac{1}{1.15} + \cdots + \frac{1}{1.95}\right) \approx 0.693$$

FIGURE 3

Notice that the integral that defines $\ln x$ is exactly the type of integral discussed in Part 1 of the Fundamental Theorem of Calculus (see Section 5.3). In fact, using that theorem, we have

$$\frac{d}{dx}\int_1^x \frac{1}{t}\, dt = \frac{1}{x}$$

and so

$$\boxed{\frac{d}{dx}(\ln x) = \frac{1}{x}} \quad \boxed{2}$$

We now use this differentiation rule to prove the following properties of the logarithm function.

> **3 Laws of Logarithms** If x and y are positive numbers and r is a rational number, then
>
> **1.** $\ln(xy) = \ln x + \ln y$ **2.** $\ln\left(\dfrac{x}{y}\right) = \ln x - \ln y$ **3.** $\ln(x^r) = r \ln x$

Proof

1. Let $f(x) = \ln(ax)$, where a is a positive constant. Then, using Equation 2 and the Chain Rule, we have

$$f'(x) = \frac{1}{ax} \frac{d}{dx}(ax) = \frac{1}{ax} \cdot a = \frac{1}{x}$$

Therefore, $f(x)$ and $\ln x$ have the same derivative and so they must differ by a constant:

$$\ln(ax) = \ln x + C$$

Putting $x = 1$ in this equation, we get $\ln a = \ln 1 + C = 0 + C = C$. Thus

$$\ln(ax) = \ln x + \ln a$$

If we now replace the constant a by any number y, we have

$$\ln(xy) = \ln x + \ln y$$

2. Using Law 1 with $x = 1/y$, we have

$$\ln \frac{1}{y} + \ln y = \ln\left(\frac{1}{y} \cdot y\right) = \ln 1 = 0$$

and so

$$\ln \frac{1}{y} = -\ln y$$

Using Law 1 again, we have

$$\ln\left(\frac{x}{y}\right) = \ln\left(x \cdot \frac{1}{y}\right) = \ln x + \ln \frac{1}{y} = \ln x - \ln y$$

The proof of Law 3 is left as an exercise.

EXAMPLE 2 Expand the expression $\ln \dfrac{(x^2 + 5)^4 \sin x}{x^3 + 1}$.

SOLUTION Using Laws 1, 2, and 3, we get

$$\ln \frac{(x^2+5)^4 \sin x}{x^3+1} = \ln(x^2+5)^4 + \ln \sin x - \ln(x^3+1)$$

$$= 4\ln(x^2+5) + \ln \sin x - \ln(x^3+1)$$

EXAMPLE 3 Express $\ln a + \frac{1}{2} \ln b$ as a single logarithm.

SOLUTION Using Laws 3 and 1 of logarithms, we have

$$\ln a + \tfrac{1}{2} \ln b = \ln a + \ln b^{1/2}$$
$$= \ln a + \ln \sqrt{b}$$
$$= \ln(a\sqrt{b})$$

In order to graph $y = \ln x$, we first determine its limits:

4 (a) $\lim_{x \to \infty} \ln x = \infty$ (b) $\lim_{x \to 0^+} \ln x = -\infty$

Proof

(a) Using Law 3 with $x = 2$ and $r = n$ (where n is any positive integer), we have $\ln(2^n) = n \ln 2$. Now $\ln 2 > 0$, so this shows that $\ln(2^n) \to \infty$ as $n \to \infty$. But $\ln x$ is an increasing function since its derivative $1/x > 0$. Therefore, $\ln x \to \infty$ as $x \to \infty$.

(b) If we let $t = 1/x$, then $t \to \infty$ as $x \to 0^+$. Thus, using (a), we have

$$\lim_{x \to 0^+} \ln x = \lim_{t \to \infty} \ln\left(\frac{1}{t}\right) = \lim_{t \to \infty} (-\ln t) = -\infty$$

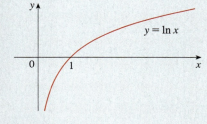

FIGURE 4

If $y = \ln x$, $x > 0$, then

$$\frac{dy}{dx} = \frac{1}{x} > 0 \quad \text{and} \quad \frac{d^2y}{dx^2} = -\frac{1}{x^2} < 0$$

which shows that $\ln x$ is increasing and concave downward on $(0, \infty)$. Putting this information together with (4), we draw the graph of $y = \ln x$ in Figure 4.

Since $\ln 1 = 0$ and $\ln x$ is an increasing continuous function that takes on arbitrarily large values, the Intermediate Value Theorem shows that there is a number where $\ln x$ takes on the value 1 (see Figure 5). This important number is denoted by e.

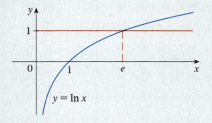

FIGURE 5

5 **Definition** e is the number such that $\ln e = 1$.

EXAMPLE 4 Use a graphing calculator or computer to estimate the value of e.

SOLUTION According to Definition 5, we estimate the value of e by graphing the curves $y = \ln x$ and $y = 1$ and determining the x-coordinate of the point of intersection. By zooming in repeatedly, as in Figure 6, we find that

$$e \approx 2.718$$

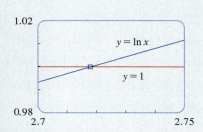

FIGURE 6

With more sophisticated methods, it can be shown that the approximate value of e, to 20 decimal places, is

$$e \approx 2.71828182845904523536$$

The decimal expansion of e is nonrepeating because e is an irrational number.

Now let's use Equation 2 to differentiate functions that involve the natural logarithmic function.

EXAMPLE 5 Differentiate $y = \ln(x^3 + 1)$.

SOLUTION To use the Chain Rule we let $u = x^3 + 1$. Then $y = \ln u$, so

$$\frac{dy}{dx} = \frac{dy}{du}\frac{du}{dx} = \frac{1}{u}\frac{du}{dx} = \frac{1}{x^3+1}(3x^2) = \frac{3x^2}{x^3+1}$$

In general, if we combine Formula 2 with the Chain Rule as in Example 5, we get

$$\boxed{6} \quad \frac{d}{dx}(\ln u) = \frac{1}{u}\frac{du}{dx} \quad \text{or} \quad \frac{d}{dx}[\ln g(x)] = \frac{g'(x)}{g(x)}$$

EXAMPLE 6 Find $\dfrac{d}{dx}\ln(\sin x)$.

SOLUTION Using (6), we have

$$\frac{d}{dx}\ln(\sin x) = \frac{1}{\sin x}\frac{d}{dx}(\sin x) = \frac{1}{\sin x}\cos x = \cot x$$

EXAMPLE 7 Differentiate $f(x) = \sqrt{\ln x}$.

SOLUTION This time the logarithm is the inner function, so the Chain Rule gives

$$f'(x) = \tfrac{1}{2}(\ln x)^{-1/2}\frac{d}{dx}(\ln x) = \frac{1}{2\sqrt{\ln x}} \cdot \frac{1}{x} = \frac{1}{2x\sqrt{\ln x}}$$

EXAMPLE 8 Find $\dfrac{d}{dx}\ln\dfrac{x+1}{\sqrt{x-2}}$.

SOLUTION 1

$$\frac{d}{dx}\ln\frac{x+1}{\sqrt{x-2}} = \frac{1}{\frac{x+1}{\sqrt{x-2}}}\frac{d}{dx}\frac{x+1}{\sqrt{x-2}}$$

$$= \frac{\sqrt{x-2}}{x+1}\,\frac{\sqrt{x-2}\cdot 1 - (x+1)(\tfrac{1}{2})(x-2)^{-1/2}}{x-2}$$

$$= \frac{x-2-\tfrac{1}{2}(x+1)}{(x+1)(x-2)} = \frac{x-5}{2(x+1)(x-2)}$$

|||| Figure 7 shows the graph of the function f of Example 8 together with the graph of its derivative. It gives a visual check on our calculation. Notice that $f'(x)$ is large negative when f is rapidly decreasing and $f'(x) = 0$ when f has a minimum.

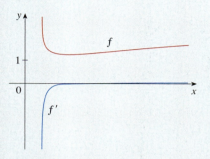

FIGURE 7

SOLUTION 2 If we first simplify the given function using the laws of logarithms, then the differentiation becomes easier:

$$\frac{d}{dx} \ln \frac{x+1}{\sqrt{x-2}} = \frac{d}{dx} \left[\ln(x+1) - \tfrac{1}{2} \ln(x-2) \right]$$

$$= \frac{1}{x+1} - \frac{1}{2}\left(\frac{1}{x-2}\right)$$

(This answer can be left as written, but if we used a common denominator we would see that it gives the same answer as in Solution 1.)

EXAMPLE 9 Discuss the curve $y = \ln(4 - x^2)$ using the guidelines of Section 4.5.

SOLUTION
A. The domain is

$$\{x \mid 4 - x^2 > 0\} = \{x \mid x^2 < 4\} = \{x \mid |x| < 2\} = (-2, 2)$$

B. The y-intercept is $f(0) = \ln 4$. To find the x-intercept we set

$$y = \ln(4 - x^2) = 0$$

We know that $\ln 1 = 0$, so we have $4 - x^2 = 1 \Rightarrow x^2 = 3$ and therefore the x-intercepts are $\pm\sqrt{3}$.

C. Since $f(-x) = f(x)$, f is even and the curve is symmetric about the y-axis.

D. We look for vertical asymptotes at the endpoints of the domain. Since $4 - x^2 \to 0^+$ as $x \to 2^-$ and also as $x \to -2^+$, we have

$$\lim_{x \to 2^-} \ln(4 - x^2) = -\infty \quad \text{and} \quad \lim_{x \to -2^+} \ln(4 - x^2) = -\infty$$

by (4). Thus, the lines $x = 2$ and $x = -2$ are vertical asymptotes.

E.
$$f'(x) = \frac{-2x}{4 - x^2}$$

Since $f'(x) > 0$ when $-2 < x < 0$ and $f'(x) < 0$ when $0 < x < 2$, f is increasing on $(-2, 0)$ and decreasing on $(0, 2)$.

F. The only critical number is $x = 0$. Since f' changes from positive to negative at 0, $f(0) = \ln 4$ is a local maximum by the First Derivative Test.

G.
$$f''(x) = \frac{(4 - x^2)(-2) + 2x(-2x)}{(4 - x^2)^2} = \frac{-8 - 2x^2}{(4 - x^2)^2}$$

Since $f''(x) < 0$ for all x, the curve is concave downward on $(-2, 2)$ and has no inflection point.

H. Using this information, we sketch the curve in Figure 8.

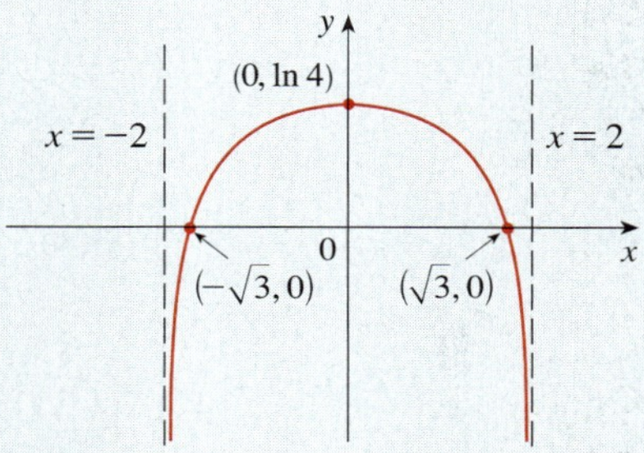

FIGURE 8
$y = \ln(4 - x^2)$

EXAMPLE 10 Find $f'(x)$ if $f(x) = \ln|x|$.

SOLUTION Since

$$f(x) = \begin{cases} \ln x & \text{if } x > 0 \\ \ln(-x) & \text{if } x < 0 \end{cases}$$

it follows that

$$f'(x) = \begin{cases} \dfrac{1}{x} & \text{if } x > 0 \\ \dfrac{1}{-x}(-1) = \dfrac{1}{x} & \text{if } x < 0 \end{cases}$$

Thus, $f'(x) = 1/x$ for all $x \neq 0$.

The result of Example 10 is worth remembering:

7
$$\frac{d}{dx}(\ln|x|) = \frac{1}{x}$$

The corresponding integration formula is

8
$$\int \frac{1}{x}\,dx = \ln|x| + C$$

Notice that this fills the gap in the rule for integrating power functions:

$$\int x^n\,dx = \frac{x^{n+1}}{n+1} + C \qquad \text{if } n \neq -1$$

The missing case ($n = -1$) is supplied by Formula 8.

EXAMPLE 11 Evaluate $\displaystyle\int \frac{x}{x^2+1}\,dx$.

SOLUTION We make the substitution $u = x^2 + 1$ because the differential $du = 2x\,dx$ occurs (except for the constant factor 2). Thus, $x\,dx = \tfrac{1}{2}\,du$ and

$$\int \frac{x}{x^2+1}\,dx = \tfrac{1}{2}\int \frac{du}{u} = \tfrac{1}{2}\ln|u| + C$$
$$= \tfrac{1}{2}\ln|x^2+1| + C = \tfrac{1}{2}\ln(x^2+1) + C$$

Notice that we removed the absolute value signs because $x^2 + 1 > 0$ for all x. We could use the properties of logarithms to write the answer as

$$\ln\sqrt{x^2+1} + C$$

but this isn't necessary.

EXAMPLE 12 Calculate $\displaystyle\int_1^e \frac{\ln x}{x}\,dx$.

SOLUTION We let $u = \ln x$ because its differential $du = dx/x$ occurs in the integral. When $x = 1$, $u = \ln 1 = 0$; when $x = e$, $u = \ln e = 1$. Thus

$$\int_1^e \frac{\ln x}{x}\,dx = \int_0^1 u\,du = \frac{u^2}{2}\bigg]_0^1 = \frac{1}{2}$$

|||| Since the function $f(x) = (\ln x)/x$ in Example 12 is positive for $x > 1$, the integral represents the area of the shaded region in Figure 9.

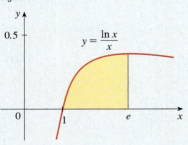

FIGURE 9

EXAMPLE 13 Calculate $\int \tan x \, dx$.

SOLUTION First we write tangent in terms of sine and cosine:

$$\int \tan x \, dx = \int \frac{\sin x}{\cos x} \, dx$$

This suggests that we should substitute $u = \cos x$ since then $du = -\sin x \, dx$ and so $\sin x \, dx = -du$:

$$\int \tan x \, dx = \int \frac{\sin x}{\cos x} \, dx = -\int \frac{du}{u}$$

$$= -\ln|u| + C = -\ln|\cos x| + C$$

Since $-\ln|\cos x| = \ln(1/|\cos x|) = \ln|\sec x|$, the result of Example 13 can also be written as

$$\int \tan x \, dx = \ln|\sec x| + C$$

Logarithmic Differentiation

The calculation of derivatives of complicated functions involving products, quotients, or powers can often be simplified by taking logarithms. The method used in the following example is called **logarithmic differentiation**.

EXAMPLE 14 Differentiate

$$y = \frac{x^{3/4}\sqrt{x^2 + 1}}{(3x + 2)^5}$$

SOLUTION We take logarithms of both sides of the equation and use the Laws of Logarithms to simplify:

$$\ln y = \tfrac{3}{4}\ln x + \tfrac{1}{2}\ln(x^2 + 1) - 5\ln(3x + 2)$$

Differentiating implicitly with respect to x gives

$$\frac{1}{y}\frac{dy}{dx} = \frac{3}{4} \cdot \frac{1}{x} + \frac{1}{2} \cdot \frac{2x}{x^2 + 1} - 5 \cdot \frac{3}{3x + 2}$$

Solving for dy/dx, we get

$$\frac{dy}{dx} = y\left(\frac{3}{4x} + \frac{x}{x^2 + 1} - \frac{15}{3x + 2}\right)$$

Because we have an explicit expression for y, we can substitute and write

$$\frac{dy}{dx} = \frac{x^{3/4}\sqrt{x^2 + 1}}{(3x + 2)^5}\left(\frac{3}{4x} + \frac{x}{x^2 + 1} - \frac{15}{3x + 2}\right)$$

|||| If we hadn't used logarithmic differentiation in Example 14, we would have had to use both the Quotient Rule and the Product Rule. The resulting calculation would have been horrendous.

Steps in Logarithmic Differentiation
1. Take logarithms of both sides of an equation $y = f(x)$ and use the Laws of Logarithms to simplify.
2. Differentiate implicitly with respect to x.
3. Solve the resulting equation for y'.

If $f(x) < 0$ for some values of x, then $\ln f(x)$ is not defined, but we can write $|y| = |f(x)|$ and use Equation 7.

7.2* Exercises

1–4 Use the Laws of Logarithms to expand the quantity.

1. $\ln \dfrac{x^3 y}{z^2}$

2. $\ln \sqrt{a(b^2 + c^2)}$

3. $\ln (uv)^{10}$

4. $\ln \dfrac{3x^2}{(x+1)^5}$

5–8 Express the quantity as a single logarithm.

5. $2 \ln 4 - \ln 2$

6. $\ln 3 + \frac{1}{3} \ln 8$

7. $\frac{1}{2} \ln x - 5 \ln(x^2 + 1)$

8. $\ln x + a \ln y - b \ln z$

9–12 Make a rough sketch of the graph of each function. Do not use a calculator. Just use the graph given in Figure 4 and, if necessary, the transformations of Section 1.3.

9. $y = -\ln x$

10. $y = \ln |x|$

11. $y = \ln(x + 3)$

12. $y = 1 + \ln(x - 2)$

13–30 Differentiate the function.

13. $f(x) = \sqrt{x} \ln x$

14. $f(x) = \ln(x^2 + 10)$

15. $f(\theta) = \ln(\cos \theta)$

16. $f(x) = \cos(\ln x)$

17. $f(x) = \sqrt[5]{\ln x}$

18. $f(x) = \ln \sqrt[5]{x}$

19. $g(x) = \ln \dfrac{a - x}{a + x}$

20. $h(x) = \ln(x + \sqrt{x^2 - 1})$

21. $f(u) = \dfrac{\ln u}{1 + \ln(2u)}$

22. $f(t) = \dfrac{1 + \ln t}{1 - \ln t}$

23. $F(t) = \ln \dfrac{(2t+1)^3}{(3t-1)^4}$

24. $y = \ln(x^4 \sin^2 x)$

25. $y = \ln |2 - x - 5x^2|$

26. $G(u) = \ln \sqrt{\dfrac{3u+2}{3u-2}}$

27. $y = \ln \left(\dfrac{x+1}{x-1}\right)^{3/5}$

28. $y = (\ln \tan x)^2$

29. $y = \tan[\ln(ax + b)]$

30. $y = \ln |\tan 2x|$

31–32 Find y' and y''.

31. $y = \ln \ln x$

32. $y = \dfrac{\ln x}{x^2}$

33–36 Differentiate f and find the domain of f.

33. $f(x) = \dfrac{x}{1 - \ln(x - 1)}$

34. $f(x) = \dfrac{1}{1 + \ln x}$

35. $f(x) = \sqrt{1 - \ln x}$

36. $f(x) = \ln \ln \ln x$

37. If $f(x) = \dfrac{x}{\ln x}$, find $f'(e)$.

38. If $f(t) = t \ln(4 + 3t)$, find $f'(-1)$.

39–40 Find $f'(x)$. Check that your answer is reasonable by comparing the graphs of f and f'.

39. $f(x) = \sin x + \ln x$

40. $f(x) = \ln(x^2 + x + 1)$

41–42 Find an equation of the tangent line to the curve at the given point.

41. $y = \sin(2 \ln x)$, $(1, 0)$

42. $y = \ln(x^3 - 7)$, $(2, 0)$

43. Find y' if $y = \ln(x^2 + y^2)$.

44. Find y' if $\ln xy = y \sin x$.

45. Find a formula for $f^{(n)}(x)$ if $f(x) = \ln(x - 1)$.

46. Find $\dfrac{d^9}{dx^9}(x^8 \ln x)$.

47–48 Use a graph to estimate the roots of the equation correct to one decimal place. Then use these estimates as the initial approximations in Newton's method to find the roots correct to six decimal places.

47. $(x-4)^2 = \ln x$ **48.** $\ln(4-x^2) = x$

49–52 Discuss the curve under the guidelines of Section 4.5.

49. $y = \ln(\sin x)$ **50.** $y = \ln(\tan^2 x)$

51. $y = \ln(1 + x^2)$ **52.** $y = \ln(x^2 - 3x + 2)$

53. If $f(x) = \ln(2x + x \sin x)$, use the graphs of f, f', and f'' to estimate the intervals of increase and the inflection points of f on the interval $(0, 15]$.

54. Investigate the family of curves $f(x) = \ln(x^2 + c)$. What happens to the inflection points and asymptotes as c changes? Graph several members of the family to illustrate what you discover.

55–58 Use logarithmic differentiation to find the derivative of the function.

55. $y = (2x + 1)^5 (x^4 - 3)^6$ **56.** $y = \dfrac{(x^3 + 1)^4 \sin^2 x}{\sqrt[3]{x}}$

57. $y = \dfrac{\sin^2 x \, \tan^4 x}{(x^2 + 1)^2}$ **58.** $y = \sqrt[4]{\dfrac{x^2 + 1}{x^2 - 1}}$

59–68 Evaluate the integral.

59. $\displaystyle\int_2^4 \dfrac{3}{x}\,dx$ **60.** $\displaystyle\int_1^2 \dfrac{4 + u^2}{u^3}\,du$

61. $\displaystyle\int_1^2 \dfrac{dt}{8 - 3t}$ **62.** $\displaystyle\int_4^9 \left(\sqrt{x} + \dfrac{1}{\sqrt{x}}\right)^2 dx$

63. $\displaystyle\int_1^e \dfrac{x^2 + x + 1}{x}\,dx$ **64.** $\displaystyle\int_e^6 \dfrac{dx}{x \ln x}$

65. $\displaystyle\int \dfrac{2 - x^2}{6x - x^3}\,dx$ **66.** $\displaystyle\int \dfrac{\cos x}{2 + \sin x}\,dx$

67. $\displaystyle\int \dfrac{(\ln x)^2}{x}\,dx$ **68.** $\displaystyle\int \dfrac{(1 + \sqrt{x})^4}{\sqrt{x}}\,dx$

69. Show that $\int \cot x \, dx = \ln|\sin x| + C$ by (a) differentiating the right side of the equation and (b) using the method of Example 13.

70. Find, correct to three decimal places, the area of the region above the hyperbola $y = 2/(x - 2)$, below the x-axis, and between the lines $x = -4$ and $x = -1$.

71. Find the volume of the solid obtained by rotating the region under the curve $y = 1/\sqrt{x+1}$ from 0 to 1 about the x-axis.

72. Find the volume of the solid obtained by rotating the region under the curve $y = 1/(x^2 + 1)$ from 0 to 3 about the y-axis.

73. The work done by a gas when it expands from volume V_1 to volume V_2 is $W = \int_{V_1}^{V_2} P \, dV$, where $P = P(V)$ is the pressure as a function of the volume V. (See Exercise 27 in Section 6.4.) Boyle's Law states that when a quantity of gas expands at constant temperature, $PV = C$, where C is a constant. If the initial volume is 600 cm^3 and the initial pressure is 150 kPa, find the work done by the gas when it expands at constant temperature to 1000 cm^3.

74. Find f if $f''(x) = x^{-2}$, $x > 0$, $f(1) = 0$, and $f(2) = 0$.

75. If g is the inverse function of $f(x) = 2x + \ln x$, find $g'(2)$.

76. (a) Find the linear approximation to $f(x) = \ln x$ near 1.
(b) Illustrate part (a) by graphing f and its linearization.
(c) For what values of x is the linear approximation accurate to within 0.1?

77. (a) By comparing areas, show that
$$\tfrac{1}{3} < \ln 1.5 < \tfrac{5}{12}$$
(b) Use the Midpoint Rule with $n = 10$ to estimate $\ln 1.5$.

78. Refer to Example 1.
(a) Find an equation of the tangent line to the curve $y = 1/t$ that is parallel to the secant line AD.
(b) Use part (a) to show that $\ln 2 > 0.66$.

79. By comparing areas, show that
$$\dfrac{1}{2} + \dfrac{1}{3} + \cdots + \dfrac{1}{n} < \ln n < 1 + \dfrac{1}{2} + \dfrac{1}{3} + \cdots + \dfrac{1}{n-1}$$

80. Prove the third law of logarithms. [*Hint:* Start by showing that both sides of the equation have the same derivative.]

81. For what values of m do the line $y = mx$ and the curve $y = x/(x^2 + 1)$ enclose a region? Find the area of the region.

82. Find $\lim_{x \to \infty} [\ln(2 + x) - \ln(1 + x)]$.

83. Use the definition of derivative to prove that
$$\lim_{x \to 0} \dfrac{\ln(1 + x)}{x} = 1$$

84. (a) Compare the rates of growth of $f(x) = x^{0.1}$ and $g(x) = \ln x$ by graphing both f and g in several viewing rectangles. When does the graph of f finally surpass the graph of g?
(b) Graph the function $h(x) = (\ln x)/x^{0.1}$ in a viewing rectangle that displays the behavior of the function as $x \to \infty$.
(c) Find a number N such that
$$\dfrac{\ln x}{x^{0.1}} < 0.1 \quad \text{whenever} \quad x > N$$

7.3* The Natural Exponential Function

Since ln is an increasing function, it is one-to-one and therefore has an inverse function, which we denote by exp. Thus, according to the definition of an inverse function,

$f^{-1}(x) = y \iff f(y) = x$

[1] $$\exp(x) = y \iff \ln y = x$$

and the cancellation equations are

$f^{-1}(f(x)) = x$
$f(f^{-1}(x)) = x$

[2] $$\exp(\ln x) = x \quad \text{and} \quad \ln(\exp x) = x$$

In particular, we have

$$\exp(0) = 1 \quad \text{since } \ln 1 = 0$$

$$\exp(1) = e \quad \text{since } \ln e = 1$$

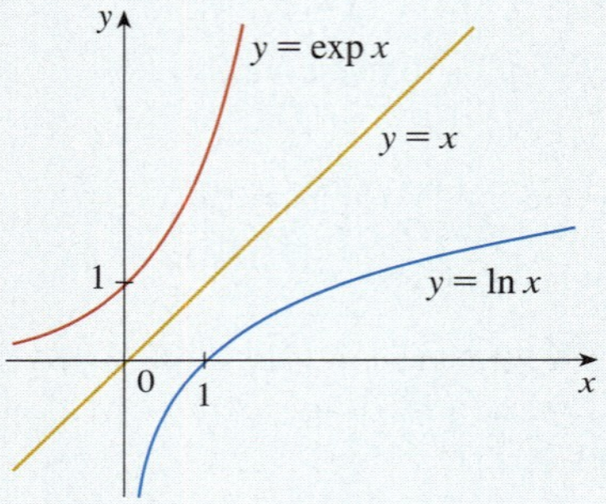

We obtain the graph of $y = \exp x$ by reflecting the graph of $y = \ln x$ about the line $y = x$. (See Figure 1.) The domain of exp is the range of ln, that is, $(-\infty, \infty)$; the range of exp is the domain of ln, that is, $(0, \infty)$.

If r is any rational number, then the third law of logarithms gives

$$\ln(e^r) = r \ln e = r$$

Therefore, by (1), $\exp(r) = e^r$

Thus, $\exp(x) = e^x$ whenever x is a rational number. This leads us to define e^x, even for irrational values of x, by the equation

FIGURE 1

$$e^x = \exp(x)$$

In other words, for the reasons given, we define e^x to be the inverse of the function $\ln x$. In this notation (1) becomes

[3] $$e^x = y \iff \ln y = x$$

and the cancellation equations (2) become

[4] $$e^{\ln x} = x \quad x > 0$$

[5] $$\ln(e^x) = x \quad \text{for all } x$$

EXAMPLE 1 Find x if $\ln x = 5$.

SOLUTION 1 From (3) we see that
$$\ln x = 5 \quad \text{means} \quad e^5 = x$$
Therefore, $x = e^5$.

SOLUTION 2 Start with the equation
$$\ln x = 5$$
and apply the exponential function to both sides of the equation:
$$e^{\ln x} = e^5$$
But (4) says that $e^{\ln x} = x$. Therefore, $x = e^5$.

EXAMPLE 2 Solve the equation $e^{5-3x} = 10$.

SOLUTION We take natural logarithms of both sides of the equation and use (5):
$$\ln(e^{5-3x}) = \ln 10$$
$$5 - 3x = \ln 10$$
$$3x = 5 - \ln 10$$
$$x = \tfrac{1}{3}(5 - \ln 10)$$

Since the natural logarithm is found on scientific calculators, we can approximate the solution to four decimal places: $x \approx 0.8991$.

The exponential function $f(x) = e^x$ is one of the most frequently occurring functions in calculus and its applications, so it is important to be familiar with its graph (Figure 2) and its properties (which follow from the fact that it is the inverse of the natural logarithmic function).

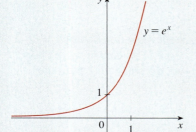

FIGURE 2
The natural exponential function

> **6 Properties of the Natural Exponential Function** The exponential function $f(x) = e^x$ is an increasing continuous function with domain $\mathbb{R}$ and range $(0, \infty)$. Thus, $e^x > 0$ for all x. Also
> $$\lim_{x \to -\infty} e^x = 0 \qquad \lim_{x \to \infty} e^x = \infty$$
> So the x-axis is a horizontal asymptote of $f(x) = e^x$.

EXAMPLE 3 Find $\displaystyle\lim_{x \to \infty} \frac{e^{2x}}{e^{2x} + 1}$.

SOLUTION We divide numerator and denominator by e^{2x}:
$$\lim_{x \to \infty} \frac{e^{2x}}{e^{2x} + 1} = \lim_{x \to \infty} \frac{1}{1 + e^{-2x}} = \frac{1}{1 + \lim_{x \to \infty} e^{-2x}}$$
$$= \frac{1}{1 + 0} = 1$$

We have used the fact that $t = -2x \to -\infty$ as $x \to \infty$ and so

$$\lim_{x \to \infty} e^{-2x} = \lim_{t \to -\infty} e^t = 0$$

We now verify that $f(x) = e^x$ has the properties expected of an exponential function.

> **7 Laws of Exponents** If x and y are real numbers and r is rational, then
>
> **1.** $e^{x+y} = e^x e^y$ **2.** $e^{x-y} = \dfrac{e^x}{e^y}$ **3.** $(e^x)^r = e^{rx}$

Proof of Law 1 Using the first law of logarithms and Equation 5, we have

$$\ln(e^x e^y) = \ln(e^x) + \ln(e^y) = x + y = \ln(e^{x+y})$$

Since ln is a one-to-one function, it follows that $e^x e^y = e^{x+y}$.

Laws 2 and 3 are proved similarly (see Exercises 85 and 86). As we will see in the next section, Law 3 actually holds when r is any real number.

Differentiation

The natural exponential function has the remarkable property that *it is its own derivative.*

$$\frac{d}{dx}(e^x) = e^x$$

Proof The function $y = e^x$ is differentiable because it is the inverse function of $y = \ln x$, which we know is differentiable with nonzero derivative. To find its derivative, we use the inverse function method. Let $y = e^x$. Then $\ln y = x$ and, differentiating this latter equation implicitly with respect to x, we get

$$\frac{1}{y}\frac{dy}{dx} = 1$$

$$\frac{dy}{dx} = y = e^x$$

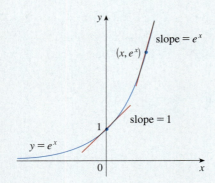

FIGURE 3

The geometric interpretation of Formula 8 is that the slope of a tangent line to the curve $y = e^x$ at any point is equal to the y-coordinate of the point (see Figure 3). This property implies that the exponential curve $y = e^x$ grows very rapidly (see Exercise 90).

EXAMPLE 4 Differentiate the function $y = e^{\tan x}$.

SOLUTION To use the Chain Rule, we let $u = \tan x$. Then we have $y = e^u$, so

$$\frac{dy}{dx} = \frac{dy}{du}\frac{du}{dx} = e^u \frac{du}{dx} = e^{\tan x} \sec^2 x$$

In general if we combine Formula 8 with the Chain Rule, as in Example 4, we get

9
$$\frac{d}{dx}(e^u) = e^u \frac{du}{dx}$$

EXAMPLE 5 Find y' if $y = e^{-4x} \sin 5x$.

SOLUTION Using Formula 9 and the Product Rule, we have

$$y' = e^{-4x}(\cos 5x)(5) + (\sin 5x)e^{-4x}(-4) = e^{-4x}(5 \cos 5x - 4 \sin 5x)$$

EXAMPLE 6 Find the absolute maximum value of the function $f(x) = xe^{-x}$.

SOLUTION We differentiate to find any critical numbers:

$$f'(x) = xe^{-x}(-1) + e^{-x}(1) = e^{-x}(1 - x)$$

Since exponential functions are always positive, we see that $f'(x) > 0$ when $1 - x > 0$, that is, when $x < 1$. Similarly, $f'(x) < 0$ when $x > 1$. By the First Derivative Test for Absolute Extreme Values, f has an absolute maximum value when $x = 1$ and the value is

$$f(1) = (1)e^{-1} = \frac{1}{e} \approx 0.37$$

EXAMPLE 7 Use the first and second derivatives of $f(x) = e^{1/x}$, together with asymptotes, to sketch its graph.

SOLUTION Notice that the domain of f is $\{x \mid x \neq 0\}$, so we check for vertical asymptotes by computing the left and right limits as $x \to 0$. As $x \to 0^+$, we know that $t = 1/x \to \infty$, so

$$\lim_{x \to 0^+} e^{1/x} = \lim_{t \to \infty} e^t = \infty$$

and this shows that $x = 0$ is a vertical asymptote. As $x \to 0^-$, we have $t = 1/x \to -\infty$, so

$$\lim_{x \to 0^-} e^{1/x} = \lim_{t \to -\infty} e^t = 0$$

As $x \to \pm\infty$, we have $1/x \to 0$ and so

$$\lim_{x \to \pm\infty} e^{1/x} = e^0 = 1$$

This shows that $y = 1$ is a horizontal asymptote.

Now let's compute the derivative. The Chain Rule gives

$$f'(x) = -\frac{e^{1/x}}{x^2}$$

Since $e^{1/x} > 0$ and $x^2 > 0$ for all $x \neq 0$, we have $f'(x) < 0$ for all $x \neq 0$. Thus, f is decreasing on $(-\infty, 0)$ and on $(0, \infty)$. There is no critical number, so the function has no maximum or minimum. The second derivative is

$$f''(x) = -\frac{x^2 e^{1/x}(-1/x^2) - e^{1/x}(2x)}{x^4} = \frac{e^{1/x}(2x + 1)}{x^4}$$

Since $e^{1/x} > 0$ and $x^4 > 0$, we have $f''(x) > 0$ when $x > -\frac{1}{2}$ ($x \neq 0$) and $f''(x) < 0$ when $x < -\frac{1}{2}$. So the curve is concave downward on $\left(-\infty, -\frac{1}{2}\right)$ and concave upward on $\left(-\frac{1}{2}, 0\right)$ and on $(0, \infty)$. The inflection point is $\left(-\frac{1}{2}, e^{-2}\right)$.

To sketch the graph of f we first draw the horizontal asymptote $y = 1$ (as a dashed line), together with the parts of the curve near the asymptotes in a preliminary sketch [Figure 4(a)]. These parts reflect the information concerning limits and the fact that f is decreasing on both $(-\infty, 0)$ and $(0, \infty)$. Notice that we have indicated that $f(x) \to 0$ as $x \to 0^-$ even though $f(0)$ does not exist. In Figure 4(b) we finish the sketch by incorporating the information concerning concavity and the inflection point. In Figure 4(c) we check our work with a graphing device.

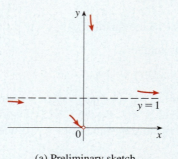
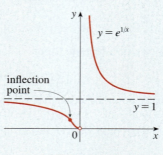
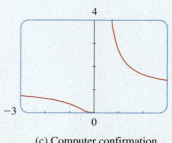

FIGURE 4 (a) Preliminary sketch (b) Finished sketch (c) Computer confirmation

Integration

Because the exponential function $y = e^x$ has a simple derivative, its integral is also simple:

$$\boxed{10 \quad \int e^x \, dx = e^x + C}$$

EXAMPLE 8 Evaluate $\int x^2 e^{x^3} \, dx$.

SOLUTION We substitute $u = x^3$. Then $du = 3x^2 \, dx$, so $x^2 \, dx = \frac{1}{3} du$ and

$$\int x^2 e^{x^3} \, dx = \frac{1}{3} \int e^u \, du = \frac{1}{3} e^u + C = \frac{1}{3} e^{x^3} + C$$

EXAMPLE 9 Find the area under the curve $y = e^{-3x}$ from 0 to 1.

SOLUTION The area is

$$A = \int_0^1 e^{-3x} \, dx = -\tfrac{1}{3} e^{-3x} \Big]_0^1 = \tfrac{1}{3}(1 - e^{-3})$$

7.3* Exercises

1. (a) How is the number e defined?
   (b) What is an approximate value for e?
   (c) Sketch, by hand, the graph of the function $f(x) = e^x$ with particular attention to how the graph crosses the y-axis. What fact allows you to do this?

2–4 Simplify each expression.

2. (a) $e^{\ln 6}$ (b) $\ln \sqrt{e}$

3. (a) $\ln e^{\sqrt{2}}$ (b) $e^{3 \ln 2}$

4. (a) $\ln e^{\sin x}$ (b) $e^{x + \ln x}$

5–12 Solve each equation for x.

5. (a) $2 \ln x = 1$ (b) $e^{-x} = 5$

6. (a) $e^{2x+3} - 7 = 0$ (b) $\ln(5 - 2x) = -3$

7. $\ln(\ln x) = 1$

8. $e^{e^x} = 10$

9. $2 \ln x = \ln 2 + \ln(3x - 4)$

10. $\ln(2x + 1) = 2 - \ln x$

11. $e^{ax} = Ce^{bx}$, where $a \neq b$

12. $7e^x - e^{2x} = 12$

13–16 Find the solution of the equation correct to four decimal places.

13. $e^{2+5x} = 100$ 14. $\ln(1 + \sqrt{x}) = 2$

15. $\ln(e^x - 2) = 3$ 16. $e^{1/(x-4)} = 7$

17–18 Solve each inequality for x.

17. (a) $e^x < 10$ (b) $\ln x > -1$

18. (a) $2 < \ln x < 9$ (b) $e^{2-3x} > 4$

19–22 Make a rough sketch of the graph of each function. Do not use a calculator. Just use the graph given in Figure 2 and, if necessary, the transformations of Section 1.3.

19. $y = e^{-x}$ 20. $y = 1 + 2e^x$

21. $y = 3 - e^x$ 22. $y = 2 + 5(1 - e^{-x})$

23. Starting with the graph of $y = e^x$, write the equation of the graph that results from
    (a) shifting 2 units downward
    (b) shifting 2 units to the right
    (c) reflecting about the x-axis
    (d) reflecting about the y-axis
    (e) reflecting about the x-axis and then about the y-axis

24. Starting with the graph of $y = e^x$, find the equation of the graph that results from
    (a) reflecting about the line $y = 4$
    (b) reflecting about the line $x = 2$

25–30 Find the limit.

25. $\lim\limits_{x \to \infty} e^{1-x^3}$ 26. $\lim\limits_{x \to (\pi/2)^+} e^{\tan x}$

27. $\lim\limits_{x \to \infty} \dfrac{e^{3x} - e^{-3x}}{e^{3x} + e^{-3x}}$ 28. $\lim\limits_{x \to -\infty} \dfrac{e^{3x} - e^{-3x}}{e^{3x} + e^{-3x}}$

29. $\lim\limits_{x \to 2^+} e^{3/(2-x)}$ 30. $\lim\limits_{x \to 2^-} e^{3/(2-x)}$

31–44 Differentiate the function.

31. $f(x) = x^2 e^x$ 32. $y = \dfrac{e^x}{1 + x}$

33. $y = e^{ax^3}$ 34. $y = e^u(\cos u + cu)$

35. $f(u) = e^{1/u}$ 36. $y = e^x \ln x$

37. $F(t) = e^{t \sin 2t}$ 38. $y = e^{k \tan \sqrt{x}}$

39. $y = \sqrt{1 + 2e^{3x}}$ 40. $y = \cos(e^{\pi x})$

41. $y = e^{e^x}$ 42. $y = \sqrt{1 + xe^{-2x}}$

43. $y = \dfrac{ae^x + b}{ce^x + d}$ 44. $y = \dfrac{e^x + e^{-x}}{e^x - e^{-x}}$

45–46 Find an equation of the tangent line to the curve at the given point.

45. $y = e^{2x} \cos \pi x$, $(0, 1)$ 46. $y = e^x/x$, $(1, e)$

47. Find y' if $e^{x^2 y} = x + y$.

48. Show that the function $y = Ae^{-x} + Bxe^{-x}$ satisfies the differential equation $y'' + 2y' + y = 0$.

49. For what values of r does the function $y = e^{rx}$ satisfy the equation $y'' + 6y' + 8y = 0$?

50. Find the values of λ for which $y = e^{\lambda x}$ satisfies the equation $y + y' = y''$.

51. If $f(x) = e^{2x}$, find a formula for $f^{(n)}(x)$.

52. Find the thousandth derivative of $f(x) = xe^{-x}$.

53. (a) Use the Intermediate Value Theorem to show that there is a root of the equation $e^x + x = 0$.
    (b) Use Newton's method to find the root of the equation in part (a) correct to six decimal places.

54. Use a graph to find an initial approximation (to one decimal place) to the root of the equation
$$e^{-x^2} = x^3 + x - 3$$
Then use Newton's method to find the root correct to six decimal places.

55. If the initial mass of a sample of ^{90}Sr is 24 mg, then the mass after t years is
$$m(t) = 24 \cdot e^{-(\ln 2)t/25}$$
    (a) Find the mass remaining after 40 years.
    (b) At what rate does the mass decay after 40 years?
    (c) How long does it take for the mass to be reduced to 5 mg?

56. For the period from 1980 to 2000, the percentage of households in the United States with at least one VCR has been modeled by the function
$$V(t) = \frac{85}{1 + 53e^{-0.5t}}$$
where the time t is measured in years since midyear 1980, so $0 \le t \le 20$. Use a graph to estimate the time at which the number of VCRs was increasing most rapidly. Then use derivatives to give a more accurate estimate.

57. Under certain circumstances a rumor spreads according to the equation
$$p(t) = \frac{1}{1 + ae^{-kt}}$$
where $p(t)$ is the proportion of the population that knows the rumor at time t and a and k are positive constants. [In Section 10.5 we will see that this is a reasonable model for $p(t)$.]
    (a) Find $\lim_{t \to \infty} p(t)$.
    (b) Find the rate of spread of the rumor.
    (c) Graph p for the case $a = 10$, $k = 0.5$ with t measured in hours. Use the graph to estimate how long it will take for 80% of the population to hear the rumor.

58. An object is attached to the end of a vibrating spring and its displacement from its equilibrium position is $y = 8e^{-t/2} \sin 4t$, where t is measured in seconds and y is measured in centimeters.
    (a) Graph the displacement function together with the functions $y = 8e^{-t/2}$ and $y = -8e^{-t/2}$. How are these graphs related? Can you explain why?
    (b) Use the graph to estimate the maximum value of the displacement. Does it occur when the graph touches the graph of $y = 8e^{-t/2}$?
    (c) What is the velocity of the object when it first returns to its equilibrium position?

    (d) Use the graph to estimate the time after which the displacement is no more than 2 cm from equilibrium.

59. Find the absolute maximum value of the function $f(x) = x - e^x$.

60. Find the absolute minimum value of the function $g(x) = e^x/x$, $x > 0$.

61. On what interval is the curve $y = xe^{3x}$ concave upward?

62. On what interval is the function $f(x) = x^2 e^{-x}$ increasing?

63–65 ▪▪▪▪ Discuss the curve using the guidelines of Section 4.5.

63. $y = e^{-1/(x+1)}$
64. $y = e^{2x} - e^x$
65. $y = e^{3x} + e^{-2x}$

66–67 ▪▪▪▪ Draw a graph of f that shows all the important aspects of the curve. Estimate the local maximum and minimum values and then use calculus to find these values exactly. Use a graph of f'' to estimate the inflection points.

66. $f(x) = e^{\cos x}$
67. $f(x) = e^{x^3 - x}$

68. The family of bell-shaped curves
$$y = \frac{1}{\sigma\sqrt{2\pi}} e^{-(x-\mu)^2/(2\sigma^2)}$$
occurs in probability and statistics, where it is called the *normal density function*. The constant μ is called the *mean* and the positive constant σ is called the *standard deviation*. For simplicity, let's scale the function so as to remove the factor $1/(\sigma\sqrt{2\pi})$ and let's analyze the special case where $\mu = 0$. So we study the function
$$f(x) = e^{-x^2/(2\sigma^2)}$$
    (a) Find the asymptote, maximum value, and inflection points of f.
    (b) What role does σ play in the shape of the curve?
    (c) Illustrate by graphing four members of this family on the same screen.

69–76 ▪▪▪▪ Evaluate the integral.

69. $\displaystyle\int_0^5 e^{-3x}\, dx$
70. $\displaystyle\int_0^1 xe^{-x^2}\, dx$
71. $\displaystyle\int e^x \sqrt{1 + e^x}\, dx$
72. $\displaystyle\int \sec^2 x\, e^{\tan x}\, dx$
73. $\displaystyle\int \frac{e^x + 1}{e^x}\, dx$
74. $\displaystyle\int \frac{e^{1/x}}{x^2}\, dx$
75. $\displaystyle\int \frac{e^{\sqrt{x}}}{\sqrt{x}}\, dx$
76. $\displaystyle\int e^x \sin(e^x)\, dx$

77. Find, correct to three decimal places, the area of the region bounded by the curves $y = e^x$, $y = e^{3x}$, and $x = 1$.

78. Find $f(x)$ if $f''(x) = 3e^x + 5 \sin x$, $f(0) = 1$, and $f'(0) = 2$.

79. Find the volume of the solid obtained by rotating about the x-axis the region bounded by the curves $y = e^x$, $y = 0$, $x = 0$, and $x = 1$.

80. Find the volume of the solid obtained by rotating about the y-axis the region bounded by the curves $y = e^{-x^2}$, $y = 0$, $x = 0$, and $x = 1$.

81–82 ⦀ Find the inverse function of f. Check your answer by graphing both f and f^{-1} on the same screen.

81. $f(x) = \ln(x + 3)$
82. $f(x) = \dfrac{1 + e^x}{1 - e^x}$

83. If $f(x) = 3 + x + e^x$, find $(f^{-1})'(4)$.

84. Evaluate $\lim\limits_{x \to \pi} \dfrac{e^{\sin x} - 1}{x - \pi}$.

85. Prove the second law of exponents [see (7)].
86. Prove the third law of exponents [see (7)].
87. (a) Show that $e^x \geq 1 + x$ if $x \geq 0$.
    [Hint: Show that $f(x) = e^x - (1 + x)$ is increasing for $x > 0$.]
    (b) Deduce that $\tfrac{4}{3} \leq \int_0^1 e^{x^2} dx \leq e$.

88. (a) Use the inequality of Exercise 87(a) to show that, for $x \geq 0$,
$$e^x \geq 1 + x + \tfrac{1}{2}x^2$$
(b) Use part (a) to improve the estimate of $\int_0^1 e^{x^2} dx$ given in Exercise 87(b).

89. (a) Use mathematical induction to prove that for $x \geq 0$ and any positive integer n,
$$e^x \geq 1 + x + \dfrac{x^2}{2!} + \cdots + \dfrac{x^n}{n!}$$
(b) Use part (a) to show that $e > 2.7$.
(c) Use part (a) to show that
$$\lim_{x \to \infty} \dfrac{e^x}{x^k} = \infty$$
for any positive integer k.

90. This exercise illustrates Exercise 89(c) for the case $k = 10$.
(a) Compare the rates of growth of $f(x) = x^{10}$ and $g(x) = e^x$ by graphing both f and g in several viewing rectangles. When does the graph of g finally surpass the graph of f?
(b) Find a viewing rectangle that shows how the function $h(x) = e^x/x^{10}$ behaves for large x.
(c) Find a number N such that
$$\dfrac{e^x}{x^{10}} > 10^{10} \quad \text{whenever} \quad x > N$$

7.4* General Logarithmic and Exponential Functions

In this section we use the natural exponential and logarithmic functions to study exponential and logarithmic functions with base $a > 0$.

General Exponential Functions

If $a > 0$ and r is any rational number, then by (4) and (7) in Section 7.3*,
$$a^r = (e^{\ln a})^r = e^{r \ln a}$$

Therefore, even for irrational numbers x, we *define*

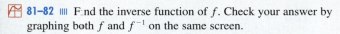

$$a^x = e^{x \ln a}$$

Thus, for instance,
$$2^{\sqrt{3}} = e^{\sqrt{3} \ln 2} \approx e^{1.20} \approx 3.32$$

The function $f(x) = a^x$ is called the **exponential function with base a**. Notice that a^x is positive for all x because e^x is positive for all x.

Definition 1 allows us to extend one of the laws of logarithms. We know that $\ln(a^r) = r \ln a$ when r is rational. But if we now let r be *any* real number we have, from Definition 1,

$$\ln a^r = \ln(e^{r \ln a}) = r \ln a$$

Thus

2 $\quad\quad\quad\quad\quad \ln a^r = r \ln a \quad$ for any real number r

The general laws of exponents follow from Definition 1 together with the laws of exponents for e^x.

3 Laws of Exponents If x and y are real numbers and $a, b > 0$, then

1. $a^{x+y} = a^x a^y \quad$ 2. $a^{x-y} = a^x/a^y \quad$ 3. $(a^x)^y = a^{xy} \quad$ 4. $(ab)^x = a^x b^x$

Proof

1. Using Definition 1 and the laws of exponents for e^x, we have

$$a^{x+y} = e^{(x+y) \ln a} = e^{x \ln a + y \ln a}$$
$$= e^{x \ln a} e^{y \ln a} = a^x a^y$$

3. Using Equation 2 we obtain

$$(a^x)^y = e^{y \ln(a^x)} = e^{yx \ln a}$$
$$= e^{xy \ln a} = a^{xy}$$

The remaining proofs are left as exercises.

The differentiation formula for exponential functions is also a consequence of Definition 1:

4 $\quad\quad\quad\quad\quad \dfrac{d}{dx}(a^x) = a^x \ln a$

Proof $\quad \dfrac{d}{dx}(a^x) = \dfrac{d}{dx}(e^{x \ln a}) = e^{x \ln a} \dfrac{d}{dx}(x \ln a)$
$$= a^x \ln a$$

Notice that if $a = e$, then $\ln e = 1$ and Formula 4 simplifies to a formula that we already know: $(d/dx) e^x = e^x$. In fact, the reason that the natural exponential function is used more often than other exponential functions is that its differentiation formula is simpler.

EXAMPLE 1 In Example 6 in Section 3.4 we considered a population of bacteria cells in a homogeneous nutrient medium. We showed that if the population doubles every hour, then the population after t hours is

$$n = n_0 2^t$$

where n_0 is the initial population. Now we can use (4) to compute the growth rate:

$$\frac{dn}{dt} = n_0 2^t \ln 2$$

For instance, if the initial population is $n_0 = 1000$ cells, then the growth rate after two hours is

$$\frac{dn}{dt}\bigg|_{t=2} = (1000)2^t \ln 2 \big|_{t=2}$$
$$= 4000 \ln 2 \approx 2773 \text{ cells/h}$$

EXAMPLE 2 Combining Formula 4 with the Chain Rule, we have

$$\frac{d}{dx}\left(10^{x^2}\right) = 10^{x^2}(\ln 10)\frac{d}{dx}(x^2) = (2\ln 10)x10^{x^2}$$

Exponential Graphs

If $a > 1$, then $\ln a > 0$, so $(d/dx)\,a^x = a^x \ln a > 0$, which shows that $y = a^x$ is increasing (see Figure 1). If $0 < a < 1$, then $\ln a < 0$ and so $y = a^x$ is decreasing (see Figure 2).

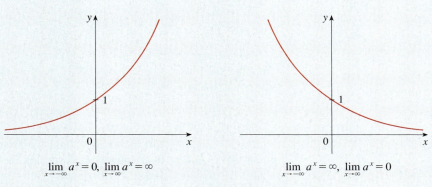

$\lim_{x \to -\infty} a^x = 0, \ \lim_{x \to \infty} a^x = \infty$ $\qquad$ $\lim_{x \to -\infty} a^x = \infty, \ \lim_{x \to \infty} a^x = 0$

FIGURE 1 $y = a^x$, $a > 1$ $\qquad$ **FIGURE 2** $y = a^x$, $0 < a < 1$

Notice from Figure 3 that as the base a gets larger, the exponential function grows more rapidly (for $x > 0$).

Figure 4 shows how the exponential function $y = 2^x$ compares with the power function $y = x^2$. The graphs intersect three times, but ultimately the exponential curve $y = 2^x$ grows far more rapidly than the parabola $y = x^2$. (See also Figure 5.)

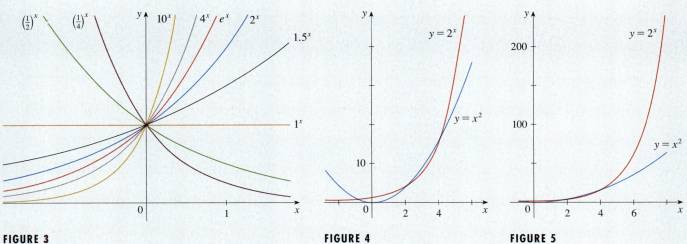

FIGURE 3 $\qquad$ **FIGURE 4** $\qquad$ **FIGURE 5**

In Chapter 10 we will show why exponential functions occur in the description of population growth and radioactive decay. Let's look at human population growth. Table 1 shows data for the population of the world in the 20th century and Figure 6 shows the corresponding scatter plot.

TABLE 1

Year	Population (millions)
1900	1650
1910	1750
1920	1860
1930	2070
1940	2300
1950	2560
1960	3040
1970	3710
1980	4450
1990	5280
2000	6080

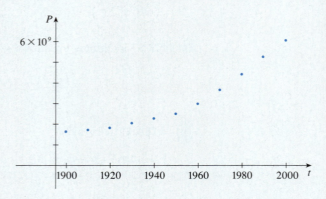

FIGURE 6 Scatter plot for world population growth

The pattern of the data points in Figure 6 suggests exponential growth, so we use a graphing calculator with exponential regression capability to apply the method of least squares and obtain the exponential model

$$P = (0.008079266) \cdot (1.013731)^t$$

Figure 7 shows the graph of this exponential function together with the original data points. We see that the exponential curve fits the data reasonably well. The period of relatively slow population growth is explained by the two world wars and the Great Depression of the 1930s.

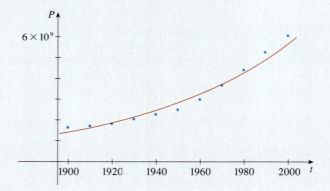

FIGURE 7
Exponential model for population growth

Exponential Integrals

The integration formula that follows from Formula 4 is

$$\int a^x \, dx = \frac{a^x}{\ln a} + C \qquad a \neq 1$$

EXAMPLE 3
$$\int_0^5 2^x \, dx = \frac{2^x}{\ln 2}\bigg]_0^5 = \frac{2^5}{\ln 2} - \frac{2^0}{\ln 2} = \frac{31}{\ln 2}$$

The Power Rule Versus the Exponential Rule

Now that we have defined arbitrary powers of numbers, we are in a position to prove the general version of the Power Rule, as promised in Section 3.3.

The Power Rule If n is any real number and $f(x) = x^n$, then
$$f'(x) = nx^{n-1}$$

|||| If $x = 0$, we can show that $f'(0) = 0$ for $n > 1$ directly from the definition of a derivative.

Proof Let $y = x^n$ and use logarithmic differentiation:
$$\ln|y| = \ln|x|^n = n \ln|x| \qquad x \neq 0$$

Therefore
$$\frac{y'}{y} = \frac{n}{x}$$

Hence
$$y' = n\frac{y}{x} = n\frac{x^n}{x} = nx^{n-1}$$

 You should distinguish carefully between the Power Rule $[(d/dx) x^n = nx^{n-1}]$, where the base is variable and the exponent is constant, and the rule for differentiating exponential functions $[(d/dx) a^x = a^x \ln a]$, where the base is constant and the exponent is variable. In general there are four cases for exponents and bases:

1. $\dfrac{d}{dx}(a^b) = 0$ (a and b are constants)

2. $\dfrac{d}{dx}[f(x)]^b = b[f(x)]^{b-1}f'(x)$

3. $\dfrac{d}{dx}[a^{g(x)}] = a^{g(x)}(\ln a)g'(x)$

4. To find $(d/dx)[f(x)]^{g(x)}$, logarithmic differentiation can be used, as in the next example.

EXAMPLE 4 Differentiate $y = x^{\sqrt{x}}$.

SOLUTION 1 Using logarithmic differentiation, we have
$$\ln y = \ln x^{\sqrt{x}} = \sqrt{x} \ln x$$
$$\frac{y'}{y} = \sqrt{x} \cdot \frac{1}{x} + (\ln x) \frac{1}{2\sqrt{x}}$$
$$y' = y\left(\frac{1}{\sqrt{x}} + \frac{\ln x}{2\sqrt{x}}\right) = x^{\sqrt{x}}\left(\frac{2 + \ln x}{2\sqrt{x}}\right)$$

|||| Figure 8 illustrates Example 4 by showing the graphs of $f(x) = x^{\sqrt{x}}$ and its derivative.

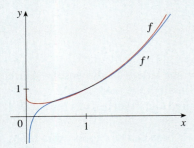

FIGURE 8

SOLUTION 2 Another method is to write $x^{\sqrt{x}} = (e^{\ln x})^{\sqrt{x}}$:

$$\frac{d}{dx}\left(x^{\sqrt{x}}\right) = \frac{d}{dx}\left(e^{\sqrt{x}\ln x}\right)$$

$$= e^{\sqrt{x}\ln x}\frac{d}{dx}\left(\sqrt{x}\ln x\right)$$

$$= x^{\sqrt{x}}\left(\frac{2 + \ln x}{2\sqrt{x}}\right) \quad \text{(as in Solution 1)}$$

General Logarithmic Functions

If $a > 0$ and $a \neq 1$, then $f(x) = a^x$ is a one-to-one function. Its inverse function is called the **logarithmic function with base a** and is denoted by $\log_a$. Thus

[5]
$$\log_a x = y \iff a^y = x$$

In particular, we see that

$$\log_e x = \ln x$$

The cancellation equations for the inverse functions $\log_a x$ and a^x are

$$a^{\log_a x} = x \quad \text{and} \quad \log_a(a^x) = x$$

Figure 9 shows the case where $a > 1$. (The most important logarithmic functions have base $a > 1$.) The fact that $y = a^x$ is a very rapidly increasing function for $x > 0$ is reflected in the fact that $y = \log_a x$ is a very slowly increasing function for $x > 1$.

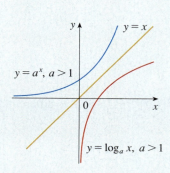

FIGURE 9

FIGURE 10

Figure 10 shows the graphs of $y = \log_a x$ with various values of the base a. Since $\log_a 1 = 0$, the graphs of all logarithmic functions pass through the point $(1, 0)$.

The laws of logarithms are similar to those for the natural logarithm and can be deduced from the laws of exponents (see Exercise 61).

The following formula shows that logarithms with any base can be expressed in terms of the natural logarithm.

6 Change of Base Formula For any positive number a ($a \neq 1$), we have

$$\log_a x = \frac{\ln x}{\ln a}$$

Proof Let $y = \log_a x$. Then, from (1), we have $a^y = x$. Taking natural logarithms of both sides of this equation, we get $y \ln a = \ln x$. Therefore

$$y = \frac{\ln x}{\ln a}$$

Scientific calculators have a key for natural logarithms, so Formula 6 enables us to use a calculator to compute a logarithm with any base (as shown in the next example). Similarly, Formula 6 allows us to graph any logarithmic function on a graphing calculator or computer (see Exercises 14–16).

EXAMPLE 5 Evaluate $\log_8 5$ correct to six decimal places.

SOLUTION Formula 6 gives

$$\log_8 5 = \frac{\ln 5}{\ln 8} \approx 0.773976$$

Formula 6 enables us to differentiate any logarithmic function. Since $\ln a$ is a constant, we can differentiate as follows:

$$\frac{d}{dx}(\log_a x) = \frac{d}{dx}\left(\frac{\ln x}{\ln a}\right)$$

$$= \frac{1}{\ln a} \frac{d}{dx}(\ln x) = \frac{1}{x \ln a}$$

$$\frac{d}{dx}(\log_a x) = \frac{1}{x \ln a}$$

|||| NOTATION FOR LOGARITHMS
Most textbooks in calculus and the sciences, as well as calculators, use the notation $\ln x$ for the natural logarithm and $\log x$ for the "common logarithm," $\log_{10} x$. In the more advanced mathematical and scientific literature and in computer languages, however, the notation $\log x$ usually denotes the natural logarithm.

EXAMPLE 6

$$\frac{d}{dx}\log_{10}(2 + \sin x) = \frac{1}{(2 + \sin x) \ln 10} \frac{d}{dx}(2 + \sin x)$$

$$= \frac{\cos x}{(2 + \sin x) \ln 10}$$

From Formula 7 we see one of the main reasons that natural logarithms (logarithms with base e) are used in calculus: The differentiation formula is simplest when $a = e$ because $\ln e = 1$.

The Number e as a Limit

We have shown that if $f(x) = \ln x$, then $f'(x) = 1/x$. Thus, $f'(1) = 1$. We now use this fact to express the number e as a limit.

From the definition of a derivative as a limit, we have

$$f'(1) = \lim_{h \to 0} \frac{f(1+h) - f(1)}{h} = \lim_{x \to 0} \frac{f(1+x) - f(1)}{x}$$

$$= \lim_{x \to 0} \frac{\ln(1+x) - \ln 1}{x} = \lim_{x \to 0} \frac{1}{x} \ln(1+x)$$

$$= \lim_{x \to 0} \ln(1+x)^{1/x}$$

Because $f'(1) = 1$, we have

$$\lim_{x \to 0} \ln(1+x)^{1/x} = 1$$

Then, by Theorem 2.5.8 and the continuity of the exponential functions, we have

$$e = e^1 = e^{\lim_{x \to 0} \ln(1+x)^{1/x}} = \lim_{x \to 0} e^{\ln(1+x)^{1/x}} = \lim_{x \to 0} (1+x)^{1/x}$$

8
$$e = \lim_{x \to 0} (1+x)^{1/x}$$

Formula 8 is illustrated by the graph of the function $y = (1+x)^{1/x}$ in Figure 11 and a table of values for small values of x.

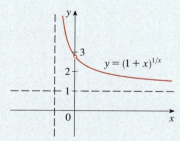

FIGURE 11

x	$(1+x)^{1/x}$
0.1	2.59374246
0.01	2.70481383
0.001	2.71692393
0.0001	2.71814593
0.00001	2.71826824
0.000001	2.71828047
0.0000001	2.71828169
0.00000001	2.71828181

If we put $n = 1/x$ in Formula 8, then $n \to \infty$ as $x \to 0^+$ and so an alternative expression for e is

9
$$e = \lim_{n \to \infty} \left(1 + \frac{1}{n}\right)^n$$

7.4* Exercises

1. (a) Write an equation that defines a^x when a is a positive number and x is a real number.
   (b) What is the domain of the function $f(x) = a^x$?
   (c) If $a \neq 1$, what is the range of this function?
   (d) Sketch the general shape of the graph of the exponential function for each of the following cases.
      (i) $a > 1$ (ii) $a = 1$ (iii) $0 < a < 1$

2. (a) If a is a positive number and $a \neq 1$, how is $\log_a x$ defined?
   (b) What is the domain of the function $f(x) = \log_a x$?
   (c) What is the range of this function?
   (d) If $a > 1$, sketch the general shapes of the graphs of $y = \log_a x$ and $y = a^x$ with a common set of axes.

3–6 ▮▮▮▮ Write the expression as a power of e.

3. $5^{\sqrt{7}}$
4. 10^{x^2}
5. $(\cos x)^x$
6. $x^{\cos x}$

7–10 ▮▮▮▮ Evaluate the expression.

7. (a) $\log_{10} 1000$ (b) $\log_2 \frac{1}{16}$
8. (a) $\log_{10} 0.1$ (b) $\log_8 320 - \log_8 5$
9. (a) $\log_{12} 3 + \log_{12} 48$ (b) $\log_5 5^{\sqrt{2}}$
10. (a) $\log_a \dfrac{1}{a}$ (b) $10^{(\log_{10} 4 + \log_{10} 7)}$

11–12 ▮▮▮▮ Graph the given functions on a common screen. How are these graphs related?

11. $y = 2^x$, $y = e^x$, $y = 5^x$, $y = 20^x$
12. $y = 3^x$, $y = 10^x$, $y = \left(\frac{1}{3}\right)^x$, $y = \left(\frac{1}{10}\right)^x$

13. Use Formula 6 to evaluate each logarithm correct to six decimal places.
    (a) $\log_{12} e$ (b) $\log_6 13.54$ (c) $\log_2 \pi$

14–16 ▮▮▮▮ Use Formula 6 to graph the given functions on a common screen. How are these graphs related?

14. $y = \log_2 x$, $y = \log_4 x$, $y = \log_6 x$, $y = \log_8 x$
15. $y = \log_{1.5} x$, $y = \ln x$, $y = \log_{10} x$, $y = \log_{50} x$
16. $y = \ln x$, $y = \log_{10} x$, $y = e^x$, $y = 10^x$

17–18 ▮▮▮▮ Find the exponential function $f(x) = Ca^x$ whose graph is given.

17.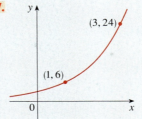

18.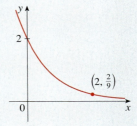

19. (a) Show that if the graphs of $f(x) = x^2$ and $g(x) = 2^x$ are drawn on a coordinate grid where the unit of measurement is 1 inch, then at a distance 2 ft to the right of the origin the height of the graph of f is 48 ft but the height of the graph of g is about 265 mi.
    (b) Suppose that the graph of $y = \log_2 x$ is drawn on a coordinate grid where the unit of measurement is an inch. How many miles to the right of the origin do we have to move before the height of the curve reaches 3 ft?

20. Compare the rates of growth of the functions $f(x) = x^5$ and $g(x) = 5^x$ by graphing both functions in several viewing rectangles. Find all points of intersection of the graphs correct to one decimal place.

21–22 ▮▮▮▮ Find the limit.

21. $\lim\limits_{t \to \infty} 2^{-t^2}$
22. $\lim\limits_{x \to 3^+} \log_{10}(x^2 - 5x + 6)$

23–38 ▮▮▮▮ Differentiate the function.

23. $h(t) = t^3 - 3^t$
24. $g(x) = x^4 4^x$
25. $y = 5^{-1/x}$
26. $y = 10^{\tan \theta}$
27. $f(u) = (2^u + 2^{-u})^{10}$
28. $y = 2^{3^{x^2}}$
29. $f(x) = \log_3(x^2 - 4)$
30. $f(x) = \log_{10}\left(\dfrac{x}{x-1}\right)$
31. $y = x^x$
32. $y = x^{1/x}$
33. $y = x^{\sin x}$
34. $y = (\sin x)^x$
35. $y = (\ln x)^x$
36. $y = x^{\ln x}$
37. $y = x^{e^x}$
38. $y = (\ln x)^{\cos x}$

39. Find an equation of the tangent line to the curve $y = 10^x$ at the point $(1, 10)$.

40. If $f(x) = x^{\cos x}$, find $f'(x)$. Check that your answer is reasonable by comparing the graphs of f and f'.

41–46 ▪ Evaluate the integral.

41. $\int_1^2 10^t \, dt$

42. $\int_0^1 4^{-2u} \, du$

43. $\int \dfrac{\log_{10} x}{x} \, dx$

44. $\int (x^5 + 5^x) \, dx$

45. $\int 3^{\sin\theta} \cos\theta \, d\theta$

46. $\int \dfrac{2^x}{2^x + 1} \, dx$

47. Find the area of the region bounded by the curves $y = 2^x$, $y = 5^x$, $x = -1$, and $x = 1$.

48. The region under the curve $y = 10^{-x}$ from $x = 0$ to $x = 1$ is rotated about the x-axis. Find the volume of the resulting solid.

49. Use a graph to find the root of the equation $2^x = 1 + 3^{-x}$ correct to one decimal place. Then use this estimate as the initial approximation in Newton's method to find the root correct to six decimal places.

50. Find y' if $x^y = y^x$.

51. Find the inverse function of $f(x) = 10^x/(10^x + 1)$.

52. Calculate $\lim_{x \to \infty} x^{-\ln x}$.

53. The geologist C. F. Richter defined the magnitude of an earthquake to be $\log_{10}(I/S)$, where I is the intensity of the quake (measured by the amplitude of a seismograph 100 km from the epicenter) and S is the intensity of a "standard" earthquake (where the amplitude is only 1 micron $= 10^{-4}$ cm). The 1989 Loma Prieta earthquake that shook San Francisco had a magnitude of 7.1 on the Richter scale. The 1906 San Francisco earthquake was 16 times as intense. What was its magnitude on the Richter scale?

54. A sound so faint that it can just be heard has intensity $I_0 = 10^{-12}$ watt/m² at a frequency of 1000 hertz (Hz). The loudness, in decibels (dB), of a sound with intensity I is then defined to be $L = 10 \log_{10}(I/I_0)$. Amplified rock music is measured at 120 dB, whereas the noise from a motor-driven lawn mower is measured at 106 dB. Find the ratio of the intensity of the rock music to that of the mower.

55. Referring to Exercise 54, find the rate of change of the loudness with respect to the intensity when the sound is measured at 50 dB (the level of ordinary conversation).

56. According to the Beer-Lambert Law, the light intensity at a depth of x meters below the surface of the ocean is $I(x) = I_0 a^x$, where I_0 is the light intensity at the surface and a is a constant such that $0 < a < 1$.
(a) Express the rate of change of $I(x)$ with respect to x in terms of $I(x)$.

(b) If $I_0 = 8$ and $a = 0.38$, find the rate of change of intensity with respect to depth at a depth of 20 m.
(c) Using the values from part (b), find the average light intensity between the surface and a depth of 20 m.

57. The flash unit on a camera operates by storing charge on a capacitor and releasing it suddenly when the flash is set off. The following data describe the charge remaining on the capacitor (measured in microcoulombs, μC) at time t (measured in seconds).

t	Q	t	Q
0.00	100.00	0.06	54.88
0.02	81.87	0.08	44.93
0.04	67.03	0.10	36.76

(a) Use a graphing calculator or computer to find an exponential model for the charge. (See Section 1.2.)
(b) The derivative $Q'(t)$ represents the electric current (measured in microamperes, μA) flowing from the capacitor to the flash bulb. Use part (a) to estimate the current when $t = 0.04$ s. Compare with the result of Example 2 in Section 2.1.

58. The table gives the U.S. population from 1790 to 1860.

Year	Population	Year	Population
1790	3,929,000	1830	12,861,000
1800	5,308,000	1840	17,063,000
1810	7,240,000	1850	23,192,000
1820	9,639,000	1860	31,443,000

(a) Use a graphing calculator or computer to fit an exponential function to the data. Graph the data points and the exponential model. How good is the fit?
(b) Estimate the rates of population growth in 1800 and 1850 by averaging slopes of secant lines.
(c) Use the exponential model in part (a) to estimate the rates of growth in 1800 and 1850. Compare these estimates with the ones in part (b).
(d) Use the exponential model to predict the population in 1870. Compare with the actual population of 38,558,000. Can you explain the discrepancy?

59. Prove the second law of exponents [see (3)].

60. Prove the fourth law of exponents [see (3)].

61. Deduce the following laws of logarithms from (3):
(a) $\log_a(xy) = \log_a x + \log_a y$
(b) $\log_a(x/y) = \log_a x - \log_a y$
(c) $\log_a(x^y) = y \log_a x$

62. Show that $\displaystyle\lim_{n \to \infty} \left(1 + \dfrac{x}{n}\right)^n = e^x$ for any $x > 0$.

7.5 Inverse Trigonometric Functions

In this section we apply the ideas of Section 7.1 to find the derivatives of the so-called inverse trigonometric functions. We have a slight difficulty in this task: Because the trigonometric functions are not one-to-one, they do not have inverse functions. The difficulty is overcome by restricting the domains of these functions so that they become one-to-one.

You can see from Figure 1 that the sine function $y = \sin x$ is not one-to-one (use the Horizontal Line Test). But the function $f(x) = \sin x$, $-\pi/2 \leq x \leq \pi/2$ (see Figure 2), is one-to-one. The inverse function of this restricted sine function f exists and is denoted by $\sin^{-1}$ or arcsin. It is called the **inverse sine function** or the **arcsine function**.

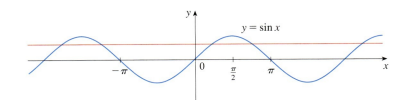

FIGURE 1

FIGURE 2 $y = \sin x$, $-\frac{\pi}{2} \leq x \leq \frac{\pi}{2}$

Since the definition of an inverse function says that

$$f^{-1}(x) = y \iff f(y) = x$$

we have

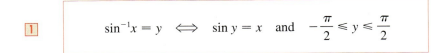

$$\boxed{1} \quad \sin^{-1} x = y \iff \sin y = x \quad \text{and} \quad -\frac{\pi}{2} \leq y \leq \frac{\pi}{2}$$

$\oslash \quad \sin^{-1} x \neq \dfrac{1}{\sin x}$

Thus, if $-1 \leq x \leq 1$, $\sin^{-1} x$ is the number between $-\pi/2$ and $\pi/2$ whose sine is x.

EXAMPLE 1 Evaluate (a) $\sin^{-1}(\frac{1}{2})$ and (b) $\tan(\arcsin \frac{1}{3})$.

SOLUTION
(a) We have

$$\sin^{-1}(\tfrac{1}{2}) = \frac{\pi}{6}$$

because $\sin(\pi/6) = \frac{1}{2}$ and $\pi/6$ lies between $-\pi/2$ and $\pi/2$.

(b) Let $\theta = \arcsin \frac{1}{3}$, so $\sin \theta = \frac{1}{3}$. Then we can draw a right triangle with angle θ as in Figure 3 and deduce from the Pythagorean Theorem that the third side has length $\sqrt{9 - 1} = 2\sqrt{2}$. This enables us to read from the triangle that

$$\tan(\arcsin \tfrac{1}{3}) = \tan \theta = \frac{1}{2\sqrt{2}}$$

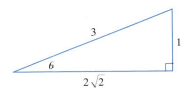

FIGURE 3

The cancellation equations for inverse functions become, in this case,

$$\boxed{\sin^{-1}(\sin x) = x \quad \text{for } -\frac{\pi}{2} \le x \le \frac{\pi}{2}}$$
$$\sin(\sin^{-1} x) = x \quad \text{for } -1 \le x \le 1$$

[2]

⊘ We must be careful when using the first cancellation equation because it is valid only when x lies in the interval $[-\pi/2, \pi/2]$. The following example shows how to proceed when x lies outside this interval.

EXAMPLE 2 Evaluate:
(a) $\sin(\sin^{-1} 0.6)$ (b) $\sin^{-1}\left(\sin \dfrac{\pi}{12}\right)$ (c) $\sin^{-1}\left(\sin \dfrac{2\pi}{3}\right)$

SOLUTION
(a) Since 0.6 lies between -1 and 1, the second cancellation equation in (2) gives

$$\sin(\sin^{-1} 0.6) = 0.6$$

(b) Since $\pi/12$ lies between $-\pi/2$ and $\pi/2$, the first cancellation equation gives

$$\sin^{-1}\left(\sin \frac{\pi}{12}\right) = \frac{\pi}{12}$$

(c) Since $2\pi/3$ does not lie in the interval $[-\pi/2, \pi/2]$, we can't use the cancellation equation. Instead we note that $\sin(2\pi/3) = \sqrt{3}/2$ and $\sin^{-1}(\sqrt{3}/2) = \pi/3$ because $\pi/3$ lies between $-\pi/2$ and $\pi/2$. Therefore

$$\sin^{-1}\left(\sin \frac{2\pi}{3}\right) = \sin^{-1}\left(\frac{\sqrt{3}}{2}\right) = \frac{\pi}{3}$$

The inverse sine function, $\sin^{-1}$, has domain $[-1, 1]$ and range $[-\pi/2, \pi/2]$, and its graph, shown in Figure 4, is obtained from that of the restricted sine function (Figure 2) by reflection about the line $y = x$.

We know that the sine function f is continuous, so the inverse sine function is also continuous. We also know from Section 3.5 that the sine function is differentiable, so the inverse sine function is also differentiable. We could calculate the derivative of $\sin^{-1}$ by the formula in Theorem 7.1.7, but since we know that $\sin^{-1}$ is differentiable, we can just as easily calculate it by implicit differentiation as follows.

Let $y = \sin^{-1} x$. Then $\sin y = x$ and $-\pi/2 \le y \le \pi/2$. Differentiating $\sin y = x$ implicitly with respect to x, we obtain

$$\cos y \frac{dy}{dx} = 1$$

and

$$\frac{dy}{dx} = \frac{1}{\cos y}$$

Now $\cos y \ge 0$ since $-\pi/2 \le y \le \pi/2$, so

$$\cos y = \sqrt{1 - \sin^2 y} = \sqrt{1 - x^2}$$

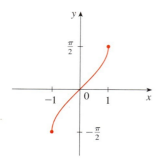

FIGURE 4 $y = \sin^{-1} x = \arcsin x$

Therefore
$$\frac{dy}{dx} = \frac{1}{\cos y} = \frac{1}{\sqrt{1-x^2}}$$

$$\boxed{3} \qquad \frac{d}{dx}(\sin^{-1}x) = \frac{1}{\sqrt{1-x^2}} \qquad -1 < x < 1$$

EXAMPLE 3 If $f(x) = \sin^{-1}(x^2 - 1)$, find (a) the domain of f, (b) $f'(x)$, and (c) the domain of f'.

SOLUTION
(a) Since the domain of the inverse sine function is $[-1, 1]$, the domain of f is
$$\{x \mid -1 \leq x^2 - 1 \leq 1\} = \{x \mid 0 \leq x^2 \leq 2\}$$
$$= \{x \mid |x| \leq \sqrt{2}\} = [-\sqrt{2}, \sqrt{2}]$$

(b) Combining Formula 3 with the Chain Rule, we have
$$f'(x) = \frac{1}{\sqrt{1-(x^2-1)^2}} \frac{d}{dx}(x^2 - 1)$$
$$= \frac{1}{\sqrt{1-(x^4-2x^2+1)}} 2x = \frac{2x}{\sqrt{2x^2-x^4}}$$

(c) The domain of f' is
$$\{x \mid -1 < x^2 - 1 < 1\} = \{x \mid 0 < x^2 < 2\}$$
$$= \{x \mid 0 < |x| < \sqrt{2}\} = (-\sqrt{2}, 0) \cup (0, \sqrt{2})$$

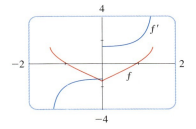

FIGURE 5

|||| The graphs of the function f of Example 3 and its derivative are shown in Figure 5. Notice that f is not differentiable at 0 and this is consistent with the fact that the graph of f' makes a sudden jump at $x = 0$.

The **inverse cosine function** is handled similarly. The restricted cosine function $f(x) = \cos x$, $0 \leq x \leq \pi$, is one-to-one (see Figure 6) and so it has an inverse function denoted by $\cos^{-1}$ or arccos.

$$\boxed{4} \qquad \cos^{-1}x = y \iff \cos y = x \text{ and } 0 \leq y \leq \pi$$

The cancellation equations are

$$\boxed{5} \qquad \cos^{-1}(\cos x) = x \quad \text{for } 0 \leq x \leq \pi$$
$$\cos(\cos^{-1}x) = x \quad \text{for } -1 \leq x \leq 1$$

The inverse cosine function, $\cos^{-1}$, has domain $[-1, 1]$ and range $[0, \pi]$ and is a continuous function whose graph is shown in Figure 7. Its derivative is given by

$$\boxed{6} \qquad \frac{d}{dx}(\cos^{-1}x) = -\frac{1}{\sqrt{1-x^2}} \qquad -1 < x < 1$$

Formula 6 can be proved by the same method as for Formula 3 and is left as Exercise 17.

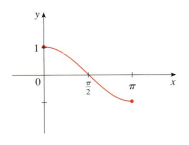

FIGURE 6
$y = \cos x$, $0 \leq x \leq \pi$

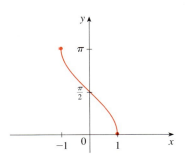

FIGURE 7
$y = \cos^{-1}x = \arccos x$

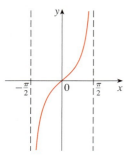

FIGURE 8
$y = \tan x$, $-\frac{\pi}{2} < x < \frac{\pi}{2}$

The tangent function can be made one-to-one by restricting it to the interval $(-\pi/2, \pi/2)$. Thus, the **inverse tangent function** is defined as the inverse of the function $f(x) = \tan x$, $-\pi/2 < x < \pi/2$. (See Figure 8.) It is denoted by $\tan^{-1}$ or arctan.

$$\boxed{7 \qquad \tan^{-1} x = y \iff \tan y = x \quad \text{and} \quad -\frac{\pi}{2} < y < \frac{\pi}{2}}$$

EXAMPLE 4 Simplify the expression $\cos(\tan^{-1} x)$.

SOLUTION 1 Let $y = \tan^{-1} x$. Then $\tan y = x$ and $-\pi/2 < y < \pi/2$. We want to find $\cos y$ but, since $\tan y$ is known, it is easier to find $\sec y$ first:

$$\sec^2 y = 1 + \tan^2 y = 1 + x^2$$

$$\sec y = \sqrt{1 + x^2} \qquad \text{(since } \sec y > 0 \text{ for } -\pi/2 < y < \pi/2\text{)}$$

Thus
$$\cos(\tan^{-1} x) = \cos y = \frac{1}{\sec y} = \frac{1}{\sqrt{1 + x^2}}$$

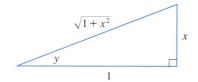

FIGURE 9

SOLUTION 2 Instead of using trigonometric identities as in Solution 1, it is perhaps easier to use a diagram. If $y = \tan^{-1} x$, then $\tan y = x$, and we can read from Figure 9 (which illustrates the case $y > 0$) that

$$\cos(\tan^{-1} x) = \cos y = \frac{1}{\sqrt{1 + x^2}}$$

The inverse tangent function, $\tan^{-1} =$ arctan, has domain $\mathbb{R}$ and range $(-\pi/2, \pi/2)$. Its graph is shown in Figure 10.

We know that
$$\lim_{x \to (\pi/2)^-} \tan x = \infty \qquad \text{and} \qquad \lim_{x \to -(\pi/2)^+} \tan x = -\infty$$

and so the lines $x = \pm\pi/2$ are vertical asymptotes of the graph of tan. Since the graph of $\tan^{-1}$ is obtained by reflecting the graph of the restricted tangent function about the line $y = x$, it follows that the lines $y = \pi/2$ and $y = -\pi/2$ are horizontal asymptotes of the graph of $\tan^{-1}$. This fact is expressed by the following limits:

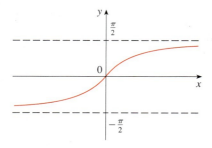

FIGURE 10
$y = \tan^{-1} x = \arctan x$

$$\boxed{8 \qquad \lim_{x \to \infty} \tan^{-1} x = \frac{\pi}{2} \qquad \lim_{x \to -\infty} \tan^{-1} x = -\frac{\pi}{2}}$$

EXAMPLE 5 Evaluate $\displaystyle\lim_{x \to 2^+} \arctan\left(\frac{1}{x - 2}\right)$.

SOLUTION Since
$$\frac{1}{x - 2} \to \infty \qquad \text{as } x \to 2^+$$

the first equation in (8) gives
$$\lim_{x \to 2^+} \arctan\left(\frac{1}{x - 2}\right) = \frac{\pi}{2}$$

Since tan is differentiable, $\tan^{-1}$ is also differentiable. To find its derivative, let $y = \tan^{-1} x$. Then $\tan y = x$. Differentiating this latter equation implicitly with respect to x, we have

$$\sec^2 y \frac{dy}{dx} = 1$$

and so

$$\frac{dy}{dx} = \frac{1}{\sec^2 y} = \frac{1}{1 + \tan^2 y} = \frac{1}{1 + x^2}$$

9
$$\frac{d}{dx}(\tan^{-1} x) = \frac{1}{1 + x^2}$$

The remaining inverse trigonometric functions are not used as frequently and are summarized here.

10
$$y = \csc^{-1} x \; (|x| \geq 1) \iff \csc y = x \text{ and } y \in (0, \pi/2] \cup (\pi, 3\pi/2]$$
$$y = \sec^{-1} x \; (|x| \geq 1) \iff \sec y = x \text{ and } y \in [0, \pi/2) \cup [\pi, 3\pi/2)$$
$$y = \cot^{-1} x \; (x \in \mathbb{R}) \iff \cot y = x \text{ and } y \in (0, \pi)$$

The choice of intervals for y in the definitions of $\csc^{-1}$ and $\sec^{-1}$ is not universally agreed upon. For instance, some authors use $y \in [0, \pi/2) \cup (\pi/2, \pi]$ in the definition of $\sec^{-1}$. [You can see from the graph of the secant function in Figure 11 that both this choice and the one in (10) will work.] The reason for the choice in (10) is that the differentiation formulas are simpler (see Exercise 79).

We collect in Table 11 the differentiation formulas for all of the inverse trigonometric functions. The proofs of the formulas for the derivatives of $\csc^{-1}$, $\sec^{-1}$, and $\cot^{-1}$ are left as Exercises 19–21.

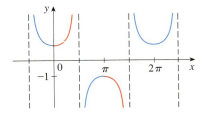

FIGURE 11
$y = \sec x$

11 Table of Derivatives of Inverse Trigonometric Functions

$$\frac{d}{dx}(\sin^{-1} x) = \frac{1}{\sqrt{1 - x^2}} \qquad \frac{d}{dx}(\csc^{-1} x) = -\frac{1}{x\sqrt{x^2 - 1}}$$

$$\frac{d}{dx}(\cos^{-1} x) = -\frac{1}{\sqrt{1 - x^2}} \qquad \frac{d}{dx}(\sec^{-1} x) = \frac{1}{x\sqrt{x^2 - 1}}$$

$$\frac{d}{dx}(\tan^{-1} x) = \frac{1}{1 + x^2} \qquad \frac{d}{dx}(\cot^{-1} x) = -\frac{1}{1 + x^2}$$

Each of these formulas can be combined with the Chain Rule. For instance, if u is a differentiable function of x, then

$$\frac{d}{dx}(\sin^{-1} u) = \frac{1}{\sqrt{1 - u^2}} \frac{du}{dx} \quad \text{and} \quad \frac{d}{dx}(\tan^{-1} u) = \frac{1}{1 + u^2} \frac{du}{dx}$$

EXAMPLE 6 Differentiate (a) $y = \dfrac{1}{\sin^{-1}x}$ and (b) $f(x) = x\tan^{-1}\sqrt{x}$.

SOLUTION

(a) $$\frac{dy}{dx} = \frac{d}{dx}(\sin^{-1}x)^{-1} = -(\sin^{-1}x)^{-2}\frac{d}{dx}(\sin^{-1}x)$$

$$= -\frac{1}{(\sin^{-1}x)^2\sqrt{1-x^2}}$$

(b) $$f'(x) = \tan^{-1}\sqrt{x} + x\,\frac{1}{1+(\sqrt{x})^2}\,\frac{1}{2}x^{-1/2}$$

$$= \tan^{-1}\sqrt{x} + \frac{\sqrt{x}}{2(1+x)}$$

EXAMPLE 7 Prove the identity $\tan^{-1}x + \cot^{-1}x = \pi/2$.

SOLUTION Although calculus is not needed to prove this identity, the proof using calculus is quite simple. If $f(x) = \tan^{-1}x + \cot^{-1}x$, then

$$f'(x) = \frac{1}{1+x^2} - \frac{1}{1+x^2} = 0$$

for all values of x. Therefore, $f(x) = C$, a constant. To determine the value of C, we put $x = 1$. Then

$$C = f(1) = \tan^{-1}1 + \cot^{-1}1 = \frac{\pi}{4} + \frac{\pi}{4} = \frac{\pi}{2}$$

Thus, $\tan^{-1}x + \cot^{-1}x = \pi/2$.

Each of the formulas in Table 11 gives rise to an integration formula. The two most useful of these are the following:

[12]
$$\int \frac{1}{\sqrt{1-x^2}}\,dx = \sin^{-1}x + C$$

[13]
$$\int \frac{1}{x^2+1}\,dx = \tan^{-1}x + C$$

EXAMPLE 8 Find $\displaystyle\int_0^{1/4} \frac{1}{\sqrt{1-4x^2}}\,dx$.

SOLUTION If we write

$$\int_0^{1/4} \frac{1}{\sqrt{1-4x^2}}\,dx = \int_0^{1/4} \frac{1}{\sqrt{1-(2x)^2}}\,dx$$

then the integral resembles Equation 12 and the substitution $u = 2x$ is suggested. This gives $du = 2\,dx$, so $dx = du/2$. When $x = 0$, $u = 0$; when $x = \frac{1}{4}$, $u = \frac{1}{2}$. So

$$\int_0^{1/4} \frac{1}{\sqrt{1-4x^2}}\,dx = \tfrac{1}{2}\int_0^{1/2} \frac{du}{\sqrt{1-u^2}} = \tfrac{1}{2}\sin^{-1}u\Big]_0^{1/2}$$

$$= \tfrac{1}{2}\left[\sin^{-1}\left(\tfrac{1}{2}\right) - \sin^{-1} 0\right] = \frac{1}{2}\cdot\frac{\pi}{6} = \frac{\pi}{12}$$

EXAMPLE 9 Evaluate $\int \dfrac{1}{x^2+a^2}\,dx$.

SOLUTION To make the given integral more like Equation 13 we write

$$\int \frac{dx}{x^2+a^2} = \int \frac{dx}{a^2\left(\dfrac{x^2}{a^2}+1\right)} = \frac{1}{a^2}\int \frac{dx}{\left(\dfrac{x}{a}\right)^2+1}$$

This suggests that we substitute $u = x/a$. Then $du = dx/a$, $dx = a\,du$, and

$$\int \frac{dx}{x^2+a^2} = \frac{1}{a^2}\int \frac{a\,du}{u^2+1} = \frac{1}{a}\int \frac{du}{u^2+1} = \frac{1}{a}\tan^{-1}u + C$$

Thus, we have the formula

|||| One of the main uses of inverse trigonometric functions is that they often arise when we integrate rational functions.

$\boxed{14}$

$$\int \frac{1}{x^2+a^2}\,dx = \frac{1}{a}\tan^{-1}\left(\frac{x}{a}\right) + C$$

EXAMPLE 10 Find $\int \dfrac{x}{x^4+9}\,dx$.

SOLUTION We substitute $u = x^2$ because then $du = 2x\,dx$ and we can use Equation 14 with $a = 3$:

$$\int \frac{x}{x^4+9}\,dx = \frac{1}{2}\int \frac{du}{u^2+9} = \frac{1}{2}\cdot\frac{1}{3}\tan^{-1}\left(\frac{u}{3}\right) + C$$

$$= \frac{1}{6}\tan^{-1}\left(\frac{x^2}{3}\right) + C$$

|||| 7.5 Exercises

1–10 |||| Find the exact value of each expression.

1. (a) $\sin^{-1}(\sqrt{3}/2)$ $\pi/3$ (b) $\cos^{-1}(-1)$
2. (a) $\arctan(-1)$ (b) $\csc^{-1} 2$
3. (a) $\tan^{-1}\sqrt{3}$ (b) $\arcsin(-1/\sqrt{2})$
4. (a) $\sec^{-1}\sqrt{2}$ (b) $\arcsin 1$
5. (a) $\arccos(\cos 2\pi)$ (b) $\tan(\tan^{-1} 5)$
6. (a) $\tan^{-1}(\tan 3\pi/4)$ (b) $\cos(\arcsin \tfrac{1}{2})$
7. $\tan(\sin^{-1}(\tfrac{2}{3}))$
8. $\csc(\arccos \tfrac{3}{5})$
9. $\sin(2\tan^{-1}\sqrt{2})$
10. $\cos(\tan^{-1} 2 + \tan^{-1} 3)$
11. Prove that $\cos(\sin^{-1}x) = \sqrt{1-x^2}$. $f(f^{-1}x) = x$

12–14 |||| Simplify the expression.

12. $\tan(\sin^{-1}x)$
13. $\sin(\tan^{-1}x)$
14. $\csc(\arctan 2x)$

15–16 Graph the given functions on the same screen. How are these graphs related?

15. $y = \sin x$, $-\pi/2 \le x \le \pi/2$; $y = \sin^{-1} x$; $y = x$

16. $y = \tan x$, $-\pi/2 < x < \pi/2$; $y = \tan^{-1} x$; $y = x$

17. Prove Formula 6 for the derivative of $\cos^{-1}$ by the same method as for Formula 3.

18. (a) Prove that $\sin^{-1} x + \cos^{-1} x = \pi/2$.
(b) Use part (a) to prove Formula 6.

19. Prove that $\dfrac{d}{dx}(\cot^{-1} x) = -\dfrac{1}{1 + x^2}$.

20. Prove that $\dfrac{d}{dx}(\sec^{-1} x) = \dfrac{1}{x\sqrt{x^2 - 1}}$.

21. Prove that $\dfrac{d}{dx}(\csc^{-1} x) = -\dfrac{1}{x\sqrt{x^2 - 1}}$.

22–35 Find the derivative of the function. Simplify where possible.

22. $y = \sqrt{\tan^{-1} x}$

23. $y = \tan^{-1}\sqrt{x}$

24. $h(x) = \sqrt{1 - x^2}\, \arcsin x$

25. $y = \sin^{-1}(2x + 1)$

26. $f(x) = x \ln(\arctan x)$

27. $H(x) = (1 + x^2) \arctan x$

28. $h(t) = e^{\sec^{-1} t}$

29. $y = \cos^{-1}(e^{2x})$

30. $y = x \cos^{-1} x - \sqrt{1 - x^2}$

31. $y = \arctan(\cos \theta)$

32. $y = \tan^{-1}(x - \sqrt{1 + x^2})$

33. $h(t) = \cot^{-1}(t) + \cot^{-1}(1/t)$

34. $y = \tan^{-1}\left(\dfrac{x}{a}\right) + \ln\sqrt{\dfrac{x - a}{x + a}}$

35. $y = \arccos\left(\dfrac{b + a \cos x}{a + b \cos x}\right)$, $0 \le x \le \pi$, $a > b > 0$

36–37 Find the derivative of the function. Find the domains of the function and its derivative.

36. $f(x) = \arcsin(e^x)$

37. $g(x) = \cos^{-1}(3 - 2x)$

38. Find y' if $\tan^{-1}(xy) = 1 + x^2 y$.

39. If $g(x) = x \sin^{-1}(x/4) + \sqrt{16 - x^2}$, find $g'(2)$.

40. Find an equation of the tangent line to the curve $y = 3 \arccos(x/2)$ at the point $(1, \pi)$.

41–42 Find $f'(x)$. Check that your answer is reasonable by comparing the graphs of f and f'.

41. $f(x) = e^{-x} \arctan x$

42. $f(x) = x \arcsin(1 - x^2)$

43–46 Find the limit.

43. $\lim\limits_{x \to -1^+} \sin^{-1} x$

44. $\lim\limits_{x \to \infty} \arccos\left(\dfrac{1 + x^2}{1 + 2x^2}\right)$

45. $\lim\limits_{x \to \infty} \arctan(e^x)$

46. $\lim\limits_{x \to 0^+} \tan^{-1}(\ln x)$

47. Where should the point P be chosen on the line segment AB so as to maximize the angle θ?

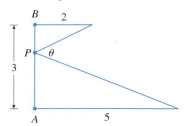

48. A painting in an art gallery has height h and is hung so that its lower edge is a distance d above the eye of an observer (as in the figure). How far from the wall should the observer stand to get the best view? (In other words, where should the observer stand so as to maximize the angle θ subtended at his eye by the painting?)

49. A ladder 10 ft long leans against a vertical wall. If the bottom of the ladder slides away from the base of the wall at a speed of 2 ft/s, how fast is the angle between the ladder and the wall changing when the bottom of the ladder is 6 ft from the base of the wall?

50. A lighthouse is located on a small island, 3 km away from the nearest point P on a straight shoreline, and its light makes four revolutions per minute. How fast is the beam of light moving along the shoreline when it is 1 km from P?

51–54 Sketch the curve using the guidelines of Section 4.5.

51. $y = \sin^{-1}\left(\dfrac{x}{x + 1}\right)$

52. $y = \tan^{-1}\left(\dfrac{x - 1}{x + 1}\right)$

53. $y = x - \tan^{-1} x$

54. $y = \tan^{-1}(\ln x)$

55. If $f(x) = \arctan(\cos(3 \arcsin x))$, use the graphs of f, f', and f'' to estimate the x-coordinates of the maximum and minimum points and inflection points of f.

56. Investigate the family of curves given by $f(x) = x - c \sin^{-1} x$. What happens to the number of maxima and minima as c changes? Graph several members of the family to illustrate what you discover.

57. Find the most general antiderivative of the function
$$f(x) = 2x + 5(1 - x^2)^{-1/2}$$

58. Find $f(x)$ if $f'(x) = 4 - 3(1 + x^2)^{-1}$ and $f(\pi/4) = 0$.

59–70 ▓ Evaluate the integral.

59. $\int_{1/2}^{\sqrt{3}/2} \dfrac{6}{\sqrt{1 - t^2}}\, dt$

60. $\int_0^1 \dfrac{4}{t^2 + 1}\, dt$

61. $\int_0^{\sqrt{3}/4} \dfrac{dx}{1 + 16x^2}$

62. $\int \dfrac{dt}{\sqrt{1 - 4t^2}}$

63. $\int_0^{1/2} \dfrac{\sin^{-1}x}{\sqrt{1 - x^2}}\, dx$

64. $\int_0^{\pi/2} \dfrac{\sin x}{1 + \cos^2 x}\, dx$

65. $\int \dfrac{x + 9}{x^2 + 9}\, dx$

66. $\int \dfrac{\tan^{-1}x}{1 + x^2}\, dx$

67. $\int \dfrac{t^2}{\sqrt{1 - t^6}}\, dt$

68. $\int \dfrac{1}{x\sqrt{x^2 - 4}}\, dx$

69. $\int \dfrac{dx}{\sqrt{x}(1 + x)}$

70. $\int \dfrac{e^{2x}}{\sqrt{1 - e^{4x}}}\, dx$

71. Use the method of Example 9 to show that, if $a > 0$,
$$\int \dfrac{1}{\sqrt{a^2 - x^2}}\, dx = \sin^{-1}\left(\dfrac{x}{a}\right) + C$$

72. The region under the curve $y = 1/\sqrt{x^2 + 4}$ from $x = 0$ to $x = 2$ is rotated about the x-axis. Find the volume of the resulting solid.

73. Evaluate $\int_0^1 \sin^{-1}x\, dx$ by interpreting it as an area and integrating with respect to y instead of x.

74. Prove that, for $xy \ne 1$,
$$\arctan x + \arctan y = \arctan \dfrac{x + y}{1 - xy}$$
if the left side lies between $-\pi/2$ and $\pi/2$.

75. Use the result of Exercise 74 to prove the following:
(a) $\arctan \tfrac{1}{2} + \arctan \tfrac{1}{3} = \pi/4$
(b) $2\arctan \tfrac{1}{3} + \arctan \tfrac{1}{7} = \pi/4$

76. (a) Sketch the graph of the function $f(x) = \sin(\sin^{-1}x)$.
(b) Sketch the graph of the function $g(x) = \sin^{-1}(\sin x)$, $x \in \mathbb{R}$.
(c) Show that $g'(x) = \dfrac{\cos x}{|\cos x|}$.
(d) Sketch the graph of $h(x) = \cos^{-1}(\sin x)$, $x \in \mathbb{R}$, and find its derivative.

77. Use the method of Example 7 to prove the identity
$$2\sin^{-1}x = \cos^{-1}(1 - 2x^2) \qquad x \ge 0$$

78. Prove the identity
$$\arcsin \dfrac{x - 1}{x + 1} = 2\arctan\sqrt{x} - \dfrac{\pi}{2}$$

79. Some authors define $y = \sec^{-1}x \iff \sec y = x$ and $y \in [0, \pi/2) \cup (\pi/2, \pi]$. Show that with this definition, we have (instead of the formula given in Exercise 20)
$$\dfrac{d}{dx}(\sec^{-1}x) = \dfrac{1}{|x|\sqrt{x^2 - 1}} \qquad |x| > 1$$

80. Let $f(x) = x\arctan(1/x)$ if $x \ne 0$ and $f(0) = 0$.
(a) Is f continuous at 0?
(b) Is f differentiable at 0?

APPLIED PROJECT

CAS Where to Sit at the Movies

A movie theater has a screen that is positioned 10 ft off the floor and is 25 ft high. The first row of seats is placed 9 ft from the screen and the rows are set 3 ft apart. The floor of the seating area is inclined at an angle of $\alpha = 20°$ above the horizontal and the distance up the incline that you sit is x. The theater has 21 rows of seats, so $0 \le x \le 60$. Suppose you decide that the best place to sit is in the row where the angle θ subtended by the screen at your eyes is a maximum. Let's also suppose that your eyes are 4 ft above the floor, as shown in the figure. (In Exercise 48 in Section 7.5 we looked at a simpler version of this problem, where the floor is horizontal, but this project involves a more complicated situation and requires technology.)

1. Show that
$$\theta = \arccos\left(\dfrac{a^2 + b^2 - 625}{2ab}\right)$$
where
$$a^2 = (9 + x\cos\alpha)^2 + (31 - x\sin\alpha)^2$$
and
$$b^2 = (9 + x\cos\alpha)^2 + (x\sin\alpha - 6)^2$$

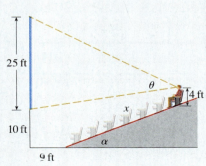

2. Use a graph of θ as a function of x to estimate the value of x that maximizes θ. In which row should you sit? What is the viewing angle θ in this row?

3. Use your computer algebra system to differentiate θ and find a numerical value for the root of the equation $d\theta/dx = 0$. Does this value confirm your result in Problem 2?

4. Use the graph of θ to estimate the average value of θ on the interval $0 \leq x \leq 60$. Then use your CAS to compute the average value. Compare with the maximum and minimum values of θ.

7.6 Hyperbolic Functions

Certain combinations of the exponential functions e^x and e^{-x} arise so frequently in mathematics and its applications that they deserve to be given special names. In many ways they are analogous to the trigonometric functions, and they have the same relationship to the hyperbola that the trigonometric functions have to the circle. For this reason they are collectively called **hyperbolic functions** and individually called **hyperbolic sine, hyperbolic cosine**, and so on.

Definition of the Hyperbolic Functions

$$\sinh x = \frac{e^x - e^{-x}}{2} \qquad \operatorname{csch} x = \frac{1}{\sinh x}$$

$$\cosh x = \frac{e^x + e^{-x}}{2} \qquad \operatorname{sech} x = \frac{1}{\cosh x}$$

$$\tanh x = \frac{\sinh x}{\cosh x} \qquad \coth x = \frac{\cosh x}{\sinh x}$$

The graphs of hyperbolic sine and cosine can be sketched using graphical addition as in Figures 1 and 2.

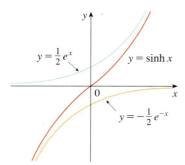

FIGURE 1
$y = \sinh x = \frac{1}{2}e^x - \frac{1}{2}e^{-x}$

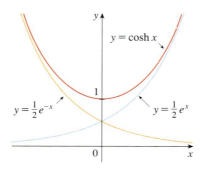

FIGURE 2
$y = \cosh x = \frac{1}{2}e^x + \frac{1}{2}e^{-x}$

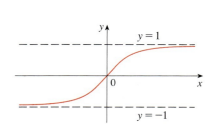

FIGURE 3
$y = \tanh x$

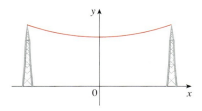

FIGURE 4
A catenary $y = c + a\cosh(x/a)$

Note that sinh has domain $\mathbb{R}$ and range $\mathbb{R}$, while cosh has domain $\mathbb{R}$ and range $[1, \infty)$. The graph of tanh is shown in Figure 3. It has the horizontal asymptotes $y = \pm 1$. (See Exercise 23.)

Some of the mathematical uses of hyperbolic functions will be seen in Chapter 8. Applications to science and engineering occur whenever an entity such as light, velocity, electricity, or radioactivity is gradually absorbed or extinguished, for the decay can be represented by hyperbolic functions. The most famous application is the use of hyperbolic cosine to describe the shape of a hanging wire. It can be proved that if a heavy flexible cable (such as a telephone or power line) is suspended between two points at the same height, then it takes the shape of a curve with equation $y = c + a\cosh(x/a)$ called a *catenary* (see Figure 4). (The Latin word *catena* means "chain.")

The hyperbolic functions satisfy a number of identities that are similar to well-known trigonometric identities. We list some of them here and leave most of the proofs to the exercises.

Hyperbolic Identities

$$\sinh(-x) = -\sinh x \qquad \cosh(-x) = \cosh x$$

$$\cosh^2 x - \sinh^2 x = 1 \qquad 1 - \tanh^2 x = \text{sech}^2 x$$

$$\sinh(x + y) = \sinh x \cosh y + \cosh x \sinh y$$

$$\cosh(x + y) = \cosh x \cosh y + \sinh x \sinh y$$

EXAMPLE 1 Prove (a) $\cosh^2 x - \sinh^2 x = 1$ and (b) $1 - \tanh^2 x = \text{sech}^2 x$.

SOLUTION

(a)
$$\cosh^2 x - \sinh^2 x = \left(\frac{e^x + e^{-x}}{2}\right)^2 - \left(\frac{e^x - e^{-x}}{2}\right)^2$$

$$= \frac{e^{2x} + 2 + e^{-2x}}{4} - \frac{e^{2x} - 2 + e^{-2x}}{4}$$

$$= \tfrac{4}{4} = 1$$

(b) We start with the identity proved in part (a):

$$\cosh^2 x - \sinh^2 x = 1$$

If we divide both sides by $\cosh^2 x$, we get

$$1 - \frac{\sinh^2 x}{\cosh^2 x} = \frac{1}{\cosh^2 x}$$

or
$$1 - \tanh^2 x = \text{sech}^2 x$$

The identity proved in Example 1(a) gives a clue to the reason for the name "hyperbolic" functions:

If t is any real number, then the point $P(\cos t, \sin t)$ lies on the unit circle $x^2 + y^2 = 1$ because $\cos^2 t + \sin^2 t = 1$. In fact, t can be interpreted as the radian measure of $\angle POQ$

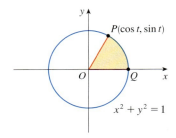

FIGURE 5

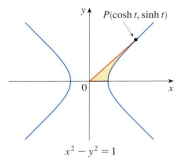

FIGURE 6

in Figure 5. For this reason the trigonometric functions are sometimes called *circular* functions.

Likewise, if t is any real number, then the point $P(\cosh t, \sinh t)$ lies on the right branch of the hyperbola $x^2 - y^2 = 1$ because $\cosh^2 t - \sinh^2 t = 1$ and $\cosh t \geq 1$. This time, t does not represent the measure of an angle. However, it turns out that t represents twice the area of the shaded hyperbolic sector in Figure 6, just as in the trigonometric case t represents twice the area of the shaded circular sector in Figure 5.

The derivatives of the hyperbolic functions are easily computed. For example,

$$\frac{d}{dx}(\sinh x) = \frac{d}{dx}\left(\frac{e^x - e^{-x}}{2}\right) = \frac{e^x + e^{-x}}{2} = \cosh x$$

We list the differentiation formulas for the hyperbolic functions as Table 1. The remaining proofs are left as exercises. Note the analogy with the differentiation formulas for trigonometric functions, but beware that the signs are different in some cases.

---

1 Derivatives of Hyperbolic Functions

$$\frac{d}{dx}(\sinh x) = \cosh x \qquad \frac{d}{dx}(\operatorname{csch} x) = -\operatorname{csch} x \coth x$$

$$\frac{d}{dx}(\cosh x) = \sinh x \qquad \frac{d}{dx}(\operatorname{sech} x) = -\operatorname{sech} x \tanh x$$

$$\frac{d}{dx}(\tanh x) = \operatorname{sech}^2 x \qquad \frac{d}{dx}(\coth x) = -\operatorname{csch}^2 x$$

EXAMPLE 2 Any of the differentiation rules in Table 1 can be combined with the Chain Rule. For instance,

$$\frac{d}{dx}(\cosh \sqrt{x}) = \sinh \sqrt{x} \cdot \frac{d}{dx}\sqrt{x} = \frac{\sinh \sqrt{x}}{2\sqrt{x}}$$

Inverse Hyperbolic Functions

You can see from Figures 1 and 3 that sinh and tanh are one-to-one functions and so they have inverse functions denoted by $\sinh^{-1}$ and $\tanh^{-1}$. Figure 2 shows that cosh is not one-to-one, but when restricted to the domain $[0, \infty)$ it becomes one-to-one. The inverse hyperbolic cosine function is defined as the inverse of this restricted function.

2
$$y = \sinh^{-1} x \iff \sinh y = x$$
$$y = \cosh^{-1} x \iff \cosh y = x \text{ and } y \geq 0$$
$$y = \tanh^{-1} x \iff \tanh y = x$$

The remaining inverse hyperbolic functions are defined similarly (see Exercise 28).

We can sketch the graphs of $\sinh^{-1}$, $\cosh^{-1}$, and $\tanh^{-1}$ in Figures 7, 8, and 9 by using Figures 1, 2, and 3.

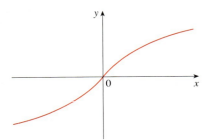

FIGURE 7 $y = \sinh^{-1} x$
domain $= \mathbb{R}$ range $= \mathbb{R}$

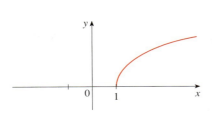

FIGURE 8 $y = \cosh^{-1} x$
domain $= [1, \infty)$ range $= [0, \infty)$

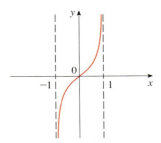

FIGURE 9 $y = \tanh^{-1} x$
domain $= (-1, 1)$ range $= \mathbb{R}$

Since the hyperbolic functions are defined in terms of exponential functions, it's not surprising to learn that the inverse hyperbolic functions can be expressed in terms of logarithms. In particular, we have:

3	$\sinh^{-1} x = \ln\left(x + \sqrt{x^2 + 1}\right)$	$x \in \mathbb{R}$
4	$\cosh^{-1} x = \ln\left(x + \sqrt{x^2 - 1}\right)$	$x \geq 1$
5	$\tanh^{-1} x = \tfrac{1}{2} \ln\left(\dfrac{1 + x}{1 - x}\right)$	$-1 < x < 1$

|||| Formula 3 is proved in Example 3. The proofs of Formulas 4 and 5 are requested in Exercises 26 and 27.

EXAMPLE 3 Show that $\sinh^{-1} x = \ln\left(x + \sqrt{x^2 + 1}\right)$.

SOLUTION Let $y = \sinh^{-1} x$. Then

$$x = \sinh y = \frac{e^y - e^{-y}}{2}$$

so

$$e^y - 2x - e^{-y} = 0$$

or, multiplying by e^y,

$$e^{2y} - 2xe^y - 1 = 0$$

This is really a quadratic equation in e^y:

$$(e^y)^2 - 2x(e^y) - 1 = 0$$

Solving by the quadratic formula, we get

$$e^y = \frac{2x \pm \sqrt{4x^2 + 4}}{2} = x \pm \sqrt{x^2 + 1}$$

Note that $e^y > 0$, but $x - \sqrt{x^2 + 1} < 0$ (because $x < \sqrt{x^2 + 1}$). Thus, the minus sign is inadmissible and we have

$$e^y = x + \sqrt{x^2 + 1}$$

Therefore $\quad y = \ln(e^y) = \ln(x + \sqrt{x^2 + 1})$

(See Exercise 25 for another method.)

6 Derivatives of Inverse Hyperbolic Functions

$$\frac{d}{dx}(\sinh^{-1}x) = \frac{1}{\sqrt{1+x^2}} \qquad \frac{d}{dx}(\operatorname{csch}^{-1}x) = -\frac{1}{|x|\sqrt{x^2+1}}$$

$$\frac{d}{dx}(\cosh^{-1}x) = \frac{1}{\sqrt{x^2-1}} \qquad \frac{d}{dx}(\operatorname{sech}^{-1}x) = -\frac{1}{x\sqrt{1-x^2}}$$

$$\frac{d}{dx}(\tanh^{-1}x) = \frac{1}{1-x^2} \qquad \frac{d}{dx}(\coth^{-1}x) = \frac{1}{1-x^2}$$

|||| Notice that the formulas for the derivatives of $\tanh^{-1}x$ and $\coth^{-1}x$ appear to be identical. But the domains of these functions have no numbers in common: $\tanh^{-1}x$ is defined for $|x| < 1$, whereas $\coth^{-1}x$ is defined for $|x| > 1$.

The inverse hyperbolic functions are all differentiable because the hyperbolic functions are differentiable. The formulas in Table 6 can be proved either by the method for inverse functions or by differentiating Formulas 3, 4, and 5.

EXAMPLE 4 Prove that $\dfrac{d}{dx}(\sinh^{-1}x) = \dfrac{1}{\sqrt{1+x^2}}$.

SOLUTION 1 Let $y = \sinh^{-1}x$. Then $\sinh y = x$. If we differentiate this equation implicitly with respect to x, we get

$$\cosh y \, \frac{dy}{dx} = 1$$

Since $\cosh^2 y - \sinh^2 y = 1$ and $\cosh y \geq 0$, we have $\cosh y = \sqrt{1 + \sinh^2 y}$, so

$$\frac{dy}{dx} = \frac{1}{\cosh y} = \frac{1}{\sqrt{1+\sinh^2 y}} = \frac{1}{\sqrt{1+x^2}}$$

SOLUTION 2 From Equation 3 (proved in Example 3), we have

$$\frac{d}{dx}(\sinh^{-1}x) = \frac{d}{dx}\ln(x + \sqrt{x^2+1})$$

$$= \frac{1}{x+\sqrt{x^2+1}} \frac{d}{dx}(x + \sqrt{x^2+1})$$

$$= \frac{1}{x+\sqrt{x^2+1}} \left(1 + \frac{x}{\sqrt{x^2+1}}\right)$$

$$= \frac{\sqrt{x^2+1} + x}{(x+\sqrt{x^2+1})\sqrt{x^2+1}}$$

$$= \frac{1}{\sqrt{x^2+1}}$$

EXAMPLE 5 Find $\dfrac{d}{dx}[\tanh^{-1}(\sin x)]$.

SOLUTION Using Table 6 and the Chain Rule, we have

$$\dfrac{d}{dx}[\tanh^{-1}(\sin x)] = \dfrac{1}{1-(\sin x)^2}\dfrac{d}{dx}(\sin x)$$

$$= \dfrac{1}{1-\sin^2 x}\cos x = \dfrac{\cos x}{\cos^2 x} = \sec x$$

EXAMPLE 6 Evaluate $\displaystyle\int_0^1 \dfrac{dx}{\sqrt{1+x^2}}$.

SOLUTION Using Table 6 (or Example 4) we know that an antiderivative of $1/\sqrt{1+x^2}$ is $\sinh^{-1}x$. Therefore

$$\int_0^1 \dfrac{dx}{\sqrt{1+x^2}} = \sinh^{-1}x\Big]_0^1$$

$$= \sinh^{-1}1$$

$$= \ln(1+\sqrt{2}) \quad \text{(from Equation 3)}$$

7.6 Exercises

1–6 ▥ Find the numerical value of each expression.

1. (a) $\sinh 0$ (b) $\cosh 0$
2. (a) $\tanh 0$ (b) $\tanh 1$
3. (a) $\sinh(\ln 2)$ (b) $\sinh 2$
4. (a) $\cosh 3$ (b) $\cosh(\ln 3)$
5. (a) $\text{sech } 0$ (b) $\cosh^{-1} 1$
6. (a) $\sinh 1$ (b) $\sinh^{-1} 1$

▫ ▫ ▫ ▫ ▫ ▫ ▫ ▫ ▫ ▫ ▫

7–19 ▥ Prove the identity.

7. $\sinh(-x) = -\sinh x$
   (This shows that sinh is an odd function.)
8. $\cosh(-x) = \cosh x$
   (This shows that cosh is an even function.)
9. $\cosh x + \sinh x = e^x$
10. $\cosh x - \sinh x = e^{-x}$
11. $\sinh(x+y) = \sinh x \cosh y + \cosh x \sinh y$
12. $\cosh(x+y) = \cosh x \cosh y + \sinh x \sinh y$
13. $\coth^2 x - 1 = \text{csch}^2 x$
14. $\tanh(x+y) = \dfrac{\tanh x + \tanh y}{1 + \tanh x \tanh y}$
15. $\sinh 2x = 2\sinh x \cosh x$
16. $\cosh 2x = \cosh^2 x + \sinh^2 x$
17. $\tanh(\ln x) = \dfrac{x^2-1}{x^2+1}$
18. $\dfrac{1+\tanh x}{1-\tanh x} = e^{2x}$
19. $(\cosh x + \sinh x)^n = \cosh nx + \sinh nx$
    (n any real number)

▫ ▫ ▫ ▫ ▫ ▫ ▫ ▫ ▫ ▫ ▫

20. If $\sinh x = \tfrac{3}{4}$, find the values of the other hyperbolic functions at x.
21. If $\tanh x = \tfrac{4}{5}$, find the values of the other hyperbolic functions at x.
22. (a) Use the graphs of sinh, cosh, and tanh in Figures 1–3 to draw the graphs of csch, sech, and coth.
    (b) Check the graphs that you sketched in part (a) by using a graphing device to produce them.

23. Use the definitions of the hyperbolic functions to find each of the following limits.
(a) $\lim_{x \to \infty} \tanh x$
(b) $\lim_{x \to -\infty} \tanh x$
(c) $\lim_{x \to \infty} \sinh x$
(d) $\lim_{x \to -\infty} \sinh x$
(e) $\lim_{x \to \infty} \operatorname{sech} x$
(f) $\lim_{x \to \infty} \coth x$
(g) $\lim_{x \to 0^+} \coth x$
(h) $\lim_{x \to 0^-} \coth x$
(i) $\lim_{x \to -\infty} \operatorname{csch} x$

24. Prove the formulas given in Table 1 for the derivatives of the functions (a) cosh, (b) tanh, (c) csch, (d) sech, and (e) coth.

25. Give an alternative solution to Example 3 by letting $y = \sinh^{-1} x$ and then using Exercise 9 and Example 1(a) with x replaced by y.

26. Prove Equation 4.

27. Prove Equation 5 using (a) the method of Example 3 and (b) Exercise 18 with x replaced by y.

28. For each of the following functions (i) give a definition like those in (2), (ii) sketch the graph, and (iii) find a formula similar to Equation 3.
(a) csch^{-1}
(b) sech^{-1}
(c) $\coth^{-1}$

29. Prove the formulas given in Table 6 for the derivatives of the following functions.
(a) $\cosh^{-1}$
(b) $\tanh^{-1}$
(c) csch^{-1}
(d) sech^{-1}
(e) $\coth^{-1}$

30–47 Find the derivative.

30. $f(x) = \tanh 4x$

31. $f(x) = x \cosh x$

32. $g(x) = \sinh^2 x$

33. $h(x) = \sinh(x^2)$

34. $F(x) = \sinh x \tanh x$

35. $G(x) = \dfrac{1 - \cosh x}{1 + \cosh x}$

36. $f(t) = e^t \operatorname{sech} t$

37. $h(t) = \coth \sqrt{1 + t^2}$

38. $f(t) = \ln(\sinh t)$

39. $H(t) = \tanh(e^t)$

40. $y = \sinh(\cosh x)$

41. $y = e^{\cosh 3x}$

42. $y = x^2 \sinh^{-1}(2x)$

43. $y = \tanh^{-1} \sqrt{x}$

44. $y = x \tanh^{-1} x + \ln \sqrt{1 - x^2}$

45. $y = x \sinh^{-1}(x/3) - \sqrt{9 + x^2}$

46. $y = \operatorname{sech}^{-1}\sqrt{1 - x^2}, \quad x > 0$

47. $y = \coth^{-1}\sqrt{x^2 + 1}$

48. A flexible cable always hangs in the shape of a catenary $y = c + a \cosh(x/a)$, where c and a are constants and $a > 0$ (see Figure 4 and Exercise 50). Graph several members of the family of functions $y = a \cosh(x/a)$. How does the graph change as a varies?

49. A telephone line hangs between two poles 14 m apart in the shape of the catenary $y = 20 \cosh(x/20) - 15$, where x and y are measured in meters.
(a) Find the slope of this curve where it meets the right pole.
(b) Find the angle θ between the line and the pole.

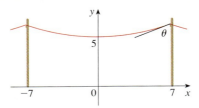

50. Using principles from physics it can be shown that when a cable is hung between two poles, it takes the shape of a curve $y = f(x)$ that satisfies the differential equation

$$\frac{d^2 y}{dx^2} = \frac{\rho g}{T} \sqrt{1 + \left(\frac{dy}{dx}\right)^2}$$

where ρ is the linear density of the cable, g is the acceleration due to gravity, and T is the tension in the cable at its lowest point, and the coordinate system is chosen appropriately. Verify that the function

$$y = f(x) = \frac{T}{\rho g} \cosh\left(\frac{\rho g x}{T}\right)$$

is a solution of this differential equation.

51. (a) Show that any function of the form

$$y = A \sinh mx + B \cosh mx$$

satisfies the differential equation $y'' = m^2 y$.
(b) Find $y = y(x)$ such that $y'' = 9y$, $y(0) = -4$, and $y'(0) = 6$.

52. Evaluate $\lim_{x \to \infty} \dfrac{\sinh x}{e^x}$.

53. At what point of the curve $y = \cosh x$ does the tangent have slope 1?

54. If $x = \ln(\sec \theta + \tan \theta)$, show that $\sec \theta = \cosh x$.

55–63 Evaluate the integral.

55. $\displaystyle\int \sinh x \cosh^2 x \, dx$

56. $\displaystyle\int \sinh(1 + 4x) \, dx$

57. $\displaystyle\int \dfrac{\sinh \sqrt{x}}{\sqrt{x}} \, dx$

58. $\displaystyle\int \tanh x \, dx$

59. $\displaystyle\int \dfrac{\cosh x}{\cosh^2 x - 1} \, dx$

60. $\displaystyle\int \dfrac{\operatorname{sech}^2 x}{2 + \tanh x} \, dx$

61. $\int_4^6 \dfrac{1}{\sqrt{t^2-9}}\,dt$

62. $\int_0^1 \dfrac{dt}{\sqrt{16t^2+1}}$

63. $\int \dfrac{e^x}{1-e^{2x}}\,dx$

64. Estimate the value of the number c such that the area under the curve $y = \sinh cx$ between $x=0$ and $x=1$ is equal to 1.

65. (a) Use Newton's method or a graphing device to find approximate solutions of the equation $\cosh 2x = 1 + \sinh x$.
(b) Estimate the area of the region bounded by the curves $y = \cosh 2x$ and $y = 1 + \sinh x$.

66. Show that the area of the shaded hyperbolic sector in Figure 6 is $A(t) = \tfrac{1}{2}t$. [*Hint:* First show that
$$A(t) = \tfrac{1}{2} \sinh t \,\cosh t - \int_1^{\cosh t} \sqrt{x^2-1}\,dx$$
and then verify that $A'(t) = \tfrac{1}{2}$.]

67. Show that if $a \neq 0$ and $b \neq 0$, then there exist numbers α and β such that $ae^x + be^{-x}$ equals either $\alpha \sinh(x+\beta)$ or $\alpha \cosh(x+\beta)$. In other words, almost every function of the form $f(x) = ae^x + be^{-x}$ is a shifted and stretched hyperbolic sine or cosine function.

7.7 Indeterminate Forms and L'Hospital's Rule

Suppose we are trying to analyze the behavior of the function
$$F(x) = \dfrac{\ln x}{x-1}$$

Although F is not defined when $x=1$, we need to know how F behaves *near* 1. In particular, we would like to know the value of the limit

$$\lim_{x \to 1} \dfrac{\ln x}{x-1}$$

In computing this limit we can't apply Law 5 of limits (the limit of a quotient is the quotient of the limits, see Section 2.3) because the limit of the denominator is 0. In fact, although the limit in (1) exists, its value is not obvious because both numerator and denominator approach 0 and $\tfrac{0}{0}$ is not defined.

In general, if we have a limit of the form
$$\lim_{x \to a} \dfrac{f(x)}{g(x)}$$

where both $f(x) \to 0$ and $g(x) \to 0$ as $x \to a$, then this limit may or may not exist and is called an **indeterminate form of type** $\tfrac{0}{0}$. We met some limits of this type in Chapter 2. For rational functions, we can cancel common factors:

$$\lim_{x \to 1} \dfrac{x^2-x}{x^2-1} = \lim_{x \to 1} \dfrac{x(x-1)}{(x+1)(x-1)} = \lim_{x \to 1} \dfrac{x}{x+1} = \dfrac{1}{2}$$

We used a geometric argument to show that
$$\lim_{x \to 0} \dfrac{\sin x}{x} = 1$$

But these methods do not work for limits such as (1), so in this section we introduce a systematic method, known as *l'Hospital's Rule*, for the evaluation of indeterminate forms.

Another situation in which a limit is not obvious occurs when we look for a horizontal asymptote of F and need to evaluate the limit

$$\boxed{2} \qquad \lim_{x \to \infty} \frac{\ln x}{x - 1}$$

It isn't obvious how to evaluate this limit because both numerator and denominator become large as $x \to \infty$. There is a struggle between numerator and denominator. If the numerator wins, the limit will be ∞; if the denominator wins, the answer will be 0. Or there may be some compromise, in which case the answer may be some finite positive number.

In general, if we have a limit of the form

$$\lim_{x \to a} \frac{f(x)}{g(x)}$$

where both $f(x) \to \infty$ (or $-\infty$) and $g(x) \to \infty$ (or $-\infty$), then the limit may or may not exist and is called an **indeterminate form of type** ∞/∞. We saw in Section 4.4 that this type of limit can be evaluated for certain functions, including rational functions, by dividing numerator and denominator by the highest power of x that occurs in the denominator. For instance,

$$\lim_{x \to \infty} \frac{x^2 - 1}{2x^2 + 1} = \lim_{x \to \infty} \frac{1 - \dfrac{1}{x^2}}{2 + \dfrac{1}{x^2}} = \frac{1 - 0}{2 + 0} = \frac{1}{2}$$

This method does not work for limits such as (2), but l'Hospital's Rule also applies to this type of indeterminate form.

> **L'Hospital's Rule** Suppose f and g are differentiable and $g'(x) \neq 0$ near a (except possibly at a). Suppose that
>
> $$\lim_{x \to a} f(x) = 0 \qquad \text{and} \qquad \lim_{x \to a} g(x) = 0$$
>
> or that
>
> $$\lim_{x \to a} f(x) = \pm\infty \qquad \text{and} \qquad \lim_{x \to a} g(x) = \pm\infty$$
>
> (In other words, we have an indeterminate form of type $\frac{0}{0}$ or ∞/∞.) Then
>
> $$\lim_{x \to a} \frac{f(x)}{g(x)} = \lim_{x \to a} \frac{f'(x)}{g'(x)}$$
>
> if the limit on the right side exists (or is ∞ or $-\infty$).

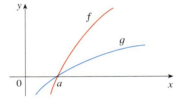

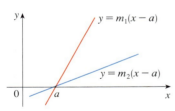

FIGURE 1

|||| Figure 1 suggests visually why l'Hospital's Rule might be true. The first graph shows two differentiable functions f and g, each of which approaches 0 as $x \to a$. If we were to zoom in toward the point $(a, 0)$, the graphs would start to look almost linear. But if the functions actually were linear, as in the second graph, then their ratio would be

$$\frac{m_1(x - a)}{m_2(x - a)} = \frac{m_1}{m_2}$$

which is the ratio of their derivatives. This suggests that

$$\lim_{x \to a} \frac{f(x)}{g(x)} = \lim_{x \to a} \frac{f'(x)}{g'(x)}$$

NOTE 1 ▫ L'Hospital's Rule says that the limit of a quotient of functions is equal to the limit of the quotient of their derivatives, provided that the given conditions are satisfied. It is especially important to verify the conditions regarding the limits of f and g before using l'Hospital's Rule.

NOTE 2 ▫ L'Hospital's Rule is also valid for one-sided limits and for limits at infinity or negative infinity; that is, "$x \to a$" can be replaced by any of the symbols $x \to a^+$, $x \to a^-$, $x \to \infty$, or $x \to -\infty$.

NOTE 3 ▫ For the special case in which $f(a) = g(a) = 0$, f' and g' are continuous, and $g'(a) \neq 0$, it is easy to see why l'Hospital's Rule is true. In fact, using the alternative form

SECTION 7.7 INDETERMINATE FORMS AND L'HOSPITAL'S RULE

■■■■ L'Hospital's Rule is named after a French nobleman, the Marquis de l'Hospital (1661–1704), but was discovered by a Swiss mathematician, John Bernoulli (1667–1748). See Exercise 78 for the example that the Marquis used to illustrate his rule. See the project on page 504 for further historical details.

of the definition of a derivative, we have

$$\lim_{x \to a} \frac{f'(x)}{g'(x)} = \frac{f'(a)}{g'(a)} = \frac{\lim_{x \to a} \dfrac{f(x) - f(a)}{x - a}}{\lim_{x \to a} \dfrac{g(x) - g(a)}{x - a}}$$

$$= \lim_{x \to a} \frac{\dfrac{f(x) - f(a)}{x - a}}{\dfrac{g(x) - g(a)}{x - a}} = \lim_{x \to a} \frac{f(x) - f(a)}{g(x) - g(a)}$$

$$= \lim_{x \to a} \frac{f(x)}{g(x)}$$

The general version of l'Hospital's Rule for the indeterminate form $\frac{0}{0}$ is somewhat more difficult and its proof is deferred to the end of this section. The proof for the indeterminate form ∞/∞ can be found in more advanced books.

EXAMPLE 1 Find $\lim\limits_{x \to 1} \dfrac{\ln x}{x - 1}$.

SOLUTION Since

$$\lim_{x \to 1} \ln x = \ln 1 = 0 \quad \text{and} \quad \lim_{x \to 1} (x - 1) = 0$$

⊘ Notice that when using l'Hospital's Rule we differentiate the numerator and denominator *separately*. We *do not* use the Quotient Rule.

we can apply l'Hospital's Rule:

$$\lim_{x \to 1} \frac{\ln x}{x - 1} = \lim_{x \to 1} \frac{\dfrac{d}{dx}(\ln x)}{\dfrac{d}{dx}(x - 1)} = \lim_{x \to 1} \frac{1/x}{1}$$

$$= \lim_{x \to 1} \frac{1}{x} = 1$$

EXAMPLE 2 Calculate $\lim\limits_{x \to \infty} \dfrac{e^x}{x^2}$.

SOLUTION We have $\lim_{x \to \infty} e^x = \infty$ and $\lim_{x \to \infty} x^2 = \infty$, so l'Hospital's Rule gives

$$\lim_{x \to \infty} \frac{e^x}{x^2} = \lim_{x \to \infty} \frac{\dfrac{d}{dx}(e^x)}{\dfrac{d}{dx}(x^2)} = \lim_{x \to \infty} \frac{e^x}{2x}$$

■■■■ The graph of the function of Example 2 is shown in Figure 2. We have noticed previously that exponential functions grow far more rapidly than power functions, so the result of Example 2 is not unexpected. See also Exercise 87.

Since $e^x \to \infty$ and $2x \to \infty$ as $x \to \infty$, the limit on the right side is also indeterminate, but a second application of l'Hospital's Rule gives

$$\lim_{x \to \infty} \frac{e^x}{x^2} = \lim_{x \to \infty} \frac{e^x}{2x} = \lim_{x \to \infty} \frac{e^x}{2} = \infty$$

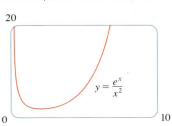

FIGURE 2

|||| The graph of the function of Example 3 is shown in Figure 3. We have discussed previously the slow growth of logarithms, so it isn't surprising that this ratio approaches 0 as $x \to \infty$. See also Exercise 88.

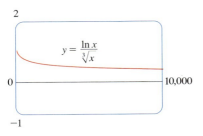

FIGURE 3

EXAMPLE 3 Calculate $\lim_{x \to \infty} \dfrac{\ln x}{\sqrt[3]{x}}$.

SOLUTION Since $\ln x \to \infty$ and $\sqrt[3]{x} \to \infty$ as $x \to \infty$, l'Hospital's Rule applies:

$$\lim_{x \to \infty} \frac{\ln x}{\sqrt[3]{x}} = \lim_{x \to \infty} \frac{1/x}{\frac{1}{3}x^{-2/3}}$$

Notice that the limit on the right side is now indeterminate of type $\frac{0}{0}$. But instead of applying l'Hospital's Rule a second time as we did in Example 2, we simplify the expression and see that a second application is unnecessary:

$$\lim_{x \to \infty} \frac{\ln x}{\sqrt[3]{x}} = \lim_{x \to \infty} \frac{1/x}{\frac{1}{3}x^{-2/3}} = \lim_{x \to \infty} \frac{3}{\sqrt[3]{x}} = 0$$

EXAMPLE 4 Find $\lim_{x \to 0} \dfrac{\tan x - x}{x^3}$. [See Exercise 36(d) in Section 2.2.]

SOLUTION Noting that both $\tan x - x \to 0$ and $x^3 \to 0$ as $x \to 0$, we use l'Hospital's Rule:

$$\lim_{x \to 0} \frac{\tan x - x}{x^3} = \lim_{x \to 0} \frac{\sec^2 x - 1}{3x^2}$$

Since the limit on the right side is still indeterminate of type $\frac{0}{0}$, we apply l'Hospital's Rule again:

$$\lim_{x \to 0} \frac{\sec^2 x - 1}{3x^2} = \lim_{x \to 0} \frac{2 \sec^2 x \tan x}{6x}$$

Because $\lim_{x \to 0} \sec^2 x = 1$, we simplify the calculation by writing

$$\lim_{x \to 0} \frac{2 \sec^2 x \tan x}{6x} = \frac{1}{3} \lim_{x \to 0} \sec^2 x \lim_{x \to 0} \frac{\tan x}{x} = \frac{1}{3} \lim_{x \to 0} \frac{\tan x}{x}$$

|||| The graph in Figure 4 gives visual confirmation of the result of Example 4. If we were to zoom in too far, however, we would get an inaccurate graph because $\tan x$ is close to x when x is small. See Exercise 36(d) in Section 2.2.

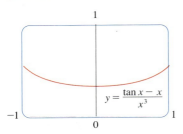

FIGURE 4

We can evaluate this last limit either by using l'Hospital's Rule a third time or by writing $\tan x$ as $(\sin x)/(\cos x)$ and making use of our knowledge of trigonometric limits. Putting together all the steps, we get

$$\lim_{x \to 0} \frac{\tan x - x}{x^3} = \lim_{x \to 0} \frac{\sec^2 x - 1}{3x^2} = \lim_{x \to 0} \frac{2 \sec^2 x \tan x}{6x}$$

$$= \frac{1}{3} \lim_{x \to 0} \frac{\tan x}{x} = \frac{1}{3} \lim_{x \to 0} \frac{\sec^2 x}{1} = \frac{1}{3}$$

EXAMPLE 5 Find $\lim_{x \to \pi^-} \dfrac{\sin x}{1 - \cos x}$.

SOLUTION If we blindly attempted to use l'Hospital's Rule, we would get

$$\lim_{x \to \pi^-} \frac{\sin x}{1 - \cos x} = \lim_{x \to \pi^-} \frac{\cos x}{\sin x} = -\infty$$

This is *wrong*! Although the numerator $\sin x \to 0$ as $x \to \pi^-$, notice that the denominator $(1 - \cos x)$ does not approach 0, so l'Hospital's Rule can't be applied here.

The required limit is, in fact, easy to find because the function is continuous and the denominator is nonzero at π:

$$\lim_{x \to \pi^-} \frac{\sin x}{1 - \cos x} = \frac{\sin \pi}{1 - \cos \pi} = \frac{0}{1 - (-1)} = 0$$

Example 5 shows what can go wrong if you use l'Hospital's Rule without thinking. Other limits *can* be found using l'Hospital's Rule but are more easily found by other methods. (See Examples 3 and 5 in Section 2.3, Example 3 in Section 2.6, and the discussion at the beginning of this section.) So when evaluating any limit, you should consider other methods before using l'Hospital's Rule.

Indeterminate Products

If $\lim_{x \to a} f(x) = 0$ and $\lim_{x \to a} g(x) = \infty$ (or $-\infty$), then it isn't clear what the value of $\lim_{x \to a} f(x)g(x)$, if any, will be. There is a struggle between f and g. If f wins, the answer will be 0; if g wins, the answer will be ∞ (or $-\infty$). Or there may be a compromise where the answer is a finite nonzero number. This kind of limit is called an **indeterminate form of type $0 \cdot \infty$**. We can deal with it by writing the product fg as a quotient:

$$fg = \frac{f}{1/g} \quad \text{or} \quad fg = \frac{g}{1/f}$$

This converts the given limit into an indeterminate form of type $\frac{0}{0}$ or ∞/∞ so that we can use l'Hospital's Rule.

EXAMPLE 6 Evaluate $\lim_{x \to 0^+} x \ln x$.

SOLUTION The given limit is indeterminate because, as $x \to 0^+$, the first factor (x) approaches 0 while the second factor ($\ln x$) approaches $-\infty$. Writing $x = 1/(1/x)$, we have $1/x \to \infty$ as $x \to 0^+$, so l'Hospital's Rule gives

$$\lim_{x \to 0^+} x \ln x = \lim_{x \to 0^+} \frac{\ln x}{1/x} = \lim_{x \to 0^+} \frac{1/x}{-1/x^2} = \lim_{x \to 0^+} (-x) = 0$$

NOTE ▫ In solving Example 6 another possible option would have been to write

$$\lim_{x \to 0^+} x \ln x = \lim_{x \to 0^+} \frac{x}{1/\ln x}$$

This gives an indeterminate form of the type 0/0, but if we apply l'Hospital's Rule we get a more complicated expression than the one we started with. In general, when we rewrite an indeterminate product, we try to choose the option that leads to the simpler limit.

EXAMPLE 7 Use l'Hospital's Rule to help sketch the graph of $f(x) = xe^x$.

SOLUTION Because both x and e^x become large as $x \to \infty$, we have $\lim_{x \to \infty} xe^x = \infty$. As $x \to -\infty$, however, $e^x \to 0$ and so we have an indeterminate product that requires the use of l'Hospital's Rule:

$$\lim_{x \to -\infty} xe^x = \lim_{x \to -\infty} \frac{x}{e^{-x}} = \lim_{x \to -\infty} \frac{1}{-e^{-x}} = \lim_{x \to -\infty} (-e^x) = 0$$

Thus, the x-axis is a horizontal asymptote.

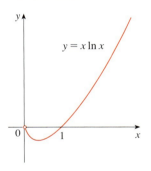

|||| Figure 5 shows the graph of the function in Example 6. Notice that the function is undefined at $x = 0$; the graph approaches the origin but never quite reaches it.

FIGURE 5

We use the methods of Chapter 4 to gather other information concerning the graph. The derivative is

$$f'(x) = xe^x + e^x = (x + 1)e^x$$

Since e^x is always positive, we see that $f'(x) > 0$ when $x + 1 > 0$, and $f'(x) < 0$ when $x + 1 < 0$. So f is increasing on $(-1, \infty)$ and decreasing on $(-\infty, -1)$. Because $f'(-1) = 0$ and f changes from negative to positive at $x = -1$, $f(-1) = -e^{-1}$ is a local (and absolute) minimum. The second derivative is

$$f''(x) = (x + 1)e^x + e^x = (x + 2)e^x$$

Since $f''(x) > 0$ if $x > -2$ and $f''(x) < 0$ if $x < -2$, f is concave upward on $(-2, \infty)$ and concave downward on $(-\infty, -2)$. The inflection point is $(-2, -2e^{-2})$.

We use this information to sketch the curve in Figure 6.

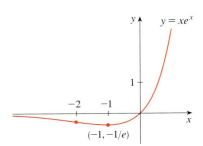

FIGURE 6

Indeterminate Differences

If $\lim_{x \to a} f(x) = \infty$ and $\lim_{x \to a} g(x) = \infty$, then the limit

$$\lim_{x \to a} [f(x) - g(x)]$$

is called an **indeterminate form of type** $\infty - \infty$. Again there is a contest between f and g. Will the answer be ∞ (f wins) or will it be $-\infty$ (g wins) or will they compromise on a finite number? To find out, we try to convert the difference into a quotient (for instance, by using a common denominator, or rationalization, or factoring out a common factor) so that we have an indeterminate form of type $\frac{0}{0}$ or ∞/∞.

EXAMPLE 8 Compute $\lim_{x \to (\pi/2)^-} (\sec x - \tan x)$.

SOLUTION First notice that $\sec x \to \infty$ and $\tan x \to \infty$ as $x \to (\pi/2)^-$, so the limit is indeterminate. Here we use a common denominator:

$$\lim_{x \to (\pi/2)^-} (\sec x - \tan x) = \lim_{x \to (\pi/2)^-} \left(\frac{1}{\cos x} - \frac{\sin x}{\cos x} \right)$$

$$= \lim_{x \to (\pi/2)^-} \frac{1 - \sin x}{\cos x} = \lim_{x \to (\pi/2)^-} \frac{-\cos x}{-\sin x} = 0$$

Note that the use of l'Hospital's Rule is justified because $1 - \sin x \to 0$ and $\cos x \to 0$ as $x \to (\pi/2)^-$.

Indeterminate Powers

Several indeterminate forms arise from the limit

$$\lim_{x \to a} [f(x)]^{g(x)}$$

1. $\lim_{x \to a} f(x) = 0$ and $\lim_{x \to a} g(x) = 0$ type 0^0

2. $\lim_{x \to a} f(x) = \infty$ and $\lim_{x \to a} g(x) = 0$ type ∞^0

3. $\lim_{x \to a} f(x) = 1$ and $\lim_{x \to a} g(x) = \pm\infty$ type 1^∞

Each of these three cases can be treated either by taking the natural logarithm:

$$\text{let} \quad y = [f(x)]^{g(x)}, \quad \text{then} \quad \ln y = g(x) \ln f(x)$$

or by writing the function as an exponential:

$$[f(x)]^{g(x)} = e^{g(x) \ln f(x)}$$

(Recall that both of these methods were used in differentiating such functions.) In either method we are led to the indeterminate product $g(x) \ln f(x)$, which is of type $0 \cdot \infty$.

EXAMPLE 9 Calculate $\lim_{x \to 0^+} (1 + \sin 4x)^{\cot x}$.

SOLUTION First notice that as $x \to 0^+$, we have $1 + \sin 4x \to 1$ and $\cot x \to \infty$, so the given limit is indeterminate. Let

$$y = (1 + \sin 4x)^{\cot x}$$

Then

$$\ln y = \ln[(1 + \sin 4x)^{\cot x}] = \cot x \ln(1 + \sin 4x)$$

so l'Hospital's Rule gives

$$\lim_{x \to 0^+} \ln y = \lim_{x \to 0^+} \frac{\ln(1 + \sin 4x)}{\tan x}$$

$$= \lim_{x \to 0^+} \frac{\frac{4 \cos 4x}{1 + \sin 4x}}{\sec^2 x} = 4$$

So far we have computed the limit of $\ln y$, but what we want is the limit of y. To find this we use the fact that $y = e^{\ln y}$:

$$\lim_{x \to 0^+} (1 + \sin 4x)^{\cot x} = \lim_{x \to 0^+} y = \lim_{x \to 0^+} e^{\ln y} = e^4$$

EXAMPLE 10 Find $\lim_{x \to 0^+} x^x$.

SOLUTION Notice that this limit is indeterminate since $0^x = 0$ for any $x > 0$ but $x^0 = 1$ for any $x \neq 0$. We could proceed as in Example 9 or by writing the function as an exponential:

$$x^x = (e^{\ln x})^x = e^{x \ln x}$$

In Example 6 we used l'Hospital's Rule to show that

$$\lim_{x \to 0^+} x \ln x = 0$$

Therefore

$$\lim_{x \to 0^+} x^x = \lim_{x \to 0^+} e^{x \ln x} = e^0 = 1$$

|||| The graph of the function $y = x^x$, $x > 0$, is shown in Figure 7. Notice that although 0^0 is not defined, the values of the function approach 1 as $x \to 0^+$. This confirms the result of Example 10.

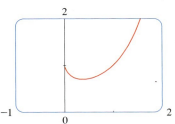

FIGURE 7

|||| See the biographical sketch of Cauchy on page 97.

In order to give the promised proof of l'Hospital's Rule we first need a generalization of the Mean Value Theorem. The following theorem is named after another French mathematician, Augustin-Louis Cauchy (1789–1857).

> **3** **Cauchy's Mean Value Theorem** Suppose that the functions f and g are continuous on $[a, b]$ and differentiable on (a, b), and $g'(x) \neq 0$ for all x in (a, b). Then there is a number c in (a, b) such that
>
> $$\frac{f'(c)}{g'(c)} = \frac{f(b) - f(a)}{g(b) - g(a)}$$

Notice that if we take the special case in which $g(x) = x$, then $g'(c) = 1$ and Theorem 3 is just the ordinary Mean Value Theorem. Furthermore, Theorem 3 can be proved in a similar manner. You can verify that all we have to do is change the function h given by Equation 4.2.4 to the function

$$h(x) = f(x) - f(a) - \frac{f(b) - f(a)}{g(b) - g(a)}[g(x) - g(a)]$$

and apply Rolle's Theorem as before.

Proof of L'Hospital's Rule We are assuming that $\lim_{x \to a} f(x) = 0$ and $\lim_{x \to a} g(x) = 0$. Let

$$L = \lim_{x \to a} \frac{f'(x)}{g'(x)}$$

We must show that $\lim_{x \to a} f(x)/g(x) = L$. Define

$$F(x) = \begin{cases} f(x) & \text{if } x \neq a \\ 0 & \text{if } x = a \end{cases} \qquad G(x) = \begin{cases} g(x) & \text{if } x \neq a \\ 0 & \text{if } x = a \end{cases}$$

Then F is continuous on I since f is continuous on $\{x \in I \mid x \neq a\}$ and

$$\lim_{x \to a} F(x) = \lim_{x \to a} f(x) = 0 = F(a)$$

Likewise, G is continuous on I. Let $x \in I$ and $x > a$. Then F and G are continuous on $[a, x]$ and differentiable on (a, x) and $G' \neq 0$ there (since $F' = f'$ and $G' = g'$). Therefore, by Cauchy's Mean Value Theorem there is a number y such that $a < y < x$ and

$$\frac{F'(y)}{G'(y)} = \frac{F(x) - F(a)}{G(x) - G(a)} = \frac{F(x)}{G(x)}$$

Here we have used the fact that, by definition, $F(a) = 0$ and $G(a) = 0$. Now, if we let $x \to a^+$, then $y \to a^+$ (since $a < y < x$), so

$$\lim_{x \to a^+} \frac{f(x)}{g(x)} = \lim_{x \to a^+} \frac{F(x)}{G(x)} = \lim_{y \to a^+} \frac{F'(y)}{G'(y)} = \lim_{y \to a^+} \frac{f'(y)}{g'(y)} = L$$

A similar argument shows that the left-hand limit is also L. Therefore

$$\lim_{x \to a} \frac{f(x)}{g(x)} = L$$

This proves l'Hospital's Rule for the case where a is finite.

If a is infinite, we let $t = 1/x$. Then $t \to 0^+$ as $x \to \infty$, so we have

$$\lim_{x \to \infty} \frac{f(x)}{g(x)} = \lim_{t \to 0^+} \frac{f(1/t)}{g(1/t)}$$

$$= \lim_{t \to 0^+} \frac{f'(1/t)(-1/t^2)}{g'(1/t)(-1/t^2)} \quad \text{(by l'Hospital's Rule for finite } a\text{)}$$

$$= \lim_{t \to 0^+} \frac{f'(1/t)}{g'(1/t)} = \lim_{x \to \infty} \frac{f'(x)}{g'(x)}$$

7.7 Exercises

1–4 Given that

$$\lim_{x \to a} f(x) = 0 \qquad \lim_{x \to a} g(x) = 0 \qquad \lim_{x \to a} h(x) = 1$$

$$\lim_{x \to a} p(x) = \infty \qquad \lim_{x \to a} q(x) = \infty$$

which of the following limits are indeterminate forms? For those that are not an indeterminate form, evaluate the limit where possible.

1. (a) $\lim_{x \to a} \dfrac{f(x)}{g(x)}$ \quad (b) $\lim_{x \to a} \dfrac{f(x)}{p(x)}$

(c) $\lim_{x \to a} \dfrac{h(x)}{p(x)}$ \quad (d) $\lim_{x \to a} \dfrac{p(x)}{f(x)}$

(e) $\lim_{x \to a} \dfrac{p(x)}{q(x)}$

2. (a) $\lim_{x \to a} [f(x)p(x)]$ \quad (b) $\lim_{x \to a} [h(x)p(x)]$

(c) $\lim_{x \to a} [p(x)q(x)]$

3. (a) $\lim_{x \to a} [f(x) - p(x)]$ \quad (b) $\lim_{x \to a} [p(x) - q(x)]$

(c) $\lim_{x \to a} [p(x) + q(x)]$

4. (a) $\lim_{x \to a} [f(x)]^{g(x)}$ \quad (b) $\lim_{x \to a} [f(x)]^{p(x)}$ \quad (c) $\lim_{x \to a} [h(x)]^{p(x)}$

(d) $\lim_{x \to a} [p(x)]^{f(x)}$ \quad (e) $\lim_{x \to a} [p(x)]^{q(x)}$ \quad (f) $\lim_{x \to a} \sqrt[q(x)]{p(x)}$

5–62 Find the limit. Use l'Hospital's Rule where appropriate. If there is a more elementary method, consider using it. If l'Hospital's Rule doesn't apply, explain why.

5. $\lim_{x \to -1} \dfrac{x^2 - 1}{x + 1}$

6. $\lim_{x \to -2} \dfrac{x + 2}{x^2 + 3x + 2}$

7. $\lim_{x \to 1} \dfrac{x^9 - 1}{x^5 - 1}$

8. $\lim_{x \to 1} \dfrac{x^a - 1}{x^b - 1}$

9. $\lim_{x \to (\pi/2)^+} \dfrac{\cos x}{1 - \sin x}$

10. $\lim_{x \to 0} \dfrac{x + \tan x}{\sin x}$

11. $\lim_{t \to 0} \dfrac{e^t - 1}{t^3}$

12. $\lim_{t \to 0} \dfrac{e^{3t} - 1}{t}$

13. $\lim_{x \to 0} \dfrac{\tan px}{\tan qx}$

14. $\lim_{\theta \to \pi/2} \dfrac{1 - \sin \theta}{\csc \theta}$

15. $\lim_{x \to \infty} \dfrac{\ln x}{x}$

16. $\lim_{x \to \infty} \dfrac{e^x}{x}$

17. $\lim_{x \to 0^+} \dfrac{\ln x}{x}$

18. $\lim_{x \to \infty} \dfrac{\ln \ln x}{x}$

19. $\lim_{t \to 0} \dfrac{5^t - 3^t}{t}$

20. $\lim_{x \to 1} \dfrac{\ln x}{\sin \pi x}$

21. $\lim_{x \to 0} \dfrac{e^x - 1 - x}{x^2}$

22. $\lim_{x \to 0} \dfrac{e^x - 1 - x - (x^2/2)}{x^3}$

23. $\lim_{x \to \infty} \dfrac{e^x}{x^3}$

24. $\lim_{x \to 0} \dfrac{\sin x}{\sinh x}$

25. $\lim_{x \to 0} \dfrac{\sin^{-1} x}{x}$

26. $\lim_{x \to 0} \dfrac{\sin x - x}{x^3}$

27. $\lim_{x \to 0} \dfrac{1 - \cos x}{x^2}$

28. $\lim_{x \to \infty} \dfrac{(\ln x)^2}{x}$

29. $\lim_{x \to 0} \dfrac{x + \sin x}{x + \cos x}$

30. $\lim_{x \to 0} \dfrac{\cos mx - \cos nx}{x^2}$

31. $\lim_{x \to \infty} \dfrac{x}{\ln(1 + 2e^x)}$

32. $\lim_{x \to 0} \dfrac{x}{\tan^{-1}(4x)}$

33. $\lim_{x \to 1} \dfrac{1 - x + \ln x}{1 + \cos \pi x}$

34. $\lim_{x \to \infty} \dfrac{\sqrt{x^2 + 2}}{\sqrt{2x^2 + 1}}$

35. $\lim_{x \to 1} \dfrac{x^a - ax + a - 1}{(x - 1)^2}$

36. $\lim_{x \to 0} \dfrac{1 - e^{-2x}}{\sec x}$

37. $\lim_{x \to 0^+} \sqrt{x} \ln x$

38. $\lim_{x \to -\infty} x^2 e^x$

39. $\lim_{x \to 0} \cot 2x \sin 6x$

40. $\lim_{x \to 0^+} \sin x \ln x$

41. $\lim_{x \to \infty} x^3 e^{-x^2}$

42. $\lim_{x \to \pi/4} (1 - \tan x) \sec x$

43. $\lim_{x \to 1^+} \ln x \tan(\pi x/2)$

44. $\lim_{x \to \infty} x \tan(1/x)$

45. $\lim_{x \to 0} \left(\frac{1}{x} - \csc x \right)$

46. $\lim_{x \to 0} (\csc x - \cot x)$

47. $\lim_{x \to \infty} (\sqrt{x^2 + x} - x)$

48. $\lim_{x \to 1} \left(\frac{1}{\ln x} - \frac{1}{x - 1} \right)$

49. $\lim_{x \to \infty} (x - \ln x)$

50. $\lim_{x \to \infty} (xe^{1/x} - x)$

51. $\lim_{x \to 0^+} x^{x^2}$

52. $\lim_{x \to 0^+} (\tan 2x)^x$

53. $\lim_{x \to 0} (1 - 2x)^{1/x}$

54. $\lim_{x \to \infty} \left(1 + \frac{a}{x} \right)^{bx}$

55. $\lim_{x \to \infty} \left(1 + \frac{3}{x} + \frac{5}{x^2} \right)^x$

56. $\lim_{x \to \infty} x^{(\ln 2)/(1 + \ln x)}$

57. $\lim_{x \to \infty} x^{1/x}$

58. $\lim_{x \to \infty} (e^x + x)^{1/x}$

59. $\lim_{x \to \infty} \left(\frac{x}{x + 1} \right)^x$

60. $\lim_{x \to 0} (\cos 3x)^{5/x}$

61. $\lim_{x \to 0^+} (\cos x)^{1/x^2}$

62. $\lim_{x \to \infty} \left(\frac{2x - 3}{2x + 5} \right)^{2x+1}$

63–64 Use a graph to estimate the value of the limit. Then use l'Hospital's Rule to find the exact value.

63. $\lim_{x \to \infty} x [\ln(x + 5) - \ln x]$

64. $\lim_{x \to \pi/4} (\tan x)^{\tan 2x}$

65–66 Illustrate l'Hospital's Rule by graphing both $f(x)/g(x)$ and $f'(x)/g'(x)$ near $x = 0$ to see that these ratios have the same limit as $x \to 0$. Also calculate the exact value of the limit.

65. $f(x) = e^x - 1$, $g(x) = x^3 + 4x$

66. $f(x) = 2x \sin x$, $g(x) = \sec x - 1$

67–72 Use l'Hospital's Rule to help sketch the curve. Use the guidelines of Section 4.5.

67. $y = xe^{-x}$

68. $y = x(\ln x)^2$

69. $y = xe^{-x^2}$

70. $y = e^x/x$

71. $y = x - \ln(1 + x)$

72. $y = e^x - 3e^{-x} - 4x$

CAS 73–75
(a) Graph the function.
(b) Use l'Hospital's Rule to explain the behavior as $x \to 0^+$ or as $x \to \infty$.
(c) Estimate the maximum and minimum values and then use calculus to find the exact values.
(d) Use a graph of f'' to estimate the x-coordinates of the inflection points.

73. $f(x) = x^{-x}$

74. $f(x) = (\sin x)^{\sin x}$

75. $f(x) = x^{1/x}$

76. Investigate the family of curves given by $f(x) = x^n e^{-x}$, where n is a positive integer. What features do these curves have in common? How do they differ from one another? In particular, what happens to the maximum and minimum points and inflection points as n increases? Illustrate by graphing several members of the family.

77. Investigate the family of curves given by $f(x) = xe^{-cx}$, where c is a real number. Start by computing the limits as $x \to \pm\infty$. Identify any transitional values of c where the basic shape changes. What happens to the maximum or minimum points and inflection points as c changes? Illustrate by graphing several members of the family.

78. The first appearance in print of l'Hospital's Rule was in the book *Analyse des Infiniment Petits* published by the Marquis de l'Hospital in 1696. This was the first calculus *textbook* ever published and the example that the Marquis used in that book to illustrate his rule was to find the limit of the function

$$y = \frac{\sqrt{2a^3x - x^4} - a\sqrt[3]{aax}}{a - \sqrt[4]{ax^3}}$$

as x approaches a, where $a > 0$. (At that time it was common to write aa instead of a^2.) Solve this problem.

79. If an initial amount A_0 of money is invested at an interest rate i compounded n times a year, the value of the investment after t years is

$$A = A_0 \left(1 + \frac{i}{n} \right)^{nt}$$

If we let $n \to \infty$, we refer to the *continuous compounding* of interest. Use l'Hospital's Rule to show that if interest is compounded continuously, then the amount after n years is

$$A = A_0 e^{it}$$

80. If an object with mass m is dropped from rest, one model for its speed v after t seconds, taking air resistance into account, is

$$v = \frac{mg}{c} (1 - e^{-ct/m})$$

where g is the acceleration due to gravity and c is a positive constant. (In Chapter 10 we will be able to deduce this

equation from the assumption that the air resistance is proportional to the speed of the object.)
(a) Calculate $\lim_{t \to \infty} v$. What is the meaning of this limit?
(b) For fixed t, use l'Hospital's Rule to calculate $\lim_{m \to \infty} v$. What can you conclude about the speed of a very heavy falling object?

81. In Section 5.3 we investigated the Fresnel function $S(x) = \int_0^x \sin(\frac{1}{2}\pi t^2)\, dt$, which arises in the study of the diffraction of light waves. Evaluate
$$\lim_{x \to 0} \frac{S(x)}{x^3}$$

82. Suppose that the temperature in a long thin rod placed along the x-axis is initially $C/(2a)$ if $|x| \le a$ and 0 if $|x| > a$. It can be shown that if the heat diffusivity of the rod is k, then the temperature of the rod at the point x at time t is
$$T(x, t) = \frac{C}{a\sqrt{4\pi kt}} \int_0^a e^{-(x-u)^2/(4kt)}\, du$$

To find the temperature distribution that results from an initial hot spot concentrated at the origin, we need to compute
$$\lim_{a \to 0} T(x, t)$$

Use l'Hospital's Rule to find this limit.

83. If f' is continuous, $f(2) = 0$, and $f'(2) = 7$, evaluate
$$\lim_{x \to 0} \frac{f(2 + 3x) + f(2 + 5x)}{x}$$

84. For what values of a and b is the following equation true?
$$\lim_{x \to 0} \left(\frac{\sin 2x}{x^3} + a + \frac{b}{x^2} \right) = 0$$

85. If f' is continuous, use l'Hospital's Rule to show that
$$\lim_{h \to 0} \frac{f(x + h) - f(x - h)}{2h} = f'(x)$$

Explain the meaning of this equation with the aid of a diagram.

86. If f'' is continuous, show that
$$\lim_{h \to 0} \frac{f(x + h) - 2f(x) + f(x - h)}{h^2} = f''(x)$$

87. Prove that
$$\lim_{x \to \infty} \frac{e^x}{x^n} = \infty$$

for any positive integer n. This shows that the exponential function approaches infinity faster than any power of x.

88. Prove that
$$\lim_{x \to \infty} \frac{\ln x}{x^p} = 0$$

for any number $p > 0$. This shows that the logarithmic function approaches ∞ more slowly than any power of x.

89. Prove that $\lim_{x \to 0^+} x^\alpha \ln x = 0$ for any $\alpha > 0$.

90. Evaluate $\lim_{x \to 0} \dfrac{1}{x^3} \int_0^x \sin(t^2)\, dt$.

91. The figure shows a sector of a circle with central angle θ. Let $A(\theta)$ be the area of the segment between the chord PR and the arc PR. Let $B(\theta)$ be the area of the triangle PQR. Find $\lim_{\theta \to 0^+} A(\theta)/B(\theta)$.

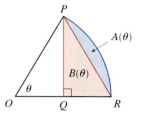

92. The figure shows two regions in the first quadrant: $A(t)$ is the area under the curve $y = \sin(x^2)$ from 0 to t, and $B(t)$ is the area of the triangle with vertices O, P, and $(t, 0)$. Find $\lim_{t \to 0^+} A(t)/B(t)$.

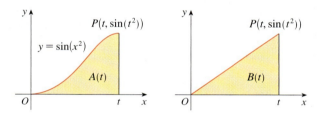

93. Let
$$f(x) = \begin{cases} e^{-1/x^2} & \text{if } x \ne 0 \\ 0 & \text{if } x = 0 \end{cases}$$

(a) Use the definition of derivative to compute $f'(0)$.
(b) Show that f has derivatives of all orders that are defined on $\mathbb{R}$. [Hint: First show by induction that there is a polynomial $p_n(x)$ and a nonnegative integer k_n such that $f^{(n)}(x) = p_n(x)f(x)/x^{k_n}$ for $x \ne 0$.]

94. Let
$$f(x) = \begin{cases} |x|^x & \text{if } x \ne 0 \\ 1 & \text{if } x = 0 \end{cases}$$

(a) Show that f is continuous at 0.
(b) Investigate graphically whether f is differentiable at 0 by zooming in several times toward the point (0, 1) on the graph of f.
(c) Show that f is not differentiable at 0. How can you reconcile this fact with the appearance of the graphs in part (b)?

WRITING PROJECT

The Origins of L'Hospital's Rule

L'Hospital's Rule was first published in 1696 in the Marquis de l'Hospital's calculus textbook *Analyse des Infiniment Petits*, but the rule was discovered in 1694 by the Swiss mathematician John (Johann) Bernoulli. The explanation is that these two mathematicians had entered into a curious business arrangement whereby the Marquis de l'Hospital bought the rights to Bernoulli's mathematical discoveries. The details, including a translation of l'Hospital's letter to Bernoulli proposing the arrangement, can be found in the book by Eves [1].

Write a report on the historical and mathematical origins of l'Hospital's Rule. Start by providing brief biographical details of both men (the dictionary edited by Gillispie [2] is a good source) and outline the business deal between them. Then give l'Hospital's statement of his rule, which is found in Struik's sourcebook [4] and more briefly in the book of Katz [3]. Notice that l'Hospital and Bernoulli formulated the rule geometrically and gave the answer in terms of differentials. Compare their statement with the version of l'Hospital's Rule given in Section 7.7 and show that the two statements are essentially the same.

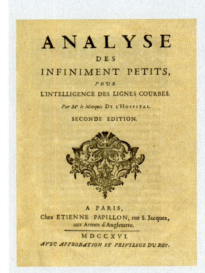

|||| The Internet is another source of information for this project. See the web site
www.stewartcalculus.com
and click on *History of Mathematics*.

1. Howard Eves, *In Mathematical Circles (Volume 2: Quadrants III and IV)* (Boston: Prindle, Weber and Schmidt, 1969), pp. 20–22.
2. C. C. Gillispie, ed., *Dictionary of Scientific Biography* (New York: Scribner's, 1974). See the article on Johann Bernoulli by E. A. Fellmann and J. O. Fleckenstein in Volume II and the article on the Marquis de l'Hospital by Abraham Robinson in Volume VIII.
3. Victor Katz, *A History of Mathematics: An Introduction* (New York: HarperCollins, 1993), p. 484.
4. D. J. Struik, ed., *A Sourcebook in Mathematics, 1200–1800* (Princeton, NJ: Princeton University Press, 1969), pp. 315–316.

7 Review

CONCEPT CHECK

1. (a) What is a one-to-one function? How can you tell if a function is one-to-one by looking at its graph?
   (b) If f is a one-to-one function, how is its inverse function f^{-1} defined? How do you obtain the graph of f^{-1} from the graph of f?
   (c) Suppose f is a one-to-one function and $g = f^{-1}$. If $f'(g(a)) \neq 0$, write a formula for $g'(a)$.

2. (a) What are the domain and range of the natural exponential function $f(x) = e^x$?
   (b) What are the domain and range of the natural logarithmic function $f(x) = \ln x$?
   (c) How are the graphs of these functions related? Sketch these graphs, by hand, using the same axes.
   (d) If a is a positive number, $a \neq 1$, write an equation that expresses $\log_a x$ in terms of $\ln x$.

3. (a) How is the inverse sine function $f(x) = \sin^{-1}x$ defined? What are its domain and range?
   (b) How is the inverse cosine function $f(x) = \cos^{-1}x$ defined? What are its domain and range?
   (c) How is the inverse tangent function $f(x) = \tan^{-1}x$ defined? What are its domain and range? Sketch its graph.

4. Write the definitions of the hyperbolic functions $\sinh x$, $\cosh x$, and $\tanh x$.

5. State the derivative of each function.
   (a) $y = e^x$ (b) $y = a^x$ (c) $y = \ln x$
   (d) $y = \log_a x$ (e) $y = \sin^{-1}x$ (f) $y = \cos^{-1}x$
   (g) $y = \tan^{-1}x$ (h) $y = \sinh x$ (i) $y = \cosh x$
   (j) $y = \tanh x$ (k) $y = \sinh^{-1}x$ (l) $y = \cosh^{-1}x$
   (m) $y = \tanh^{-1}x$

6. (a) How is the number e defined?
   (b) Express e as a limit.
   (c) Why is the natural exponential function $y = e^x$ used more often in calculus than the other exponential functions $y = a^x$?
   (d) Why is the natural logarithmic function $y = \ln x$ used more often in calculus than the other logarithmic functions $y = \log_a x$?

7. (a) What does l'Hospital's Rule say?
   (b) How can you use l'Hospital's Rule if you have a product $f(x)g(x)$ where $f(x) \to 0$ and $g(x) \to \infty$ as $x \to a$?
   (c) How can you use l'Hospital's Rule if you have a difference $f(x) - g(x)$ where $f(x) \to \infty$ and $g(x) \to \infty$ as $x \to a$?
   (d) How can you use l'Hospital's Rule if you have a power $[f(x)]^{g(x)}$ where $f(x) \to 0$ and $g(x) \to 0$ as $x \to a$?

TRUE-FALSE QUIZ

Determine whether the statement is true or false. If it is true, explain why. If it is false, explain why or give an example that disproves the statement.

1. If f is one-to-one, with domain $\mathbb{R}$, then $f^{-1}(f(6)) = 6$.
2. If f is one-to-one and differentiable, with domain $\mathbb{R}$, then $(f^{-1})'(6) = 1/f'(6)$.
3. The function $f(x) = \cos x$, $-\pi/2 \le x \le \pi/2$, is one-to-one.
4. $\tan^{-1}(-1) = 3\pi/4$
5. If $0 < a < b$, then $\ln a < \ln b$.
6. $\pi^{\sqrt{5}} = e^{\sqrt{5} \ln \pi}$
7. You can always divide by e^x.
8. If $a > 0$ and $b > 0$, then $\ln(a + b) = \ln a + \ln b$.
9. If $x > 0$, then $(\ln x)^6 = 6 \ln x$.
10. $\dfrac{d}{dx}(10^x) = x 10^{x-1}$
11. $\dfrac{d}{dx}(\ln 10) = \dfrac{1}{10}$
12. The inverse function of $y = e^{3x}$ is $y = \tfrac{1}{3} \ln x$.
13. $\cos^{-1} x = \dfrac{1}{\cos x}$
14. $\tan^{-1} x = \dfrac{\sin^{-1} x}{\cos^{-1} x}$
15. $\cosh x \ge 1$ for all x
16. $\ln \dfrac{1}{10} = -\displaystyle\int_1^{10} \dfrac{dx}{x}$
17. $\displaystyle\int_2^{16} \dfrac{dx}{x} = 3 \ln 2$
18. $\displaystyle\lim_{x \to \pi^-} \dfrac{\tan x}{1 - \cos x} = \lim_{x \to \pi^-} \dfrac{\sec^2 x}{\sin x} = \infty$

EXERCISES

1. The graph of f is shown. Is f one-to-one? Explain.

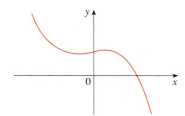

2. The graph of g is given.
   (a) Why is g one-to-one?
   (b) Estimate the value of $g^{-1}(2)$.
   (c) Estimate the domain of g^{-1}.
   (d) Sketch the graph of g^{-1}.

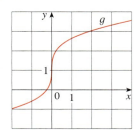

3. Suppose f is one-to-one, $f(7) = 3$, and $f'(7) = 8$. Find
   (a) $f^{-1}(3)$ and (b) $(f^{-1})'(3)$.

4. Find the inverse function of $f(x) = \dfrac{x+1}{2x+1}$.

5–9 ∎ Sketch a rough graph of the function without using a calculator.

5. $y = 5^x - 1$
6. $y = -e^{-x}$
7. $y = -\ln x$
8. $y = \ln(x - 1)$
9. $y = 2 \arctan x$

10. Let $a > 1$. For large values of x, which of the functions $y = x^a$, $y = a^x$, and $y = \log_a x$ has the largest values and which has the smallest values?

11–12 ∎ Find the exact value of each expression.

11. (a) $e^{2 \ln 3}$ (b) $\log_{10} 25 + \log_{10} 4$
12. (a) $\ln e^\pi$ (b) $\tan(\arcsin \tfrac{1}{2})$

13–20 ∎ Solve the equation for x.

13. $\ln x = \tfrac{1}{3}$
14. $e^x = \tfrac{1}{3}$
15. $e^{e^x} = 17$
16. $\ln(1 + e^{-x}) = 3$
17. $\ln(x + 1) + \ln(x - 1) = 1$
18. $\log_5(c^x) = d$
19. $\tan^{-1} x = 1$
20. $\sin x = 0.3$

21–47 ∎ Differentiate.

21. $f(t) = t^2 \ln t$
22. $g(t) = \dfrac{e^t}{1 + e^t}$
23. $h(\theta) = e^{\tan 2\theta}$
24. $h(u) = 10^{\sqrt{u}}$
25. $y = \ln|\sec 5x + \tan 5x|$
26. $y = e^{-t}(t^2 - 2t + 2)$
27. $y = e^{cx}(c \sin x - \cos x)$
28. $y = \sin^{-1}(e^x)$
29. $y = \ln(\sec^2 x)$
30. $y = \ln(x^2 e^x)$
31. $y = xe^{-1/x}$
32. $y = x' e^{sx}$
33. $y = 2^{-t^2}$
34. $y = e^{\cos x} + \cos(e^x)$

35. $H(v) = v \tan^{-1} v$

36. $F(z) = \log_{10}(1 + z^2)$

37. $y = x \sinh(x^2)$

38. $y = (\cos x)^x$

39. $y = \ln \sin x - \tfrac{1}{2} \sin^2 x$

40. $y = \arctan(\arcsin \sqrt{x})$

41. $y = \ln\left(\dfrac{1}{x}\right) + \dfrac{1}{\ln x}$

42. $xe^y = y - 1$

43. $y = \ln(\cosh 3x)$

44. $y = \dfrac{(x^2 + 1)^4}{(2x + 1)^3(3x - 1)^5}$

45. $y = \cosh^{-1}(\sinh x)$

46. $y = x \tanh^{-1} \sqrt{x}$

47. $f(x) = e^{\sin^3(\ln(x^2+1))}$

48. Show that
$$\dfrac{d}{dx}\left(\tfrac{1}{2} \tan^{-1} x + \tfrac{1}{4} \ln \dfrac{(x + 1)^2}{x^2 + 1}\right) = \dfrac{1}{(1 + x)(1 + x^2)}$$

49–52 ■ Find f' in terms of g'.

49. $f(x) = e^{g(x)}$

50. $f(x) = g(e^x)$

51. $f(x) = \ln |g(x)|$

52. $f(x) = g(\ln x)$

53–54 ■ Find $f^{(n)}(x)$.

53. $f(x) = 2^x$

54. $f(x) = \ln(2x)$

55. Use mathematical induction to show that if $f(x) = xe^x$, then $f^{(n)}(x) = (x + n)e^x$.

56. Find y' if $y = x + \arctan y$.

57–58 ■ Find an equation of the tangent to the curve at the given point.

57. $y = (2 + x)e^{-x}$, $(0, 2)$

58. $y = x \ln x$, (e, e)

59. At what point on the curve $y = [\ln(x + 4)]^2$ is the tangent horizontal?

60. If $f(x) = xe^{\sin x}$, find $f'(x)$. Graph f and f' on the same screen and comment.

61. (a) Find an equation of the tangent to the curve $y = e^x$ that is parallel to the line $x - 4y = 1$.
(b) Find an equation of the tangent to the curve $y = e^x$ that passes through the origin.

62. The function $C(t) = K(e^{-at} - e^{-bt})$, where a, b, and K are positive constants and $b > a$, is used to model the concentration at time t of a drug injected into the bloodstream.
(a) Show that $\lim_{t \to \infty} C(t) = 0$.
(b) Find $C'(t)$, the rate at which the drug is cleared from circulation.
(c) When is this rate equal to 0?

63–78 ■ Evaluate the limit.

63. $\lim\limits_{x \to \infty} e^{-3x}$

64. $\lim\limits_{x \to 10^-} \ln(100 - x^2)$

65. $\lim\limits_{x \to 3^-} e^{2/(x-3)}$

66. $\lim\limits_{x \to \infty} \arctan(x^3 - x)$

67. $\lim\limits_{x \to 0^+} \ln(\sinh x)$

68. $\lim\limits_{x \to \infty} e^{-x} \sin x$

69. $\lim\limits_{x \to \infty} \dfrac{1 + 2^x}{1 - 2^x}$

70. $\lim\limits_{x \to \infty} \left(1 + \dfrac{4}{x}\right)^x$

71. $\lim\limits_{x \to 0} \dfrac{\tan \pi x}{\ln(1 + x)}$

72. $\lim\limits_{x \to 0} \dfrac{1 - \cos x}{x^2 + x}$

73. $\lim\limits_{x \to 0} \dfrac{e^{4x} - 1 - 4x}{x^2}$

74. $\lim\limits_{x \to \infty} \dfrac{e^{4x} - 1 - 4x}{x^2}$

75. $\lim\limits_{x \to \infty} x^3 e^{-x}$

76. $\lim\limits_{x \to 0^+} x^2 \ln x$

77. $\lim\limits_{x \to 1^+} \left(\dfrac{x}{x - 1} - \dfrac{1}{\ln x}\right)$

78. $\lim\limits_{x \to (\pi/2)^-} (\tan x)^{\cos x}$

79–84 ■ Sketch the curve using the guidelines of Section 4.5.

79. $y = \tan^{-1}(1/x)$

80. $y = \sin^{-1}(1/x)$

81. $y = x \ln x$

82. $y = e^{2x - x^2}$

83. $y = e^x + e^{-3x}$

84. $y = \ln(x^2 - 1)$

85. Graph $f(x) = e^{-1/x^2}$ in a viewing rectangle that shows all the main aspects of this function. Estimate the inflection points. Then use calculus to find them exactly.

86. Investigate the family of functions $f(x) = cxe^{-cx^2}$. What happens to the maximum and minimum points and the inflection points as c changes? Illustrate your conclusions by graphing several members of the family.

87. An equation of motion of the form $s = Ae^{-ct} \cos(\omega t + \delta)$ represents damped oscillation of an object. Find the velocity and acceleration of the object.

88. (a) Show that there is exactly one root of the equation $\ln x = 3 - x$ and that it lies between 2 and e.
(b) Find the root of the equation in part (a) correct to four decimal places.

89. The biologist G. F. Gause conducted an experiment in the 1930s with the protozoan *Paramecium* and used the population function
$$P(t) = \dfrac{64}{1 + 31e^{-0.7944t}}$$
to model his data, where t was measured in days. Use this model to determine when the population was increasing most rapidly.

90–103 ▪ Evaluate the integral.

90. $\int_0^4 \dfrac{1}{16 + t^2}\, dt$

91. $\int_0^1 y e^{-2y^2}\, dy$

92. $\int_2^5 \dfrac{dr}{1 + 2r}$

93. $\int_2^4 \dfrac{1 + x - x^2}{x^2}\, dx$

94. $\int_0^{\pi/2} \dfrac{\cos x}{1 + \sin^2 x}\, dx$

95. $\int \dfrac{e^{\sqrt{x}}}{\sqrt{x}}\, dx$

96. $\int \dfrac{\cos(\ln x)}{x}\, dx$

97. $\int \dfrac{x + 1}{x^2 + 2x}\, dx$

98. $\int \dfrac{e^{-x}}{1 + e^{-2x}}\, dx$

99. $\int \tan x \ln(\cos x)\, dx$

100. $\int \dfrac{x}{\sqrt{1 - x^4}}\, dx$

101. $\int 2^{\tan \theta} \sec^2 \theta\, d\theta$

102. $\int \sinh au\, du$

103. $\int \dfrac{\sec \theta \tan \theta}{1 + \sec \theta}\, d\theta$

104–106 ▪ Use properties of integrals to prove the inequality.

104. $\int_0^1 \sqrt{1 + e^{2x}}\, dx \geq e - 1$

105. $\int_0^1 e^x \cos x\, dx \leq e - 1$

106. $\int_0^1 x \sin^{-1} x\, dx \leq \pi/4$

107–108 ▪ Find $f'(x)$.

107. $f(x) = \int_1^{\sqrt{x}} \dfrac{e^s}{s}\, ds$

108. $f(x) = \int_{\ln x}^{2x} e^{-t^2}\, dt$

109. Find the average value of the function $f(x) = 1/x$ on the interval $[1, 4]$.

110. Find the area of the region bounded by the curves $y = e^x$, $y = e^{-x}$, $x = -2$, and $x = 1$.

111. Find the volume of the solid obtained by rotating about the y-axis the region under the curve $y = 1/(1 + x^4)$ from $x = 0$ to $x = 1$.

112. If $f(x) = x + x^2 + e^x$ and $g(x) = f^{-1}(x)$, find $g'(1)$.

113. If g is the inverse function of $f(x) = \ln x + \tan^{-1} x$, find $g'(\pi/4)$.

114. What is the area of the largest rectangle in the first quadrant with two sides on the axes and one vertex on the curve $y = e^{-x}$?

115. What is the area of the largest triangle in the first quadrant with two sides on the axes and the third side tangent to the curve $y = e^{-x}$?

116. Evaluate $\int_0^1 e^x\, dx$ without using the Fundamental Theorem of Calculus. [*Hint:* Use the definition of a definite integral with right endpoints, sum a geometric series, and then use l'Hospital's Rule.]

117. If $F(x) = \int_a^b t^x\, dt$, where $a, b > 0$, then, by the Fundamental Theorem,

$$F(x) = \dfrac{b^{x+1} - a^{x+1}}{x + 1} \qquad x \neq -1$$

$$F(-1) = \ln b - \ln a$$

Use l'Hospital's Rule to show that F is continuous at -1.

118. Show that

$$\cos\{\arctan[\sin(\text{arccot } x)]\} = \sqrt{\dfrac{x^2 + 1}{x^2 + 2}}$$

119. If f is a continuous function such that

$$\int_0^x f(t)\, dt = xe^{2x} + \int_0^x e^{-t} f(t)\, dt$$

for all x, find an explicit formula for $f(x)$.

120. (a) Show that $\ln x < x - 1$ for $x > 0$, $x \neq 1$.
(b) Show that, for $x > 0$, $x \neq 1$,

$$\dfrac{x - 1}{x} < \ln x$$

(c) Deduce *Napier's Inequality:*

$$\dfrac{1}{b} < \dfrac{\ln b - \ln a}{b - a} < \dfrac{1}{a}$$

if $b > a > 0$.

(d) Give a geometric proof of Napier's Inequality by comparing the slopes of the three lines shown in the figure.

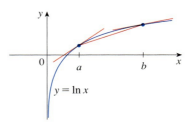

(e) Give another proof of Napier's Inequality by applying Property 8 of integrals (see Section 5.2) to $\int_a^b (1/x)\, dx$.

PROBLEMS PLUS

Cover up the solution to the example and try it yourself.

EXAMPLE For what values of c does the equation $\ln x = cx^2$ have exactly one solution?

SOLUTION One of the most important principles of problem solving is to draw a diagram, even if the problem as stated doesn't explicitly mention a geometric situation. Our present problem can be reformulated geometrically as follows: For what values of c does the curve $y = \ln x$ intersect the curve $y = cx^2$ in exactly one point?

Let's start by graphing $y = \ln x$ and $y = cx^2$ for various values of c. We know that, for $c \neq 0$, $y = cx^2$ is a parabola that opens upward if $c > 0$ and downward if $c < 0$. Figure 1 shows the parabolas $y = cx^2$ for several positive values of c. Most of them don't intersect $y = \ln x$ at all and one intersects twice. We have the feeling that there must be a value of c (somewhere between 0.1 and 0.3) for which the curves intersect exactly once, as in Figure 2.

To find that particular value of c, we let a be the x-coordinate of the single point of intersection. In other words, $\ln a = ca^2$, so a is the unique solution of the given equation. We see from Figure 2 that the curves just touch, so they have a common tangent line when $x = a$. That means the curves $y = \ln x$ and $y = cx^2$ have the same slope when $x = a$. Therefore

$$\frac{1}{a} = 2ca$$

FIGURE 1

Solving the equations $\ln a = ca^2$ and $1/a = 2ca$, we get

$$\ln a = ca^2 = c \cdot \frac{1}{2c} = \frac{1}{2}$$

Thus, $a = e^{1/2}$ and

$$c = \frac{\ln a}{a^2} = \frac{\ln e^{1/2}}{e} = \frac{1}{2e}$$

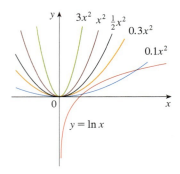

FIGURE 2

For negative values of c we have the situation illustrated in Figure 3: All parabolas $y = cx^2$ with negative values of c intersect $y = \ln x$ exactly once. And let's not forget about $c = 0$: The curve $y = 0x^2 = 0$ is just the x-axis, which intersects $y = \ln x$ exactly once.

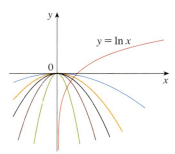

FIGURE 3

To summarize, the required values of c are $c = 1/(2e)$ and $c \leq 0$.

PROBLEMS

1. If a rectangle has its base on the x-axis and two vertices on the curve $y = e^{-x^2}$, show that the rectangle has the largest possible area when the two vertices are at the points of inflection of the curve.

2. Prove that $\log_2 5$ is an irrational number.

3. Show that
$$\frac{d^n}{dx^n}(e^{ax}\sin bx) = r^n e^{ax}\sin(bx + n\theta)$$
where a and b are positive numbers, $r^2 = a^2 + b^2$, and $\theta = \tan^{-1}(b/a)$.

4. Show that $\sin^{-1}(\tanh x) = \tan^{-1}(\sinh x)$.

5. Show that, for $x > 0$,
$$\frac{x}{1+x^2} < \tan^{-1}x < x$$

6. Suppose f is continuous, $f(0) = 0$, $f(1) = 1$, $f'(x) > 0$, and $\int_0^1 f(x)\,dx = \frac{1}{3}$. Find the value of the integral $\int_0^1 f^{-1}(y)\,dy$.

7. Show that $f(x) = \int_1^x \sqrt{1+t^3}\,dt$ is one-to-one and find $(f^{-1})'(0)$.

8. If
$$y = \frac{x}{\sqrt{a^2-1}} - \frac{2}{\sqrt{a^2-1}}\arctan\frac{\sin x}{a + \sqrt{a^2-1} + \cos x}$$
show that $y' = \dfrac{1}{a + \cos x}$.

9. For what value of a is the following equation true?
$$\lim_{x\to\infty}\left(\frac{x+a}{x-a}\right)^x = e$$

10. Sketch the set of all points (x, y) such that $|x + y| \leq e^x$.

11. Prove that $\cosh(\sinh x) < \sinh(\cosh x)$ for all x.

12. Show that, for all positive values of x and y,
$$\frac{e^{x+y}}{xy} \geq e^2$$

13. For what value of k does the equation $e^{2x} = k\sqrt{x}$ have exactly one solution?

14. For which positive numbers a is it true that $a^x \geq 1 + x$ for all x?

15. For which positive numbers a does the curve $y = a^x$ intersect the line $y = x$?

Chapter 8

The techniques of this chapter enable us to find the height of a rocket a minute after liftoff and to compute the escape velocity of the rocket.

Techniques of Integration

Because of the Fundamental Theorem of Calculus, we can integrate a function if we know an antiderivative, that is, an indefinite integral. We summarize here the most important integrals that we have learned so far.

$$\int x^n \, dx = \frac{x^{n+1}}{n+1} + C \quad (n \neq -1) \qquad \int \frac{1}{x} \, dx = \ln|x| + C$$

$$\int e^x \, dx = e^x + C \qquad \int a^x \, dx = \frac{a^x}{\ln a} + C$$

$$\int \sin x \, dx = -\cos x + C \qquad \int \cos x \, dx = \sin x + C$$

$$\int \sec^2 x \, dx = \tan x + C \qquad \int \csc^2 x \, dx = -\cot x + C$$

$$\int \sec x \tan x \, dx = \sec x + C \qquad \int \csc x \cot x \, dx = -\csc x + C$$

$$\int \sinh x \, dx = \cosh x + C \qquad \int \cosh x \, dx = \sinh x + C$$

$$\int \tan x \, dx = \ln|\sec x| + C \qquad \int \cot x \, dx = \ln|\sin x| + C$$

$$\int \frac{1}{x^2 + a^2} \, dx = \frac{1}{a} \tan^{-1}\left(\frac{x}{a}\right) + C \qquad \int \frac{1}{\sqrt{a^2 - x^2}} \, dx = \sin^{-1}\left(\frac{x}{a}\right) + C$$

In this chapter we develop techniques for using these basic integration formulas to obtain indefinite integrals of more complicated functions. We learned the most important method of integration, the Substitution Rule, in Section 5.5. The other general technique, integration by parts, is presented in Section 8.1. Then we learn methods that are special to particular classes of functions such as trigonometric functions and rational functions.

Integration is not as straightforward as differentiation; there are no rules that absolutely guarantee obtaining an indefinite integral of a function. Therefore, in Section 8.5 we discuss a strategy for integration.

8.1 Integration by Parts

Every differentiation rule has a corresponding integration rule. For instance, the Substitution Rule for integration corresponds to the Chain Rule for differentiation. The rule that corresponds to the Product Rule for differentiation is called the rule for *integration by parts*.

The Product Rule states that if f and g are differentiable functions, then

$$\frac{d}{dx}[f(x)g(x)] = f(x)g'(x) + g(x)f'(x)$$

In the notation for indefinite integrals this equation becomes

$$\int [f(x)g'(x) + g(x)f'(x)]\,dx = f(x)g(x)$$

or

$$\int f(x)g'(x)\,dx + \int g(x)f'(x)\,dx = f(x)g(x)$$

We can rearrange this equation as

[1]
$$\int f(x)g'(x)\,dx = f(x)g(x) - \int g(x)f'(x)\,dx$$

Formula 1 is called the **formula for integration by parts**. It is perhaps easier to remember in the following notation. Let $u = f(x)$ and $v = g(x)$. Then the differentials are $du = f'(x)\,dx$ and $dv = g'(x)\,dx$, so, by the Substitution Rule, the formula for integration by parts becomes

[2]
$$\int u\,dv = uv - \int v\,du$$

EXAMPLE 1 Find $\int x \sin x\,dx$.

SOLUTION USING FORMULA 1 Suppose we choose $f(x) = x$ and $g'(x) = \sin x$. Then $f'(x) = 1$ and $g(x) = -\cos x$. (For g we can choose *any* antiderivative of g'.) Thus, using Formula 1, we have

$$\int x \sin x\,dx = f(x)g(x) - \int g(x)f'(x)\,dx$$

$$= x(-\cos x) - \int (-\cos x)\,dx$$

$$= -x \cos x + \int \cos x\,dx$$

$$= -x \cos x + \sin x + C$$

It's wise to check the answer by differentiating it. If we do so, we get $x \sin x$, as expected.

SOLUTION USING FORMULA 2 Let

$$u = x \qquad dv = \sin x\,dx$$

|||| It is helpful to use the pattern:
$u = \square \qquad dv = \square$
$du = \square \qquad v = \square$

Then

$$du = dx \qquad v = -\cos x$$

and so

$$\int x \sin x\,dx = \int \underbrace{x}_{u}\,\underbrace{\sin x\,dx}_{dv} = \underbrace{x}_{u}\,\underbrace{(-\cos x)}_{v} - \int \underbrace{(-\cos x)}_{v}\,\underbrace{dx}_{du}$$

$$= -x \cos x + \int \cos x\,dx$$

$$= -x \cos x + \sin x + C$$

NOTE ▫ Our aim in using integration by parts is to obtain a simpler integral than the one we started with. Thus, in Example 1 we started with $\int x \sin x \, dx$ and expressed it in terms of the simpler integral $\int \cos x \, dx$. If we had chosen $u = \sin x$ and $dv = x \, dx$, then $du = \cos x \, dx$ and $v = x^2/2$, so integration by parts gives

$$\int x \sin x \, dx = (\sin x)\frac{x^2}{2} - \frac{1}{2}\int x^2 \cos x \, dx$$

Although this is true, $\int x^2 \cos x \, dx$ is a more difficult integral than the one we started with. In general, when deciding on a choice for u and dv, we usually try to choose $u = f(x)$ to be a function that becomes simpler when differentiated (or at least not more complicated) as long as $dv = g'(x) \, dx$ can be readily integrated to give v.

EXAMPLE 2 Evaluate $\int \ln x \, dx$.

SOLUTION Here we don't have much choice for u and dv. Let

$$u = \ln x \qquad dv = dx$$

Then

$$du = \frac{1}{x} dx \qquad v = x$$

Integrating by parts, we get

$$\int \ln x \, dx = x \ln x - \int x \, \frac{dx}{x}$$

|||| It's customary to write $\int 1 \, dx$ as $\int dx$.

$$= x \ln x - \int dx$$

|||| Check the answer by differentiating it.

$$= x \ln x - x + C$$

Integration by parts is effective in this example because the derivative of the function $f(x) = \ln x$ is simpler than f.

EXAMPLE 3 Find $\int t^2 e^t \, dt$.

SOLUTION Notice that t^2 becomes simpler when differentiated (whereas e^t is unchanged when differentiated or integrated), so we choose

$$u = t^2 \qquad dv = e^t \, dt$$

Then

$$du = 2t \, dt \qquad v = e^t$$

Integration by parts gives

$$\boxed{3} \qquad \int t^2 e^t \, dt = t^2 e^t - 2 \int t e^t \, dt$$

The integral that we obtained, $\int t e^t \, dt$, is simpler than the original integral but is still not obvious. Therefore, we use integration by parts a second time, this time with $u = t$ and

$dv = e^t\, dt$. Then $du = dt$, $v = e^t$, and

$$\int te^t\, dt = te^t - \int e^t\, dt$$
$$= te^t - e^t + C$$

Putting this in Equation 3, we get

$$\int t^2 e^t\, dt = t^2 e^t - 2 \int te^t\, dt$$
$$= t^2 e^t - 2(te^t - e^t + C)$$
$$= t^2 e^t - 2te^t + 2e^t + C_1 \qquad \text{where } C_1 = -2C$$

EXAMPLE 4 Evaluate $\int e^x \sin x\, dx$.

SOLUTION Neither e^x nor $\sin x$ becomes simpler when differentiated, but we try choosing $u = e^x$ and $dv = \sin x\, dx$ anyway. Then $du = e^x\, dx$ and $v = -\cos x$, so integration by parts gives

|||| An easier method, using complex numbers, is given in Exercise 48 in Appendix G.

$$\boxed{4} \qquad \int e^x \sin x\, dx = -e^x \cos x + \int e^x \cos x\, dx$$

The integral that we have obtained, $\int e^x \cos x\, dx$, is no simpler than the original one, but at least it's no more difficult. Having had success in the preceding example integrating by parts twice, we persevere and integrate by parts again. This time we use $u = e^x$ and $dv = \cos x\, dx$. Then $du = e^x\, dx$, $v = \sin x$, and

$$\boxed{5} \qquad \int e^x \cos x\, dx = e^x \sin x - \int e^x \sin x\, dx$$

At first glance, it appears as if we have accomplished nothing because we have arrived at $\int e^x \sin x\, dx$, which is where we started. However, if we put Equation 5 into Equation 4 we get

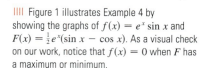

|||| Figure 1 illustrates Example 4 by showing the graphs of $f(x) = e^x \sin x$ and $F(x) = \tfrac{1}{2} e^x (\sin x - \cos x)$. As a visual check on our work, notice that $f(x) = 0$ when F has a maximum or minimum.

$$\int e^x \sin x\, dx = -e^x \cos x + e^x \sin x - \int e^x \sin x\, dx$$

This can be regarded as an equation to be solved for the unknown integral. Adding $\int e^x \sin x\, dx$ to both sides, we obtain

$$2 \int e^x \sin x\, dx = -e^x \cos x + e^x \sin x$$

Dividing by 2 and adding the constant of integration, we get

$$\int e^x \sin x\, dx = \tfrac{1}{2} e^x (\sin x - \cos x) + C$$

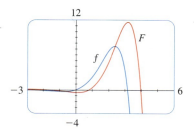

FIGURE 1

If we combine the formula for integration by parts with Part 2 of the Fundamental Theorem of Calculus, we can evaluate definite integrals by parts. Evaluating both sides of Formula 1 between a and b, assuming f' and g' are continuous, and using the Fundamental

Theorem, we obtain

$$\boxed{6} \qquad \int_a^b f(x)g'(x)\,dx = f(x)g(x)\Big]_a^b - \int_a^b g(x)f'(x)\,dx$$

EXAMPLE 5 Calculate $\int_0^1 \tan^{-1}x\,dx$.

SOLUTION Let

$$u = \tan^{-1}x \qquad dv = dx$$

Then

$$du = \frac{dx}{1+x^2} \qquad v = x$$

So Formula 6 gives

$$\int_0^1 \tan^{-1}x\,dx = x\tan^{-1}x\Big]_0^1 - \int_0^1 \frac{x}{1+x^2}\,dx$$

$$= 1 \cdot \tan^{-1}1 - 0 \cdot \tan^{-1}0 - \int_0^1 \frac{x}{1+x^2}\,dx$$

$$= \frac{\pi}{4} - \int_0^1 \frac{x}{1+x^2}\,dx$$

|||| Since $\tan^{-1}x \geq 0$ for $x \geq 0$, the integral in Example 5 can be interpreted as the area of the region shown in Figure 2.

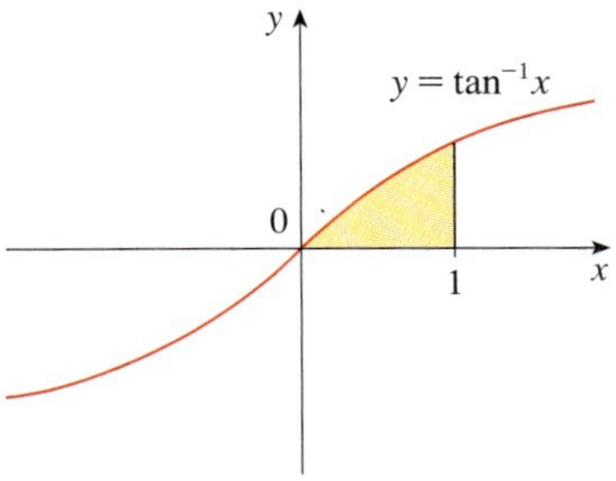

To evaluate this integral we use the substitution $t = 1 + x^2$ (since u has another meaning in this example). Then $dt = 2x\,dx$, so $x\,dx = dt/2$. When $x = 0$, $t = 1$; when $x = 1$, $t = 2$; so

$$\int_0^1 \frac{x}{1+x^2}\,dx = \frac{1}{2}\int_1^2 \frac{dt}{t} = \frac{1}{2}\ln|t|\Big]_1^2$$

$$= \frac{1}{2}(\ln 2 - \ln 1) = \frac{1}{2}\ln 2$$

Therefore

$$\int_0^1 \tan^{-1}x\,dx = \frac{\pi}{4} - \int_0^1 \frac{x}{1+x^2}\,dx = \frac{\pi}{4} - \frac{\ln 2}{2}$$

FIGURE 2

EXAMPLE 6 Prove the reduction formula

|||| Equation 7 is called a *reduction formula* because the exponent n has been *reduced* to $n - 1$ and $n - 2$.

$$\boxed{7} \qquad \int \sin^n x\,dx = -\frac{1}{n}\cos x \sin^{n-1}x + \frac{n-1}{n}\int \sin^{n-2}x\,dx$$

where $n \geq 2$ is an integer.

SOLUTION Let $u = \sin^{n-1}x \qquad\qquad dv = \sin x\,dx$

Then $du = (n-1)\sin^{n-2}x \cos x\,dx \qquad v = -\cos x$

so integration by parts gives

$$\int \sin^n x\,dx = -\cos x \sin^{n-1}x + (n-1)\int \sin^{n-2}x \cos^2 x\,dx$$

Since $\cos^2 x = 1 - \sin^2 x$, we have

$$\int \sin^n x \, dx = -\cos x \sin^{n-1} x + (n-1) \int \sin^{n-2} x \, dx - (n-1) \int \sin^n x \, dx$$

As in Example 4, we solve this equation for the desired integral by taking the last term on the right side to the left side. Thus, we have

$$n \int \sin^n x \, dx = -\cos x \sin^{n-1} x + (n-1) \int \sin^{n-2} x \, dx$$

or

$$\int \sin^n x \, dx = -\frac{1}{n} \cos x \sin^{n-1} x + \frac{n-1}{n} \int \sin^{n-2} x \, dx$$

The reduction formula (7) is useful because by using it repeatedly we could eventually express $\int \sin^n x \, dx$ in terms of $\int \sin x \, dx$ (if n is odd) or $\int (\sin x)^0 \, dx = \int dx$ (if n is even).

8.1 Exercises

1–2 Evaluate the integral using integration by parts with the indicated choices of u and dv.

1. $\int x \ln x \, dx$; $\quad u = \ln x, \ dv = x \, dx$

2. $\int \theta \sec^2 \theta \, d\theta$; $\quad u = \theta, \ dv = \sec^2 \theta \, d\theta$

3–32 Evaluate the integral.

3. $\int x \cos 5x \, dx$

4. $\int xe^{-x} \, dx$

5. $\int re^{r/2} \, dr$

6. $\int t \sin 2t \, dt$

7. $\int x^2 \sin \pi x \, dx$

8. $\int x^2 \cos mx \, dx$

9. $\int \ln(2x+1) \, dx$

10. $\int \sin^{-1} x \, dx$

11. $\int \arctan 4t \, dt$

12. $\int p^5 \ln p \, dp$

13. $\int (\ln x)^2 \, dx$

14. $\int t^3 e^t \, dt$

15. $\int e^{2\theta} \sin 3\theta \, d\theta$

16. $\int e^{-\theta} \cos 2\theta \, d\theta$

17. $\int y \sinh y \, dy$

18. $\int y \cosh ay \, dy$

19. $\int_0^\pi t \sin 3t \, dt$

20. $\int_0^1 (x^2+1)e^{-x} \, dx$

21. $\int_1^2 \frac{\ln x}{x^2} \, dx$

22. $\int_1^4 \sqrt{t} \ln t \, dt$

23. $\int_0^1 \frac{y}{e^{2y}} \, dy$

24. $\int_{\pi/4}^{\pi/2} x \csc^2 x \, dx$

25. $\int_0^{1/2} \cos^{-1} x \, dx$

26. $\int_0^1 x 5^x \, dx$

27. $\int \cos x \ln(\sin x) \, dx$

28. $\int_1^{\sqrt{3}} \arctan(1/x) \, dx$

29. $\int \cos(\ln x) \, dx$

30. $\int_0^1 \frac{r^3}{\sqrt{4+r^2}} \, dr$

31. $\int_1^2 x^4 (\ln x)^2 \, dx$

32. $\int_0^t e^s \sin(t-s) \, ds$

33–36 First make a substitution and then use integration by parts to evaluate the integral.

33. $\int \sin \sqrt{x} \, dx$

34. $\int_1^4 e^{\sqrt{x}} \, dx$

35. $\int_{\sqrt{\pi/2}}^{\sqrt{\pi}} \theta^3 \cos(\theta^2) \, d\theta$

36. $\int x^5 e^{x^2} \, dx$

37–40 Evaluate the indefinite integral. Illustrate, and check that your answer is reasonable, by graphing both the function and its antiderivative (take $C = 0$).

37. $\int x \cos \pi x \, dx$

38. $\int x^{3/2} \ln x \, dx$

39. $\int (2x+3) e^x \, dx$

40. $\int x^3 e^{x^2} \, dx$

41. (a) Use the reduction formula in Example 6 to show that

$$\int \sin^2 x \, dx = \frac{x}{2} - \frac{\sin 2x}{4} + C$$

(b) Use part (a) and the reduction formula to evaluate $\int \sin^4 x \, dx$.

42. (a) Prove the reduction formula

$$\int \cos^n x \, dx = \frac{1}{n} \cos^{n-1} x \sin x + \frac{n-1}{n} \int \cos^{n-2} x \, dx$$

(b) Use part (a) to evaluate $\int \cos^2 x \, dx$.
(c) Use parts (a) and (b) to evaluate $\int \cos^4 x \, dx$.

43. (a) Use the reduction formula in Example 6 to show that

$$\int_0^{\pi/2} \sin^n x \, dx = \frac{n-1}{n} \int_0^{\pi/2} \sin^{n-2} x \, dx$$

where $n \geq 2$ is an integer.
(b) Use part (a) to evaluate $\int_0^{\pi/2} \sin^3 x \, dx$ and $\int_0^{\pi/2} \sin^5 x \, dx$.
(c) Use part (a) to show that, for odd powers of sine,

$$\int_0^{\pi/2} \sin^{2n+1} x \, dx = \frac{2 \cdot 4 \cdot 6 \cdot \cdots \cdot 2n}{3 \cdot 5 \cdot 7 \cdot \cdots \cdot (2n+1)}$$

44. Prove that, for even powers of sine,

$$\int_0^{\pi/2} \sin^{2n} x \, dx = \frac{1 \cdot 3 \cdot 5 \cdot \cdots \cdot (2n-1)}{2 \cdot 4 \cdot 6 \cdot \cdots \cdot 2n} \frac{\pi}{2}$$

45–48 |||| Use integration by parts to prove the reduction formula.

45. $\int (\ln x)^n \, dx = x(\ln x)^n - n \int (\ln x)^{n-1} \, dx$

46. $\int x^n e^x \, dx = x^n e^x - n \int x^{n-1} e^x \, dx$

47. $\int (x^2 + a^2)^n \, dx$

$= \frac{x(x^2 + a^2)^n}{2n+1} + \frac{2na^2}{2n+1} \int (x^2 + a^2)^{n-1} \, dx \quad (n \neq -\tfrac{1}{2})$

48. $\int \sec^n x \, dx = \frac{\tan x \sec^{n-2} x}{n-1} + \frac{n-2}{n-1} \int \sec^{n-2} x \, dx \quad (n \neq 1)$

49. Use Exercise 45 to find $\int (\ln x)^3 \, dx$.

50. Use Exercise 46 to find $\int x^4 e^x \, dx$.

51–52 |||| Find the area of the region bounded by the given curves.

51. $y = xe^{-0.4x}$, $y = 0$, $x = 5$

52. $y = 5 \ln x$, $y = x \ln x$

53–54 |||| Use a graph to find approximate x-coordinates of the points of intersection of the given curves. Then find (approximately) the area of the region bounded by the curves.

53. $y = x \sin x$, $y = (x - 2)^2$

54. $y = \arctan 3x$, $y = x/2$

55–58 |||| Use the method of cylindrical shells to find the volume generated by rotating the region bounded by the given curves about the specified axis.

55. $y = \cos(\pi x/2)$, $y = 0$, $0 \leq x \leq 1$; about the y-axis

56. $y = e^x$, $y = e^{-x}$, $x = 1$; about the y-axis

57. $y = e^{-x}$, $y = 0$, $x = -1$, $x = 0$; about $x = 1$

58. $y = e^x$, $x = 0$, $y = \pi$; about the x-axis

59. Find the average value of $f(x) = x^2 \ln x$ on the interval $[1, 3]$.

60. A rocket accelerates by burning its onboard fuel, so its mass decreases with time. Suppose the initial mass of the rocket at liftoff (including its fuel) is m, the fuel is consumed at rate r, and the exhaust gases are ejected with constant velocity v_e (relative to the rocket). A model for the velocity of the rocket at time t is given by the equation

$$v(t) = -gt - v_e \ln \frac{m - rt}{m}$$

where g is the acceleration due to gravity and t is not too large. If $g = 9.8$ m/s^2, $m = 30{,}000$ kg, $r = 160$ kg/s, and $v_e = 3000$ m/s, find the height of the rocket one minute after liftoff.

61. A particle that moves along a straight line has velocity $v(t) = t^2 e^{-t}$ meters per second after t seconds. How far will it travel during the first t seconds?

62. If $f(0) = g(0) = 0$ and f'' and g'' are continuous, show that

$$\int_0^a f(x) g''(x) \, dx = f(a) g'(a) - f'(a) g(a) + \int_0^a f''(x) g(x) \, dx$$

63. Suppose that $f(1) = 2$, $f(4) = 7$, $f'(1) = 5$, $f'(4) = 3$, and f'' is continuous. Find the value of $\int_1^4 x f''(x) \, dx$.

64. (a) Use integration by parts to show that

$$\int f(x) \, dx = x f(x) - \int x f'(x) \, dx$$

(b) If f and g are inverse functions and f' is continuous, prove that

$$\int_a^b f(x) \, dx = bf(b) - af(a) - \int_{f(a)}^{f(b)} g(y) \, dy$$

[*Hint:* Use part (a) and make the substitution $y = f(x)$.]

(c) In the case where f and g are positive functions and $b > a > 0$, draw a diagram to give a geometric interpretation of part (b).

(d) Use part (b) to evaluate $\int_1^e \ln x \, dx$.

65. We arrived at Formula 6.3.2, $V = \int_a^b 2\pi x f(x) \, dx$, by using cylindrical shells, but now we can use integration by parts to prove it using the slicing method of Section 6.2, at least for the case where f is one-to-one and therefore has an inverse function g. Use the figure to show that

$$V = \pi b^2 d - \pi a^2 c - \int_c^d \pi [g(y)]^2 \, dy$$

Make the substitution $y = f(x)$ and then use integration by parts on the resulting integral to prove that $V = \int_a^b 2\pi x f(x) \, dx$.

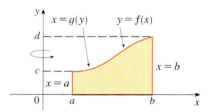

66. Let $I_n = \int_0^{\pi/2} \sin^n x \, dx$.

(a) Show that $I_{2n+2} \leq I_{2n+1} \leq I_{2n}$.

(b) Use Exercise 44 to show that

$$\frac{I_{2n+2}}{I_{2n}} = \frac{2n+1}{2n+2}$$

(c) Use parts (a) and (b) to show that

$$\frac{2n+1}{2n+2} \leq \frac{I_{2n+1}}{I_{2n}} \leq 1$$

and deduce that $\lim_{n \to \infty} I_{2n+1}/I_{2n} = 1$.

(d) Use part (c) and Exercises 43 and 44 to show that

$$\lim_{n \to \infty} \frac{2}{1} \cdot \frac{2}{3} \cdot \frac{4}{3} \cdot \frac{4}{5} \cdot \frac{6}{5} \cdot \frac{6}{7} \cdot \ldots \cdot \frac{2n}{2n-1} \cdot \frac{2n}{2n+1} = \frac{\pi}{2}$$

This formula is usually written as an infinite product:

$$\frac{\pi}{2} = \frac{2}{1} \cdot \frac{2}{3} \cdot \frac{4}{3} \cdot \frac{4}{5} \cdot \frac{6}{5} \cdot \frac{6}{7} \cdot \ldots$$

and is called the *Wallis product*.

(e) We construct rectangles as follows. Start with a square of area 1 and attach rectangles of area 1 alternately beside or on top of the previous rectangle (see the figure). Find the limit of the ratios of width to height of these rectangles.

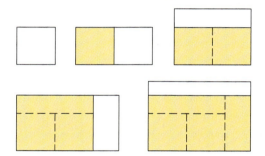

8.2 Trigonometric Integrals

In this section we use trigonometric identities to integrate certain combinations of trigonometric functions. We start with powers of sine and cosine.

EXAMPLE 1 Evaluate $\int \cos^3 x \, dx$.

SOLUTION Simply substituting $u = \cos x$ isn't helpful, since then $du = -\sin x \, dx$. In order to integrate powers of cosine, we would need an extra $\sin x$ factor. Similarly, a power of sine would require an extra $\cos x$ factor. Thus, here we can separate one cosine factor and convert the remaining $\cos^2 x$ factor to an expression involving sine using the identity $\sin^2 x + \cos^2 x = 1$:

$$\cos^3 x = \cos^2 x \cdot \cos x = (1 - \sin^2 x) \cos x$$

We can then evaluate the integral by substituting $u = \sin x$, so $du = \cos x \, dx$ and

$$\int \cos^3 x \, dx = \int \cos^2 x \cdot \cos x \, dx = \int (1 - \sin^2 x) \cos x \, dx$$

$$= \int (1 - u^2) \, du = u - \tfrac{1}{3} u^3 + C$$

$$= \sin x - \tfrac{1}{3} \sin^3 x + C$$

In general, we try to write an integrand involving powers of sine and cosine in a form where we have only one sine factor (and the remainder of the expression in terms of cosine) or only one cosine factor (and the remainder of the expression in terms of sine). The identity $\sin^2 x + \cos^2 x = 1$ enables us to convert back and forth between even powers of sine and cosine.

EXAMPLE 2 Find $\int \sin^5 x \cos^2 x \, dx$

SOLUTION We could convert $\cos^2 x$ to $1 - \sin^2 x$, but we would be left with an expression in terms of $\sin x$ with no extra $\cos x$ factor. Instead, we separate a single sine factor and rewrite the remaining $\sin^4 x$ factor in terms of $\cos x$:

$$\sin^5 x \cos^2 x = (\sin^2 x)^2 \cos^2 x \sin x = (1 - \cos^2 x)^2 \cos^2 x \sin x$$

Substituting $u = \cos x$, we have $du = -\sin x \, dx$ and so

$$\int \sin^5 x \cos^2 x \, dx = \int (\sin^2 x)^2 \cos^2 x \sin x \, dx$$

$$= \int (1 - \cos^2 x)^2 \cos^2 x \sin x \, dx$$

$$= \int (1 - u^2)^2 u^2 (-du) = -\int (u^2 - 2u^4 + u^6) \, du$$

$$= -\left(\frac{u^3}{3} - 2\frac{u^5}{5} + \frac{u^7}{7}\right) + C$$

$$= -\tfrac{1}{3}\cos^3 x + \tfrac{2}{5}\cos^5 x - \tfrac{1}{7}\cos^7 x + C$$

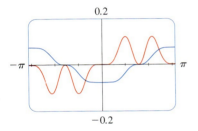

|||| Figure 1 shows the graphs of the integrand $\sin^5 x \cos^2 x$ in Example 2 and its indefinite integral (with $C = 0$). Which is which?

FIGURE 1

In the preceding examples, an odd power of sine or cosine enabled us to separate a single factor and convert the remaining even power. If the integrand contains even powers of both sine and cosine, this strategy fails. In this case, we can take advantage of the following half-angle identities (see Equations 17b and 17a in Appendix D):

$$\sin^2 x = \tfrac{1}{2}(1 - \cos 2x) \quad \text{and} \quad \cos^2 x = \tfrac{1}{2}(1 + \cos 2x)$$

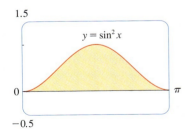

|||| Example 3 shows that the area of the region shown in Figure 2 is $\pi/2$.

FIGURE 2

EXAMPLE 3 Evaluate $\int_0^\pi \sin^2 x \, dx$.

SOLUTION If we write $\sin^2 x = 1 - \cos^2 x$, the integral is no simpler to evaluate. Using the half-angle formula for $\sin^2 x$, however, we have

$$\int_0^\pi \sin^2 x \, dx = \tfrac{1}{2} \int_0^\pi (1 - \cos 2x) \, dx = \left[\tfrac{1}{2}\left(x - \tfrac{1}{2}\sin 2x\right)\right]_0^\pi$$

$$= \tfrac{1}{2}\left(\pi - \tfrac{1}{2}\sin 2\pi\right) - \tfrac{1}{2}\left(0 - \tfrac{1}{2}\sin 0\right) = \tfrac{1}{2}\pi$$

Notice that we mentally made the substitution $u = 2x$ when integrating $\cos 2x$. Another method for evaluating this integral was given in Exercise 41 in Section 8.1.

EXAMPLE 4 Find $\int \sin^4 x \, dx$.

SOLUTION We could evaluate this integral using the reduction formula for $\int \sin^n x \, dx$ (Equation 8.1.7) together with Example 1 (as in Exercise 41 in Section 8.1), but a better method is to write $\sin^4 x = (\sin^2 x)^2$ and use a half-angle formula:

$$\int \sin^4 x \, dx = \int (\sin^2 x)^2 \, dx$$

$$= \int \left(\frac{1 - \cos 2x}{2} \right)^2 dx$$

$$= \tfrac{1}{4} \int (1 - 2 \cos 2x + \cos^2 2x) \, dx$$

Since $\cos^2 2x$ occurs, we must use another half-angle formula

$$\cos^2 2x = \tfrac{1}{2}(1 + \cos 4x)$$

This gives

$$\int \sin^4 x \, dx = \tfrac{1}{4} \int [1 - 2 \cos 2x + \tfrac{1}{2}(1 + \cos 4x)] \, dx$$

$$= \tfrac{1}{4} \int \left(\tfrac{3}{2} - 2 \cos 2x + \tfrac{1}{2} \cos 4x \right) dx$$

$$= \tfrac{1}{4} \left(\tfrac{3}{2} x - \sin 2x + \tfrac{1}{8} \sin 4x \right) + C$$

To summarize, we list guidelines to follow when evaluating integrals of the form $\int \sin^m x \cos^n x \, dx$, where $m \geq 0$ and $n \geq 0$ are integers.

Strategy for Evaluating $\int \sin^m x \cos^n x \, dx$

(a) If the power of cosine is odd ($n = 2k + 1$), save one cosine factor and use $\cos^2 x = 1 - \sin^2 x$ to express the remaining factors in terms of sine:

$$\int \sin^m x \cos^{2k+1} x \, dx = \int \sin^m x \, (\cos^2 x)^k \cos x \, dx$$

$$= \int \sin^m x \, (1 - \sin^2 x)^k \cos x \, dx$$

Then substitute $u = \sin x$.

(b) If the power of sine is odd ($m = 2k + 1$), save one sine factor and use $\sin^2 x = 1 - \cos^2 x$ to express the remaining factors in terms of cosine:

$$\int \sin^{2k+1} x \cos^n x \, dx = \int (\sin^2 x)^k \cos^n x \sin x \, dx$$

$$= \int (1 - \cos^2 x)^k \cos^n x \sin x \, dx$$

Then substitute $u = \cos x$. [Note that if the powers of both sine and cosine are odd, either (a) or (b) can be used.]

(c) If the powers of both sine and cosine are even, use the half-angle identities

$$\sin^2 x = \tfrac{1}{2}(1 - \cos 2x) \qquad \cos^2 x = \tfrac{1}{2}(1 + \cos 2x)$$

It is sometimes helpful to use the identity

$$\sin x \cos x = \tfrac{1}{2} \sin 2x$$

We can use a similar strategy to evaluate integrals of the form $\int \tan^m x \sec^n x \, dx$. Since $(d/dx) \tan x = \sec^2 x$, we can separate a $\sec^2 x$ factor and convert the remaining (even) power of secant to an expression involving tangent using the identity $\sec^2 x = 1 + \tan^2 x$. Or, since $(d/dx) \sec x = \sec x \tan x$, we can separate a $\sec x \tan x$ factor and convert the remaining (even) power of tangent to secant.

EXAMPLE 5 Evaluate $\int \tan^6 x \sec^4 x \, dx$.

SOLUTION If we separate one $\sec^2 x$ factor, we can express the remaining $\sec^2 x$ factor in terms of tangent using the identity $\sec^2 x = 1 + \tan^2 x$. We can then evaluate the integral by substituting $u = \tan x$ with $du = \sec^2 x \, dx$:

$$\int \tan^6 x \sec^4 x \, dx = \int \tan^6 x \sec^2 x \sec^2 x \, dx$$

$$= \int \tan^6 x \, (1 + \tan^2 x) \sec^2 x \, dx$$

$$= \int u^6 (1 + u^2) \, du = \int (u^6 + u^8) \, du$$

$$= \frac{u^7}{7} + \frac{u^9}{9} + C$$

$$= \tfrac{1}{7} \tan^7 x + \tfrac{1}{9} \tan^9 x + C$$

EXAMPLE 6 Find $\int \tan^5 \theta \sec^7 \theta \, d\theta$.

SOLUTION If we separate a $\sec^2 \theta$ factor, as in the preceding example, we are left with a $\sec^5 \theta$ factor, which isn't easily converted to tangent. However, if we separate a $\sec \theta \tan \theta$ factor, we can convert the remaining power of tangent to an expression involving only secant using the identity $\tan^2 \theta = \sec^2 \theta - 1$. We can then evaluate the integral by substituting $u = \sec \theta$, so $du = \sec \theta \tan \theta \, d\theta$:

$$\int \tan^5 \theta \sec^7 \theta \, d\theta = \int \tan^4 \theta \sec^6 \theta \sec \theta \tan \theta \, d\theta$$

$$= \int (\sec^2 \theta - 1)^2 \sec^6 \theta \sec \theta \tan \theta \, d\theta$$

$$= \int (u^2 - 1)^2 u^6 \, du = \int (u^{10} - 2u^8 + u^6) \, du$$

$$= \frac{u^{11}}{11} - 2\frac{u^9}{9} + \frac{u^7}{7} + C$$

$$= \tfrac{1}{11} \sec^{11} \theta - \tfrac{2}{9} \sec^9 \theta + \tfrac{1}{7} \sec^7 \theta + C$$

The preceding examples demonstrate strategies for evaluating integrals of the form $\int \tan^m x \sec^n x \, dx$ for two cases, which we summarize here.

> **Strategy for Evaluating $\int \tan^m x \sec^n x \, dx$**
>
> (a) If the power of secant is even ($n = 2k$, $k \geq 2$), save a factor of $\sec^2 x$ and use $\sec^2 x = 1 + \tan^2 x$ to express the remaining factors in terms of $\tan x$:
>
> $$\int \tan^m x \sec^{2k} x \, dx = \int \tan^m x \, (\sec^2 x)^{k-1} \sec^2 x \, dx$$
>
> $$= \int \tan^m x \, (1 + \tan^2 x)^{k-1} \sec^2 x \, dx$$
>
> Then substitute $u = \tan x$.
>
> (b) If the power of tangent is odd ($m = 2k + 1$), save a factor of $\sec x \tan x$ and use $\tan^2 x = \sec^2 x - 1$ to express the remaining factors in terms of $\sec x$:
>
> $$\int \tan^{2k+1} x \sec^n x \, dx = \int (\tan^2 x)^k \sec^{n-1} x \, \sec x \tan x \, dx$$
>
> $$= \int (\sec^2 x - 1)^k \sec^{n-1} x \, \sec x \tan x \, dx$$
>
> Then substitute $u = \sec x$.

For other cases, the guidelines are not as clear-cut. We may need to use identities, integration by parts, and occasionally a little ingenuity. We will sometimes need to be able to integrate $\tan x$ by using the formula established in (5.5.5):

$$\int \tan x \, dx = \ln|\sec x| + C$$

We will also need the indefinite integral of secant:

$$\boxed{1} \quad \int \sec x \, dx = \ln|\sec x + \tan x| + C$$

We could verify Formula 1 by differentiating the right side, or as follows. First we multiply numerator and denominator by $\sec x + \tan x$:

$$\int \sec x \, dx = \int \sec x \, \frac{\sec x + \tan x}{\sec x + \tan x} \, dx$$

$$= \int \frac{\sec^2 x + \sec x \tan x}{\sec x + \tan x} \, dx$$

If we substitute $u = \sec x + \tan x$, then $du = (\sec x \tan x + \sec^2 x) \, dx$, so the integral becomes $\int (1/u) \, du = \ln|u| + C$. Thus, we have

$$\int \sec x \, dx = \ln|\sec x + \tan x| + C$$

EXAMPLE 7 Find $\int \tan^3 x \, dx$.

SOLUTION Here only $\tan x$ occurs, so we use $\tan^2 x = \sec^2 x - 1$ to rewrite a $\tan^2 x$ factor in terms of $\sec^2 x$:

$$\int \tan^3 x \, dx = \int \tan x \tan^2 x \, dx$$
$$= \int \tan x \, (\sec^2 x - 1) \, dx$$
$$= \int \tan x \sec^2 x \, dx - \int \tan x \, dx$$
$$= \frac{\tan^2 x}{2} - \ln|\sec x| + C$$

In the first integral we mentally substituted $u = \tan x$ so that $du = \sec^2 x \, dx$.

If an even power of tangent appears with an odd power of secant, it is helpful to express the integrand completely in terms of $\sec x$. Powers of $\sec x$ may require integration by parts, as shown in the following example.

EXAMPLE 8 Find $\int \sec^3 x \, dx$.

SOLUTION Here we integrate by parts with

$$u = \sec x \qquad dv = \sec^2 x \, dx$$
$$du = \sec x \tan x \, dx \qquad v = \tan x$$

Then

$$\int \sec^3 x \, dx = \sec x \tan x - \int \sec x \tan^2 x \, dx$$
$$= \sec x \tan x - \int \sec x \, (\sec^2 x - 1) \, dx$$
$$= \sec x \tan x - \int \sec^3 x \, dx + \int \sec x \, dx$$

Using Formula 1 and solving for the required integral, we get

$$\int \sec^3 x \, dx = \tfrac{1}{2}\bigl(\sec x \tan x + \ln|\sec x + \tan x|\bigr) + C$$

Integrals such as the one in the preceding example may seem very special but they occur frequently in applications of integration, as we will see in Chapter 9. Integrals of the form $\int \cot^m x \csc^n x \, dx$ can be found by similar methods because of the identity $1 + \cot^2 x = \csc^2 x$.

Finally, we can make use of another set of trigonometric identities:

> **2** To evaluate the integrals (a) $\int \sin mx \cos nx \, dx$, (b) $\int \sin mx \sin nx \, dx$, or (c) $\int \cos mx \cos nx \, dx$, use the corresponding identity:
>
> (a) $\sin A \cos B = \tfrac{1}{2}[\sin(A - B) + \sin(A + B)]$
>
> (b) $\sin A \sin B = \tfrac{1}{2}[\cos(A - B) - \cos(A + B)]$
>
> (c) $\cos A \cos B = \tfrac{1}{2}[\cos(A - B) + \cos(A + B)]$

|||| These product identities are discussed in Appendix D.

EXAMPLE 9 Evaluate $\int \sin 4x \cos 5x \, dx$.

SOLUTION This integral could be evaluated using integration by parts, but it's easier to use the identity in Equation 2(a) as follows:

$$\int \sin 4x \cos 5x \, dx = \int \tfrac{1}{2}[\sin(-x) + \sin 9x] \, dx$$

$$= \tfrac{1}{2} \int (-\sin x + \sin 9x) \, dx$$

$$= \tfrac{1}{2}(\cos x - \tfrac{1}{9} \cos 9x) + C$$

8.2 Exercises

1–47 Evaluate the integral.

1. $\int \sin^3 x \cos^2 x \, dx$
2. $\int \sin^6 x \cos^3 x \, dx$
3. $\int_{\pi/2}^{3\pi/4} \sin^5 x \cos^3 x \, dx$
4. $\int_0^{\pi/2} \cos^5 x \, dx$
5. $\int \cos^5 x \sin^4 x \, dx$
6. $\int \sin^3(mx) \, dx$
7. $\int_0^{\pi/2} \cos^2 \theta \, d\theta$
8. $\int_0^{\pi/2} \sin^2(2\theta) \, d\theta$
9. $\int_0^{\pi} \sin^4(3t) \, dt$
10. $\int_0^{\pi} \cos^6 \theta \, d\theta$
11. $\int (1 + \cos \theta)^2 \, d\theta$
12. $\int x \cos^2 x \, dx$
13. $\int_0^{\pi/4} \sin^4 x \cos^2 x \, dx$
14. $\int_0^{\pi/2} \sin^2 x \cos^2 x \, dx$
15. $\int \sin^3 x \sqrt{\cos x} \, dx$
16. $\int \cos \theta \cos^5(\sin \theta) \, d\theta$
17. $\int \cos^2 x \tan^3 x \, dx$
18. $\int \cot^5 \theta \sin^4 \theta \, d\theta$
19. $\int \dfrac{1 - \sin x}{\cos x} \, dx$
20. $\int \cos^2 x \sin 2x \, dx$
21. $\int \sec^2 x \tan x \, dx$
22. $\int_0^{\pi/2} \sec^4(t/2) \, dt$
23. $\int \tan^2 x \, dx$
24. $\int \tan^4 x \, dx$
25. $\int \sec^6 t \, dt$
26. $\int_0^{\pi/4} \sec^4 \theta \tan^4 \theta \, d\theta$
27. $\int_0^{\pi/3} \tan^5 x \sec^4 x \, dx$
28. $\int \tan^3(2x) \sec^5(2x) \, dx$
29. $\int \tan^3 x \sec x \, dx$
30. $\int_0^{\pi/3} \tan^5 x \sec^6 x \, dx$
31. $\int \tan^5 x \, dx$
32. $\int \tan^6(ay) \, dy$
33. $\int \dfrac{\tan^3 \theta}{\cos^4 \theta} \, d\theta$
34. $\int \tan^2 x \sec x \, dx$
35. $\int_{\pi/6}^{\pi/2} \cot^2 x \, dx$
36. $\int_{\pi/4}^{\pi/2} \cot^3 x \, dx$
37. $\int \cot^3 \alpha \csc^3 \alpha \, d\alpha$
38. $\int \csc^4 x \cot^6 x \, dx$
39. $\int \csc x \, dx$
40. $\int_{\pi/6}^{\pi/3} \csc^3 x \, dx$
41. $\int \sin 5x \sin 2x \, dx$
42. $\int \sin 3x \cos x \, dx$
43. $\int \cos 7\theta \cos 5\theta \, d\theta$
44. $\int \dfrac{\cos x + \sin x}{\sin 2x} \, dx$
45. $\int \dfrac{1 - \tan^2 x}{\sec^2 x} \, dx$
46. $\int \dfrac{dx}{\cos x - 1}$
47. $\int t \sec^2(t^2) \tan^4(t^2) \, dt$

48. If $\int_0^{\pi/4} \tan^6 x \sec x \, dx = I$, express the value of $\int_0^{\pi/4} \tan^8 x \sec x \, dx$ in terms of I.

49–52 Evaluate the indefinite integral. Illustrate, and check that your answer is reasonable, by graphing both the integrand and its antiderivative (taking $C = 0$).

49. $\int \sin^5 x \, dx$
50. $\int \sin^4 x \cos^4 x \, dx$
51. $\int \sin 3x \sin 6x \, dx$
52. $\int \sec^4 \dfrac{x}{2} \, dx$

53. Find the average value of the function $f(x) = \sin^2 x \cos^3 x$ on the interval $[-\pi, \pi]$.

54. Evaluate $\int \sin x \cos x \, dx$ by four methods: (a) the substitution $u = \cos x$, (b) the substitution $u = \sin x$, (c) the identity $\sin 2x = 2 \sin x \cos x$, and (d) integration by parts. Explain the different appearances of the answers.

55–56 Find the area of the region bounded by the given curves.

55. $y = \sin x$, $y = \sin^3 x$, $x = 0$, $x = \pi/2$

56. $y = \sin x$, $y = 2 \sin^2 x$, $x = 0$, $x = \pi/2$

57–58 Use a graph of the integrand to guess the value of the integral. Then use the methods of this section to prove that your guess is correct.

57. $\int_0^{2\pi} \cos^3 x \, dx$

58. $\int_0^2 \sin 2\pi x \cos 5\pi x \, dx$

59–62 Find the volume obtained by rotating the region bounded by the given curves about the specified axis.

59. $y = \sin x$, $x = \pi/2$, $x = \pi$, $y = 0$; about the x-axis

60. $y = \tan^2 x$, $y = 0$, $x = 0$, $x = \pi/4$; about the x-axis

61. $y = \cos x$, $y = 0$, $x = 0$, $x = \pi/2$; about $y = -1$

62. $y = \cos x$, $y = 0$, $x = 0$, $x = \pi/2$; about $y = 1$

63. A particle moves on a straight line with velocity function $v(t) = \sin \omega t \cos^2 \omega t$. Find its position function $s = f(t)$ if $f(0) = 0$.

64. Household electricity is supplied in the form of alternating current that varies from 155 V to -155 V with a frequency of 60 cycles per second (Hz). The voltage is thus given by the equation

$$E(t) = 155 \sin(120 \pi t)$$

where t is the time in seconds. Voltmeters read the RMS (root-mean-square) voltage, which is the square root of the average value of $[E(t)]^2$ over one cycle.
(a) Calculate the RMS voltage of household current.
(b) Many electric stoves require an RMS voltage of 220 V. Find the corresponding amplitude A needed for the voltage $E(t) = A \sin(120 \pi t)$.

65–67 Prove the formula, where m and n are positive integers.

65. $\int_{-\pi}^{\pi} \sin mx \cos nx \, dx = 0$

66. $\int_{-\pi}^{\pi} \sin mx \sin nx \, dx = \begin{cases} 0 & \text{if } m \neq n \\ \pi & \text{if } m = n \end{cases}$

67. $\int_{-\pi}^{\pi} \cos mx \cos nx \, dx = \begin{cases} 0 & \text{if } m \neq n \\ \pi & \text{if } m = n \end{cases}$

68. A *finite Fourier series* is given by the sum

$$f(x) = \sum_{n=1}^{N} a_n \sin nx$$
$$= a_1 \sin x + a_2 \sin 2x + \cdots + a_N \sin Nx$$

Show that the mth coefficient a_m is given by the formula

$$a_m = \frac{1}{\pi} \int_{-\pi}^{\pi} f(x) \sin mx \, dx$$

8.3 Trigonometric Substitution

In finding the area of a circle or an ellipse, an integral of the form $\int \sqrt{a^2 - x^2} \, dx$ arises, where $a > 0$. If it were $\int x\sqrt{a^2 - x^2} \, dx$, the substitution $u = a^2 - x^2$ would be effective but, as it stands, $\int \sqrt{a^2 - x^2} \, dx$ is more difficult. If we change the variable from x to θ by the substitution $x = a \sin \theta$, then the identity $1 - \sin^2 \theta = \cos^2 \theta$ allows us to get rid of the root sign because

$$\sqrt{a^2 - x^2} = \sqrt{a^2 - a^2 \sin^2 \theta} = \sqrt{a^2(1 - \sin^2 \theta)} = \sqrt{a^2 \cos^2 \theta} = a|\cos \theta|$$

Notice the difference between the substitution $u = a^2 - x^2$ (in which the new variable is a function of the old one) and the substitution $x = a \sin \theta$ (the old variable is a function of the new one).

In general we can make a substitution of the form $x = g(t)$ by using the Substitution Rule in reverse. To make our calculations simpler, we assume that g has an inverse function; that is, g is one-to-one. In this case, if we replace u by x and x by t in the Substitution Rule (Equation 5.5.4), we obtain

$$\int f(x) \, dx = \int f(g(t)) g'(t) \, dt$$

This kind of substitution is called *inverse substitution*.

We can make the inverse substitution $x = a \sin \theta$ provided that it defines a one-to-one function. This can be accomplished by restricting θ to lie in the interval $[-\pi/2, \pi/2]$.

In the following table we list trigonometric substitutions that are effective for the given radical expressions because of the specified trigonometric identities. In each case the restriction on θ is imposed to ensure that the function that defines the substitution is one-to-one. (These are the same intervals used in Appendix D in defining the inverse functions.)

Table of Trigonometric Substitutions

Expression	Substitution	Identity
$\sqrt{a^2 - x^2}$	$x = a \sin \theta,\ -\dfrac{\pi}{2} \leq \theta \leq \dfrac{\pi}{2}$	$1 - \sin^2\theta = \cos^2\theta$
$\sqrt{a^2 + x^2}$	$x = a \tan \theta,\ -\dfrac{\pi}{2} < \theta < \dfrac{\pi}{2}$	$1 + \tan^2\theta = \sec^2\theta$
$\sqrt{x^2 - a^2}$	$x = a \sec \theta,\ 0 \leq \theta < \dfrac{\pi}{2}\ \text{or}\ \pi \leq \theta < \dfrac{3\pi}{2}$	$\sec^2\theta - 1 = \tan^2\theta$

EXAMPLE 1 Evaluate $\displaystyle\int \frac{\sqrt{9 - x^2}}{x^2}\, dx$.

SOLUTION Let $x = 3 \sin \theta$, where $-\pi/2 \leq \theta \leq \pi/2$. Then $dx = 3 \cos \theta\, d\theta$ and

$$\sqrt{9 - x^2} = \sqrt{9 - 9\sin^2\theta} = \sqrt{9\cos^2\theta} = 3|\cos \theta| = 3\cos \theta$$

(Note that $\cos \theta \geq 0$ because $-\pi/2 \leq \theta \leq \pi/2$.) Thus, the Inverse Substitution Rule gives

$$\int \frac{\sqrt{9 - x^2}}{x^2}\, dx = \int \frac{3\cos \theta}{9 \sin^2\theta}\, 3\cos \theta\, d\theta$$

$$= \int \frac{\cos^2\theta}{\sin^2\theta}\, d\theta = \int \cot^2\theta\, d\theta$$

$$= \int (\csc^2\theta - 1)\, d\theta$$

$$= -\cot \theta - \theta + C$$

Since this is an indefinite integral, we must return to the original variable x. This can be done either by using trigonometric identities to express $\cot \theta$ in terms of $\sin \theta = x/3$ or by drawing a diagram, as in Figure 1, where θ is interpreted as an angle of a right triangle. Since $\sin \theta = x/3$, we label the opposite side and the hypotenuse as having lengths x and 3. Then the Pythagorean Theorem gives the length of the adjacent side as $\sqrt{9 - x^2}$, so we can simply read the value of $\cot \theta$ from the figure:

$$\cot \theta = \frac{\sqrt{9 - x^2}}{x}$$

(Although $\theta > 0$ in the diagram, this expression for $\cot \theta$ is valid even when $\theta < 0$.) Since $\sin \theta = x/3$, we have $\theta = \sin^{-1}(x/3)$ and so

$$\int \frac{\sqrt{9 - x^2}}{x^2}\, dx = -\frac{\sqrt{9 - x^2}}{x} - \sin^{-1}\!\left(\frac{x}{3}\right) + C$$

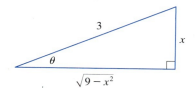

FIGURE 1
$\sin \theta = \dfrac{x}{3}$

EXAMPLE 2 Find the area enclosed by the ellipse

$$\frac{x^2}{a^2} + \frac{y^2}{b^2} = 1$$

SOLUTION Solving the equation of the ellipse for y, we get

$$\frac{y^2}{b^2} = 1 - \frac{x^2}{a^2} = \frac{a^2 - x^2}{a^2} \quad \text{or} \quad y = \pm \frac{b}{a}\sqrt{a^2 - x^2}$$

Because the ellipse is symmetric with respect to both axes, the total area A is four times the area in the first quadrant (see Figure 2). The part of the ellipse in the first quadrant is given by the function

$$y = \frac{b}{a}\sqrt{a^2 - x^2} \qquad 0 \le x \le a$$

and so

$$\tfrac{1}{4}A = \int_0^a \frac{b}{a}\sqrt{a^2 - x^2}\, dx$$

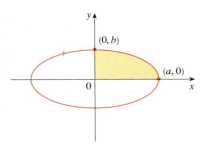

FIGURE 2
$\dfrac{x^2}{a^2} + \dfrac{y^2}{b^2} = 1$

To evaluate this integral we substitute $x = a \sin\theta$. Then $dx = a \cos\theta\, d\theta$. To change the limits of integration we note that when $x = 0$, $\sin\theta = 0$, so $\theta = 0$; when $x = a$, $\sin\theta = 1$, so $\theta = \pi/2$. Also

$$\sqrt{a^2 - x^2} = \sqrt{a^2 - a^2 \sin^2\theta} = \sqrt{a^2 \cos^2\theta} = a|\cos\theta| = a\cos\theta$$

since $0 \le \theta \le \pi/2$. Therefore

$$A = 4\frac{b}{a}\int_0^a \sqrt{a^2 - x^2}\, dx = 4\frac{b}{a}\int_0^{\pi/2} a\cos\theta \cdot a\cos\theta\, d\theta$$

$$= 4ab \int_0^{\pi/2} \cos^2\theta\, d\theta = 4ab \int_0^{\pi/2} \tfrac{1}{2}(1 + \cos 2\theta)\, d\theta$$

$$= 2ab\bigl[\theta + \tfrac{1}{2}\sin 2\theta\bigr]_0^{\pi/2} = 2ab\left(\frac{\pi}{2} + 0 - 0\right)$$

$$= \pi ab$$

We have shown that the area of an ellipse with semiaxes a and b is πab. In particular, taking $a = b = r$, we have proved the famous formula that the area of a circle with radius r is πr^2.

NOTE ▫ Since the integral in Example 2 was a definite integral, we changed the limits of integration and did not have to convert back to the original variable x.

EXAMPLE 3 Find $\displaystyle\int \frac{1}{x^2\sqrt{x^2 + 4}}\, dx$.

SOLUTION Let $x = 2\tan\theta$, $-\pi/2 < \theta < \pi/2$. Then $dx = 2\sec^2\theta\, d\theta$ and

$$\sqrt{x^2 + 4} = \sqrt{4(\tan^2\theta + 1)} = \sqrt{4\sec^2\theta} = 2|\sec\theta| = 2\sec\theta$$

Thus, we have

$$\int \frac{dx}{x^2\sqrt{x^2 + 4}} = \int \frac{2\sec^2\theta\, d\theta}{4\tan^2\theta \cdot 2\sec\theta} = \frac{1}{4}\int \frac{\sec\theta}{\tan^2\theta}\, d\theta$$

To evaluate this trigonometric integral we put everything in terms of $\sin\theta$ and $\cos\theta$:

$$\frac{\sec\theta}{\tan^2\theta} = \frac{1}{\cos\theta} \cdot \frac{\cos^2\theta}{\sin^2\theta} = \frac{\cos\theta}{\sin^2\theta}$$

Therefore, making the substitution $u = \sin\theta$, we have

$$\int \frac{dx}{x^2\sqrt{x^2+4}} = \frac{1}{4}\int \frac{\cos\theta}{\sin^2\theta}\,d\theta = \frac{1}{4}\int \frac{du}{u^2}$$

$$= \frac{1}{4}\left(-\frac{1}{u}\right) + C = -\frac{1}{4\sin\theta} + C$$

$$= -\frac{\csc\theta}{4} + C$$

We use Figure 3 to determine that $\csc\theta = \sqrt{x^2+4}/x$ and so

$$\int \frac{dx}{x^2\sqrt{x^2+4}} = -\frac{\sqrt{x^2+4}}{4x} + C$$

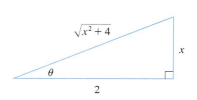

FIGURE 3
$\tan\theta = \dfrac{x}{2}$

EXAMPLE 4 Find $\displaystyle\int \frac{x}{\sqrt{x^2+4}}\,dx$.

SOLUTION It would be possible to use the trigonometric substitution $x = 2\tan\theta$ here (as in Example 3). But the direct substitution $u = x^2 + 4$ is simpler, because then $du = 2x\,dx$ and

$$\int \frac{x}{\sqrt{x^2+4}}\,dx = \frac{1}{2}\int \frac{du}{\sqrt{u}} = \sqrt{u} + C = \sqrt{x^2+4} + C$$

NOTE ▪ Example 4 illustrates the fact that even when trigonometric substitutions are possible, they may not give the easiest solution. You should look for a simpler method first.

EXAMPLE 5 Evaluate $\displaystyle\int \frac{dx}{\sqrt{x^2-a^2}}$, where $a > 0$.

SOLUTION 1 We let $x = a\sec\theta$, where $0 < \theta < \pi/2$ or $\pi < \theta < 3\pi/2$. Then $dx = a\sec\theta\tan\theta\,d\theta$ and

$$\sqrt{x^2-a^2} = \sqrt{a^2(\sec^2\theta - 1)} = \sqrt{a^2\tan^2\theta} = a|\tan\theta| = a\tan\theta$$

Therefore

$$\int \frac{dx}{\sqrt{x^2-a^2}} = \int \frac{a\sec\theta\tan\theta}{a\tan\theta}\,d\theta$$

$$= \int \sec\theta\,d\theta = \ln|\sec\theta + \tan\theta| + C$$

The triangle in Figure 4 gives $\tan\theta = \sqrt{x^2-a^2}/a$, so we have

$$\int \frac{dx}{\sqrt{x^2-a^2}} = \ln\left|\frac{x}{a} + \frac{\sqrt{x^2-a^2}}{a}\right| + C$$

$$= \ln\left|x + \sqrt{x^2-a^2}\right| - \ln a + C$$

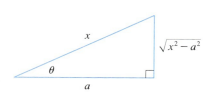

FIGURE 4
$\sec\theta = \dfrac{x}{a}$

Writing $C_1 = C - \ln a$, we have

$$\boxed{1} \qquad \int \frac{dx}{\sqrt{x^2 - a^2}} = \ln|x + \sqrt{x^2 - a^2}| + C_1$$

SOLUTION 2 For $x > 0$ the hyperbolic substitution $x = a \cosh t$ can also be used. Using the identity $\cosh^2 y - \sinh^2 y = 1$, we have

$$\sqrt{x^2 - a^2} = \sqrt{a^2(\cosh^2 t - 1)} = \sqrt{a^2 \sinh^2 t} = a \sinh t$$

Since $dx = a \sinh t \, dt$, we obtain

$$\int \frac{dx}{\sqrt{x^2 - a^2}} = \int \frac{a \sinh t \, dt}{a \sinh t} = \int dt = t + C$$

Since $\cosh t = x/a$, we have $t = \cosh^{-1}(x/a)$ and

$$\boxed{2} \qquad \int \frac{dx}{\sqrt{x^2 - a^2}} = \cosh^{-1}\left(\frac{x}{a}\right) + C$$

Although Formulas 1 and 2 look quite different, they are actually equivalent by Formula 7.6.4.

NOTE ▪ As Example 5 illustrates, hyperbolic substitutions can be used in place of trigonometric substitutions and sometimes they lead to simpler answers. But we usually use trigonometric substitutions because trigonometric identities are more familiar than hyperbolic identities.

EXAMPLE 6 Find $\int_0^{3\sqrt{3}/2} \frac{x^3}{(4x^2 + 9)^{3/2}} \, dx$.

SOLUTION First we note that $(4x^2 + 9)^{3/2} = (\sqrt{4x^2 + 9})^3$ so trigonometric substitution is appropriate. Although $\sqrt{4x^2 + 9}$ is not quite one of the expressions in the table of trigonometric substitutions, it becomes one of them if we make the preliminary substitution $u = 2x$. When we combine this with the tangent substitution, we have $x = \frac{3}{2} \tan \theta$, which gives $dx = \frac{3}{2} \sec^2 \theta \, d\theta$ and

$$\sqrt{4x^2 + 9} = \sqrt{9 \tan^2 \theta + 9} = 3 \sec \theta$$

When $x = 0$, $\tan \theta = 0$, so $\theta = 0$; when $x = 3\sqrt{3}/2$, $\tan \theta = \sqrt{3}$, so $\theta = \pi/3$.

$$\int_0^{3\sqrt{3}/2} \frac{x^3}{(4x^2 + 9)^{3/2}} \, dx = \int_0^{\pi/3} \frac{\frac{27}{8} \tan^3 \theta}{27 \sec^3 \theta} \, \frac{3}{2} \sec^2 \theta \, d\theta$$

$$= \frac{3}{16} \int_0^{\pi/3} \frac{\tan^3 \theta}{\sec \theta} \, d\theta = \frac{3}{16} \int_0^{\pi/3} \frac{\sin^3 \theta}{\cos^2 \theta} \, d\theta$$

$$= \frac{3}{16} \int_0^{\pi/3} \frac{1 - \cos^2 \theta}{\cos^2 \theta} \sin \theta \, d\theta$$

Now we substitute $u = \cos \theta$ so that $du = -\sin \theta \, d\theta$. When $\theta = 0$, $u = 1$; when $\theta = \pi/3$, $u = \frac{1}{2}$.

Therefore

$$\int_0^{3\sqrt{3}/2} \frac{x^3}{(4x^2+9)^{3/2}}\,dx = -\tfrac{3}{16}\int_1^{1/2} \frac{1-u^2}{u^2}\,du = \tfrac{3}{16}\int_1^{1/2}(1-u^{-2})\,du$$

$$= \tfrac{3}{16}\left[u+\frac{1}{u}\right]_1^{1/2} = \tfrac{3}{16}\left[(\tfrac{1}{2}+2)-(1+1)\right] = \tfrac{3}{32}$$

EXAMPLE 7 Evaluate $\displaystyle\int \frac{x}{\sqrt{3-2x-x^2}}\,dx$.

SOLUTION We can transform the integrand into a function for which trigonometric substitution is appropriate by first completing the square under the root sign:

$$3-2x-x^2 = 3-(x^2+2x) = 3+1-(x^2+2x+1)$$
$$= 4-(x+1)^2$$

This suggests that we make the substitution $u = x+1$. Then $du = dx$ and $x = u-1$, so

$$\int \frac{x}{\sqrt{3-2x-x^2}}\,dx = \int \frac{u-1}{\sqrt{4-u^2}}\,du$$

We now substitute $u = 2\sin\theta$, giving $du = 2\cos\theta\,d\theta$ and $\sqrt{4-u^2} = 2\cos\theta$, so

$$\int \frac{x}{\sqrt{3-2x-x^2}}\,dx = \int \frac{2\sin\theta - 1}{2\cos\theta}\,2\cos\theta\,d\theta$$

$$= \int (2\sin\theta - 1)\,d\theta$$

$$= -2\cos\theta - \theta + C$$

$$= -\sqrt{4-u^2} - \sin^{-1}\!\left(\frac{u}{2}\right) + C$$

$$= -\sqrt{3-2x-x^2} - \sin^{-1}\!\left(\frac{x+1}{2}\right) + C$$

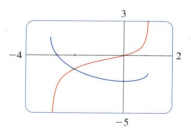

Figure 5 shows the graphs of the integrand in Example 7 and its indefinite integral (with $C = 0$). Which is which?

FIGURE 5

8.3 Exercises

1–3 Evaluate the integral using the indicated trigonometric substitution. Sketch and label the associated right triangle.

1. $\displaystyle\int \frac{1}{x^2\sqrt{x^2-9}}\,dx;\quad x = 3\sec\theta$

2. $\displaystyle\int x^3\sqrt{9-x^2}\,dx;\quad x = 3\sin\theta$

3. $\displaystyle\int \frac{x^3}{\sqrt{x^2+9}}\,dx;\quad x = 3\tan\theta$

4–30 Evaluate the integral.

4. $\displaystyle\int_0^{2\sqrt{3}} \frac{x^3}{\sqrt{16-x^2}}\,dx$

5. $\displaystyle\int_{\sqrt{2}}^2 \frac{1}{t^3\sqrt{t^2-1}}\,dt$

6. $\displaystyle\int_0^2 x^3\sqrt{x^2+4}\,dx$

7. $\displaystyle\int \frac{1}{x^2\sqrt{25-x^2}}\,dx$

8. $\displaystyle\int \frac{\sqrt{x^2-a^2}}{x^4}\,dx$

9. $\displaystyle\int \frac{dx}{\sqrt{x^2+16}}$

10. $\displaystyle\int \frac{t^5}{\sqrt{t^2+2}}\,dt$

11. $\int \sqrt{1 - 4x^2}\, dx$

12. $\int_0^1 x\sqrt{x^2 + 4}\, dx$

13. $\int \dfrac{\sqrt{x^2 - 9}}{x^3}\, dx$

14. $\int \dfrac{du}{u\sqrt{5 - u^2}}$

15. $\int \dfrac{x^2}{(a^2 - x^2)^{3/2}}\, dx$

16. $\int \dfrac{dx}{x^2\sqrt{16x^2 - 9}}$

17. $\int \dfrac{x}{\sqrt{x^2 - 7}}\, dx$

18. $\int \dfrac{dx}{[(ax)^2 - b^2]^{3/2}}$

19. $\int \dfrac{\sqrt{1 + x^2}}{x}\, dx$

20. $\int \dfrac{t}{\sqrt{25 - t^2}}\, dt$

21. $\int_0^{2/3} x^3\sqrt{4 - 9x^2}\, dx$

22. $\int_0^1 \sqrt{x^2 + 1}\, dx$

23. $\int \sqrt{5 + 4x - x^2}\, dx$

24. $\int \dfrac{dt}{\sqrt{t^2 - 6t + 13}}$

25. $\int \dfrac{1}{\sqrt{9x^2 + 6x - 8}}\, dx$

26. $\int \dfrac{x^2}{\sqrt{4x - x^2}}\, dx$

27. $\int \dfrac{dx}{(x^2 + 2x + 2)^2}$

28. $\int \dfrac{dx}{(5 - 4x - x^2)^{5/2}}$

29. $\int x\sqrt{1 - x^4}\, dx$

30. $\int_0^{\pi/2} \dfrac{\cos t}{\sqrt{1 + \sin^2 t}}\, dt$

31. (a) Use trigonometric substitution to show that

$$\int \dfrac{dx}{\sqrt{x^2 + a^2}} = \ln(x + \sqrt{x^2 + a^2}) + C$$

(b) Use the hyperbolic substitution $x = a \sinh t$ to show that

$$\int \dfrac{dx}{\sqrt{x^2 + a^2}} = \sinh^{-1}\left(\dfrac{x}{a}\right) + C$$

These formulas are connected by Formula 7.6.3.

32. Evaluate

$$\int \dfrac{x^2}{(x^2 + a^2)^{3/2}}\, dx$$

(a) by trigonometric substitution.
(b) by the hyperbolic substitution $x = a \sinh t$.

33. Find the average value of $f(x) = \sqrt{x^2 - 1}/x$, $1 \leq x \leq 7$.

34. Find the area of the region bounded by the hyperbola $9x^2 - 4y^2 = 36$ and the line $x = 3$.

35. Prove the formula $A = \tfrac{1}{2}r^2\theta$ for the area of a sector of a circle with radius r and central angle θ. [*Hint:* Assume $0 < \theta < \pi/2$ and place the center of the circle at the origin so it has the equation $x^2 + y^2 = r^2$. Then A is the sum of the area of the triangle POQ and the area of the region PQR in the figure.]

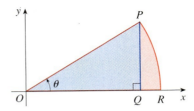

36. Evaluate the integral

$$\int \dfrac{dx}{x^4\sqrt{x^2 - 2}}$$

Graph the integrand and its indefinite integral on the same screen and check that your answer is reasonable.

37. Use a graph to approximate the roots of the equation $x^2\sqrt{4 - x^2} = 2 - x$. Then approximate the area bounded by the curve $y = x^2\sqrt{4 - x^2}$ and the line $y = 2 - x$.

38. A charged rod of length L produces an electric field at point $P(a, b)$ given by

$$E(P) = \int_{-a}^{L-a} \dfrac{\lambda b}{4\pi\varepsilon_0(x^2 + b^2)^{3/2}}\, dx$$

where λ is the charge density per unit length on the rod and ε_0 is the free space permittivity (see the figure). Evaluate the integral to determine an expression for the electric field $E(P)$.

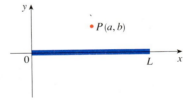

39. Find the area of the crescent-shaped region (called a *lune*) bounded by arcs of circles with radii r and R. (See the figure.)

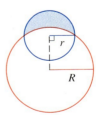

40. A water storage tank has the shape of a cylinder with diameter 10 ft. It is mounted so that the circular cross-sections are vertical. If the depth of the water is 7 ft, what percentage of the total capacity is being used?

41. A torus is generated by rotating the circle $x^2 + (y - R)^2 = r^2$ about the x-axis. Find the volume enclosed by the torus.

8.4 Integration of Rational Functions by Partial Fractions

In this section we show how to integrate any rational function (a ratio of polynomials) by expressing it as a sum of simpler fractions, called *partial fractions*, that we already know how to integrate. To illustrate the method, observe that by taking the fractions $2/(x-1)$ and $1/(x+2)$ to a common denominator we obtain

$$\frac{2}{x-1} - \frac{1}{x+2} = \frac{2(x+2) - (x-1)}{(x-1)(x+2)} = \frac{x+5}{x^2+x-2}$$

If we now reverse the procedure, we see how to integrate the function on the right side of this equation:

$$\int \frac{x+5}{x^2+x-2}\,dx = \int \left(\frac{2}{x-1} - \frac{1}{x+2}\right) dx$$

$$= 2\ln|x-1| - \ln|x+2| + C$$

To see how the method of partial fractions works in general, let's consider a rational function

$$f(x) = \frac{P(x)}{Q(x)}$$

where P and Q are polynomials. It's possible to express f as a sum of simpler fractions provided that the degree of P is less than the degree of Q. Such a rational function is called *proper*. Recall that if

$$P(x) = a_n x^n + a_{n-1} x^{n-1} + \cdots + a_1 x + a_0$$

where $a_n \neq 0$, then the degree of P is n and we write $\deg(P) = n$.

If f is improper, that is, $\deg(P) \geq \deg(Q)$, then we must take the preliminary step of dividing Q into P (by long division) until a remainder $R(x)$ is obtained such that $\deg(R) < \deg(Q)$. The division statement is

$$\boxed{1} \qquad f(x) = \frac{P(x)}{Q(x)} = S(x) + \frac{R(x)}{Q(x)}$$

where S and R are also polynomials.

As the following example illustrates, sometimes this preliminary step is all that is required.

EXAMPLE 1 Find $\int \dfrac{x^3 + x}{x - 1}\,dx$.

SOLUTION Since the degree of the numerator is greater than the degree of the denominator, we first perform the long division. This enables us to write

$$\int \frac{x^3 + x}{x-1}\,dx = \int \left(x^2 + x + 2 + \frac{2}{x-1}\right) dx$$

$$= \frac{x^3}{3} + \frac{x^2}{2} + 2x + 2\ln|x-1| + C$$

```
              x² + x + 2
        _____
x - 1 ) x³         + x
        x³ - x²
        _____
             x² + x
             x² - x
             _____
                  2x
                  2x - 2
                  _____
                       2
```

The next step is to factor the denominator $Q(x)$ as far as possible. It can be shown that any polynomial Q can be factored as a product of linear factors (of the form $ax + b$) and irreducible quadratic factors (of the form $ax^2 + bx + c$, where $b^2 - 4ac < 0$). For instance, if $Q(x) = x^4 - 16$, we could factor it as

$$Q(x) = (x^2 - 4)(x^2 + 4) = (x - 2)(x + 2)(x^2 + 4)$$

The third step is to express the proper rational function $R(x)/Q(x)$ (from Equation 1) as a sum of **partial fractions** of the form

$$\frac{A}{(ax+b)^i} \quad \text{or} \quad \frac{Ax+B}{(ax^2+bx+c)^j}$$

A theorem in algebra guarantees that it is always possible to do this. We explain the details for the four cases that occur.

CASE I ▫ The denominator $Q(x)$ is a product of distinct linear factors.
This means that we can write

$$Q(x) = (a_1 x + b_1)(a_2 x + b_2) \cdots (a_k x + b_k)$$

where no factor is repeated (and no factor is a constant multiple of another). In this case the partial fraction theorem states that there exist constants $A_1, A_2, \ldots, A_k$ such that

$$\boxed{2} \qquad \frac{R(x)}{Q(x)} = \frac{A_1}{a_1 x + b_1} + \frac{A_2}{a_2 x + b_2} + \cdots + \frac{A_k}{a_k x + b_k}$$

These constants can be determined as in the following example.

EXAMPLE 2 Evaluate $\displaystyle \int \frac{x^2 + 2x - 1}{2x^3 + 3x^2 - 2x} \, dx$.

SOLUTION Since the degree of the numerator is less than the degree of the denominator, we don't need to divide. We factor the denominator as

$$2x^3 + 3x^2 - 2x = x(2x^2 + 3x - 2) = x(2x - 1)(x + 2)$$

Since the denominator has three distinct linear factors, the partial fraction decomposition of the integrand (2) has the form

$$\boxed{3} \qquad \frac{x^2 + 2x - 1}{x(2x-1)(x+2)} = \frac{A}{x} + \frac{B}{2x-1} + \frac{C}{x+2}$$

|||| Another method for finding A, B, and C is given in the note after this example.

To determine the values of A, B, and C, we multiply both sides of this equation by the product of the denominators, $x(2x - 1)(x + 2)$, obtaining

$$\boxed{4} \qquad x^2 + 2x - 1 = A(2x - 1)(x + 2) + Bx(x + 2) + Cx(2x - 1)$$

Expanding the right side of Equation 4 and writing it in the standard form for polynomials, we get

$$\boxed{5} \qquad x^2 + 2x - 1 = (2A + B + 2C)x^2 + (3A + 2B - C)x - 2A$$

|||| Figure 1 shows the graphs of the integrand in Example 2 and its indefinite integral (with $K = 0$). Which is which?

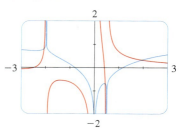

FIGURE 1

|||| We could check our work by taking the terms to a common denominator and adding them.

The polynomials in Equation 5 are identical, so their coefficients must be equal. The coefficient of x^2 on the right side, $2A + B + 2C$, must equal the coefficient of x^2 on the left side—namely, 1. Likewise, the coefficients of x are equal and the constant terms are equal. This gives the following system of equations for A, B, and C:

$$2A + B + 2C = 1$$
$$3A + 2B - C = 2$$
$$-2A = -1$$

Solving, we get $A = \frac{1}{2}$, $B = \frac{1}{5}$, and $C = -\frac{1}{10}$, and so

$$\int \frac{x^2 + 2x - 1}{2x^3 + 3x^2 - 2x}\, dx = \int \left(\frac{1}{2}\frac{1}{x} + \frac{1}{5}\frac{1}{2x-1} - \frac{1}{10}\frac{1}{x+2}\right) dx$$

$$= \tfrac{1}{2}\ln|x| + \tfrac{1}{10}\ln|2x - 1| - \tfrac{1}{10}\ln|x + 2| + K$$

In integrating the middle term we have made the mental substitution $u = 2x - 1$, which gives $du = 2\, dx$ and $dx = du/2$.

NOTE We can use an alternative method to find the coefficients A, B, and C in Example 2. Equation 4 is an identity; it is true for every value of x. Let's choose values of x that simplify the equation. If we put $x = 0$ in Equation 4, then the second and third terms on the right side vanish and the equation then becomes $-2A = -1$, or $A = \frac{1}{2}$. Likewise, $x = \frac{1}{2}$ gives $5B/4 = \frac{1}{4}$ and $x = -2$ gives $10C = -1$, so $B = \frac{1}{5}$ and $C = -\frac{1}{10}$. (You may object that Equation 3 is not valid for $x = 0$, $\frac{1}{2}$, or -2, so why should Equation 4 be valid for those values? In fact, Equation 4 is true for all values of x, even $x = 0$, $\frac{1}{2}$, and -2. See Exercise 67 for the reason.)

EXAMPLE 3 Find $\int \dfrac{dx}{x^2 - a^2}$, where $a \neq 0$.

SOLUTION The method of partial fractions gives

$$\frac{1}{x^2 - a^2} = \frac{1}{(x-a)(x+a)} = \frac{A}{x-a} + \frac{B}{x+a}$$

and therefore

$$A(x + a) + B(x - a) = 1$$

Using the method of the preceding note, we put $x = a$ in this equation and get $A(2a) = 1$, so $A = 1/(2a)$. If we put $x = -a$, we get $B(-2a) = 1$, so $B = -1/(2a)$. Thus

$$\int \frac{dx}{x^2 - a^2} = \frac{1}{2a}\int \left(\frac{1}{x-a} - \frac{1}{x+a}\right) dx$$

$$= \frac{1}{2a}\left(\ln|x - a| - \ln|x + a|\right) + C$$

Since $\ln x - \ln y = \ln(x/y)$, we can write the integral as

$$\boxed{6} \quad \int \frac{dx}{x^2 - a^2} = \frac{1}{2a} \ln \left| \frac{x-a}{x+a} \right| + C$$

See Exercises 53–54 for ways of using Formula 6.

CASE II ▪ $Q(x)$ **is a product of linear factors, some of which are repeated.**
Suppose the first linear factor $(a_1 x + b_1)$ is repeated r times; that is, $(a_1 x + b_1)^r$ occurs in the factorization of $Q(x)$. Then instead of the single term $A_1/(a_1 x + b_1)$ in Equation 2, we would use

$$\boxed{7} \quad \frac{A_1}{a_1 x + b_1} + \frac{A_2}{(a_1 x + b_1)^2} + \cdots + \frac{A_r}{(a_1 x + b_1)^r}$$

By way of illustration, we could write

$$\frac{x^3 - x + 1}{x^2(x-1)^3} = \frac{A}{x} + \frac{B}{x^2} + \frac{C}{x-1} + \frac{D}{(x-1)^2} + \frac{E}{(x-1)^3}$$

but we prefer to work out in detail a simpler example.

EXAMPLE 4 Find $\displaystyle\int \frac{x^4 - 2x^2 + 4x + 1}{x^3 - x^2 - x + 1} dx$.

SOLUTION The first step is to divide. The result of long division is

$$\frac{x^4 - 2x^2 + 4x + 1}{x^3 - x^2 - x + 1} = x + 1 + \frac{4x}{x^3 - x^2 - x + 1}$$

The second step is to factor the denominator $Q(x) = x^3 - x^2 - x + 1$. Since $Q(1) = 0$, we know that $x - 1$ is a factor and we obtain

$$x^3 - x^2 - x + 1 = (x-1)(x^2 - 1) = (x-1)(x-1)(x+1)$$
$$= (x-1)^2(x+1)$$

Since the linear factor $x - 1$ occurs twice, the partial fraction decomposition is

$$\frac{4x}{(x-1)^2(x+1)} = \frac{A}{x-1} + \frac{B}{(x-1)^2} + \frac{C}{x+1}$$

Multiplying by the least common denominator, $(x-1)^2(x+1)$, we get

$$\boxed{8} \quad 4x = A(x-1)(x+1) + B(x+1) + C(x-1)^2$$
$$= (A+C)x^2 + (B-2C)x + (-A+B+C)$$

|||| Another method for finding the coefficients:
Put $x = 1$ in (8): $B = 2$.
Put $x = -1$: $C = -1$.
Put $x = 0$: $A = B + C = 1$.

Now we equate coefficients:

$$A + C = 0$$
$$B - 2C = 4$$
$$-A + B + C = 0$$

Solving, we obtain $A = 1$, $B = 2$, and $C = -1$, so

$$\int \frac{x^4 - 2x^2 + 4x + 1}{x^3 - x^2 - x + 1} \, dx = \int \left[x + 1 + \frac{1}{x-1} + \frac{2}{(x-1)^2} - \frac{1}{x+1} \right] dx$$

$$= \frac{x^2}{2} + x + \ln|x - 1| - \frac{2}{x-1} - \ln|x+1| + K$$

$$= \frac{x^2}{2} + x - \frac{2}{x-1} + \ln \left| \frac{x-1}{x+1} \right| + K$$

CASE III ▫ $Q(x)$ **contains irreducible quadratic factors, none of which is repeated.**

If $Q(x)$ has the factor $ax^2 + bx + c$, where $b^2 - 4ac < 0$, then, in addition to the partial fractions in Equations 2 and 7, the expression for $R(x)/Q(x)$ will have a term of the form

$$\boxed{9} \qquad \frac{Ax + B}{ax^2 + bx + c}$$

where A and B are constants to be determined. For instance, the function given by $f(x) = x/[(x - 2)(x^2 + 1)(x^2 + 4)]$ has a partial fraction decomposition of the form

$$\frac{x}{(x - 2)(x^2 + 1)(x^2 + 4)} = \frac{A}{x - 2} + \frac{Bx + C}{x^2 + 1} + \frac{Dx + E}{x^2 + 4}$$

The term given in (9) can be integrated by completing the square and using the formula

$$\boxed{10} \qquad \int \frac{dx}{x^2 + a^2} = \frac{1}{a} \tan^{-1}\left(\frac{x}{a}\right) + C$$

EXAMPLE 5 Evaluate $\displaystyle\int \frac{2x^2 - x + 4}{x^3 + 4x} \, dx$.

SOLUTION Since $x^3 + 4x = x(x^2 + 4)$ can't be factored further, we write

$$\frac{2x^2 - x + 4}{x(x^2 + 4)} = \frac{A}{x} + \frac{Bx + C}{x^2 + 4}$$

Multiplying by $x(x^2 + 4)$, we have

$$2x^2 - x + 4 = A(x^2 + 4) + (Bx + C)x$$

$$= (A + B)x^2 + Cx + 4A$$

Equating coefficients, we obtain

$$A + B = 2 \qquad C = -1 \qquad 4A = 4$$

Thus $A = 1$, $B = 1$, and $C = -1$ and so

$$\int \frac{2x^2 - x + 4}{x^3 + 4x} \, dx = \int \left(\frac{1}{x} + \frac{x - 1}{x^2 + 4} \right) dx$$

In order to integrate the second term we split it into two parts:

$$\int \frac{x-1}{x^2+4}\,dx = \int \frac{x}{x^2+4}\,dx - \int \frac{1}{x^2+4}\,dx$$

We make the substitution $u = x^2 + 4$ in the first of these integrals so that $du = 2x\,dx$. We evaluate the second integral by means of Formula 10 with $a = 2$:

$$\int \frac{2x^2 - x + 4}{x(x^2+4)}\,dx = \int \frac{1}{x}\,dx + \int \frac{x}{x^2+4}\,dx - \int \frac{1}{x^2+4}\,dx$$

$$= \ln|x| + \tfrac{1}{2}\ln(x^2+4) - \tfrac{1}{2}\tan^{-1}(x/2) + K$$

EXAMPLE 6 Evaluate $\displaystyle\int \frac{4x^2 - 3x + 2}{4x^2 - 4x + 3}\,dx$.

SOLUTION Since the degree of the numerator is *not less than* the degree of the denominator, we first divide and obtain

$$\frac{4x^2 - 3x + 2}{4x^2 - 4x + 3} = 1 + \frac{x-1}{4x^2 - 4x + 3}$$

Notice that the quadratic $4x^2 - 4x + 3$ is irreducible because its discriminant is $b^2 - 4ac = -32 < 0$. This means it can't be factored, so we don't need to use the partial fraction technique.

To integrate the given function we complete the square in the denominator:

$$4x^2 - 4x + 3 = (2x-1)^2 + 2$$

This suggests that we make the substitution $u = 2x - 1$. Then, $du = 2\,dx$ and $x = (u+1)/2$, so

$$\int \frac{4x^2 - 3x + 2}{4x^2 - 4x + 3}\,dx = \int \left(1 + \frac{x-1}{4x^2 - 4x + 3}\right) dx$$

$$= x + \tfrac{1}{2}\int \frac{\tfrac{1}{2}(u+1) - 1}{u^2 + 2}\,du = x + \tfrac{1}{4}\int \frac{u - 1}{u^2 + 2}\,du$$

$$= x + \tfrac{1}{4}\int \frac{u}{u^2+2}\,du - \tfrac{1}{4}\int \frac{1}{u^2+2}\,du$$

$$= x + \tfrac{1}{8}\ln(u^2+2) - \tfrac{1}{4}\cdot\tfrac{1}{\sqrt{2}}\tan^{-1}\!\left(\tfrac{u}{\sqrt{2}}\right) + C$$

$$= x + \tfrac{1}{8}\ln(4x^2 - 4x + 3) - \tfrac{1}{4\sqrt{2}}\tan^{-1}\!\left(\tfrac{2x-1}{\sqrt{2}}\right) + C$$

NOTE ▫ Example 6 illustrates the general procedure for integrating a partial fraction of the form

$$\frac{Ax + B}{ax^2 + bx + c} \qquad \text{where } b^2 - 4ac < 0$$

We complete the square in the denominator and then make a substitution that brings the integral into the form

$$\int \frac{Cu + D}{u^2 + a^2}\, du = C \int \frac{u}{u^2 + a^2}\, du + D \int \frac{1}{u^2 + a^2}\, du$$

Then the first integral is a logarithm and the second is expressed in terms of $\tan^{-1}$.

CASE IV ▫ **$Q(x)$ contains a repeated irreducible quadratic factor.**
If $Q(x)$ has the factor $(ax^2 + bx + c)^r$, where $b^2 - 4ac < 0$, then instead of the single partial fraction (9), the sum

$$\boxed{11} \qquad \frac{A_1 x + B_1}{ax^2 + bx + c} + \frac{A_2 x + B_2}{(ax^2 + bx + c)^2} + \cdots + \frac{A_r x + B_r}{(ax^2 + bx + c)^r}$$

occurs in the partial fraction decomposition of $R(x)/Q(x)$. Each of the terms in (11) can be integrated by first completing the square.

|||| It would be extremely tedious to work out by hand the numerical values of the coefficients in Example 7. Most computer algebra systems, however, can find the numerical values very quickly. For instance, the Maple command

    convert(f, parfrac, x)

or the Mathematica command

    Apart[f]

gives the following values:

$A = -1$, $B = \frac{1}{8}$, $C = D = -1$,
$E = \frac{15}{8}$, $F = -\frac{1}{8}$, $G = H = \frac{3}{4}$,
$I = -\frac{1}{2}$, $J = \frac{1}{2}$

EXAMPLE 7 Write out the form of the partial fraction decomposition of the function

$$\frac{x^3 + x^2 + 1}{x(x-1)(x^2 + x + 1)(x^2 + 1)^3}$$

SOLUTION

$$\frac{x^3 + x^2 + 1}{x(x-1)(x^2 + x + 1)(x^2 + 1)^3}$$

$$= \frac{A}{x} + \frac{B}{x-1} + \frac{Cx + D}{x^2 + x + 1} + \frac{Ex + F}{x^2 + 1} + \frac{Gx + H}{(x^2 + 1)^2} + \frac{Ix + J}{(x^2 + 1)^3}$$

EXAMPLE 8 Evaluate $\displaystyle\int \frac{1 - x + 2x^2 - x^3}{x(x^2 + 1)^2}\, dx$.

SOLUTION The form of the partial fraction decomposition is

$$\frac{1 - x + 2x^2 - x^3}{x(x^2 + 1)^2} = \frac{A}{x} + \frac{Bx + C}{x^2 + 1} + \frac{Dx + E}{(x^2 + 1)^2}$$

Multiplying by $x(x^2 + 1)^2$, we have

$$-x^3 + 2x^2 - x + 1 = A(x^2 + 1)^2 + (Bx + C)x(x^2 + 1) + (Dx + E)x$$

$$= A(x^4 + 2x^2 + 1) + B(x^4 + x^2) + C(x^3 + x) + Dx^2 + Ex$$

$$= (A + B)x^4 + Cx^3 + (2A + B + D)x^2 + (C + E)x + A$$

If we equate coefficients, we get the system

$$A + B = 0 \qquad C = -1 \qquad 2A + B + D = 2 \qquad C + E = -1 \qquad A = 1$$

which has the solution $A = 1$, $B = -1$, $C = -1$, $D = 1$, and $E = 0$. Thus

$$\int \frac{1 - x + 2x^2 - x^3}{x(x^2 + 1)^2} \, dx = \int \left(\frac{1}{x} - \frac{x + 1}{x^2 + 1} + \frac{x}{(x^2 + 1)^2} \right) dx$$

$$= \int \frac{dx}{x} - \int \frac{x}{x^2 + 1} \, dx - \int \frac{dx}{x^2 + 1} + \int \frac{x \, dx}{(x^2 + 1)^2}$$

$$= \ln|x| - \tfrac{1}{2} \ln(x^2 + 1) - \tan^{-1} x - \frac{1}{2(x^2 + 1)} + K$$

|||| In the second and fourth terms we made the mental substitution $u = x^2 + 1$.

We note that sometimes partial fractions can be avoided when integrating a rational function. For instance, although the integral

$$\int \frac{x^2 + 1}{x(x^2 + 3)} \, dx$$

could be evaluated by the method of Case III, it's much easier to observe that if $u = x(x^2 + 3) = x^3 + 3x$, then $du = (3x^2 + 3) \, dx$ and so

$$\int \frac{x^2 + 1}{x(x^2 + 3)} \, dx = \tfrac{1}{3} \ln|x^3 + 3x| + C$$

Rationalizing Substitutions

Some nonrational functions can be changed into rational functions by means of appropriate substitutions. In particular, when an integrand contains an expression of the form $\sqrt[n]{g(x)}$, then the substitution $u = \sqrt[n]{g(x)}$ may be effective. Other instances appear in the exercises.

EXAMPLE 9 Evaluate $\displaystyle\int \frac{\sqrt{x + 4}}{x} \, dx$.

SOLUTION Let $u = \sqrt{x + 4}$. Then $u^2 = x + 4$, so $x = u^2 - 4$ and $dx = 2u \, du$. Therefore

$$\int \frac{\sqrt{x + 4}}{x} \, dx = \int \frac{u}{u^2 - 4} 2u \, du = 2 \int \frac{u^2}{u^2 - 4} \, du$$

$$= 2 \int \left(1 + \frac{4}{u^2 - 4} \right) du$$

We can evaluate this integral either by factoring $u^2 - 4$ as $(u - 2)(u + 2)$ and using partial fractions or by using Formula 6 with $a = 2$:

$$\int \frac{\sqrt{x + 4}}{x} \, dx = 2 \int du + 8 \int \frac{du}{u^2 - 4}$$

$$= 2u + 8 \cdot \frac{1}{2 \cdot 2} \ln \left| \frac{u - 2}{u + 2} \right| + C$$

$$= 2\sqrt{x + 4} + 2 \ln \left| \frac{\sqrt{x + 4} - 2}{\sqrt{x + 4} + 2} \right| + C$$

8.4 Exercises

1–6 Write out the form of the partial fraction decomposition of the function (as in Example 7). Do not determine the numerical values of the coefficients.

1. (a) $\dfrac{2x}{(x+3)(3x+1)}$ (b) $\dfrac{1}{x^3+2x^2+x}$

2. (a) $\dfrac{x-1}{x^3+x^2}$ (b) $\dfrac{x-1}{x^3+x}$

3. (a) $\dfrac{2}{x^2+3x-4}$ (b) $\dfrac{x^2}{(x-1)(x^2+x+1)}$

4. (a) $\dfrac{x^3}{x^2+4x+3}$ (b) $\dfrac{2x+1}{(x+1)^3(x^2+4)^2}$

5. (a) $\dfrac{x^4}{x^4-1}$ (b) $\dfrac{t^4+t^2+1}{(t^2+1)(t^2+4)^2}$

6. (a) $\dfrac{x^4}{(x^3+x)(x^2-x+3)}$ (b) $\dfrac{1}{x^6-x^3}$

7–38 Evaluate the integral.

7. $\displaystyle\int \dfrac{x}{x-6}\,dx$

8. $\displaystyle\int \dfrac{r^2}{r+4}\,dr$

9. $\displaystyle\int \dfrac{x-9}{(x+5)(x-2)}\,dx$

10. $\displaystyle\int \dfrac{1}{(t+4)(t-1)}\,dt$

11. $\displaystyle\int_2^3 \dfrac{1}{x^2-1}\,dx$

12. $\displaystyle\int_0^1 \dfrac{x-1}{x^2+3x+2}\,dx$

13. $\displaystyle\int \dfrac{ax}{x^2-bx}\,dx$

14. $\displaystyle\int \dfrac{1}{(x+a)(x+b)}\,dx$

15. $\displaystyle\int_0^1 \dfrac{2x+3}{(x+1)^2}\,dx$

16. $\displaystyle\int_0^1 \dfrac{x^3-4x-10}{x^2-x-6}\,dx$

17. $\displaystyle\int_1^2 \dfrac{4y^2-7y-12}{y(y+2)(y-3)}\,dy$

18. $\displaystyle\int \dfrac{x^2+2x-1}{x^3-x}\,dx$

19. $\displaystyle\int \dfrac{1}{(x+5)^2(x-1)}\,dx$

20. $\displaystyle\int \dfrac{x^2}{(x-3)(x+2)^2}\,dx$

21. $\displaystyle\int \dfrac{5x^2+3x-2}{x^3+2x^2}\,dx$

22. $\displaystyle\int \dfrac{ds}{s^2(s-1)^2}$

23. $\displaystyle\int \dfrac{x^2}{(x+1)^3}\,dx$

24. $\displaystyle\int \dfrac{x^3}{(x+1)^3}\,dx$

25. $\displaystyle\int \dfrac{10}{(x-1)(x^2+9)}\,dx$

26. $\displaystyle\int \dfrac{x^2-x+6}{x^3+3x}\,dx$

27. $\displaystyle\int \dfrac{x^3+x^2+2x+1}{(x^2+1)(x^2+2)}\,dx$

28. $\displaystyle\int \dfrac{x^2-2x-1}{(x-1)^2(x^2+1)}\,dx$

29. $\displaystyle\int \dfrac{x+4}{x^2+2x+5}\,dx$

30. $\displaystyle\int \dfrac{x^3-2x^2+x+1}{x^4+5x^2+4}\,dx$

31. $\displaystyle\int \dfrac{1}{x^3-1}\,dx$

32. $\displaystyle\int_0^1 \dfrac{x}{x^2+4x+13}\,dx$

33. $\displaystyle\int_2^5 \dfrac{x^2+2x}{x^3+3x^2+4}\,dx$

34. $\displaystyle\int \dfrac{x^3}{x^3+1}\,dx$

35. $\displaystyle\int \dfrac{dx}{x^4-x^2}$

36. $\displaystyle\int_0^1 \dfrac{2x^3+5x}{x^4+5x^2+4}\,dx$

37. $\displaystyle\int \dfrac{x-3}{(x^2+2x+4)^2}\,dx$

38. $\displaystyle\int \dfrac{x^4+1}{x(x^2+1)^2}\,dx$

39–48 Make a substitution to express the integrand as a rational function and then evaluate the integral.

39. $\displaystyle\int \dfrac{1}{x\sqrt{x+1}}\,dx$

40. $\displaystyle\int \dfrac{1}{x-\sqrt{x+2}}\,dx$

41. $\displaystyle\int_9^{16} \dfrac{\sqrt{x}}{x-4}\,dx$

42. $\displaystyle\int_0^1 \dfrac{1}{1+\sqrt[3]{x}}\,dx$

43. $\displaystyle\int \dfrac{x^3}{\sqrt[3]{x^2+1}}\,dx$

44. $\displaystyle\int_{1/3}^3 \dfrac{\sqrt{x}}{x^2+x}\,dx$

45. $\displaystyle\int \dfrac{1}{\sqrt{x}-\sqrt[3]{x}}\,dx$ [Hint: Substitute $u=\sqrt[6]{x}$.]

46. $\displaystyle\int \dfrac{1}{\sqrt[3]{x}+\sqrt[4]{x}}\,dx$ [Hint: Substitute $u=\sqrt[12]{x}$.]

47. $\displaystyle\int \dfrac{e^{2x}}{e^{2x}+3e^x+2}\,dx$

48. $\displaystyle\int \dfrac{\cos x}{\sin^2 x+\sin x}\,dx$

49–50 Use integration by parts, together with the techniques of this section, to evaluate the integral.

49. $\displaystyle\int \ln(x^2-x+2)\,dx$

50. $\displaystyle\int x\tan^{-1}x\,dx$

51. Use a graph of $f(x)=1/(x^2-2x-3)$ to decide whether $\int_0^2 f(x)\,dx$ is positive or negative. Use the graph to give a rough estimate of the value of the integral and then use partial fractions to find the exact value.

52. Graph both $y=1/(x^3-2x^2)$ and an antiderivative on the same screen.

53–54 Evaluate the integral by completing the square and using Formula 6.

53. $\displaystyle\int \dfrac{dx}{x^2-2x}$

54. $\displaystyle\int \dfrac{2x+1}{4x^2+12x-7}\,dx$

55. The German mathematician Karl Weierstrass (1815–1897) noticed that the substitution $t=\tan(x/2)$ will convert any

rational function of sin x and cos x into an ordinary rational function of t.

(a) If $t = \tan(x/2)$, $-\pi < x < \pi$, sketch a right triangle or use trigonometric identities to show that

$$\cos\left(\frac{x}{2}\right) = \frac{1}{\sqrt{1+t^2}} \quad \text{and} \quad \sin\left(\frac{x}{2}\right) = \frac{t}{\sqrt{1+t^2}}$$

(b) Show that

$$\cos x = \frac{1-t^2}{1+t^2} \quad \text{and} \quad \sin x = \frac{2t}{1+t^2}$$

(c) Show that

$$dx = \frac{2}{1+t^2}\, dt$$

56–59 Use the substitution in Exercise 55 to transform the integrand into a rational function of t and then evaluate the integral.

56. $\displaystyle\int \frac{dx}{3 - 5\sin x}$

57. $\displaystyle\int \frac{1}{3\sin x - 4\cos x}\, dx$

58. $\displaystyle\int_{\pi/3}^{\pi/2} \frac{1}{1 + \sin x - \cos x}\, dx$

59. $\displaystyle\int \frac{1}{2\sin x + \sin 2x}\, dx$

60–61 Find the area of the region under the given curve from a to b.

60. $y = \dfrac{1}{x^2 - 6x + 8}$, $a = 5$, $b = 10$

61. $y = \dfrac{x+1}{x-1}$, $a = 2$, $b = 3$

62. Find the volume of the resulting solid if the region under the curve $y = 1/(x^2 + 3x + 2)$ from $x = 0$ to $x = 1$ is rotated about (a) the x-axis and (b) the y-axis.

63. One method of slowing the growth of an insect population without using pesticides is to introduce into the population a number of sterile males that mate with fertile females but produce no offspring. If P represents the number of female insects in a population, S the number of sterile males introduced each generation, and r the population's natural growth rate, then the female population is related to time t by

$$t = \int \frac{P+S}{P[(r-1)P - S]}\, dP$$

Suppose an insect population with 10,000 females grows at a rate of $r = 0.10$ and 900 sterile males are added. Evaluate the integral to give an equation relating the female population to time. (Note that the resulting equation can't be solved explicitly for P.)

64. Factor $x^4 + 1$ as a difference of squares by first adding and subtracting the same quantity. Use this factorization to evaluate $\int 1/(x^4 + 1)\, dx$.

CAS 65. (a) Use a computer algebra system to find the partial fraction decomposition of the function

$$f(x) = \frac{4x^3 - 27x^2 + 5x - 32}{30x^5 - 13x^4 + 50x^3 - 286x^2 - 299x - 70}$$

(b) Use part (a) to find $\int f(x)\, dx$ (by hand) and compare with the result of using the CAS to integrate f directly. Comment on any discrepancy.

CAS 66. (a) Find the partial fraction decomposition of the function

$$f(x) = \frac{12x^5 - 7x^3 - 13x^2 + 8}{100x^6 - 80x^5 + 116x^4 - 80x^3 + 41x^2 - 20x + 4}$$

(b) Use part (a) to find $\int f(x)\, dx$ and graph f and its indefinite integral on the same screen.

(c) Use the graph of f to discover the main features of the graph of $\int f(x)\, dx$.

67. Suppose that F, G, and Q are polynomials and

$$\frac{F(x)}{Q(x)} = \frac{G(x)}{Q(x)}$$

for all x except when $Q(x) = 0$. Prove that $F(x) = G(x)$ for all x. [*Hint:* Use continuity.]

68. If f is a quadratic function such that $f(0) = 1$ and

$$\int \frac{f(x)}{x^2(x+1)^3}\, dx$$

is a rational function, find the value of $f'(0)$.

8.5 Strategy for Integration

As we have seen, integration is more challenging than differentiation. In finding the derivative of a function it is obvious which differentiation formula we should apply. But it may not be obvious which technique we should use to integrate a given function.

Until now individual techniques have been applied in each section. For instance, we usually used substitution in Exercises 5.5, integration by parts in Exercises 8.1, and partial fractions in Exercises 8.4. But in this section we present a collection of miscellaneous integrals in random order and the main challenge is to recognize which technique or formula to use. No hard and fast rules can be given as to which method applies in a given situation, but we give some advice on strategy that you may find useful.

A prerequisite for strategy selection is a knowledge of the basic integration formulas. In the following table we have collected the integrals from our previous list together with several additional formulas that we have learned in this chapter. Most of them should be memorized. It is useful to know them all, but the ones marked with an asterisk need not be memorized since they are easily derived. Formula 19 can be avoided by using partial fractions, and trigonometric substitutions can be used in place of Formula 20.

Table of Integration Formulas Constants of integration have been omitted.

1. $\int x^n \, dx = \dfrac{x^{n+1}}{n+1} \quad (n \neq -1)$
2. $\int \dfrac{1}{x} \, dx = \ln|x|$
3. $\int e^x \, dx = e^x$
4. $\int a^x \, dx = \dfrac{a^x}{\ln a}$
5. $\int \sin x \, dx = -\cos x$
6. $\int \cos x \, dx = \sin x$
7. $\int \sec^2 x \, dx = \tan x$
8. $\int \csc^2 x \, dx = -\cot x$
9. $\int \sec x \tan x \, dx = \sec x$
10. $\int \csc x \cot x \, dx = -\csc x$
11. $\int \sec x \, dx = \ln|\sec x + \tan x|$
12. $\int \csc x \, dx = \ln|\csc x - \cot x|$
13. $\int \tan x \, dx = \ln|\sec x|$
14. $\int \cot x \, dx = \ln|\sin x|$
15. $\int \sinh x \, dx = \cosh x$
16. $\int \cosh x \, dx = \sinh x$
17. $\int \dfrac{dx}{x^2 + a^2} = \dfrac{1}{a} \tan^{-1}\left(\dfrac{x}{a}\right)$
18. $\int \dfrac{dx}{\sqrt{a^2 - x^2}} = \sin^{-1}\left(\dfrac{x}{a}\right)$
*19. $\int \dfrac{dx}{x^2 - a^2} = \dfrac{1}{2a} \ln\left|\dfrac{x-a}{x+a}\right|$
*20. $\int \dfrac{dx}{\sqrt{x^2 \pm a^2}} = \ln\left|x + \sqrt{x^2 \pm a^2}\right|$

Once you are armed with these basic integration formulas, if you don't immediately see how to attack a given integral, you might try the following four-step strategy.

1. Simplify the Integrand if Possible Sometimes the use of algebraic manipulation or trigonometric identities will simplify the integrand and make the method of integration obvious. Here are some examples:

$$\int \sqrt{x}\,(1 + \sqrt{x})\, dx = \int (\sqrt{x} + x)\, dx$$

$$\int \dfrac{\tan \theta}{\sec^2 \theta}\, d\theta = \int \dfrac{\sin \theta}{\cos \theta} \cos^2 \theta \, d\theta$$

$$= \int \sin \theta \cos \theta \, d\theta = \tfrac{1}{2} \int \sin 2\theta \, d\theta$$

$$\int (\sin x + \cos x)^2 \, dx = \int (\sin^2 x + 2\sin x \cos x + \cos^2 x) \, dx$$

$$= \int (1 + 2\sin x \cos x) \, dx$$

2. Look for an Obvious Substitution Try to find some function $u = g(x)$ in the integrand whose differential $du = g'(x) \, dx$ also occurs, apart from a constant factor. For instance, in the integral

$$\int \frac{x}{x^2 - 1} \, dx$$

we notice that if $u = x^2 - 1$, then $du = 2x \, dx$. Therefore, we use the substitution $u = x^2 - 1$ instead of the method of partial fractions.

3. Classify the Integrand According to Its Form If Steps 1 and 2 have not led to the solution, then we take a look at the form of the integrand $f(x)$.
  (a) *Trigonometric functions.* If $f(x)$ is a product of powers of $\sin x$ and $\cos x$, of $\tan x$ and $\sec x$, or of $\cot x$ and $\csc x$, then we use the substitutions recommended in Section 8.2.
  (b) *Rational functions.* If f is a rational function, we use the procedure of Section 8.4 involving partial fractions.
  (c) *Integration by parts.* If $f(x)$ is a product of a power of x (or a polynomial) and a transcendental function (such as a trigonometric, exponential, or logarithmic function), then we try integration by parts, choosing u and dv according to the advice given in Section 8.1. If you look at the functions in Exercises 8.1, you will see that most of them are the type just described.
  (d) *Radicals.* Particular kinds of substitutions are recommended when certain radicals appear.
      (i) If $\sqrt{\pm x^2 \pm a^2}$ occurs, we use a trigonometric substitution according to the table in Section 8.3.
      (ii) If $\sqrt[n]{ax + b}$ occurs, we use the rationalizing substitution $u = \sqrt[n]{ax + b}$. More generally, this sometimes works for $\sqrt[n]{g(x)}$.

4. Try Again If the first three steps have not produced the answer, remember that there are basically only two methods of integration: substitution and parts.
  (a) *Try substitution.* Even if no substitution is obvious (Step 2), some inspiration or ingenuity (or even desperation) may suggest an appropriate substitution.
  (b) *Try parts.* Although integration by parts is used most of the time on products of the form described in Step 3(c), it is sometimes effective on single functions. Looking at Section 8.1, we see that it works on $\tan^{-1} x$, $\sin^{-1} x$, and $\ln x$, and these are all inverse functions.
  (c) *Manipulate the integrand.* Algebraic manipulations (perhaps rationalizing the denominator or using trigonometric identities) may be useful in transforming the integral into an easier form. These manipulations may be more substantial than in Step 1 and may involve some ingenuity. Here is an example:

$$\int \frac{dx}{1 - \cos x} = \int \frac{1}{1 - \cos x} \cdot \frac{1 + \cos x}{1 + \cos x} \, dx = \int \frac{1 + \cos x}{1 - \cos^2 x} \, dx$$

$$= \int \frac{1 + \cos x}{\sin^2 x} \, dx = \int \left(\csc^2 x + \frac{\cos x}{\sin^2 x} \right) dx$$

(d) *Relate the problem to previous problems.* When you have built up some experience in integration, you may be able to use a method on a given integral that is similar to a method you have already used on a previous integral. Or you may even be able to express the given integral in terms of a previous one. For instance, $\int \tan^2 x \sec x \, dx$ is a challenging integral, but if we make use of the identity $\tan^2 x = \sec^2 x - 1$, we can write

$$\int \tan^2 x \sec x \, dx = \int \sec^3 x \, dx - \int \sec x \, dx$$

and if $\int \sec^3 x \, dx$ has previously been evaluated (see Example 8 in Section 8.2), then that calculation can be used in the present problem.

(e) *Use several methods.* Sometimes two or three methods are required to evaluate an integral. The evaluation could involve several successive substitutions of different types, or it might combine integration by parts with one or more substitutions.

In the following examples we indicate a method of attack but do not fully work out the integral.

EXAMPLE 1 $\int \dfrac{\tan^3 x}{\cos^3 x} \, dx$

In Step 1 we rewrite the integral:

$$\int \frac{\tan^3 x}{\cos^3 x} \, dx = \int \tan^3 x \sec^3 x \, dx$$

The integral is now of the form $\int \tan^m x \sec^n x \, dx$ with m odd, so we can use the advice in Section 7.2.

Alternatively, if in Step 1 we had written

$$\int \frac{\tan^3 x}{\cos^3 x} \, dx = \int \frac{\sin^3 x}{\cos^3 x} \frac{1}{\cos^3 x} \, dx = \int \frac{\sin^3 x}{\cos^6 x} \, dx$$

then we could have continued as follows with the substitution $u = \cos x$:

$$\int \frac{\sin^3 x}{\cos^6 x} \, dx = \int \frac{1 - \cos^2 x}{\cos^6 x} \sin x \, dx = \int \frac{1 - u^2}{u^6} (-du)$$

$$= \int \frac{u^2 - 1}{u^6} \, du = \int (u^{-4} - u^{-6}) \, du$$

EXAMPLE 2 $\int e^{\sqrt{x}} \, dx$

According to Step 3(d)(ii) we substitute $u = \sqrt{x}$. Then $x = u^2$, so $dx = 2u \, du$ and

$$\int e^{\sqrt{x}} \, dx = 2 \int u e^u \, du$$

The integrand is now a product of u and the transcendental function e^u so it can be integrated by parts.

EXAMPLE 3 $\int \dfrac{x^5 + 1}{x^3 - 3x^2 - 10x}\, dx$

No algebraic simplification or substitution is obvious, so Steps 1 and 2 don't apply here. The integrand is a rational function so we apply the procedure of Section 8.4, remembering that the first step is to divide.

EXAMPLE 4 $\int \dfrac{dx}{x\sqrt{\ln x}}$

Here Step 2 is all that is needed. We substitute $u = \ln x$ because its differential is $du = dx/x$, which occurs in the integral.

EXAMPLE 5 $\int \sqrt{\dfrac{1-x}{1+x}}\, dx$

Although the rationalizing substitution

$$u = \sqrt{\dfrac{1-x}{1+x}}$$

works here [Step 3(d)(ii)], it leads to a very complicated rational function. An easier method is to do some algebraic manipulation [either as Step 1 or as Step 4(c)]. Multiplying numerator and denominator by $\sqrt{1-x}$, we have

$$\int \sqrt{\dfrac{1-x}{1+x}}\, dx = \int \dfrac{1-x}{\sqrt{1-x^2}}\, dx$$

$$= \int \dfrac{1}{\sqrt{1-x^2}}\, dx - \int \dfrac{x}{\sqrt{1-x^2}}\, dx$$

$$= \sin^{-1} x + \sqrt{1-x^2} + C$$

Can We Integrate All Continuous Functions?

The question arises: Will our strategy for integration enable us to find the integral of every continuous function? For example, can we use it to evaluate $\int e^{x^2}\, dx$? The answer is no, at least not in terms of the functions that we are familiar with.

The functions that we have been dealing with in this book are called **elementary functions**. These are the polynomials, rational functions, power functions (x^a), exponential functions (a^x), logarithmic functions, trigonometric and inverse trigonometric functions, hyperbolic and inverse hyperbolic functions, and all functions that can be obtained from these by the five operations of addition, subtraction, multiplication, division, and composition. For instance, the function

$$f(x) = \sqrt{\dfrac{x^2 - 1}{x^3 + 2x - 1}} + \ln(\cosh x) - xe^{\sin 2x}$$

is an elementary function.

If f is an elementary function, then f' is an elementary function but $\int f(x)\, dx$ need not be an elementary function. Consider $f(x) = e^{x^2}$. Since f is continuous, its integral exists, and if we define the function F by

$$F(x) = \int_0^x e^{t^2}\, dt$$

then we know from Part 1 of the Fundamental Theorem of Calculus that

$$F'(x) = e^{x^2}$$

Thus, $f(x) = e^{x^2}$ has an antiderivative F, but it has been proved that F is not an elementary function. This means that no matter how hard we try, we will never succeed in evaluating $\int e^{x^2} dx$ in terms of the functions we know. (In Chapter 11, however, we will see how to express $\int e^{x^2} dx$ as an infinite series.) The same can be said of the following integrals:

$$\int \frac{e^x}{x} dx \qquad \int \sin(x^2) dx \qquad \int \cos(e^x) dx$$

$$\int \sqrt{x^3 + 1} \, dx \qquad \int \frac{1}{\ln x} dx \qquad \int \frac{\sin x}{x} dx$$

In fact, the majority of elementary functions don't have elementary antiderivatives. You may be assured, though, that the integrals in the following exercises are all elementary functions.

8.5 Exercises

1–80 ▪ Evaluate the integral.

1. $\int \dfrac{\sin x + \sec x}{\tan x} dx$

2. $\int \tan^3 \theta \, d\theta$

3. $\int_0^2 \dfrac{2t}{(t-3)^2} dt$

4. $\int \dfrac{x}{\sqrt{3-x^4}} dx$

5. $\int_{-1}^{1} \dfrac{e^{\arctan y}}{1+y^2} dy$

6. $\int x \csc x \cot x \, dx$

7. $\int_1^3 r^4 \ln r \, dr$

8. $\int_0^4 \dfrac{x-1}{x^2 - 4x - 5} dx$

9. $\int \dfrac{x-1}{x^2 - 4x + 5} dx$

10. $\int \dfrac{x}{x^4 + x^2 + 1} dx$

11. $\int \sin^3 \theta \cos^5 \theta \, d\theta$

12. $\int \sin x \cos(\cos x) \, dx$

13. $\int \dfrac{dx}{(1-x^2)^{3/2}}$

14. $\int \dfrac{\sqrt{1 + \ln x}}{x \ln x} dx$

15. $\int_0^{1/2} \dfrac{x}{\sqrt{1-x^2}} dx$

16. $\int_0^{\sqrt{2}/2} \dfrac{x^2}{\sqrt{1-x^2}} dx$

17. $\int x \sin^2 x \, dx$

18. $\int \dfrac{e^{2t}}{1 + e^{4t}} dt$

19. $\int e^{x + e^x} dx$

20. $\int e^{\sqrt[3]{x}} dx$

21. $\int t^3 e^{-2t} dt$

22. $\int x \sin^{-1} x \, dx$

23. $\int_0^1 (1 + \sqrt{x})^8 dx$

24. $\int \ln(x^2 - 1) \, dx$

25. $\int \dfrac{3x^2 - 2}{x^2 - 2x - 8} dx$

26. $\int \dfrac{3x^2 - 2}{x^3 - 2x - 8} dx$

27. $\int \cot x \ln(\sin x) \, dx$

28. $\int \sin \sqrt{at} \, dt$

29. $\int_0^5 \dfrac{3w - 1}{w + 2} dw$

30. $\int_{-2}^{2} |x^2 - 4x| \, dx$

31. $\int \sqrt{\dfrac{1+x}{1-x}} dx$

32. $\int \dfrac{\sqrt{2x-1}}{2x+3} dx$

33. $\int \sqrt{3 - 2x - x^2} \, dx$

34. $\int_{\pi/4}^{\pi/2} \dfrac{1 + 4 \cot x}{4 - \cot x} dx$

35. $\int_{-1}^{1} x^8 \sin x \, dx$

36. $\int \sin 4x \cos 3x \, dx$

37. $\int_0^{\pi/4} \cos^2 \theta \tan^2 \theta \, d\theta$

38. $\int_0^{\pi/4} \tan^5 \theta \sec^3 \theta \, d\theta$

39. $\int \dfrac{x}{1 - x^2 + \sqrt{1-x^2}} dx$

40. $\int \dfrac{1}{\sqrt{4y^2 - 4y - 3}} dy$

41. $\int \theta \tan^2 \theta \, d\theta$

42. $\int x^2 \tan^{-1} x \, dx$

43. $\int e^x \sqrt{1 + e^x} \, dx$

44. $\int \sqrt{1 + e^x} \, dx$

45. $\int x^5 e^{-x^3} dx$

46. $\int \dfrac{1 + e^x}{1 - e^x} dx$

47. $\int \dfrac{x + a}{x^2 + a^2} dx$

48. $\int \dfrac{x}{x^4 - a^4} dx$

49. $\displaystyle\int \frac{1}{x\sqrt{4x+1}}\,dx$

50. $\displaystyle\int \frac{1}{x^2\sqrt{4x+1}}\,dx$

51. $\displaystyle\int \frac{1}{x\sqrt{4x^2+1}}\,dx$

52. $\displaystyle\int \frac{dx}{x(x^4+1)}$

53. $\displaystyle\int x^2 \sinh mx\,dx$

54. $\displaystyle\int (x+\sin x)^2\,dx$

55. $\displaystyle\int \frac{1}{x+4+4\sqrt{x+1}}\,dx$

56. $\displaystyle\int \frac{x\ln x}{\sqrt{x^2-1}}\,dx$

57. $\displaystyle\int x\sqrt[3]{x+c}\,dx$

58. $\displaystyle\int x^2 \ln(1+x)\,dx$

59. $\displaystyle\int \frac{1}{e^{3x}-e^x}\,dx$

60. $\displaystyle\int \frac{1}{x+\sqrt[3]{x}}\,dx$

61. $\displaystyle\int \frac{x^4}{x^{10}+16}\,dx$

62. $\displaystyle\int \frac{x^3}{(x+1)^{10}}\,dx$

63. $\displaystyle\int \sqrt{x}\,e^{\sqrt{x}}\,dx$

64. $\displaystyle\int_{\pi/4}^{\pi/3} \frac{\ln(\tan x)}{\sin x \cos x}\,dx$

65. $\displaystyle\int \frac{1}{\sqrt{x+1}+\sqrt{x}}\,dx$

66. $\displaystyle\int_{2}^{3} \frac{u^3+1}{u^3-u^2}\,du$

67. $\displaystyle\int_{1}^{3} \frac{\arctan\sqrt{t}}{\sqrt{t}}\,dt$

68. $\displaystyle\int \frac{1}{1+2e^x-e^{-x}}\,dx$

69. $\displaystyle\int \frac{e^{2x}}{1+e^x}\,dx$

70. $\displaystyle\int \frac{\ln(x+1)}{x^2}\,dx$

71. $\displaystyle\int \frac{x}{x^4+4x^2+3}\,dx$

72. $\displaystyle\int \frac{\sqrt{t}}{1+\sqrt[3]{t}}\,dt$

73. $\displaystyle\int \frac{1}{(x-2)(x^2+4)}\,dx$

74. $\displaystyle\int \frac{dx}{e^x-e^{-x}}$

75. $\displaystyle\int \sin x \sin 2x \sin 3x\,dx$

76. $\displaystyle\int (x^2-bx)\sin 2x\,dx$

77. $\displaystyle\int \frac{\sqrt{x}}{1+x^3}\,dx$

78. $\displaystyle\int \frac{\sec x \cos 2x}{\sin x + \sec x}\,dx$

79. $\displaystyle\int x\sin^2 x \cos x\,dx$

80. $\displaystyle\int \frac{\sin x \cos x}{\sin^4 x + \cos^4 x}\,dx$

81. The functions $y=e^{x^2}$ and $y=x^2 e^{x^2}$ don't have elementary antiderivatives, but $y=(2x^2+1)e^{x^2}$ does. Evaluate $\int (2x^2+1)e^{x^2}\,dx$.

8.6 Integration Using Tables and Computer Algebra Systems

In this section we describe how to use tables and computer algebra systems to integrate functions that have elementary antiderivatives. You should bear in mind, though, that even the most powerful computer algebra systems can't find explicit formulas for the antiderivatives of functions like e^{x^2} or the other functions described at the end of Section 8.5.

Tables of Integrals

Tables of indefinite integrals are very useful when we are confronted by an integral that is difficult to evaluate by hand and we don't have access to a computer algebra system. A relatively brief table of 120 integrals, categorized by form, is provided on the Reference Pages at the back of the book. More extensive tables are available in *CRC Standard Mathematical Tables and Formulae*, 30th ed. by Daniel Zwillinger (Boca Raton, FL: CRC Press, 1995) (581 entries) or in Gradshteyn and Ryzhik's *Table of Integrals, Series, and Products*, 6e (New York: Academic Press, 2000), which contains hundreds of pages of integrals. It should be remembered, however, that integrals do not often occur in exactly the form listed in a table. Usually we need to use substitution or algebraic manipulation to transform a given integral into one of the forms in the table.

EXAMPLE 1 The region bounded by the curves $y=\arctan x$, $y=0$, and $x=1$ is rotated about the y-axis. Find the volume of the resulting solid.

SOLUTION Using the method of cylindrical shells, we see that the volume is

$$V = \int_{0}^{1} 2\pi x \arctan x\,dx$$

In the section of the Table of Integrals entitled *Inverse Trigonometric Forms* we locate Formula 92:

$$\int u \tan^{-1} u \, du = \frac{u^2 + 1}{2} \tan^{-1} u - \frac{u}{2} + C$$

> The Table of Integrals appears on the Reference Pages at the back of the book.

Thus, the volume is

$$V = 2\pi \int_0^1 x \tan^{-1} x \, dx = 2\pi \left[\frac{x^2 + 1}{2} \tan^{-1} x - \frac{x}{2} \right]_0^1$$

$$= \pi [(x^2 + 1) \tan^{-1} x - x]_0^1 = \pi (2 \tan^{-1} 1 - 1)$$

$$= \pi [2(\pi/4) - 1] = \tfrac{1}{2}\pi^2 - \pi$$

EXAMPLE 2 Use the Table of Integrals to find $\int \dfrac{x^2}{\sqrt{5 - 4x^2}} \, dx$.

SOLUTION If we look at the section of the table entitled *Forms involving* $\sqrt{a^2 - u^2}$, we see that the closest entry is number 34:

$$\int \frac{u^2}{\sqrt{a^2 - u^2}} \, du = -\frac{u}{2} \sqrt{a^2 - u^2} + \frac{a^2}{2} \sin^{-1}\left(\frac{u}{a}\right) + C$$

This is not exactly what we have, but we will be able to use it if we first make the substitution $u = 2x$:

$$\int \frac{x^2}{\sqrt{5 - 4x^2}} \, dx = \int \frac{(u/2)^2}{\sqrt{5 - u^2}} \frac{du}{2} = \frac{1}{8} \int \frac{u^2}{\sqrt{5 - u^2}} \, du$$

Then we use Formula 34 with $a^2 = 5$ (so $a = \sqrt{5}$):

$$\int \frac{x^2}{\sqrt{5 - 4x^2}} \, dx = \frac{1}{8} \int \frac{u^2}{\sqrt{5 - u^2}} \, du = \frac{1}{8}\left(-\frac{u}{2} \sqrt{5 - u^2} + \frac{5}{2} \sin^{-1} \frac{u}{\sqrt{5}} \right) + C$$

$$= -\frac{x}{8} \sqrt{5 - 4x^2} + \frac{5}{16} \sin^{-1}\left(\frac{2x}{\sqrt{5}}\right) + C$$

EXAMPLE 3 Use the Table of Integrals to find $\int x^3 \sin x \, dx$.

SOLUTION If we look in the section called *Trigonometric Forms*, we see that none of the entries explicitly includes a u^3 factor. However, we can use the reduction formula in entry 84 with $n = 3$:

$$\int x^3 \sin x \, dx = -x^3 \cos x + 3 \int x^2 \cos x \, dx$$

We now need to evaluate $\int x^2 \cos x \, dx$. We can use the reduction formula in entry 85 with $n = 2$, followed by entry 82:

$$\int x^2 \cos x \, dx = x^2 \sin x - 2 \int x \sin x \, dx$$

$$= x^2 \sin x - 2(\sin x - x \cos x) + K$$

> 85. $\int u^n \cos u \, du$
> $= u^n \sin u - n \int u^{n-1} \sin u \, du$

Combining these calculations, we get

$$\int x^3 \sin x \, dx = -x^3 \cos x + 3x^2 \sin x + 6x \cos x - 6 \sin x + C$$

where $C = 3K$.

EXAMPLE 4 Use the Table of Integrals to find $\int x\sqrt{x^2 + 2x + 4} \, dx$.

SOLUTION Since the table gives forms involving $\sqrt{a^2 + x^2}$, $\sqrt{a^2 - x^2}$, and $\sqrt{x^2 - a^2}$, but not $\sqrt{ax^2 + bx + c}$, we first complete the square:

$$x^2 + 2x + 4 = (x + 1)^2 + 3$$

If we make the substitution $u = x + 1$ (so $x = u - 1$), the integrand will involve the pattern $\sqrt{a^2 + u^2}$:

$$\int x\sqrt{x^2 + 2x + 4} \, dx = \int (u - 1)\sqrt{u^2 + 3} \, du$$

$$= \int u\sqrt{u^2 + 3} \, du - \int \sqrt{u^2 + 3} \, du$$

The first integral is evaluated using the substitution $t = u^2 + 3$:

$$\int u\sqrt{u^2 + 3} \, du = \tfrac{1}{2} \int \sqrt{t} \, dt = \tfrac{1}{2} \cdot \tfrac{2}{3} t^{3/2} = \tfrac{1}{3}(u^2 + 3)^{3/2}$$

21. $\int \sqrt{a^2 + u^2} \, du = \dfrac{u}{2} \sqrt{a^2 + u^2}$
$+ \dfrac{a^2}{2} \ln(u + \sqrt{a^2 + u^2}) + C$

For the second integral we use Formula 21 with $a = \sqrt{3}$:

$$\int \sqrt{u^2 + 3} \, du = \dfrac{u}{2} \sqrt{u^2 + 3} + \tfrac{3}{2} \ln(u + \sqrt{u^2 + 3})$$

Thus

$$\int x\sqrt{x^2 + 2x + 4} \, dx$$

$$= \tfrac{1}{3}(x^2 + 2x + 4)^{3/2} - \dfrac{x + 1}{2} \sqrt{x^2 + 2x + 4} - \tfrac{3}{2} \ln(x + 1 + \sqrt{x^2 + 2x + 4}) + C$$

Computer Algebra Systems

We have seen that the use of tables involves matching the form of the given integrand with the forms of the integrands in the tables. Computers are particularly good at matching patterns. And just as we used substitutions in conjunction with tables, a CAS can perform substitutions that transform a given integral into one that occurs in its stored formulas. So it isn't surprising that computer algebra systems excel at integration. That doesn't mean that integration by hand is an obsolete skill. We will see that a hand computation sometimes produces an indefinite integral in a form that is more convenient than a machine answer.

To begin, let's see what happens when we ask a machine to integrate the relatively simple function $y = 1/(3x - 2)$. Using the substitution $u = 3x - 2$, an easy calculation by hand gives

$$\int \dfrac{1}{3x - 2} \, dx = \tfrac{1}{3} \ln |3x - 2| + C$$

whereas Derive, Mathematica, and Maple all return the answer

$$\tfrac{1}{3}\ln(3x-2)$$

The first thing to notice is that computer algebra systems omit the constant of integration. In other words, they produce a *particular* antiderivative, not the most general one. Therefore, when making use of a machine integration, we might have to add a constant. Second, the absolute value signs are omitted in the machine answer. That is fine if our problem is concerned only with values of x greater than $\tfrac{2}{3}$. But if we are interested in other values of x, then we need to insert the absolute value symbol.

In the next example we reconsider the integral of Example 4, but this time we ask a machine for the answer.

EXAMPLE 5 Use a computer algebra system to find $\int x\sqrt{x^2+2x+4}\,dx$.

SOLUTION Maple responds with the answer

$$\tfrac{1}{3}(x^2+2x+4)^{3/2} - \tfrac{1}{4}(2x+2)\sqrt{x^2+2x+4} - \frac{3}{2}\operatorname{arcsinh}\frac{\sqrt{3}}{3}(1+x)$$

This looks different from the answer we found in Example 4, but it is equivalent because the third term can be rewritten using the identity

$$\operatorname{arcsinh} x = \ln\!\left(x + \sqrt{x^2+1}\right)$$

|||| This is Equation 7.6.3.

Thus

$$\operatorname{arcsinh}\frac{\sqrt{3}}{3}(1+x) = \ln\!\left[\frac{\sqrt{3}}{3}(1+x) + \sqrt{\tfrac{1}{3}(1+x)^2+1}\right]$$

$$= \ln\frac{1}{\sqrt{3}}\left[1 + x + \sqrt{(1+x)^2+3}\right]$$

$$= \ln\frac{1}{\sqrt{3}} + \ln\!\left(x + 1 + \sqrt{x^2+2x+4}\right)$$

The resulting extra term $-\tfrac{3}{2}\ln(1/\sqrt{3})$ can be absorbed into the constant of integration.

Mathematica gives the answer

$$\left(\frac{5}{6} + \frac{x}{6} + \frac{x^2}{3}\right)\sqrt{x^2+2x+4} - \frac{3}{2}\operatorname{arcsinh}\!\left(\frac{1+x}{\sqrt{3}}\right)$$

Mathematica combined the first two terms of Example 4 (and the Maple result) into a single term by factoring.

Derive gives the answer

$$\tfrac{1}{6}\sqrt{x^2+2x+4}\,(2x^2+x+5) - \tfrac{3}{2}\ln\!\left(\sqrt{x^2+2x+4} + x + 1\right)$$

The first term is like the first term in the Mathematica answer, and the second term is identical to the last term in Example 4.

EXAMPLE 6 Use a CAS to evaluate $\int x(x^2+5)^8\,dx$.

SOLUTION Maple and Mathematica give the same answer:

$$\tfrac{1}{18}x^{18} + \tfrac{5}{2}x^{16} + 50x^{14} + \tfrac{1750}{3}x^{12} + 4375x^{10} + 21875x^8 + \tfrac{218750}{3}x^6 + 156250x^4 + \tfrac{390625}{2}x^2$$

It's clear that both systems must have expanded $(x^2 + 5)^8$ by the Binomial Theorem and then integrated each term.

If we integrate by hand instead, using the substitution $u = x^2 + 5$, we get

$$\int x(x^2 + 5)^8 \, dx = \tfrac{1}{18}(x^2 + 5)^9 + C$$

▌ Derive and the TI-89/92 also give this answer.

For most purposes, this is a more convenient form of the answer.

EXAMPLE 7 Use a CAS to find $\int \sin^5 x \cos^2 x \, dx$.

SOLUTION In Example 2 in Section 8.2 we found that

$$\boxed{1} \qquad \int \sin^5 x \cos^2 x \, dx = -\tfrac{1}{3}\cos^3 x + \tfrac{2}{5}\cos^5 x - \tfrac{1}{7}\cos^7 x + C$$

Derive and Maple report the answer

$$-\tfrac{1}{7}\sin^4 x \cos^3 x - \tfrac{4}{35}\sin^2 x \cos^3 x - \tfrac{8}{105}\cos^3 x$$

whereas Mathematica produces

$$-\tfrac{5}{64}\cos x - \tfrac{1}{192}\cos 3x + \tfrac{3}{320}\cos 5x - \tfrac{1}{448}\cos 7x$$

We suspect that there are trigonometric identities which show these three answers are equivalent. Indeed, if we ask Derive, Maple, and Mathematica to simplify their expressions using trigonometric identities, they ultimately produce the same form of the answer as in Equation 1.

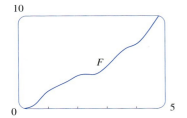

FIGURE 1

EXAMPLE 8 If $f(x) = x + 60 \sin^4 x \cos^5 x$, find the antiderivative F of f such that $F(0) = 0$. Graph F for $0 \le x \le 5$. Where does F have extreme values and inflection points?

SOLUTION The antiderivative of f produced by Maple is

$$F(x) = \tfrac{1}{2}x^2 - \tfrac{20}{3}\sin^3 x \cos^6 x - \tfrac{20}{7}\sin x \cos^6 x + \tfrac{4}{7}\cos^4 x \sin x + \tfrac{16}{21}\cos^2 x \sin x + \tfrac{32}{21}\sin x$$

and we note that $F(0) = 0$. This expression could probably be simplified, but there's no need to do so because a computer algebra system can graph this version of F as easily as any other version. A graph of F is shown in Figure 1. To locate the extreme values of F, we graph its derivative $F' = f$ in Figure 2 and observe that F has a local maximum when $x \approx 2.3$ and a local minimum when $x \approx 2.5$. The graph of $F'' = f'$ in Figure 2 shows that F has inflection points when $x \approx 0.7, 1.3, 1.8, 2.4, 3.3,$ and 3.9.

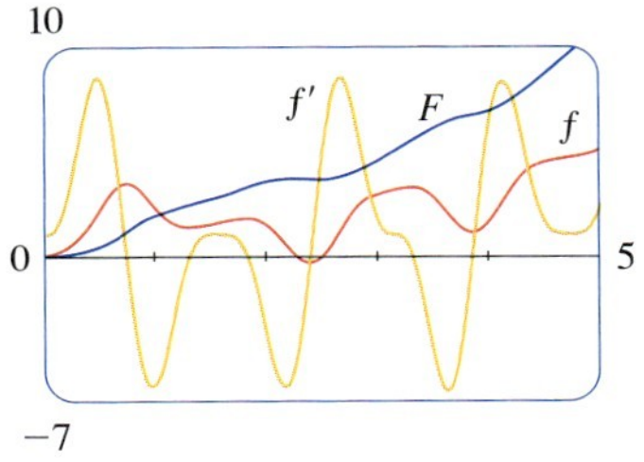

FIGURE 2

8.6 Exercises

1–4 ▌ Use the indicated entry in the Table of Integrals on the Reference Pages to evaluate the integral.

1. $\displaystyle\int \frac{\sqrt{7 - 2x^2}}{x^2} \, dx;$ entry 33

2. $\displaystyle\int \frac{3x}{\sqrt{3 - 2x}} \, dx;$ entry 55

3. $\displaystyle\int \sec^3(\pi x) \, dx;$ entry 71

4. $\displaystyle\int e^{2\theta} \sin 3\theta \, d\theta;$ entry 98

5–30 Use the Table of Integrals on the Reference Pages to evaluate the integral.

5. $\int_0^1 2x \cos^{-1}x \, dx$

6. $\int_2^3 \dfrac{1}{x^2\sqrt{4x^2-7}} \, dx$

7. $\int e^{-3x} \cos 4x \, dx$

8. $\int \csc^3(x/2) \, dx$

9. $\int \dfrac{dx}{x^2\sqrt{4x^2+9}}$

10. $\int \dfrac{\sqrt{2y^2-3}}{y^2} \, dy$

11. $\int_{-1}^0 t^2 e^{-t} \, dt$

12. $\int_0^\pi x^2 \cos 3x \, dx$

13. $\int \dfrac{\tan^3(1/z)}{z^2} \, dz$

14. $\int \sin^{-1}\sqrt{x} \, dx$

15. $\int e^x \operatorname{sech}(e^x) \, dx$

16. $\int x \sin(x^2) \cos(3x^2) \, dx$

17. $\int y\sqrt{6+4y-4y^2} \, dy$

18. $\int \dfrac{x^5}{x^2+\sqrt{2}} \, dx$

19. $\int \sin^2 x \cos x \ln(\sin x) \, dx$

20. $\int \dfrac{dx}{e^x(1+2e^x)}$

21. $\int \dfrac{e^x}{3-e^{2x}} \, dx$

22. $\int_0^2 x^3\sqrt{4x^2-x^4} \, dx$

23. $\int \sec^5 x \, dx$

24. $\int \sin^6 2x \, dx$

25. $\int \dfrac{\sqrt{4+(\ln x)^2}}{x} \, dx$

26. $\int_0^1 x^4 e^{-x} \, dx$

27. $\int \sqrt{e^{2x}-1} \, dx$

28. $\int e^t \sin(\alpha t - 3) \, dt$

29. $\int \dfrac{x^4 \, dx}{\sqrt{x^{10}-2}}$

30. $\int \dfrac{\sec^2\theta \tan^2\theta}{\sqrt{9-\tan^2\theta}} \, d\theta$

31. Find the volume of the solid obtained when the region under the curve $y = x\sqrt{4-x^2}$, $0 \leq x \leq 2$, is rotated about the y-axis.

32. The region under the curve $y = \tan^2 x$ from 0 to $\pi/4$ is rotated about the x-axis. Find the volume of the resulting solid.

33. Verify Formula 53 in the Table of Integrals (a) by differentiation and (b) by using the substitution $t = a + bu$.

34. Verify Formula 31 (a) by differentiation and (b) by substituting $u = a \sin \theta$.

CAS **35–42** Use a computer algebra system to evaluate the integral. Compare the answer with the result of using tables. If the answers are not the same, show that they are equivalent.

35. $\int x^2\sqrt{5-x^2} \, dx$

36. $\int x^2(1+x^3)^4 \, dx$

37. $\int \sin^3 x \cos^2 x \, dx$

38. $\int \tan^2 x \sec^4 x \, dx$

39. $\int x\sqrt{1+2x} \, dx$

40. $\int \sin^4 x \, dx$

41. $\int \tan^5 x \, dx$

42. $\int x^5\sqrt{x^2+1} \, dx$

CAS **43.** Computer algebra systems sometimes need a helping hand from human beings. Ask your CAS to evaluate

$$\int 2^x \sqrt{4^x - 1} \, dx$$

If it doesn't return an answer, ask it to try

$$\int 2^x \sqrt{2^{2x} - 1} \, dx$$

instead. Why do you think it was successful with this form of the integrand?

CAS **44.** Try to evaluate

$$\int (1 + \ln x)\sqrt{1 + (x \ln x)^2} \, dx$$

with a computer algebra system. If it doesn't return an answer, make a substitution that changes the integral into one that the CAS can evaluate.

CAS **45–48** Use a CAS to find an antiderivative F of f such that $F(0) = 0$. Graph f and F and locate approximately the x-coordinates of the extreme points and inflection points of F.

45. $f(x) = \dfrac{x^2 - 1}{x^4 + x^2 + 1}$

46. $f(x) = xe^{-x} \sin x$, $-5 \leq x \leq 5$

47. $f(x) = \sin^4 x \cos^6 x$, $0 \leq x \leq \pi$

48. $f(x) = \dfrac{x^3 - x}{x^6 + 1}$

DISCOVERY PROJECT

Patterns in Integrals

In this project a computer algebra system is used to investigate indefinite integrals of families of functions. By observing the patterns that occur in the integrals of several members of the family, you will first guess, and then prove, a general formula for the integral of any member of the family.

1. (a) Use a computer algebra system to evaluate the following integrals.

    (i) $\displaystyle\int \frac{1}{(x+2)(x+3)}\,dx$ (ii) $\displaystyle\int \frac{1}{(x+1)(x+5)}\,dx$

    (iii) $\displaystyle\int \frac{1}{(x+2)(x-5)}\,dx$ (iv) $\displaystyle\int \frac{1}{(x+2)^2}\,dx$

   (b) Based on the pattern of your responses in part (a), guess the value of the integral
   $$\int \frac{1}{(x+a)(x+b)}\,dx$$
   if $a \neq b$. What if $a = b$?

   (c) Check your guess by asking your CAS to evaluate the integral in part (b). Then prove it using partial fractions.

2. (a) Use a computer algebra system to evaluate the following integrals.

    (i) $\displaystyle\int \sin x \cos 2x\,dx$ (ii) $\displaystyle\int \sin 3x \cos 7x\,dx$ (iii) $\displaystyle\int \sin 8x \cos 3x\,dx$

   (b) Based on the pattern of your responses in part (a), guess the value of the integral
   $$\int \sin ax \cos bx\,dx$$

   (c) Check your guess with a CAS. Then prove it using the techniques of Section 7.2. For what values of a and b is it valid?

3. (a) Use a computer algebra system to evaluate the following integrals.

    (i) $\displaystyle\int \ln x\,dx$ (ii) $\displaystyle\int x \ln x\,dx$ (iii) $\displaystyle\int x^2 \ln x\,dx$

    (iv) $\displaystyle\int x^3 \ln x\,dx$ (v) $\displaystyle\int x^7 \ln x\,dx$

   (b) Based on the pattern of your responses in part (a), guess the value of
   $$\int x^n \ln x\,dx$$

   (c) Use integration by parts to prove the conjecture that you made in part (b). For what values of n is it valid?

4. (a) Use a computer algebra system to evaluate the following integrals.

    (i) $\displaystyle\int xe^x\,dx$ (ii) $\displaystyle\int x^2 e^x\,dx$ (iii) $\displaystyle\int x^3 e^x\,dx$

    (iv) $\displaystyle\int x^4 e^x\,dx$ (v) $\displaystyle\int x^5 e^x\,dx$

   (b) Based on the pattern of your responses in part (a), guess the value of $\int x^6 e^x\,dx$. Then use your CAS to check your guess.

   (c) Based on the patterns in parts (a) and (b), make a conjecture as to the value of the integral
   $$\int x^n e^x\,dx$$
   when n is a positive integer.

   (d) Use mathematical induction to prove the conjecture you made in part (c).

8.7 Approximate Integration

There are two situations in which it is impossible to find the exact value of a definite integral.

The first situation arises from the fact that in order to evaluate $\int_a^b f(x)\,dx$ using the Fundamental Theorem of Calculus we need to know an antiderivative of f. Sometimes, however, it is difficult, or even impossible, to find an antiderivative (see Section 8.5). For example, it is impossible to evaluate the following integrals exactly:

$$\int_0^1 e^{x^2}\,dx \qquad \int_{-1}^1 \sqrt{1+x^3}\,dx$$

The second situation arises when the function is determined from a scientific experiment through instrument readings or collected data. There may be no formula for the function (see Example 5).

In both cases we need to find approximate values of definite integrals. We already know one such method. Recall that the definite integral is defined as a limit of Riemann sums, so any Riemann sum could be used as an approximation to the integral: If we divide $[a, b]$ into n subintervals of equal length $\Delta x = (b-a)/n$, then we have

$$\int_a^b f(x)\,dx \approx \sum_{i=1}^n f(x_i^*)\,\Delta x$$

where x_i^* is any point in the ith subinterval $[x_{i-1}, x_i]$. If x_i^* is chosen to be the left endpoint of the interval, then $x_i^* = x_{i-1}$ and we have

$$\boxed{1} \qquad \int_a^b f(x)\,dx \approx L_n = \sum_{i=1}^n f(x_{i-1})\,\Delta x$$

If $f(x) \geq 0$, then the integral represents an area and (1) represents an approximation of this area by the rectangles shown in Figure 1(a). If we choose x_i^* to be the right endpoint, then $x_i^* = x_i$ and we have

$$\boxed{2} \qquad \int_a^b f(x)\,dx \approx R_n = \sum_{i=1}^n f(x_i)\,\Delta x$$

[See Figure 1(b).] The approximations L_n and R_n defined by Equations 1 and 2 are called the **left endpoint approximation** and **right endpoint approximation**, respectively.

In Section 5.2 we also considered the case where x_i^* is chosen to be the midpoint $\bar{x}_i$ of the subinterval $[x_{i-1}, x_i]$. Figure 1(c) shows the midpoint approximation M_n, which appears to be better than either L_n or R_n.

Midpoint Rule

$$\int_a^b f(x)\,dx \approx M_n = \Delta x\,[f(\bar{x}_1) + f(\bar{x}_2) + \cdots + f(\bar{x}_n)]$$

where $\qquad \Delta x = \dfrac{b-a}{n}$

and $\qquad \bar{x}_i = \tfrac{1}{2}(x_{i-1} + x_i) = $ midpoint of $[x_{i-1}, x_i]$

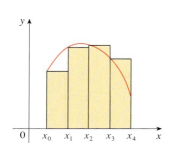

(a) Left endpoint approximation

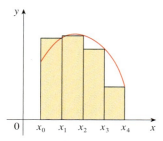

(b) Right endpoint approximation

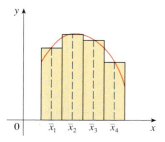

(c) Midpoint approximation

FIGURE 1

Another approximation, called the Trapezoidal Rule, results from averaging the approximations in Equations 1 and 2:

$$\int_a^b f(x)\,dx \approx \frac{1}{2}\left[\sum_{i=1}^n f(x_{i-1})\,\Delta x + \sum_{i=1}^n f(x_i)\,\Delta x\right] = \frac{\Delta x}{2}\left[\sum_{i=1}^n (f(x_{i-1}) + f(x_i))\right]$$

$$= \frac{\Delta x}{2}[(f(x_0) + f(x_1)) + (f(x_1) + f(x_2)) + \cdots + (f(x_{n-1}) + f(x_n))]$$

$$= \frac{\Delta x}{2}[f(x_0) + 2f(x_1) + 2f(x_2) + \cdots + 2f(x_{n-1}) + f(x_n)]$$

Trapezoidal Rule

$$\int_a^b f(x)\,dx \approx T_n = \frac{\Delta x}{2}[f(x_0) + 2f(x_1) + 2f(x_2) + \cdots + 2f(x_{n-1}) + f(x_n)]$$

where $\Delta x = (b - a)/n$ and $x_i = a + i\,\Delta x$.

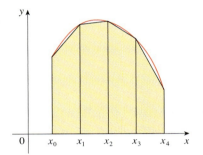

FIGURE 2
Trapezoidal approximation

The reason for the name Trapezoidal Rule can be seen from Figure 2, which illustrates the case $f(x) \geq 0$. The area of the trapezoid that lies above the ith subinterval is

$$\Delta x \left(\frac{f(x_{i-1}) + f(x_i)}{2}\right) = \frac{\Delta x}{2}[f(x_{i-1}) + f(x_i)]$$

and if we add the areas of all these trapezoids, we get the right side of the Trapezoidal Rule.

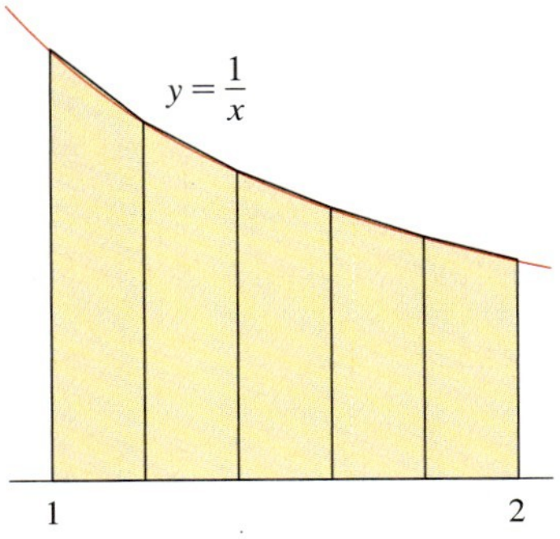

EXAMPLE 1 Use (a) the Trapezoidal Rule and (b) the Midpoint Rule with $n = 5$ to approximate the integral $\int_1^2 (1/x)\,dx$.

SOLUTION
(a) With $n = 5$, $a = 1$, and $b = 2$, we have $\Delta x = (2 - 1)/5 = 0.2$, and so the Trapezoidal Rule gives

$$\int_1^2 \frac{1}{x}\,dx \approx T_5 = \frac{0.2}{2}[f(1) + 2f(1.2) + 2f(1.4) + 2f(1.6) + 2f(1.8) + f(2)]$$

$$= 0.1\left(\frac{1}{1} + \frac{2}{1.2} + \frac{2}{1.4} + \frac{2}{1.6} + \frac{2}{1.8} + \frac{1}{2}\right)$$

$$\approx 0.695635$$

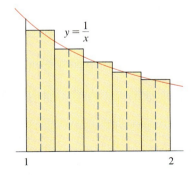

FIGURE 3

This approximation is illustrated in Figure 3.

(b) The midpoints of the five subintervals are 1.1, 1.3, 1.5, 1.7, and 1.9, so the Midpoint Rule gives

$$\int_1^2 \frac{1}{x}\,dx \approx \Delta x\,[f(1.1) + f(1.3) + f(1.5) + f(1.7) + f(1.9)]$$

$$= \frac{1}{5}\left(\frac{1}{1.1} + \frac{1}{1.3} + \frac{1}{1.5} + \frac{1}{1.7} + \frac{1}{1.9}\right)$$

$$\approx 0.691908$$

FIGURE 4

This approximation is illustrated in Figure 4.

In Example 1 we deliberately chose an integral whose value can be computed explicitly so that we can see how accurate the Trapezoidal and Midpoint Rules are. By the Fundamental Theorem of Calculus,

$$\int_1^2 \frac{1}{x}\, dx = \ln x \Big]_1^2 = \ln 2 = 0.693147\ldots$$

$\int_a^b f(x)\, dx = \text{approximation} + \text{error}$

The **error** in using an approximation is defined to be the amount that needs to be added to the approximation to make it exact. From the values in Example 1 we see that the errors in the Trapezoidal and Midpoint Rule approximations for $n = 5$ are

$$E_T \approx -0.002488 \quad \text{and} \quad E_M \approx 0.001239$$

In general, we have

$$E_T = \int_a^b f(x)\, dx - T_n \quad \text{and} \quad E_M = \int_a^b f(x)\, dx - M_n$$

Module 5.1/5.2/8.7 allows you to compare approximation methods.

The following tables show the results of calculations similar to those in Example 1, but for $n = 5$, 10, and 20 and for the left and right endpoint approximations as well as the Trapezoidal and Midpoint Rules.

Approximations to $\int_1^2 \frac{1}{x}\, dx$

n	L_n	R_n	T_n	M_n
5	0.745635	0.645635	0.695635	0.691908
10	0.718771	0.668771	0.693771	0.692835
20	0.705803	0.680803	0.693303	0.693069

Corresponding errors

n	E_L	E_R	E_T	E_M
5	−0.052488	0.047512	−0.002488	0.001239
10	−0.025624	0.024376	−0.000624	0.000312
20	−0.012656	0.012344	−0.000156	0.000078

We can make several observations from these tables:

1. In all of the methods we get more accurate approximations when we increase the value of n. (But very large values of n result in so many arithmetic operations that we have to beware of accumulated round-off error.)

2. The errors in the left and right endpoint approximations are opposite in sign and appear to decrease by a factor of about 2 when we double the value of n.

▌ It turns out that these observations are true in most cases.

3. The Trapezoidal and Midpoint Rules are much more accurate than the endpoint approximations.

4. The errors in the Trapezoidal and Midpoint Rules are opposite in sign and appear to decrease by a factor of about 4 when we double the value of n.

5. The size of the error in the Midpoint Rule is about half the size of the error in the Trapezoidal Rule.

Figure 5 shows why we can usually expect the Midpoint Rule to be more accurate than the Trapezoidal Rule. The area of a typical rectangle in the Midpoint Rule is the same as the trapezoid $ABCD$ whose upper side is tangent to the graph at P. The area of this trapezoid is closer to the area under the graph than is the area of the trapezoid $AQRD$ used in the Trapezoidal Rule. [The midpoint error (shaded red) is smaller than the trapezoidal error (shaded blue).]

These observations are corroborated in the following error estimates, which are proved in books on numerical analysis. Notice that Observation 4 corresponds to the n^2 in each denominator because $(2n)^2 = 4n^2$. The fact that the estimates depend on the size of the second derivative is not surprising if you look at Figure 5, because $f''(x)$ measures how much the graph is curved. [Recall that $f''(x)$ measures how fast the slope of $y = f(x)$ changes.]

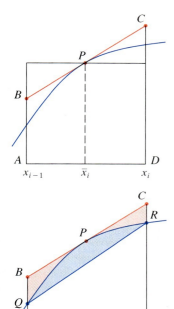

FIGURE 5

|||| K can be any number larger than all the values of $|f''(x)|$, but smaller values of K give better error bounds.

3 Error Bounds Suppose $|f''(x)| \le K$ for $a \le x \le b$. If E_T and E_M are the errors in the Trapezoidal and Midpoint Rules, then

$$|E_T| \le \frac{K(b-a)^3}{12n^2} \quad \text{and} \quad |E_M| \le \frac{K(b-a)^3}{24n^2}$$

Let's apply this error estimate to the Trapezoidal Rule approximation in Example 1. If $f(x) = 1/x$, then $f'(x) = -1/x^2$ and $f''(x) = 2/x^3$. Since $1 \le x \le 2$, we have $1/x \le 1$, so

$$|f''(x)| = \left|\frac{2}{x^3}\right| \le \frac{2}{1^3} = 2$$

Therefore, taking $K = 2$, $a = 1$, $b = 2$, and $n = 5$ in the error estimate (3), we see that

$$|E_T| \le \frac{2(2-1)^3}{12(5)^2} = \frac{1}{150} \approx 0.006667$$

Comparing this error estimate of 0.006667 with the actual error of about 0.002488, we see that it can happen that the actual error is substantially less than the upper bound for the error given by (3).

EXAMPLE 2 How large should we take n in order to guarantee that the Trapezoidal and Midpoint Rule approximations for $\int_1^2 (1/x)\,dx$ are accurate to within 0.0001?

SOLUTION We saw in the preceding calculation that $|f''(x)| \le 2$ for $1 \le x \le 2$, so we can take $K = 2$, $a = 1$, and $b = 2$ in (3). Accuracy to within 0.0001 means that the size of the error should be less than 0.0001. Therefore, we choose n so that

$$\frac{2(1)^3}{12n^2} < 0.0001$$

Solving the inequality for n, we get

$$n^2 > \frac{2}{12(0.0001)}$$

or

$$n > \frac{1}{\sqrt{0.0006}} \approx 40.8$$

|||| It's quite possible that a lower value for n would suffice, but 41 is the smallest value for which the error bound formula can *guarantee* us accuracy to within 0.0001.

Thus, $n = 41$ will ensure the desired accuracy.

For the same accuracy with the Midpoint Rule we choose n so that

$$\frac{2(1)^3}{24n^2} < 0.0001$$

which gives

$$n > \frac{1}{\sqrt{0.0012}} \approx 29$$

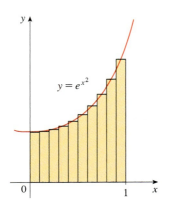

FIGURE 6

|||| Error estimates are upper bounds for the error. They give theoretical, worst-case scenarios. The actual error in this case turns out to be about 0.0023.

EXAMPLE 3
(a) Use the Midpoint Rule with $n = 10$ to approximate the integral $\int_0^1 e^{x^2} dx$.
(b) Give an upper bound for the error involved in this approximation.

SOLUTION
(a) Since $a = 0$, $b = 1$, and $n = 10$, the Midpoint Rule gives

$$\int_0^1 e^{x^2} dx \approx \Delta x \left[f(0.05) + f(0.15) + \cdots + f(0.85) + f(0.95) \right]$$

$$= 0.1[e^{0.0025} + e^{0.0225} + e^{0.0625} + e^{0.1225} + e^{0.2025} + e^{0.3025}$$
$$+ e^{0.4225} + e^{0.5625} + e^{0.7225} + e^{0.9025}]$$

$$\approx 1.460393$$

Figure 6 illustrates this approximation.

(b) Since $f(x) = e^{x^2}$, we have $f'(x) = 2xe^{x^2}$ and $f''(x) = (2 + 4x^2)e^{x^2}$. Also, since $0 \leq x \leq 1$, we have $x^2 \leq 1$ and so

$$0 \leq f''(x) = (2 + 4x^2)e^{x^2} \leq 6e$$

Taking $K = 6e$, $a = 0$, $b = 1$, and $n = 10$ in the error estimate (3), we see that an upper bound for the error is

$$\frac{6e(1)^3}{24(10)^2} = \frac{e}{400} \approx 0.007$$

|||| Simpson's Rule

Another rule for approximate integration results from using parabolas instead of straight line segments to approximate a curve. As before, we divide $[a, b]$ into n subintervals of equal length $h = \Delta x = (b - a)/n$, but this time we assume that n is an *even* number. Then on each consecutive pair of intervals we approximate the curve $y = f(x) \geq 0$ by a parabola as shown in Figure 7. If $y_i = f(x_i)$, then $P_i(x_i, y_i)$ is the point on the curve lying above x_i. A typical parabola passes through three consecutive points P_i, P_{i+1}, and P_{i+2}.

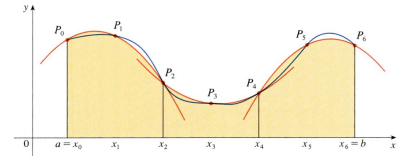

FIGURE 7

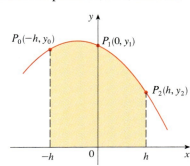

FIGURE 8

To simplify our calculations, we first consider the case where $x_0 = -h$, $x_1 = 0$, and $x_2 = h$. (See Figure 8.) We know that the equation of the parabola through P_0, P_1, and P_2 is of the form $y = Ax^2 + Bx + C$ and so the area under the parabola from $x = -h$ to $x = h$ is

|||| Here we have used Theorem 5.5.6. Notice that $Ax^2 + C$ is even and Bx is odd.

$$\int_{-h}^{h} (Ax^2 + Bx + C)\, dx = 2 \int_{0}^{h} (Ax^2 + C)\, dx$$

$$= 2 \left[A \frac{x^3}{3} + Cx \right]_0^h$$

$$= 2 \left(A \frac{h^3}{3} + Ch \right) = \frac{h}{3}(2Ah^2 + 6C)$$

But, since the parabola passes through $P_0(-h, y_0)$, $P_1(0, y_1)$, and $P_2(h, y_2)$, we have

$$y_0 = A(-h)^2 + B(-h) + C = Ah^2 - Bh + C$$

$$y_1 = C$$

$$y_2 = Ah^2 + Bh + C$$

and therefore
$$y_0 + 4y_1 + y_2 = 2Ah^2 + 6C$$

Thus, we can rewrite the area under the parabola as

$$\frac{h}{3}(y_0 + 4y_1 + y_2)$$

Now, by shifting this parabola horizontally we do not change the area under it. This means that the area under the parabola through P_0, P_1, and P_2 from $x = x_0$ to $x = x_2$ in Figure 7 is still

$$\frac{h}{3}(y_0 + 4y_1 + y_2)$$

Similarly, the area under the parabola through P_2, P_3, and P_4 from $x = x_2$ to $x = x_4$ is

$$\frac{h}{3}(y_2 + 4y_3 + y_4)$$

If we compute the areas under all the parabolas in this manner and add the results, we get

$$\int_a^b f(x)\, dx \approx \frac{h}{3}(y_0 + 4y_1 + y_2) + \frac{h}{3}(y_2 + 4y_3 + y_4) + \cdots + \frac{h}{3}(y_{n-2} + 4y_{n-1} + y_n)$$

$$= \frac{h}{3}(y_0 + 4y_1 + 2y_2 + 4y_3 + 2y_4 + \cdots + 2y_{n-2} + 4y_{n-1} + y_n)$$

Although we have derived this approximation for the case in which $f(x) \geq 0$, it is a reasonable approximation for any continuous function f and is called Simpson's Rule after the English mathematician Thomas Simpson (1710–1761). Note the pattern of coefficients: 1, 4, 2, 4, 2, 4, 2, ..., 4, 2, 4, 1.

IIII Thomas Simpson was a weaver who taught himself mathematics and went on to become one of the best English mathematicians of the 18th century. What we call Simpson's Rule was actually known to Cavalieri and Gregory in the 17th century, but Simpson popularized it in his best-selling calculus textbook, entitled *A New Treatise of Fluxions.*

Simpson's Rule

$$\int_a^b f(x)\,dx \approx S_n = \frac{\Delta x}{3}[f(x_0) + 4f(x_1) + 2f(x_2) + 4f(x_3) + \cdots + 2f(x_{n-2}) + 4f(x_{n-1}) + f(x_n)]$$

where n is even and $\Delta x = (b - a)/n$.

EXAMPLE 4 Use Simpson's Rule with $n = 10$ to approximate $\int_1^2 (1/x)\,dx$.

SOLUTION Putting $f(x) = 1/x$, $n = 10$, and $\Delta x = 0.1$ in Simpson's Rule, we obtain

$$\int_1^2 \frac{1}{x}\,dx \approx S_{10}$$

$$= \frac{\Delta x}{3}[f(1) + 4f(1.1) + 2f(1.2) + 4f(1.3) + \cdots + 2f(1.8) + 4f(1.9) + f(2)]$$

$$= \frac{0.1}{3}\left(\frac{1}{1} + \frac{4}{1.1} + \frac{2}{1.2} + \frac{4}{1.3} + \frac{2}{1.4} + \frac{4}{1.5} + \frac{2}{1.6} + \frac{4}{1.7} + \frac{2}{1.8} + \frac{4}{1.9} + \frac{1}{2}\right)$$

$$\approx 0.693150$$

Notice that, in Example 4, Simpson's Rule gives us a *much* better approximation ($S_{10} \approx 0.693150$) to the true value of the integral ($\ln 2 \approx 0.693147\ldots$) than does the Trapezoidal Rule ($T_{10} \approx 0.693771$) or the Midpoint Rule ($M_{10} \approx 0.692835$). It turns out (see Exercise 48) that the approximations in Simpson's Rule are weighted averages of those in the Trapezoidal and Midpoint Rules:

$$S_{2n} = \tfrac{1}{3}T_n + \tfrac{2}{3}M_n$$

(Recall that E_T and E_M usually have opposite signs and $|E_M|$ is about half the size of $|E_T|$.)

In many applications of calculus we need to evaluate an integral even if no explicit formula is known for y as a function of x. A function may be given graphically or as a table of values of collected data. If there is evidence that the values are not changing rapidly, then the Trapezoidal Rule or Simpson's Rule can still be used to find an approximate value for $\int_a^b y\,dx$, the integral of y with respect to x.

EXAMPLE 5 Figure 9 shows data traffic on the link from the United States to SWITCH, the Swiss academic and research network, on February 10, 1998. $D(t)$ is the data throughput, measured in megabits per second (Mb/s). Use Simpson's Rule to estimate the total amount of data transmitted on the link up to noon on that day.

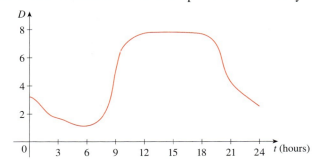

FIGURE 9

SOLUTION Because we want the units to be consistent and $D(t)$ is measured in megabits per second, we convert the units for t from hours to seconds. If we let $A(t)$ be the amount of data (in megabits) transmitted by time t, where t is measured in seconds, then $A'(t) = D(t)$. So, by the Net Change Theorem (see Section 5.4), the total amount of data transmitted by noon (when $t = 12 \times 60^2 = 43{,}200$) is

$$A(43{,}200) = \int_0^{43{,}200} D(t)\, dt$$

We estimate the values of $D(t)$ at hourly intervals from the graph and compile them in the table.

t (hours)	t (seconds)	$D(t)$	t (hours)	t (seconds)	$D(t)$
0	0	3.2	7	25,200	1.3
1	3,600	2.7	8	28,800	2.8
2	7,200	1.9	9	32,400	5.7
3	10,800	1.7	10	36,000	7.1
4	14,400	1.3	11	39,600	7.7
5	18,000	1.0	12	43,200	7.9
6	21,600	1.1			

Then we use Simpson's Rule with $n = 12$ and $\Delta t = 3600$ to estimate the integral:

$$\int_0^{43{,}200} A(t)\, dt \approx \frac{\Delta t}{3}[D(0) + 4D(3600) + 2D(7200) + \cdots + 4D(39{,}600) + D(43{,}200)]$$

$$\approx \frac{3600}{3}[3.2 + 4(2.7) + 2(1.9) + 4(1.7) + 2(1.3) + 4(1.0)$$
$$+ 2(1.1) + 4(1.3) + 2(2.8) + 4(5.7) + 2(7.1) + 4(7.7) + 7.9]$$

$$= 143{,}880$$

Thus, the total amount of data transmitted up to noon is about 144,000 megabits, or 144 gigabits.

In Exercises 27 and 28 you are asked to demonstrate, in particular cases, that the error in Simpson's Rule decreases by a factor of about 16 when n is doubled. That is consistent with the appearance of n^4 in the denominator of the following error estimate for Simpson's Rule. It is similar to the estimates given in (3) for the Trapezoidal and Midpoint Rules, but it uses the fourth derivative of f.

4 Error Bound for Simpson's Rule Suppose that $|f^{(4)}(x)| \leq K$ for $a \leq x \leq b$. If E_S is the error involved in using Simpson's Rule, then

$$|E_S| \leq \frac{K(b-a)^5}{180n^4}$$

EXAMPLE 6 How large should we take n in order to guarantee that the Simpson's Rule approximation for $\int_1^2 (1/x)\, dx$ is accurate to within 0.0001?

SOLUTION If $f(x) = 1/x$, then $f^{(4)}(x) = 24/x^5$. Since $x \geq 1$, we have $1/x \leq 1$ and so

$$|f^{(4)}(x)| = \left|\frac{24}{x^5}\right| \leq 24$$

IIII Many calculators and computer algebra systems have a built-in algorithm that computes an approximation of a definite integral. Some of these machines use Simpson's Rule; others use more sophisticated techniques such as *adaptive* numerical integration. This means that if a function fluctuates much more on a certain part of the interval than it does elsewhere, then that part gets divided into more subintervals. This strategy reduces the number of calculations required to achieve a prescribed accuracy.

Therefore, we can take $K = 24$ in (4). Thus, for an error less than 0.0001 we should choose n so that

$$\frac{24(1)^5}{180n^4} < 0.0001$$

This gives

$$n^4 > \frac{24}{180(0.0001)}$$

or

$$n > \frac{1}{\sqrt[4]{0.00075}} \approx 6.04$$

Therefore, $n = 8$ (n must be even) gives the desired accuracy. (Compare this with Example 2, where we obtained $n = 41$ for the Trapezoidal Rule and $n = 29$ for the Midpoint Rule.)

EXAMPLE 7
(a) Use Simpson's Rule with $n = 10$ to approximate the integral $\int_0^1 e^{x^2}\, dx$.
(b) Estimate the error involved in this approximation.

SOLUTION
(a) If $n = 10$, then $\Delta x = 0.1$ and Simpson's Rule gives

$$\int_0^1 e^{x^2}\, dx \approx \frac{\Delta x}{3}[f(0) + 4f(0.1) + 2f(0.2) + \cdots + 2f(0.8) + 4f(0.9) + f(1)]$$

$$= \frac{0.1}{3}[e^0 + 4e^{0.01} + 2e^{0.04} + 4e^{0.09} + 2e^{0.16} + 4e^{0.25} + 2e^{0.36}$$

$$+ 4e^{0.49} + 2e^{0.64} + 4e^{0.81} + e^1]$$

$$\approx 1.462681$$

IIII Figure 10 illustrates the calculation in Example 7. Notice that the parabolic arcs are so close to the graph of $y = e^{x^2}$ that they are practically indistinguishable from it.

(b) The fourth derivative of $f(x) = e^{x^2}$ is

$$f^{(4)}(x) = (12 + 48x^2 + 16x^4)e^{x^2}$$

and so, since $0 \leq x \leq 1$, we have

$$0 \leq f^{(4)}(x) \leq (12 + 48 + 16)e^1 = 76e$$

Therefore, putting $K = 76e$, $a = 0$, $b = 1$, and $n = 10$ in (4), we see that the error is at most

$$\frac{76e(1)^5}{180(10)^4} \approx 0.000115$$

(Compare this with Example 3.) Thus, correct to three decimal places, we have

$$\int_0^1 e^{x^2}\, dx \approx 1.463$$

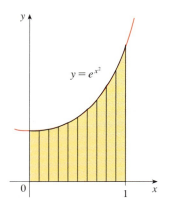

FIGURE 10

8.7 Exercises

1. Let $I = \int_0^4 f(x)\,dx$, where f is the function whose graph is shown.
(a) Use the graph to find L_2, R_2, and M_2.
(b) Are these underestimates or overestimates of I?
(c) Use the graph to find T_2. How does it compare with I?
(d) For any value of n, list the numbers L_n, R_n, M_n, T_n, and I in increasing order.

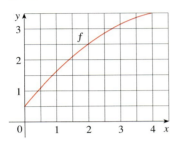

2. The left, right, Trapezoidal, and Midpoint Rule approximations were used to estimate $\int_0^2 f(x)\,dx$, where f is the function whose graph is shown. The estimates were 0.7811, 0.8675, 0.8632, and 0.9540, and the same number of subintervals were used in each case.
(a) Which rule produced which estimate?
(b) Between which two approximations does the true value of $\int_0^2 f(x)\,dx$ lie?

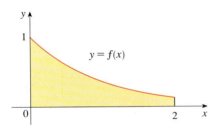

3. Estimate $\int_0^1 \cos(x^2)\,dx$ using (a) the Trapezoidal Rule and (b) the Midpoint Rule, each with $n = 4$. From a graph of the integrand, decide whether your answers are underestimates or overestimates. What can you conclude about the true value of the integral?

4. Draw the graph of $f(x) = \sin(x^2/2)$ in the viewing rectangle $[0, 1]$ by $[0, 0.5]$ and let $I = \int_0^1 f(x)\,dx$.
(a) Use the graph to decide whether L_2, R_2, M_2, and T_2 underestimate or overestimate I.
(b) For any value of n, list the numbers L_n, R_n, M_n, T_n, and I in increasing order.
(c) Compute L_5, R_5, M_5, and T_5. From the graph, which do you think gives the best estimate of I?

5–6 Use (a) the Midpoint Rule and (b) Simpson's Rule to approximate the given integral with the specified value of n. (Round your answers to six decimal places.) Compare your results to the actual value to determine the error in each approximation.

5. $\int_0^\pi x^2 \sin x\,dx$, $n = 8$

6. $\int_0^1 e^{-\sqrt{x}}\,dx$, $n = 6$

7–18 Use (a) the Trapezoidal Rule, (b) the Midpoint Rule, and (c) Simpson's Rule to approximate the given integral with the specified value of n. (Round your answers to six decimal places.)

7. $\int_0^2 \sqrt[4]{1 + x^2}\,dx$, $n = 8$

8. $\int_0^{1/2} \sin(x^2)\,dx$, $n = 4$

9. $\int_1^2 \dfrac{\ln x}{1 + x}\,dx$, $n = 10$

10. $\int_0^3 \dfrac{dt}{1 + t^2 + t^4}$, $n = 6$

11. $\int_0^{1/2} \sin(e^{t/2})\,dt$, $n = 8$

12. $\int_0^4 \sqrt{1 + \sqrt{x}}\,dx$, $n = 8$

13. $\int_1^2 e^{1/x}\,dx$, $n = 4$

14. $\int_0^4 \sqrt{x}\, \sin x\,dx$, $n = 8$

15. $\int_1^5 \dfrac{\cos x}{x}\,dx$, $n = 8$

16. $\int_4^6 \ln(x^3 + 2)\,dx$, $n = 10$

17. $\int_0^3 \dfrac{1}{1 + y^5}\,dy$, $n = 6$

18. $\int_2^4 \dfrac{e^x}{x}\,dx$, $n = 10$

19. (a) Find the approximations T_{10} and M_{10} for the integral $\int_0^2 e^{-x^2}\,dx$.
(b) Estimate the errors in the approximations of part (a).
(c) How large do we have to choose n so that the approximations T_n and M_n to the integral in part (a) are accurate to within 0.00001?

20. (a) Find the approximations T_8 and M_8 for $\int_0^1 \cos(x^2)\,dx$.
(b) Estimate the errors involved in the approximations of part (a).
(c) How large do we have to choose n so that the approximations T_n and M_n to the integral in part (a) are accurate to within 0.00001?

21. (a) Find the approximations T_{10} and S_{10} for $\int_0^1 e^x\,dx$ and the corresponding errors E_T and E_S.
(b) Compare the actual errors in part (a) with the error estimates given by (3) and (4).
(c) How large do we have to choose n so that the approximations T_n, M_n, and S_n to the integral in part (a) are accurate to within 0.00001?

22. How large should n be to guarantee that the Simpson's Rule approximation to $\int_0^1 e^{x^2}\,dx$ is accurate to within 0.00001?

CAS 23. The trouble with the error estimates is that it is often very difficult to compute four derivatives and obtain a good upper bound K for $|f^{(4)}(x)|$ by hand. But computer algebra systems have no

problem computing $f^{(4)}$ and graphing it, so we can easily find a value for K from a machine graph. This exercise deals with approximations to the integral $I = \int_0^{2\pi} f(x)\,dx$, where $f(x) = e^{\cos x}$.
(a) Use a graph to get a good upper bound for $|f''(x)|$.
(b) Use M_{10} to approximate I.
(c) Use part (a) to estimate the error in part (b).
(d) Use the built-in numerical integration capability of your CAS to approximate I.
(e) How does the actual error compare with the error estimate in part (c)?
(f) Use a graph to get a good upper bound for $|f^{(4)}(x)|$.
(g) Use S_{10} to approximate I.
(h) Use part (f) to estimate the error in part (g).
(i) How does the actual error compare with the error estimate in part (h)?
(j) How large should n be to guarantee that the size of the error in using S_n is less than 0.0001?

CAS **24.** Repeat Exercise 23 for the integral $\int_{-1}^{1} \sqrt{4 - x^3}\,dx$.

25–26 Find the approximations L_n, R_n, T_n, and M_n for $n = 4, 8$, and 16. Then compute the corresponding errors E_L, E_R, E_T, and E_M. (Round your answers to six decimal places. You may wish to use the sum command on a computer algebra system.) What observations can you make? In particular, what happens to the errors when n is doubled?

25. $\int_0^1 x^3\,dx$ **26.** $\int_0^2 e^x\,dx$

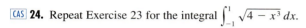

27–28 Find the approximations T_n, M_n, and S_n for $n = 6$ and 12. Then compute the corresponding errors E_T, E_M, and E_S. (Round your answers to six decimal places. You may wish to use the sum command on a computer algebra system.) What observations can you make? In particular, what happens to the errors when n is doubled?

27. $\int_1^4 \sqrt{x}\,dx$ **28.** $\int_{-1}^2 xe^x\,dx$

29. Estimate the area under the graph in the figure by using (a) the Trapezoidal Rule, (b) the Midpoint Rule, and (c) Simpson's Rule, each with $n = 4$.

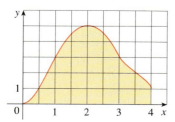

30. The widths (in meters) of a kidney-shaped swimming pool were measured at 2-meter intervals as indicated in the figure. Use Simpson's Rule to estimate the area of the pool.

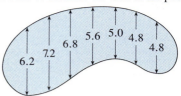

31. (a) Use the Midpoint Rule and the given data to estimate the value of the integral $\int_0^{3.2} f(x)\,dx$.

x	$f(x)$	x	$f(x)$
0.0	6.8	2.0	7.6
0.4	6.5	2.4	8.4
0.8	6.3	2.8	8.8
1.2	6.4	3.2	9.0
1.6	6.9		

(b) If it is known that $-4 \leq f''(x) \leq 1$ for all x, estimate the error involved in the approximation in part (a).

32. A radar gun was used to record the speed of a runner during the first 5 seconds of a race (see the table). Use Simpson's Rule to estimate the distance the runner covered during those 5 seconds.

t (s)	v (m/s)	t (s)	v (m/s)
0	0	3.0	10.51
0.5	4.67	3.5	10.67
1.0	7.34	4.0	10.76
1.5	8.86	4.5	10.81
2.0	9.73	5.0	10.81
2.5	10.22		

33. The graph of the acceleration $a(t)$ of a car measured in ft/s² is shown. Use Simpson's Rule to estimate the increase in the velocity of the car during the 6-second time interval.

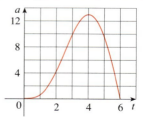

34. Water leaked from a tank at a rate of $r(t)$ liters per hour, where the graph of r is as shown. Use Simpson's Rule to estimate the total amount of water that leaked out during the first six hours.

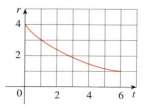

35. The table (supplied by San Diego Gas and Electric) gives the power consumption in megawatts in San Diego County from midnight to 6:00 A.M. on December 8, 1999. Use Simpson's Rule to estimate the energy used during that time period. (Use the fact that power is the derivative of energy.)

t	P	t	P
0:00	1814	3:30	1611
0:30	1735	4:00	1621
1:00	1686	4:30	1666
1:30	1646	5:00	1745
2:00	1637	5:30	1886
2:30	1609	6:00	2052
3:00	1604		

36. Shown is the graph of traffic on an Internet service provider's T1 data line from midnight to 8:00 A.M. D is the data throughput, measured in megabits per second. Use Simpson's Rule to estimate the total amount of data transmitted during that time period.

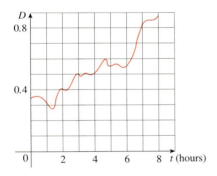

37. If the region shown in the figure is rotated about the y-axis to form a solid, use Simpson's Rule with $n = 8$ to estimate the volume of the solid.

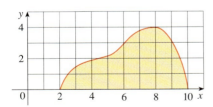

38. The table shows values of a force function $f(x)$ where x is measured in meters and $f(x)$ in newtons. Use Simpson's Rule to estimate the work done by the force in moving an object a distance of 18 m.

x	0	3	6	9	12	15	18
$f(x)$	9.8	9.1	8.5	8.0	7.7	7.5	7.4

39. The region bounded by the curves $y = \sqrt[3]{1 + x^3}$, $y = 0$, $x = 0$, and $x = 2$ is rotated about the x-axis. Use Simpson's Rule with $n = 10$ to estimate the volume of the resulting solid.

CAS 40. The figure shows a pendulum with length L that makes a maximum angle θ_0 with the vertical. Using Newton's Second Law it can be shown that the period T (the time for one complete swing) is given by

$$T = 4\sqrt{\frac{L}{g}} \int_0^{\pi/2} \frac{dx}{\sqrt{1 - k^2 \sin^2 x}}$$

where $k = \sin(\frac{1}{2}\theta_0)$ and g is the acceleration due to gravity. If $L = 1$ m and $\theta_0 = 42°$, use Simpson's Rule with $n = 10$ to find the period.

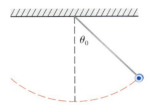

41. The intensity of light with wavelength λ traveling through a diffraction grating with N slits at an angle θ is given by $I(\theta) = N^2 \sin^2 k/k^2$, where $k = (\pi Nd \sin \theta)/\lambda$ and d is the distance between adjacent slits. A helium-neon laser with wavelength $\lambda = 632.8 \times 10^{-9}$ m is emitting a narrow band of light, given by $-10^{-6} < \theta < 10^{-6}$, through a grating with 10,000 slits spaced 10^{-4} m apart. Use the Midpoint Rule with $n = 10$ to estimate the total light intensity $\int_{-10^{-6}}^{10^{-6}} I(\theta)\, d\theta$ emerging from the grating.

42. Use the Trapezoidal Rule with $n = 10$ to approximate $\int_0^{20} \cos(\pi x)\, dx$. Compare your result to the actual value. Can you explain the discrepancy?

43. Sketch the graph of a continuous function on $[0, 2]$ for which the Trapezoidal Rule with $n = 2$ is more accurate than the Midpoint Rule.

44. Sketch the graph of a continuous function on $[0, 2]$ for which the right endpoint approximation with $n = 2$ is more accurate than Simpson's Rule.

45. If f is a positive function and $f''(x) < 0$ for $a \leq x \leq b$, show that

$$T_n < \int_a^b f(x)\, dx < M_n$$

46. Show that if f is a polynomial of degree 3 or lower, then Simpson's Rule gives the exact value of $\int_a^b f(x)\, dx$.

47. Show that $\frac{1}{2}(T_n + M_n) = T_{2n}$.

48. Show that $\frac{1}{3}T_n + \frac{2}{3}M_n = S_{2n}$.

8.8 Improper Integrals

In defining a definite integral $\int_a^b f(x)\,dx$ we dealt with a function f defined on a finite interval $[a, b]$ and we assumed that f does not have an infinite discontinuity (see Section 5.2). In this section we extend the concept of a definite integral to the case where the interval is infinite and also to the case where f has an infinite discontinuity in $[a, b]$. In either case the integral is called an *improper* integral. One of the most important applications of this idea, probability distributions, will be studied in Section 9.5.

Type I: Infinite Intervals

Try painting a fence that never ends.
Resources / Module 6
/ How To Calculate
/ Start of Improper Integrals

Consider the infinite region S that lies under the curve $y = 1/x^2$, above the x-axis, and to the right of the line $x = 1$. You might think that, since S is infinite in extent, its area must be infinite, but let's take a closer look. The area of the part of S that lies to the left of the line $x = t$ (shaded in Figure 1) is

$$A(t) = \int_1^t \frac{1}{x^2}\,dx = -\frac{1}{x}\Big]_1^t = 1 - \frac{1}{t}$$

Notice that $A(t) < 1$ no matter how large t is chosen.

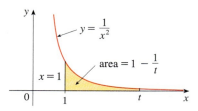

FIGURE 1

We also observe that

$$\lim_{t \to \infty} A(t) = \lim_{t \to \infty}\left(1 - \frac{1}{t}\right) = 1$$

The area of the shaded region approaches 1 as $t \to \infty$ (see Figure 2), so we say that the area of the infinite region S is equal to 1 and we write

$$\int_1^\infty \frac{1}{x^2}\,dx = \lim_{t \to \infty} \int_1^t \frac{1}{x^2}\,dx = 1$$

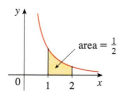 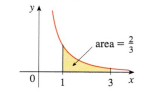 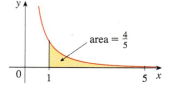 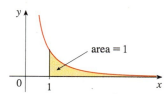

FIGURE 2

Using this example as a guide, we define the integral of f (not necessarily a positive function) over an infinite interval as the limit of integrals over finite intervals.

> **1 Definition of an Improper Integral of Type 1**
>
> (a) If $\int_a^t f(x)\,dx$ exists for every number $t \geq a$, then
>
> $$\int_a^\infty f(x)\,dx = \lim_{t \to \infty} \int_a^t f(x)\,dx$$
>
> provided this limit exists (as a finite number).
>
> (b) If $\int_t^b f(x)\,dx$ exists for every number $t \leq b$, then
>
> $$\int_{-\infty}^b f(x)\,dx = \lim_{t \to -\infty} \int_t^b f(x)\,dx$$
>
> provided this limit exists (as a finite number).
>
> The improper integrals $\int_a^\infty f(x)\,dx$ and $\int_{-\infty}^b f(x)\,dx$ are called **convergent** if the corresponding limit exists and **divergent** if the limit does not exist.
>
> (c) If both $\int_a^\infty f(x)\,dx$ and $\int_{-\infty}^a f(x)\,dx$ are convergent, then we define
>
> $$\int_{-\infty}^\infty f(x)\,dx = \int_{-\infty}^a f(x)\,dx + \int_a^\infty f(x)\,dx$$
>
> In part (c) any real number a can be used (see Exercise 74).

Any of the improper integrals in Definition 1 can be interpreted as an area provided that f is a positive function. For instance, in case (a) if $f(x) \geq 0$ and the integral $\int_a^\infty f(x)\,dx$ is convergent, then we define the area of the region $S = \{(x, y) \mid x \geq a, 0 \leq y \leq f(x)\}$ in Figure 3 to be

$$A(S) = \int_a^\infty f(x)\,dx$$

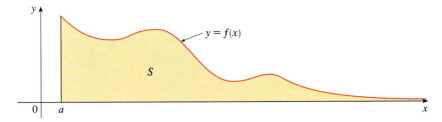

FIGURE 3

This is appropriate because $\int_a^\infty f(x)\,dx$ is the limit as $t \to \infty$ of the area under the graph of f from a to t.

EXAMPLE 1 Determine whether the integral $\int_1^\infty (1/x)\,dx$ is convergent or divergent.

SOLUTION According to part (a) of Definition 1, we have

$$\int_1^\infty \frac{1}{x}\,dx = \lim_{t \to \infty} \int_1^t \frac{1}{x}\,dx = \lim_{t \to \infty} \ln|x|\Big]_1^t$$
$$= \lim_{t \to \infty} (\ln t - \ln 1) = \lim_{t \to \infty} \ln t = \infty$$

The limit does not exist as a finite number and so the improper integral $\int_1^\infty (1/x)\,dx$ is divergent.

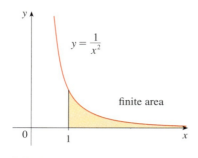

FIGURE 4

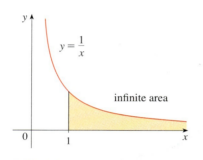

FIGURE 5

Let's compare the result of Example 1 with the example given at the beginning of this section:

$$\int_1^\infty \frac{1}{x^2}\,dx \text{ converges} \qquad \int_1^\infty \frac{1}{x}\,dx \text{ diverges}$$

Geometrically, this says that although the curves $y = 1/x^2$ and $y = 1/x$ look very similar for $x > 0$, the region under $y = 1/x^2$ to the right of $x = 1$ (the shaded region in Figure 4) has finite area whereas the corresponding region under $y = 1/x$ (in Figure 5) has infinite area. Note that both $1/x^2$ and $1/x$ approach 0 as $x \to \infty$ but $1/x^2$ approaches 0 faster than $1/x$. The values of $1/x$ don't decrease fast enough for its integral to have a finite value.

EXAMPLE 2 Evaluate $\int_{-\infty}^0 xe^x\,dx$.

SOLUTION Using part (b) of Definition 1, we have

$$\int_{-\infty}^0 xe^x\,dx = \lim_{t \to -\infty} \int_t^0 xe^x\,dx$$

We integrate by parts with $u = x$, $dv = e^x\,dx$ so that $du = dx$, $v = e^x$:

$$\int_t^0 xe^x\,dx = xe^x\Big]_t^0 - \int_t^0 e^x\,dx$$

$$= -te^t - 1 + e^t$$

We know that $e^t \to 0$ as $t \to -\infty$, and by l'Hospital's Rule we have

$$\lim_{t \to -\infty} te^t = \lim_{t \to -\infty} \frac{t}{e^{-t}} = \lim_{t \to -\infty} \frac{1}{-e^{-t}}$$

$$= \lim_{t \to -\infty} (-e^t) = 0$$

Therefore

$$\int_{-\infty}^0 xe^x\,dx = \lim_{t \to -\infty} (-te^t - 1 + e^t)$$

$$= -0 - 1 + 0 = -1$$

EXAMPLE 3 Evaluate $\int_{-\infty}^\infty \frac{1}{1+x^2}\,dx$.

SOLUTION It's convenient to choose $a = 0$ in Definition 1(c):

$$\int_{-\infty}^\infty \frac{1}{1+x^2}\,dx = \int_{-\infty}^0 \frac{1}{1+x^2}\,dx + \int_0^\infty \frac{1}{1+x^2}\,dx$$

We must now evaluate the integrals on the right side separately:

$$\int_0^\infty \frac{1}{1+x^2}\,dx = \lim_{t \to \infty} \int_0^t \frac{dx}{1+x^2} = \lim_{t \to \infty} \tan^{-1}x\Big]_0^t$$

$$= \lim_{t \to \infty} (\tan^{-1}t - \tan^{-1}0) = \lim_{t \to \infty} \tan^{-1}t = \frac{\pi}{2}$$

$$\int_{-\infty}^{0} \frac{1}{1+x^2}\,dx = \lim_{t\to-\infty} \int_{t}^{0} \frac{dx}{1+x^2} = \lim_{t\to-\infty} \tan^{-1}x \Big]_{t}^{0}$$

$$= \lim_{t\to-\infty} (\tan^{-1} 0 - \tan^{-1} t)$$

$$= 0 - \left(-\frac{\pi}{2}\right) = \frac{\pi}{2}$$

Since both of these integrals are convergent, the given integral is convergent and

$$\int_{-\infty}^{\infty} \frac{1}{1+x^2}\,dx = \frac{\pi}{2} + \frac{\pi}{2} = \pi$$

Since $1/(1+x^2) > 0$, the given improper integral can be interpreted as the area of the infinite region that lies under the curve $y = 1/(1+x^2)$ and above the x-axis (see Figure 6).

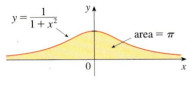

FIGURE 6

EXAMPLE 4 For what values of p is the integral

$$\int_{1}^{\infty} \frac{1}{x^p}\,dx$$

convergent?

SOLUTION We know from Example 1 that if $p = 1$, then the integral is divergent, so let's assume that $p \neq 1$. Then

$$\int_{1}^{\infty} \frac{1}{x^p}\,dx = \lim_{t\to\infty} \int_{1}^{t} x^{-p}\,dx$$

$$= \lim_{t\to\infty} \frac{x^{-p+1}}{-p+1} \Big]_{x=1}^{x=t}$$

$$= \lim_{t\to\infty} \frac{1}{1-p} \left[\frac{1}{t^{p-1}} - 1\right]$$

If $p > 1$, then $p - 1 > 0$, so as $t \to \infty$, $t^{p-1} \to \infty$ and $1/t^{p-1} \to 0$. Therefore

$$\int_{1}^{\infty} \frac{1}{x^p}\,dx = \frac{1}{p-1} \quad \text{if } p > 1$$

and so the integral converges. But if $p < 1$, then $p - 1 < 0$ and so

$$\frac{1}{t^{p-1}} = t^{1-p} \to \infty \qquad \text{as } t \to \infty$$

and the integral diverges.

We summarize the result of Example 4 for future reference:

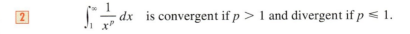

$\boxed{2}$ $\int_{1}^{\infty} \dfrac{1}{x^p}\,dx$ is convergent if $p > 1$ and divergent if $p \leq 1$.

Type 2: Discontinuous Integrands

Suppose that f is a positive continuous function defined on a finite interval $[a, b)$ but has a vertical asymptote at b. Let S be the unbounded region under the graph of f and above the x-axis between a and b. (For Type 1 integrals, the regions extended indefinitely in a

570 ||| **CHAPTER 8** TECHNIQUES OF INTEGRATION

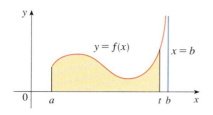

FIGURE 7

horizontal direction. Here the region is infinite in a vertical direction.) The area of the part of S between a and t (the shaded region in Figure 7) is

$$A(t) = \int_a^t f(x)\,dx$$

If it happens that $A(t)$ approaches a definite number A as $t \to b^-$, then we say that the area of the region S is A and we write

$$\int_a^b f(x)\,dx = \lim_{t \to b^-} \int_a^t f(x)\,dx$$

We use this equation to define an improper integral of Type 2 even when f is not a positive function, no matter what type of discontinuity f has at b.

|||| Parts (b) and (c) of Definition 3 are illustrated in Figures 8 and 9 for the case where $f(x) \geq 0$ and f has vertical asymptotes at a and c, respectively.

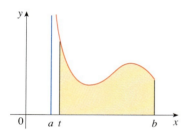

FIGURE 8

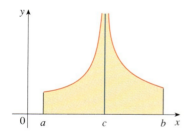

FIGURE 9

3 **Definition of an Improper Integral of Type 2**

(a) If f is continuous on $[a, b)$ and is discontinuous at b, then

$$\int_a^b f(x)\,dx = \lim_{t \to b^-} \int_a^t f(x)\,dx$$

if this limit exists (as a finite number).

(b) If f is continuous on $(a, b]$ and is discontinuous at a, then

$$\int_a^b f(x)\,dx = \lim_{t \to a^+} \int_t^b f(x)\,dx$$

if this limit exists (as a finite number).

The improper integral $\int_a^b f(x)\,dx$ is called **convergent** if the corresponding limit exists and **divergent** if the limit does not exist.

(c) If f has a discontinuity at c, where $a < c < b$, and both $\int_a^c f(x)\,dx$ and $\int_c^b f(x)\,dx$ are convergent, then we define

$$\int_a^b f(x)\,dx = \int_a^c f(x)\,dx + \int_c^b f(x)\,dx$$

EXAMPLE 5 Find $\displaystyle\int_2^5 \frac{1}{\sqrt{x-2}}\,dx$.

SOLUTION We note first that the given integral is improper because $f(x) = 1/\sqrt{x-2}$ has the vertical asymptote $x = 2$. Since the infinite discontinuity occurs at the left endpoint of $[2, 5]$, we use part (b) of Definition 3:

$$\int_2^5 \frac{dx}{\sqrt{x-2}} = \lim_{t \to 2^+} \int_t^5 \frac{dx}{\sqrt{x-2}}$$
$$= \lim_{t \to 2^+} 2\sqrt{x-2}\Big]_t^5$$
$$= \lim_{t \to 2^+} 2(\sqrt{3} - \sqrt{t-2})$$
$$= 2\sqrt{3}$$

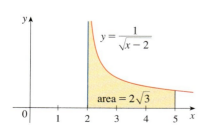

FIGURE 10

Thus, the given improper integral is convergent and, since the integrand is positive, we can interpret the value of the integral as the area of the shaded region in Figure 10.

EXAMPLE 6 Determine whether $\int_0^{\pi/2} \sec x \, dx$ converges or diverges.

SOLUTION Note that the given integral is improper because $\lim_{x \to (\pi/2)^-} \sec x = \infty$. Using part (a) of Definition 3 and Formula 14 from the Table of Integrals, we have

$$\int_0^{\pi/2} \sec x \, dx = \lim_{t \to (\pi/2)^-} \int_0^t \sec x \, dx$$

$$= \lim_{t \to (\pi/2)^-} \ln|\sec x + \tan x|\Big]_0^t$$

$$= \lim_{t \to (\pi/2)^-} [\ln(\sec t + \tan t) - \ln 1]$$

$$= \infty$$

because $\sec t \to \infty$ and $\tan t \to \infty$ as $t \to (\pi/2)^-$. Thus, the given improper integral is divergent.

EXAMPLE 7 Evaluate $\int_0^3 \dfrac{dx}{x-1}$ if possible.

SOLUTION Observe that the line $x = 1$ is a vertical asymptote of the integrand. Since it occurs in the middle of the interval $[0, 3]$, we must use part (c) of Definition 3 with $c = 1$:

$$\int_0^3 \frac{dx}{x-1} = \int_0^1 \frac{dx}{x-1} + \int_1^3 \frac{dx}{x-1}$$

where

$$\int_0^1 \frac{dx}{x-1} = \lim_{t \to 1^-} \int_0^t \frac{dx}{x-1} = \lim_{t \to 1^-} \ln|x-1|\Big]_0^t$$

$$= \lim_{t \to 1^-} (\ln|t-1| - \ln|-1|)$$

$$= \lim_{t \to 1^-} \ln(1-t) = -\infty$$

because $1 - t \to 0^+$ as $t \to 1^-$. Thus, $\int_0^1 dx/(x-1)$ is divergent. This implies that $\int_0^3 dx/(x-1)$ is divergent. [We do not need to evaluate $\int_1^3 dx/(x-1)$.]

⊘ **WARNING** ▫ If we had not noticed the asymptote $x = 1$ in Example 7 and had instead confused the integral with an ordinary integral, then we might have made the following erroneous calculation:

$$\int_0^3 \frac{dx}{x-1} = \ln|x-1|\Big]_0^3 = \ln 2 - \ln 1 = \ln 2$$

This is wrong because the integral is improper and must be calculated in terms of limits.

From now on, whenever you meet the symbol $\int_a^b f(x) \, dx$ you must decide, by looking at the function f on $[a, b]$, whether it is an ordinary definite integral or an improper integral.

EXAMPLE 8 Evaluate $\int_0^1 \ln x \, dx$.

SOLUTION We know that the function $f(x) = \ln x$ has a vertical asymptote at 0 since $\lim_{x \to 0^+} \ln x = -\infty$. Thus, the given integral is improper and we have

$$\int_0^1 \ln x \, dx = \lim_{t \to 0^+} \int_t^1 \ln x \, dx$$

Now we integrate by parts with $u = \ln x$, $dv = dx$, $du = dx/x$, and $v = x$:

$$\int_t^1 \ln x \, dx = x \ln x \Big]_t^1 - \int_t^1 dx$$

$$= 1 \ln 1 - t \ln t - (1 - t)$$

$$= -t \ln t - 1 + t$$

To find the limit of the first term we use l'Hospital's Rule:

$$\lim_{t \to 0^+} t \ln t = \lim_{t \to 0^+} \frac{\ln t}{1/t}$$

$$= \lim_{t \to 0^+} \frac{1/t}{-1/t^2}$$

$$= \lim_{t \to 0^+} (-t) = 0$$

Therefore

$$\int_0^1 \ln x \, dx = \lim_{t \to 0^+} (-t \ln t - 1 + t)$$

$$= -0 - 1 + 0 = -1$$

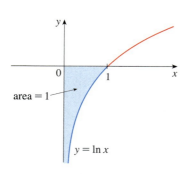

FIGURE 11

Figure 11 shows the geometric interpretation of this result. The area of the shaded region above $y = \ln x$ and below the x-axis is 1.

A Comparison Test for Improper Integrals

Sometimes it is impossible to find the exact value of an improper integral and yet it is important to know whether it is convergent or divergent. In such cases the following theorem is useful. Although we state it for Type 1 integrals, a similar theorem is true for Type 2 integrals.

Comparison Theorem Suppose that f and g are continuous functions with $f(x) \geq g(x) \geq 0$ for $x \geq a$.

(a) If $\int_a^\infty f(x) \, dx$ is convergent, then $\int_a^\infty g(x) \, dx$ is convergent.

(b) If $\int_a^\infty g(x) \, dx$ is divergent, then $\int_a^\infty f(x) \, dx$ is divergent.

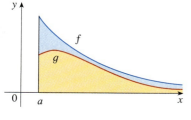

FIGURE 12

We omit the proof of the Comparison Theorem, but Figure 12 makes it seem plausible. If the area under the top curve $y = f(x)$ is finite, then so is the area under the bottom curve $y = g(x)$. And if the area under $y = g(x)$ is infinite, then so is the area under $y = f(x)$. [Note that the reverse is not necessarily true: If $\int_a^\infty g(x) \, dx$ is convergent, $\int_a^\infty f(x) \, dx$ may or may not be convergent, and if $\int_a^\infty f(x) \, dx$ is divergent, $\int_a^\infty g(x) \, dx$ may or may not be divergent.]

EXAMPLE 9 Show that $\int_0^\infty e^{-x^2} \, dx$ is convergent.

SOLUTION We can't evaluate the integral directly because the antiderivative of e^{-x^2} is not an elementary function (as explained in Section 8.5). We write

$$\int_0^\infty e^{-x^2} \, dx = \int_0^1 e^{-x^2} \, dx + \int_1^\infty e^{-x^2} \, dx$$

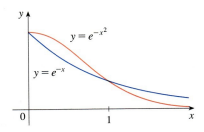

FIGURE 13

and observe that the first integral on the right-hand side is just an ordinary definite integral. In the second integral we use the fact that for $x \geq 1$ we have $x^2 \geq x$, so $-x^2 \leq -x$ and therefore $e^{-x^2} \leq e^{-x}$. (See Figure 13.) The integral of e^{-x} is easy to evaluate:

$$\int_1^\infty e^{-x}\,dx = \lim_{t \to \infty} \int_1^t e^{-x}\,dx$$

$$= \lim_{t \to \infty}(e^{-1} - e^{-t}) = e^{-1}$$

Thus, taking $f(x) = e^{-x}$ and $g(x) = e^{-x^2}$ in the Comparison Theorem, we see that $\int_1^\infty e^{-x^2}\,dx$ is convergent. It follows that $\int_0^\infty e^{-x^2}\,dx$ is convergent.

In Example 9 we showed that $\int_0^\infty e^{-x^2}\,dx$ is convergent without computing its value. In Exercise 70 we indicate how to show that its value is approximately 0.8862. In probability theory it is important to know the exact value of this improper integral, as we will see in Section 9.5; using the methods of multivariable calculus it can be shown that the exact value is $\sqrt{\pi}/2$. Table 1 illustrates the definition of an improper integral by showing how the (computer-generated) values of $\int_0^t e^{-x^2}\,dx$ approach $\sqrt{\pi}/2$ as t becomes large. In fact, these values converge quite quickly because $e^{-x^2} \to 0$ very rapidly as $x \to \infty$.

TABLE 1

t	$\int_0^t e^{-x^2}\,dx$
1	0.7468241328
2	0.8820813908
3	0.8862073483
4	0.8862269118
5	0.8862269255
6	0.8862269255

EXAMPLE 10 The integral $\int_1^\infty \dfrac{1 + e^{-x}}{x}\,dx$ is divergent by the Comparison Theorem because

$$\frac{1 + e^{-x}}{x} > \frac{1}{x}$$

and $\int_1^\infty (1/x)\,dx$ is divergent by Example 1 [or by (2) with $p = 1$].

Table 2 illustrates the divergence of the integral in Example 10. It appears that the values are not approaching any fixed number.

TABLE 2

t	$\int_1^t [(1 + e^{-x})/x]\,dx$
2	0.8636306042
5	1.8276735512
10	2.5219648704
100	4.8245541204
1000	7.1271392134
10000	9.4297243064

8.8 Exercises

1. Explain why each of the following integrals is improper.

(a) $\int_1^\infty x^4 e^{-x^4}\,dx$ (b) $\int_0^{\pi/2} \sec x\,dx$

(c) $\int_0^2 \dfrac{x}{x^2 - 5x + 6}\,dx$ (d) $\int_{-\infty}^0 \dfrac{1}{x^2 + 5}\,dx$

2. Which of the following integrals are improper? Why?

(a) $\int_1^2 \dfrac{1}{2x - 1}\,dx$ (b) $\int_0^1 \dfrac{1}{2x - 1}\,dx$

(c) $\int_{-\infty}^\infty \dfrac{\sin x}{1 + x^2}\,dx$ (d) $\int_1^2 \ln(x - 1)\,dx$

3. Find the area under the curve $y = 1/x^3$ from $x = 1$ to $x = t$ and evaluate it for $t = 10$, 100, and 1000. Then find the total area under this curve for $x \geq 1$.

4. (a) Graph the functions $f(x) = 1/x^{1.1}$ and $g(x) = 1/x^{0.9}$ in the viewing rectangles $[0, 10]$ by $[0, 1]$ and $[0, 100]$ by $[0, 1]$.
   (b) Find the areas under the graphs of f and g from $x = 1$ to $x = t$ and evaluate for $t = 10$, 100, 10^4, 10^6, 10^{10}, and 10^{20}.
   (c) Find the total area under each curve for $x \geq 1$, if it exists.

5–40 Determine whether each integral is convergent or divergent. Evaluate those that are convergent.

5. $\int_1^\infty \dfrac{1}{(3x+1)^2}\, dx$

6. $\int_{-\infty}^0 \dfrac{1}{2x-5}\, dx$

7. $\int_{-\infty}^{-1} \dfrac{1}{\sqrt{2-w}}\, dw$

8. $\int_0^\infty \dfrac{x}{(x^2+2)^2}\, dx$

9. $\int_4^\infty e^{-y/2}\, dy$

10. $\int_{-\infty}^{-1} e^{-2t}\, dt$

11. $\int_{-\infty}^\infty \dfrac{x}{1+x^2}\, dx$

12. $\int_{-\infty}^\infty (2-v^4)\, dv$

13. $\int_{-\infty}^\infty xe^{-x^2}\, dx$

14. $\int_{-\infty}^\infty x^2 e^{-x^3}\, dx$

15. $\int_{2\pi}^\infty \sin\theta\, d\theta$

16. $\int_0^\infty \cos^2\alpha\, d\alpha$

17. $\int_1^\infty \dfrac{x+1}{x^2+2x}\, dx$

18. $\int_0^\infty \dfrac{dz}{z^2+3z+2}$

19. $\int_0^\infty se^{-5s}\, ds$

20. $\int_{-\infty}^6 re^{r/3}\, dr$

21. $\int_1^\infty \dfrac{\ln x}{x}\, dx$

22. $\int_{-\infty}^\infty e^{-|x|}\, dx$

23. $\int_{-\infty}^\infty \dfrac{x^2}{9+x^6}\, dx$

24. $\int_1^\infty \dfrac{\ln x}{x^3}\, dx$

25. $\int_1^\infty \dfrac{\ln x}{x^2}\, dx$

26. $\int_0^\infty \dfrac{x \arctan x}{(1+x^2)^2}\, dx$

27. $\int_0^3 \dfrac{1}{\sqrt{x}}\, dx$

28. $\int_0^3 \dfrac{1}{x\sqrt{x}}\, dx$

29. $\int_{-1}^0 \dfrac{1}{x^2}\, dx$

30. $\int_1^9 \dfrac{1}{\sqrt[3]{x-9}}\, dx$

31. $\int_{-2}^3 \dfrac{1}{x^4}\, dx$

32. $\int_0^1 \dfrac{dx}{\sqrt{1-x^2}}$

33. $\int_0^{33} (x-1)^{-1/5}\, dx$

34. $\int_0^1 \dfrac{1}{4y-1}\, dy$

35. $\int_0^\pi \sec x\, dx$

36. $\int_0^4 \dfrac{1}{x^2+x-6}\, dx$

37. $\int_{-1}^1 \dfrac{e^x}{e^x-1}\, dx$

38. $\int_0^2 \dfrac{x-3}{2x-3}\, dx$

39. $\int_0^2 z^2 \ln z\, dz$

40. $\int_0^1 \dfrac{\ln x}{\sqrt{x}}\, dx$

41–46 Sketch the region and find its area (if the area is finite).

41. $S = \{(x, y) \mid x \leq 1,\ 0 \leq y \leq e^x\}$

42. $S = \{(x, y) \mid x \geq -2,\ 0 \leq y \leq e^{-x/2}\}$

43. $S = \{(x, y) \mid 0 \leq y \leq 2/(x^2+9)\}$

44. $S = \{(x, y) \mid x \geq 0,\ 0 \leq y \leq x/(x^2+9)\}$

45. $S = \{(x, y) \mid 0 \leq x < \pi/2,\ 0 \leq y \leq \sec^2 x\}$

46. $S = \{(x, y) \mid -2 < x \leq 0,\ 0 \leq y \leq 1/\sqrt{x+2}\}$

47. (a) If $g(x) = (\sin^2 x)/x^2$, use your calculator or computer to make a table of approximate values of $\int_1^t g(x)\, dx$ for $t = 2$, 5, 10, 100, 1000, and $10{,}000$. Does it appear that $\int_1^\infty g(x)\, dx$ is convergent?
    (b) Use the Comparison Theorem with $f(x) = 1/x^2$ to show that $\int_1^\infty g(x)\, dx$ is convergent.
    (c) Illustrate part (b) by graphing f and g on the same screen for $1 \leq x \leq 10$. Use your graph to explain intuitively why $\int_1^\infty g(x)\, dx$ is convergent.

48. (a) If $g(x) = 1/(\sqrt{x}-1)$, use your calculator or computer to make a table of approximate values of $\int_2^t g(x)\, dx$ for $t = 5$, 10, 100, 1000, and $10{,}000$. Does it appear that $\int_2^\infty g(x)\, dx$ is convergent or divergent?
    (b) Use the Comparison Theorem with $f(x) = 1/\sqrt{x}$ to show that $\int_2^\infty g(x)\, dx$ is divergent.
    (c) Illustrate part (b) by graphing f and g on the same screen for $2 \leq x \leq 20$. Use your graph to explain intuitively why $\int_2^\infty g(x)\, dx$ is divergent.

49–54 Use the Comparison Theorem to determine whether the integral is convergent or divergent.

49. $\int_1^\infty \dfrac{\cos^2 x}{1+x^2}\, dx$

50. $\int_1^\infty \dfrac{2+e^{-x}}{x}\, dx$

51. $\int_1^\infty \dfrac{dx}{x+e^{2x}}$

52. $\int_1^\infty \dfrac{x}{\sqrt{1+x^6}}\, dx$

53. $\int_0^{\pi/2} \dfrac{dx}{x \sin x}$

54. $\int_0^1 \dfrac{e^{-x}}{\sqrt{x}}\, dx$

55. The integral

$$\int_0^\infty \dfrac{1}{\sqrt{x}\,(1+x)}\, dx$$

is improper for two reasons: The interval $[0, \infty)$ is infinite and the integrand has an infinite discontinuity at 0. Evaluate it by expressing it as a sum of improper integrals of Type 2 and Type 1 as follows:

$$\int_0^\infty \dfrac{1}{\sqrt{x}\,(1+x)}\, dx = \int_0^1 \dfrac{1}{\sqrt{x}\,(1+x)}\, dx + \int_1^\infty \dfrac{1}{\sqrt{x}\,(1+x)}\, dx$$

56. Evaluate
$$\int_2^\infty \frac{1}{x\sqrt{x^2-4}}\,dx$$
by the same method as in Exercise 55.

57–59 Find the values of p for which the integral converges and evaluate the integral for those values of p.

57. $\int_0^1 \frac{1}{x^p}\,dx$

58. $\int_e^\infty \frac{1}{x(\ln x)^p}\,dx$

59. $\int_0^1 x^p \ln x\,dx$

60. (a) Evaluate the integral $\int_0^\infty x^n e^{-x}\,dx$ for $n = 0, 1, 2,$ and 3.
(b) Guess the value of $\int_0^\infty x^n e^{-x}\,dx$ when n is an arbitrary positive integer.
(c) Prove your guess using mathematical induction.

61. (a) Show that $\int_{-\infty}^\infty x\,dx$ is divergent.
(b) Show that
$$\lim_{t\to\infty}\int_{-t}^t x\,dx = 0$$
This shows that we can't define
$$\int_{-\infty}^\infty f(x)\,dx = \lim_{t\to\infty}\int_{-t}^t f(x)\,dx$$

62. The *average speed* of molecules in an ideal gas is
$$\bar{v} = \frac{4}{\sqrt{\pi}}\left(\frac{M}{2RT}\right)^{3/2}\int_0^\infty v^3 e^{-Mv^2/(2RT)}\,dv$$
where M is the molecular weight of the gas, R is the gas constant, T is the gas temperature, and v is the molecular speed. Show that
$$\bar{v} = \sqrt{\frac{8RT}{\pi M}}$$

63. We know from Example 1 that the region $\mathcal{R} = \{(x, y) \mid x \geq 1, 0 \leq y \leq 1/x\}$ has infinite area. Show that by rotating $\mathcal{R}$ about the x-axis we obtain a solid with finite volume.

64. Use the information and data in Exercises 29 and 30 of Section 6.4 to find the work required to propel a 1000-kg satellite out of Earth's gravitational field.

65. Find the *escape velocity* v_0 that is needed to propel a rocket of mass m out of the gravitational field of a planet with mass M and radius R. Use Newton's Law of Gravitation (see Exercise 29 in Section 6.4) and the fact that the initial kinetic energy of $\frac{1}{2}mv_0^2$ supplies the needed work.

66. Astronomers use a technique called *stellar stereography* to determine the density of stars in a star cluster from the observed (two-dimensional) density that can be analyzed from a photograph. Suppose that in a spherical cluster of radius R the density of stars depends only on the distance r from the center of the cluster. If the perceived star density is given by $y(s)$, where s is the observed planar distance from the center of the cluster, and $x(r)$ is the actual density, it can be shown that
$$y(s) = \int_s^R \frac{2r}{\sqrt{r^2-s^2}}\,x(r)\,dr$$
If the actual density of stars in a cluster is $x(r) = \frac{1}{2}(R-r)^2$, find the perceived density $y(s)$.

67. A manufacturer of lightbulbs wants to produce bulbs that last about 700 hours but, of course, some bulbs burn out faster than others. Let $F(t)$ be the fraction of the company's bulbs that burn out before t hours, so $F(t)$ always lies between 0 and 1.
(a) Make a rough sketch of what you think the graph of F might look like.
(b) What is the meaning of the derivative $r(t) = F'(t)$?
(c) What is the value of $\int_0^\infty r(t)\,dt$? Why?

68. As we will see in Section 10.4, a radioactive substance decays exponentially: The mass at time t is $m(t) = m(0)e^{kt}$, where $m(0)$ is the initial mass and k is a negative constant. The *mean life* M of an atom in the substance is
$$M = -k\int_0^\infty te^{kt}\,dt$$
For the radioactive carbon isotope, ^{14}C, used in radiocarbon dating, the value of k is -0.000121. Find the mean life of a ^{14}C atom.

69. Determine how large the number a has to be so that
$$\int_a^\infty \frac{1}{x^2+1}\,dx < 0.001$$

70. Estimate the numerical value of $\int_0^\infty e^{-x^2}\,dx$ by writing it as the sum of $\int_0^4 e^{-x^2}\,dx$ and $\int_4^\infty e^{-x^2}\,dx$. Approximate the first integral by using Simpson's Rule with $n = 8$ and show that the second integral is smaller than $\int_4^\infty e^{-4x}\,dx$, which is less than 0.0000001.

71. If $f(t)$ is continuous for $t \geq 0$, the *Laplace transform* of f is the function F defined by
$$F(s) = \int_0^\infty f(t)e^{-st}\,dt$$
and the domain of F is the set consisting of all numbers s for which the integral converges. Find the Laplace transforms of the following functions.
(a) $f(t) = 1$ (b) $f(t) = e^t$ (c) $f(t) = t$

72. Show that if $0 \leq f(t) \leq Me^{at}$ for $t \geq 0$, where M and a are constants, then the Laplace transform $F(s)$ exists for $s > a$.

73. Suppose that $0 \leq f(t) \leq Me^{at}$ and $0 \leq f'(t) \leq Ke^{at}$ for $t \geq 0$, where f' is continuous. If the Laplace transform of $f(t)$ is $F(s)$

and the Laplace transform of $f'(t)$ is $G(s)$, show that
$$G(s) = sF(s) - f(0) \quad s > a$$

74. If $\int_{-\infty}^{\infty} f(x)\, dx$ is convergent and a and b are real numbers, show that
$$\int_{-\infty}^{a} f(x)\, dx + \int_{a}^{\infty} f(x)\, dx = \int_{-\infty}^{b} f(x)\, dx + \int_{b}^{\infty} f(x)\, dx$$

75. Show that $\int_{0}^{\infty} x^2 e^{-x^2}\, dx = \tfrac{1}{2} \int_{0}^{\infty} e^{-x^2}\, dx$.

76. Show that $\int_{0}^{\infty} e^{-x^2}\, dx = \int_{0}^{1} \sqrt{-\ln y}\, dy$ by interpreting the integrals as areas.

77. Find the value of the constant C for which the integral
$$\int_{0}^{\infty} \left(\frac{1}{\sqrt{x^2 + 4}} - \frac{C}{x + 2} \right) dx$$
converges. Evaluate the integral for this value of C.

78. Find the value of the constant C for which the integral
$$\int_{0}^{\infty} \left(\frac{x}{x^2 + 1} - \frac{C}{3x + 1} \right) dx$$
converges. Evaluate the integral for this value of C.

8 Review

CONCEPT CHECK

1. State the rule for integration by parts. In practice, how do you use it?

2. How do you evaluate $\int \sin^m x \cos^n x\, dx$ if m is odd? What if n is odd? What if m and n are both even?

3. If the expression $\sqrt{a^2 - x^2}$ occurs in an integral, what substitution might you try? What if $\sqrt{a^2 + x^2}$ occurs? What if $\sqrt{x^2 - a^2}$ occurs?

4. What is the form of the partial fraction expansion of a rational function $P(x)/Q(x)$ if the degree of P is less than the degree of Q and $Q(x)$ has only distinct linear factors? What if a linear factor is repeated? What if $Q(x)$ has an irreducible quadratic factor (not repeated)? What if the quadratic factor is repeated?

5. State the rules for approximating the definite integral $\int_a^b f(x)\, dx$ with the Midpoint Rule, the Trapezoidal Rule, and Simpson's Rule. Which would you expect to give the best estimate? How do you approximate the error for each rule?

6. Define the following improper integrals.
   (a) $\int_a^{\infty} f(x)\, dx$ (b) $\int_{-\infty}^{b} f(x)\, dx$ (c) $\int_{-\infty}^{\infty} f(x)\, dx$

7. Define the improper integral $\int_a^b f(x)\, dx$ for each of the following cases.
   (a) f has an infinite discontinuity at a.
   (b) f has an infinite discontinuity at b.
   (c) f has an infinite discontinuity at c, where $a < c < b$.

8. State the Comparison Theorem for improper integrals.

TRUE-FALSE QUIZ

Determine whether the statement is true or false. If it is true, explain why. If it is false, explain why or give an example that disproves the statement.

1. $\dfrac{x(x^2 + 4)}{x^2 - 4}$ can be put in the form $\dfrac{A}{x + 2} + \dfrac{B}{x - 2}$.

2. $\dfrac{x^2 + 4}{x(x^2 - 4)}$ can be put in the form $\dfrac{A}{x} + \dfrac{B}{x + 2} + \dfrac{C}{x - 2}$.

3. $\dfrac{x^2 + 4}{x^2(x - 4)}$ can be put in the form $\dfrac{A}{x^2} + \dfrac{B}{x - 4}$.

4. $\dfrac{x^2 - 4}{x(x^2 + 4)}$ can be put in the form $\dfrac{A}{x} + \dfrac{B}{x^2 + 4}$.

5. $\int_0^4 \dfrac{x}{x^2 - 1}\, dx = \tfrac{1}{2} \ln 15$

6. $\int_1^{\infty} \dfrac{1}{x^{\sqrt{2}}}\, dx$ is convergent.

7. If f is continuous, then $\int_{-\infty}^{\infty} f(x)\, dx = \lim_{t \to \infty} \int_{-t}^{t} f(x)\, dx$.

8. The Midpoint Rule is always more accurate than the Trapezoidal Rule.

9. (a) Every elementary function has an elementary derivative.
   (b) Every elementary function has an elementary antiderivative.

10. If f is continuous on $[0, \infty)$ and $\int_1^{\infty} f(x)\, dx$ is convergent, then $\int_0^{\infty} f(x)\, dx$ is convergent.

11. If f is a continuous, decreasing function on $[1, \infty)$ and $\lim_{x \to \infty} f(x) = 0$, then $\int_1^{\infty} f(x)\, dx$ is convergent.

12. If $\int_a^{\infty} f(x)\, dx$ and $\int_a^{\infty} g(x)\, dx$ are both convergent, then $\int_a^{\infty} [f(x) + g(x)]\, dx$ is convergent.

13. If $\int_a^{\infty} f(x)\, dx$ and $\int_a^{\infty} g(x)\, dx$ are both divergent, then $\int_a^{\infty} [f(x) + g(x)]\, dx$ is divergent.

14. If $f(x) \leq g(x)$ and $\int_0^{\infty} g(x)\, dx$ diverges, then $\int_0^{\infty} f(x)\, dx$ also diverges.

EXERCISES

Note: Additional practice in techniques of integration is provided in Exercises 7.5.

1–40 ▮ Evaluate the integral.

1. $\displaystyle\int_0^5 \frac{x}{x+10}\,dx$

2. $\displaystyle\int_0^5 y e^{-0.6y}\,dy$

3. $\displaystyle\int_0^{\pi/2} \frac{\cos\theta}{1+\sin\theta}\,d\theta$

4. $\displaystyle\int_1^4 \frac{dt}{(2t+1)^3}$

5. $\displaystyle\int \tan^7 x \sec^3 x\,dx$

6. $\displaystyle\int \frac{1}{y^2 - 4y - 12}\,dy$

7. $\displaystyle\int \frac{\sin(\ln t)}{t}\,dt$

8. $\displaystyle\int \frac{dx}{x^2\sqrt{1+x^2}}$

9. $\displaystyle\int_1^4 x^{3/2}\ln x\,dx$

10. $\displaystyle\int_0^1 \frac{\sqrt{\arctan x}}{1+x^2}\,dx$

11. $\displaystyle\int_1^2 \frac{\sqrt{x^2-1}}{x}\,dx$

12. $\displaystyle\int_{-1}^1 \frac{\sin x}{1+x^2}\,dx$

13. $\displaystyle\int \frac{dx}{x^3+x}$

14. $\displaystyle\int \frac{x^2+2}{x+2}\,dx$

15. $\displaystyle\int \sin^2\theta \cos^5\theta\,d\theta$

16. $\displaystyle\int \frac{\sec^6\theta}{\tan^2\theta}\,d\theta$

17. $\displaystyle\int x\sec x \tan x\,dx$

18. $\displaystyle\int \frac{x^2+8x-3}{x^3+3x^2}\,dx$

19. $\displaystyle\int \frac{x+1}{9x^2+6x+5}\,dx$

20. $\displaystyle\int \frac{dt}{\sin^2 t + \cos 2t}$

21. $\displaystyle\int \frac{dx}{\sqrt{x^2-4x}}$

22. $\displaystyle\int \frac{x^3}{(x+1)^{10}}\,dx$

23. $\displaystyle\int \csc^4 4x\,dx$

24. $\displaystyle\int e^x \cos x\,dx$

25. $\displaystyle\int \frac{3x^3-x^2+6x-4}{(x^2+1)(x^2+2)}\,dx$

26. $\displaystyle\int \frac{dx}{1+e^x}$

27. $\displaystyle\int_0^{\pi/2} \cos^3 x \sin 2x\,dx$

28. $\displaystyle\int \frac{\sqrt[3]{x}+1}{\sqrt[3]{x}-1}\,dx$

29. $\displaystyle\int_{-1}^1 x^5 \sec x\,dx$

30. $\displaystyle\int \frac{dx}{e^x\sqrt{1-e^{-2x}}}$

31. $\displaystyle\int_0^{\ln 10} \frac{e^x\sqrt{e^x-1}}{e^x+8}\,dx$

32. $\displaystyle\int_0^{\pi/4} \frac{x\sin x}{\cos^3 x}\,dx$

33. $\displaystyle\int \frac{x^2}{(4-x^2)^{3/2}}\,dx$

34. $\displaystyle\int (\arcsin x)^2\,dx$

35. $\displaystyle\int \frac{1}{\sqrt{x}+x^{3/2}}\,dx$

36. $\displaystyle\int \frac{1-\tan\theta}{1+\tan\theta}\,d\theta$

37. $\displaystyle\int (\cos x + \sin x)^2 \cos 2x\,dx$

38. $\displaystyle\int x(\tan^{-1} x)^2\,dx$

39. $\displaystyle\int_0^{1/2} \frac{xe^{2x}}{(1+2x)^2}\,dx$

40. $\displaystyle\int_{\pi/4}^{\pi/3} \frac{\sqrt{\tan\theta}}{\sin 2\theta}\,d\theta$

41–50 ▮ Evaluate the integral or show that it is divergent.

41. $\displaystyle\int_1^\infty \frac{1}{(2x+1)^3}\,dx$

42. $\displaystyle\int_0^1 \frac{t^2+1}{t^2-1}\,dt$

43. $\displaystyle\int_2^\infty \frac{dx}{x\ln x}$

44. $\displaystyle\int_2^6 \frac{y}{\sqrt{y-2}}\,dy$

45. $\displaystyle\int_0^4 \frac{\ln x}{\sqrt{x}}\,dx$

46. $\displaystyle\int_0^1 \frac{1}{2-3x}\,dx$

47. $\displaystyle\int_0^3 \frac{dx}{x^2-x-2}$

48. $\displaystyle\int_{-1}^1 \frac{x+1}{\sqrt[3]{x^4}}\,dx$

49. $\displaystyle\int_{-\infty}^\infty \frac{dx}{4x^2+4x+5}$

50. $\displaystyle\int_1^\infty \frac{\tan^{-1}x}{x^2}\,dx$

51–52 ▮ Evaluate the indefinite integral. Illustrate and check that your answer is reasonable by graphing both the function and its antiderivative (take $C = 0$).

51. $\displaystyle\int \ln(x^2+2x+2)\,dx$

52. $\displaystyle\int \frac{x^3}{\sqrt{x^2+1}}\,dx$

53. Graph the function $f(x) = \cos^2 x \sin^3 x$ and use the graph to guess the value of the integral $\int_0^{2\pi} f(x)\,dx$. Then evaluate the integral to confirm your guess.

CAS 54. (a) How would you evaluate $\int x^5 e^{-2x}\,dx$ by hand? (Don't actually carry out the integration.)
  (b) How would you evaluate $\int x^5 e^{-2x}\,dx$ using tables? (Don't actually do it.)
  (c) Use a CAS to evaluate $\int x^5 e^{-2x}\,dx$.
  (d) Graph the integrand and the indefinite integral on the same screen.

55–58 ▮ Use the Table of Integrals on the Reference Pages to evaluate the integral.

55. $\displaystyle\int e^x\sqrt{1-e^{2x}}\,dx$

56. $\displaystyle\int \csc^5 t\,dt$

57. $\displaystyle\int \sqrt{x^2+x+1}\,dx$

58. $\displaystyle\int \frac{\cot x}{\sqrt{1+2\sin x}}\,dx$

59. Verify Formula 33 in the Table of Integrals (a) by differentiation and (b) by using a trigonometric substitution.

60. Verify Formula 62 in the Table of Integrals.

61. Is it possible to find a number n such that $\int_0^\infty x^n\,dx$ is convergent?

62. For what values of a is $\int_0^\infty e^{ax} \cos x \, dx$ convergent? Evaluate the integral for those values of a.

63–64 Use (a) the Trapezoidal Rule, (b) the Midpoint Rule, and (c) Simpson's Rule with $n = 10$ to approximate the given integral. Round your answers to six decimal places.

63. $\int_0^1 \sqrt{1 + x^4} \, dx$

64. $\int_0^{\pi/2} \sqrt{\sin x} \, dx$

65. Estimate the errors involved in Exercise 63, parts (a) and (b). How large should n be in each case to guarantee an error of less than 0.00001?

66. Use Simpson's Rule with $n = 6$ to estimate the area under the curve $y = e^x/x$ from $x = 1$ to $x = 4$.

67. The speedometer reading (v) on a car was observed at 1-minute intervals and recorded in the chart. Use Simpson's Rule to estimate the distance traveled by the car.

t (min)	v (mi/h)	t (min)	v (mi/h)
0	40	6	56
1	42	7	57
2	45	8	57
3	49	9	55
4	52	10	56
5	54		

68. A population of honeybees increased at a rate of $r(t)$ bees per week, where the graph of r is as shown. Use Simpson's Rule with six subintervals to estimate the increase in the bee population during the first 24 weeks.

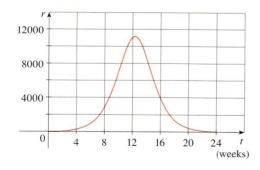

CAS 69. (a) If $f(x) = \sin(\sin x)$, use a graph to find an upper bound for $|f^{(4)}(x)|$.
(b) Use Simpson's Rule with $n = 10$ to approximate $\int_0^\pi f(x) \, dx$ and use part (a) to estimate the error.
(c) How large should n be to guarantee that the size of the error in using S_n is less than 0.00001?

70. Suppose you are asked to estimate the volume of a football. You measure and find that a football is 28 cm long. You use a piece of string and measure the circumference at its widest point to be 53 cm. The circumference 7 cm from each end is 45 cm. Use Simpson's Rule to make your estimate.

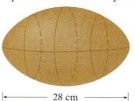

71. Use the Comparison Theorem to determine whether the integral
$$\int_1^\infty \frac{x^3}{x^5 + 2} \, dx$$
is convergent or divergent.

72. Find the area of the region bounded by the hyperbola $y^2 - x^2 = 1$ and the line $y = 3$.

73. Find the area bounded by the curves $y = \cos x$ and $y = \cos^2 x$ between $x = 0$ and $x = \pi$.

74. Find the area of the region bounded by the curves $y = 1/(2 + \sqrt{x})$, $y = 1/(2 - \sqrt{x})$, and $x = 1$.

75. The region under the curve $y = \cos^2 x$, $0 \leq x \leq \pi/2$, is rotated about the x-axis. Find the volume of the resulting solid.

76. The region in Exercise 75 is rotated about the y-axis. Find the volume of the resulting solid.

77. If f' is continuous on $[0, \infty)$ and $\lim_{x \to \infty} f(x) = 0$, show that
$$\int_0^\infty f'(x) \, dx = -f(0)$$

78. We can extend our definition of average value of a continuous function to an infinite interval by defining the average value of f on the interval $[a, \infty)$ to be
$$\lim_{t \to \infty} \frac{1}{t - a} \int_a^t f(x) \, dx$$
(a) Find the average value of $y = \tan^{-1} x$ on the interval $[0, \infty)$.
(b) If $f(x) \geq 0$ and $\int_a^\infty f(x) \, dx$ is divergent, show that the average value of f on the interval $[a, \infty)$ is $\lim_{x \to \infty} f(x)$, if this limit exists.
(c) If $\int_a^\infty f(x) \, dx$ is convergent, what is the average value of f on the interval $[a, \infty)$?
(d) Find the average value of $y = \sin x$ on the interval $[0, \infty)$.

79. Use the substitution $u = 1/x$ to show that
$$\int_0^\infty \frac{\ln x}{1 + x^2} \, dx = 0$$

80. The magnitude of the repulsive force between two point charges with the same sign, one of size 1 and the other of size q, is
$$F = \frac{q}{4\pi\varepsilon_0 r^2}$$
where r is the distance between the charges and ε_0 is a constant. The *potential* V at a point P due to the charge q is defined to be the work expended in bringing a unit charge to P from infinity along the straight line that joins q and P. Find a formula for V.

PROBLEMS PLUS

Cover up the solution to the example and try it yourself first.

EXAMPLE
(a) Prove that if f is a continuous function, then
$$\int_0^a f(x)\, dx = \int_0^a f(a - x)\, dx$$

(b) Use part (a) to show that
$$\int_0^{\pi/2} \frac{\sin^n x}{\sin^n x + \cos^n x}\, dx = \frac{\pi}{4}$$

for all positive numbers n.

SOLUTION
(a) At first sight, the given equation may appear somewhat baffling. How is it possible to connect the left side to the right side? Connections can often be made through one of the principles of problem solving: *introduce something extra*. Here the extra ingredient is a new variable. We often think of introducing a new variable when we use the Substitution Rule to integrate a specific function. But that technique is still useful in the present circumstance in which we have a general function f.

Once we think of making a substitution, the form of the right side suggests that it should be $u = a - x$. Then $du = -dx$. When $x = 0$, $u = a$; when $x = a$, $u = 0$. So

$$\int_0^a f(a - x)\, dx = -\int_a^0 f(u)\, du = \int_0^a f(u)\, du$$

But this integral on the right side is just another way of writing $\int_0^a f(x)\, dx$. So the given equation is proved.

(b) If we let the given integral be I and apply part (a) with $a = \pi/2$, we get

$$I = \int_0^{\pi/2} \frac{\sin^n x}{\sin^n x + \cos^n x}\, dx = \int_0^{\pi/2} \frac{\sin^n(\pi/2 - x)}{\sin^n(\pi/2 - x) + \cos^n(\pi/2 - x)}\, dx$$

A well-known trigonometric identity tells us that $\sin(\pi/2 - x) = \cos x$ and $\cos(\pi/2 - x) = \sin x$, so we get

$$I = \int_0^{\pi/2} \frac{\cos^n x}{\cos^n x + \sin^n x}\, dx$$

Notice that the two expressions for I are very similar. In fact, the integrands have the same denominator. This suggests that we should add the two expressions. If we do so, we get

$$2I = \int_0^{\pi/2} \frac{\sin^n x + \cos^n x}{\sin^n x + \cos^n x}\, dx = \int_0^{\pi/2} 1\, dx = \frac{\pi}{2}$$

Therefore, $I = \pi/4$.

||| The principles of problem solving are discussed on page 58.

||| The computer graphs in Figure 1 make it seem plausible that all of the integrals in the example have the same value. The graph of each integrand is labeled with the corresponding value of n.

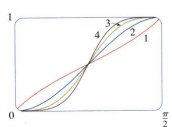

FIGURE 1

PROBLEMS

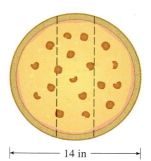

FIGURE FOR PROBLEM 1

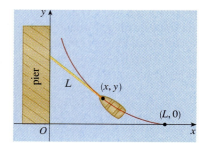

FIGURE FOR PROBLEM 4

1. Three mathematics students have ordered a 14-inch pizza. Instead of slicing it in the traditional way, they decide to slice it by parallel cuts, as shown in the figure. Being mathematics majors, they are able to determine where to slice so that each gets the same amount of pizza. Where are the cuts made?

2. Evaluate $\int \dfrac{1}{x^7 - x}\, dx$.

   The straightforward approach would be to start with partial fractions, but that would be brutal. Try a substitution.

3. Evaluate $\int_0^1 \left(\sqrt[3]{1 - x^7} - \sqrt[7]{1 - x^3}\right) dx$.

4. A man initially standing at the point O walks along a pier pulling a rowboat by a rope of length L. The man keeps the rope straight and taut. The path followed by the boat is a curve called a *tractrix* and it has the property that the rope is always tangent to the curve (see the figure).
   (a) Show that if the path followed by the boat is the graph of the function $y = f(x)$, then
   $$f'(x) = \frac{dy}{dx} = \frac{-\sqrt{L^2 - x^2}}{x}$$
   (b) Determine the function $y = f(x)$.

5. A function f is defined by
   $$f(x) = \int_0^\pi \cos t \, \cos(x - t)\, dt \qquad 0 \le x \le 2\pi$$
   Find the minimum value of f.

6. If n is a positive integer, prove that
   $$\int_0^1 (\ln x)^n \, dx = (-1)^n n!$$

7. Show that
   $$\int_0^1 (1 - x^2)^n \, dx = \frac{2^{2n}(n!)^2}{(2n + 1)!}$$
   Hint: Start by showing that if I_n denotes the integral, then
   $$I_{k+1} = \frac{2k + 2}{2k + 3} I_k$$

8. Suppose that f is a positive function such that f' is continuous.
   (a) How is the graph of $y = f(x) \sin nx$ related to the graph of $y = f(x)$? What happens as $n \to \infty$?
   (b) Make a guess as to the value of the limit
   $$\lim_{n \to \infty} \int_0^1 f(x) \sin nx \, dx$$
   based on graphs of the integrand.
   (c) Using integration by parts, confirm the guess that you made in part (b). [Use the fact that, since f' is continuous, there is a constant M such that $|f'(x)| \le M$ for $0 \le x \le 1$.]

9. If $0 < a < b$, find $\displaystyle\lim_{t \to 0} \left\{\int_0^1 [bx + a(1 - x)]^t \, dx\right\}^{1/t}$.

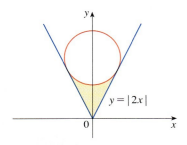

FIGURE FOR PROBLEM 11

10. Graph $f(x) = \sin(e^x)$ and use the graph to estimate the value of t such that $\int_t^{t+1} f(x)\, dx$ is a maximum. Then find the exact value of t that maximizes this integral.

11. The circle with radius 1 shown in the figure touches the curve $y = |2x|$ twice. Find the area of the region that lies between the two curves.

12. A rocket is fired straight up, burning fuel at the constant rate of b kilograms per second. Let $v = v(t)$ be the velocity of the rocket at time t and suppose that the velocity u of the exhaust gas is constant. Let $M = M(t)$ be the mass of the rocket at time t and note that M decreases as the fuel burns. If we neglect air resistance, it follows from Newton's Second Law that

$$F = M\frac{dv}{dt} - ub$$

where the force $F = -Mg$. Thus

$$\boxed{1} \qquad M\frac{dv}{dt} - ub = -Mg$$

Let M_1 be the mass of the rocket without fuel, M_2 the initial mass of the fuel, and $M_0 = M_1 + M_2$. Then, until the fuel runs out at time $t = M_2/b$, the mass is $M = M_0 - bt$.
(a) Substitute $M = M_0 - bt$ into Equation 1 and solve the resulting equation for v. Use the initial condition $v(0) = 0$ to evaluate the constant.
(b) Determine the velocity of the rocket at time $t = M_2/b$. This is called the *burnout velocity*.
(c) Determine the height of the rocket $y = y(t)$ at the burnout time.
(d) Find the height of the rocket at any time t.

13. Use integration by parts to show that, for all $x > 0$,

$$0 < \int_0^\infty \frac{\sin t}{\ln(1 + x + t)}\, dt < \frac{2}{\ln(1 + x)}$$

14. The **Chebyshev polynomials** T_n are defined by

$$T_n(x) = \cos(n \arccos x) \qquad n = 0, 1, 2, 3, \ldots$$

(a) What are the domain and range of these functions?
(b) We know that $T_0(x) = 1$ and $T_1(x) = x$. Express T_2 explicitly as a quadratic polynomial and T_3 as a cubic polynomial.
(c) Show that, for $n \geq 1$,

$$T_{n+1}(x) = 2x T_n(x) - T_{n-1}(x)$$

(d) Use part (c) to show that T_n is a polynomial of degree n.
(e) Use parts (b) and (c) to express T_4, T_5, T_6, and T_7 explicitly as polynomials.
(f) What are the zeros of T_n? At what numbers does T_n have local maximum and minimum values?
(g) Graph T_2, T_3, T_4, and T_5 on a common screen.
(h) Graph T_5, T_6, and T_7 on a common screen.
(i) Based on your observations from parts (g) and (h), how are the zeros of T_n related to the zeros of T_{n+1}? What about the x-coordinates of the maximum and minimum values?
(j) Based on your graphs in parts (g) and (h), what can you say about $\int_{-1}^1 T_n(x)\, dx$ when n is odd and when n is even?
(k) Use the substitution $u = \arccos x$ to evaluate the integral in part (j).
(l) The family of functions $f(x) = \cos(c \arccos x)$ are defined even when c is not an integer (but then f is not a polynomial). Describe how the graph of f changes as c increases.

Chapter 9

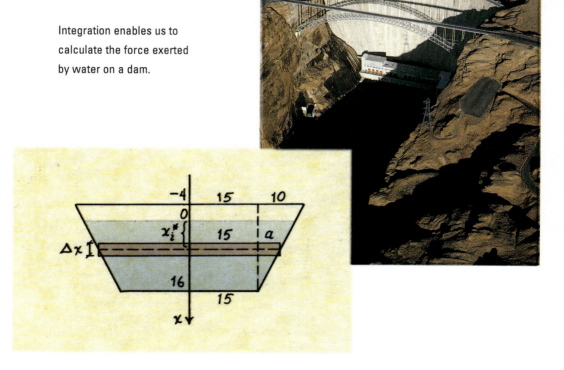

Integration enables us to calculate the force exerted by water on a dam.

Further Applications of Integration

We looked at some applications of integrals in Chapter 6: areas, volumes, work, and average values. Here we explore some of the many other geometric applications of integration—the length of a curve, the area of a surface— as well as quantities of interest in physics, engineering, biology, economics, and statistics. For instance, we will investigate the center of gravity of a plate, the force exerted by water pressure on a dam, the flow of blood from the human heart, and the average time spent on hold during a telephone call.

9.1 Arc Length

FIGURE 1

What do we mean by the length of a curve? We might think of fitting a piece of string to the curve in Figure 1 and then measuring the string against a ruler. But that might be difficult to do with much accuracy if we have a complicated curve. We need a precise definition for the length of an arc of a curve, in the same spirit as the definitions we developed for the concepts of area and volume.

If the curve is a polygon, we can easily find its length; we just add the lengths of the line segments that form the polygon. (We can use the distance formula to find the distance between the endpoints of each segment.) We are going to define the length of a general curve by first approximating it by a polygon and then taking a limit as the number of segments of the polygon is increased. This process is familiar for the case of a circle, where the circumference is the limit of lengths of inscribed polygons (see Figure 2).

Now suppose that a curve C is defined by the equation $y = f(x)$, where f is continuous and $a \leq x \leq b$. We obtain a polygonal approximation to C by dividing the interval $[a, b]$ into n subintervals with endpoints $x_0, x_1, \ldots, x_n$ and equal width Δx. If $y_i = f(x_i)$, then the point $P_i(x_i, y_i)$ lies on C and the polygon with vertices $P_0, P_1, \ldots, P_n$, illustrated in Figure 3, is an approximation to C. The length L of C is approximately the length of this polygon and the approximation gets better as we let n increase. (See Figure 4, where the arc of the curve between P_{i-1} and P_i has been magnified and approximations with successively smaller values of Δx are shown.) Therefore, we define the **length** L of the curve C

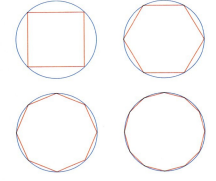

FIGURE 2

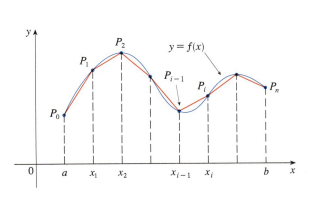

FIGURE 3

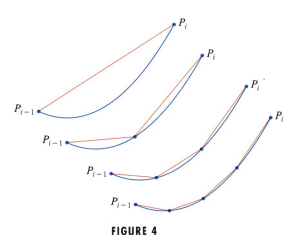

FIGURE 4

with equation $y = f(x)$, $a \leq x \leq b$, as the limit of the lengths of these inscribed polygons (if the limit exists):

$$\boxed{1} \qquad L = \lim_{n \to \infty} \sum_{i=1}^{n} |P_{i-1}P_i|$$

Notice that the procedure for defining arc length is very similar to the procedure we used for defining area and volume: We divided the curve into a large number of small parts. We then found the approximate lengths of the small parts and added them. Finally, we took the limit as $n \to \infty$.

The definition of arc length given by Equation 1 is not very convenient for computational purposes, but we can derive an integral formula for L in the case where f has a continuous derivative. [Such a function f is called **smooth** because a small change in x produces a small change in $f'(x)$.]

If we let $\Delta y_i = y_i - y_{i-1}$, then

$$|P_{i-1}P_i| = \sqrt{(x_i - x_{i-1})^2 + (y_i - y_{i-1})^2} = \sqrt{(\Delta x)^2 + (\Delta y_i)^2}$$

By applying the Mean Value Theorem to f on the interval $[x_{i-1}, x_i]$, we find that there is a number x_i^* between x_{i-1} and x_i such that

$$f(x_i) - f(x_{i-1}) = f'(x_i^*)(x_i - x_{i-1})$$

that is,
$$\Delta y_i = f'(x_i^*) \, \Delta x$$

Thus, we have

$$|P_{i-1}P_i| = \sqrt{(\Delta x)^2 + (\Delta y_i)^2}$$
$$= \sqrt{(\Delta x)^2 + [f'(x_i^*) \, \Delta x]^2}$$
$$= \sqrt{1 + [f'(x_i^*)]^2} \sqrt{(\Delta x)^2}$$
$$= \sqrt{1 + [f'(x_i^*)]^2} \, \Delta x \qquad \text{(since } \Delta x > 0\text{)}$$

Therefore, by Definition 1,

$$L = \lim_{n \to \infty} \sum_{i=1}^{n} |P_{i-1}P_i| = \lim_{n \to \infty} \sum_{i=1}^{n} \sqrt{1 + [f'(x_i^*)]^2} \, \Delta x$$

We recognize this expression as being equal to

$$\int_a^b \sqrt{1 + [f'(x)]^2} \, dx$$

by the definition of a definite integral. This integral exists because the function $g(x) = \sqrt{1 + [f'(x)]^2}$ is continuous. Thus, we have proved the following theorem:

---

2 The Arc Length Formula If f' is continuous on $[a, b]$, then the length of the curve $y = f(x)$, $a \leq x \leq b$, is

$$L = \int_a^b \sqrt{1 + [f'(x)]^2} \, dx$$

If we use Leibniz notation for derivatives, we can write the arc length formula as follows:

$$\boxed{3 \quad L = \int_a^b \sqrt{1 + \left(\frac{dy}{dx}\right)^2}\, dx}$$

EXAMPLE 1 Find the length of the arc of the semicubical parabola $y^2 = x^3$ between the points $(1, 1)$ and $(4, 8)$. (See Figure 5.)

SOLUTION For the top half of the curve we have

$$y = x^{3/2} \qquad \frac{dy}{dx} = \tfrac{3}{2} x^{1/2}$$

and so the arc length formula gives

$$L = \int_1^4 \sqrt{1 + \left(\frac{dy}{dx}\right)^2}\, dx = \int_1^4 \sqrt{1 + \tfrac{9}{4}x}\, dx$$

If we substitute $u = 1 + 9x/4$, then $du = 9\, dx/4$. When $x = 1$, $u = \tfrac{13}{4}$; when $x = 4$, $u = 10$. Therefore

$$L = \tfrac{4}{9} \int_{13/4}^{10} \sqrt{u}\, du = \tfrac{4}{9} \cdot \tfrac{2}{3} u^{3/2} \Big]_{13/4}^{10}$$

$$= \tfrac{8}{27} \left[10^{3/2} - \left(\tfrac{13}{4}\right)^{3/2} \right]$$

$$= \tfrac{1}{27} \left(80\sqrt{10} - 13\sqrt{13} \right)$$

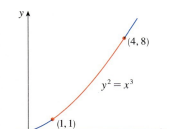

FIGURE 5

|||| As a check on our answer to Example 1, notice from Figure 5 that it ought to be slightly larger than the distance from $(1, 1)$ to $(4, 8)$, which is

$$\sqrt{58} \approx 7.615773$$

According to our calculation in Example 1, we have

$$L = \tfrac{1}{27}(80\sqrt{10} - 13\sqrt{13}) \approx 7.633705$$

Sure enough, this is a bit greater than the length of the line segment.

If a curve has the equation $x = g(y)$, $c \leq y \leq d$, and $g'(y)$ is continuous, then by interchanging the roles of x and y in Formula 2 or Equation 3, we obtain the following formula for its length:

$$\boxed{4 \quad L = \int_c^d \sqrt{1 + [g'(y)]^2}\, dy = \int_c^d \sqrt{1 + \left(\frac{dx}{dy}\right)^2}\, dy}$$

EXAMPLE 2 Find the length of the arc of the parabola $y^2 = x$ from $(0, 0)$ to $(1, 1)$.

SOLUTION Since $x = y^2$, we have $dx/dy = 2y$, and Formula 4 gives

$$L = \int_0^1 \sqrt{1 + \left(\frac{dx}{dy}\right)^2}\, dy = \int_0^1 \sqrt{1 + 4y^2}\, dy$$

We make the trigonometric substitution $y = \tfrac{1}{2} \tan\theta$, which gives $dy = \tfrac{1}{2} \sec^2\theta\, d\theta$ and $\sqrt{1 + 4y^2} = \sqrt{1 + \tan^2\theta} = \sec\theta$. When $y = 0$, $\tan\theta = 0$, so $\theta = 0$; when $y = 1$, $\tan\theta = 2$, so $\theta = \tan^{-1} 2 = \alpha$, say. Thus

$$L = \int_0^\alpha \sec\theta \cdot \tfrac{1}{2} \sec^2\theta\, d\theta = \tfrac{1}{2} \int_0^\alpha \sec^3\theta\, d\theta$$

$$= \tfrac{1}{2} \cdot \tfrac{1}{2} \big[\sec\theta \tan\theta + \ln|\sec\theta + \tan\theta| \big]_0^\alpha \qquad \text{(from Example 8 in Section 8.2)}$$

$$= \tfrac{1}{4} \big(\sec\alpha \tan\alpha + \ln|\sec\alpha + \tan\alpha| \big)$$

(We could have used Formula 21 in the Table of Integrals.) Since $\tan \alpha = 2$, we have $\sec^2 \alpha = 1 + \tan^2 \alpha = 5$, so $\sec \alpha = \sqrt{5}$ and

$$L = \frac{\sqrt{5}}{2} + \frac{\ln(\sqrt{5} + 2)}{4}$$

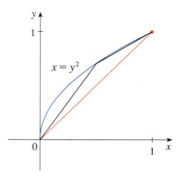

FIGURE 6

n	L_n
1	1.414
2	1.445
4	1.464
8	1.472
16	1.476
32	1.478
64	1.479

|||| Figure 6 shows the arc of the parabola whose length is computed in Example 2, together with polygonal approximations having $n = 1$ and $n = 2$ line segments, respectively. For $n = 1$ the approximate length is $L_1 = \sqrt{2}$, the diagonal of a square. The table shows the approximations L_n that we get by dividing $[0, 1]$ into n equal subintervals. Notice that each time we double the number of sides of the polygon, we get closer to the exact length, which is

$$L = \frac{\sqrt{5}}{2} + \frac{\ln(\sqrt{5} + 2)}{4} \approx 1.478943$$

Because of the presence of the square root sign in Formulas 2 and 4, the calculation of an arc length often leads to an integral that is very difficult or even impossible to evaluate explicitly. Thus, we sometimes have to be content with finding an approximation to the length of a curve as in the following example.

EXAMPLE 3
(a) Set up an integral for the length of the arc of the hyperbola $xy = 1$ from the point $(1, 1)$ to the point $(2, \frac{1}{2})$.
(b) Use Simpson's Rule with $n = 10$ to estimate the arc length.

SOLUTION
(a) We have

$$y = \frac{1}{x} \qquad \frac{dy}{dx} = -\frac{1}{x^2}$$

and so the arc length is

$$L = \int_1^2 \sqrt{1 + \left(\frac{dy}{dx}\right)^2}\, dx = \int_1^2 \sqrt{1 + \frac{1}{x^4}}\, dx = \int_1^2 \frac{\sqrt{x^4 + 1}}{x^2}\, dx$$

(b) Using Simpson's Rule (see Section 8.7) with $a = 1$, $b = 2$, $n = 10$, $\Delta x = 0.1$, and $f(x) = \sqrt{1 + 1/x^4}$, we have

$$L = \int_1^2 \sqrt{1 + \frac{1}{x^4}}\, dx$$

|||| Checking the value of the definite integral with a more accurate approximation produced by a computer algebra system, we see that the approximation using Simpson's Rule is accurate to four decimal places.

$$\approx \frac{\Delta x}{3}[f(1) + 4f(1.1) + 2f(1.2) + 4f(1.3) + \cdots + 2f(1.8) + 4f(1.9) + f(2)]$$

$$\approx 1.1321$$

The Arc Length Function

We will find it useful to have a function that measures the arc length of a curve from a particular starting point to any other point on the curve. Thus, if a smooth curve C has the equation $y = f(x)$, $a \leq x \leq b$, let $s(x)$ be the distance along C from the initial point $P_0(a, f(a))$ to the point $Q(x, f(x))$. Then s is a function, called the **arc length function**, and, by Formula 2,

$$\boxed{5} \qquad s(x) = \int_a^x \sqrt{1 + [f'(t)]^2}\, dt$$

(We have replaced the variable of integration by t so that x does not have two meanings.) We can use Part 1 of the Fundamental Theorem of Calculus to differentiate Equation 5 (since the integrand is continuous):

$$\boxed{6} \qquad \frac{ds}{dx} = \sqrt{1 + [f'(x)]^2} = \sqrt{1 + \left(\frac{dy}{dx}\right)^2}$$

Equation 6 shows that the rate of change of s with respect to x is always at least 1 and is equal to 1 when $f'(x)$, the slope of the curve, is 0. The differential of arc length is

$$\boxed{7} \qquad ds = \sqrt{1 + \left(\frac{dy}{dx}\right)^2}\, dx$$

and this equation is sometimes written in the symmetric form

$$\boxed{8} \qquad (ds)^2 = (dx)^2 + (dy)^2$$

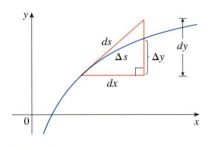

FIGURE 7

The geometric interpretation of Equation 8 is shown in Figure 7. It can be used as a mnemonic device for remembering both of the Formulas 3 and 4. If we write $L = \int ds$, then from Equation 8 either we can solve to get (7), which gives (3), or we can solve to get

$$ds = \sqrt{1 + \left(\frac{dx}{dy}\right)^2}\, dy$$

which gives (4).

EXAMPLE 4 Find the arc length function for the curve $y = x^2 - \frac{1}{8} \ln x$ taking $P_0(1, 1)$ as the starting point.

SOLUTION If $f(x) = x^2 - \frac{1}{8} \ln x$, then

$$f'(x) = 2x - \frac{1}{8x}$$

$$1 + [f'(x)]^2 = 1 + \left(2x - \frac{1}{8x}\right)^2 = 1 + 4x^2 - \frac{1}{2} + \frac{1}{64x^2}$$

$$= 4x^2 + \frac{1}{2} + \frac{1}{64x^2} = \left(2x + \frac{1}{8x}\right)^2$$

$$\sqrt{1 + [f'(x)]^2} = 2x + \frac{1}{8x}$$

Thus, the arc length function is given by

$$s(x) = \int_1^x \sqrt{1 + [f'(t)]^2}\, dt$$

$$= \int_1^x \left(2t - \frac{1}{8t}\right) dt = t^2 + \tfrac{1}{8}\ln t \Big]_1^x$$

$$= x^2 + \tfrac{1}{8}\ln x - 1$$

For instance, the arc length along the curve from $(1, 1)$ to $(3, f(3))$ is

$$s(3) = 3^2 + \tfrac{1}{8}\ln 3 - 1 = 8 + \frac{\ln 3}{8} \approx 8.1373$$

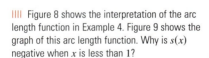

|||| Figure 8 shows the interpretation of the arc length function in Example 4. Figure 9 shows the graph of this arc length function. Why is $s(x)$ negative when x is less than 1?

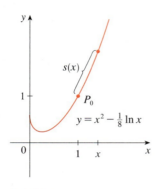

FIGURE 8

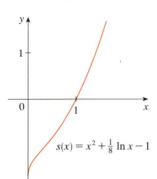

FIGURE 9

9.1 Exercises

1. Use the arc length formula (3) to find the length of the curve $y = 2 - 3x$, $-2 \leq x \leq 1$. Check your answer by noting that the curve is a line segment and calculating its length by the distance formula.

2. Use the arc length formula to find the length of the curve $y = \sqrt{4 - x^2}$, $0 \leq x \leq 2$. Check your answer by noting that the curve is a quarter-circle.

3–4 |||| Graph the curve and visually estimate its length. Then find its exact length.

3. $y = \tfrac{2}{3}(x^2 - 1)^{3/2}$, $1 \leq x \leq 3$

4. $y = \dfrac{x^3}{6} + \dfrac{1}{2x}$, $\tfrac{1}{2} \leq x \leq 1$

5–16 |||| Find the length of the curve.

5. $y = 1 + 6x^{3/2}$, $0 \leq x \leq 1$

6. $y^2 = 4(x + 4)^3$, $0 \leq x \leq 2$, $y > 0$

7. $y = \dfrac{x^5}{6} + \dfrac{1}{10x^3}$, $1 \leq x \leq 2$

8. $y = \dfrac{x^2}{2} - \dfrac{\ln x}{4}$, $2 \leq x \leq 4$

9. $x = \tfrac{1}{3}\sqrt{y}\,(y - 3)$, $1 \leq y \leq 9$

10. $y = \ln(\cos x)$, $0 \leq x \leq \pi/3$

11. $y = \ln(\sec x)$, $0 \leq x \leq \pi/4$

12. $y = \ln x$, $1 \leq x \leq \sqrt{3}$

13. $y = \cosh x$, $0 \leq x \leq 1$

14. $y^2 = 4x$, $0 \leq y \leq 2$

15. $y = e^x$, $0 \leq x \leq 1$

16. $y = \ln\left(\dfrac{e^x + 1}{e^x - 1}\right)$, $a \leq x \leq b$, $a > 0$

17–20 |||| Set up, but do not evaluate, an integral for the length of the curve.

17. $y = \cos x$, $0 \leq x \leq 2\pi$

18. $y = 2^x$, $0 \leq x \leq 3$

19. $x = y + y^3$, $1 \leq y \leq 4$ **20.** $\dfrac{x^2}{a^2} + \dfrac{y^2}{b^2} = 1$

21–24 |||| Use Simpson's rule with $n = 10$ to estimate the arc length of the curve. Compare your answer with the value of the integral produced by your calculator.

21. $y = xe^{-x}$, $0 \leq x \leq 5$ **22.** $x = y + \sqrt{y}$, $1 \leq y \leq 2$

23. $y = \sec x$, $0 \leq x \leq \pi/3$ **24.** $y = x \ln x$, $1 \leq x \leq 3$

25. (a) Graph the curve $y = x\sqrt[3]{4-x}$, $0 \leq x \leq 4$.
(b) Compute the lengths of inscribed polygons with $n = 1, 2$, and 4 sides. (Divide the interval into equal subintervals.) Illustrate by sketching these polygons (as in Figure 6).
(c) Set up an integral for the length of the curve.
(d) Use your calculator to find the length of the curve to four decimal places. Compare with the approximations in part (b).

26. Repeat Exercise 25 for the curve
$$y = x + \sin x \qquad 0 \leq x \leq 2\pi$$

CAS 27. Use either a computer algebra system or a table of integrals to find the *exact* length of the arc of the curve $x = \ln(1 - y^2)$ that lies between the points $(0, 0)$ and $\left(\ln \tfrac{3}{4}, \tfrac{1}{2}\right)$.

CAS 28. Use either a computer algebra system or a table of integrals to find the *exact* length of the arc of the curve $y = x^{4/3}$ that lies between the points $(0, 0)$ and $(1, 1)$. If your CAS has trouble evaluating the integral, make a substitution that changes the integral into one that the CAS can evaluate.

29. Sketch the curve with equation $x^{2/3} + y^{2/3} = 1$ and use symmetry to find its length.

30. (a) Sketch the curve $y^3 = x^2$.
(b) Use Formulas 3 and 4 to set up two integrals for the arc length from $(0, 0)$ to $(1, 1)$. Observe that one of these is an improper integral and evaluate both of them.
(c) Find the length of the arc of this curve from $(-1, 1)$ to $(8, 4)$.

31. Find the arc length function for the curve $y = 2x^{3/2}$ with starting point $P_0(1, 2)$.

32. (a) Graph the curve $y = \tfrac{1}{3}x^3 + 1/(4x)$, $x > 0$.
(b) Find the arc length function for this curve with starting point $P_0\left(1, \tfrac{7}{12}\right)$.
(c) Graph the arc length function.

33. A hawk flying at 15 m/s at an altitude of 180 m accidentally drops its prey. The parabolic trajectory of the falling prey is described by the equation
$$y = 180 - \dfrac{x^2}{45}$$
until it hits the ground, where y is its height above the ground and x is the horizontal distance traveled in meters. Calculate the distance traveled by the prey from the time it is dropped until the time it hits the ground. Express your answer correct to the nearest tenth of a meter.

34. A steady wind blows a kite due west. The kite's height above ground from horizontal position $x = 0$ to $x = 80$ ft is given by
$$y = 150 - \tfrac{1}{40}(x - 50)^2$$
Find the distance traveled by the kite.

35. A manufacturer of corrugated metal roofing wants to produce panels that are 28 in. wide and 2 in. thick by processing flat sheets of metal as shown in the figure. The profile of the roofing takes the shape of a sine wave. Verify that the sine curve has equation $y = \sin(\pi x/7)$ and find the width w of a flat metal sheet that is needed to make a 28-inch panel. (Use your calculator to evaluate the integral correct to four significant digits.)

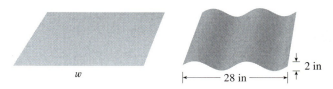

36. (a) The figure shows a telephone wire hanging between two poles at $x = -b$ and $x = b$. It takes the shape of a catenary with equation $y = c + a \cosh(x/a)$. Find the length of the wire.
(b) Suppose two telephone poles are 50 ft apart and the length of the wire between the poles is 51 ft. If the lowest point of the wire must be 20 ft above the ground, how high up on each pole should the wire be attached?

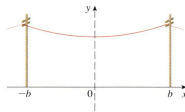

37. Find the length of the curve $y = \int_1^x \sqrt{t^3 - 1}\, dt$, $1 \leq x \leq 4$.

38. The curves with equations $x^n + y^n = 1$, $n = 4, 6, 8, \ldots$, are called **fat circles**. Graph the curves with $n = 2, 4, 6, 8$, and 10 to see why. Set up an integral for the length L_{2k} of the fat circle with $n = 2k$. Without attempting to evaluate this integral, state the value of
$$\lim_{k \to \infty} L_{2k}$$

DISCOVERY PROJECT

Arc Length Contest

The curves shown are all examples of graphs of continuous functions f that have the following properties.

1. $f(0) = 0$ and $f(1) = 0$
2. $f(x) \geq 0$ for $0 \leq x \leq 1$
3. The area under the graph of f from 0 to 1 is equal to 1.

The lengths L of these curves, however, are different.

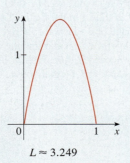
$L \approx 3.249$

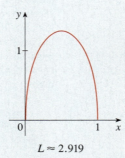
$L \approx 2.919$

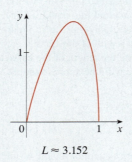
$L \approx 3.152$

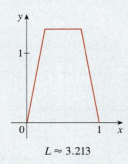
$L \approx 3.213$

Try to discover formulas for two functions that satisfy the given conditions 1, 2, and 3. (Your graphs might be similar to the ones shown or could look quite different.) Then calculate the arc length of each graph. The winning entry will be the one with the smallest arc length.

9.2 Area of a Surface of Revolution

A surface of revolution is formed when a curve is rotated about a line. Such a surface is the lateral boundary of a solid of revolution of the type discussed in Sections 6.2 and 6.3.

We want to define the area of a surface of revolution in such a way that it corresponds to our intuition. If the surface area is A, we can imagine that painting the surface would require the same amount of paint as does a flat region with area A.

Let's start with some simple surfaces. The lateral surface area of a circular cylinder with radius r and height h is taken to be $A = 2\pi rh$ because we can imagine cutting the cylinder and unrolling it (as in Figure 1) to obtain a rectangle with dimensions $2\pi r$ and h.

Likewise, we can take a circular cone with base radius r and slant height l, cut it along the dashed line in Figure 2, and flatten it to form a sector of a circle with radius l and central angle $\theta = 2\pi r/l$. We know that, in general, the area of a sector of a circle with radius l and angle θ is $\frac{1}{2}l^2\theta$ (see Exercise 35 in Section 8.3) and so in this case it is

$$A = \tfrac{1}{2}l^2\theta = \tfrac{1}{2}l^2\left(\frac{2\pi r}{l}\right) = \pi rl$$

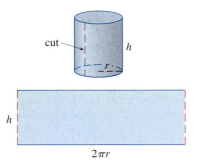

FIGURE 1

Therefore, we define the lateral surface area of a cone to be $A = \pi rl$.

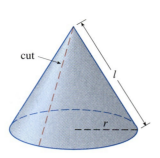
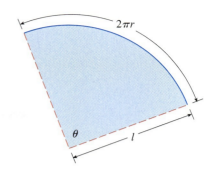

FIGURE 2

What about more complicated surfaces of revolution? If we follow the strategy we used with arc length, we can approximate the original curve by a polygon. When this polygon is rotated about an axis, it creates a simpler surface whose surface area approximates the actual surface area. By taking a limit, we can determine the exact surface area.

The approximating surface, then, consists of a number of *bands*, each formed by rotating a line segment about an axis. To find the surface area, each of these bands can be considered a portion of a circular cone, as shown in Figure 3. The area of the band (or frustum of a cone) with slant height l and upper and lower radii r_1 and r_2 is found by subtracting the areas of two cones:

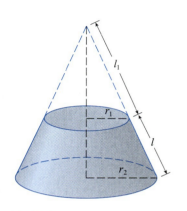

FIGURE 3

$$\boxed{1} \qquad A = \pi r_2(l_1 + l) - \pi r_1 l_1 = \pi[(r_2 - r_1)l_1 + r_2 l]$$

From similar triangles we have

$$\frac{l_1}{r_1} = \frac{l_1 + l}{r_2}$$

which gives

$$r_2 l_1 = r_1 l_1 + r_1 l \qquad \text{or} \qquad (r_2 - r_1)l_1 = r_1 l$$

Putting this in Equation 1, we get

$$A = \pi(r_1 l + r_2 l)$$

or

$$\boxed{2} \qquad \boxed{A = 2\pi r l}$$

where $r = \frac{1}{2}(r_1 + r_2)$ is the average radius of the band.

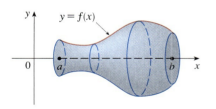

(a) Surface of revolution

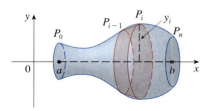

(b) Approximating band

FIGURE 4

Now we apply this formula to our strategy. Consider the surface shown in Figure 4, which is obtained by rotating the curve $y = f(x)$, $a \leq x \leq b$, about the x-axis, where f is positive and has a continuous derivative. In order to define its surface area, we divide the interval $[a, b]$ into n subintervals with endpoints $x_0, x_1, \ldots, x_n$ and equal width Δx, as we did in determining arc length. If $y_i = f(x_i)$, then the point $P_i(x_i, y_i)$ lies on the curve. The part of the surface between x_{i-1} and x_i is approximated by taking the line segment $P_{i-1}P_i$ and rotating it about the x-axis. The result is a band with slant height $l = |P_{i-1}P_i|$ and average radius $r = \frac{1}{2}(y_{i-1} + y_i)$ so, by Formula 2, its surface area is

$$2\pi \frac{y_{i-1} + y_i}{2} |P_{i-1}P_i|$$

As in the proof of Theorem 9.1.2, we have

$$|P_{i-1}P_i| = \sqrt{1 + [f'(x_i^*)]^2}\,\Delta x$$

where x_i^* is some number in $[x_{i-1}, x_i]$. When Δx is small, we have $y_i = f(x_i) \approx f(x_i^*)$ and also $y_{i-1} = f(x_{i-1}) \approx f(x_i^*)$, since f is continuous. Therefore

$$2\pi \frac{y_{i-1} + y_i}{2}|P_{i-1}P_i| \approx 2\pi f(x_i^*)\sqrt{1 + [f'(x_i^*)]^2}\,\Delta x$$

and so an approximation to what we think of as the area of the complete surface of revolution is

3
$$\sum_{i=1}^{n} 2\pi f(x_i^*)\sqrt{1 + [f'(x_i^*)]^2}\,\Delta x$$

This approximation appears to become better as $n \to \infty$ and, recognizing (3) as a Riemann sum for the function $g(x) = 2\pi f(x)\sqrt{1 + [f'(x)]^2}$, we have

$$\lim_{n \to \infty} \sum_{i=1}^{n} 2\pi f(x_i^*)\sqrt{1 + [f'(x_i^*)]^2}\,\Delta x = \int_a^b 2\pi f(x)\sqrt{1 + [f'(x)]^2}\,dx$$

Therefore, in the case where f is positive and has a continuous derivative, we define the **surface area** of the surface obtained by rotating the curve $y = f(x)$, $a \leq x \leq b$, about the x-axis as

4
$$S = \int_a^b 2\pi f(x)\sqrt{1 + [f'(x)]^2}\,dx$$

With the Leibniz notation for derivatives, this formula becomes

5
$$S = \int_a^b 2\pi y\sqrt{1 + \left(\frac{dy}{dx}\right)^2}\,dx$$

If the curve is described as $x = g(y)$, $c \leq y \leq d$, then the formula for surface area becomes

6
$$S = \int_c^d 2\pi y\sqrt{1 + \left(\frac{dx}{dy}\right)^2}\,dy$$

and both Formulas 5 and 6 can be summarized symbolically, using the notation for arc length given in Section 9.1, as

7
$$S = \int 2\pi y\,ds$$

For rotation about the y-axis, the surface area formula becomes

$$\boxed{S = \int 2\pi x \, ds} \qquad \boxed{8}$$

where, as before, we can use either

$$ds = \sqrt{1 + \left(\frac{dy}{dx}\right)^2} \, dx \quad \text{or} \quad ds = \sqrt{1 + \left(\frac{dx}{dy}\right)^2} \, dy$$

These formulas can be remembered by thinking of $2\pi y$ or $2\pi x$ as the circumference of a circle traced out by the point (x, y) on the curve as it is rotated about the x-axis or y-axis, respectively (see Figure 5).

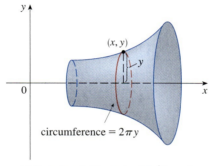

FIGURE 5   (a) Rotation about x-axis: $S = \int 2\pi y \, ds$ (b) Rotation about y-axis: $S = \int 2\pi x \, ds$

EXAMPLE 1 The curve $y = \sqrt{4 - x^2}$, $-1 \leq x \leq 1$, is an arc of the circle $x^2 + y^2 = 4$. Find the area of the surface obtained by rotating this arc about the x-axis. (The surface is a portion of a sphere of radius 2. See Figure 6.)

SOLUTION We have

$$\frac{dy}{dx} = \tfrac{1}{2}(4 - x^2)^{-1/2}(-2x) = \frac{-x}{\sqrt{4 - x^2}}$$

and so, by Formula 5, the surface area is

$$S = \int_{-1}^{1} 2\pi y \sqrt{1 + \left(\frac{dy}{dx}\right)^2} \, dx$$

$$= 2\pi \int_{-1}^{1} \sqrt{4 - x^2} \sqrt{1 + \frac{x^2}{4 - x^2}} \, dx$$

$$= 2\pi \int_{-1}^{1} \sqrt{4 - x^2} \, \frac{2}{\sqrt{4 - x^2}} \, dx$$

$$= 4\pi \int_{-1}^{1} 1 \, dx = 4\pi(2) = 8\pi$$

FIGURE 6

Figure 6 shows the portion of the sphere whose surface area is computed in Example 1.

||| Figure 7 shows the surface of revolution whose area is computed in Example 2.

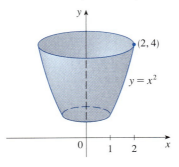

FIGURE 7

EXAMPLE 2 The arc of the parabola $y = x^2$ from $(1, 1)$ to $(2, 4)$ is rotated about the y-axis. Find the area of the resulting surface.

SOLUTION 1 Using

$$y = x^2 \quad \text{and} \quad \frac{dy}{dx} = 2x$$

we have, from Formula 8,

$$S = \int 2\pi x \, ds$$

$$= \int_1^2 2\pi x \sqrt{1 + \left(\frac{dy}{dx}\right)^2} \, dx$$

$$= 2\pi \int_1^2 x \sqrt{1 + 4x^2} \, dx$$

Substituting $u = 1 + 4x^2$, we have $du = 8x \, dx$. Remembering to change the limits of integration, we have

$$S = \frac{\pi}{4} \int_5^{17} \sqrt{u} \, du = \frac{\pi}{4} \left[\tfrac{2}{3} u^{3/2}\right]_5^{17}$$

$$= \frac{\pi}{6} \left(17\sqrt{17} - 5\sqrt{5}\right)$$

SOLUTION 2 Using

$$x = \sqrt{y} \quad \text{and} \quad \frac{dx}{dy} = \frac{1}{2\sqrt{y}}$$

||| As a check on our answer to Example 2, notice from Figure 7 that the surface area should be close to that of a circular cylinder with the same height and radius halfway between the upper and lower radius of the surface: $2\pi(1.5)(3) \approx 28.27$. We computed that the surface area was

$$\frac{\pi}{6}\left(17\sqrt{17} - 5\sqrt{5}\right) \approx 30.85$$

which seems reasonable. Alternatively, the surface area should be slightly larger than the area of a frustum of a cone with the same top and bottom edges. From Equation 2, this is $2\pi(1.5)(\sqrt{10}) \approx 29.80$.

we have

$$S = \int 2\pi x \, ds = \int_1^4 2\pi x \sqrt{1 + \left(\frac{dx}{dy}\right)^2} \, dy$$

$$= 2\pi \int_1^4 \sqrt{y} \sqrt{1 + \frac{1}{4y}} \, dy = \pi \int_1^4 \sqrt{4y + 1} \, dy$$

$$= \frac{\pi}{4} \int_5^{17} \sqrt{u} \, du \qquad \text{(where } u = 1 + 4y\text{)}$$

$$= \frac{\pi}{6}\left(17\sqrt{17} - 5\sqrt{5}\right) \qquad \text{(as in Solution 1)}$$

EXAMPLE 3 Find the area of the surface generated by rotating the curve $y = e^x$, $0 \le x \le 1$, about the x-axis.

||| Another method: Use Formula 6 with $x = \ln y$.

SOLUTION Using Formula 5 with

$$y = e^x \quad \text{and} \quad \frac{dy}{dx} = e^x$$

we have

$$S = \int_0^1 2\pi y \sqrt{1 + \left(\frac{dy}{dx}\right)^2}\, dx = 2\pi \int_0^1 e^x \sqrt{1 + e^{2x}}\, dx$$

$$= 2\pi \int_1^e \sqrt{1 + u^2}\, du \qquad \text{(where } u = e^x\text{)}$$

$$= 2\pi \int_{\pi/4}^{\alpha} \sec^3\theta\, d\theta \qquad \text{(where } u = \tan\theta \text{ and } \alpha = \tan^{-1}e\text{)}$$

|||| Or use Formula 21 in the Table of Integrals.

$$= 2\pi \cdot \tfrac{1}{2}\bigl[\sec\theta \tan\theta + \ln|\sec\theta + \tan\theta|\bigr]_{\pi/4}^{\alpha} \qquad \text{(by Example 8 in Section 8.2)}$$

$$= \pi\bigl[\sec\alpha \tan\alpha + \ln(\sec\alpha + \tan\alpha) - \sqrt{2} - \ln(\sqrt{2} + 1)\bigr]$$

Since $\tan\alpha = e$, we have $\sec^2\alpha = 1 + \tan^2\alpha = 1 + e^2$ and

$$S = \pi\bigl[e\sqrt{1 + e^2} + \ln(e + \sqrt{1 + e^2}) - \sqrt{2} - \ln(\sqrt{2} + 1)\bigr]$$

9.2 Exercises

1–4 |||| Set up, but do not evaluate, an integral for the area of the surface obtained by rotating the curve about the given axis.

1. $y = \ln x,\ 1 \le x \le 3;\ x$-axis

2. $y = \sin^2 x,\ 0 \le x \le \pi/2;\ x$-axis

3. $y = \sec x,\ 0 \le x \le \pi/4;\ y$-axis

4. $y = e^x,\ 1 \le y \le 2;\ y$-axis

5–12 |||| Find the area of the surface obtained by rotating the curve about the x-axis.

5. $y = x^3,\ 0 \le x \le 2$

6. $9x = y^2 + 18,\ 2 \le x \le 6$

7. $y = \sqrt{x},\ 4 \le x \le 9$

8. $y = \cos 2x,\ 0 \le x \le \pi/6$

9. $y = \cosh x,\ 0 \le x \le 1$

10. $y = \dfrac{x^3}{6} + \dfrac{1}{2x},\ \tfrac{1}{2} \le x \le 1$

11. $x = \tfrac{1}{3}(y^2 + 2)^{3/2},\ 1 \le y \le 2$

12. $x = 1 + 2y^2,\ 1 \le y \le 2$

13–16 |||| The given curve is rotated about the y-axis. Find the area of the resulting surface.

13. $y = \sqrt[3]{x},\ 1 \le y \le 2$

14. $y = 1 - x^2,\ 0 \le x \le 1$

15. $x = \sqrt{a^2 - y^2},\ 0 \le y \le a/2$

16. $x = a\cosh(y/a),\ -a \le y \le a$

17–20 |||| Use Simpson's Rule with $n = 10$ to approximate the area of the surface obtained by rotating the curve about the x-axis. Compare your answer with the value of the integral produced by your calculator.

17. $y = \ln x,\ 1 \le x \le 3$

18. $y = x + \sqrt{x},\ 1 \le x \le 2$

19. $y = \sec x,\ 0 \le x \le \pi/3$

20. $y = \sqrt{1 + e^x},\ 0 \le x \le 1$

[CAS] **21–22** |||| Use either a CAS or a table of integrals to find the exact area of the surface obtained by rotating the given curve about the x-axis.

21. $y = 1/x,\ 1 \le x \le 2$

22. $y = \sqrt{x^2 + 1},\ 0 \le x \le 3$

[CAS] **23–24** |||| Use a CAS to find the exact area of the surface obtained by rotating the curve about the y-axis. If your CAS has trouble evaluating the integral, express the surface area as an integral in the other variable.

23. $y = x^3,\ 0 \le y \le 1$

24. $y = \ln(x + 1),\ 0 \le x \le 1$

25. If the region $\mathcal{R} = \{(x, y)\,|\,x \ge 1,\ 0 \le y \le 1/x\}$ is rotated about the x-axis, the volume of the resulting solid is finite (see Exercise 63 in Section 8.8). Show that the surface area is infinite. (The surface is shown in the figure and is known as **Gabriel's horn**.)

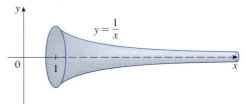

26. If the infinite curve $y = e^{-x}$, $x \geq 0$, is rotated about the x-axis, find the area of the resulting surface.

27. (a) If $a > 0$, find the area of the surface generated by rotating the loop of the curve $3ay^2 = x(a - x)^2$ about the x-axis.
(b) Find the surface area if the loop is rotated about the y-axis.

28. A group of engineers is building a parabolic satellite dish whose shape will be formed by rotating the curve $y = ax^2$ about the y-axis. If the dish is to have a 10-ft diameter and a maximum depth of 2 ft, find the value of a and the surface area of the dish.

29. The ellipse
$$\frac{x^2}{a^2} + \frac{y^2}{b^2} = 1 \qquad a > b$$
is rotated about the x-axis to form a surface called an *ellipsoid*. Find the surface area of this ellipsoid.

30. Find the surface area of the torus in Exercise 61 in Section 6.2.

31. If the curve $y = f(x)$, $a \leq x \leq b$, is rotated about the horizontal line $y = c$, where $f(x) \leq c$, find a formula for the area of the resulting surface.

[CAS] **32.** Use the result of Exercise 31 to set up an integral to find the area of the surface generated by rotating the curve $y = \sqrt{x}$, $0 \leq x \leq 4$, about the line $y = 4$. Then use a CAS to evaluate the integral.

33. Find the area of the surface obtained by rotating the circle $x^2 + y^2 = r^2$ about the line $y = r$.

34. Show that the surface area of a zone of a sphere that lies between two parallel planes is $S = \pi dh$, where d is the diameter of the sphere and h is the distance between the planes. (Notice that S depends only on the distance between the planes and not on their location, provided that both planes intersect the sphere.)

35. Formula 4 is valid only when $f(x) \geq 0$. Show that when $f(x)$ is not necessarily positive, the formula for surface area becomes
$$S = \int_a^b 2\pi |f(x)| \sqrt{1 + [f'(x)]^2}\, dx$$

36. Let L be the length of the curve $y = f(x)$, $a \leq x \leq b$, where f is positive and has a continuous derivative. Let S_f be the surface area generated by rotating the curve about the x-axis. If c is a positive constant, define $g(x) = f(x) + c$ and let S_g be the corresponding surface area generated by the curve $y = g(x)$, $a \leq x \leq b$. Express S_g in terms of S_f and L.

DISCOVERY PROJECT

Rotating on a Slant

We know how to find the volume of a solid of revolution obtained by rotating a region about a horizontal or vertical line (see Section 6.2). We also know how to find the surface area of a surface of revolution if we rotate a curve about a horizontal or vertical line (see Section 9.2). But what if we rotate about a slanted line, that is, a line that is neither horizontal nor vertical? In this project you are asked to discover formulas for the volume of a solid of revolution and for the area of a surface of revolution when the axis of rotation is a slanted line.

Let C be the arc of the curve $y = f(x)$ between the points $P(p, f(p))$ and $Q(q, f(q))$ and let $\mathcal{R}$ be the region bounded by C, by the line $y = mx + b$ (which lies entirely below C), and by the perpendiculars to the line from P and Q.

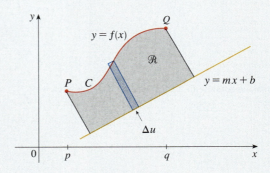

1. Show that the area of $\mathcal{R}$ is

$$\frac{1}{1+m^2}\int_p^q [f(x) - mx - b][1 + mf'(x)]\, dx$$

[*Hint:* This formula can be verified by subtracting areas, but it will be helpful throughout the project to derive it by first approximating the area using rectangles perpendicular to the line, as shown in the figure. Use the figure to help express Δu in terms of Δx.]

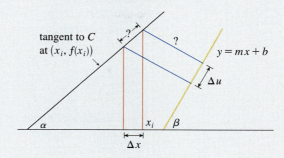

2. Find the area of the region shown in the figure at the left.

3. Find a formula similar to the one in Problem 1 for the volume of the solid obtained by rotating $\mathcal{R}$ about the line $y = mx + b$.

4. Find the volume of the solid obtained by rotating the region of Problem 2 about the line $y = x - 2$.

5. Find a formula for the area of the surface obtained by rotating C about the line $y = mx + b$.

6. Use a computer algebra system to find the exact area of the surface obtained by rotating the curve $y = \sqrt{x}$, $0 \le x \le 4$, about the line $y = \frac{1}{2}x$. Then approximate your result to three decimal places.

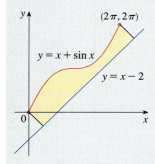

FIGURE FOR PROBLEM 2

9.3 Applications to Physics and Engineering

Among the many applications of integral calculus to physics and engineering, we consider two here: force due to water pressure and centers of mass. As with our previous applications to geometry (areas, volumes, and lengths) and to work, our strategy is to break up the physical quantity into a large number of small parts, approximate each small part, add the results, take the limit, and then evaluate the resulting integral.

Hydrostatic Pressure and Force

Deep-sea divers realize that water pressure increases as they dive deeper. This is because the weight of the water above them increases.

In general, suppose that a thin horizontal plate with area A square meters is submerged in a fluid of density ρ kilograms per cubic meter at a depth d meters below the surface of the fluid as in Figure 1. The fluid directly above the plate has volume $V = Ad$, so its mass is $m = \rho V = \rho A d$. The force exerted by the fluid on the plate is therefore

$$F = mg = \rho g A d$$

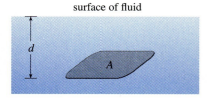

FIGURE 1

where g is the acceleration due to gravity. The **pressure** P on the plate is defined to be the force per unit area:

$$P = \frac{F}{A} = \rho g d$$

|||| When using U.S. Customary units, we write $P = \rho g d = \delta d$, where $\delta = \rho g$ is the weight density (as opposed to ρ, which is the mass density). For instance, the weight density of water is $\delta = 62.5$ lb/ft^3.

The SI unit for measuring pressure is newtons per square meter, which is called a pascal (abbreviation: 1 N/m^2 = 1 Pa). Since this is a small unit, the kilopascal (kPa) is often used. For instance, because the density of water is $\rho = 1000$ kg/m^3, the pressure at the bottom of a swimming pool 2 m deep is

$$P = \rho g d = 1000 \text{ kg/m}^3 \times 9.8 \text{ m/s}^2 \times 2 \text{ m}$$

$$= 19{,}600 \text{ Pa} = 19.6 \text{ kPa}$$

An important principle of fluid pressure is the experimentally verified fact that at *any point in a liquid the pressure is the same in all directions.* (A diver feels the same pressure on nose and both ears.) Thus, the pressure in *any* direction at a depth d in a fluid with mass density ρ is given by

$$\boxed{1} \qquad P = \rho g d = \delta d$$

This helps us determine the hydrostatic force against a *vertical* plate or wall or dam in a fluid. This is not a straightforward problem because the pressure is not constant but increases as the depth increases.

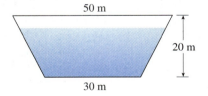

FIGURE 2

EXAMPLE 1 A dam has the shape of the trapezoid shown in Figure 2. The height is 20 m, and the width is 50 m at the top and 30 m at the bottom. Find the force on the dam due to hydrostatic pressure if the water level is 4 m from the top of the dam.

SOLUTION We choose a vertical x-axis with origin at the surface of the water as in Figure 3(a). The depth of the water is 16 m, so we divide the interval $[0, 16]$ into subintervals of equal length with endpoints x_i and we choose $x_i^* \in [x_{i-1}, x_i]$. The ith horizontal strip of the dam is approximated by a rectangle with height Δx and width w_i, where, from similar triangles in Figure 3(b),

$$\frac{a}{16 - x_i^*} = \frac{10}{20} \quad \text{or} \quad a = \frac{16 - x_i^*}{2} = 8 - \frac{x_i^*}{2}$$

and so
$$w_i = 2(15 + a) = 2\left(15 + 8 - \tfrac{1}{2}x_i^*\right) = 46 - x_i^*$$

If A_i is the area of the ith strip, then

$$A_i \approx w_i \, \Delta x = (46 - x_i^*) \, \Delta x$$

If Δx is small, then the pressure P_i on the ith strip is almost constant and we can use Equation 1 to write

$$P_i \approx 1000 g x_i^*$$

The hydrostatic force F_i acting on the ith strip is the product of the pressure and the area:

$$F_i = P_i A_i \approx 1000 g x_i^* (46 - x_i^*) \, \Delta x$$

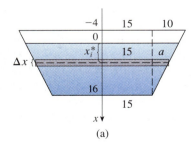

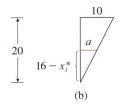

FIGURE 3

Adding these forces and taking the limit as $n \to \infty$, we obtain the total hydrostatic force on the dam:

$$F = \lim_{n \to \infty} \sum_{i=1}^{n} 1000 g x_i^* (46 - x_i^*) \Delta x$$

$$= \int_0^{16} 1000 g x (46 - x)\, dx$$

$$= 1000(9.8) \int_0^{16} (46x - x^2)\, dx$$

$$= 9800 \left[23x^2 - \frac{x^3}{3} \right]_0^{16}$$

$$\approx 4.43 \times 10^7 \text{ N}$$

EXAMPLE 2 Find the hydrostatic force on one end of a cylindrical drum with radius 3 ft if the drum is submerged in water 10 ft deep.

SOLUTION In this example it is convenient to choose the axes as in Figure 4 so that the origin is placed at the center of the drum. Then the circle has a simple equation, $x^2 + y^2 = 9$. As in Example 1 we divide the circular region into horizontal strips of equal width. From the equation of the circle, we see that the length of the ith strip is $2\sqrt{9 - (y_i^*)^2}$ and so its area is

$$A_i = 2\sqrt{9 - (y_i^*)^2}\, \Delta y$$

The pressure on this strip is approximately

$$\delta d_i = 62.5(7 - y_i^*)$$

and so the force on the strip is approximately

$$\delta d_i A_i = 62.5(7 - y_i^*) 2\sqrt{9 - (y_i^*)^2}\, \Delta y$$

The total force is obtained by adding the forces on all the strips and taking the limit:

$$F = \lim_{n \to \infty} \sum_{i=1}^{n} 62.5(7 - y_i^*) 2\sqrt{9 - (y_i^*)^2}\, \Delta y$$

$$= 125 \int_{-3}^{3} (7 - y)\sqrt{9 - y^2}\, dy$$

$$= 125 \cdot 7 \int_{-3}^{3} \sqrt{9 - y^2}\, dy - 125 \int_{-3}^{3} y\sqrt{9 - y^2}\, dy$$

The second integral is 0 because the integrand is an odd function (see Theorem 5.5.6). The first integral can be evaluated using the trigonometric substitution $y = 3 \sin \theta$, but it's simpler to observe that it is the area of a semicircular disk with radius 3. Thus

$$F = 875 \int_{-3}^{3} \sqrt{9 - y^2}\, dy = 875 \cdot \tfrac{1}{2} \pi (3)^2$$

$$= \frac{7875\pi}{2} \approx 12{,}370 \text{ lb}$$

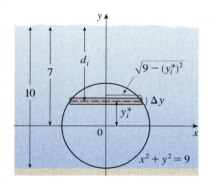

FIGURE 4

Moments and Centers of Mass

Our main objective here is to find the point P on which a thin plate of any given shape balances horizontally as in Figure 5. This point is called the **center of mass** (or center of gravity) of the plate.

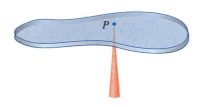

FIGURE 5

We first consider the simpler situation illustrated in Figure 6, where two masses m_1 and m_2 are attached to a rod of negligible mass on opposite sides of a fulcrum and at distances d_1 and d_2 from the fulcrum. The rod will balance if

$$\boxed{2} \qquad m_1 d_1 = m_2 d_2$$

This is an experimental fact discovered by Archimedes and called the Law of the Lever. (Think of a lighter person balancing a heavier one on a seesaw by sitting farther away from the center.)

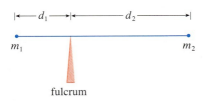

FIGURE 6

Now suppose that the rod lies along the x-axis with m_1 at x_1 and m_2 at x_2 and the center of mass at $\bar{x}$. If we compare Figures 6 and 7, we see that $d_1 = \bar{x} - x_1$ and $d_2 = x_2 - \bar{x}$ and so Equation 2 gives

$$m_1(\bar{x} - x_1) = m_2(x_2 - \bar{x})$$

$$m_1 \bar{x} + m_2 \bar{x} = m_1 x_1 + m_2 x_2$$

$$\boxed{3} \qquad \bar{x} = \frac{m_1 x_1 + m_2 x_2}{m_1 + m_2}$$

The numbers $m_1 x_1$ and $m_2 x_2$ are called the **moments** of the masses m_1 and m_2 (with respect to the origin), and Equation 3 says that the center of mass $\bar{x}$ is obtained by adding the moments of the masses and dividing by the total mass $m = m_1 + m_2$.

FIGURE 7

In general, if we have a system of n particles with masses $m_1, m_2, \ldots, m_n$ located at the points $x_1, x_2, \ldots, x_n$ on the x-axis, it can be shown similarly that the center of mass of the system is located at

$$\boxed{4} \qquad \bar{x} = \frac{\sum_{i=1}^{n} m_i x_i}{\sum_{i=1}^{n} m_i} = \frac{\sum_{i=1}^{n} m_i x_i}{m}$$

where $m = \sum m_i$ is the total mass of the system, and the sum of the individual moments

$$M = \sum_{i=1}^{n} m_i x_i$$

is called the **moment of the system about the origin.** Then Equation 4 could be rewritten as $m\bar{x} = M$, which says that if the total mass were considered as being concentrated at the center of mass $\bar{x}$, then its moment would be the same as the moment of the system.

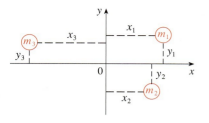

FIGURE 8

Now we consider a system of n particles with masses $m_1, m_2, \ldots, m_n$ located at the points $(x_1, y_1), (x_2, y_2), \ldots, (x_n, y_n)$ in the xy-plane as shown in Figure 8. By analogy with the one-dimensional case, we define the **moment of the system about the y-axis** to be

$$\boxed{5} \qquad M_y = \sum_{i=1}^{n} m_i x_i$$

and the **moment of the system about the x-axis** as

$$\boxed{6} \qquad M_x = \sum_{i=1}^{n} m_i y_i$$

Then M_y measures the tendency of the system to rotate about the y-axis and M_x measures the tendency to rotate about the x-axis.

As in the one-dimensional case, the coordinates $(\bar{x}, \bar{y})$ of the center of mass are given in terms of the moments by the formulas

$$\boxed{7} \qquad \bar{x} = \frac{M_y}{m} \qquad \bar{y} = \frac{M_x}{m}$$

where $m = \sum m_i$ is the total mass. Since $m\bar{x} = M_y$ and $m\bar{y} = M_x$, the center of mass $(\bar{x}, \bar{y})$ is the point where a single particle of mass m would have the same moments as the system.

EXAMPLE 3 Find the moments and center of mass of the system of objects that have masses 3, 4, and 8 at the points $(-1, 1)$, $(2, -1)$, and $(3, 2)$.

SOLUTION We use Equations 5 and 6 to compute the moments:

$$M_y = 3(-1) + 4(2) + 8(3) = 29$$

$$M_x = 3(1) + 4(-1) + 8(2) = 15$$

Since $m = 3 + 4 + 8 = 15$, we use Equations 7 to obtain

$$\bar{x} = \frac{M_y}{m} = \frac{29}{15} \qquad \bar{y} = \frac{M_x}{m} = \frac{15}{15} = 1$$

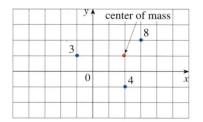

FIGURE 9

Thus, the center of mass is $\left(1\frac{14}{15}, 1\right)$. (See Figure 9.)

Next we consider a flat plate (called a *lamina*) with uniform density ρ that occupies a region $\mathcal{R}$ of the plane. We wish to locate the center of mass of the plate, which is called the **centroid** of $\mathcal{R}$. In doing so we use the following physical principles: The **symmetry principle** says that if $\mathcal{R}$ is symmetric about a line l, then the centroid of $\mathcal{R}$ lies on l. (If $\mathcal{R}$ is reflected about l, then $\mathcal{R}$ remains the same so its centroid remains fixed. But the only fixed points lie on l.) Thus, the centroid of a rectangle is its center. Moments should be defined so that if the entire mass of a region is concentrated at the center of mass, then its moments remain unchanged. Also, the moment of the union of two nonoverlapping regions should be the sum of the moments of the individual regions.

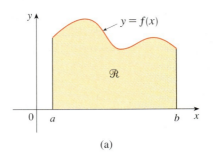

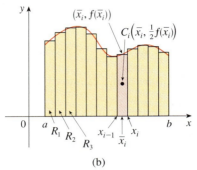

FIGURE 10

Suppose that the region $\mathcal{R}$ is of the type shown in Figure 10(a); that is, $\mathcal{R}$ lies between the lines $x = a$ and $x = b$, above the x-axis, and beneath the graph of f, where f is a continuous function. We divide the interval $[a, b]$ into n subintervals with endpoints $x_0, x_1, \ldots, x_n$ and equal width Δx. We choose the sample point x_i^* to be the midpoint $\bar{x}_i$ of the ith subinterval, that is, $\bar{x}_i = (x_{i-1} + x_i)/2$. This determines the polygonal approximation to $\mathcal{R}$ shown in Figure 10(b). The centroid of the ith approximating rectangle R_i is its center $C_i(\bar{x}_i, \frac{1}{2}f(\bar{x}_i))$. Its area is $f(\bar{x}_i)\,\Delta x$, so its mass is

$$\rho f(\bar{x}_i)\,\Delta x$$

The moment of R_i about the y-axis is the product of its mass and the distance from C_i to the y-axis, which is $\bar{x}_i$. Thus

$$M_y(R_i) = [\rho f(\bar{x}_i)\,\Delta x]\,\bar{x}_i = \rho \bar{x}_i f(\bar{x}_i)\,\Delta x$$

Adding these moments, we obtain the moment of the polygonal approximation to $\mathcal{R}$, and then by taking the limit as $n \to \infty$ we obtain the moment of $\mathcal{R}$ itself about the y-axis:

$$M_y = \lim_{n \to \infty} \sum_{i=1}^{n} \rho \bar{x}_i f(\bar{x}_i)\,\Delta x = \rho \int_a^b x f(x)\,dx$$

In a similar fashion we compute the moment of R_i about the x-axis as the product of its mass and the distance from C_i to the x-axis:

$$M_x(R_i) = [\rho f(\bar{x}_i)\,\Delta x]\tfrac{1}{2}f(\bar{x}_i) = \rho \cdot \tfrac{1}{2}[f(\bar{x}_i)]^2\,\Delta x$$

Again we add these moments and take the limit to obtain the moment of $\mathcal{R}$ about the x-axis:

$$M_x = \lim_{n \to \infty} \sum_{i=1}^{n} \rho \cdot \tfrac{1}{2}[f(\bar{x}_i)]^2\,\Delta x = \rho \int_a^b \tfrac{1}{2}[f(x)]^2\,dx$$

Just as for systems of particles, the center of mass of the plate is defined so that $m\bar{x} = M_y$ and $m\bar{y} = M_x$. But the mass of the plate is the product of its density and its area:

$$m = \rho A = \rho \int_a^b f(x)\,dx$$

and so

$$\bar{x} = \frac{M_y}{m} = \frac{\rho \int_a^b x f(x)\,dx}{\rho \int_a^b f(x)\,dx} = \frac{\int_a^b x f(x)\,dx}{\int_a^b f(x)\,dx}$$

$$\bar{y} = \frac{M_x}{m} = \frac{\rho \int_a^b \tfrac{1}{2}[f(x)]^2\,dx}{\rho \int_a^b f(x)\,dx} = \frac{\int_a^b \tfrac{1}{2}[f(x)]^2\,dx}{\int_a^b f(x)\,dx}$$

Notice the cancellation of the ρ's. The location of the center of mass is independent of the density.

In summary, the center of mass of the plate (or the centroid of $\mathcal{R}$) is located at the point $(\bar{x}, \bar{y})$, where

$$\boxed{8} \quad \bar{x} = \frac{1}{A} \int_a^b x f(x)\, dx \qquad \bar{y} = \frac{1}{A} \int_a^b \tfrac{1}{2} [f(x)]^2\, dx$$

EXAMPLE 4 Find the center of mass of a semicircular plate of radius r.

SOLUTION In order to use (8) we place the semicircle as in Figure 11 so that $f(x) = \sqrt{r^2 - x^2}$ and $a = -r$, $b = r$. Here there is no need to use the formula to calculate $\bar{x}$ because, by the symmetry principle, the center of mass must lie on the y-axis, so $\bar{x} = 0$. The area of the semicircle is $A = \pi r^2/2$, so

$$\bar{y} = \frac{1}{A} \int_{-r}^r \tfrac{1}{2} [f(x)]^2\, dx$$

$$= \frac{1}{\pi r^2/2} \cdot \tfrac{1}{2} \int_{-r}^r \left(\sqrt{r^2 - x^2}\right)^2 dx$$

$$= \frac{2}{\pi r^2} \int_0^r (r^2 - x^2)\, dx = \frac{2}{\pi r^2} \left[r^2 x - \frac{x^3}{3} \right]_0^r$$

$$= \frac{2}{\pi r^2} \cdot \frac{2r^3}{3} = \frac{4r}{3\pi}$$

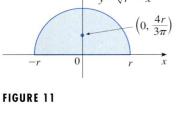

FIGURE 11

The center of mass is located at the point $(0, 4r/(3\pi))$.

EXAMPLE 5 Find the centroid of the region bounded by the curves $y = \cos x$, $y = 0$, $x = 0$, and $x = \pi/2$.

SOLUTION The area of the region is

$$A = \int_0^{\pi/2} \cos x\, dx = \sin x \Big]_0^{\pi/2} = 1$$

so Formulas 8 give

$$\bar{x} = \frac{1}{A} \int_0^{\pi/2} x f(x)\, dx = \int_0^{\pi/2} x \cos x\, dx$$

$$= x \sin x \Big]_0^{\pi/2} - \int_0^{\pi/2} \sin x\, dx \qquad \text{(by integration by parts)}$$

$$= \frac{\pi}{2} - 1$$

$$\bar{y} = \frac{1}{A} \int_0^{\pi/2} \tfrac{1}{2} [f(x)]^2\, dx = \tfrac{1}{2} \int_0^{\pi/2} \cos^2 x\, dx$$

$$= \tfrac{1}{4} \int_0^{\pi/2} (1 + \cos 2x)\, dx = \tfrac{1}{4} \left[x + \tfrac{1}{2} \sin 2x \right]_0^{\pi/2}$$

$$= \frac{\pi}{8}$$

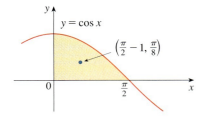

FIGURE 12

The centroid is $((\pi/2) - 1, \pi/8)$ and is shown in Figure 12.

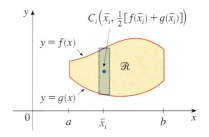

FIGURE 13

If the region $\mathcal{R}$ lies between two curves $y = f(x)$ and $y = g(x)$, where $f(x) \geq g(x)$, as illustrated in Figure 13, then the same sort of argument that led to Formulas 8 can be used to show that the centroid of $\mathcal{R}$ is $(\bar{x}, \bar{y})$, where

$$\boxed{9} \quad \bar{x} = \frac{1}{A} \int_a^b x[f(x) - g(x)] \, dx \qquad \bar{y} = \frac{1}{A} \int_a^b \tfrac{1}{2}\{[f(x)]^2 - [g(x)]^2\} \, dx$$

(See Exercise 43.)

EXAMPLE 6 Find the centroid of the region bounded by the line $y = x$ and the parabola $y = x^2$.

SOLUTION The region is sketched in Figure 14. We take $f(x) = x$, $g(x) = x^2$, $a = 0$, and $b = 1$ in Formulas 9. First we note that the area of the region is

$$A = \int_0^1 (x - x^2) \, dx = \frac{x^2}{2} - \frac{x^3}{3} \bigg]_0^1 = \frac{1}{6}$$

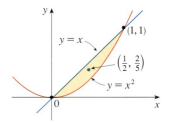

FIGURE 14

Therefore

$$\bar{x} = \frac{1}{A} \int_0^1 x[f(x) - g(x)] \, dx = \frac{1}{\frac{1}{6}} \int_0^1 x(x - x^2) \, dx$$

$$= 6 \int_0^1 (x^2 - x^3) \, dx = 6 \left[\frac{x^3}{3} - \frac{x^4}{4} \right]_0^1 = \frac{1}{2}$$

$$\bar{y} = \frac{1}{A} \int_0^1 \tfrac{1}{2}\{[f(x)]^2 - [g(x)]^2\} \, dx = \frac{1}{\frac{1}{6}} \int_0^1 \tfrac{1}{2}(x^2 - x^4) \, dx$$

$$= 3 \left[\frac{x^3}{3} - \frac{x^5}{5} \right]_0^1 = \frac{2}{5}$$

The centroid is $\left(\frac{1}{2}, \frac{2}{5}\right)$.

We end this section by showing a surprising connection between centroids and volumes of revolution.

|||| This theorem is named after the Greek mathematician Pappus of Alexandria, who lived in the fourth century A.D.

Theorem of Pappus Let $\mathcal{R}$ be a plane region that lies entirely on one side of a line l in the plane. If $\mathcal{R}$ is rotated about l, then the volume of the resulting solid is the product of the area A of $\mathcal{R}$ and the distance d traveled by the centroid of $\mathcal{R}$.

Proof We give the proof for the special case in which the region lies between $y = f(x)$ and $y = g(x)$ as in Figure 13 and the line l is the y-axis. Using the method of cylindrical shells (see Section 6.3), we have

$$V = \int_a^b 2\pi x [f(x) - g(x)] \, dx$$

$$= 2\pi \int_a^b x[f(x) - g(x)] \, dx$$

$$= 2\pi(\bar{x}A) \qquad \text{(by Formulas 9)}$$

$$= (2\pi\bar{x})A = Ad$$

where $d = 2\pi \bar{x}$ is the distance traveled by the centroid during one rotation about the y-axis.

EXAMPLE 7 A torus is formed by rotating a circle of radius r about a line in the plane of the circle that is a distance $R\ (> r)$ from the center of the circle. Find the volume of the torus.

SOLUTION The circle has area $A = \pi r^2$. By the symmetry principle, its centroid is its center and so the distance traveled by the centroid during a rotation is $d = 2\pi R$. Therefore, by the Theorem of Pappus, the volume of the torus is

$$V = Ad = (2\pi R)(\pi r^2) = 2\pi^2 r^2 R$$

The method of Example 7 should be compared with the method of Exercise 61 in Section 6.2.

9.3 Exercises

1. An aquarium 5 ft long, 2 ft wide, and 3 ft deep is full of water. Find (a) the hydrostatic pressure on the bottom of the aquarium, (b) the hydrostatic force on the bottom, and (c) the hydrostatic force on one end of the aquarium.

2. A swimming pool 5 m wide, 10 m long, and 3 m deep is filled with seawater of density 1030 kg/m³ to a depth of 2.5 m. Find (a) the hydrostatic pressure at the bottom of the pool, (b) the hydrostatic force on the bottom, and (c) the hydrostatic force on one end of the pool.

3–9 A vertical plate is submerged in water and has the indicated shape. Explain how to approximate the hydrostatic force against one side of the plate by a Riemann sum. Then express the force as an integral and evaluate it.

3.
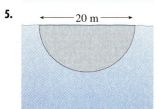

4.
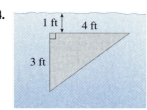

5.

6.
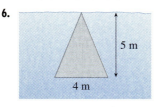

7.
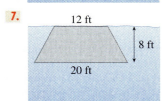

8.
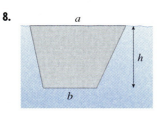

9.
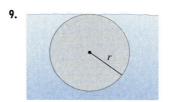

10. A large tank is designed with ends in the shape of the region between the curves $y = x^2/2$ and $y = 12$, measured in feet. Find the hydrostatic force on one end of the tank if it is filled to a depth of 8 ft with gasoline. (Assume the gasoline's density is 42.0 lb/ft³.)

11. A trough is filled with a liquid of density 840 kg/m³. The ends of the trough are equilateral triangles with sides 8 m long and vertex at the bottom. Find the hydrostatic force on one end of the trough.

12. A vertical dam has a semicircular gate as shown in the figure. Find the hydrostatic force against the gate.

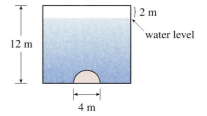

13. A cube with 20-cm-long sides is sitting on the bottom of an aquarium in which the water is one meter deep. Find the hydrostatic force on (a) the top of the cube and (b) one of the sides of the cube.

14. A dam is inclined at an angle of 30° from the vertical and has the shape of an isosceles trapezoid 100 ft wide at the top and

50 ft wide at the bottom and with a slant height of 70 ft. Find the hydrostatic force on the dam when it is full of water.

15. A swimming pool is 20 ft wide and 40 ft long and its bottom is an inclined plane, the shallow end having a depth of 3 ft and the deep end, 9 ft. If the pool is full of water, find the hydrostatic force on (a) the shallow end, (b) the deep end, (c) one of the sides, and (d) the bottom of the pool.

16. Suppose that a plate is immersed vertically in a fluid with density ρ and the width of the plate is $w(x)$ at a depth of x meters beneath the surface of the fluid. If the top of the plate is at depth a and the bottom is at depth b, show that the hydrostatic force on one side of the plate is

$$F = \int_a^b \rho g x w(x)\, dx$$

17. A vertical, irregularly shaped plate is submerged in water. The table shows measurements of its width, taken at the indicated depths. Use Simpson's rule to estimate the force of the water against the plate.

Depth (m)	2.0	2.5	3.0	3.5	4.0	4.5	5.0
Plate width (m)	0	0.8	1.7	2.4	2.9	3.3	3.6

18. (a) Use the formula of Exercise 16 to show that

$$F = (\rho g \bar{x}) A$$

where $\bar{x}$ is the x-coordinate of the centroid of the plate and A is its area. This equation shows that the hydrostatic force against a vertical plane region is the same as if the region were horizontal at the depth of the centroid of the region.
(b) Use the result of part (a) to give another solution to Exercise 9.

19–20 Point-masses m_i are located on the x-axis as shown. Find the moment M of the system about the origin and the center of mass $\bar{x}$.

19. $m_1 = 40$, $m_2 = 30$ at $x = 2, 5$

20. $m_1 = 25$, $m_2 = 20$, $m_3 = 10$ at $x = -2, 3, 7$

21–22 The masses m_i are located at the points P_i. Find the moments M_x and M_y and the center of mass of the system.

21. $m_1 = 6$, $m_2 = 5$, $m_3 = 10$;
$P_1(1, 5)$, $P_2(3, -2)$, $P_3(-2, -1)$

22. $m_1 = 6$, $m_2 = 5$, $m_3 = 1$, $m_4 = 4$;
$P_1(1, -2)$, $P_2(3, 4)$, $P_3(-3, -7)$, $P_4(6, -1)$

23–26 Sketch the region bounded by the curves, and visually estimate the location of the centroid. Then find the exact coordinates of the centroid.

23. $y = 4 - x^2$, $y = 0$

24. $3x + 2y = 6$, $y = 0$, $x = 0$

25. $y = e^x$, $y = 0$, $x = 0$, $x = 1$

26. $y = 1/x$, $y = 0$, $x = 1$, $x = 2$

27–31 Find the centroid of the region bounded by the given curves.

27. $y = \sqrt{x}$, $y = x$

28. $y = x + 2$, $y = x^2$

29. $y = \sin x$, $y = \cos x$, $x = 0$, $x = \pi/4$

30. $y = x$, $y = 0$, $y = 1/x$, $x = 2$

31. $x = 5 - y^2$, $x = 0$

32–34 Calculate the moments M_x and M_y and the center of mass of a lamina with the given density and shape.

32. $\rho = 5$

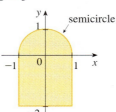

33. $\rho = 1$

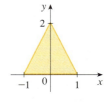

34. $\rho = 2$

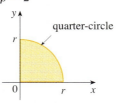

35. Find the centroid of the region bounded by the curves $y = 2^x$ and $y = x^2$, $0 \leq x \leq 2$, to three decimal places. Sketch the region and plot the centroid to see if your answer is reasonable.

36. Use a graph to find approximate x-coordinates of the points of intersection of the curves $y = x + \ln x$ and $y = x^3 - x$. Then find (approximately) the centroid of the region bounded by these curves.

37. Prove that the centroid of any triangle is located at the point of intersection of the medians. [*Hints:* Place the axes so that the vertices are $(a, 0)$, $(0, b)$, and $(c, 0)$. Recall that a median is a line segment from a vertex to the midpoint of the oppo-

site side. Recall also that the medians intersect at a point two-thirds of the way from each vertex (along the median) to the opposite side.]

38–39 |||| Find the centroid of the region shown, not by integration, but by locating the centroids of the rectangles and triangles (from Exercise 37) and using additivity of moments.

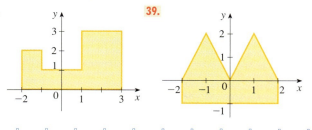

38.   **39.**

40–42 |||| Use the Theorem of Pappus to find the volume of the given solid.

40. A sphere of radius r (Use Example 4.)

41. A cone with height h and base radius r

42. The solid obtained by rotating the triangle with vertices $(2, 3)$, $(2, 5)$, and $(5, 4)$ about the x-axis

43. Prove Formulas 9.

44. Let $\mathcal{R}$ be the region that lies between the curves $y = x^m$ and $y = x^n$, $0 \le x \le 1$, where m and n are integers with $0 \le n < m$.
(a) Sketch the region $\mathcal{R}$.
(b) Find the coordinates of the centroid of $\mathcal{R}$.
(c) Try to find values of m and n such that the centroid lies outside $\mathcal{R}$.

9.4 Applications to Economics and Biology

In this section we consider some applications of integration to economics (consumer surplus) and biology (blood flow, cardiac output). Others are described in the exercises.

Consumer Surplus

Recall from Section 4.8 that the demand function $p(x)$ is the price that a company has to charge in order to sell x units of a commodity. Usually, selling larger quantities requires lowering prices, so the demand function is a decreasing function. The graph of a typical demand function, called a **demand curve**, is shown in Figure 1. If X is the amount of the commodity that is currently available, then $P = p(X)$ is the current selling price.

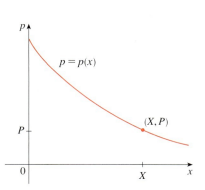

FIGURE 1
A typical demand curve

We divide the interval $[0, X]$ into n subintervals, each of length $\Delta x = X/n$, and let $x_i^* = x_i$ be the right endpoint of the ith subinterval, as in Figure 2. If, after the first x_{i-1} units were sold, a total of only x_i units had been available and the price per unit had been set at $p(x_i)$ dollars, then the additional Δx units could have been sold (but no more). The consumers who would have paid $p(x_i)$ dollars placed a high value on the product; they would have paid what it was worth to them. So, in paying only P dollars they have saved an amount of

(savings per unit)(number of units) $= [p(x_i) - P]\,\Delta x$

Considering similar groups of willing consumers for each of the subintervals and adding the savings, we get the total savings:

$$\sum_{i=1}^{n} [p(x_i) - P]\,\Delta x$$

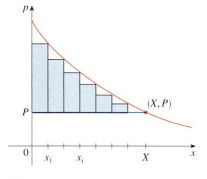

FIGURE 2

(This sum corresponds to the area enclosed by the rectangles in Figure 2.) If we let $n \to \infty$, this Riemann sum approaches the integral

$$\boxed{1} \qquad \int_0^X [p(x) - P]\,dx$$

which economists call the **consumer surplus** for the commodity.

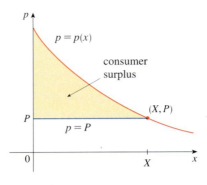

FIGURE 3

The consumer surplus represents the amount of money saved by consumers in purchasing the commodity at price P, corresponding to an amount demanded of X. Figure 3 shows the interpretation of the consumer surplus as the area under the demand curve and above the line $p = P$.

EXAMPLE 1 The demand for a product, in dollars, is

$$p = 1200 - 0.2x - 0.0001x^2$$

Find the consumer surplus when the sales level is 500.

SOLUTION Since the number of products sold is $X = 500$, the corresponding price is

$$P = 1200 - (0.2)(500) - (0.0001)(500)^2 = 1075$$

Therefore, from Definition 1, the consumer surplus is

$$\int_0^{500} [p(x) - P]\,dx = \int_0^{500} (1200 - 0.2x - 0.0001x^2 - 1075)\,dx$$

$$= \int_0^{500} (125 - 0.2x - 0.0001x^2)\,dx$$

$$= 125x - 0.1x^2 - (0.0001)\left(\frac{x^3}{3}\right)\Big]_0^{500}$$

$$= (125)(500) - (0.1)(500)^2 - \frac{(0.0001)(500)^3}{3}$$

$$= \$33{,}333.33$$

Blood Flow

In Example 7 in Section 3.4 we discussed the law of laminar flow:

$$v(r) = \frac{P}{4\eta l}(R^2 - r^2)$$

which gives the velocity v of blood that flows along a blood vessel with radius R and length l at a distance r from the central axis, where P is the pressure difference between the ends of the vessel and η is the viscosity of the blood. Now, in order to compute the rate of blood flow, or *flux* (volume per unit time), we consider smaller, equally spaced radii $r_1, r_2, \ldots$. The approximate area of the ring (or washer) with inner radius r_{i-1} and outer radius r_i is

$$2\pi r_i\,\Delta r \qquad \text{where} \qquad \Delta r = r_i - r_{i-1}$$

(See Figure 4.) If Δr is small, then the velocity is almost constant throughout this ring and can be approximated by $v(r_i)$. Thus, the volume of blood per unit time that flows across the ring is approximately

$$(2\pi r_i\,\Delta r)v(r_i) = 2\pi r_i v(r_i)\,\Delta r$$

and the total volume of blood that flows across a cross-section per unit time is approximately

$$\sum_{i=1}^{n} 2\pi r_i v(r_i)\,\Delta r$$

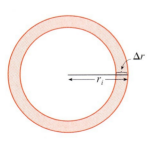

FIGURE 4

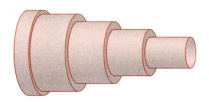

FIGURE 5

This approximation is illustrated in Figure 5. Notice that the velocity (and hence the volume per unit time) increases toward the center of the blood vessel. The approximation gets better as n increases. When we take the limit we get the exact value of the **flux** (or *discharge*), which is the volume of blood that passes a cross-section per unit time:

$$F = \lim_{n \to \infty} \sum_{i=1}^{n} 2\pi r_i v(r_i)\, \Delta r = \int_0^R 2\pi r v(r)\, dr$$

$$= \int_0^R 2\pi r \frac{P}{4\eta l}(R^2 - r^2)\, dr$$

$$= \frac{\pi P}{2\eta l} \int_0^R (R^2 r - r^3)\, dr = \frac{\pi P}{2\eta l}\left[R^2 \frac{r^2}{2} - \frac{r^4}{4}\right]_{r=0}^{r=R}$$

$$= \frac{\pi P}{2\eta l}\left[\frac{R^4}{2} - \frac{R^4}{4}\right] = \frac{\pi P R^4}{8\eta l}$$

The resulting equation

$$\boxed{2} \qquad F = \frac{\pi P R^4}{8\eta l}$$

is called **Poiseuille's Law**; it shows that the flux is proportional to the fourth power of the radius of the blood vessel.

Cardiac Output

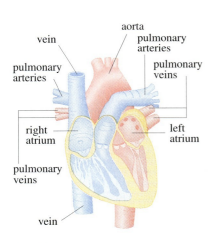

FIGURE 6

Figure 6 shows the human cardiovascular system. Blood returns from the body through the veins, enters the right atrium of the heart, and is pumped to the lungs through the pulmonary arteries for oxygenation. It then flows back into the left atrium through the pulmonary veins and then out to the rest of the body through the aorta. The **cardiac output** of the heart is the volume of blood pumped by the heart per unit time, that is, the rate of flow into the aorta.

The *dye dilution method* is used to measure the cardiac output. Dye is injected into the right atrium and flows through the heart into the aorta. A probe inserted into the aorta measures the concentration of the dye leaving the heart at equally spaced times over a time interval $[0, T]$ until the dye has cleared. Let $c(t)$ be the concentration of the dye at time t. If we divide $[0, T]$ into subintervals of equal length Δt, then the amount of dye that flows past the measuring point during the subinterval from $t = t_{i-1}$ to $t = t_i$ is approximately

$$(\text{concentration})(\text{volume}) = c(t_i)(F\, \Delta t)$$

where F is the rate of flow that we are trying to determine. Thus, the total amount of dye is approximately

$$\sum_{i=1}^{n} c(t_i) F\, \Delta t = F \sum_{i=1}^{n} c(t_i)\, \Delta t$$

and, letting $n \to \infty$, we find that the amount of dye is

$$A = F \int_0^T c(t)\, dt$$

Thus, the cardiac output is given by

$$\boxed{3} \qquad F = \frac{A}{\int_0^T c(t)\, dt}$$

where the amount of dye A is known and the integral can be approximated from the concentration readings.

EXAMPLE 2 A 5-mg bolus of dye is injected into a right atrium. The concentration of the dye (in milligrams per liter) is measured in the aorta at one-second intervals as shown in the chart. Estimate the cardiac output.

SOLUTION Here $A = 5$, $\Delta t = 1$, and $T = 10$. We use Simpson's Rule to approximate the integral of the concentration:

$$\int_0^{10} c(t)\, dt \approx \tfrac{1}{3}[0 + 4(0.4) + 2(2.8) + 4(6.5) + 2(9.8) + 4(8.9)$$

$$+ 2(6.1) + 4(4.0) + 2(2.3) + 4(1.1) + 0]$$

$$\approx 41.87$$

Thus, Formula 3 gives the cardiac output to be

$$F = \frac{A}{\int_0^{10} c(t)\, dt} \approx \frac{5}{41.87}$$

$$\approx 0.12 \text{ L/s} = 7.2 \text{ L/min}$$

t	$c(t)$	t	$c(t)$
0	0	6	6.1
1	0.4	7	4.0
2	2.8	8	2.3
3	6.5	9	1.1
4	9.8	10	0
5	8.9		

9.4 Exercises

1. The marginal cost function $C'(x)$ was defined to be the derivative of the cost function. (See Sections 3.4 and 4.8.) If the marginal cost of maufacturing x meters of a fabric is $C'(x) = 5 - 0.008x + 0.000009x^2$ (measured in dollars per meter) and the fixed start-up cost is $C(0) = \$20{,}000$, use the Net Change Theorem to find the cost of producing the first 2000 units.

2. The marginal revenue from the sale of x units of a product is $12 - 0.0004x$. If the revenue from the sale of the first 1000 units is $\$12{,}400$, find the revenue from the sale of the first 5000 units.

3. The marginal cost of producing x units of a certain product is $74 + 1.1x - 0.002x^2 + 0.00004x^3$ (in dollars per unit). Find the increase in cost if the production level is raised from 1200 units to 1600 units.

4. The demand function for a certain commodity is $p = 5 - x/10$. Find the consumer surplus when the sales level is 30. Illustrate by drawing the demand curve and identifying the consumer surplus as an area.

5. A demand curve is given by $p = 450/(x + 8)$. Find the consumer surplus when the selling price is $10.

6. The **supply function** $p_S(x)$ for a commodity gives the relation between the selling price and the number of units that manufacturers will produce at that price. For a higher price, manufacturers will produce more units, so p_S is an increasing function of x. Let X be the amount of the commodity currently produced and let $P = p_S(X)$ be the current price. Some producers would be willing to make and sell the commodity for a lower selling price and are therefore receiving more than their minimal price. The excess is called the **producer surplus**. An argument similar to that for consumer surplus shows that the surplus is given by the integral

$$\int_0^X [P - p_S(x)]\, dx$$

Calculate the producer surplus for the supply function $p_S(x) = 3 + 0.01x^2$ at the sales level $X = 10$. Illustrate by drawing the supply curve and identifying the producer surplus as an area.

7. If a supply curve is modeled by the equation $p = 200 + 0.2x^{3/2}$, find the producer surplus when the selling price is $400.

8. For a given commodity and pure competition, the number of units produced and the price per unit are determined as the coordinates of the point of intersection of the supply and demand curves. Given the demand curve $p = 50 - x/20$ and the supply curve $p = 20 + x/10$, find the consumer surplus and the producer surplus. Illustrate by sketching the supply and demand curves and identifying the surpluses as areas.

9. A company modeled the demand curve for its product (in dollars) by

$$p = \frac{800{,}000e^{-x/5000}}{x + 20{,}000}$$

Use a graph to estimate the sales level when the selling price is $16. Then find (approximately) the consumer surplus for this sales level.

10. A movie theater has been charging $7.50 per person and selling about 400 tickets on a typical weeknight. After surveying their customers, the theater estimates that for every 50 cents that they lower the price, the number of moviegoers will increase by 35 per night. Find the demand function and calculate the consumer surplus when the tickets are priced at $6.00.

11. If the amount of capital that a company has at time t is $f(t)$, then the derivative, $f'(t)$, is called the *net investment flow*. Suppose that the net investment flow is $\sqrt{t}$ million dollars per year (where t is measured in years). Find the increase in capital (the *capital formation*) from the fourth year to the eighth year.

12. A hot, wet summer is causing a mosquito population explosion in a lake resort area. The number of mosquitos is increasing at an estimated rate of $2200 + 10e^{0.8t}$ per week (where t is measured in weeks). By how much does the mosquito population increase between the fifth and ninth weeks of summer?

13. Use Poiseuille's Law to calculate the rate of flow in a small human artery where we can take $\eta = 0.027$, $R = 0.008$ cm, $l = 2$ cm, and $P = 4000$ dynes/cm^2.

14. High blood pressure results from constriction of the arteries. To maintain a normal flow rate (flux), the heart has to pump harder, thus increasing the blood pressure. Use Poiseuille's Law to show that if R_0 and P_0 are normal values of the radius and pressure in an artery and the constricted values are R and P, then for the flux to remain constant, P and R are related by the equation

$$\frac{P}{P_0} = \left(\frac{R_0}{R}\right)^4$$

Deduce that if the radius of an artery is reduced to three-fourths of its former value, then the pressure is more than tripled.

15. The dye dilution method is used to measure cardiac output with 8 mg of dye. The dye concentrations, in mg/L, are modeled by $c(t) = \frac{1}{4}t(12 - t)$, $0 \le t \le 12$, where t is measured in seconds. Find the cardiac output.

16. After an 8-mg injection of dye, the readings of dye concentration at two-second intervals are as shown in the table. Use Simpson's Rule to estimate the cardiac output.

t	$c(t)$	t	$c(t)$
0	0	12	3.9
2	2.4	14	2.3
4	5.1	16	1.6
6	7.8	18	0.7
8	7.6	20	0
10	5.4		

9.5 Probability

Calculus plays a role in the analysis of random behavior. Suppose we consider the cholesterol level of a person chosen at random from a certain age group, or the height of an adult female chosen at random, or the lifetime of a randomly chosen battery of a certain type. Such quantities are called **continuous random variables** because their values actually range over an interval of real numbers, although they might be measured or recorded only to the nearest integer. We might want to know the probability that a blood cholesterol level is greater than 250, or the probability that the height of an adult female is between 60 and 70 inches, or the probability that the battery we are buying lasts between 100 and 200 hours. If X represents the lifetime of that type of battery, we denote this last probability as follows:

$$P(100 \le X \le 200)$$

According to the frequency interpretation of probability, this number is the long-run proportion of all batteries of the specified type whose lifetimes are between 100 and 200 hours. Since it represents a proportion, the probability naturally falls between 0 and 1.

Every continuous random variable X has a **probability density function** f. This means that the probability that X lies between a and b is found by integrating f from a to b:

$$\boxed{1} \qquad P(a \leq X \leq b) = \int_a^b f(x)\, dx$$

For example, Figure 1 shows the graph of a model of the probability density function f for a random variable X defined to be the height in inches of an adult female in the United States (according to data from the National Health Survey). The probability that the height of a woman chosen at random from this population is between 60 and 70 inches is equal to the area under the graph of f from 60 to 70.

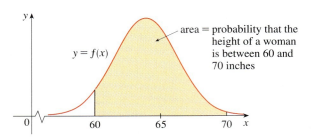

FIGURE 1
Probability density function for the height of an adult female

In general, the probability density function f of a random variable X satisfies the condition $f(x) \geq 0$ for all x. Because probabilities are measured on a scale from 0 to 1, it follows that

$$\boxed{2} \qquad \int_{-\infty}^{\infty} f(x)\, dx = 1$$

EXAMPLE 1 Let $f(x) = 0.006x(10 - x)$ for $0 \leq x \leq 10$ and $f(x) = 0$ for all other values of x.
(a) Verify that f is a probability density function.
(b) Find $P(4 \leq X \leq 8)$.

SOLUTION
(a) For $0 \leq x \leq 10$ we have $0.006x(10 - x) \geq 0$, so $f(x) \geq 0$ for all x. We also need to check that Equation 2 is satisfied:

$$\int_{-\infty}^{\infty} f(x)\, dx = \int_0^{10} 0.006x(10 - x)\, dx = 0.006 \int_0^{10} (10x - x^2)\, dx$$

$$= 0.006\left[5x^2 - \tfrac{1}{3}x^3\right]_0^{10} = 0.006\left(500 - \tfrac{1000}{3}\right) = 1$$

Therefore, f is a probability density function.
(b) The probability that X lies between 4 and 8 is

$$P(4 \leq X \leq 8) = \int_4^8 f(x)\, dx = 0.006 \int_4^8 (10x - x^2)\, dx$$

$$= 0.006\left[5x^2 - \tfrac{1}{3}x^3\right]_4^8 = 0.544$$

EXAMPLE 2 Phenomena such as waiting times and equipment failure times are commonly modeled by exponentially decreasing probability density functions. Find the exact form of such a function.

SOLUTION Think of the random variable as being the time you wait on hold before an agent of a company you're telephoning answers your call. So instead of x, let's use t to represent time, in minutes. If f is the probability density function and you call at time $t = 0$, then, from Definition 1, $\int_0^2 f(t)\,dt$ represents the probability that an agent answers within the first two minutes and $\int_4^5 f(t)\,dt$ is the probability that your call is answered during the fifth minute.

It's clear that $f(t) = 0$ for $t < 0$ (the agent can't answer before you place the call). For $t > 0$ we are told to use an exponentially decreasing function, that is, a function of the form $f(t) = Ae^{-ct}$, where A and c are positive constants. Thus

$$f(t) = \begin{cases} 0 & \text{if } t < 0 \\ Ae^{-ct} & \text{if } t \geq 0 \end{cases}$$

We use condition 2 to determine the value of A:

$$1 = \int_{-\infty}^{\infty} f(t)\,dt = \int_{-\infty}^{0} f(t)\,dt + \int_{0}^{\infty} f(t)\,dt$$

$$= \int_{0}^{\infty} Ae^{-ct}\,dt = \lim_{x \to \infty} \int_0^x Ae^{-ct}\,dt$$

$$= \lim_{x \to \infty} \left[-\frac{A}{c}e^{-ct}\right]_0^x = \lim_{x \to \infty} \frac{A}{c}(1 - e^{-cx})$$

$$= \frac{A}{c}$$

Therefore, $A/c = 1$ and so $A = c$. Thus, every exponential density function has the form

$$f(t) = \begin{cases} 0 & \text{if } t < 0 \\ ce^{-ct} & \text{if } t \geq 0 \end{cases}$$

A typical graph is shown in Figure 2.

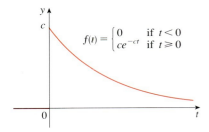

FIGURE 2
An exponential density function

Average Values

Suppose you're waiting for a company to answer your phone call and you wonder how long, on average, you can expect to wait. Let $f(t)$ be the corresponding density function, where t is measured in minutes, and think of a sample of N people who have called this company. Most likely, none of them had to wait more than an hour, so let's restrict our attention to the interval $0 \leq t \leq 60$. Let's divide that interval into n intervals of length Δt and endpoints $0, t_1, t_2, \ldots$. (Think of Δt as lasting a minute, or half a minute, or 10 seconds, or even a second.) The probability that somebody's call gets answered during the time period from t_{i-1} to t_i is the area under the curve $y = f(t)$ from t_{i-1} to t_i, which is approximately equal to $f(\bar{t}_i)\,\Delta t$. (This is the area of the approximating rectangle in Figure 3, where $\bar{t}_i$ is the midpoint of the interval.)

Since the long-run proportion of calls that get answered in the time period from t_{i-1} to t_i is $f(\bar{t}_i)\,\Delta t$, we expect that, out of our sample of N callers, the number whose call was answered in that time period is approximately $Nf(\bar{t}_i)\,\Delta t$ and the time that each waited is about $\bar{t}_i$. Therefore, the total time they waited is the product of these numbers: approximately $\bar{t}_i[Nf(\bar{t}_i)\,\Delta t]$. Adding over all such intervals, we get the approximate total of everybody's waiting times:

$$\sum_{i=1}^{n} N\bar{t}_i f(\bar{t}_i)\,\Delta t$$

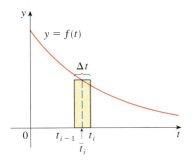

FIGURE 3

If we now divide by the number of callers N, we get the approximate *average* waiting time:

$$\sum_{i=1}^{n} \bar{t}_i f(\bar{t}_i) \, \Delta t$$

We recognize this as a Riemann sum for the function $tf(t)$. As the time interval shrinks (that is, $\Delta t \to 0$ and $n \to \infty$), this Riemann sum approaches the integral

$$\int_0^{60} t f(t) \, dt$$

This integral is called the *mean waiting time*.

In general, the **mean** of any probability density function f is defined to be

$$\mu = \int_{-\infty}^{\infty} x f(x) \, dx$$

▌ It is traditional to denote the mean by the Greek letter μ (mu).

The mean can be interpreted as the long-run average value of the random variable X. It can also be interpreted as a measure of centrality of the probability density function.

The expression for the mean resembles an integral we have seen before. If $\mathcal{R}$ is the region that lies under the graph of f, we know from Formula 9.3.8 that the x-coordinate of the centroid of $\mathcal{R}$ is

$$\bar{x} = \frac{\int_{-\infty}^{\infty} x f(x) \, dx}{\int_{-\infty}^{\infty} f(x) \, dx} = \int_{-\infty}^{\infty} x f(x) \, dx = \mu$$

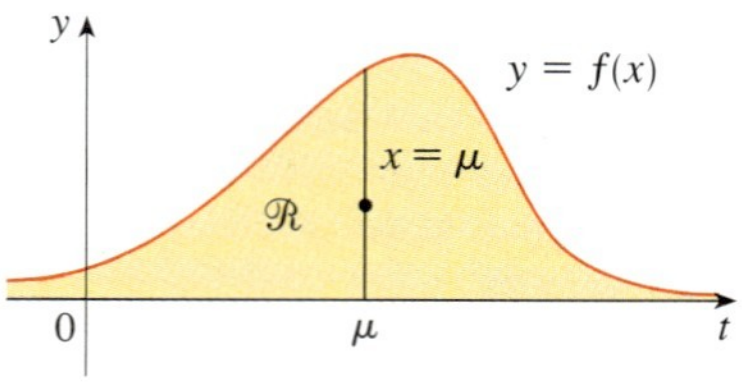

because of Equation 2. So a thin plate in the shape of $\mathcal{R}$ balances at a point on the vertical line $x = \mu$. (See Figure 4.)

FIGURE 4
$\mathcal{R}$ balances at a point on the line $x = \mu$

EXAMPLE 3 Find the mean of the exponential distribution of Example 2:

$$f(t) = \begin{cases} 0 & \text{if } t < 0 \\ c e^{-ct} & \text{if } t \geq 0 \end{cases}$$

SOLUTION According to the definition of a mean, we have

$$\mu = \int_{-\infty}^{\infty} t f(t) \, dt = \int_0^{\infty} t c e^{-ct} \, dt$$

To evaluate this integral we use integration by parts, with $u = t$ and $dv = c e^{-ct} \, dt$:

$$\int_0^{\infty} t c e^{-ct} \, dt = \lim_{x \to \infty} \int_0^x t c e^{-ct} \, dt = \lim_{x \to \infty} \left(-t e^{-ct} \Big]_0^x + \int_0^x e^{-ct} \, dt \right)$$

$$= \lim_{x \to \infty} \left(-x e^{-cx} + \frac{1}{c} - \frac{e^{-cx}}{c} \right)$$

$$= \frac{1}{c}$$

▌ The limit of the first term is 0 by l'Hospital's Rule.

The mean is $\mu = 1/c$, so we can rewrite the probability density function as

$$f(t) = \begin{cases} 0 & \text{if } t < 0 \\ \mu^{-1} e^{-t/\mu} & \text{if } t \geq 0 \end{cases}$$

EXAMPLE 4 Suppose the average waiting time for a customer's call to be answered by a company representative is five minutes.
(a) Find the probability that a call is answered during the first minute.
(b) Find the probability that a customer waits more than five minutes to be answered.

SOLUTION
(a) We are given that the mean of the exponential distribution is $\mu = 5$ min and so, from the result of Example 3, we know that the probability density function is

$$f(t) = \begin{cases} 0 & \text{if } t < 0 \\ 0.2 e^{-t/5} & \text{if } t \geq 0 \end{cases}$$

Thus, the probability that a call is answered during the first minute is

$$P(0 \leq T \leq 1) = \int_0^1 f(t)\, dt$$

$$= \int_0^1 0.2 e^{-t/5}\, dt$$

$$= 0.2(-5) e^{-t/5} \Big]_0^1$$

$$= 1 - e^{-1/5} \approx 0.1813$$

So about 18% of customers' calls are answered during the first minute.

(b) The probability that a customer waits more than five minutes is

$$P(T > 5) = \int_5^\infty f(t)\, dt = \int_5^\infty 0.2 e^{-t/5}\, dt$$

$$= \lim_{x \to \infty} \int_5^x 0.2 e^{-t/5}\, dt = \lim_{x \to \infty} (e^{-1} - e^{-x/5})$$

$$= \frac{1}{e} \approx 0.368$$

About 37% of customers wait more than five minutes before their calls are answered.

Notice the result of Example 4(b): Even though the mean waiting time is 5 minutes, only 37% of callers wait more than 5 minutes. The reason is that some callers have to wait much longer (maybe 10 or 15 minutes), and this brings up the average.

Another measure of centrality of a probability density function is the *median*. That is a number m such that half the callers have a waiting time less than m and the other callers have a waiting time longer than m. In general, the **median** of a probability density function is the number m such that

$$\int_m^\infty f(x)\, dx = \tfrac{1}{2}$$

This means that half the area under the graph of f lies to the right of m. In Exercise 7 you are asked to show that the median waiting time for the company described in Example 4 is approximately 3.5 minutes.

Normal Distributions

Many important random phenomena—such as test scores on aptitude tests, heights and weights of individuals from a homogeneous population, annual rainfall in a given location—are modeled by a **normal distribution**. This means that the probability density function of the random variable X is a member of the family of functions

$$\boxed{3} \qquad f(x) = \frac{1}{\sigma\sqrt{2\pi}} e^{-(x-\mu)^2/(2\sigma^2)}$$

|||| The standard deviation is denoted by the lowercase Greek letter σ (sigma).

You can verify that the mean for this function is μ. The positive constant σ is called the **standard deviation**; it measures how spread out the values of X are. From the bell-shaped graphs of members of the family in Figure 5, we see that for small values of σ the values of X are clustered about the mean, whereas for larger values of σ the values of X are more spread out. Statisticians have methods for using sets of data to estimate μ and σ.

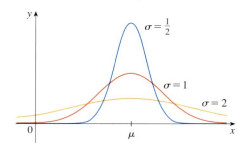

FIGURE 5
Normal distributions

The factor $1/(\sigma\sqrt{2\pi})$ is needed to make f a probability density function. In fact, it can be verified using the methods of multivariable calculus that

$$\int_{-\infty}^{\infty} \frac{1}{\sigma\sqrt{2\pi}} e^{-(x-\mu)^2/(2\sigma^2)} dx = 1$$

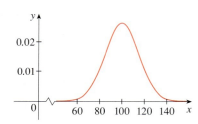

FIGURE 6
Distribution of IQ scores

EXAMPLE 5 Intelligence Quotient (IQ) scores are distributed normally with mean 100 and standard deviation 15. (Figure 6 shows the corresponding probability density function.)
(a) What percentage of the population has an IQ score between 85 and 115?
(b) What percentage of the population has an IQ above 140?

SOLUTION
(a) Since IQ scores are normally distributed, we use the probability density function given by Equation 3 with $\mu = 100$ and $\sigma = 15$:

$$P(85 \leq X \leq 115) = \int_{85}^{115} \frac{1}{15\sqrt{2\pi}} e^{-(x-100)^2/(2 \cdot 15^2)} dx$$

Recall from Section 8.5 that the function $y = e^{-x^2}$ doesn't have an elementary antiderivative, so we can't evaluate the integral exactly. But we can use the numerical integration capability of a calculator or computer (or the Midpoint Rule or Simpson's Rule) to estimate the integral. Doing so, we find that

$$P(85 \leq X \leq 115) \approx 0.68$$

So about 68% of the population has an IQ between 85 and 115, that is, within one standard deviation of the mean.

(b) The probability that the IQ score of a person chosen at random is more than 140 is

$$P(X > 140) = \int_{140}^{\infty} \frac{1}{15\sqrt{2\pi}} e^{-(x-100)^2/450} \, dx$$

To avoid the improper integral we could approximate it by the integral from 140 to 200. (It's quite safe to say that people with an IQ over 200 are extremely rare.) Then

$$P(X > 140) \approx \int_{140}^{200} \frac{1}{15\sqrt{2\pi}} e^{-(x-100)^2/450} \, dx \approx 0.0038$$

Therefore, about 0.4% of the population has an IQ over 140.

9.5 Exercises

1. Let $f(x)$ be the probability density function for the lifetime of a manufacturer's highest quality car tire, where x is measured in miles. Explain the meaning of each integral.

(a) $\int_{30,000}^{40,000} f(x) \, dx$
(b) $\int_{25,000}^{\infty} f(x) \, dx$

2. Let $f(t)$ be the probability density function for the time it takes you to drive to school in the morning, where t is measured in minutes. Express the following probabilities as integrals.
(a) The probability that you drive to school in less than 15 minutes
(b) The probability that it takes you more than half an hour to get to school

3. Let $f(x) = \frac{3}{64} x\sqrt{16 - x^2}$ for $0 \leq x \leq 4$ and $f(x) = 0$ for all other values of x.
(a) Verify that f is a probability density function.
(b) Find $P(X < 2)$.

4. Let $f(x) = kx^2(1 - x)$ if $0 \leq x \leq 1$ and $f(x) = 0$ if $x < 0$ or $x > 1$.
(a) For what value of k is f a probability density function?
(b) For that value of k, find $P(X \geq \frac{1}{2})$.
(c) Find the mean.

5. A spinner from a board game randomly indicates a real number between 0 and 10. The spinner is fair in the sense that it indicates a number in a given interval with the same probability as it indicates a number in any other interval of the same length.
(a) Explain why the function

$$f(x) = \begin{cases} 0.1 & \text{if } 0 \leq x \leq 10 \\ 0 & \text{if } x < 0 \text{ or } x > 10 \end{cases}$$

is a probability density function for the spinner's values.
(b) What does your intuition tell you about the value of the mean? Check your guess by evaluating an integral.

6. (a) Explain why the function whose graph is shown is a probability density function.

(b) Use the graph to find the following probabilities:
(i) $P(X < 3)$ (ii) $P(3 \leq X \leq 8)$
(c) Calculate the mean.

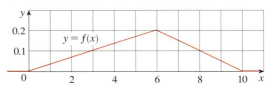

7. Show that the median waiting time for a phone call to the company described in Example 4 is about 3.5 minutes.

8. (a) A type of lightbulb is labeled as having an average lifetime of 1000 hours. It's reasonable to model the probability of failure of these bulbs by an exponential density function with mean $\mu = 1000$. Use this model to find the probability that a bulb
(i) fails within the first 200 hours,
(ii) burns for more than 800 hours.
(b) What is the median lifetime of these lightbulbs?

9. The manager of a fast-food restaurant determines that the average time that her customers wait for service is 2.5 minutes.
(a) Find the probability that a customer has to wait for more than 4 minutes.
(b) Find the probability that a customer is served within the first 2 minutes.
(c) The manager wants to advertise that anybody who isn't served within a certain number of minutes gets a free hamburger. But she doesn't want to give away free hamburgers to more than 2% of her customers. What should the advertisement say?

10. According to the National Health Survey, the heights of adult males in the United States are normally distributed with mean 69.0 inches and standard deviation 2.8 inches.
(a) What is the probability that an adult male chosen at random is between 65 inches and 73 inches tall?

(b) What percentage of the adult male population is more than 6 feet tall?

11. The "Garbage Project" at the University of Arizona reports that the amount of paper discarded by households per week is normally distributed with mean 9.4 lb and standard deviation 4.2 lb. What percentage of households throw out at least 10 lb of paper a week?

12. Boxes are labeled as containing 500 g of cereal. The machine filling the boxes produces weights that are normally distributed with standard deviation 12 g.
  (a) If the target weight is 500 g, what is the probability that the machine produces a box with less than 480 g of cereal?
  (b) Suppose a law states that no more than 5% of a manufacturer's cereal boxes can contain less than the stated weight of 500 g. At what target weight should the manufacturer set its filling machine?

13. For any normal distribution, find the probability that the random variable lies within two standard deviations of the mean.

14. The standard deviation for a random variable with probability density function f and mean μ is defined by

$$\sigma = \left[\int_{-\infty}^{\infty} (x - \mu)^2 f(x)\, dx \right]^{1/2}$$

Find the standard deviation for an exponential density function with mean μ.

15. The hydrogen atom is composed of one proton in the nucleus and one electron, which moves about the nucleus. In the quantum theory of atomic structure, it is assumed that the electron does not move in a well-defined orbit. Instead, it occupies a state known as an *orbital*, which may be thought of as a "cloud" of negative charge surrounding the nucleus. At the state of lowest energy, called the *ground state*, or *1s-orbital*, the shape of this cloud is assumed to be a sphere centered at the nucleus. This sphere is described in terms of the probability density function

$$p(r) = \frac{4}{a_0^3} r^2 e^{-2r/a_0} \qquad r \geq 0$$

where a_0 is the *Bohr radius* ($a_0 \approx 5.59 \times 10^{-11}$ m). The integral

$$P(r) = \int_0^r \frac{4}{a_0^3} s^2 e^{-2s/a_0}\, ds$$

gives the probability that the electron will be found within the sphere of radius r meters centered at the nucleus.
  (a) Verify that $p(r)$ is a probability density function.
  (b) Find $\lim_{r \to \infty} p(r)$. For what value of r does $p(r)$ have its maximum value?
  (c) Graph the density function.
  (d) Find the probability that the electron will be within the sphere of radius $4a_0$ centered at the nucleus.
  (e) Calculate the mean distance of the electron from the nucleus in the ground state of the hydrogen atom.

9 Review

CONCEPT CHECK

1. (a) How is the length of a curve defined?
  (b) Write an expression for the length of a smooth curve given by $y = f(x)$, $a \leq x \leq b$.
  (c) What if x is given as a function of y?

2. (a) Write an expression for the surface area of the surface obtained by rotating the curve $y = f(x)$, $a \leq x \leq b$, about the x-axis.
  (b) What if x is given as a function of y?
  (c) What if the curve is rotated about the y-axis?

3. Describe how we can find the hydrostatic force against a vertical wall submersed in a fluid.

4. (a) What is the physical significance of the center of mass of a thin plate?
  (b) If the plate lies between $y = f(x)$ and $y = 0$, where $a \leq x \leq b$, write expressions for the coordinates of the center of mass.

5. What does the Theorem of Pappus say?

6. Given a demand function $p(x)$, explain what is meant by the consumer surplus when the amount of a commodity currently available is X and the current selling price is P. Illustrate with a sketch.

7. (a) What is the cardiac output of the heart?
  (b) Explain how the cardiac output can be measured by the dye dilution method.

8. What is a probability density function? What properties does such a function have?

9. Suppose $f(x)$ is the probability density function for the weight of a female college student, where x is measured in pounds.
  (a) What is the meaning of the integral $\int_0^{100} f(x)\, dx$?
  (b) Write an expression for the mean of this density function.
  (c) How can we find the median of this density function?

10. What is a normal distribution? What is the significance of the standard deviation?

EXERCISES

1–2 Find the length of the curve.

1. $y = \frac{1}{6}(x^2 + 4)^{3/2}, \quad 0 \leq x \leq 3$

2. $y = 2 \ln \sin(\frac{1}{2}x), \quad \pi/3 \leq x \leq \pi$

3. (a) Find the length of the curve
$$y = \frac{x^4}{16} + \frac{1}{2x^2} \quad 1 \leq x \leq 2$$
   (b) Find the area of the surface obtained by rotating the curve in part (a) about the y-axis.

4. (a) The curve $y = x^2$, $0 \leq x \leq 1$, is rotated about the y-axis. Find the area of the resulting surface.
   (b) Find the area of the surface obtained by rotating the curve in part (a) about the x-axis.

5. Use Simpson's Rule with $n = 6$ to estimate the length of the curve $y = e^{-x^2}, 0 \leq x \leq 3$.

6. Use Simpson's Rule with $n = 6$ to estimate the area of the surface obtained by rotating the curve in Exercise 5 about the x-axis.

7. Find the length of the curve
$$y = \int_1^x \sqrt{\sqrt{t} - 1} \, dt \quad 1 \leq x \leq 16$$

8. Find the area of the surface obtained by rotating the curve in Exercise 7 about the y-axis.

9. A gate in an irrigation canal is constructed in the form of a trapezoid 3 ft wide at the bottom, 5 ft wide at the top, and 2 ft high. It is placed vertically in the canal, with the water extending to its top. Find the hydrostatic force on one side of the gate.

10. A trough is filled with water and its vertical ends have the shape of the parabolic region in the figure. Find the hydrostatic force on one end of the trough.

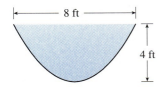

11–12 Find the centroid of the region bounded by the given curves.

11. $y = 4 - x^2, \quad y = x + 2$

12. $y = \sin x, \quad y = 0, \quad x = \pi/4, \quad x = 3\pi/4$

13–14 Find the centroid of the region shown.

13.

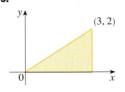

14.

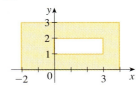

15. Find the volume obtained when the circle of radius 1 with center $(1, 0)$ is rotated about the y-axis.

16. Use the Theorem of Pappus and the fact that the volume of a sphere of radius r is $\frac{4}{3}\pi r^3$ to find the centroid of the semicircular region bounded by the curve $y = \sqrt{r^2 - x^2}$ and the x-axis.

17. The demand function for a commodity is given by $p = 2000 - 0.1x - 0.01x^2$. Find the consumer surplus when the sales level is 100.

18. After a 6-mg injection of dye into a heart, the readings of dye concentration at two-second intervals are as shown in the table. Use Simpson's Rule to estimate the cardiac output.

t	$c(t)$	t	$c(t)$
0	0	14	4.7
2	1.9	16	3.3
4	3.3	18	2.1
6	5.1	20	1.1
8	7.6	22	0.5
10	7.1	24	0
12	5.8		

19. (a) Explain why the function
$$f(x) = \begin{cases} \frac{\pi}{20} \sin\left(\frac{\pi x}{10}\right) & \text{if } 0 \leq x \leq 10 \\ 0 & \text{if } x < 0 \text{ or } x > 10 \end{cases}$$
is a probability density function.
   (b) Find $P(X < 4)$.
   (c) Calculate the mean. Is the value what you would expect?

20. Lengths of human pregnancies are normally distributed with mean 268 days and standard deviation 15 days. What percentage of pregnancies last between 250 days and 280 days?

21. The length of time spent waiting in line at a certain bank is modeled by an exponential density function with mean 8 minutes.
   (a) What is the probability that a customer is served in the first 3 minutes?
   (b) What is the probability that a customer has to wait more than 10 minutes?
   (c) What is the median waiting time?

PROBLEMS PLUS

1. Find the area of the region $S = \{(x, y) \mid x \geq 0, \ y \leq 1, \ x^2 + y^2 \leq 4y\}$.

2. Find the centroid of the region enclosed by the loop of the curve $y^2 = x^3 - x^4$.

3. If a sphere of radius r is sliced by a plane whose distance from the center of the sphere is d, then the sphere is divided into two pieces called *segments of one base*. The corresponding surfaces are called *spherical zones of one base*.
   (a) Determine the surface areas of the two spherical zones indicated in the figure.
   (b) Determine the approximate area of the Arctic Ocean by assuming that it is approximately circular in shape, with center at the North Pole and "circumference" at 75° north latitude. Use $r = 3960$ mi for the radius of Earth.
   (c) A sphere of radius r is inscribed in a right circular cylinder of radius r. Two planes perpendicular to the central axis of the cylinder and a distance h apart cut off a *spherical zone of two bases* on the sphere. Show that the surface area of the spherical zone equals the surface area of the region that the two planes cut off on the cylinder.
   (d) The *Torrid Zone* is the region on the surface of Earth that is between the Tropic of Cancer (23.45° north latitude) and the Tropic of Capricorn (23.45° south latitude). What is the area of the Torrid Zone?

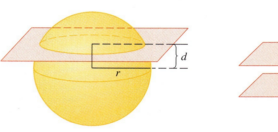
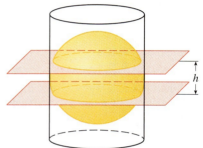

4. (a) Show that an observer at height H above the north pole of a sphere of radius r can see a part of the sphere that has area
$$\frac{2\pi r^2 H}{r + H}$$
   (b) Two spheres with radii r and R are placed so that the distance between their centers is d, where $d > r + R$. Where should a light be placed on the line joining the centers of the spheres in order to illuminate the largest total surface?

5. Suppose that the density of seawater, $\rho = \rho(z)$, varies with the depth z below the surface.
   (a) Show that the hydrostatic pressure is governed by the differential equation
$$\frac{dP}{dz} = \rho(z)g$$
   where g is the acceleration due to gravity. Let P_0 and ρ_0 be the pressure and density at $z = 0$. Express the pressure at depth z as an integral.
   (b) Suppose the density of seawater at depth z is given by $\rho = \rho_0 e^{z/H}$, where H is a positive constant. Find the total force, expressed as an integral, exerted on a vertical circular porthole of radius r whose center is located at a distance $L > r$ below the surface.

6. The figure shows a semicircle with radius 1, horizontal diameter PQ, and tangent lines at P and Q. At what height above the diameter should the horizontal line be placed so as to minimize the shaded area?

7. Let P be a pyramid with a square base of side $2b$ and suppose that S is a sphere with its center on the base of P and is tangent to all eight edges of P. Find the height of P. Then find the volume of the intersection of S and P.

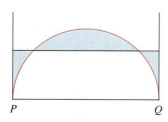

FIGURE FOR PROBLEM 6

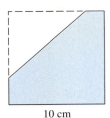

FIGURE FOR PROBLEM 10

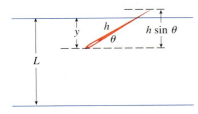

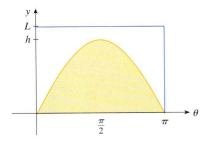

FIGURE FOR PROBLEM 11

8. Consider a flat metal plate to be placed vertically under water with its top 2 m below the surface of the water. Determine a shape for the plate so that if the plate is divided into any number of horizontal strips of equal height, the hydrostatic force on each strip is the same.

9. A uniform disk with radius 1 m is to be cut by a line so that the center of mass of the smaller piece lies halfway along a radius. How close to the center of the disk should the cut be made? (Express your answer correct to two decimal places.)

10. A triangle with area 30 cm^2 is cut from a corner of a square with side 10 cm, as shown in the figure. If the centroid of the remaining region is 4 cm from the right side of the square, how far is it from the bottom of the square?

11. In a famous 18th-century problem, known as *Buffon's needle problem*, a needle of length h is dropped onto a flat surface (for example, a table) on which parallel lines L units apart, $L \geq h$, have been drawn. The problem is to determine the probability that the needle will come to rest intersecting one of the lines. Assume that the lines run east-west, parallel to the x-axis in a rectangular coordinate system (as in the figure). Let y be the distance from the "southern" end of the needle to the nearest line to the north. (If the needle's southern end lies on a line, let $y = 0$. If the needle happens to lie east-west, let the "western" end be the "southern" end.) Let θ be the angle that the needle makes with a ray extending eastward from the "southern" end. Then $0 \leq y \leq L$ and $0 \leq \theta \leq \pi$. Note that the needle intersects one of the lines only when $y < h \sin \theta$. Now, the total set of possibilities for the needle can be identified with the rectangular region $0 \leq y \leq L$, $0 \leq \theta \leq \pi$, and the proportion of times that the needle intersects a line is the ratio

$$\frac{\text{area under } y = h \sin \theta}{\text{area of rectangle}}$$

This ratio is the probability that the needle intersects a line. Find the probability that the needle will intersect a line if $h = L$. What if $h = L/2$?

12. If the needle in Problem 11 has length $h > L$, it's possible for the needle to intersect more than one line.
    (a) If $L = 4$, find the probability that a needle of length 7 will intersect at least one line.
    [*Hint:* Proceed as in Problem 11. Define y as before; then the total set of possibilities for the needle can be identified with the same rectangular region $0 \leq y \leq L$, $0 \leq \theta \leq \pi$. What portion of the rectangle corresponds to the needle intersecting a line?]
    (b) If $L = 4$, find the probability that a needle of length 7 will intersect *two* lines.
    (c) If $2L < h \leq 3L$, find a general formula for the probability that the needle intersects three lines.

Chapter 10

By analyzing pairs of differential equations we gain insight into population cycles of predators and prey, such as the Canada lynx and snowshoe hare.

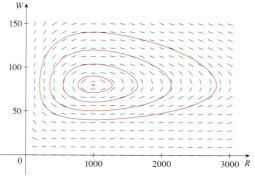

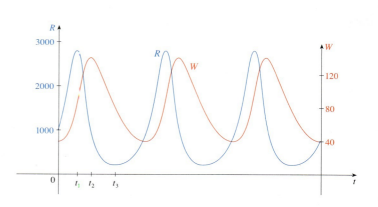

Differential Equations

Perhaps the most important of all the applications of calculus is to differential equations. When physical scientists or social scientists use calculus, more often than not it is to analyze a differential equation that has arisen in the process of modeling some phenomenon that they are studying.

Although it is often impossible to find an explicit formula for the solution of a differential equation, we will see that graphical and numerical approaches provide the needed information.

10.1 Modeling with Differential Equations

|||| Now is a good time to read (or reread) the discussion of mathematical modeling on page 25.

In describing the process of modeling in Section 1.2, we talked about formulating a mathematical model of a real-world problem either through intuitive reasoning about the phenomenon or from a physical law based on evidence from experiments. The mathematical model often takes the form of a *differential equation*, that is, an equation that contains an unknown function and some of its derivatives. This is not surprising because in a real-world problem we often notice that changes occur and we want to predict future behavior on the basis of how current values change. Let's begin by examining several examples of how differential equations arise when we model physical phenomena.

Models of Population Growth

One model for the growth of a population is based on the assumption that the population grows at a rate proportional to the size of the population. That is a reasonable assumption for a population of bacteria or animals under ideal conditions (unlimited environment, adequate nutrition, absence of predators, immunity from disease).

Let's identify and name the variables in this model:

$t =$ time (the independent variable)

$P =$ the number of individuals in the population (the dependent variable)

The rate of growth of the population is the derivative dP/dt. So our assumption that the rate of growth of the population is proportional to the population size is written as the equation

$$\frac{dP}{dt} = kP$$

where k is the proportionality constant. Equation 1 is our first model for population growth; it is a differential equation because it contains an unknown function P and its derivative dP/dt.

Having formulated a model, let's look at its consequences. If we rule out a population of 0, then $P(t) > 0$ for all t. So, if $k > 0$, then Equation 1 shows that $P'(t) > 0$ for all t. This means that the population is always increasing. In fact, as $P(t)$ increases, Equation 1 shows that dP/dt becomes larger. In other words, the growth rate increases as the population increases.

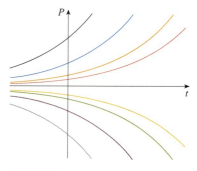

FIGURE 1
The family of solutions of $dP/dt = kP$

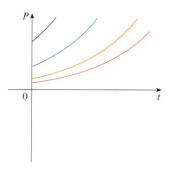

FIGURE 2
The family of solutions $P(t) = Ce^{kt}$
with $C > 0$ and $t \geq 0$

Let's try to think of a solution of Equation 1. This equation asks us to find a function whose derivative is a constant multiple of itself. We know that exponential functions have that property. In fact, if we let $P(t) = Ce^{kt}$, then

$$P'(t) = C(ke^{kt}) = k(Ce^{kt}) = kP(t)$$

Thus, any exponential function of the form $P(t) = Ce^{kt}$ is a solution of Equation 1. When we study this equation in detail in Section 10.4, we will see that there is no other solution.

Allowing C to vary through all the real numbers, we get the *family* of solutions $P(t) = Ce^{kt}$ whose graphs are shown in Figure 1. But populations have only positive values and so we are interested only in the solutions with $C > 0$. And we are probably concerned only with values of t greater than the initial time $t = 0$. Figure 2 shows the physically meaningful solutions. Putting $t = 0$, we get $P(0) = Ce^{k(0)} = C$, so the constant C turns out to be the initial population, $P(0)$.

Equation 1 is appropriate for modeling population growth under ideal conditions, but we have to recognize that a more realistic model must reflect the fact that a given environment has limited resources. Many populations start by increasing in an exponential manner, but the population levels off when it approaches its *carrying capacity K* (or decreases toward K if it ever exceeds K). For a model to take into account both trends, we make two assumptions:

- $\dfrac{dP}{dt} \approx kP$ if P is small   (Initially, the growth rate is proportional to P.)

- $\dfrac{dP}{dt} < 0$ if $P > K$ (P decreases if it ever exceeds K.)

A simple expression that incorporates both assumptions is given by the equation

$$\boxed{2} \qquad \frac{dP}{dt} = kP\left(1 - \frac{P}{K}\right)$$

Notice that if P is small compared with K, then P/K is close to 0 and so $dP/dt \approx kP$. If $P > K$, then $1 - P/K$ is negative and so $dP/dt < 0$.

Equation 2 is called the *logistic differential equation* and was proposed by the Dutch mathematical biologist Pierre-François Verhulst in the 1840s as a model for world population growth. We will develop techniques that enable us to find explicit solutions of the logistic equation in Section 10.5, but for now we can deduce qualitative characteristics of the solutions directly from Equation 2. We first observe that the constant functions $P(t) = 0$ and $P(t) = K$ are solutions because, in either case, one of the factors on the right side of Equation 2 is zero. (This certainly makes physical sense: If the population is ever either 0 or at the carrying capacity, it stays that way.) These two constant solutions are called *equilibrium solutions*.

If the initial population $P(0)$ lies between 0 and K, then the right side of Equation 2 is positive, so $dP/dt > 0$ and the population increases. But if the population exceeds the carrying capacity ($P > K$), then $1 - P/K$ is negative, so $dP/dt < 0$ and the population decreases. Notice that, in either case, if the population approaches the carrying capacity ($P \to K$), then $dP/dt \to 0$, which means the population levels off. So we expect that the solutions of the logistic differential equation have graphs that look something like the ones in Figure 3. Notice that the graphs move away from the equilibrium solution $P = 0$ and move toward the equilibrium solution $P = K$.

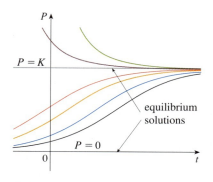

FIGURE 3
Solutions of the logistic equation

A Model for the Motion of a Spring

Let's now look at an example of a model from the physical sciences. We consider the motion of an object with mass m at the end of a vertical spring (as in Figure 4). In Section 6.4 we discussed Hooke's Law, which says that if the spring is stretched (or compressed) x units from its natural length, then it exerts a force that is proportional to x:

$$\text{restoring force} = -kx$$

where k is a positive constant (called the *spring constant*). If we ignore any external resisting forces (due to air resistance or friction) then, by Newton's Second Law (force equals mass times acceleration), we have

$$\boxed{3} \qquad m\frac{d^2x}{dt^2} = -kx$$

This is an example of what is called a *second-order differential equation* because it involves second derivatives. Let's see what we can guess about the form of the solution directly from the equation. We can rewrite Equation 3 in the form

$$\frac{d^2x}{dt^2} = -\frac{k}{m}x$$

which says that the second derivative of x is proportional to x but has the opposite sign. We know two functions with this property, the sine and cosine functions. In fact, it turns out that all solutions of Equation 3 can be written as combinations of certain sine and cosine functions (see Exercise 3). This is not surprising; we expect the spring to oscillate about its equilibrium position and so it is natural to think that trigonometric functions are involved.

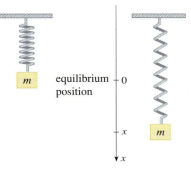

FIGURE 4

General Differential Equations

In general, a **differential equation** is an equation that contains an unknown function and one or more of its derivatives. The **order** of a differential equation is the order of the highest derivative that occurs in the equation. Thus, Equations 1 and 2 are first-order equations and Equation 3 is a second-order equation. In all three of those equations the independent variable is called t and represents time, but in general the independent variable doesn't have to represent time. For example, when we consider the differential equation

$$\boxed{4} \qquad y' = xy$$

it is understood that y is an unknown function of x.

A function f is called a **solution** of a differential equation if the equation is satisfied when $y = f(x)$ and its derivatives are substituted into the equation. Thus, f is a solution of Equation 4 if

$$f'(x) = xf(x)$$

for all values of x in some interval.

When we are asked to *solve* a differential equation we are expected to find all possible solutions of the equation. We have already solved some particularly simple differential equations, namely, those of the form

$$y' = f(x)$$

For instance, we know that the general solution of the differential equation

$$y' = x^3$$

is given by

$$y = \frac{x^4}{4} + C$$

where C is an arbitrary constant.

But, in general, solving a differential equation is not an easy matter. There is no systematic technique that enables us to solve all differential equations. In Section 10.2, however, we will see how to draw rough graphs of solutions even when we have no explicit formula. We will also learn how to find numerical approximations to solutions.

EXAMPLE 1 Show that every member of the family of functions

$$y = \frac{1 + ce^t}{1 - ce^t}$$

is a solution of the differential equation $y' = \frac{1}{2}(y^2 - 1)$.

SOLUTION We use the Quotient Rule to differentiate the expression for y:

$$y' = \frac{(1 - ce^t)(ce^t) - (1 + ce^t)(-ce^t)}{(1 - ce^t)^2}$$

$$= \frac{ce^t - c^2 e^{2t} + ce^t + c^2 e^{2t}}{(1 - ce^t)^2} = \frac{2ce^t}{(1 - ce^t)^2}$$

The right side of the differential equation becomes

$$\tfrac{1}{2}(y^2 - 1) = \frac{1}{2}\left[\left(\frac{1 + ce^t}{1 - ce^t}\right)^2 - 1\right] = \frac{1}{2}\left[\frac{(1 + ce^t)^2 - (1 - ce^t)^2}{(1 - ce^t)^2}\right]$$

$$= \frac{1}{2}\frac{4ce^t}{(1 - ce^t)^2} = \frac{2ce^t}{(1 - ce^t)^2}$$

Therefore, for every value of c, the given function is a solution of the differential equation.

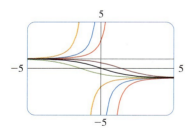

|||| Figure 5 shows graphs of seven members of the family in Example 1. The differential equation shows that if $y \approx \pm 1$, then $y' \approx 0$. That is borne out by the flatness of the graphs near $y = 1$ and $y = -1$.

FIGURE 5

When applying differential equations, we are usually not as interested in finding a family of solutions (the *general solution*) as we are in finding a solution that satisfies some additional requirement. In many physical problems we need to find the particular solution that satisfies a condition of the form $y(t_0) = y_0$. This is called an **initial condition**, and the problem of finding a solution of the differential equation that satisfies the initial condition is called an **initial-value problem**.

Geometrically, when we impose an initial condition, we look at the family of solution curves and pick the one that passes through the point (t_0, y_0). Physically, this corresponds to measuring the state of a system at time t_0 and using the solution of the initial-value problem to predict the future behavior of the system.

EXAMPLE 2 Find a solution of the differential equation $y' = \frac{1}{2}(y^2 - 1)$ that satisfies the initial condition $y(0) = 2$.

SOLUTION Substituting the values $t = 0$ and $y = 2$ into the formula

$$y = \frac{1 + ce^t}{1 - ce^t}$$

from Example 1, we get

$$2 = \frac{1 + ce^0}{1 - ce^0} = \frac{1 + c}{1 - c}$$

Solving this equation for c, we get $2 - 2c = 1 + c$, which gives $c = \frac{1}{3}$. So the solution of the initial-value problem is

$$y = \frac{1 + \frac{1}{3}e^t}{1 - \frac{1}{3}e^t} = \frac{3 + e^t}{3 - e^t}$$

10.1 Exercises

1. Show that $y = x - x^{-1}$ is a solution of the differential equation $xy' + y = 2x$.

2. Verify that $y = \sin x \cos x - \cos x$ is a solution of the initial-value problem
$$y' + (\tan x)y = \cos^2 x \qquad y(0) = -1$$
on the interval $-\pi/2 < x < \pi/2$.

3. (a) For what nonzero values of k does the function $y = \sin kt$ satisfy the differential equation $y'' + 9y = 0$?
   (b) For those values of k, verify that every member of the family of functions
   $$y = A \sin kt + B \cos kt$$
   is also a solution.

4. For what values of r does the function $y = e^{rt}$ satisfy the differential equation $y'' + y' - 6y = 0$?

5. Which of the following functions are solutions of the differential equation $y'' + 2y' + y = 0$?
   (a) $y = e^t$ (b) $y = e^{-t}$
   (c) $y = te^{-t}$ (d) $y = t^2 e^{-t}$

6. (a) Show that every member of the family of functions $y = Ce^{x^2/2}$ is a solution of the differential equation $y' = xy$.
   (b) Illustrate part (a) by graphing several members of the family of solutions on a common screen.
   (c) Find a solution of the differential equation $y' = xy$ that satisfies the initial condition $y(0) = 5$.
   (d) Find a solution of the differential equation $y' = xy$ that satisfies the initial condition $y(1) = 2$.

7. (a) What can you say about a solution of the equation $y' = -y^2$ just by looking at the differential equation?
   (b) Verify that all members of the family $y = 1/(x + C)$ are solutions of the equation in part (a).
   (c) Can you think of a solution of the differential equation $y' = -y^2$ that is not a member of the family in part (b)?
   (d) Find a solution of the initial-value problem
   $$y' = -y^2 \qquad y(0) = 0.5$$

8. (a) What can you say about the graph of a solution of the equation $y' = xy^3$ when x is close to 0? What if x is large?
   (b) Verify that all members of the family $y = (c - x^2)^{-1/2}$ are solutions of the differential equation $y' = xy^3$.
   (c) Graph several members of the family of solutions on a common screen. Do the graphs confirm what you predicted in part (a)?
   (d) Find a solution of the initial-value problem
   $$y' = xy^3 \qquad y(0) = 2$$

9. A population is modeled by the differential equation
$$\frac{dP}{dt} = 1.2P\left(1 - \frac{P}{4200}\right)$$
   (a) For what values of P is the population increasing?
   (b) For what values of P is the population decreasing?
   (c) What are the equilibrium solutions?

10. A function $y(t)$ satisfies the differential equation

$$\frac{dy}{dt} = y^4 - 6y^3 + 5y^2$$

(a) What are the constant solutions of the equation?
(b) For what values of y is y increasing?
(c) For what values of y is y decreasing?

11. Explain why the functions with the given graphs *can't* be solutions of the differential equation

$$\frac{dy}{dt} = e^t(y-1)^2$$

(a) 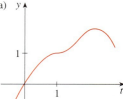 (b)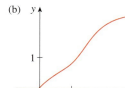

12. The function with the given graph is a solution of one of the following differential equations. Decide which is the correct equation and justify your answer.

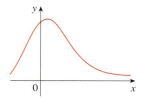

A. $y' = 1 + xy$ B. $y' = -2xy$ C. $y' = 1 - 2xy$

13. Psychologists interested in learning theory study **learning curves**. A learning curve is the graph of a function $P(t)$, the performance of someone learning a skill as a function of the training time t. The derivative dP/dt represents the rate at which performance improves.

(a) When do you think P increases most rapidly? What happens to dP/dt as t increases? Explain.
(b) If M is the maximum level of performance of which the learner is capable, explain why the differential equation

$$\frac{dP}{dt} = k(M - P) \qquad k \text{ a positive constant}$$

is a reasonable model for learning.
(c) Make a rough sketch of a possible solution of this differential equation.

14. Suppose you have just poured a cup of freshly brewed coffee with temperature 95°C in a room where the temperature is 20°C.

(a) When do you think the coffee cools most quickly? What happens to the rate of cooling as time goes by? Explain.
(b) **Newton's Law of Cooling** states that the rate of cooling of an object is proportional to the temperature difference between the object and its surroundings, provided that this difference is not too large. Write a differential equation that expresses Newton's Law of Cooling for this particular situation. What is the initial condition? In view of your answer to part (a), do you think this differential equation is an appropriate model for cooling?
(c) Make a rough sketch of the graph of the solution of the initial-value problem in part (b).

10.2 Direction Fields and Euler's Method

Unfortunately, it's impossible to solve most differential equations in the sense of obtaining an explicit formula for the solution. In this section we show that, despite the absence of an explicit solution, we can still learn a lot about the solution through a graphical approach (direction fields) or a numerical approach (Euler's method).

Direction Fields

Suppose we are asked to sketch the graph of the solution of the initial-value problem

$$y' = x + y \qquad y(0) = 1$$

We don't know a formula for the solution, so how can we possibly sketch its graph? Let's think about what the differential equation means. The equation $y' = x + y$ tells us that the slope at any point (x, y) on the graph (called the *solution curve*) is equal to the sum of the

x- and y-coordinates of the point (see Figure 1). In particular, because the curve passes through the point $(0, 1)$, its slope there must be $0 + 1 = 1$. So a small portion of the solution curve near the point $(0, 1)$ looks like a short line segment through $(0, 1)$ with slope 1 (see Figure 2).

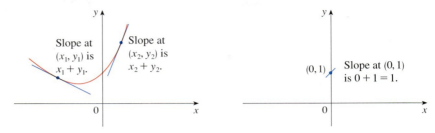

FIGURE 1
A solution of $y' = x + y$

FIGURE 2
Beginning of the solution curve through $(0, 1)$

As a guide to sketching the rest of the curve, let's draw short line segments at a number of points (x, y) with slope $x + y$. The result is called a *direction field* and is shown in Figure 3. For instance, the line segment at the point $(1, 2)$ has slope $1 + 2 = 3$. The direction field allows us to visualize the general shape of the solution curves by indicating the direction in which the curves proceed at each point.

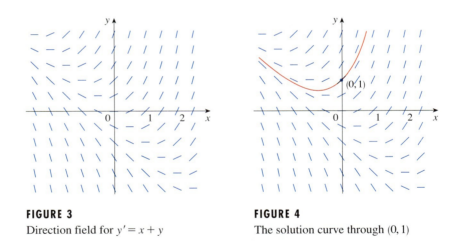

FIGURE 3
Direction field for $y' = x + y$

FIGURE 4
The solution curve through $(0, 1)$

Now we can sketch the solution curve through the point $(0, 1)$ by following the direction field as in Figure 4. Notice that we have drawn the curve so that it is parallel to nearby line segments.

In general, suppose we have a first-order differential equation of the form

$$y' = F(x, y)$$

where $F(x, y)$ is some expression in x and y. The differential equation says that the slope of a solution curve at a point (x, y) on the curve is $F(x, y)$. If we draw short line segments with slope $F(x, y)$ at several points (x, y), the result is called a **direction field** (or **slope field**). These line segments indicate the direction in which a solution curve is heading, so the direction field helps us visualize the general shape of these curves.

630 |||| CHAPTER 10 DIFFERENTIAL EQUATIONS

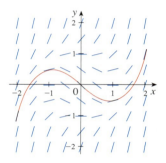

FIGURE 5

Module 10.2A shows direction fields and solution curves for a variety of differential equations.

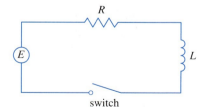

FIGURE 9

EXAMPLE 1
(a) Sketch the direction field for the differential equation $y' = x^2 + y^2 - 1$.
(b) Use part (a) to sketch the solution curve that passes through the origin.

SOLUTION
(a) We start by computing the slope at several points in the following chart:

x	-2	-1	0	1	2	-2	-1	0	1	2	...
y	0	0	0	0	0	1	1	1	1	1	...
$y' = x^2 + y^2 - 1$	3	0	-1	0	3	4	1	0	1	4	...

Now we draw short line segments with these slopes at these points. The result is the direction field shown in Figure 5.

(b) We start at the origin and move to the right in the direction of the line segment (which has slope -1). We continue to draw the solution curve so that it moves parallel to the nearby line segments. The resulting solution curve is shown in Figure 6. Returning to the origin, we draw the solution curve to the left as well.

The more line segments we draw in a direction field, the clearer the picture becomes. Of course, it's tedious to compute slopes and draw line segments for a huge number of points by hand, but computers are well suited for this task. Figure 7 shows a more detailed, computer-drawn direction field for the differential equation in Example 1. It enables us to draw, with reasonable accuracy, the solution curves shown in Figure 8 with y-intercepts $-2, -1, 0, 1,$ and 2.

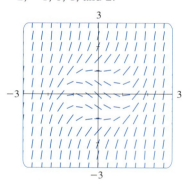

FIGURE 7

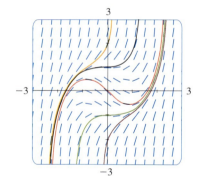

FIGURE 8

Now let's see how direction fields give insight into physical situations. The simple electric circuit shown in Figure 9 contains an electromotive force (usually a battery or generator) that produces a voltage of $E(t)$ volts (V) and a current of $I(t)$ amperes (A) at time t. The circuit also contains a resistor with a resistance of R ohms (Ω) and an inductor with an inductance of L henries (H).

Ohm's Law gives the drop in voltage due to the resistor as RI. The voltage drop due to the inductor is $L(dI/dt)$. One of Kirchhoff's laws says that the sum of the voltage drops is equal to the supplied voltage $E(t)$. Thus, we have

$$\boxed{1} \qquad L\frac{dI}{dt} + RI = E(t)$$

which is a first-order differential equation that models the current I at time t.

EXAMPLE 2 Suppose that in the simple circuit of Figure 9 the resistance is 12 Ω, the inductance is 4 H, and a battery gives a constant voltage of 60 V.
(a) Draw a direction field for Equation 1 with these values.
(b) What can you say about the limiting value of the current?
(c) Identify any equilibrium solutions.
(d) If the switch is closed when $t = 0$ so the current starts with $I(0) = 0$, use the direction field to sketch the solution curve.

SOLUTION
(a) If we put $L = 4$, $R = 12$, and $E(t) = 60$ in Equation 1, we get

$$4\frac{dI}{dt} + 12I = 60 \quad \text{or} \quad \frac{dI}{dt} = 15 - 3I$$

The direction field for this differential equation is shown in Figure 10.

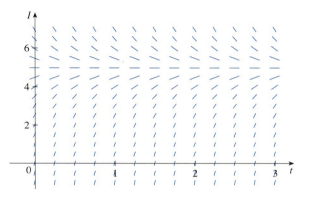

FIGURE 10

(b) It appears from the direction field that all solutions approach the value 5 A, that is,

$$\lim_{t \to \infty} I(t) = 5$$

$\frac{dI}{dt} = 15 - 3I$

(c) It appears that the constant function $I(t) = 5$ is an equilibrium solution. Indeed, we can verify this directly from the differential equation. If $I(t) = 5$, then the left side is $dI/dt = 0$ and the right side is $15 - 3(5) = 0$.

(d) We use the direction field to sketch the solution curve that passes through $(0, 0)$, as shown in red in Figure 11.

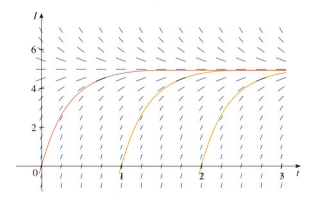

FIGURE 11

Notice from Figure 10 that the line segments along any horizontal line are parallel. That is because the independent variable t does not occur on the right side of the equation $I' = 15 - 3I$. In general, a differential equation of the form

$$y' = f(y)$$

in which the independent variable is missing from the right side, is called **autonomous**. For such an equation, the slopes corresponding to two different points with the same y-coordinate must be equal. This means that if we know one solution to an autonomous differential equation, then we can obtain infinitely many others just by shifting the graph of the known solution to the right or left. In Figure 11 we have shown the solutions that result from shifting the solution curve of Example 2 one and two time units (namely, seconds) to the right. They correspond to closing the switch when $t = 1$ or $t = 2$.

Euler's Method

The basic idea behind direction fields can be used to find numerical approximations to solutions of differential equations. We illustrate the method on the initial-value problem that we used to introduce direction fields:

$$y' = x + y \qquad y(0) = 1$$

The differential equation tells us that $y'(0) = 0 + 1 = 1$, so the solution curve has slope 1 at the point $(0, 1)$. As a first approximation to the solution we could use the linear approximation $L(x) = x + 1$. In other words, we could use the tangent line at $(0, 1)$ as a rough approximation to the solution curve (see Figure 12).

Euler's idea was to improve on this approximation by proceeding only a short distance along this tangent line and then making a midcourse correction by changing direction as indicated by the direction field. Figure 13 shows what happens if we start out along the tangent line but stop when $x = 0.5$. (This horizontal distance traveled is called the *step size*.) Since $L(0.5) = 1.5$, we have $y(0.5) \approx 1.5$ and we take $(0.5, 1.5)$ as the starting point for a new line segment. The differential equation tells us that $y'(0.5) = 0.5 + 1.5 = 2$, so we use the linear function

$$y = 1.5 + 2(x - 0.5) = 2x + 0.5$$

as an approximation to the solution for $x > 0.5$ (the orange segment in Figure 13). If we decrease the step size from 0.5 to 0.25, we get the better Euler approximation shown in Figure 14.

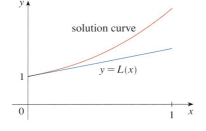

FIGURE 12
First Euler approximation

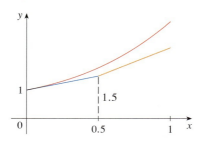

FIGURE 13
Euler approximation with step size 0.5

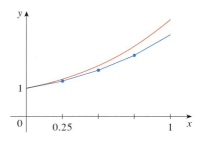

FIGURE 14
Euler approximation with step size 0.25

In general, Euler's method says to start at the point given by the initial value and proceed in the direction indicated by the direction field. Stop after a short time, look at the slope at the new location, and proceed in that direction. Keep stopping and changing direc-

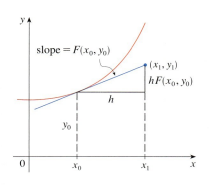

FIGURE 15

 Module 10.2B shows how Euler's method works numerically and visually for a variety of differential equations and step sizes.

|||| Computer software packages that produce numerical approximations to solutions of differential equations use methods that are refinements of Euler's method. Although Euler's method is simple and not as accurate, it is the basic idea on which the more accurate methods are based.

tion according to the direction field. Euler's method does not produce the exact solution to an initial-value problem—it gives approximations. But by decreasing the step size (and therefore increasing the number of midcourse corrections), we obtain successively better approximations to the exact solution. (Compare Figures 12, 13, and 14.)

For the general first-order initial-value problem $y' = F(x, y)$, $y(x_0) = y_0$, our aim is to find approximate values for the solution at equally spaced numbers x_0, $x_1 = x_0 + h$, $x_2 = x_1 + h, \ldots$, where h is the step size. The differential equation tells us that the slope at (x_0, y_0) is $y' = F(x_0, y_0)$, so Figure 15 shows that the approximate value of the solution when $x = x_1$ is

$$y_1 = y_0 + hF(x_0, y_0)$$

Similarly,
$$y_2 = y_1 + hF(x_1, y_1)$$

In general,
$$y_n = y_{n-1} + hF(x_{n-1}, y_{n-1})$$

EXAMPLE 3 Use Euler's method with step size 0.1 to construct a table of approximate values for the solution of the initial-value problem

$$y' = x + y \qquad y(0) = 1$$

SOLUTION We are given that $h = 0.1$, $x_0 = 0$, $y_0 = 1$, and $F(x, y) = x + y$. So we have

$$y_1 = y_0 + hF(x_0, y_0) = 1 + 0.1(0 + 1) = 1.1$$

$$y_2 = y_1 + hF(x_1, y_1) = 1.1 + 0.1(0.1 + 1.1) = 1.22$$

$$y_3 = y_2 + hF(x_2, y_2) = 1.22 + 0.1(0.2 + 1.22) = 1.362$$

This means that if $y(x)$ is the exact solution, then $y(0.3) \approx 1.362$.

Proceeding with similar calculations, we get the values in the table:

n	x_n	y_n	n	x_n	y_n
1	0.1	1.100000	6	0.6	1.943122
2	0.2	1.220000	7	0.7	2.197434
3	0.3	1.362000	8	0.8	2.487178
4	0.4	1.528200	9	0.9	2.815895
5	0.5	1.721020	10	1.0	3.187485

For a more accurate table of values in Example 3 we could decrease the step size. But for a large number of small steps the amount of computation is considerable and so we need to program a calculator or computer to carry out these calculations. The following table shows the results of applying Euler's method with decreasing step size to the initial-value problem of Example 3.

Step size	Euler estimate of $y(0.5)$	Euler estimate of $y(1)$
0.500	1.500000	2.500000
0.250	1.625000	2.882813
0.100	1.721020	3.187485
0.050	1.757789	3.306595
0.020	1.781212	3.383176
0.010	1.789264	3.409628
0.005	1.793337	3.423034
0.001	1.796619	3.433848

Notice that the Euler estimates in the table seem to be approaching limits, namely, the true values of $y(0.5)$ and $y(1)$. Figure 16 shows graphs of the Euler approximations with step sizes 0.5, 0.25, 0.1, 0.05, 0.02, 0.01, and 0.005. They are approaching the exact solution curve as the step size h approaches 0.

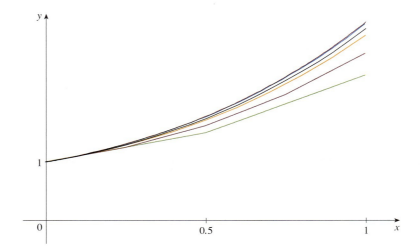

FIGURE 16
Euler approximations
approaching the exact solution

EXAMPLE 4 In Example 2 we discussed a simple electric circuit with resistance 12 Ω, inductance 4 H, and a battery with voltage 60 V. If the switch is closed when $t = 0$, we modeled the current I at time t by the initial-value problem

$$\frac{dI}{dt} = 15 - 3I \qquad I(0) = 0$$

Estimate the current in the circuit half a second after the switch is closed.

SOLUTION We use Euler's method with $F(t, I) = 15 - 3I$, $t_0 = 0$, $I_0 = 0$, and step size $h = 0.1$ second:

$$I_1 = 0 + 0.1(15 - 3 \cdot 0) = 1.5$$

$$I_2 = 1.5 + 0.1(15 - 3 \cdot 1.5) = 2.55$$

$$I_3 = 2.55 + 0.1(15 - 3 \cdot 2.55) = 3.285$$

$$I_4 = 3.285 + 0.1(15 - 3 \cdot 3.285) = 3.7995$$

$$I_5 = 3.7995 + 0.1(15 - 3 \cdot 3.7995) = 4.15965$$

So the current after 0.5 s is

$$I(0.5) \approx 4.16 \text{ A}$$

10.2 Exercises

1. A direction field for the differential equation $y' = y(1 - \frac{1}{4}y^2)$ is shown.
   (a) Sketch the graphs of the solutions that satisfy the given initial conditions.
      (i) $y(0) = 1$ (ii) $y(0) = -1$
      (iii) $y(0) = -3$ (iv) $y(0) = 3$
   (b) Find all the equilibrium solutions.

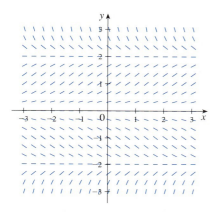

2. A direction field for the differential equation $y' = x \sin y$ is shown.
   (a) Sketch the graphs of the solutions that satisfy the given initial conditions.
      (i) $y(0) = 1$ (ii) $y(0) = 2$ (iii) $y(0) = \pi$
      (iv) $y(0) = 4$ (v) $y(0) = 5$
   (b) Find all the equilibrium solutions.

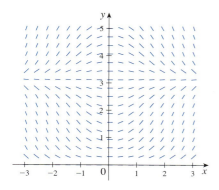

3–6 ⫼ Match the differential equation with its direction field (labeled I–IV). Give reasons for your answer.

3. $y' = y - 1$
4. $y' = y - x$
5. $y' = y^2 - x^2$
6. $y' = y^3 - x^3$

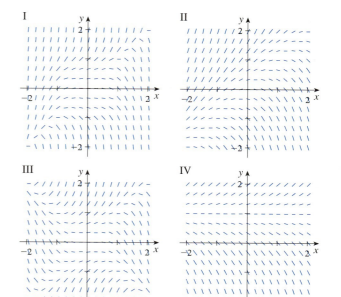

7. Use the direction field labeled I (for Exercises 3–6) to sketch the graphs of the solutions that satisfy the given initial conditions.
   (a) $y(0) = 1$ (b) $y(0) = 0$ (c) $y(0) = -1$

8. Repeat Exercise 7 for the direction field labeled III.

9–10 ⫼ Sketch a direction field for the differential equation. Then use it to sketch three solution curves.

9. $y' = 1 + y$
10. $y' = x^2 - y^2$

11–14 ⫼ Sketch the direction field of the differential equation. Then use it to sketch a solution curve that passes through the given point.

11. $y' = y - 2x$, $(1, 0)$
12. $y' = 1 - xy$, $(0, 0)$
13. $y' = y + xy$, $(0, 1)$
14. $y' = x - xy$, $(1, 0)$

CAS **15–16** ⫼ Use a computer algebra system to draw a direction field for the given differential equation. Get a printout and sketch on it the solution curve that passes through $(0, 1)$. Then use the CAS to draw the solution curve and compare it with your sketch.

15. $y' = y \sin 2x$
16. $y' = \sin(x + y)$

CAS **17.** Use a computer algebra system to draw a direction field for the differential equation $y' = y^3 - 4y$. Get a printout and sketch on it solutions that satisfy the initial condition $y(0) = c$ for various values of c. For what values of c does $\lim_{t \to \infty} y(t)$ exist? What are the possible values for this limit?

18. Make a rough sketch of a direction field for the autonomous differential equation $y' = f(y)$, where the graph of f is as shown. How does the limiting behavior of solutions depend on the value of $y(0)$?

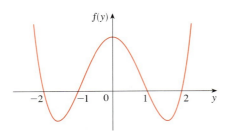

19. (a) Use Euler's method with each of the following step sizes to estimate the value of $y(0.4)$, where y is the solution of the initial-value problem $y' = y$, $y(0) = 1$.
 (i) $h = 0.4$ (ii) $h = 0.2$ (iii) $h = 0.1$
 (b) We know that the exact solution of the initial-value problem in part (a) is $y = e^x$. Draw, as accurately as you can, the graph of $y = e^x$, $0 \leq x \leq 0.4$, together with the Euler approximations using the step sizes in part (a). (Your sketches should resemble Figures 12, 13, and 14.) Use your sketches to decide whether your estimates in part (a) are underestimates or overestimates.
 (c) The error in Euler's method is the difference between the exact value and the approximate value. Find the errors made in part (a) in using Euler's method to estimate the true value of $y(0.4)$, namely $e^{0.4}$. What happens to the error each time the step size is halved?

20. A direction field for a differential equation is shown. Draw, with a ruler, the graphs of the Euler approximations to the solution curve that passes through the origin. Use step sizes $h = 1$ and $h = 0.5$. Will the Euler estimates be underestimates or overestimates? Explain.

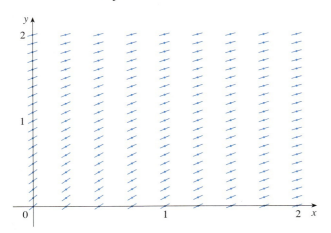

21. Use Euler's method with step size 0.5 to compute the approximate y-values y_1, y_2, y_3, and y_4 of the solution of the initial-value problem $y' = y - 2x$, $y(1) = 0$.

22. Use Euler's method with step size 0.2 to estimate $y(1)$, where $y(x)$ is the solution of the initial-value problem $y' = 1 - xy$, $y(0) = 0$.

23. Use Euler's method with step size 0.1 to estimate $y(0.5)$, where $y(x)$ is the solution of the initial-value problem $y' = y + xy$, $y(0) = 1$.

24. (a) Use Euler's method with step size 0.2 to estimate $y(1.4)$, where $y(x)$ is the solution of the initial-value problem $y' = x - xy$, $y(1) = 0$.
 (b) Repeat part (a) with step size 0.1.

25. (a) Program a calculator or computer to use Euler's method to compute $y(1)$, where $y(x)$ is the solution of the initial-value problem

$$\frac{dy}{dx} + 3x^2 y = 6x^2 \qquad y(0) = 3$$

 (i) $h = 1$ (ii) $h = 0.1$
 (iii) $h = 0.01$ (iv) $h = 0.001$
 (b) Verify that $y = 2 + e^{-x^3}$ is the exact solution of the differential equation.
 (c) Find the errors in using Euler's method to compute $y(1)$ with the step sizes in part (a). What happens to the error when the step size is divided by 10?

CAS 26. (a) Program your computer algebra system, using Euler's method with step size 0.01, to calculate $y(2)$, where y is the solution of the initial-value problem

$$y' = x^3 - y^3 \qquad y(0) = 1$$

 (b) Check your work by using the CAS to draw the solution curve.

27. The figure shows a circuit containing an electromotive force, a capacitor with a capacitance of C farads (F), and a resistor with a resistance of R ohms (Ω). The voltage drop across the capacitor is Q/C, where Q is the charge (in coulombs), so in this case Kirchhoff's Law gives

$$RI + \frac{Q}{C} = E(t)$$

But $I = dQ/dt$, so we have

$$R\frac{dQ}{dt} + \frac{1}{C}Q = E(t)$$

Suppose the resistance is 5 Ω, the capacitance is 0.05 F, and a battery gives a constant voltage of 60 V.
 (a) Draw a direction field for this differential equation.
 (b) What is the limiting value of the charge?

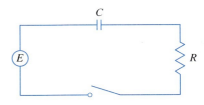

(c) Is there an equilibrium solution?
(d) If the initial charge is $Q(0) = 0$ C, use the direction field to sketch the solution curve.
(e) If the initial charge is $Q(0) = 0$ C, use Euler's method with step size 0.1 to estimate the charge after half a second.

28. In Exercise 14 in Section 10.1 we considered a 95°C cup of coffee in a 20°C room. Suppose it is known that the coffee cools at a rate of 1°C per minute when its temperature is 70°C.
(a) What does the differential equation become in this case?
(b) Sketch a direction field and use it to sketch the solution curve for the initial-value problem. What is the limiting value of the temperature?
(c) Use Euler's method with step size $h = 2$ minutes to estimate the temperature of the coffee after 10 minutes.

10.3 Separable Equations

We have looked at first-order differential equations from a geometric point of view (direction fields) and from a numerical point of view (Euler's method). What about the symbolic point of view? It would be nice to have an explicit formula for a solution of a differential equation. Unfortunately, that is not always possible. But in this section we examine a certain type of differential equation that *can* be solved explicitly.

A **separable equation** is a first-order differential equation in which the expression for dy/dx can be factored as a function of x times a function of y. In other words, it can be written in the form

$$\frac{dy}{dx} = g(x)f(y)$$

The name *separable* comes from the fact that the expression on the right side can be "separated" into a function of x and a function of y. Equivalently, if $f(y) \neq 0$, we could write

$$\boxed{1} \qquad \frac{dy}{dx} = \frac{g(x)}{h(y)}$$

where $h(y) = 1/f(y)$. To solve this equation we rewrite it in the differential form

$$h(y)\,dy = g(x)\,dx$$

so that all y's are on one side of the equation and all x's are on the other side. Then we integrate both sides of the equation:

|||| The technique for solving separable differential equations was first used by James Bernoulli (in 1690) in solving a problem about pendulums and by Leibniz (in a letter to Huygens in 1691). John Bernoulli explained the general method in a paper published in 1694.

$$\boxed{2} \qquad \int h(y)\,dy = \int g(x)\,dx$$

Equation 2 defines y implicitly as a function of x. In some cases we may be able to solve for y in terms of x.

We use the Chain Rule to justify this procedure: If h and g satisfy (2), then

$$\frac{d}{dx}\left(\int h(y)\,dy\right) = \frac{d}{dx}\left(\int g(x)\,dx\right)$$

so

$$\frac{d}{dy}\left(\int h(y)\,dy\right)\frac{dy}{dx} = g(x)$$

and

$$h(y)\frac{dy}{dx} = g(x)$$

Thus, Equation 1 is satisfied.

EXAMPLE 1

(a) Solve the differential equation $\dfrac{dy}{dx} = \dfrac{x^2}{y^2}$.

(b) Find the solution of this equation that satisfies the initial condition $y(0) = 2$.

SOLUTION

(a) We write the equation in terms of differentials and integrate both sides:

$$y^2\,dy = x^2\,dx$$

$$\int y^2\,dy = \int x^2\,dx$$

$$\tfrac{1}{3}y^3 = \tfrac{1}{3}x^3 + C$$

where C is an arbitrary constant. (We could have used a constant C_1 on the left side and another constant C_2 on the right side. But then we could combine these constants by writing $C = C_2 - C_1$.)

Solving for y, we get

$$y = \sqrt[3]{x^3 + 3C}$$

We could leave the solution like this or we could write it in the form

$$y = \sqrt[3]{x^3 + K}$$

where $K = 3C$. (Since C is an arbitrary constant, so is K.)

(b) If we put $x = 0$ in the general solution in part (a), we get $y(0) = \sqrt[3]{K}$. To satisfy the initial condition $y(0) = 2$, we must have $\sqrt[3]{K} = 2$ and so $K = 8$.

Thus, the solution of the initial-value problem is

$$y = \sqrt[3]{x^3 + 8}$$

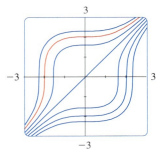

|||| Figure 1 shows graphs of several members of the family of solutions of the differential equation in Example 1. The solution of the initial-value problem in part (b) is shown in red.

FIGURE 1

EXAMPLE 2 Solve the differential equation $\dfrac{dy}{dx} = \dfrac{6x^2}{2y + \cos y}$.

SOLUTION Writing the equation in differential form and integrating both sides, we have

$$(2y + \cos y)\,dy = 6x^2\,dx$$

$$\int (2y + \cos y)\,dy = \int 6x^2\,dx$$

$$\boxed{3} \qquad y^2 + \sin y = 2x^3 + C$$

where C is a constant. Equation 3 gives the general solution implicitly. In this case it's impossible to solve the equation to express y explicitly as a function of x.

EXAMPLE 3 Solve the equation $y' = x^2 y$.

SOLUTION First we rewrite the equation using Leibniz notation:

$$\dfrac{dy}{dx} = x^2 y$$

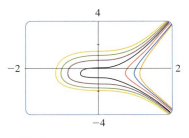

|||| Some computer algebra systems can plot curves defined by implicit equations. Figure 2 shows the graphs of several members of the family of solutions of the differential equation in Example 2. As we look at the curves from left to right, the values of C are 3, 2, 1, 0, -1, -2, and -3.

FIGURE 2

SECTION 10.3 SEPARABLE EQUATIONS ❙❙❙❙ 639

|||| If a solution y is a function that satisfies $y(x) \neq 0$ for some x, it follows from a uniqueness theorem for solutions of differential equations that $y(x) \neq 0$ for all x.

If $y \neq 0$, we can rewrite it in differential notation and integrate:

$$\frac{dy}{y} = x^2\, dx \qquad y \neq 0$$

$$\int \frac{dy}{y} = \int x^2\, dx$$

$$\ln|y| = \frac{x^3}{3} + C$$

This equation defines y implicitly as a function of x. But in this case we can solve explicitly for y as follows:

$$|y| = e^{\ln|y|} = e^{(x^3/3)+C} = e^C e^{x^3/3}$$

so

$$y = \pm e^C e^{x^3/3}$$

We can easily verify that the function $y = 0$ is also a solution of the given differential equation. So we can write the general solution in the form

$$y = A e^{x^3/3}$$

where A is an arbitrary constant ($A = e^C$, or $A = -e^C$, or $A = 0$).

|||| Figure 3 shows a direction field for the differential equation in Example 3. Compare it with Figure 4, in which we use the equation $y = Ae^{x^3/3}$ to graph solutions for several values of A. If you use the direction field to sketch solution curves with y-intercepts 5, 2, 1, -1, and -2, they will resemble the curves in Figure 4.

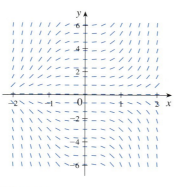

FIGURE 3

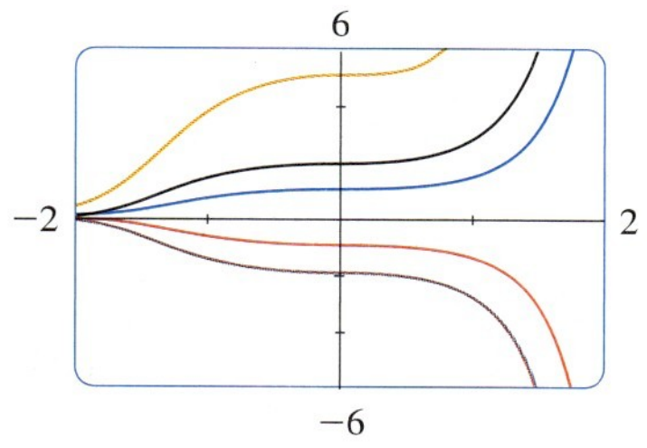

FIGURE 4

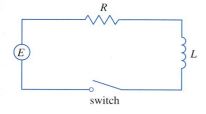

FIGURE 5

EXAMPLE 4 In Section 10.2 we modeled the current $I(t)$ in the electric circuit shown in Figure 5 by the differential equation

$$L \frac{dI}{dt} + RI = E(t)$$

Find an expression for the current in a circuit where the resistance is 12 Ω, the inductance is 4 H, a battery gives a constant voltage of 60 V, and the switch is turned on when $t = 0$. What is the limiting value of the current?

SOLUTION With $L = 4$, $R = 12$, and $E(t) = 60$, the equation becomes

$$4 \frac{dI}{dt} + 12I = 60$$

or
$$\frac{dI}{dt} = 15 - 3I$$

and the initial-value problem is

$$\frac{dI}{dt} = 15 - 3I \qquad I(0) = 0$$

We recognize this equation as being separable, and we solve it as follows:

$$\int \frac{dI}{15 - 3I} = \int dt \qquad (15 - 3I \neq 0)$$

$$-\tfrac{1}{3} \ln |15 - 3I| = t + C$$

$$|15 - 3I| = e^{-3(t+C)}$$

$$15 - 3I = \pm e^{-3C} e^{-3t} = A e^{-3t}$$

$$I = 5 - \tfrac{1}{3} A e^{-3t}$$

Since $I(0) = 0$, we have $5 - \tfrac{1}{3} A = 0$, so $A = 15$ and the solution is

$$I(t) = 5 - 5e^{-3t}$$

The limiting current, in amperes, is

$$\lim_{t \to \infty} I(t) = \lim_{t \to \infty} (5 - 5e^{-3t}) = 5 - 5 \lim_{t \to \infty} e^{-3t}$$

$$= 5 - 0 = 5$$

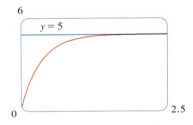

Figure 6 shows how the solution in Example 4 (the current) approaches its limiting value. Comparison with Figure 11 in Section 10.2 shows that we were able to draw a fairly accurate solution curve from the direction field.

FIGURE 6

Orthogonal Trajectories

An **orthogonal trajectory** of a family of curves is a curve that intersects each curve of the family orthogonally, that is, at right angles (see Figure 7). For instance, each member of the family $y = mx$ of straight lines through the origin is an orthogonal trajectory of the family $x^2 + y^2 = r^2$ of concentric circles with center the origin (see Figure 8). We say that the two families are orthogonal trajectories of each other.

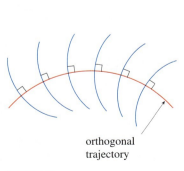

orthogonal trajectory

FIGURE 7

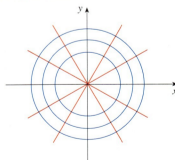

FIGURE 8

EXAMPLE 5 Find the orthogonal trajectories of the family of curves $x = ky^2$, where k is an arbitrary constant.

SOLUTION The curves $x = ky^2$ form a family of parabolas whose axis of symmetry is the x-axis. The first step is to find a single differential equation that is satisfied by all members of the family. If we differentiate $x = ky^2$, we get

$$1 = 2ky \frac{dy}{dx} \quad \text{or} \quad \frac{dy}{dx} = \frac{1}{2ky}$$

This differential equation depends on k, but we need an equation that is valid for all values of k simultaneously. To eliminate k we note that, from the equation of the given general parabola $x = ky^2$, we have $k = x/y^2$ and so the differential equation can be written as

$$\frac{dy}{dx} = \frac{1}{2ky} = \frac{1}{2\frac{x}{y^2}y}$$

or

$$\frac{dy}{dx} = \frac{y}{2x}$$

This means that the slope of the tangent line at any point (x, y) on one of the parabolas is $y' = y/(2x)$. On an orthogonal trajectory the slope of the tangent line must be the negative reciprocal of this slope. Therefore, the orthogonal trajectories must satisfy the differential equation

$$\frac{dy}{dx} = -\frac{2x}{y}$$

This differential equation is separable, and we solve it as follows:

$$\int y \, dy = -\int 2x \, dx$$

$$\frac{y^2}{2} = -x^2 + C$$

$$\boxed{4} \qquad x^2 + \frac{y^2}{2} = C$$

where C is an arbitrary positive constant. Thus, the orthogonal trajectories are the family of ellipses given by Equation 4 and sketched in Figure 9.

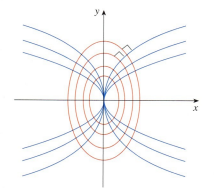

FIGURE 9

Orthogonal trajectories occur in various branches of physics. For example, in an electrostatic field the lines of force are orthogonal to the lines of constant potential. Also, the streamlines in aerodynamics are orthogonal trajectories of the velocity-equipotential curves.

Mixing Problems

A typical mixing problem involves a tank of fixed capacity filled with a thoroughly mixed solution of some substance, such as salt. A solution of a given concentration enters the tank at a fixed rate and the mixture, thoroughly stirred, leaves at a fixed rate, which may differ from the entering rate. If $y(t)$ denotes the amount of substance in the tank at time t, then $y'(t)$ is the rate at which the substance is being added minus the rate at which it is being removed. The mathematical description of this situation often leads to a first-order separable differential equation. We can use the same type of reasoning to model a variety of phenomena: chemical reactions, discharge of pollutants into a lake, injection of a drug into the bloodstream.

EXAMPLE 6 A tank contains 20 kg of salt dissolved in 5000 L of water. Brine that contains 0.03 kg of salt per liter of water enters the tank at a rate of 25 L/min. The solution is kept thoroughly mixed and drains from the tank at the same rate. How much salt remains in the tank after half an hour?

SOLUTION Let $y(t)$ be the amount of salt (in kilograms) after t minutes. We are given that $y(0) = 20$ and we want to find $y(30)$. We do this by finding a differential equation satisfied by $y(t)$. Note that dy/dt is the rate of change of the amount of salt, so

$$\boxed{5} \qquad \frac{dy}{dt} = (\text{rate in}) - (\text{rate out})$$

where (rate in) is the rate at which salt enters the tank and (rate out) is the rate at which salt leaves the tank. We have

$$\text{rate in} = \left(0.03 \, \frac{\text{kg}}{\text{L}}\right)\left(25 \, \frac{\text{L}}{\text{min}}\right) = 0.75 \, \frac{\text{kg}}{\text{min}}$$

The tank always contains 5000 L of liquid, so the concentration at time t is $y(t)/5000$ (measured in kilograms per liter). Since the brine flows out at a rate of 25 L/min, we have

$$\text{rate out} = \left(\frac{y(t)}{5000} \, \frac{\text{kg}}{\text{L}}\right)\left(25 \, \frac{\text{L}}{\text{min}}\right) = \frac{y(t)}{200} \, \frac{\text{kg}}{\text{min}}$$

Thus, from Equation 5 we get

$$\frac{dy}{dt} = 0.75 - \frac{y(t)}{200} = \frac{150 - y(t)}{200}$$

Solving this separable differential equation, we obtain

$$\int \frac{dy}{150 - y} = \int \frac{dt}{200}$$

$$-\ln|150 - y| = \frac{t}{200} + C$$

Since $y(0) = 20$, we have $-\ln 130 = C$, so

$$-\ln|150 - y| = \frac{t}{200} - \ln 130$$

Therefore

$$|150 - y| = 130 e^{-t/200}$$

Since $y(t)$ is continuous and $y(0) = 20$ and the right side is never 0, we deduce that $150 - y(t)$ is always positive. Thus, $|150 - y| = 150 - y$ and so

$$y(t) = 150 - 130 e^{-t/200}$$

The amount of salt after 30 min is

$$y(30) = 150 - 130 e^{-30/200} \approx 38.1 \text{ kg}$$

IIII Figure 10 shows the graph of the function $y(t)$ of Example 6. Notice that, as time goes by, the amount of salt approaches 150 kg.

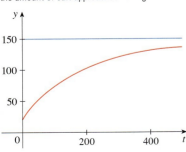

FIGURE 10

10.3 Exercises

1–10 Solve the differential equation.

1. $\dfrac{dy}{dx} = \dfrac{y}{x}$

2. $\dfrac{dy}{dx} = \dfrac{e^{2x}}{4y^3}$

3. $(x^2 + 1)y' = xy$

4. $y' = y^2 \sin x$

5. $(1 + \tan y)y' = x^2 + 1$

6. $\dfrac{du}{dr} = \dfrac{1 + \sqrt{r}}{1 + \sqrt{u}}$

7. $\dfrac{dy}{dt} = \dfrac{te^t}{y\sqrt{1 + y^2}}$

8. $y' = \dfrac{xy}{2 \ln y}$

9. $\dfrac{du}{dt} = 2 + 2u + t + tu$

10. $\dfrac{dz}{dt} + e^{t+z} = 0$

11–18 Find the solution of the differential equation that satisfies the given initial condition.

11. $\dfrac{dy}{dx} = y^2 + 1$, $y(1) = 0$

12. $\dfrac{dy}{dx} = \dfrac{y \cos x}{1 + y^2}$, $y(0) = 1$

13. $x \cos x = (2y + e^{3y})y'$, $y(0) = 0$

14. $\dfrac{dP}{dt} = \sqrt{Pt}$, $P(1) = 2$

15. $\dfrac{du}{dt} = \dfrac{2t + \sec^2 t}{2u}$, $u(0) = -5$

16. $\dfrac{dy}{dt} = te^y$, $y(1) = 0$

17. $y' \tan x = a + y$, $y(\pi/3) = a$, $0 < x < \pi/2$

18. $xy' + y = y^2$, $y(1) = -1$

19. Find an equation of the curve that satisfies $dy/dx = 4x^3 y$ and whose y-intercept is 7.

20. Find an equation of the curve that passes through the point $(1, 1)$ and whose slope at (x, y) is y^2/x^3.

21. (a) Solve the differential equation $y' = 2x\sqrt{1 - y^2}$.
    (b) Solve the initial-value problem $y' = 2x\sqrt{1 - y^2}$, $y(0) = 0$, and graph the solution.
    (c) Does the initial-value problem $y' = 2x\sqrt{1 - y^2}$, $y(0) = 2$, have a solution? Explain.

22. Solve the equation $e^{-y}y' + \cos x = 0$ and graph several members of the family of solutions. How does the solution curve change as the constant C varies?

CAS 23. Solve the initial-value problem $y' = (\sin x)/\sin y$, $y(0) = \pi/2$, and graph the solution (if your CAS does implicit plots).

CAS 24. Solve the equation $y' = x\sqrt{x^2 + 1}/(ye^y)$ and graph several members of the family of solutions (if your CAS does implicit plots). How does the solution curve change as the constant C varies?

CAS 25–26
(a) Use a computer algebra system to draw a direction field for the differential equation. Get a printout and use it to sketch some solution curves without solving the differential equation.
(b) Solve the differential equation.
(c) Use the CAS to draw several members of the family of solutions obtained in part (b). Compare with the curves from part (a).

25. $y' = 1/y$

26. $y' = x^2/y$

27–30 Find the orthogonal trajectories of the family of curves. Use a graphing device to draw several members of each family on a common screen.

27. $y = kx^2$

28. $x^2 - y^2 = k$

29. $y = (x + k)^{-1}$

30. $y = ke^{-x}$

31. Solve the initial-value problem in Exercise 27 in Section 10.2 to find an expression for the charge at time t. Find the limiting value of the charge.

32. In Exercise 28 in Section 10.2 we discussed a differential equation that models the temperature of a 95°C cup of coffee in a 20°C room. Solve the differential equation to find an expression for the temperature of the coffee at time t.

33. In Exercise 13 in Section 10.1 we formulated a model for learning in the form of the differential equation

$$\dfrac{dP}{dt} = k(M - P)$$

where $P(t)$ measures the performance of someone learning a skill after a training time t, M is the maximum level of performance, and k is a positive constant. Solve this differential equation to find an expression for $P(t)$. What is the limit of this expression?

34. In an elementary chemical reaction, single molecules of two reactants A and B form a molecule of the product C: $A + B \to C$. The law of mass action states that the rate of reaction is proportional to the product of the concentrations of A and B:

$$\dfrac{d[C]}{dt} = k[A][B]$$

(See Example 4 in Section 3.4.) Thus, if the initial concentrations are $[A] = a$ moles/L and $[B] = b$ moles/L and we write

$x = [C]$, then we have

$$\frac{dx}{dt} = k(a - x)(b - x)$$

(a) Assuming that $a \neq b$, find x as a function of t. Use the fact that the initial concentration of C is 0.
(b) Find $x(t)$ assuming that $a = b$. How does this expression for $x(t)$ simplify if it is known that $[C] = a/2$ after 20 seconds?

35. In contrast to the situation of Exercise 34, experiments show that the reaction $H_2 + Br_2 \rightarrow 2HBr$ satisfies the rate law

$$\frac{d[HBr]}{dt} = k[H_2][Br_2]^{1/2}$$

and so for this reaction the differential equation becomes

$$\frac{dx}{dt} = k(a - x)(b - x)^{1/2}$$

where $x = [HBr]$ and a and b are the initial concentrations of hydrogen and bromine.
(a) Find x as a function of t in the case where $a = b$. Use the fact that $x(0) = 0$.
(b) If $a > b$, find t as a function of x. [*Hint:* In performing the integration, make the substitution $u = \sqrt{b - x}$.]

36. A sphere with radius 1 m has temperature 15°C. It lies inside a concentric sphere with radius 2 m and temperature 25°C. The temperature $T(r)$ at a distance r from the common center of the spheres satisfies the differential equation

$$\frac{d^2T}{dr^2} + \frac{2}{r}\frac{dT}{dr} = 0$$

If we let $S = dT/dr$, then S satisfies a first-order differential equation. Solve it to find an expression for the temperature $T(r)$ between the spheres.

37. A glucose solution is administered intravenously into the bloodstream at a constant rate r. As the glucose is added, it is converted into other substances and removed from the bloodstream at a rate that is proportional to the concentration at that time. Thus, a model for the concentration $C = C(t)$ of the glucose solution in the bloodstream is

$$\frac{dC}{dt} = r - kC$$

where k is a positive constant.
(a) Suppose that the concentration at time $t = 0$ is C_0. Determine the concentration at any time t by solving the differential equation.
(b) Assuming that $C_0 < r/k$, find $\lim_{t \to \infty} C(t)$ and interpret your answer.

38. A certain small country has $10 billion in paper currency in circulation, and each day $50 million comes into the country's banks. The government decides to introduce new currency by having the banks replace old bills with new ones whenever old currency comes into the banks. Let $x = x(t)$ denote the amount of new currency in circulation at time t, with $x(0) = 0$.
(a) Formulate a mathematical model in the form of an initial-value problem that represents the "flow" of the new currency into circulation.
(b) Solve the initial-value problem found in part (a).
(c) How long will it take for the new bills to account for 90% of the currency in circulation?

39. A tank contains 1000 L of brine with 15 kg of dissolved salt. Pure water enters the tank at a rate of 10 L/min. The solution is kept thoroughly mixed and drains from the tank at the same rate. How much salt is in the tank (a) after t minutes and (b) after 20 minutes?

40. A tank contains 1000 L of pure water. Brine that contains 0.05 kg of salt per liter of water enters the tank at a rate of 5 L/min. Brine that contains 0.04 kg of salt per liter of water enters the tank at a rate of 10 L/min. The solution is kept thoroughly mixed and drains from the tank at a rate of 15 L/min. How much salt is in the tank (a) after t minutes and (b) after one hour?

41. When a raindrop falls, it increases in size and so its mass at time t is a function of t, $m(t)$. The rate of growth of the mass is $km(t)$ for some positive constant k. When we apply Newton's Law of Motion to the raindrop, we get $(mv)' = gm$, where v is the velocity of the raindrop (directed downward) and g is the acceleration due to gravity. The *terminal velocity* of the raindrop is $\lim_{t \to \infty} v(t)$. Find an expression for the terminal velocity in terms of g and k.

42. An object of mass m is moving horizontally through a medium which resists the motion with a force that is a function of the velocity; that is,

$$m\frac{d^2s}{dt^2} = m\frac{dv}{dt} = f(v)$$

where $v = v(t)$ and $s = s(t)$ represent the velocity and position of the object at time t, respectively. For example, think of a boat moving through the water.
(a) Suppose that the resisting force is proportional to the velocity, that is, $f(v) = -kv$, k a positive constant. (This model is appropriate for small values of v.) Let $v(0) = v_0$ and $s(0) = s_0$ be the initial values of v and s. Determine v and s at any time t. What is the total distance that the object travels from time $t = 0$?
(b) For larger values of v a better model is obtained by supposing that the resisting force is proportional to the square of the velocity, that is, $f(v) = -kv^2$, $k > 0$. (This model was first proposed by Newton.) Let v_0 and s_0 be the initial values of v and s. Determine v and s at any time t. What is the total distance that the object travels in this case?

43. Let $A(t)$ be the area of a tissue culture at time t and let M be the final area of the tissue when growth is complete. Most cell divisions occur on the periphery of the tissue and the number of cells on the periphery is proportional to $\sqrt{A(t)}$. So a reason-

able model for the growth of tissue is obtained by assuming that the rate of growth of the area is jointly proportional to $\sqrt{A(t)}$ and $M - A(t)$.
(a) Formulate a differential equation and use it to show that the tissue grows fastest when $A(t) = M/3$.
(b) Solve the differential equation to find an expression for $A(t)$. Use a computer algebra system to perform the integration.

44. According to Newton's Law of Universal Gravitation, the gravitational force on an object of mass m that has been projected vertically upward from Earth's surface is

$$F = \frac{mgR^2}{(x+R)^2}$$

where $x = x(t)$ is the object's distance above the surface at time t, R is Earth's radius, and g is the acceleration due to gravity. Also, by Newton's Second Law, $F = ma = m(dv/dt)$

and so

$$m\frac{dv}{dt} = -\frac{mgR^2}{(x+R)^2}$$

(a) Suppose a rocket is fired vertically upward with an initial velocity v_0. Let h be the maximum height above the surface reached by the object. Show that

$$v_0 = \sqrt{\frac{2gRh}{R+h}}$$

[*Hint:* By the Chain Rule, $m(dv/dt) = mv(dv/dx)$.]
(b) Calculate $v_e = \lim_{h \to \infty} v_0$. This limit is called the *escape velocity* for Earth.
(c) Use $R = 3960$ mi and $g = 32$ ft/s^2 to calculate v_e in feet per second and in miles per second.

APPLIED PROJECT

How Fast Does a Tank Drain?

If water (or other liquid) drains from a tank, we expect that the flow will be greatest at first (when the water depth is greatest) and will gradually decrease as the water level decreases. But we need a more precise mathematical description of how the flow decreases in order to answer the kinds of questions that engineers ask: How long does it take for a tank to drain completely? How much water should a tank hold in order to guarantee a certain minimum water pressure for a sprinkler system?

Let $h(t)$ and $V(t)$ be the height and volume of water in a tank at time t. If water leaks through a hole with area a at the bottom of the tank, then Torricelli's Law says that

$$\frac{dV}{dt} = -a\sqrt{2gh}$$

where g is the acceleration due to gravity. So the rate at which water flows from the tank is proportional to the square root of the water height.

1. (a) Suppose the tank is cylindrical with height 6 ft and radius 2 ft and the hole is circular with radius 1 in. If we take $g = 32$ ft/s^2, show that y satisfies the differential equation

$$\frac{dh}{dt} = -\frac{1}{72}\sqrt{h}$$

(b) Solve this equation to find the height of the water at time t, assuming the tank is full at time $t = 0$.
(c) How long will it take for the water to drain completely?

2. Because of the rotation and viscosity of the liquid, the theoretical model given by Equation 1 isn't quite accurate. Instead, the model

$$\frac{dh}{dt} = k\sqrt{h}$$

is often used and the constant k (which depends on the physical properties of the liquid) is determined from data concerning the draining of the tank.

(a) Suppose that a hole is drilled in the side of a cylindrical bottle and the height h of the water (above the hole) decreases from 10 cm to 3 cm in 68 seconds. Use Equation 2 to find an expression for $h(t)$. Evaluate $h(t)$ for $t = 10, 20, 30, 40, 50, 60$.

(b) Drill a 4-mm hole near the bottom of the cylindrical part of a two-liter plastic soft-drink bottle. Attach a strip of masking tape marked in centimeters from 0 to 10, with 0 corresponding to the top of the hole. With one finger over the hole, fill the bottle with water to the 10-cm mark. Then take your finger off the hole and record the values of $h(t)$ for $t = 10, 20, 30, 40, 50, 60$ seconds. (You will probably find that it takes 68 seconds for the level to decrease to $h = 3$ cm.) Compare your data with the values of $h(t)$ from part (a). How well did the model predict the actual values?

|||| This part of the project is best done as a classroom demonstration or as a group project with three students in each group: a timekeeper to call out seconds, a bottle keeper to estimate the height every 10 seconds, and a record keeper to record these values.

3. In many parts of the world, the water for sprinkler systems in large hotels and hospitals is supplied by gravity from cylindrical tanks on or near the roofs of the buildings. Suppose such a tank has radius 10 ft and the diameter of the outlet is 2.5 inches. An engineer has to guarantee that the water pressure will be at least 2160 lb/ft² for a period of 10 minutes. (When a fire happens, the electrical system might fail and it could take up to 10 minutes for the emergency generator and fire pump to be activated.) What height should the engineer specify for the tank in order to make such a guarantee? (Use the fact that the water pressure at a depth of d feet is $P = 62.5d$. See Section 9.3.)

4. Not all water tanks are shaped like cylinders. Suppose a tank has cross-sectional area $A(h)$ at height h. Then the volume of water up to height h is $V = \int_0^h A(u)\,du$ and so the Fundamental Theorem of Calculus gives $dV/dh = A(h)$. It follows that

$$\frac{dV}{dt} = \frac{dV}{dh}\frac{dh}{dt} = A(h)\frac{dh}{dt}$$

and so Torricelli's Law becomes

$$A(h)\frac{dh}{dt} = -a\sqrt{2gh}$$

(a) Suppose the tank has the shape of a sphere with radius 2 m and is initially half full of water. If the radius of the circular hole is 1 cm and we take $g = 10$ m/s², show that h satisfies the differential equation

$$(4h - h^2)\frac{dh}{dt} = -0.0001\sqrt{20h}$$

(b) How long will it take for the water to drain completely?

|||| APPLIED PROJECT

Which Is Faster, Going Up or Coming Down?

Suppose you throw a ball into the air. Do you think it takes longer to reach its maximum height or to fall back to Earth from its maximum height? We will solve the problem in this project but, before getting started, think about that situation and make a guess based on your physical intuition.

1. A ball with mass m is projected vertically upward from Earth's surface with a positive initial velocity v_0. We assume the forces acting on the ball are the force of gravity and a retarding force of air resistance with direction opposite to the direction of motion and with magnitude $p\,|v(t)|$, where p is a positive constant and $v(t)$ is the velocity of the ball at time t. In both

■■■ In modeling force due to air resistance, various functions have been used, depending on the physical characteristics and speed of the ball. Here we use a linear model, $-pv$, but a quadratic model ($-pv^2$ on the way up and pv^2 on the way down) is another possibility for higher speeds (see Exercise 42 in Section 10.3). For a golf ball, experiments have shown that a good model is $-pv^{1.3}$ going up and $p|v|^{1.3}$ coming down. But no matter which force function $-f(v)$ is used [where $f(v) > 0$ for $v > 0$ and $f(v) < 0$ for $v < 0$], the answer to the question remains the same. See F. Brauer, "What Goes Up Must Come Down, Eventually," *Amer. Math. Monthly* 108 (2001), pp. 437–440.

the ascent and the descent, the total force acting on the ball is $-pv - mg$. [During ascent, $v(t)$ is positive and the resistance acts downward; during descent, $v(t)$ is negative and the resistance acts upward.] So, by Newton's Second Law, the equation of motion is

$$mv' = -pv - mg$$

Solve this differential equation to show that the velocity is

$$v(t) = \left(v_0 + \frac{mg}{p}\right)e^{-pt/m} - \frac{mg}{p}$$

2. Show that the height of the ball, until it hits the ground, is

$$y(t) = \left(v_0 + \frac{mg}{p}\right)\frac{m}{p}(1 - e^{-pt/m}) - \frac{mgt}{p}$$

3. Let t_1 be the time that the ball takes to reach its maximum height. Show that

$$t_1 = \frac{m}{p}\ln\left(\frac{mg + pv_0}{mg}\right)$$

Find this time for a ball with mass 1 kg and initial velocity 20 m/s. Assume the air resistance is $\frac{1}{10}$ of the speed.

4. Let t_2 be the time at which the ball falls back to Earth. For the particular ball in Problem 3, estimate t_2 by using a graph of the height function $y(t)$. Which is faster, going up or coming down?

5. In general, it's not easy to find t_2 because it's impossible to solve the equation $y(t) = 0$ explicitly. We can, however, use an indirect method to determine whether ascent or descent is faster; we determine whether $y(2t_1)$ is positive or negative. Show that

$$y(2t_1) = \frac{m^2 g}{p^2}\left(x - \frac{1}{x} - 2\ln x\right)$$

where $x = e^{pt_1/m}$. Then show that $x > 1$ and the function

$$f(x) = x - \frac{1}{x} - 2\ln x$$

is increasing for $x > 1$. Use this result to decide whether $y(2t_1)$ is positive or negative. What can you conclude? Is ascent or descent faster?

10.4 Exponential Growth and Decay

One of the models for population growth that we considered in Section 10.1 was based on the assumption that the population grows at a rate proportional to the size of the population:

$$\frac{dP}{dt} = kP$$

Is that a reasonable assumption? Suppose we have a population (of bacteria, for instance) with size $P = 1000$ and at a certain time it is growing at a rate of $P' = 300$ bacteria per hour. Now let's take another 1000 bacteria of the same type and put them with the first population. Each half of the new population was growing at a rate of 300 bacteria per hour. We would expect the total population of 2000 to increase at a rate of 600 bacteria per hour initially (provided there's enough room and nutrition). So if we double the size, we double the growth rate. In general, it seems reasonable that the growth rate should be proportional to the size.

The same assumption applies in other situations as well. In nuclear physics, the mass of a radioactive substance decays at a rate proportional to the mass. In chemistry, the rate of a unimolecular first-order reaction is proportional to the concentration of the substance. In finance, the value of a savings account with continuously compounded interest increases at a rate proportional to that value.

In general, if $y(t)$ is the value of a quantity y at time t and if the rate of change of y with respect to t is proportional to its size $y(t)$ at any time, then

$$\boxed{1} \qquad \frac{dy}{dt} = ky$$

where k is a constant. Equation 1 is sometimes called the **law of natural growth** (if $k > 0$) or the **law of natural decay** (if $k < 0$). Because it is a separable differential equation we can solve it by the methods of Section 10.3:

$$\int \frac{dy}{y} = \int k\, dt$$

$$\ln|y| = kt + C$$

$$|y| = e^{kt+C} = e^C e^{kt}$$

$$y = Ae^{kt}$$

where A ($= \pm e^C$ or 0) is an arbitrary constant. To see the significance of the constant A, we observe that

$$y(0) = Ae^{k \cdot 0} = A$$

Therefore, A is the initial value of the function.

Because Equation 1 occurs so frequently in nature, we summarize what we have just proved for future use.

$\boxed{2}$ The solution of the initial-value problem

$$\frac{dy}{dt} = ky \qquad y(0) = y_0$$

is

$$y(t) = y_0 e^{kt}$$

Population Growth

What is the significance of the proportionality constant k? In the context of population growth, we can write

$$\boxed{3} \qquad \frac{dP}{dt} = kP \qquad \text{or} \qquad \frac{1}{P}\frac{dP}{dt} = k$$

The quantity

$$\frac{1}{P}\frac{dP}{dt}$$

is the growth rate divided by the population size; it is called the **relative growth rate**. According to (3), instead of saying "the growth rate is proportional to population size" we could say "the relative growth rate is constant." Then (2) says that a population with constant relative growth rate must grow exponentially. Notice that the relative growth rate k appears as the coefficient of t in the exponential function $y_0 e^{kt}$. For instance, if

$$\frac{dP}{dt} = 0.02P$$

and t is measured in years, then the relative growth rate is $k = 0.02$ and the population grows at a rate of 2% per year. If the population at time 0 is P_0, then the expression for the population is

$$P(t) = P_0 e^{0.02t}$$

TABLE 1

Year	Population (millions)
1900	1650
1910	1750
1920	1860
1930	2070
1940	2300
1950	2560
1960	3040
1970	3710
1980	4450
1990	5280
2000	6080

EXAMPLE 1 Assuming that the growth rate is proportional to population size, use the data in Table 1 to model the population of the world in the 20th century. What is the relative growth rate? How well does the model fit the data?

SOLUTION We measure the time t in years and let $t = 0$ in the year 1900. We measure the population $P(t)$ in millions of people. Then the initial condition is $P(0) = 1650$. We are assuming that the growth rate is proportional to population size, so the initial-value problem is

$$\frac{dP}{dt} = kP \qquad P(0) = 1650$$

From (2) we know that the solution is

$$P(t) = 1650 e^{kt}$$

One way to estimate the relative growth rate k is to use the fact that the population in 1910 was 1750 million. Therefore

$$P(10) = 1650 e^{k(10)} = 1750$$

We solve this equation for k:

$$e^{10k} = \frac{1750}{1650}$$

$$k = \frac{1}{10} \ln \frac{1750}{1650} \approx 0.005884$$

Thus, the relative growth rate is about 0.6% per year and the model becomes

$$P(t) = 1650 e^{0.005884t}$$

Table 2 and Figure 1 allow us to compare the predictions of this model with the actual data. You can see that the predictions become quite inaccurate after about 30 years and they underestimate by a factor of more than 2 in 2000.

TABLE 2

Year	Model	Population
1900	1650	1650
1910	1750	1750
1920	1856	1860
1930	1969	2070
1940	2088	2300
1950	2214	2560
1960	2349	3040
1970	2491	3710
1980	2642	4450
1990	2802	5280
2000	2972	6080

FIGURE 1 A possible model for world population growth

In Sections 7.2 and 7.4* we modeled the same data with an exponential function, but there we used the method of least squares.

Another possibility for estimating k would be to use the given population for 1950, for instance, instead of 1910. Then

$$P(50) = 1650e^{50k} = 2560$$

$$k = \frac{1}{50} \ln \frac{2560}{1650} \approx 0.0087846$$

The estimate for the relative growth rate is now 0.88% per year and the model is

$$P(t) = 1650e^{0.0087846t}$$

The predictions with this second model are shown in Table 3 and Figure 2. This exponential model is more accurate over a longer period of time, but it too lags behind reality in recent years.

TABLE 3

Year	Model	Population
1900	1650	1650
1910	1802	1750
1920	1967	1860
1930	2148	2070
1940	2345	2300
1950	2560	2560
1960	2795	3040
1970	3052	3710
1980	3332	4450
1990	3638	5280
2000	3972	6080

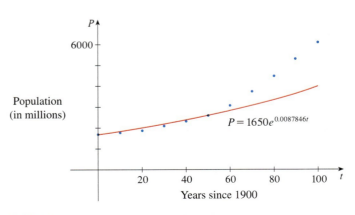

FIGURE 2 Another model for world population growth

EXAMPLE 2 Use the data in Table 1 to model the population of the world in the second half of the 20th century. Use the model to estimate the population in 1993 and to predict the population in the year 2010.

SOLUTION Here we let $t = 0$ in the year 1950. Then the initial-value problem is

$$\frac{dP}{dt} = kP \qquad P(0) = 2560$$

and the solution is

$$P(t) = 2560e^{kt}$$

Let's estimate k by using the population in 1960:

$$P(10) = 2560e^{10k} = 3040$$

$$k = \frac{1}{10} \ln \frac{3040}{2560} \approx 0.017185$$

The relative growth rate is about 1.7% per year and the model is

$$P(t) = 2560e^{0.017185t}$$

We estimate that the world population in 1993 was

$$P(43) = 2560e^{0.017185(43)} \approx 5360 \text{ million}$$

The model predicts that the population in 2010 will be

$$P(60) = 2560e^{0.017185(60)} \approx 7179 \text{ million}$$

The graph in Figure 3 shows that the model is fairly accurate to date, so the estimate for 1993 is quite reliable. But the prediction for 2010 is riskier.

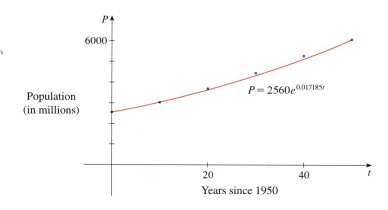

FIGURE 3
A model for world population growth in the second half of the 20th century

Radioactive Decay

Radioactive substances decay by spontaneously emitting radiation. If $m(t)$ is the mass remaining from an initial mass m_0 of the substance after time t, then the relative decay rate

$$-\frac{1}{m}\frac{dm}{dt}$$

has been found experimentally to be constant. (Since dm/dt is negative, the relative decay rate is positive.) It follows that

$$\frac{dm}{dt} = km$$

where k is a negative constant. In other words, radioactive substances decay at a rate proportional to the remaining mass. This means that we can use (2) to show that the mass decays exponentially:

$$m(t) = m_0 e^{kt}$$

Physicists express the rate of decay in terms of **half-life**, the time required for half of any given quantity to decay.

EXAMPLE 3 The half-life of radium-226 ($^{226}_{88}$Ra) is 1590 years.
(a) A sample of radium-226 has a mass of 100 mg. Find a formula for the mass of $^{226}_{88}$Ra that remains after t years.
(b) Find the mass after 1000 years correct to the nearest milligram.
(c) When will the mass be reduced to 30 mg?

SOLUTION
(a) Let $m(t)$ be the mass of radium-226 (in milligrams) that remains after t years. Then $dm/dt = km$ and $y(0) = 100$, so (2) gives

$$m(t) = m(0)e^{kt} = 100e^{kt}$$

In order to determine the value of k, we use the fact that $y(1590) = \tfrac{1}{2}(100)$. Thus

$$100e^{1590k} = 50 \quad \text{so} \quad e^{1590k} = \tfrac{1}{2}$$

and

$$1590k = \ln \tfrac{1}{2} = -\ln 2$$

$$k = -\frac{\ln 2}{1590}$$

Therefore

$$m(t) = 100e^{-(\ln 2/1590)t}$$

We could use the fact that $e^{\ln 2} = 2$ to write the expression for $m(t)$ in the alternative form

$$m(t) = 100 \times 2^{-t/1590}$$

(b) The mass after 1000 years is

$$m(1000) = 100e^{-(\ln 2/1590)1000} \approx 65 \text{ mg}$$

(c) We want to find the value of t such that $m(t) = 30$, that is,

$$100e^{-(\ln 2/1590)t} = 30 \quad \text{or} \quad e^{-(\ln 2/1590)t} = 0.3$$

We solve this equation for t by taking the natural logarithm of both sides:

$$-\frac{\ln 2}{1590} t = \ln 0.3$$

Thus
$$t = -1590 \frac{\ln 0.3}{\ln 2} \approx 2762 \text{ years}$$

As a check on our work in Example 3, we use a graphing device to draw the graph of $m(t)$ in Figure 4 together with the horizontal line $m = 30$. These curves intersect when $t \approx 2800$, and this agrees with the answer to part (c).

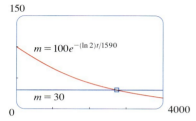

FIGURE 4

Newton's Law of Cooling

Newton's Law of Cooling states that the rate of cooling of an object is proportional to the temperature difference between the object and its surroundings, provided that this difference is not too large. (This law also applies to warming.) If we let $T(t)$ be the temperature of the object at time t and T_s be the temperature of the surroundings, then we can formulate Newton's Law of Cooling as a differential equation:

$$\frac{dT}{dt} = k(T - T_s)$$

where k is a constant. We could solve this equation as a separable differential equation by the method of Section 10.3, but an easier method is to make the change of variable $y(t) = T(t) - T_s$. Because T_s is constant, we have $y'(t) = T'(t)$ and so the equation becomes

$$\frac{dy}{dt} = ky$$

We can then use (2) to find an expression for y, from which we can find T.

EXAMPLE 4 A bottle of soda pop at room temperature (72°F) is placed in a refrigerator where the temperature is 44°F. After half an hour the soda pop has cooled to 61°F.
(a) What is the temperature of the soda pop after another half hour?
(b) How long does it take for the soda pop to cool to 50°F?

SOLUTION
(a) Let $T(t)$ be the temperature of the soda after t minutes. The surrounding temperature is $T_s = 44°F$, so Newton's Law of Cooling states that

$$\frac{dT}{dt} = k(T - 44)$$

If we let $y = T - 44$, then $y(0) = T(0) - 44 = 72 - 44 = 28$, so y is a solution of the initial-value problem

$$\frac{dy}{dt} = ky \qquad y(0) = 28$$

and by (2) we have

$$y(t) = y(0)e^{kt} = 28e^{kt}$$

We are given that $T(30) = 61$, so $y(30) = 61 - 44 = 17$ and

$$28e^{30k} = 17 \qquad e^{30k} = \tfrac{17}{28}$$

Taking logarithms, we have

$$k = \frac{\ln\left(\tfrac{17}{28}\right)}{30} \approx -0.01663$$

Thus

$$y(t) = 28e^{-0.01663t}$$

$$T(t) = 44 + 28e^{-0.01663t}$$

$$T(60) = 44 + 28e^{-0.01663(60)} \approx 54.3$$

So after another half hour the pop has cooled to about 54°F.
(b) We have $T(t) = 50$ when

$$44 + 28e^{-0.01663t} = 50$$

$$e^{-0.01663t} = \tfrac{6}{28}$$

$$t = \frac{\ln\left(\tfrac{6}{28}\right)}{-0.01663} \approx 92.6$$

The pop cools to 50°F after about 1 hour 33 minutes.

Notice that in Example 4, we have

$$\lim_{t \to \infty} T(t) = \lim_{t \to \infty} (44 + 28e^{-0.01663t}) = 44 + 28 \cdot 0 = 44$$

which is to be expected. The graph of the temperature function is shown in Figure 5.

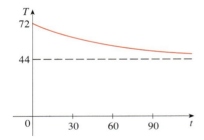

FIGURE 5

Continuously Compounded Interest

EXAMPLE 5 If $1000 is invested at 6% interest, compounded annually, then after 1 year the investment is worth $1000(1.06) = $1060, after 2 years it's worth $[1000(1.06)]1.06 = $1123.60, and after t years it's worth $1000(1.06)^t$. In general, if an amount A_0 is invested at an interest rate r ($r = 0.06$ in this example), then after t years it's worth $A_0(1 + r)^t$. Usually, however, interest is compounded more frequently, say, n times a year. Then in each compounding period the interest rate is r/n and there are nt compounding periods in t years, so the value of the investment is

$$A_0\left(1 + \frac{r}{n}\right)^{nt}$$

For instance, after 3 years at 6% interest a $1000 investment will be worth

$$\$1000(1.06)^3 = \$1191.02 \quad \text{with annual compounding}$$

$$\$1000(1.03)^6 = \$1194.05 \quad \text{with semiannual compounding}$$

$$\$1000(1.015)^{12} = \$1195.62 \quad \text{with quarterly compounding}$$

$$\$1000(1.005)^{36} = \$1196.68 \quad \text{with monthly compounding}$$

$$\$1000\left(1 + \frac{0.06}{365}\right)^{365 \cdot 3} = \$1197.20 \quad \text{with daily compounding}$$

You can see that the interest paid increases as the number of compounding periods (n) increases. If we let $n \to \infty$, then we will be compounding the interest **continuously** and the value of the investment will be

$$A(t) = \lim_{n \to \infty} A_0\left(1 + \frac{r}{n}\right)^{nt} = \lim_{n \to \infty} A_0\left[\left(1 + \frac{r}{n}\right)^{n/r}\right]^{rt}$$

$$= A_0\left[\lim_{n \to \infty} \left(1 + \frac{r}{n}\right)^{n/r}\right]^{rt}$$

$$= A_0\left[\lim_{m \to \infty} \left(1 + \frac{1}{m}\right)^{m}\right]^{rt} \quad \text{(where } m = n/r\text{)}$$

But the limit in this expression is equal to the number e (see Equation 7.4.9 or 7.4*.9). So with continuous compounding of interest at interest rate r, the amount after t years is

$$A(t) = A_0 e^{rt}$$

If we differentiate this equation, we get

$$\frac{dA}{dt} = rA_0 e^{rt} = rA(t)$$

which says that, with continuous compounding of interest, the rate of increase of an investment is proportional to its size.

Returning to the example of $1000 invested for 3 years at 6% interest, we see that with continuous compounding of interest the value of the investment will be

$$A(3) = \$1000 e^{(0.06)3}$$

$$= \$1000 e^{0.18} = \$1197.22$$

Notice how close this is to the amount we calculated for daily compounding, $1197.20. But the amount is easier to compute if we use continuous compounding.

10.4 Exercises

1. A population of protozoa develops with a constant relative growth rate of 0.7944 per member per day. On day zero the population consists of two members. Find the population size after six days.

2. A common inhabitant of human intestines is the bacterium *Escherichia coli*. A cell of this bacterium in a nutrient-broth medium divides into two cells every 20 minutes. The initial population of a culture is 60 cells.
   (a) Find the relative growth rate.
   (b) Find an expression for the number of cells after t hours.
   (c) Find the number of cells after 8 hours.
   (d) Find the rate of growth after 8 hours.
   (e) When will the population reach 20,000 cells?

3. A bacteria culture starts with 500 bacteria and grows at a rate proportional to its size. After 3 hours there are 8000 bacteria.
   (a) Find an expression for the number of bacteria after t hours.
   (b) Find the number of bacteria after 4 hours.
   (c) Find the rate of growth after 4 hours.
   (d) When will the population reach 30,000?

4. A bacteria culture grows with constant relative growth rate. After 2 hours there are 600 bacteria and after 8 hours the count is 75,000.
   (a) Find the initial population.
   (b) Find an expression for the population after t hours.
   (c) Find the number of cells after 5 hours.
   (d) Find the rate of growth after 5 hours.
   (e) When will the population reach 200,000?

5. The table gives estimates of the world population, in millions, from 1750 to 2000:

Year	Population	Year	Population
1750	790	1900	1650
1800	980	1950	2560
1850	1260	2000	6080

   (a) Use the exponential model and the population figures for 1750 and 1800 to predict the world population in 1900 and 1950. Compare with the actual figures.
   (b) Use the exponential model and the population figures for 1850 and 1900 to predict the world population in 1950. Compare with the actual population.
   (c) Use the exponential model and the population figures for 1900 and 1950 to predict the world population in 2000. Compare with the actual population and try to explain the discrepancy.

6. The table gives the population of the United States, in millions, for the years 1900–2000.

Year	Population	Year	Population
1900	76	1960	179
1910	92	1970	203
1920	106	1980	227
1930	123	1990	250
1940	131	2000	275
1950	150		

   (a) Use the exponential model and the census figures for 1900 and 1910 to predict the population in 2000. Compare with the actual figure and try to explain the discrepancy.
   (b) Use the exponential model and the census figures for 1980 and 1990 to predict the population in 2000. Compare with the actual population. Then use this model to predict the population in the years 2010 and 2020.
   (c) Graph both of the exponential functions in parts (a) and (b) together with a plot of the actual population. Are these models reasonable ones?

7. Experiments show that if the chemical reaction

$$N_2O_5 \rightarrow 2NO_2 + \tfrac{1}{2}O_2$$

takes place at 45°C, the rate of reaction of dinitrogen pentoxide is proportional to its concentration as follows:

$$-\frac{d[N_2O_5]}{dt} = 0.0005[N_2O_5]$$

(See Example 4 in Section 3.4.)
   (a) Find an expression for the concentration $[N_2O_5]$ after t seconds if the initial concentration is C.
   (b) How long will the reaction take to reduce the concentration of N_2O_5 to 90% of its original value?

8. Bismuth-210 has a half-life of 5.0 days.
   (a) A sample originally has a mass of 800 mg. Find a formula for the mass remaining after t days.
   (b) Find the mass remaining after 30 days.
   (c) When is the mass reduced to 1 mg?
   (d) Sketch the graph of the mass function.

9. The half-life of cesium-137 is 30 years. Suppose we have a 100-mg sample.
   (a) Find the mass that remains after t years.
   (b) How much of the sample remains after 100 years?
   (c) After how long will only 1 mg remain?

10. After 3 days a sample of radon-222 decayed to 58% of its original amount.
    (a) What is the half-life of radon-222?
    (b) How long would it take the sample to decay to 10% of its original amount?

11. Scientists can determine the age of ancient objects by a method called *radiocarbon dating*. The bombardment of the upper atmosphere by cosmic rays converts nitrogen to a radioactive isotope of carbon, ^{14}C, with a half-life of about 5730 years. Vegetation absorbs carbon dioxide through the atmosphere and animal life assimilates ^{14}C through food chains. When a plant or animal dies, it stops replacing its carbon and the amount of ^{14}C begins to decrease through radioactive decay. Therefore, the level of radioactivity must also decay exponentially. A parchment fragment was discovered that had about 74% as much ^{14}C radioactivity as does plant material on Earth today. Estimate the age of the parchment.

12. A curve passes through the point (0, 5) and has the property that the slope of the curve at every point P is twice the y-coordinate of P. What is the equation of the curve?

13. A roast turkey is taken from an oven when its temperature has reached 185°F and is placed on a table in a room where the temperature is 75°F.
    (a) If the temperature of the turkey is 150°F after half an hour, what is the temperature after 45 min?
    (b) When will the turkey have cooled to 100°F?

14. A thermometer is taken from a room where the temperature is 20°C to the outdoors, where the temperature is 5°C. After one minute the thermometer reads 12°C.
    (a) What will the reading on the thermometer be after one more minute?
    (b) When will the thermometer read 6°C?

15. When a cold drink is taken from a refrigerator, its temperature is 5°C. After 25 minutes in a 20°C room its temperature has increased to 10°C.
    (a) What is the temperature of the drink after 50 minutes?
    (b) When will its temperature be 15°C?

16. A freshly brewed cup of coffee has temperature 95°C in a 20°C room. When its temperature is 70°C, it is cooling at a rate of 1°C per minute. When does this occur?

17. The rate of change of atmospheric pressure P with respect to altitude h is proportional to P, provided that the temperature is constant. At 15°C the pressure is 101.3 kPa at sea level and 87.14 kPa at $h = 1000$ m.
    (a) What is the pressure at an altitude of 3000 m?
    (b) What is the pressure at the top of Mount McKinley, at an altitude of 6187 m?

18. (a) If $500 is borrowed at 14% interest, find the amounts due at the end of 2 years if the interest is compounded (i) annually, (ii) quarterly, (iii) monthly, (iv) daily, (v) hourly, and (vi) continuously.
    (b) Suppose $500 is borrowed and the interest is compounded continuously. If $A(t)$ is the amount due after t years, where $0 \leq t \leq 2$, graph $A(t)$ for each of the interest rates 14%, 10%, and 6% on a common screen.

19. (a) If $3000 is invested at 5% interest, find the value of the investment at the end of 5 years if the interest is compounded (i) annually, (ii) semiannually, (iii) monthly, (iv) weekly, (v) daily, and (vi) continuously.
    (b) If $A(t)$ is the amount of the investment at time t for the case of continuous compounding, write a differential equation and an initial condition satisfied by $A(t)$.

20. (a) How long will it take an investment to double in value if the interest rate is 6% compounded continuously?
    (b) What is the equivalent annual interest rate?

21. Consider a population $P = P(t)$ with constant relative birth and death rates α and β, respectively, and a constant emigration rate m, where α, β, and m are positive constants. Assume that $\alpha > \beta$. Then the rate of change of the population at time t is modeled by the differential equation

$$\frac{dP}{dt} = kP - m \qquad \text{where } k = \alpha - \beta$$

(a) Find the solution of this equation that satisfies the initial condition $P(0) = P_0$.
(b) What condition on m will lead to an exponential expansion of the population?
(c) What condition on m will result in a constant population? A population decline?
(d) In 1847, the population of Ireland was about 8 million and the difference between the relative birth and death rates was 1.6% of the population. Because of the potato famine in the 1840s and 1850s, about 210,000 inhabitants per year emigrated from Ireland. Was the population expanding or declining at that time?

22. Let c be a positive number. A differential equation of the form

$$\frac{dy}{dt} = ky^{1+c}$$

where k is a positive constant, is called a *doomsday equation* because the exponent in the expression ky^{1+c} is larger than that for natural growth (that is, ky).
(a) Determine the solution that satisfies the initial condition $y(0) = y_0$.
(b) Show that there is a finite time $t = T$ (doomsday) such that $\lim_{t \to T^-} y(t) = \infty$.
(c) An especially prolific breed of rabbits has the growth term $ky^{1.01}$. If 2 such rabbits breed initially and the warren has 16 rabbits after three months, then when is doomsday?

APPLIED PROJECT

Calculus and Baseball

In this project we explore three of the many applications of calculus to baseball. The physical interactions of the game, especially the collision of ball and bat, are quite complex and their models are discussed in detail in a book by Robert Adair, *The Physics of Baseball,* 3d ed. (New York: HarperPerennial, 2002).

1. It may surprise you to learn that the collision of baseball and bat lasts only about a thousandth of a second. Here we calculate the average force on the bat during this collision by first computing the change in the ball's momentum.

    The *momentum p* of an object is the product of its mass m and its velocity v, that is, $p = mv$. Suppose an object, moving along a straight line, is acted on by a force $F = F(t)$ that is a continuous function of time.

    (a) Show that the change in momentum over a time interval $[t_0, t_1]$ is equal to the integral of F from t_0 to t_1; that is, show that

    $$p(t_1) - p(t_0) = \int_{t_0}^{t_1} F(t)\, dt$$

    This integral is called the *impulse* of the force over the time interval.

    (b) A pitcher throws a 90-mi/h fastball to a batter, who hits a line drive directly back to the pitcher. The ball is in contact with the bat for 0.001 s and leaves the bat with velocity 110 mi/h. A baseball weighs 5 oz and, in U.S. Customary units, its mass is measured in slugs: $m = w/g$ where $g = 32$ ft/s^2.
      (i) Find the change in the ball's momentum.
      (ii) Find the average force on the bat.

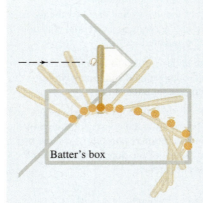

Batter's box

An overhead view of the position of a baseball bat, shown every fiftieth of a second during a typical swing. (Adapted from *The Physics of Baseball*)

2. In this problem we calculate the work required for a pitcher to throw a 90-mi/h fastball by first considering kinetic energy.

    The *kinetic energy K* of an object of mass m and velocity v is given by $K = \frac{1}{2}mv^2$. Suppose an object of mass m, moving in a straight line, is acted on by a force $F = F(s)$ that depends on its position s. According to Newton's Second Law

    $$F(s) = ma = m\frac{dv}{dt}$$

    where a and v denote the acceleration and velocity of the object.

    (a) Show that the work done in moving the object from a position s_0 to a position s_1 is equal to the change in the object's kinetic energy; that is, show that

    $$W = \int_{s_0}^{s_1} F(s)\, ds = \tfrac{1}{2}mv_1^2 - \tfrac{1}{2}mv_0^2$$

    where $v_0 = v(s_0)$ and $v_1 = v(s_1)$ are the velocities of the object at the positions s_0 and s_1. *Hint:* By the Chain Rule,

    $$m\frac{dv}{dt} = m\frac{dv}{ds}\frac{ds}{dt} = mv\frac{dv}{ds}$$

    (b) How many foot-pounds of work does it take to throw a baseball at a speed of 90 mi/h?

3. (a) An outfielder fields a baseball 280 ft away from home plate and throws it directly to the catcher with an initial velocity of 100 ft/s. Assume that the velocity $v(t)$ of the ball after t seconds satisfies the differential equation $dv/dt = -v/10$ because of air resistance. How long does it take for the ball to reach home plate? (Ignore any vertical motion of the ball.)

    (b) The manager of the team wonders whether the ball will reach home plate sooner if it is relayed by an infielder. The shortstop can position himself directly between the outfielder and home plate, catch the ball thrown by the outfielder, turn, and throw the ball to the catcher with an initial velocity of 105 ft/s. The manager clocks the relay time of the shortstop (catching, turning, throwing) at half a second. How far from home plate should the shortstop position himself to minimize the total time for the ball to reach the plate? Should the manager encourage a direct throw or a relayed throw? What if the shortstop can throw at 115 ft/s?

(c) For what throwing velocity of the shortstop does a relayed throw take the same time as a direct throw?

10.5 The Logistic Equation

In this section we discuss in detail a model for population growth, the logistic model, that is more sophisticated than exponential growth. In doing so we use all the tools at our disposal—direction fields and Euler's method from Section 10.2 and the explicit solution of separable differential equations from Section 10.3. In the exercises we investigate other possible models for population growth, some of which take into account harvesting and seasonal growth.

The Logistic Model

As we discussed in Section 10.1, a population often increases exponentially in its early stages but levels off eventually and approaches its carrying capacity because of limited resources. If $P(t)$ is the size of the population at time t, we assume that

$$\frac{dP}{dt} \approx kP \qquad \text{if } P \text{ is small}$$

This says that the growth rate is initially close to being proportional to size. In other words, the relative growth rate is almost constant when the population is small. But we also want to reflect the fact that the relative growth rate decreases as the population P increases and becomes negative if P ever exceeds its **carrying capacity** K, the maximum population that the environment is capable of sustaining in the long run. The simplest expression for the relative growth rate that incorporates these assumptions is

$$\frac{1}{P}\frac{dP}{dt} = k\left(1 - \frac{P}{K}\right)$$

Multiplying by P, we obtain the model for population growth known as the **logistic differential equation**:

$$\frac{dP}{dt} = kP\left(1 - \frac{P}{K}\right)$$

Notice from Equation 1 that if P is small compared with K, then P/K is close to 0 and so $dP/dt \approx kP$. However, if $P \to K$ (the population approaches its carrying capacity), then $P/K \to 1$, so $dP/dt \to 0$. We can deduce information about whether solutions increase or decrease directly from Equation 1. If the population P lies between 0 and K, then the right side of the equation is positive, so $dP/dt > 0$ and the population increases. But if the population exceeds the carrying capacity ($P > K$), then $1 - P/K$ is negative, so $dP/dt < 0$ and the population decreases.

Direction Fields

Let's start our more detailed analysis of the logistic differential equation by looking at a direction field.

EXAMPLE 1 Draw a direction field for the logistic equation with $k = 0.08$ and carrying capacity $K = 1000$. What can you deduce about the solutions?

SOLUTION In this case the logistic differential equation is

$$\frac{dP}{dt} = 0.08P\left(1 - \frac{P}{1000}\right)$$

A direction field for this equation is shown in Figure 1. We show only the first quadrant because negative populations aren't meaningful and we are interested only in what happens after $t = 0$.

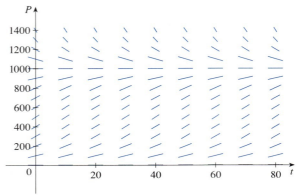

FIGURE 1
Direction field for the logistic equation in Example 1

The logistic equation is autonomous (dP/dt depends only on P, not on t), so the slopes are the same along any horizontal line. As expected, the slopes are positive for $0 < P < 1000$ and negative for $P > 1000$.

The slopes are small when P is close to 0 or 1000 (the carrying capacity). Notice that the solutions move away from the equilibrium solution $P = 0$ and move toward the equilibrium solution $P = 1000$.

In Figure 2 we use the direction field to sketch solution curves with initial populations $P(0) = 100$, $P(0) = 400$, and $P(0) = 1300$. Notice that solution curves that start below $P = 1000$ are increasing and those that start above $P = 1000$ are decreasing. The slopes are greatest when $P \approx 500$ and, therefore, the solution curves that start below $P = 1000$ have inflection points when $P \approx 500$. In fact we can prove that all solution curves that start below $P = 500$ have an inflection point when P is exactly 500 (see Exercise 9).

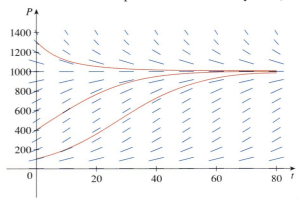

FIGURE 2
Solution curves for the logistic equation in Example 1

Euler's Method

Next let's use Euler's method to obtain numerical estimates for solutions of the logistic differential equation at specific times.

EXAMPLE 2 Use Euler's method with step sizes 20, 10, 5, 1, and 0.1 to estimate the population sizes $P(40)$ and $P(80)$, where P is the solution of the initial-value problem

$$\frac{dP}{dt} = 0.08P\left(1 - \frac{P}{1000}\right) \qquad P(0) = 100$$

SOLUTION With step size $h = 20$, $t_0 = 0$, $P_0 = 100$, and

$$F(t, P) = 0.08P\left(1 - \frac{P}{1000}\right)$$

we get, using the notation of Section 10.2,

$$P_1 = 100 + 20F(0, 100) = 244$$
$$P_2 = 244 + 20F(20, 244) \approx 539.14$$
$$P_3 = 539.14 + 20F(40, 539.14) \approx 936.69$$
$$P_4 = 936.69 + 20F(60, 936.69) \approx 1031.57$$

Thus, our estimates for the population sizes at times $t = 40$ and $t = 80$ are

$$P(40) \approx 539 \qquad P(80) \approx 1032$$

For smaller step sizes we need to program a calculator or computer. The table gives the results.

Step size	Euler estimate of $P(40)$	Euler estimate of $P(80)$
20	539	1032
10	647	997
5	695	991
1	725	986
0.1	731	985

Figure 3 shows a graph of the Euler approximations with step sizes $h = 10$ and $h = 1$. We see that the Euler approximation with $h = 1$ looks very much like the lower solution curve that we drew using a direction field in Figure 2.

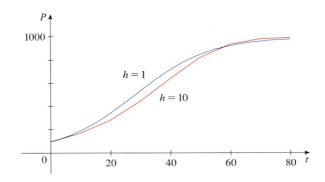

FIGURE 3
Euler approximations of the solution curve in Example 2

The Analytic Solution

The logistic equation (1) is separable and so we can solve it explicitly using the method of Section 10.3. Since

$$\frac{dP}{dt} = kP\left(1 - \frac{P}{K}\right)$$

we have

2 $$\int \frac{dP}{P(1 - P/K)} = \int k\,dt$$

To evaluate the integral on the left side, we write

$$\frac{1}{P(1 - P/K)} = \frac{K}{P(K - P)}$$

Using partial fractions (see Section 8.4), we get

$$\frac{K}{P(K - P)} = \frac{1}{P} + \frac{1}{K - P}$$

This enables us to rewrite Equation 2:

$$\int \left(\frac{1}{P} + \frac{1}{K - P}\right) dP = \int k\,dt$$

$$\ln|P| - \ln|K - P| = kt + C$$

$$\ln\left|\frac{K - P}{P}\right| = -kt - C$$

$$\left|\frac{K - P}{P}\right| = e^{-kt-C} = e^{-C}e^{-kt}$$

3 $$\frac{K - P}{P} = Ae^{-kt}$$

where $A = \pm e^{-C}$. Solving Equation 3 for P, we get

$$\frac{K}{P} - 1 = Ae^{-kt} \quad\Rightarrow\quad \frac{P}{K} = \frac{1}{1 + Ae^{-kt}}$$

so

$$P = \frac{K}{1 + Ae^{-kt}}$$

We find the value of A by putting $t = 0$ in Equation 3. If $t = 0$, then $P = P_0$ (the initial population), so

$$\frac{K - P_0}{P_0} = Ae^0 = A$$

Thus, the solution to the logistic equation is

$$\boxed{4} \qquad P(t) = \frac{K}{1 + Ae^{-kt}} \qquad \text{where } A = \frac{K - P_0}{P_0}$$

Using the expression for $P(t)$ in Equation 4, we see that

$$\lim_{t \to \infty} P(t) = K$$

which is to be expected.

EXAMPLE 3 Write the solution of the initial-value problem

$$\frac{dP}{dt} = 0.08P\left(1 - \frac{P}{1000}\right) \qquad P(0) = 100$$

and use it to find the population sizes $P(40)$ and $P(80)$. At what time does the population reach 900?

SOLUTION The differential equation is a logistic equation with $k = 0.08$, carrying capacity $K = 1000$, and initial population $P_0 = 100$. So Equation 4 gives the population at time t as

$$P(t) = \frac{1000}{1 + Ae^{-0.08t}} \qquad \text{where } A = \frac{1000 - 100}{100} = 9$$

Thus $$P(t) = \frac{1000}{1 + 9e^{-0.08t}}$$

|||| Compare these values with the Euler estimates from Example 2:
$P(40) \approx 731 \qquad P(80) \approx 985$

So the population sizes when $t = 40$ and 80 are

$$P(40) = \frac{1000}{1 + 9e^{-3.2}} \approx 731.6 \qquad P(80) = \frac{1000}{1 + 9e^{-6.4}} \approx 985.3$$

The population reaches 900 when

$$\frac{1000}{1 + 9e^{-0.08t}} = 900$$

Solving this equation for t, we get

$$1 + 9e^{-0.08t} = \tfrac{10}{9}$$

$$e^{-0.08t} = \tfrac{1}{81}$$

$$-0.08t = \ln \tfrac{1}{81} = -\ln 81$$

$$t = \frac{\ln 81}{0.08} \approx 54.9$$

|||| Compare the solution curve in Figure 4 with the lowest solution curve we drew from the direction field in Figure 2.

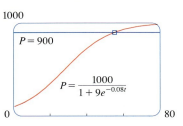

FIGURE 4

So the population reaches 900 when t is approximately 55. As a check on our work, we graph the population curve in Figure 4 and observe where it intersects the line $P = 900$. The cursor indicates that $t \approx 55$.

Comparison of the Natural Growth and Logistic Models

In the 1930s the biologist G. F. Gause conducted an experiment with the protozoan *Paramecium* and used a logistic equation to model his data. The table gives his daily count of the population of protozoa. He estimated the initial relative growth rate to be 0.7944 and the carrying capacity to be 64.

t (days)	0	1	2	3	4	5	6	7	8	9	10	11	12	13	14	15	16
P (observed)	2	3	22	16	39	52	54	47	50	76	69	51	57	70	53	59	57

EXAMPLE 4 Find the exponential and logistic models for Gause's data. Compare the predicted values with the observed values and comment on the fit.

SOLUTION Given the relative growth rate $k = 0.7944$ and the initial population $P_0 = 2$, the exponential model is

$$P(t) = P_0 e^{kt} = 2e^{0.7944t}$$

Gause used the same value of k for his logistic model. [This is reasonable because $P_0 = 2$ is small compared with the carrying capacity ($K = 64$). The equation

$$\frac{1}{P_0}\frac{dP}{dt}\bigg|_{t=0} = k\left(1 - \frac{2}{64}\right) \approx k$$

shows that the value of k for the logistic model is very close to the value for the exponential model.]

Then the solution of the logistic equation in Equation 4 gives

$$P(t) = \frac{K}{1 + Ae^{-kt}} = \frac{64}{1 + Ae^{-0.7944t}}$$

where

$$A = \frac{K - P_0}{P_0} = \frac{64 - 2}{2} = 31$$

So

$$P(t) = \frac{64}{1 + 31e^{-0.7944t}}$$

We use these equations to calculate the predicted values (rounded to the nearest integer) and compare them in the table.

t (days)	0	1	2	3	4	5	6	7	8	9	10	11	12	13	14	15	16
P (observed)	2	3	22	16	39	52	54	47	50	76	69	51	57	70	53	59	57
P (logistic model)	2	4	9	17	28	40	51	57	61	62	63	64	64	64	64	64	64
P (exponential model)	2	4	10	22	48	106	...										

We notice from the table and from the graph in Figure 5 that for the first three or four days the exponential model gives results comparable to those of the more sophisticated logistic model. For $t \geq 5$, however, the exponential model is hopelessly inaccurate, but the logistic model fits the observations reasonably well.

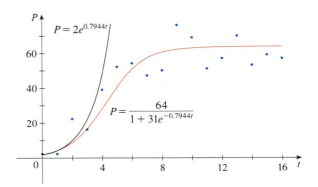

FIGURE 5
The exponential and logistic models for the *Paramecium* data

Other Models for Population Growth

The Law of Natural Growth and the logistic differential equation are not the only equations that have been proposed to model population growth. In Exercise 14 we look at the Gompertz growth function and in Exercises 15 and 16 we investigate seasonal-growth models.

Two of the other models are modifications of the logistic model. The differential equation

$$\frac{dP}{dt} = kP\left(1 - \frac{P}{K}\right) - c$$

has been used to model populations that are subject to "harvesting" of one sort or another. (Think of a population of fish being caught at a constant rate). This equation is explored in Exercises 11 and 12.

For some species there is a minimum population level m below which the species tends to become extinct. (Adults may not be able to find suitable mates.) Such populations have been modeled by the differential equation

$$\frac{dP}{dt} = kP\left(1 - \frac{P}{K}\right)\left(1 - \frac{m}{P}\right)$$

where the extra factor, $1 - m/P$, takes into account the consequences of a sparse population (see Exercise 13).

10.5 Exercises

1. Suppose that a population develops according to the logistic equation

$$\frac{dP}{dt} = 0.05P - 0.0005P^2$$

where t is measured in weeks.
(a) What is the carrying capacity? What is the value of k?
(b) A direction field for this equation is shown at the right. Where are the slopes close to 0? Where are they largest? Which solutions are increasing? Which solutions are decreasing?
(c) Use the direction field to sketch solutions for initial populations of 20, 40, 60, 80, 120, and 140. What do these

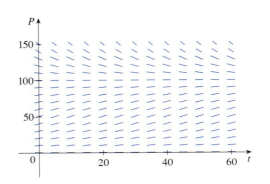

solutions have in common? How do they differ? Which solutions have inflection points? At what population levels do they occur?

(d) What are the equilibrium solutions? How are the other solutions related to these solutions?

2. Suppose that a population grows according to a logistic model with carrying capacity 6000 and $k = 0.0015$ per year.
   (a) Write the logistic differential equation for these data.
   (b) Draw a direction field (either by hand or with a computer algebra system). What does it tell you about the solution curves?
   (c) Use the direction field to sketch the solution curves for initial populations of 1000, 2000, 4000, and 8000. What can you say about the concavity of these curves? What is the significance of the inflection points?
   (d) Program a calculator or computer to use Euler's method with step size $h = 1$ to estimate the population after 50 years if the initial population is 1000.
   (e) If the initial population is 1000, write a formula for the population after t years. Use it to find the population after 50 years and compare with your estimate in part (d).
   (f) Graph the solution in part (e) and compare with the solution curve you sketched in part (c).

3. The Pacific halibut fishery has been modeled by the differential equation
$$\frac{dy}{dt} = ky\left(1 - \frac{y}{K}\right)$$
where $y(t)$ is the biomass (the total mass of the members of the population) in kilograms at time t (measured in years), the carrying capacity is estimated to be $K = 8 \times 10^7$ kg, and $k = 0.71$ per year.
   (a) If $y(0) = 2 \times 10^7$ kg, find the biomass a year later.
   (b) How long will it take for the biomass to reach 4×10^7 kg?

4. The table gives the number of yeast cells in a new laboratory culture.

Time (hours)	Yeast cells	Time (hours)	Yeast cells
0	18	10	509
2	39	12	597
4	80	14	640
6	171	16	664
8	336	18	672

   (a) Plot the data and use the plot to estimate the carrying capacity for the yeast population.
   (b) Use the data to estimate the initial relative growth rate.
   (c) Find both an exponential model and a logistic model for these data.
   (d) Compare the predicted values with the observed values, both in a table and with graphs. Comment on how well your models fit the data.
   (e) Use your logistic model to estimate the number of yeast cells after 7 hours.

5. The population of the world was about 5.3 billion in 1990. Birth rates in the 1990s ranged from 35 to 40 million per year and death rates ranged from 15 to 20 million per year. Let's assume that the carrying capacity for world population is 100 billion.
   (a) Write the logistic differential equation for these data. (Because the initial population is small compared to the carrying capacity, you can take k to be an estimate of the initial relative growth rate.)
   (b) Use the logistic model to estimate the world population in the year 2000 and compare with the actual population of 6.1 billion.
   (c) Use the logistic model to predict the world population in the years 2100 and 2500.
   (d) What are your predictions if the carrying capacity is 50 billion?

6. (a) Make a guess as to the carrying capacity for the U.S. population. Use it and the fact that the population was 250 million in 1990 to formulate a logistic model for the U.S. population.
   (b) Determine the value of k in your model by using the fact that the population in 2000 was 275 million.
   (c) Use your model to predict the U.S. population in the years 2100 and 2200.
   (d) Use your model to predict the year in which the U.S. population will exceed 300 million.

7. One model for the spread of a rumor is that the rate of spread is proportional to the product of the fraction y of the population who have heard the rumor and the fraction who have not heard the rumor.
   (a) Write a differential equation that is satisfied by y.
   (b) Solve the differential equation.
   (c) A small town has 1000 inhabitants. At 8 A.M., 80 people have heard a rumor. By noon half the town has heard it. At what time will 90% of the population have heard the rumor?

8. Biologists stocked a lake with 400 fish and estimated the carrying capacity (the maximal population for the fish of that species in that lake) to be 10,000. The number of fish tripled in the first year.
   (a) Assuming that the size of the fish population satisfies the logistic equation, find an expression for the size of the population after t years.
   (b) How long will it take for the population to increase to 5000?

9. (a) Show that if P satisfies the logistic equation (1), then
$$\frac{d^2P}{dt^2} = k^2 P\left(1 - \frac{P}{K}\right)\left(1 - \frac{2P}{K}\right)$$
   (b) Deduce that a population grows fastest when it reaches half its carrying capacity.

10. For a fixed value of K (say $K = 10$), the family of logistic functions given by Equation 4 depends on the initial value P_0 and the proportionality constant k. Graph several members of

this family. How does the graph change when P_0 varies? How does it change when k varies?

11. Let's modify the logistic differential equation of Example 1 as follows:

$$\frac{dP}{dt} = 0.08P\left(1 - \frac{P}{1000}\right) - 15$$

(a) Suppose $P(t)$ represents a fish population at time t, where t is measured in weeks. Explain the meaning of the term -15.
(b) Draw a direction field for this differential equation.
(c) What are the equilibrium solutions?
(d) Use the direction field to sketch several solution curves. Describe what happens to the fish population for various initial populations.
(e) Solve this differential equation explicitly, either by using partial fractions or with a computer algebra system. Use the initial populations 200 and 300. Graph the solutions and compare with your sketches in part (d).

12. Consider the differential equation

$$\frac{dP}{dt} = 0.08P\left(1 - \frac{P}{1000}\right) - c$$

as a model for a fish population, where t is measured in weeks and c is a constant.
(a) Use a CAS to draw direction fields for various values of c.
(b) From your direction fields in part (a), determine the values of c for which there is at least one equilibrium solution. For what values of c does the fish population always die out?
(c) Use the differential equation to prove what you discovered graphically in part (b).
(d) What would you recommend for a limit to the weekly catch of this fish population?

13. There is considerable evidence to support the theory that for some species there is a minimum population m such that the species will become extinct if the size of the population falls below m. This condition can be incorporated into the logistic equation by introducing the factor $(1 - m/P)$. Thus, the modified logistic model is given by the differential equation

$$\frac{dP}{dt} = kP\left(1 - \frac{P}{K}\right)\left(1 - \frac{m}{P}\right)$$

(a) Use the differential equation to show that any solution is increasing if $m < P < K$ and decreasing if $0 < P < m$.
(b) For the case where $k = 0.08$, $K = 1000$, and $m = 200$, draw a direction field and use it to sketch several solution curves. Describe what happens to the population for various initial populations. What are the equilibrium solutions?
(c) Solve the differential equation explicitly, either by using partial fractions or with a computer algebra system. Use the initial population P_0.

(d) Use the solution in part (c) to show that if $P_0 < m$, then the species will become extinct. [*Hint:* Show that the numerator in your expression for $P(t)$ is 0 for some value of t.]

14. Another model for a growth function for a limited population is given by the **Gompertz function**, which is a solution of the differential equation

$$\frac{dP}{dt} = c \ln\left(\frac{K}{P}\right)P$$

where c is a constant and K is the carrying capacity.
(a) Solve this differential equation.
(b) Compute $\lim_{t \to \infty} P(t)$.
(c) Graph the Gompertz growth function for $K = 1000$, $P_0 = 100$, and $c = 0.05$, and compare it with the logistic function in Example 3. What are the similarities? What are the differences?
(d) We know from Exercise 9 that the logistic function grows fastest when $P = K/2$. Use the Gompertz differential equation to show that the Gompertz function grows fastest when $P = K/e$.

15. In a **seasonal-growth model**, a periodic function of time is introduced to account for seasonal variations in the rate of growth. Such variations could, for example, be caused by seasonal changes in the availability of food.
(a) Find the solution of the seasonal-growth model

$$\frac{dP}{dt} = kP \cos(rt - \phi) \qquad P(0) = P_0$$

where k, r, and ϕ are positive constants.
(b) By graphing the solution for several values of k, r, and ϕ, explain how the values of k, r, and ϕ affect the solution. What can you say about $\lim_{t \to \infty} P(t)$?

16. Suppose we alter the differential equation in Exercise 15 as follows:

$$\frac{dP}{dt} = kP \cos^2(rt - \phi) \qquad P(0) = P_0$$

(a) Solve this differential equation with the help of a table of integrals or a CAS.
(b) Graph the solution for several values of k, r, and ϕ. How do the values of k, r, and ϕ affect the solution? What can you say about $\lim_{t \to \infty} P(t)$ in this case?

17. Graphs of logistic functions (Figures 2 and 4) look suspiciously similar to the graph of the hyperbolic tangent function (Figure 3 in Section 7.6). Explain the similarity by showing that the logistic function given by Equation 4 can be written as

$$P(t) = \tfrac{1}{2}K\left[1 + \tanh\left(\tfrac{1}{2}k(t - c)\right)\right]$$

where $c = (\ln A)/k$. Thus, the logistic function is really just a shifted hyperbolic tangent.

10.6 Linear Equations

A first-order **linear** differential equation is one that can be put into the form

$$\boxed{1} \qquad \frac{dy}{dx} + P(x)y = Q(x)$$

where P and Q are continuous functions on a given interval. This type of equation occurs frequently in various sciences, as we will see.

An example of a linear equation is $xy' + y = 2x$ because, for $x \neq 0$, it can be written in the form

$$\boxed{2} \qquad y' + \frac{1}{x}y = 2$$

Notice that this differential equation is not separable because it's impossible to factor the expression for y' as a function of x times a function of y. But we can still solve the equation by noticing, by the Product Rule, that

$$xy' + y = (xy)'$$

and so we can rewrite the equation as

$$(xy)' = 2x$$

If we now integrate both sides of this equation, we get

$$xy = x^2 + C \qquad \text{or} \qquad y = x + \frac{C}{x}$$

If we had been given the differential equation in the form of Equation 2, we would have had to take the preliminary step of multiplying each side of the equation by x.

It turns out that every first-order linear differential equation can be solved in a similar fashion by multiplying both sides of Equation 1 by a suitable function $I(x)$ called an *integrating factor*. We try to find I so that the left side of Equation 1, when multiplied by $I(x)$, becomes the derivative of the product $I(x)y$:

$$\boxed{3} \qquad I(x)(y' + P(x)y) = (I(x)y)'$$

If we can find such a function I, then Equation 1 becomes

$$(I(x)y)' = I(x)Q(x)$$

Integrating both sides, we would have

$$I(x)y = \int I(x)Q(x)\, dx + C$$

so the solution would be

$$\boxed{4} \qquad y(x) = \frac{1}{I(x)}\left[\int I(x)Q(x)\, dx + C\right]$$

To find such an I, we expand Equation 3 and cancel terms:

$$I(x)y' + I(x)P(x)y = (I(x)y)' = I'(x)y + I(x)y'$$

$$I(x)P(x) = I'(x)$$

This is a separable differential equation for I, which we solve as follows:

$$\int \frac{dI}{I} = \int P(x)\,dx$$

$$\ln|I| = \int P(x)\,dx$$

$$I = Ae^{\int P(x)\,dx}$$

where $A = \pm e^C$. We are looking for a particular integrating factor, not the most general one, so we take $A = 1$ and use

$$\boxed{5} \qquad I(x) = e^{\int P(x)\,dx}$$

Thus, a formula for the general solution to Equation 1 is provided by Equation 4, where I is given by Equation 5. Instead of memorizing this formula, however, we just remember the form of the integrating factor.

> To solve the linear differential equation $y' + P(x)y = Q(x)$, multiply both sides by the **integrating factor** $I(x) = e^{\int P(x)\,dx}$ and integrate both sides.

EXAMPLE 1 Solve the differential equation $\dfrac{dy}{dx} + 3x^2 y = 6x^2$.

SOLUTION The given equation is linear since it has the form of Equation 1 with $P(x) = 3x^2$ and $Q(x) = 6x^2$. An integrating factor is

$$I(x) = e^{\int 3x^2\,dx} = e^{x^3}$$

Multiplying both sides of the differential equation by e^{x^3}, we get

$$e^{x^3}\frac{dy}{dx} + 3x^2 e^{x^3} y = 6x^2 e^{x^3}$$

or

$$\frac{d}{dx}(e^{x^3} y) = 6x^2 e^{x^3}$$

Integrating both sides, we have

$$e^{x^3} y = \int 6x^2 e^{x^3}\,dx = 2e^{x^3} + C$$

$$y = 2 + Ce^{-x^3}$$

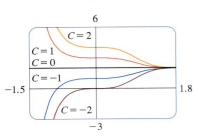

Figure 1 shows the graphs of several members of the family of solutions in Example 1. Notice that they all approach 2 as $x \to \infty$.

FIGURE 1

EXAMPLE 2 Find the solution of the initial-value problem

$$x^2 y' + xy = 1 \qquad x > 0 \qquad y(1) = 2$$

SOLUTION We must first divide both sides by the coefficient of y' to put the differential equation into standard form:

$$\boxed{6} \qquad y' + \frac{1}{x} y = \frac{1}{x^2} \qquad x > 0$$

The integrating factor is

$$I(x) = e^{\int (1/x)\,dx} = e^{\ln x} = x$$

Multiplication of Equation 6 by x gives

$$xy' + y = \frac{1}{x} \quad \text{or} \quad (xy)' = \frac{1}{x}$$

Then

$$xy = \int \frac{1}{x}\,dx = \ln x + C$$

and so

$$y = \frac{\ln x + C}{x}$$

Since $y(1) = 2$, we have

$$2 = \frac{\ln 1 + C}{1} = C$$

Therefore, the solution to the initial-value problem is

$$y = \frac{\ln x + 2}{x}$$

|||| The solution of the initial-value problem in Example 2 is shown in Figure 2.

FIGURE 2

EXAMPLE 3 Solve $y' + 2xy = 1$.

SOLUTION The given equation is in the standard form for a linear equation. Multiplying by the integrating factor

$$e^{\int 2x\,dx} = e^{x^2}$$

we get

$$e^{x^2} y' + 2xe^{x^2} y = e^{x^2}$$

or

$$\left(e^{x^2} y\right)' = e^{x^2}$$

Therefore

$$e^{x^2} y = \int e^{x^2}\,dx + C$$

Recall from Section 8.5 that $\int e^{x^2}\,dx$ can't be expressed in terms of elementary functions. Nonetheless, it's a perfectly good function and we can leave the answer as

$$y = e^{-x^2} \int e^{x^2}\,dx + Ce^{-x^2}$$

|||| Even though the solutions of the differential equation in Example 3 are expressed in terms of an integral, they can still be graphed by a computer algebra system (Figure 3).

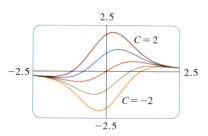

FIGURE 3

Another way of writing the solution is

$$y = e^{-x^2} \int_0^x e^{t^2}\, dt + Ce^{-x^2}$$

(Any number can be chosen for the lower limit of integration.)

Application to Electric Circuits

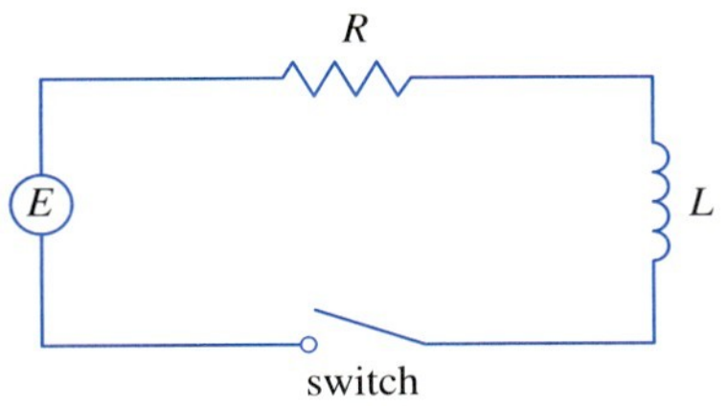

In Section 10.2 we considered the simple electric circuit shown in Figure 4: An electromotive force (usually a battery or generator) produces a voltage of $E(t)$ volts (V) and a current of $I(t)$ amperes (A) at time t. The circuit also contains a resistor with a resistance of R ohms (Ω) and an inductor with an inductance of L henries (H).

Ohm's Law gives the drop in voltage due to the resistor as RI. The voltage drop due to the inductor is $L(dI/dt)$. One of Kirchhoff's laws says that the sum of the voltage drops is equal to the supplied voltage $E(t)$. Thus, we have

$$\boxed{7} \qquad L\frac{dI}{dt} + RI = E(t)$$

FIGURE 4

which is a first-order linear differential equation. The solution gives the current I at time t.

EXAMPLE 4 Suppose that in the simple circuit of Figure 4 the resistance is 12 Ω and the inductance is 4 H. If a battery gives a constant voltage of 60 V and the switch is closed when $t = 0$ so the current starts with $I(0) = 0$, find (a) $I(t)$, (b) the current after 1 s, and (c) the limiting value of the current.

SOLUTION

(a) If we put $L = 4$, $R = 12$, and $E(t) = 60$ in Equation 7, we obtain the initial-value problem

$$4\frac{dI}{dt} + 12I = 60 \qquad I(0) = 0$$

or

$$\frac{dI}{dt} + 3I = 15 \qquad I(0) = 0$$

Multiplying by the integrating factor $e^{\int 3\, dt} = e^{3t}$, we get

$$e^{3t}\frac{dI}{dt} + 3e^{3t}I = 15e^{3t}$$

$$\frac{d}{dt}(e^{3t}I) = 15e^{3t}$$

$$e^{3t}I = \int 15e^{3t}\, dt = 5e^{3t} + C$$

$$I(t) = 5 + Ce^{-3t}$$

Since $I(0) = 0$, we have $5 + C = 0$, so $C = -5$ and

$$I(t) = 5(1 - e^{-3t})$$

||| The differential equation in Example 4 is both linear and separable, so an alternative method is to solve it as a separable equation (Example 4 in Section 10.3). If we replace the battery by a generator, however, we get an equation that is linear but not separable (Example 5).

Figure 5 shows how the current in Example 4 approaches its limiting value.

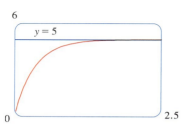

FIGURE 5

(b) After 1 second the current is

$$I(1) = 5(1 - e^{-3}) \approx 4.75 \text{ A}$$

(c)
$$\lim_{t \to \infty} I(t) = \lim_{t \to \infty} 5(1 - e^{-3t})$$
$$= 5 - 5 \lim_{t \to \infty} e^{-3t}$$
$$= 5 - 0 = 5$$

EXAMPLE 5 Suppose that the resistance and inductance remain as in Example 4 but, instead of the battery, we use a generator that produces a variable voltage of $E(t) = 60 \sin 30t$ volts. Find $I(t)$.

SOLUTION This time the differential equation becomes

$$4\frac{dI}{dt} + 12I = 60 \sin 30t \quad \text{or} \quad \frac{dI}{dt} + 3I = 15 \sin 30t$$

The same integrating factor e^{3t} gives

$$\frac{d}{dt}(e^{3t}I) = e^{3t}\frac{dI}{dt} + 3e^{3t}I = 15e^{3t} \sin 30t$$

Using Formula 98 in the Table of Integrals, we have

$$e^{3t}I = \int 15e^{3t} \sin 30t \, dt = 15\frac{e^{3t}}{909}(3 \sin 30t - 30 \cos 30t) + C$$

$$I = \tfrac{5}{101}(\sin 30t - 10 \cos 30t) + Ce^{-3t}$$

Since $I(0) = 0$, we get

$$-\tfrac{50}{101} + C = 0$$

so
$$I(t) = \tfrac{5}{101}(\sin 30t - 10 \cos 30t) + \tfrac{50}{101}e^{-3t}$$

Figure 6 shows the graph of the current when the battery is replaced by a generator.

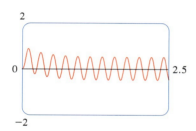

FIGURE 6

10.6 Exercises

1–4 Determine whether the differential equation is linear.

1. $y' + e^x y = x^2 y^2$
2. $y + \sin x = x^3 y'$
3. $xy' + \ln x - x^2 y = 0$
4. $y' + \cos y = \tan x$

5–14 Solve the differential equation.

5. $y' + 2y = 2e^x$
6. $y' = x + 5y$
7. $xy' - 2y = x^2$
8. $x^2 y' + 2xy = \cos^2 x$
9. $xy' + y = \sqrt{x}$
10. $1 + xy = xy'$
11. $\dfrac{dy}{dx} + 2xy = x^2$
12. $\dfrac{dy}{dx} = x \sin 2x + y \tan x, \quad -\pi/2 < x < \pi/2$
13. $(1 + t)\dfrac{du}{dt} + u = 1 + t, \quad t > 0$
14. $t \ln t \dfrac{dr}{dt} + r = te^t$

15–20 Solve the initial-value problem.

15. $y' = x + y, \quad y(0) = 2$
16. $t\dfrac{dy}{dt} + 2y = t^3, \quad t > 0, \quad y(1) = 0$
17. $\dfrac{dv}{dt} - 2tv = 3t^2 e^{t^2}, \quad v(0) = 5$

18. $2xy' + y = 6x$, $x > 0$, $y(4) = 20$

19. $xy' = y + x^2 \sin x$, $y(\pi) = 0$

20. $x \dfrac{dy}{dx} - \dfrac{y}{x+1} = x$, $y(1) = 0$, $x > 0$

21–22 Solve the differential equation and use a graphing calculator or computer to graph several members of the family of solutions. How does the solution curve change as C varies?

21. $xy' + y = x \cos x$, $x > 0$

22. $y' + (\cos x)y = \cos x$

23. A **Bernoulli differential equation** (named after James Bernoulli) is of the form

$$\dfrac{dy}{dx} + P(x)y = Q(x)y^n$$

Observe that, if $n = 0$ or 1, the Bernoulli equation is linear. For other values of n, show that the substitution $u = y^{1-n}$ transforms the Bernoulli equation into the linear equation

$$\dfrac{du}{dx} + (1 - n)P(x)u = (1 - n)Q(x)$$

24–26 Use the method of Exercise 23 to solve the differential equation.

24. $xy' + y = -xy^2$

25. $y' + \dfrac{2}{x} y = \dfrac{y^3}{x^2}$

26. $y' + y = xy^3$

27. In the circuit shown in Figure 4, a battery supplies a constant voltage of 40 V, the inductance is 2 H, the resistance is 10 Ω, and $I(0) = 0$.
(a) Find $I(t)$.
(b) Find the current after 0.1 s.

28. In the circuit shown in Figure 4, a generator supplies a voltage of $E(t) = 40 \sin 60t$ volts, the inductance is 1 H, the resistance is 20 Ω, and $I(0) = 1$ A.
(a) Find $I(t)$.
(b) Find the current after 0.1 s.
(c) Use a graphing device to draw the graph of the current function.

29. The figure shows a circuit containing an electromotive force, a capacitor with a capacitance of C farads (F), and a resistor with a resistance of R ohms (Ω). The voltage drop across the capacitor is Q/C, where Q is the charge (in coulombs), so in this case Kirchhoff's Law gives

$$RI + \dfrac{Q}{C} = E(t)$$

But $I = dQ/dt$ (see Example 3 in Section 3.4), so we have

$$R \dfrac{dQ}{dt} + \dfrac{1}{C} Q = E(t)$$

Suppose the resistance is 5 Ω, the capacitance is 0.05 F, a battery gives a constant voltage of 60 V, and the initial charge is $Q(0) = 0$ C. Find the charge and the current at time t.

30. In the circuit of Exercise 29, $R = 2\ \Omega$, $C = 0.01$ F, $Q(0) = 0$, and $E(t) = 10 \sin 60t$. Find the charge and the current at time t.

31. Let $P(t)$ be the performance level of someone learning a skill as a function of the training time t. The graph of P is called a *learning curve*. In Exercise 13 in Section 10.1 we proposed the differential equation

$$\dfrac{dP}{dt} = k[M - P(t)]$$

as a reasonable model for learning, where k is a positive constant. Solve it as a linear differential equation and use your solution to graph the learning curve.

32. Two new workers were hired for an assembly line. Jim processed 25 units during the first hour and 45 units during the second hour. Mark processed 35 units during the first hour and 50 units the second hour. Using the model of Exercise 31 and assuming that $P(0) = 0$, estimate the maximum number of units per hour that each worker is capable of processing.

33. In Section 10.3 we looked at mixing problems in which the volume of fluid remained constant and saw that such problems give rise to separable equations. (See Example 6 in that section.) If the rates of flow into and out of the system are different, then the volume is not constant and the resulting differential equation is linear but not separable.

A tank contains 100 L of water. A solution with a salt concentration of 0.4 kg/L is added at a rate of 5 L/min. The solution is kept mixed and is drained from the tank at a rate of 3 L/min. If $y(t)$ is the amount of salt (in kilograms) after t minutes, show that y satisfies the differential equation

$$\dfrac{dy}{dt} = 2 - \dfrac{3y}{100 + 2t}$$

Solve this equation and find the concentration after 20 minutes.

34. A tank with a capacity of 400 L is full of a mixture of water and chlorine with a concentration of 0.05 g of chlorine per liter. In order to reduce the concentration of chlorine, fresh water is pumped into the tank at a rate of 4 L/s. The mixture is kept

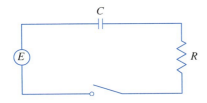

stirred and is pumped out at a rate of 10 L/s. Find the amount of chlorine in the tank as a function of time.

35. An object with mass m is dropped from rest and we assume that the air resistance is proportional to the speed of the object. If $s(t)$ is the distance dropped after t seconds, then the speed is $v = s'(t)$ and the acceleration is $a = v'(t)$. If g is the acceleration due to gravity, then the downward force on the object is $mg - cv$, where c is a positive constant, and Newton's Second Law gives

$$m \frac{dv}{dt} = mg - cv$$

(a) Solve this as a linear equation to show that

$$v = \frac{mg}{c}(1 - e^{-ct/m})$$

(b) What is the limiting velocity?
(c) Find the distance the object has fallen after t seconds.

36. If we ignore air resistance, we can conclude that heavier objects fall no faster than lighter objects. But if we take air resistance into account, our conclusion changes. Use the expression for the velocity of a falling object in Exercise 35(a) to find dv/dm and show that heavier objects *do* fall faster than lighter ones.

10.7 Predator-Prey Systems

We have looked at a variety of models for the growth of a single species that lives alone in an environment. In this section we consider more realistic models that take into account the interaction of two species in the same habitat. We will see that these models take the form of a pair of linked differential equations.

We first consider the situation in which one species, called the *prey,* has an ample food supply and the second species, called the *predators,* feeds on the prey. Examples of prey and predators include rabbits and wolves in an isolated forest, food fish and sharks, aphids and ladybugs, and bacteria and amoebas. Our model will have two dependent variables and both are functions of time. We let $R(t)$ be the number of prey (using R for rabbits) and $W(t)$ be the number of predators (with W for wolves) at time t.

In the absence of predators, the ample food supply would support exponential growth of the prey, that is,

$$\frac{dR}{dt} = kR \qquad \text{where } k \text{ is a positive constant}$$

In the absence of prey, we assume that the predator population would decline at a rate proportional to itself, that is,

$$\frac{dW}{dt} = -rW \qquad \text{where } r \text{ is a positive constant}$$

With both species present, however, we assume that the principal cause of death among the prey is being eaten by a predator, and the birth and survival rates of the predators depend on their available food supply, namely, the prey. We also assume that the two species encounter each other at a rate that is proportional to both populations and is therefore proportional to the product RW. (The more there are of either population, the more encounters there are likely to be.) A system of two differential equations that incorporates these assumptions is as follows:

W represents the predator.
R represents the prey.

$$\boxed{1} \qquad \frac{dR}{dt} = kR - aRW \qquad \frac{dW}{dt} = -rW + bRW$$

where k, r, a, and b are positive constants. Notice that the term $-aRW$ decreases the natural growth rate of the prey and the term bRW increases the natural growth rate of the predators.

▐▐▐▐ The Lotka-Volterra equations were proposed as a model to explain the variations in the shark and food-fish populations in the Adriatic Sea by the Italian mathematician Vito Volterra (1860–1940).

The equations in (1) are known as the **predator-prey equations**, or the **Lotka-Volterra equations**. A **solution** of this system of equations is a pair of functions $R(t)$ and $W(t)$ that describe the populations of prey and predator as functions of time. Because the system is coupled (R and W occur in both equations), we can't solve one equation and then the other; we have to solve them simultaneously. Unfortunately, it is usually impossible to find explicit formulas for R and W as functions of t. We can, however, use graphical methods to analyze the equations.

EXAMPLE 1 Suppose that populations of rabbits and wolves are described by the Lotka-Volterra equations (1) with $k = 0.08$, $a = 0.001$, $r = 0.02$, and $b = 0.00002$. The time t is measured in months.
(a) Find the constant solutions (called the **equilibrium solutions**) and interpret the answer.
(b) Use the system of differential equations to find an expression for dW/dR.
(c) Draw a direction field for the resulting differential equation in the RW-plane. Then use that direction field to sketch some solution curves.
(d) Suppose that, at some point in time, there are 1000 rabbits and 40 wolves. Draw the corresponding solution curve and use it to describe the changes in both population levels.
(e) Use part (d) to make sketches of R and W as functions of t.

SOLUTION
(a) With the given values of k, a, r, and b, the Lotka-Volterra equations become

$$\frac{dR}{dt} = 0.08R - 0.001RW$$

$$\frac{dW}{dt} = -0.02W + 0.00002RW$$

Both R and W will be constant if both derivatives are 0, that is,

$$R' = R(0.08 - 0.001W) = 0$$

$$W' = W(-0.02 + 0.00002R) = 0$$

One solution is given by $R = 0$ and $W = 0$. (This makes sense: If there are no rabbits or wolves, the populations are certainly not going to increase.) The other constant solution is

$$W = \frac{0.08}{0.001} = 80 \qquad R = \frac{0.02}{0.00002} = 1000$$

So the equilibrium populations consist of 80 wolves and 1000 rabbits. This means that 1000 rabbits are just enough to support a constant wolf population of 80. There are neither too many wolves (which would result in fewer rabbits) nor too few wolves (which would result in more rabbits).

(b) We use the Chain Rule to eliminate t:

$$\frac{dW}{dt} = \frac{dW}{dR} \frac{dR}{dt}$$

so

$$\frac{dW}{dR} = \frac{\dfrac{dW}{dt}}{\dfrac{dR}{dt}} = \frac{-0.02W + 0.00002RW}{0.08R - 0.001RW}$$

(c) If we think of W as a function of R, we have the differential equation

$$\frac{dW}{dR} = \frac{-0.02W + 0.00002RW}{0.08R - 0.001RW}$$

We draw the direction field for this differential equation in Figure 1 and we use it to sketch several solution curves in Figure 2. If we move along a solution curve, we observe how the relationship between R and W changes as time passes. Notice that the curves appear to be closed in the sense that if we travel along a curve, we always return to the same point. Notice also that the point $(1000, 80)$ is inside all the solution curves. That point is called an *equilibrium point* because it corresponds to the equilibrium solution $R = 1000$, $W = 80$.

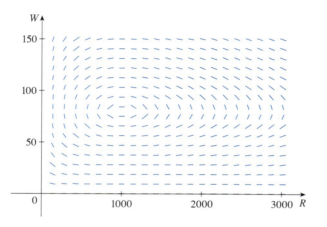

FIGURE 1 Direction field for the predator-prey system

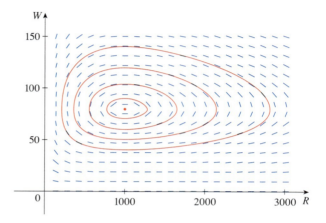

FIGURE 2 Phase portrait of the system

When we represent solutions of a system of differential equations as in Figure 2, we refer to the RW-plane as the **phase plane**, and we call the solution curves **phase trajectories**. So a phase trajectory is a path traced out by solutions (R, W) as time goes by. A **phase portrait** consists of equilibrium points and typical phase trajectories, as shown in Figure 2.

(d) Starting with 1000 rabbits and 40 wolves corresponds to drawing the solution curve through the point $P_0(1000, 40)$. Figure 3 shows this phase trajectory with the direction field removed. Starting at the point P_0 at time $t = 0$ and letting t increase, do we move clockwise or counterclockwise around the phase trajectory? If we put $R = 1000$ and

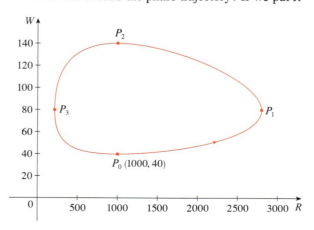

FIGURE 3
Phase trajectory through $(1000, 40)$

$W = 40$ in the first differential equation, we get

$$\frac{dR}{dt} = 0.08(1000) - 0.001(1000)(40) = 80 - 40 = 40$$

Since $dR/dt > 0$, we conclude that R is increasing at P_0 and so we move counterclockwise around the phase trajectory.

We see that at P_0 there aren't enough wolves to maintain a balance between the populations, so the rabbit population increases. That results in more wolves and eventually there are so many wolves that the rabbits have a hard time avoiding them. So the number of rabbits begins to decline (at P_1, where we estimate that R reaches its maximum population of about 2800). This means that at some later time the wolf population starts to fall (at P_2, where $R = 1000$ and $W \approx 140$). But this benefits the rabbits, so their population later starts to increase (at P_3, where $W = 80$ and $R \approx 210$). As a consequence, the wolf population eventually starts to increase as well. This happens when the populations return to their initial values of $R = 1000$ and $W = 40$, and the entire cycle begins again.

(e) From the description in part (d) of how the rabbit and wolf populations rise and fall, we can sketch the graphs of $R(t)$ and $W(t)$. Suppose the points P_1, P_2, and P_3 in Figure 3 are reached at times t_1, t_2, and t_3. Then we can sketch graphs of R and W as in Figure 4.

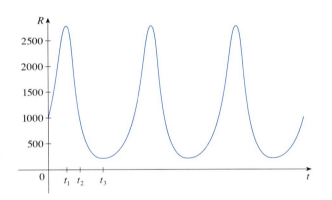
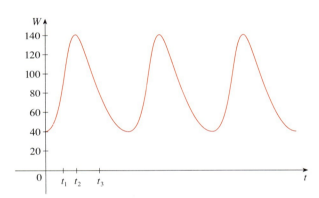

FIGURE 4
Graphs of the rabbit and wolf populations as functions of time

To make the graphs easier to compare, we draw the graphs on the same axes but with different scales for R and W, as in Figure 5. Notice that the rabbits reach their maximum populations about a quarter of a cycle before the wolves.

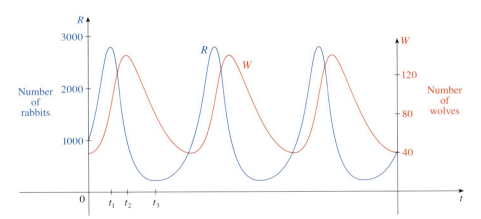

FIGURE 5
Comparison of the rabbit and wolf populations

An important part of the modeling process, as we discussed in Section 1.2, is to interpret our mathematical conclusions as real-world predictions and to test the predictions against real data. The Hudson's Bay Company, which started trading in animal furs in Canada in 1670, has kept records that date back to the 1840s. Figure 6 shows graphs of the number of pelts of the snowshoe hare and its predator, the Canada lynx, traded by the company over a 90-year period. You can see that the coupled oscillations in the hare and lynx populations predicted by the Lotka-Volterra model do actually occur and the period of these cycles is roughly 10 years.

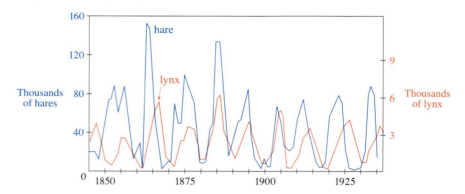

FIGURE 6
Relative abundance of hare and lynx from Hudson's Bay Company records

Although the relatively simple Lotka-Volterra model has had some success in explaining and predicting coupled populations, more sophisticated models have also been proposed. One way to modify the Lotka-Volterra equations is to assume that, in the absence of predators, the prey grow according to a logistic model with carrying capacity K. Then the Lotka-Volterra equations (1) are replaced by the system of differential equations

$$\frac{dR}{dt} = kR\left(1 - \frac{R}{K}\right) - aRW \qquad \frac{dW}{dt} = -rW + bRW$$

This model is investigated in Exercises 9 and 10.

Models have also been proposed to describe and predict population levels of two species that compete for the same resources or cooperate for mutual benefit. Such models are explored in Exercise 2.

10.7 Exercises

1. For each predator-prey system, determine which of the variables, x or y, represents the prey population and which represents the predator population. Is the growth of the prey restricted just by the predators or by other factors as well? Do the predators feed only on the prey or do they have additional food sources? Explain.

   (a) $\dfrac{dx}{dt} = -0.05x + 0.0001xy$

   $\dfrac{dy}{dt} = 0.1y - 0.005xy$

   (b) $\dfrac{dx}{dt} = 0.2x - 0.0002x^2 - 0.006xy$

   $\dfrac{dy}{dt} = -0.015y + 0.00008xy$

2. Each system of differential equations is a model for two species that either compete for the same resources or cooperate for mutual benefit (flowering plants and insect pollinators, for instance). Decide whether each system describes competition or cooperation and explain why it is a reasonable model. (Ask yourself what effect an increase in one species has on the growth rate of the other.)

   (a) $\dfrac{dx}{dt} = 0.12x - 0.0006x^2 + 0.00001xy$

   $\dfrac{dy}{dt} = 0.08x + 0.00004xy$

   (b) $\dfrac{dx}{dt} = 0.15x - 0.0002x^2 - 0.0006xy$

   $\dfrac{dy}{dt} = 0.2y - 0.00008y^2 - 0.0002xy$

3–4 ▮▮▮ A phase trajectory is shown for populations of rabbits (R) and foxes (F).
(a) Describe how each population changes as time goes by.
(b) Use your description to make a rough sketch of the graphs of R and F as functions of time.

3.

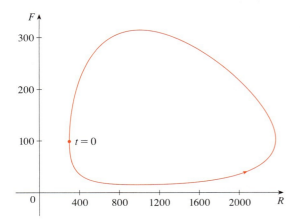

4.

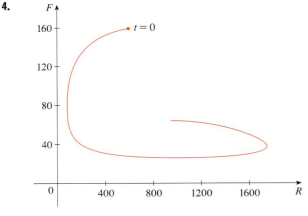

5–6 ▮▮▮ Graphs of populations of two species are shown. Use them to sketch the corresponding phase trajectory.

5.

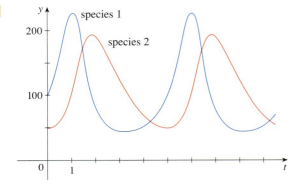

6.

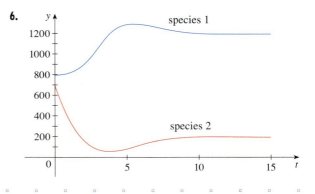

7. In Example 1(b) we showed that the rabbit and wolf populations satisfy the differential equation

$$\frac{dW}{dR} = \frac{-0.02W + 0.00002RW}{0.08R - 0.001RW}$$

By solving this separable differential equation, show that

$$\frac{R^{0.02}W^{0.08}}{e^{0.00002R}e^{0.001W}} = C$$

where C is a constant.

It is impossible to solve this equation for W as an explicit function of R (or vice versa). If you have a computer algebra system that graphs implicitly defined curves, use this equation and your CAS to draw the solution curve that passes through the point $(1000, 40)$ and compare with Figure 3.

8. Populations of aphids and ladybugs are modeled by the equations

$$\frac{dA}{dt} = 2A - 0.01AL$$

$$\frac{dL}{dt} = -0.5L + 0.0001AL$$

(a) Find the equilibrium solutions and explain their significance.
(b) Find an expression for dL/dA.
(c) The direction field for the differential equation in part (b) is shown. Use it to sketch a phase portrait. What do the phase trajectories have in common?

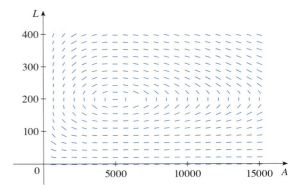

(d) Suppose that at time $t = 0$ there are 1000 aphids and 200 ladybugs. Draw the corresponding phase trajectory and use it to describe how both populations change.

(e) Use part (d) to make rough sketches of the aphid and ladybug populations as functions of t. How are the graphs related to each other?

9. In Example 1 we used Lotka-Volterra equations to model populations of rabbits and wolves. Let's modify those equations as follows:

$$\frac{dR}{dt} = 0.08R(1 - 0.0002R) - 0.001RW$$

$$\frac{dW}{dt} = -0.02W + 0.00002RW$$

(a) According to these equations, what happens to the rabbit population in the absence of wolves?

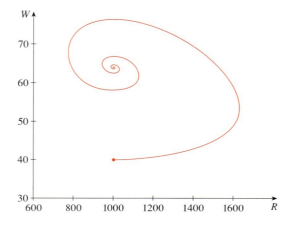

(b) Find all the equilibrium solutions and explain their significance.

(c) The figure shows the phase trajectory that starts at the point (1000, 40). Describe what eventually happens to the rabbit and wolf populations.

(d) Sketch graphs of the rabbit and wolf populations as functions of time.

CAS 10. In Exercise 8 we modeled populations of aphids and ladybugs with a Lotka-Volterra system. Suppose we modify those equations as follows:

$$\frac{dA}{dt} = 2A(1 - 0.0001A) - 0.01AL$$

$$\frac{dL}{dt} = -0.5L + 0.0001AL$$

(a) In the absence of ladybugs, what does the model predict about the aphids?
(b) Find the equilibrium solutions.
(c) Find an expression for dL/dA.
(d) Use a computer algebra system to draw a direction field for the differential equation in part (c). Then use the direction field to sketch a phase portrait. What do the phase trajectories have in common?
(e) Suppose that at time $t = 0$ there are 1000 aphids and 200 ladybugs. Draw the corresponding phase trajectory and use it to describe how both populations change.
(f) Use part (e) to make rough sketches of the aphid and ladybug populations as functions of t. How are the graphs related to each other?

10 Review

CONCEPT CHECK

1. (a) What is a differential equation?
(b) What is the order of a differential equation?
(c) What is an initial condition?

2. What can you say about the solutions of the equation $y' = x^2 + y^2$ just by looking at the differential equation?

3. What is a direction field for the differential equation $y' = F(x, y)$?

4. Explain how Euler's method works.

5. What is a separable differential equation? How do you solve it?

6. What is a first-order linear differential equation? How do you solve it?

7. (a) Write a differential equation that expresses the law of natural growth. What does it say in terms of relative growth rate?
(b) Under what circumstances is this an appropriate model for population growth?
(c) What are the solutions of this equation?

8. (a) Write the logistic equation.
(b) Under what circumstances is this an appropriate model for population growth?

9. (a) Write Lotka-Volterra equations to model populations of food fish (F) and sharks (S).
(b) What do these equations say about each population in the absence of the other?

TRUE-FALSE QUIZ

Determine whether the statement is true or false. If it is true, explain why. If it is false, explain why or give an example that disproves the statement.

1. All solutions of the differential equation $y' = -1 - y^4$ are decreasing functions.

2. The function $f(x) = (\ln x)/x$ is a solution of the differential equation $x^2 y' + xy = 1$.

3. The equation $y' = x + y$ is separable.

4. The equation $y' = 3y - 2x + 6xy - 1$ is separable.

5. The equation $e^x y' = y$ is linear.

6. The equation $y' + xy = e^y$ is linear.

7. If y is the solution of the initial-value problem
$$\frac{dy}{dt} = 2y\left(1 - \frac{y}{5}\right) \qquad y(0) = 1$$
then $\lim_{t \to \infty} y = 5$.

EXERCISES

1. (a) A direction field for the differential equation $y' = y(y-2)(y-4)$ is shown. Sketch the graphs of the solutions that satisfy the given initial conditions.
   (i) $y(0) = -0.3$ (ii) $y(0) = 1$
   (iii) $y(0) = 3$ (iv) $y(0) = 4.3$
   (b) If the initial condition is $y(0) = c$, for what values of c is $\lim_{t \to \infty} y(t)$ finite? What are the equilibrium solutions?

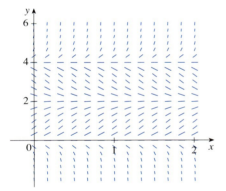

2. (a) Sketch a direction field for the differential equation $y' = x/y$. Then use it to sketch the four solutions that satisfy the initial conditions $y(0) = 1$, $y(0) = -1$, $y(2) = 1$, and $y(-2) = 1$.
   (b) Check your work in part (a) by solving the differential equation explicitly. What type of curve is each solution curve?

3. (a) A direction field for the differential equation $y' = x^2 - y^2$ is shown. Sketch the solution of the initial-value problem
$$y' = x^2 - y^2 \qquad y(0) = 1$$
Use your graph to estimate the value of $y(0.3)$.

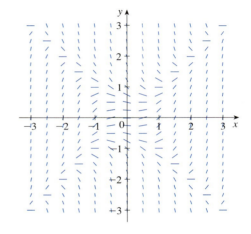

(b) Use Euler's method with step size 0.1 to estimate $y(0.3)$ where $y(x)$ is the solution of the initial-value problem in part (a). Compare with your estimate from part (a).
(c) On what lines are the centers of the horizontal line segments of the direction field in part (a) located? What happens when a solution curve crosses these lines?

4. (a) Use Euler's method with step size 0.2 to estimate $y(0.4)$, where $y(x)$ is the solution of the initial-value problem
$$y' = 2xy^2 \qquad y(0) = 1$$
(b) Repeat part (a) with step size 0.1.
(c) Find the exact solution of the differential equation and compare the value at 0.4 with the approximations in parts (a) and (b).

5–8 Solve the differential equation.

5. $y' = xe^{-\sin x} - y \cos x$

6. $\dfrac{dx}{dt} = 1 - t + x - tx$

7. $(3y^2 + 2y)y' = x \cos x$

8. $x^2 y' - y = 2x^3 e^{-1/x}$

9–11 ▪ Solve the initial-value problem.

9. $xyy' = \ln x$, $y(1) = 2$

10. $1 + x = 2xyy'$, $x > 0$, $y(1) = -2$

11. $y' + y = \sqrt{x}e^{-x}$, $y(0) = 3$

12. Solve the initial-value problem $2yy' = xe^x$, $y(0) = 1$, and graph the solution.

13–14 ▪ Find the orthogonal trajectories of the family of curves.

13. $kx^2 + y^2 = 1$
14. $y = \dfrac{k}{1+x^2}$

15. A bacteria culture starts with 1000 bacteria and the growth rate is proportional to the number of bacteria. After 2 hours the population is 9000.
(a) Find an expression for the number of bacteria after t hours.
(b) Find the number of bacteria after 3 h.
(c) Find the rate of growth after 3 h.
(d) How long does it take for the number of bacteria to double?

16. An isotope of strontium, ^{90}Sr, has a half-life of 25 years.
(a) Find the mass of ^{90}Sr that remains from a sample of 18 mg after t years.
(b) How long would it take for the mass to decay to 2 mg?

17. Let $C(t)$ be the concentration of a drug in the bloodstream. As the body eliminates the drug, $C(t)$ decreases at a rate that is proportional to the amount of the drug that is present at the time. Thus, $C'(t) = -kC(t)$, where k is a positive number called the *elimination constant* of the drug.
(a) If C_0 is the concentration at time $t = 0$, find the concentration at time t.
(b) If the body eliminates half the drug in 30 h, how long does it take to eliminate 90% of the drug?

18. (a) The population of the world was 5.28 billion in 1990 and 6.07 billion in 2000. Find an exponential model for these data and use the model to predict the world population in the year 2020.
(b) According to the model in part (a), when will the world population exceed 10 billion?
(c) Use the data in part (a) to find a logistic model for the population. Assume a carrying capacity of 100 billion. Then use the logistic model to predict the population in 2020. Compare with your prediction from the exponential model.
(d) According to the logistic model, when will the world population exceed 10 billion? Compare with your prediction in part (b).

19. The von Bertalanffy growth model is used to predict the length $L(t)$ of a fish over a period of time. If L_∞ is the largest length for a species, then the hypothesis is that the rate of growth in length is proportional to $L_\infty - L$, the length yet to be achieved.
(a) Formulate and solve a differential equation to find an expression for $L(t)$.
(b) For the North Sea haddock it has been determined that $L_\infty = 53$ cm, $L(0) = 10$ cm, and the constant of proportionality is 0.2. What does the expression for $L(t)$ become with these data?

20. A tank contains 100 L of pure water. Brine that contains 0.1 kg of salt per liter enters the tank at a rate of 10 L/min. The solution is kept thoroughly mixed and drains from the tank at the same rate. How much salt is in the tank after 6 minutes?

21. One model for the spread of an epidemic is that the rate of spread is jointly proportional to the number of infected people and the number of uninfected people. In an isolated town of 5000 inhabitants, 160 people have a disease at the beginning of the week and 1200 have it at the end of the week. How long does it take for 80% of the population to become infected?

22. The Brentano-Stevens Law in psychology models the way that a subject reacts to a stimulus. It states that if R represents the reaction to an amount S of stimulus, then the relative rates of increase are proportional:

$$\frac{1}{R}\frac{dR}{dt} = \frac{k}{S}\frac{dS}{dt}$$

where k is a positive constant. Find R as a function of S.

23. The transport of a substance across a capillary wall in lung physiology has been modeled by the differential equation

$$\frac{dh}{dt} = -\frac{R}{V}\left(\frac{h}{k+h}\right)$$

where h is the hormone concentration in the bloodstream, t is time, R is the maximum transport rate, V is the volume of the capillary, and k is a positive constant that measures the affinity between the hormones and the enzymes that assist the process. Solve this differential equation to find a relationship between h and t.

24. Populations of birds and insects are modeled by the equations

$$\frac{dx}{dt} = 0.4x - 0.002xy$$

$$\frac{dy}{dt} = -0.2y + 0.000008xy$$

(a) Which of the variables, x or y, represents the bird population and which represents the insect population? Explain.

(b) Find the equilibrium solutions and explain their significance.
(c) Find an expression for dy/dx.
(d) The direction field for the differential equation in part (c) is shown. Use it to sketch the phase trajectory corresponding to initial populations of 100 birds and 40,000 insects. Then use the phase trajectory to describe how both populations change.

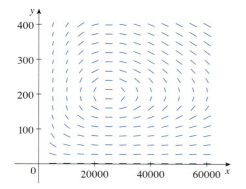

(e) Use part (d) to make rough sketches of the bird and insect populations as functions of time. How are these graphs related to each other?

25. Suppose the model of Exercise 24 is replaced by the equations

$$\frac{dx}{dt} = 0.4x(1 - 0.000005x) - 0.002xy$$

$$\frac{dy}{dt} = -0.2y + 0.000008xy$$

(a) According to these equations, what happens to the insect population in the absence of birds?
(b) Find the equilibrium solutions and explain their significance.
(c) The figure shows the phase trajectory that starts with 100 birds and 40,000 insects. Describe what eventually happens to the bird and insect populations.

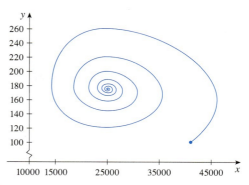

(d) Sketch graphs of the bird and insect populations as functions of time.

26. Barbara weighs 60 kg and is on a diet of 1600 calories per day, of which 850 are used automatically by basal metabolism. She spends about 15 cal/kg/day times her weight doing exercise. If 1 kg of fat contains 10,000 cal and we assume that the storage of calories in the form of fat is 100% efficient, formulate a differential equation and solve it to find her weight as a function of time. Does her weight ultimately approach an equilibrium weight?

27. When a flexible cable of uniform density is suspended between two fixed points and hangs of its own weight, the shape $y = f(x)$ of the cable must satisfy a differential equation of the form

$$\frac{d^2y}{dx^2} = k\sqrt{1 + \left(\frac{dy}{dx}\right)^2}$$

where k is a positive constant. Consider the cable shown in the figure.
(a) Let $z = dy/dx$ in the differential equation. Solve the resulting first-order differential equation (in z), and then integrate to find y.
(b) Determine the length of the cable.

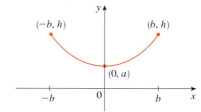

PROBLEMS PLUS

1. Find all functions f such that f' is continuous and

$$[f(x)]^2 = 100 + \int_0^x \{[f(t)]^2 + [f'(t)]^2\}\, dt \qquad \text{for all real } x$$

2. A student forgot the Product Rule for differentiation and made the mistake of thinking that $(fg)' = f'g'$. However, he was lucky and got the correct answer. The function f that he used was $f(x) = e^{x^2}$ and the domain of his problem was the interval $(\tfrac{1}{2}, \infty)$. What was the function g?

3. Let f be a function with the property that $f(0) = 1$, $f'(0) = 1$, and $f(a + b) = f(a)f(b)$ for all real numbers a and b. Show that $f'(x) = f(x)$ for all x and deduce that $f(x) = e^x$.

4. Find all functions f that satisfy the equation

$$\left(\int f(x)\, dx \right)\left(\int \frac{1}{f(x)}\, dx \right) = -1$$

5. A peach pie is taken out of the oven at 5:00 P.M. At that time it is piping hot: 100°C. At 5:10 P.M. its temperature is 80°C; at 5:20 P.M. it is 65°C. What is the temperature of the room?

6. Snow began to fall during the morning of February 2 and continued steadily into the afternoon. At noon a snowplow began removing snow from a road at a constant rate. The plow traveled 6 km from noon to 1 P.M. but only 3 km from 1 P.M. to 2 P.M. When did the snow begin to fall? [*Hints:* To get started, let t be the time measured in hours after noon; let $x(t)$ be the distance traveled by the plow at time t; then the speed of the plow is dx/dt. Let b be the number of hours before noon that it began to snow. Find an expression for the height of the snow at time t. Then use the given information that the rate of removal R (in m³/h) is constant.]

7. A dog sees a rabbit running in a straight line across an open field and gives chase. In a rectangular coordinate system (as shown in the figure), assume:
   (i) The rabbit is at the origin and the dog is at the point $(L, 0)$ at the instant the dog first sees the rabbit.
   (ii) The rabbit runs up the y-axis and the dog always runs straight for the rabbit.
   (iii) The dog runs at the same speed as the rabbit.
   (a) Show that the dog's path is the graph of the function $y = f(x)$, where y satisfies the differential equation

$$x\frac{d^2y}{dx^2} = \sqrt{1 + \left(\frac{dy}{dx}\right)^2}$$

   (b) Determine the solution of the equation in part (a) that satisfies the initial conditions $y = y' = 0$ when $x = L$. [*Hint:* Let $z = dy/dx$ in the differential equation and solve the resulting first-order equation to find z; then integrate z to find y.]
   (c) Does the dog ever catch the rabbit?

8. (a) Suppose that the dog in Problem 7 runs twice as fast as the rabbit. Find a differential equation for the path of the dog. Then solve it to find the point where the dog catches the rabbit.
   (b) Suppose the dog runs half as fast as the rabbit. How close does the dog get to the rabbit? What are their positions when they are closest?

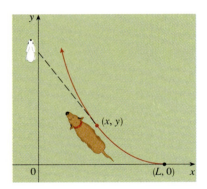

FIGURE FOR PROBLEM 7

9. A planning engineer for a new alum plant must present some estimates to his company regarding the capacity of a silo designed to contain bauxite ore until it is processed into alum. The ore resembles pink talcum powder and is poured from a conveyor at the top of the silo. The silo is a cylinder 100 ft high with a radius of 200 ft. The conveyor carries $60,000\pi$ ft³/h and the ore maintains a conical shape whose radius is 1.5 times its height.
   (a) If, at a certain time t, the pile is 60 ft high, how long will it take for the pile to reach the top of the silo?
   (b) Management wants to know how much room will be left in the floor area of the silo when the pile is 60 ft high. How fast is the floor area of the pile growing at that height?
   (c) Suppose a loader starts removing the ore at the rate of $20,000\pi$ ft³/h when the height of the pile reaches 90 ft. Suppose, also, that the pile continues to maintain its shape. How long will it take for the pile to reach the top of the silo under these conditions?

10. Find the curve that passes through the point (3, 2) and has the property that if the tangent line is drawn at any point P on the curve, then the part of the tangent line that lies in the first quadrant is bisected at P.

11. Recall that the normal line to a curve at a point P on the curve is the line that passes through P and is perpendicular to the tangent line at P. Find the curve that passes through the point (3, 2) and has the property that if the normal line is drawn at any point on the curve, then the y-intercept of the normal line is always 6.

12. Find all curves with the property that if the normal line is drawn at any point P on the curve, then the part of the normal line between P and the x-axis is bisected by the y-axis.

Chapter 11

Parametric curves are used to represent letters and other symbols on laser printers. See the Laboratory Project on page 705.

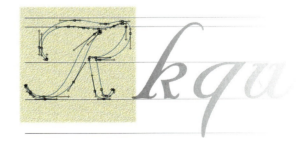

Parametric Equations and Polar Coordinates

So far we have described plane curves by giving y as a function of x $[y = f(x)]$ or x as a function of y $[x = g(y)]$ or by giving a relation between x and y that defines y implicitly as a function of x $[f(x, y) = 0]$. In this chapter we discuss two new methods for describing curves.

Some curves, such as the cycloid, are best handled when both x and y are given in terms of a third variable t called a parameter $[x = f(t), y = g(t)]$. Other curves, such as the cardioid, have their most convenient description when we use a new coordinate system, called the polar coordinate system.

11.1 Curves Defined by Parametric Equations

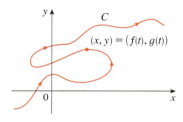

FIGURE 1

Imagine that a particle moves along the curve C shown in Figure 1. It is impossible to describe C by an equation of the form $y = f(x)$ because C fails the Vertical Line Test. But the x- and y-coordinates of the particle are functions of time and so we can write $x = f(t)$ and $y = g(t)$. Such a pair of equations is often a convenient way of describing a curve and gives rise to the following definition.

Suppose that x and y are both given as functions of a third variable t (called a **parameter**) by the equations

$$x = f(t) \qquad y = g(t)$$

(called **parametric equations**). Each value of t determines a point (x, y), which we can plot in a coordinate plane. As t varies, the point $(x, y) = (f(t), g(t))$ varies and traces out a curve C, which we call a **parametric curve**. The parameter t does not necessarily represent time and, in fact, we could use a letter other than t for the parameter. But in many applications of parametric curves, t does denote time and therefore we can interpret $(x, y) = (f(t), g(t))$ as the position of a particle at time t.

EXAMPLE 1 Sketch and identify the curve defined by the parametric equations

$$x = t^2 - 2t \qquad y = t + 1$$

SOLUTION Each value of t gives a point on the curve, as shown in the table. For instance, if $t = 0$, then $x = 0$, $y = 1$ and so the corresponding point is $(0, 1)$. In Figure 2 we plot the points (x, y) determined by several values of the parameter and we join them to produce a curve.

t	x	y
-2	8	-1
-1	3	0
0	0	1
1	-1	2
2	0	3
3	3	4
4	8	5

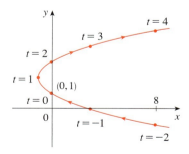

FIGURE 2

A particle whose position is given by the parametric equations moves along the curve in the direction of the arrows as t increases. Notice that the consecutive points marked on the curve appear at equal time intervals but not at equal distances. That is because the particle slows down and then speeds up as t increases.

It appears from Figure 2 that the curve traced out by the particle may be a parabola. This can be confirmed by eliminating the parameter t as follows. We obtain $t = y - 1$ from the second equation and substitute into the first equation. This gives

$$x = t^2 - 2t = (y-1)^2 - 2(y-1) = y^2 - 4y + 3$$

and so the curve represented by the given parametric equations is the parabola $x = y^2 - 4y + 3$.

No restriction was placed on the parameter t in Example 1, so we assumed that t could be any real number. But sometimes we restrict t to lie in a finite interval. For instance, the parametric curve

$$x = t^2 - 2t \qquad y = t + 1 \qquad 0 \leq t \leq 4$$

shown in Figure 3 is the part of the parabola in Example 1 that starts at the point $(0, 1)$ and ends at the point $(8, 5)$. The arrowhead indicates the direction in which the curve is traced as t increases from 0 to 4.

In general, the curve with parametric equations

$$x = f(t) \qquad y = g(t) \qquad a \leq t \leq b$$

has **initial point** $(f(a), g(a))$ and **terminal point** $(f(b), g(b))$.

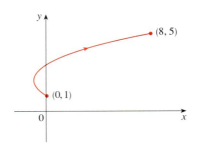

FIGURE 3

EXAMPLE 2 What curve is represented by the parametric equations $x = \cos t$, $y = \sin t$, $0 \leq t \leq 2\pi$?

SOLUTION If we plot points, it appears that the curve is a circle. We can confirm this impression by eliminating t. Observe that

$$x^2 + y^2 = \cos^2 t + \sin^2 t = 1$$

Thus, the point (x, y) moves on the unit circle $x^2 + y^2 = 1$. Notice that in this example the parameter t can be interpreted as the angle (in radians) shown in Figure 4. As t increases from 0 to 2π, the point $(x, y) = (\cos t, \sin t)$ moves once around the circle in the counterclockwise direction starting from the point $(1, 0)$.

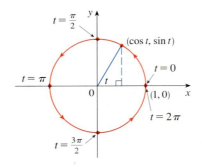

FIGURE 4

EXAMPLE 3 What curve is represented by the parametric equations $x = \sin 2t$, $y = \cos 2t$, $0 \leq t \leq 2\pi$?

SOLUTION Again we have

$$x^2 + y^2 = \sin^2 2t + \cos^2 2t = 1$$

so the parametric equations again represent the unit circle $x^2 + y^2 = 1$. But as t increases from 0 to 2π, the point $(x, y) = (\sin 2t, \cos 2t)$ starts at $(0, 1)$ and moves *twice* around the circle in the clockwise direction as indicated in Figure 5.

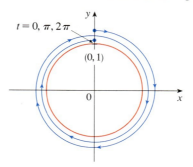

FIGURE 5

Examples 2 and 3 show that different sets of parametric equations can represent the same curve. Thus, we distinguish between a *curve*, which is a set of points, and a *parametric curve*, in which the points are traced in a particular way.

EXAMPLE 4 Sketch the curve with parametric equations $x = \sin t$, $y = \sin^2 t$.

SOLUTION Observe that $y = (\sin t)^2 = x^2$ and so the point (x, y) moves on the parabola $y = x^2$. But note also that, since $-1 \leq \sin t \leq 1$, we have $-1 \leq x \leq 1$, so the parametric equations represent only the part of the parabola for which $-1 \leq x \leq 1$. Since $\sin t$ is periodic, the point $(x, y) = (\sin t, \sin^2 t)$ moves back and forth infinitely often along the parabola from $(-1, 1)$ to $(1, 1)$. (See Figure 6.)

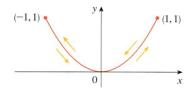

FIGURE 6

Module 11.1A gives an animation of the relationship between motion along a parametric curve $x = f(t)$, $y = g(t)$ and motion along the graphs of f and g as functions of t. Clicking on TRIG gives you the family of parametric curves

$$x = a \cos bt \quad y = c \sin dt$$

If you choose $a = b = c = d = 1$ and click START, you will see how the graphs of $x = \cos t$ and $y = \sin t$ relate to the circle in Example 2. If you choose $a = b = c = 1, d = 2$, you will see graphs as in Figure 7. By clicking on PAUSE and then repeatedly on STEP, you can see from the color coding how motion along the graphs of $x = \cos t$ and $y = \sin 2t$ corresponds to motion along the parametric curve, which is called a **Lissajous figure**.

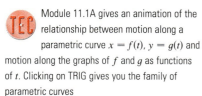

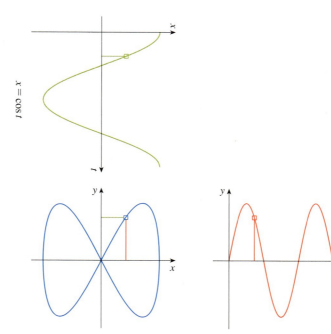

FIGURE 7 $x = \cos t$ $y = \sin 2t$ $y = \sin 2t$

Graphing Devices

Most graphing calculators and computer graphing programs can be used to graph curves defined by parametric equations. In fact, it's instructive to watch a parametric curve being drawn by a graphing calculator because the points are plotted in order as the corresponding parameter values increase.

EXAMPLE 5 Use a graphing device to graph the curve $x = y^4 - 3y^2$.

SOLUTION If we let the parameter be $t = y$, then we have the equations

$$x = t^4 - 3t^2 \qquad y = t$$

Using these parametric equations to graph the curve, we obtain Figure 8. It would be possible to solve the given equation ($x = y^4 - 3y^2$) for y as four functions of x and graph them individually, but the parametric equations provide a much easier method.

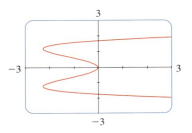

FIGURE 8

In general, if we need to graph an equation of the form $x = g(y)$, we can use the parametric equations

$$x = g(t) \qquad y = t$$

Notice also that curves with equations $y = f(x)$ (the ones we are most familiar with—graphs of functions) can also be regarded as curves with parametric equations

$$x = t \qquad y = f(t)$$

Graphing devices are particularly useful when sketching complicated curves. For instance, the curves shown in Figures 9 and 10 would be virtually impossible to produce by hand.

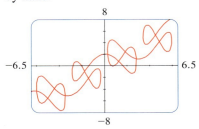

FIGURE 9
$x = t + 2 \sin 2t, \ y = t + 2 \cos 5t,$
$-2\pi \le t \le 2\pi$

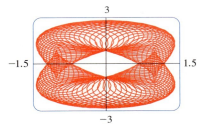

FIGURE 10
$x = \cos t - \cos 80t \sin t,$
$y = 2 \sin t - \sin 80t, \ 0 \le t \le 2\pi$

One of the most important uses of parametric curves is in computer-aided design (CAD). In the Laboratory Project after Section 10.2 we will investigate special parametric curves, called **Bézier curves**, that are used extensively in manufacturing, especially in the automotive industry. These curves are also employed in specifying the shapes of letters and other symbols in laser printers.

The Cycloid

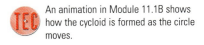

An animation in Module 11.1B shows how the cycloid is formed as the circle moves.

EXAMPLE 6 The curve traced out by a point P on the circumference of a circle as the circle rolls along a straight line is called a **cycloid** (see Figure 11). If the circle has radius r and rolls along the x-axis and if one position of P is the origin, find parametric equations for the cycloid.

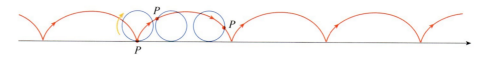

FIGURE 11

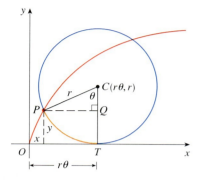

FIGURE 12

SOLUTION   We choose as parameter the angle of rotation θ of the circle ($\theta = 0$ when P is at the origin). Suppose the circle has rotated through θ radians. Because the circle has been in contact with the line, we see from Figure 12 that the distance it has rolled from the origin is

$$|OT| = \text{arc } PT = r\theta$$

Therefore, the center of the circle is $C(r\theta, r)$. Let the coordinates of P be (x, y). Then from Figure 12 we see that

$$x = |OT| - |PQ| = r\theta - r \sin \theta = r(\theta - \sin \theta)$$

$$y = |TC| - |QC| = r - r \cos \theta = r(1 - \cos \theta)$$

Therefore, parametric equations of the cycloid are

$$\boxed{1} \qquad x = r(\theta - \sin \theta) \qquad y = r(1 - \cos \theta) \qquad \theta \in \mathbb{R}$$

One arch of the cycloid comes from one rotation of the circle and so is described by $0 \leq \theta \leq 2\pi$. Although Equations 1 were derived from Figure 12, which illustrates the case where $0 < \theta < \pi/2$, it can be seen that these equations are still valid for other values of θ (see Exercise 37).

Although it is possible to eliminate the parameter θ from Equations 1, the resulting Cartesian equation in x and y is very complicated and not as convenient to work with as the parametric equations.

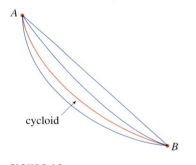

FIGURE 13

FIGURE 14

One of the first people to study the cycloid was Galileo, who proposed that bridges be built in the shape of cycloids and who tried to find the area under one arch of a cycloid. Later this curve arose in connection with the *brachistochrone problem:* Find the curve along which a particle will slide in the shortest time (under the influence of gravity) from a point A to a lower point B not directly beneath A. The Swiss mathematician John Bernoulli, who posed this problem in 1696, showed that among all possible curves that join A to B, as in Figure 13, the particle will take the least time sliding from A to B if the curve is part of an inverted arch of a cycloid.

The Dutch physicist Huygens had already shown that the cycloid is also the solution to the *tautochrone problem;* that is, no matter where a particle P is placed on an inverted cycloid, it takes the same time to slide to the bottom (see Figure 14). Huygens proposed that pendulum clocks (which he invented) swing in cycloidal arcs because then the pendulum takes the same time to make a complete oscillation whether it swings through a wide or a small arc.

Families of Parametric Curves

EXAMPLE 7   Investigate the family of curves with parametric equations

$$x = a + \cos t \qquad y = a \tan t + \sin t$$

What do these curves have in common? How does the shape change as a increases?

SOLUTION We use a graphing device to produce the graphs for the cases $a = -2, -1, -0.5, -0.2, 0, 0.5, 1,$ and 2 shown in Figure 15. Notice that all of these curves (except the case $a = 0$) have two branches, and both branches approach the vertical asymptote $x = a$ as x approaches a from the left or right.

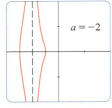
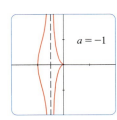
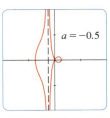

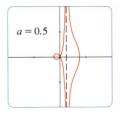
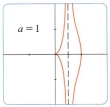
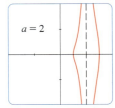

FIGURE 15 Members of the family $x = a + \cos t$, $y = a \tan t + \sin t$, all graphed in the viewing rectangle $[-4, 4]$ by $[-4, 4]$

When $a < -1$, both branches are smooth; but when a reaches -1, the right branch acquires a sharp point, called a *cusp*. For a between -1 and 0 the cusp turns into a loop, which becomes larger as a approaches 0. When $a = 0$, both branches come together and form a circle (see Example 2). For a between 0 and 1, the left branch has a loop, which shrinks to become a cusp when $a = 1$. For $a > 1$, the branches become smooth again, and as a increases further, they become less curved. Notice that the curves with a positive are reflections about the y-axis of the corresponding curves with a negative.

These curves are called **conchoids of Nicomedes** after the ancient Greek scholar Nicomedes. He called them conchoids because the shape of their outer branches resembles that of a conch shell or mussel shell.

11.1 Exercises

1–4 Sketch the curve by using the parametric equations to plot points. Indicate with an arrow the direction in which the curve is traced as t increases.

1. $x = 1 + \sqrt{t}$, $y = t^2 - 4t$, $0 \leq t \leq 5$
2. $x = 2 \cos t$, $y = t - \cos t$, $0 \leq t \leq 2\pi$
3. $x = 5 \sin t$, $y = t^2$, $-\pi \leq t \leq \pi$
4. $x = e^{-t} + t$, $y = e^t - t$, $-2 \leq t \leq 2$

5–10
(a) Sketch the curve by using the parametric equations to plot points. Indicate with an arrow the direction in which the curve is traced as t increases.
(b) Eliminate the parameter to find a Cartesian equation of the curve.

5. $x = 3t - 5$, $y = 2t + 1$
6. $x = 1 + t$, $y = 5 - 2t$, $-2 \leq t \leq 3$
7. $x = t^2 - 2$, $y = 5 - 2t$, $-3 \leq t \leq 4$
8. $x = 1 + 3t$, $y = 2 - t^2$
9. $x = \sqrt{t}$, $y = 1 - t$
10. $x = t^2$, $y = t^3$

11–18
(a) Eliminate the parameter to find a Cartesian equation of the curve.
(b) Sketch the curve and indicate with an arrow the direction in which the curve is traced as the parameter increases.

11. $x = \sin \theta$, $y = \cos \theta$, $0 \leq \theta \leq \pi$
12. $x = 4 \cos \theta$, $y = 5 \sin \theta$, $-\pi/2 \leq \theta \leq \pi/2$
13. $x = \sin^2 \theta$, $y = \cos^2 \theta$
14. $x = \sec \theta$, $y = \tan \theta$, $-\pi/2 < \theta < \pi/2$
15. $x = e^t$, $y = e^{-t}$
16. $x = \ln t$, $y = \sqrt{t}$, $t \geq 1$

17. $x = \cosh t$, $y = \sinh t$

18. $x = 1 + \cos\theta$, $y = 2\cos\theta - 1$

19–22 Describe the motion of a particle with position (x, y) as t varies in the given interval.

19. $x = \cos\pi t$, $y = \sin\pi t$, $1 \le t \le 2$

20. $x = 2 + \cos t$, $y = 3 + \sin t$, $0 \le t \le 2\pi$

21. $x = 2\sin t$, $y = 3\cos t$, $0 \le t \le 2\pi$

22. $x = \cos^2 t$, $y = \cos t$, $0 \le t \le 4\pi$

23. Suppose a curve is given by the parametric equations $x = f(t)$, $y = g(t)$, where the range of f is $[1, 4]$ and the range of g is $[2, 3]$. What can you say about the curve?

24. Match the graphs of the parametric equations $x = f(t)$ and $y = g(t)$ in (a)–(d) with the parametric curves labeled I–IV. Give reasons for your choices.

(a)
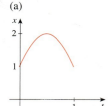

I
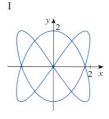

(b)
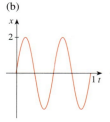

II
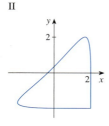

(c)
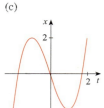

III
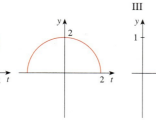

(d)
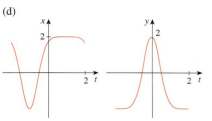

IV
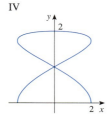

25–27 Use the graphs of $x = f(t)$ and $y = g(t)$ to sketch the parametric curve $x = f(t)$, $y = g(t)$. Indicate with arrows the direction in which the curve is traced as t increases.

25.
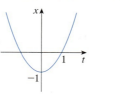

26.

27.
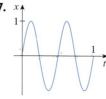
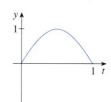

28. Match the parametric equations with the graphs labeled I–VI. Give reasons for your choices. (Do not use a graphing device.)

(a) $x = t^3 - 2t$, $y = t^2 - t$
(b) $x = t^3 - 1$, $y = 2 - t^2$
(c) $x = \sin 3t$, $y = \sin 4t$
(d) $x = t + \sin 2t$, $y = t + \sin 3t$
(e) $x = \sin(t + \sin t)$, $y = \cos(t + \cos t)$
(f) $x = \cos t$, $y = \sin(t + \sin 5t)$

I
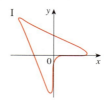

II
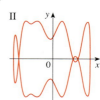

III

IV
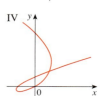

V

VI
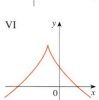

29. Graph the curve $x = y - 3y^3 + y^5$.

30. Graph the curves $y = x^5$ and $x = y(y-1)^2$ and find their points of intersection correct to one decimal place.

31. (a) Show that the parametric equations
$$x = x_1 + (x_2 - x_1)t \qquad y = y_1 + (y_2 - y_1)t$$
where $0 \le t \le 1$, describe the line segment that joins the points $P_1(x_1, y_1)$ and $P_2(x_2, y_2)$.
(b) Find parametric equations to represent the line segment from $(-2, 7)$ to $(3, -1)$.

32. Use a graphing device and the result of Exercise 31(a) to draw the triangle with vertices $A(1, 1)$, $B(4, 2)$, and $C(1, 5)$.

33. Find parametric equations for the path of a particle that moves along the circle $x^2 + (y - 1)^2 = 4$ in the manner described.
(a) Once around clockwise, starting at $(2, 1)$
(b) Three times around counterclockwise, starting at $(2, 1)$
(c) Halfway around counterclockwise, starting at $(0, 3)$

34. Graph the semicircle traced by the particle in Exercise 33(c).

35. (a) Find parametric equations for the ellipse $x^2/a^2 + y^2/b^2 = 1$. [Hint: Modify the equations of a circle in Example 2.]
(b) Use these parametric equations to graph the ellipse when $a = 3$ and $b = 1, 2, 4$, and 8.
(c) How does the shape of the ellipse change as b varies?

36. Find three different sets of parametric equations to represent the curve $y = x^3$, $x \in \mathbb{R}$.

37. Derive Equations 1 for the case $\pi/2 < \theta < \pi$.

38. Let P be a point at a distance d from the center of a circle of radius r. The curve traced out by P as the circle rolls along a straight line is called a **trochoid.** (Think of the motion of a point on a spoke of a bicycle wheel.) The cycloid is the special case of a trochoid with $d = r$. Using the same parameter θ as for the cycloid and assuming the line is the x-axis and $\theta = 0$ when P is at one of its lowest points, show that parametric equations of the trochoid are
$$x = r\theta - d \sin \theta \qquad y = r - d \cos \theta$$
Sketch the trochoid for the cases $d < r$ and $d > r$.

39. If a and b are fixed numbers, find parametric equations for the curve that consists of all possible positions of the point P in the figure, using the angle θ as the parameter. Then eliminate the parameter and identify the curve.

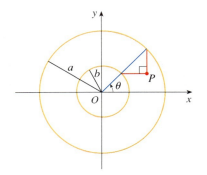

40. If a and b are fixed numbers, find parametric equations for the curve that consists of all possible positions of the point P in the figure, using the angle θ as the parameter. The line segment AB is tangent to the larger circle.

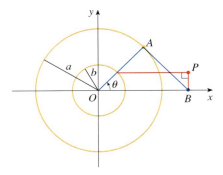

41. A curve, called a **witch of Maria Agnesi**, consists of all possible positions of the point P in the figure. Show that parametric equations for this curve can be written as
$$x = 2a \cot \theta \qquad y = 2a \sin^2 \theta$$
Sketch the curve.

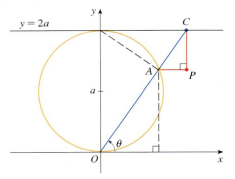

42. Find parametric equations for the curve that consists of all possible positions of the point P in the figure, where $|OP| = |AB|$. Sketch the curve. (This curve is called the **cissoid of Diocles** after the Greek scholar Diocles, who introduced the cissoid as a graphical method for constructing the edge of a cube whose volume is twice that of a given cube.)

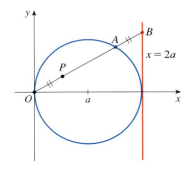

43. Suppose that the position of one particle at time t is given by

$$x_1 = 3 \sin t \qquad y_1 = 2 \cos t \qquad 0 \leq t \leq 2\pi$$

and the position of a second particle is given by

$$x_2 = -3 + \cos t \qquad y_2 = 1 + \sin t \qquad 0 \leq t \leq 2\pi$$

(a) Graph the paths of both particles. How many points of intersection are there?
(b) Are any of these points of intersection *collision points*? In other words, are the particles ever at the same place at the same time? If so, find the collision points.
(c) Describe what happens if the path of the second particle is given by

$$x_2 = 3 + \cos t \qquad y_2 = 1 + \sin t \qquad 0 \leq t \leq 2\pi$$

44. If a projectile is fired with an initial velocity of v_0 meters per second at an angle α above the horizontal and air resistance is assumed to be negligible, then its position after t seconds is given by the parametric equations

$$x = (v_0 \cos \alpha)t \qquad y = (v_0 \sin \alpha)t - \tfrac{1}{2}gt^2$$

where g is the acceleration due to gravity (9.8 m/s^2).
(a) If a gun is fired with $\alpha = 30°$ and $v_0 = 500$ m/s, when will the bullet hit the ground? How far from the gun will it hit the ground? What is the maximum height reached by the bullet?

(b) Use a graphing device to check your answers to part (a). Then graph the path of the projectile for several other values of the angle α to see where it hits the ground. Summarize your findings.
(c) Show that the path is parabolic by eliminating the parameter.

45. Investigate the family of curves defined by the parametric equations $x = t^2$, $y = t^3 - ct$. How does the shape change as c increases? Illustrate by graphing several members of the family.

46. The **swallowtail catastrophe curves** are defined by the parametric equations $x = 2ct - 4t^3$, $y = -ct^2 + 3t^4$. Graph several of these curves. What features do the curves have in common? How do they change when c increases?

47. The curves with equations $x = a \sin nt$, $y = b \cos t$ are called **Lissajous figures**. Investigate how these curves vary when a, b, and n vary. (Take n to be a positive integer.)

48. Investigate the family of curves defined by the parametric equations

$$x = \sin t\,(c - \sin t) \qquad y = \cos t\,(c - \sin t)$$

How does the shape change as c changes? In particular, you should identify the transitional values of c for which the basic shape of the curve changes.

LABORATORY PROJECT

Running Circles around Circles

In this project we investigate families of curves, called *hypocycloids* and *epicycloids*, that are generated by the motion of a point on a circle that rolls inside or outside another circle.

1. A **hypocycloid** is a curve traced out by a fixed point P on a circle C of radius b as C rolls on the inside of a circle with center O and radius a. Show that if the initial position of P is $(a, 0)$ and the parameter θ is chosen as in the figure, then parametric equations of the hypocycloid are

$$x = (a - b) \cos \theta + b \cos\left(\frac{a - b}{b}\theta\right) \qquad y = (a - b) \sin \theta - b \sin\left(\frac{a - b}{b}\theta\right)$$

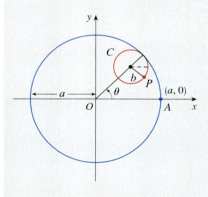

2. Use a graphing device (or the interactive graphic in TEC Module 11.1B) to draw the graphs of hypocycloids with a a positive integer and $b = 1$. How does the value of a affect the graph? Show that if we take $a = 4$, then the parametric equations of the hypocycloid reduce to

$$x = 4 \cos^3\theta \qquad y = 4 \sin^3\theta$$

This curve is called a **hypocycloid of four cusps**, or an **astroid**.

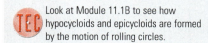

Look at Module 11.1B to see how hypocycloids and epicycloids are formed by the motion of rolling circles.

3. Now try $b = 1$ and $a = n/d$, a fraction where n and d have no common factor. First let $n = 1$ and try to determine graphically the effect of the denominator d on the shape of the graph. Then let n vary while keeping d constant. What happens when $n = d + 1$?

4. What happens if $b = 1$ and a is irrational? Experiment with an irrational number like $\sqrt{2}$ or $e - 2$. Take larger and larger values for θ and speculate on what would happen if we were to graph the hypocycloid for all real values of θ.

5. If the circle C rolls on the *outside* of the fixed circle, the curve traced out by P is called an **epicycloid**. Find parametric equations for the epicycloid.

6. Investigate the possible shapes for epicycloids. Use methods similar to Problems 2–4.

11.2 Calculus with Parametric Curves

Having seen how to represent curves by parametric equations, we now apply the methods of calculus to these parametric curves. In particular, we solve problems involving tangents, area, arc length, and surface area.

Tangents

In the preceding section we saw that some curves defined by parametric equations $x = f(t)$ and $y = g(t)$ can also be expressed, by eliminating the parameter, in the form $y = F(x)$. (See Exercise 67 for general conditions under which this is possible.) If we substitute $x = f(t)$ and $y = g(t)$ in the equation $y = F(x)$, we get

$$g(t) = F(f(t))$$

and so, if g, F, and f are differentiable, the Chain Rule gives

$$g'(t) = F'(f(t))f'(t) = F'(x)f'(t)$$

If $f'(t) \neq 0$, we can solve for $F'(x)$:

$$\boxed{1} \qquad F'(x) = \frac{g'(t)}{f'(t)}$$

Since the slope of the tangent to the curve $y = F(x)$ at $(x, F(x))$ is $F'(x)$, Equation 1 enables us to find tangents to parametric curves without having to eliminate the parameter. Using Leibniz notation, we can rewrite Equation 1 in an easily remembered form:

|||| If we think of a parametric curve as being traced out by a moving particle, then dy/dt and dx/dt are the vertical and horizontal velocities of the particle and Formula 2 says that the slope of the tangent is the ratio of these velocities.

$$\boxed{2} \qquad \frac{dy}{dx} = \frac{\dfrac{dy}{dt}}{\dfrac{dx}{dt}} \quad \text{if} \quad \frac{dx}{dt} \neq 0$$

It can be seen from Equation 2 that the curve has a horizontal tangent when $dy/dt = 0$ (provided that $dx/dt \neq 0$) and it has a vertical tangent when $dx/dt = 0$ (provided that $dy/dt \neq 0$). This information is useful for sketching parametric curves.

As we know from Chapter 4, it is also useful to consider d^2y/dx^2. This can be found by replacing y by dy/dx in Equation 2:

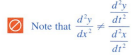
Note that $\dfrac{d^2y}{dx^2} \neq \dfrac{\frac{d^2y}{dt^2}}{\frac{d^2x}{dt^2}}$

$$\frac{d^2y}{dx^2} = \frac{d}{dx}\left(\frac{dy}{dx}\right) = \frac{\frac{d}{dt}\left(\frac{dy}{dx}\right)}{\frac{dx}{dt}}$$

EXAMPLE 1
A curve C is defined by the parametric equations $x = t^2$, $y = t^3 - 3t$.
(a) Show that C has two tangents at the point $(3, 0)$ and find their equations.
(b) Find the points on C where the tangent is horizontal or vertical.
(c) Determine where the curve is concave upward or downward.
(d) Sketch the curve.

SOLUTION
(a) Notice that $y = t^3 - 3t = t(t^2 - 3) = 0$ when $t = 0$ or $t = \pm\sqrt{3}$. Therefore, the point $(3, 0)$ on C arises from two values of the parameter, $t = \sqrt{3}$ and $t = -\sqrt{3}$. This indicates that C crosses itself at $(3, 0)$. Since

$$\frac{dy}{dx} = \frac{dy/dt}{dx/dt} = \frac{3t^2 - 3}{2t} = \frac{3}{2}\left(t - \frac{1}{t}\right)$$

the slope of the tangent when $t = \pm\sqrt{3}$ is $dy/dx = \pm 6/(2\sqrt{3}) = \pm\sqrt{3}$, so the equations of the tangents at $(3, 0)$ are

$$y = \sqrt{3}\,(x - 3) \quad \text{and} \quad y = -\sqrt{3}\,(x - 3)$$

(b) C has a horizontal tangent when $dy/dx = 0$, that is, when $dy/dt = 0$ and $dx/dt \neq 0$. Since $dy/dt = 3t^2 - 3$, this happens when $t^2 = 1$, that is, $t = \pm 1$. The corresponding points on C are $(1, -2)$ and $(1, 2)$. C has a vertical tangent when $dx/dt = 2t = 0$, that is, $t = 0$. (Note that $dy/dt \neq 0$ there.) The corresponding point on C is $(0, 0)$.

(c) To determine concavity we calculate the second derivative:

$$\frac{d^2y}{dx^2} = \frac{\frac{d}{dt}\left(\frac{dy}{dx}\right)}{\frac{dx}{dt}} = \frac{\frac{3}{2}\left(1 + \frac{1}{t^2}\right)}{2t} = \frac{3(t^2 + 1)}{4t^3}$$

Thus, the curve is concave upward when $t > 0$ and concave downward when $t < 0$.

(d) Using the information from parts (b) and (c), we sketch C in Figure 1.

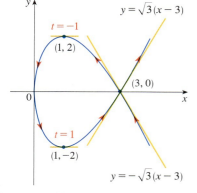

FIGURE 1

EXAMPLE 2
(a) Find the tangent to the cycloid $x = r(\theta - \sin\theta)$, $y = r(1 - \cos\theta)$ at the point where $\theta = \pi/3$. (See Example 6 in Section 11.1.)
(b) At what points is the tangent horizontal? When is it vertical?

SOLUTION
(a) The slope of the tangent line is

$$\frac{dy}{dx} = \frac{dy/d\theta}{dx/d\theta} = \frac{r\sin\theta}{r(1 - \cos\theta)} = \frac{\sin\theta}{1 - \cos\theta}$$

When $\theta = \pi/3$, we have

$$x = r\left(\frac{\pi}{3} - \sin\frac{\pi}{3}\right) = r\left(\frac{\pi}{3} - \frac{\sqrt{3}}{2}\right) \qquad y = r\left(1 - \cos\frac{\pi}{3}\right) = \frac{r}{2}$$

and

$$\frac{dy}{dx} = \frac{\sin(\pi/3)}{1 - \cos(\pi/3)} = \frac{\sqrt{3}/2}{1 - \frac{1}{2}} = \sqrt{3}$$

Therefore, the slope of the tangent is $\sqrt{3}$ and its equation is

$$y - \frac{r}{2} = \sqrt{3}\left(x - \frac{r\pi}{3} + \frac{r\sqrt{3}}{2}\right) \qquad \text{or} \qquad \sqrt{3}\,x - y = r\left(\frac{\pi}{\sqrt{3}} - 2\right)$$

The tangent is sketched in Figure 2.

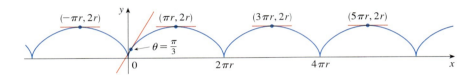

FIGURE 2

(b) The tangent is horizontal when $dy/dx = 0$, which occurs when $\sin\theta = 0$ and $1 - \cos\theta \neq 0$, that is, $\theta = (2n - 1)\pi$, n an integer. The corresponding point on the cycloid is $((2n - 1)\pi r, 2r)$.

When $\theta = 2n\pi$, both $dx/d\theta$ and $dy/d\theta$ are 0. It appears from the graph that there are vertical tangents at those points. We can verify this by using l'Hospital's Rule as follows:

$$\lim_{\theta \to 2n\pi^+} \frac{dy}{dx} = \lim_{\theta \to 2n\pi^+} \frac{\sin\theta}{1 - \cos\theta} = \lim_{\theta \to 2n\pi^+} \frac{\cos\theta}{\sin\theta} = \infty$$

A similar computation shows that $dy/dx \to -\infty$ as $\theta \to 2n\pi^-$, so indeed there are vertical tangents when $\theta = 2n\pi$, that is, when $x = 2n\pi r$.

Areas

We know that the area under a curve $y = F(x)$ from a to b is $A = \int_a^b F(x)\,dx$, where $F(x) \geq 0$. If the curve is given by parametric equations $x = f(t)$, $y = g(t)$ and is traversed once as t increases from α to β, then we can adapt the earlier formula by using the Substitution Rule for Definite Integrals as follows:

$$A = \int_a^b y\,dx = \int_\alpha^\beta g(t)f'(t)\,dt$$

$$\left[\text{or } \int_\beta^\alpha g(t)f'(t)\,dt \quad \text{if } (f(\beta), g(\beta)) \text{ is the leftmost endpoint}\right]$$

EXAMPLE 3 Find the area under one arch of the cycloid $x = r(\theta - \sin\theta)$, $y = r(1 - \cos\theta)$. (See Figure 3.)

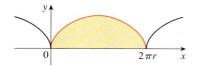

FIGURE 3

|||| The result of Example 3 says that the area under one arch of the cycloid is three times the area of the rolling circle that generates the cycloid (see Example 6 in Section 11.1). Galileo guessed this result but it was first proved by the French mathematician Roberval and the Italian mathematician Torricelli.

SOLUTION One arch of the cycloid is given by $0 \leq \theta \leq 2\pi$. Using the Substitution Rule with $y = r(1 - \cos\theta)$ and $dx = r(1 - \cos\theta)\, d\theta$, we have

$$A = \int_0^{2\pi r} y\, dx = \int_0^{2\pi} r(1 - \cos\theta)\, r(1 - \cos\theta)\, d\theta$$

$$= r^2 \int_0^{2\pi} (1 - \cos\theta)^2\, d\theta = r^2 \int_0^{2\pi} (1 - 2\cos\theta + \cos^2\theta)\, d\theta$$

$$= r^2 \int_0^{2\pi} \left[1 - 2\cos\theta + \tfrac{1}{2}(1 + \cos 2\theta)\right] d\theta$$

$$= r^2 \left[\tfrac{3}{2}\theta - 2\sin\theta + \tfrac{1}{4}\sin 2\theta\right]_0^{2\pi}$$

$$= r^2 (\tfrac{3}{2} \cdot 2\pi) = 3\pi r^2$$

Arc Length

We already know how to find the length L of a curve C given in the form $y = F(x)$, $a \leq x \leq b$. Formula 9.1.3 says that if F' is continuous, then

$$\boxed{3} \qquad L = \int_a^b \sqrt{1 + \left(\frac{dy}{dx}\right)^2}\, dx$$

Suppose that C can also be described by the parametric equations $x = f(t)$ and $y = g(t)$, $\alpha \leq t \leq \beta$, where $dx/dt = f'(t) > 0$. This means that C is traversed once, from left to right, as t increases from α to β and $f(\alpha) = a$, $f(\beta) = b$. Putting Formula 2 into Formula 3 and using the Substitution Rule, we obtain

$$L = \int_a^b \sqrt{1 + \left(\frac{dy}{dx}\right)^2}\, dx = \int_\alpha^\beta \sqrt{1 + \left(\frac{dy/dt}{dx/dt}\right)^2}\, \frac{dx}{dt}\, dt$$

Since $dx/dt > 0$, we have

$$\boxed{4} \qquad L = \int_\alpha^\beta \sqrt{\left(\frac{dx}{dt}\right)^2 + \left(\frac{dy}{dt}\right)^2}\, dt$$

Even if C can't be expressed in the form $y = F(x)$, Formula 4 is still valid but we obtain it by polygonal approximations. We divide the parameter interval $[\alpha, \beta]$ into n subintervals of equal width Δt. If $t_0, t_1, t_2, \ldots, t_n$ are the endpoints of these subintervals, then $x_i = f(t_i)$ and $y_i = g(t_i)$ are the coordinates of points $P_i(x_i, y_i)$ that lie on C and the polygon with vertices $P_0, P_1, \ldots, P_n$ approximates C (see Figure 4).

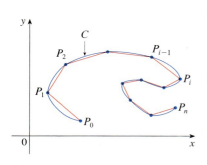

FIGURE 4

As in Section 9.1, we define the length L of C to be the limit of the lengths of these approximating polygons as $n \to \infty$:

$$L = \lim_{n \to \infty} \sum_{i=1}^n |P_{i-1}P_i|$$

The Mean Value Theorem, when applied to f on the interval $[t_{i-1}, t_i]$, gives a number t_i^* in (t_{i-1}, t_i) such that

$$f(t_i) - f(t_{i-1}) = f'(t_i^*)(t_i - t_{i-1})$$

If we let $\Delta x_i = x_i - x_{i-1}$ and $\Delta y_i = y_i - y_{i-1}$, this equation becomes

$$\Delta x_i = f'(t_i^*)\, \Delta t$$

Similarly, when applied to g, the Mean Value Theorem gives a number t_i^{**} in (t_{i-1}, t_i) such that

$$\Delta y_i = g'(t_i^{**}) \, \Delta t$$

Therefore

$$|P_{i-1}P_i| = \sqrt{(\Delta x_i)^2 + (\Delta y_i)^2}$$
$$= \sqrt{[f'(t_i^*) \, \Delta t]^2 + [g'(t_i^{**}) \, \Delta t]^2}$$
$$= \sqrt{[f'(t_i^*)]^2 + [g'(t_i^{**})]^2} \, \Delta t$$

and so

$$\boxed{5} \qquad L = \lim_{n \to \infty} \sum_{i=1}^{n} \sqrt{[f'(t_i^*)]^2 + [g'(t_i^{**})]^2} \, \Delta t$$

The sum in (5) resembles a Riemann sum for the function $\sqrt{[f'(t)]^2 + [g'(t)]^2}$ but it is not exactly a Riemann sum because $t_i^* \neq t_i^{**}$ in general. Nevertheless, if f' and g' are continuous, it can be shown that the limit in (5) is the same as if t_i^* and t_i^{**} were equal, namely,

$$L = \int_{\alpha}^{\beta} \sqrt{[f'(t)]^2 + [g'(t)]^2} \, dt$$

Thus, using Leibniz notation, we have the following result, which has the same form as (4).

$\boxed{6}$ Theorem If a curve C is described by the parametric equations $x = f(t)$, $y = g(t)$, $\alpha \leq t \leq \beta$, where f' and g' are continuous on $[\alpha, \beta]$ and C is traversed exactly once as t increases from α to β, then the length of C is

$$L = \int_{\alpha}^{\beta} \sqrt{\left(\frac{dx}{dt}\right)^2 + \left(\frac{dy}{dt}\right)^2} \, dt$$

Notice that the formula in Theorem 6 is consistent with the general formulas $L = \int ds$ and $(ds)^2 = (dx)^2 + (dy)^2$ of Section 9.1.

EXAMPLE 4 If we use the representation of the unit circle given in Example 2 in Section 11.1,

$$x = \cos t \qquad y = \sin t \qquad 0 \leq t \leq 2\pi$$

then $dx/dt = -\sin t$ and $dy/dt = \cos t$, so Theorem 6 gives

$$L = \int_0^{2\pi} \sqrt{\left(\frac{dx}{dt}\right)^2 + \left(\frac{dy}{dt}\right)^2} \, dt = \int_0^{2\pi} \sqrt{\sin^2 t + \cos^2 t} \, dt$$
$$= \int_0^{2\pi} dt = 2\pi$$

as expected. If, on the other hand, we use the representation given in Example 3 in Section 11.1,

$$x = \sin 2t \qquad y = \cos 2t \qquad 0 \leq t \leq 2\pi$$

then $dx/dt = 2\cos 2t$, $dy/dt = -2\sin 2t$, and the integral in Theorem 6 gives

$$\int_0^{2\pi} \sqrt{\left(\frac{dx}{dt}\right)^2 + \left(\frac{dy}{dt}\right)^2}\,dt = \int_0^{2\pi} \sqrt{4\cos^2 2t + 4\sin^2 2t}\,dt = \int_0^{2\pi} 2\,dt = 4\pi$$

⊘ Notice that the integral gives twice the arc length of the circle because as t increases from 0 to 2π, the point $(\sin 2t, \cos 2t)$ traverses the circle twice. In general, when finding the length of a curve C from a parametric representation, we have to be careful to ensure that C is traversed only once as t increases from α to β.

EXAMPLE 5 Find the length of one arch of the cycloid $x = r(\theta - \sin\theta)$, $y = r(1 - \cos\theta)$.

SOLUTION From Example 3 we see that one arch is described by the parameter interval $0 \le \theta \le 2\pi$. Since

$$\frac{dx}{d\theta} = r(1 - \cos\theta) \qquad \text{and} \qquad \frac{dy}{d\theta} = r\sin\theta$$

we have

$$L = \int_0^{2\pi} \sqrt{\left(\frac{dx}{d\theta}\right)^2 + \left(\frac{dy}{d\theta}\right)^2}\,d\theta = \int_0^{2\pi} \sqrt{r^2(1-\cos\theta)^2 + r^2\sin^2\theta}\,d\theta$$

$$= \int_0^{2\pi} \sqrt{r^2(1 - 2\cos\theta + \cos^2\theta + \sin^2\theta)}\,d\theta = r\int_0^{2\pi} \sqrt{2(1-\cos\theta)}\,d\theta$$

To evaluate this integral we use the identity $\sin^2 x = \frac{1}{2}(1 - \cos 2x)$ with $\theta = 2x$, which gives $1 - \cos\theta = 2\sin^2(\theta/2)$. Since $0 \le \theta \le 2\pi$, we have $0 \le \theta/2 \le \pi$ and so $\sin(\theta/2) \ge 0$. Therefore

$$\sqrt{2(1-\cos\theta)} = \sqrt{4\sin^2(\theta/2)} = 2|\sin(\theta/2)| = 2\sin(\theta/2)$$

and so

$$L = 2r \int_0^{2\pi} \sin(\theta/2)\,d\theta = 2r[-2\cos(\theta/2)]_0^{2\pi}$$

$$= 2r[2+2] = 8r$$

|||| The result of Example 5 says that the length of one arch of a cycloid is eight times the radius of the generating circle (see Figure 5). This was first proved in 1658 by Sir Christopher Wren, who later became the architect of St. Paul's Cathedral in London.

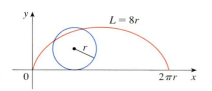

FIGURE 5

|||| **Surface Area**

In the same way as for arc length, we can adapt Formula 9.2.5 to obtain a formula for surface area. If the curve given by the parametric equations $x = f(t)$, $y = g(t)$, $\alpha \le t \le \beta$, is rotated about the x-axis, where f', g' are continuous and $g(t) \ge 0$, then the area of the resulting surface is given by

$$\boxed{7} \qquad S = \int_\alpha^\beta 2\pi y \sqrt{\left(\frac{dx}{dt}\right)^2 + \left(\frac{dy}{dt}\right)^2}\,dt$$

The general symbolic formulas $S = \int 2\pi y\,ds$ and $S = \int 2\pi x\,ds$ (Formulas 9.2.7 and 9.2.8) are still valid, but for parametric curves we use

$$ds = \sqrt{\left(\frac{dx}{dt}\right)^2 + \left(\frac{dy}{dt}\right)^2}\,dt$$

EXAMPLE 6 Show that the surface area of a sphere of radius r is $4\pi r^2$.

SOLUTION The sphere is obtained by rotating the semicircle
$$x = r\cos t \qquad y = r\sin t \qquad 0 \le t \le \pi$$
about the x-axis. Therefore, from Formula 7, we get
$$S = \int_0^\pi 2\pi r \sin t \sqrt{(-r\sin t)^2 + (r\cos t)^2}\, dt$$
$$= 2\pi \int_0^\pi r \sin t \sqrt{r^2(\sin^2 t + \cos^2 t)}\, dt$$
$$= 2\pi \int_0^\pi r \sin t \cdot r\, dt = 2\pi r^2 \int_0^\pi \sin t\, dt$$
$$= 2\pi r^2 (-\cos t)\Big]_0^\pi = 4\pi r^2$$

11.2 Exercises

1–2 Find dy/dx.
1. $x = t - t^3$, $y = 2 - 5t$
2. $x = te^t$, $y = t + e^t$

3–6 Find an equation of the tangent to the curve at the point corresponding to the given value of the parameter.
3. $x = t^4 + 1$, $y = t^3 + t$; $t = -1$
4. $x = 2t^2 + 1$, $y = \frac{1}{3}t^3 - t$; $t = 3$
5. $x = e^{\sqrt{t}}$, $y = t - \ln t^2$; $t = 1$
6. $x = \cos\theta + \sin 2\theta$, $y = \sin\theta + \cos 2\theta$; $\theta = 0$

7–8 Find an equation of the tangent to the curve at the given point by two methods: (a) without eliminating the parameter and (b) by first eliminating the parameter.
7. $x = e^t$, $y = (t - 1)^2$; $(1, 1)$
8. $x = \tan\theta$, $y = \sec\theta$; $(1, \sqrt{2})$

9–10 Find an equation of the tangent(s) to the curve at the given point. Then graph the curve and the tangent(s).
9. $x = 2\sin 2t$, $y = 2\sin t$; $(\sqrt{3}, 1)$
10. $x = \sin t$, $y = \sin(t + \sin t)$; $(0, 0)$

11–16 Find dy/dx and d^2y/dx^2. For which values of t is the curve concave upward?
11. $x = 4 + t^2$, $y = t^2 + t^3$
12. $x = t^3 - 12t$, $y = t^2 - 1$
13. $x = t - e^t$, $y = t + e^{-t}$
14. $x = t + \ln t$, $y = t - \ln t$
15. $x = 2\sin t$, $y = 3\cos t$, $0 < t < 2\pi$
16. $x = \cos 2t$, $y = \cos t$, $0 < t < \pi$

17–20 Find the points on the curve where the tangent is horizontal or vertical. If you have a graphing device, graph the curve to check your work.
17. $x = 10 - t^2$, $y = t^3 - 12t$
18. $x = 2t^3 + 3t^2 - 12t$, $y = 2t^3 + 3t^2 + 1$
19. $x = 2\cos\theta$, $y = \sin 2\theta$
20. $x = \cos 3\theta$, $y = 2\sin\theta$

21. Use a graph to estimate the coordinates of the leftmost point on the curve $x = t^4 - t^2$, $y = t + \ln t$. Then use calculus to find the exact coordinates.

22. Try to estimate the coordinates of the highest point and the leftmost point on the curve $x = te^t$, $y = te^{-t}$. Then find the exact coordinates. What are the asymptotes of this curve?

23–24 Graph the curve in a viewing rectangle that displays all the important aspects of the curve.
23. $x = t^4 - 2t^3 - 2t^2$, $y = t^3 - t$
24. $x = t^4 + 4t^3 - 8t^2$, $y = 2t^2 - t$

25. Show that the curve $x = \cos t$, $y = \sin t \cos t$ has two tangents at $(0, 0)$ and find their equations. Sketch the curve.

26. At what point does the curve $x = 1 - 2\cos^2 t$, $y = (\tan t)(1 - 2\cos^2 t)$ cross itself? Find the equations of both tangents at that point.

27. (a) Find the slope of the tangent line to the trochoid $x = r\theta - d\sin\theta$, $y = r - d\cos\theta$ in terms of θ. (See Exercise 38 in Section 11.1.)
    (b) Show that if $d < r$, then the trochoid does not have a vertical tangent.

28. (a) Find the slope of the tangent to the astroid $x = a\cos^3\theta$, $y = a\sin^3\theta$ in terms of θ. (Astroids are explored in the Laboratory Project on page 695.)

(b) At what points is the tangent horizontal or vertical?
(c) At what points does the tangent have slope 1 or -1?

29. At what points on the curve $x = t^3 + 4t$, $y = 6t^2$ is the tangent parallel to the line with equations $x = -7t$, $y = 12t - 5$?

30. Find equations of the tangents to the curve $x = 3t^2 + 1$, $y = 2t^3 + 1$ that pass through the point $(4, 3)$.

31. Use the parametric equations of an ellipse, $x = a \cos \theta$, $y = b \sin \theta$, $0 \leq \theta \leq 2\pi$, to find the area that it encloses.

32. Find the area bounded by the curve $x = t - 1/t$, $y = t + 1/t$ and the line $y = 2.5$.

33. Find the area bounded by the curve $x = \cos t$, $y = e^t$, $0 \leq t \leq \pi/2$, and the lines $y = 1$ and $x = 0$.

34. Find the area of the region enclosed by the astroid $x = a \cos^3 \theta$, $y = a \sin^3 \theta$. (Astroids are explored in the Laboratory Project on page 695.)

35. Find the area under one arch of the trochoid of Exercise 38 in Section 11.1 for the case $d < r$.

36. Let $\mathcal{R}$ be the region enclosed by the loop of the curve in Example 1.
(a) Find the area of $\mathcal{R}$.
(b) If $\mathcal{R}$ is rotated about the x-axis, find the volume of the resulting solid.
(c) Find the centroid of $\mathcal{R}$.

37–40 ■ Set up, but do not evaluate, an integral that represents the length of the curve.

37. $x = t - t^2$, $y = \frac{4}{3}t^{3/2}$, $1 \leq t \leq 2$

38. $x = 1 + e^t$, $y = t^2$, $-3 \leq t \leq 3$

39. $x = t + \cos t$, $y = t - \sin t$, $0 \leq t \leq 2\pi$

40. $x = \ln t$, $y = \sqrt{t + 1}$, $1 \leq t \leq 5$

41–44 ■ Find the length of the curve.

41. $x = 1 + 3t^2$, $y = 4 + 2t^3$, $0 \leq t \leq 1$

42. $x = a(\cos \theta + \theta \sin \theta)$, $y = a(\sin \theta - \theta \cos \theta)$, $0 \leq \theta \leq \pi$

43. $x = \dfrac{t}{1 + t}$, $y = \ln(1 + t)$, $0 \leq t \leq 2$

44. $x = e^t + e^{-t}$, $y = 5 - 2t$, $0 \leq t \leq 3$

45–47 ■ Graph the curve and find its length.

45. $x = e^t \cos t$, $y = e^t \sin t$, $0 \leq t \leq \pi$

46. $x = \cos t + \ln(\tan \frac{1}{2} t)$, $y = \sin t$, $\pi/4 \leq t \leq 3\pi/4$

47. $x = e^t - t$, $y = 4e^{t/2}$, $-8 \leq t \leq 3$

48. Find the length of the loop of the curve $x = 3t - t^3$, $y = 3t^2$.

49. Use Simpson's Rule with $n = 6$ to estimate the length of the curve $x = t - e^t$, $y = t + e^t$, $-6 \leq t \leq 6$.

50. In Exercise 41 in Section 11.1 you were asked to derive the parametric equations $x = 2a \cot \theta$, $y = 2a \sin^2 \theta$ for the curve called the witch of Maria Agnesi. Use Simpson's Rule with $n = 4$ to estimate the length of the arc of this curve given by $\pi/4 \leq \theta \leq \pi/2$.

51–52 ■ Find the distance traveled by a particle with position (x, y) as t varies in the given time interval. Compare with the length of the curve.

51. $x = \sin^2 t$, $y = \cos^2 t$, $0 \leq t \leq 3\pi$

52. $x = \cos^2 t$, $y = \cos t$, $0 \leq t \leq 4\pi$

53. Show that the total length of the ellipse $x = a \sin \theta$, $y = b \cos \theta$, $a > b > 0$, is

$$L = 4a \int_0^{\pi/2} \sqrt{1 - e^2 \sin^2 \theta}\, d\theta$$

where e is the eccentricity of the ellipse $(e = c/a$, where $c = \sqrt{a^2 - b^2})$.

54. Find the total length of the astroid $x = a \cos^3 \theta$, $y = a \sin^3 \theta$, where $a > 0$.

CAS 55. (a) Graph the **epitrochoid** with equations
$$x = 11 \cos t - 4 \cos(11t/2)$$
$$y = 11 \sin t - 4 \sin(11t/2)$$
What parameter interval gives the complete curve?
(b) Use your CAS to find the approximate length of this curve.

CAS 56. A curve called **Cornu's spiral** is defined by the parametric equations
$$x = C(t) = \int_0^t \cos(\pi u^2/2)\, du$$
$$y = S(t) = \int_0^t \sin(\pi u^2/2)\, du$$
where C and S are the Fresnel functions that were introduced in Chapter 5.
(a) Graph this curve. What happens as $t \to \infty$ and as $t \to -\infty$?
(b) Find the length of Cornu's spiral from the origin to the point with parameter value t.

57–58 ■ Set up, but do not evaluate, an integral that represents the area of the surface obtained by rotating the given curve about the x-axis.

57. $x = t - t^2$, $y = \frac{4}{3}t^{3/2}$, $1 \leq t \leq 2$

58. $x = \sin^2 t$, $y = \sin 3t$, $0 \leq t \leq \pi/3$

59–61 Find the area of the surface obtained by rotating the given curve about the x-axis.

59. $x = t^3$, $y = t^2$, $0 \leq t \leq 1$

60. $x = 3t - t^3$, $y = 3t^2$, $0 \leq t \leq 1$

61. $x = a\cos^3\theta$, $y = a\sin^3\theta$, $0 \leq \theta \leq \pi/2$

62. Graph the curve
$$x = 2\cos\theta - \cos 2\theta \qquad y = 2\sin\theta - \sin 2\theta$$
If this curve is rotated about the x-axis, find the area of the resulting surface. (Use your graph to help find the correct parameter interval.)

63. If the curve
$$x = t + t^3 \qquad y = t - \frac{1}{t^2} \qquad 1 \leq t \leq 2$$
is rotated about the x-axis, use your calculator to estimate the area of the resulting surface to three decimal places.

64. If the arc of the curve in Exercise 50 is rotated about the x-axis, estimate the area of the resulting surface using Simpson's Rule with $n = 4$.

65–66 Find the surface area generated by rotating the given curve about the y-axis.

65. $x = 3t^2$, $y = 2t^3$, $0 \leq t \leq 5$

66. $x = e^t - t$, $y = 4e^{t/2}$, $0 \leq t \leq 1$

67. If f' is continuous and $f'(t) \neq 0$ for $a \leq t \leq b$, show that the parametric curve $x = f(t)$, $y = g(t)$, $a \leq t \leq b$, can be put in the form $y = F(x)$. [*Hint:* Show that f^{-1} exists.]

68. Use Formula 2 to derive Formula 7 from Formula 9.2.5 for the case in which the curve can be represented in the form $y = F(x)$, $a \leq x \leq b$.

69. The **curvature** at a point P of a curve is defined as
$$\kappa = \left| \frac{d\phi}{ds} \right|$$
where ϕ is the angle of inclination of the tangent line at P, as shown in the figure. Thus, the curvature is the absolute value of the rate of change of ϕ with respect to arc length. It can be regarded as a measure of the rate of change of direction of the curve at P and will be studied in greater detail in Chapter 14.
(a) For a parametric curve $x = x(t)$, $y = y(t)$, derive the formula
$$\kappa = \frac{|\dot{x}\ddot{y} - \ddot{x}\dot{y}|}{[\dot{x}^2 + \dot{y}^2]^{3/2}}$$
where the dots indicate derivatives with respect to t, so $\dot{x} = dx/dt$. [*Hint:* Use $\phi = \tan^{-1}(dy/dx)$ and Equation 2 to find $d\phi/dt$. Then use the Chain Rule to find $d\phi/ds$.]

(b) By regarding a curve $y = f(x)$ as the parametric curve $x = x$, $y = f(x)$, with parameter x, show that the formula in part (a) becomes
$$\kappa = \frac{|d^2y/dx^2|}{[1 + (dy/dx)^2]^{3/2}}$$

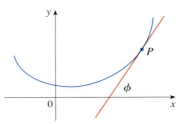

70. (a) Use the formula in Exercise 69(b) to find the curvature of the parabola $y = x^2$ at the point $(1, 1)$.
(b) At what point does this parabola have maximum curvature?

71. Use the formula in Exercise 69(a) to find the curvature of the cycloid $x = \theta - \sin\theta$, $y = 1 - \cos\theta$ at the top of one of its arches.

72. (a) Show that the curvature at each point of a straight line is $\kappa = 0$.
(b) Show that the curvature at each point of a circle of radius r is $\kappa = 1/r$.

73. A string is wound around a circle and then unwound while being held taut. The curve traced by the point P at the end of the string is called the **involute** of the circle. If the circle has radius r and center O and the initial position of P is $(r, 0)$, and if the parameter θ is chosen as in the figure, show that parametric equations of the involute are
$$x = r(\cos\theta + \theta\sin\theta) \qquad y = r(\sin\theta - \theta\cos\theta)$$

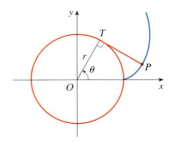

74. A cow is tied to a silo with radius r by a rope just long enough to reach the opposite side of the silo. Find the area available for grazing by the cow.

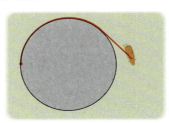

LABORATORY PROJECT

Bézier Curves

The **Bézier curves** are used in computer-aided design and are named after the French mathematician Pierre Bézier (1910–1999), who worked in the automotive industry. A cubic Bézier curve is determined by four **control points**, $P_0(x_0, y_0)$, $P_1(x_1, y_1)$, $P_2(x_2, y_2)$, and $P_3(x_3, y_3)$, and is defined by the parametric equations

$$x = x_0(1-t)^3 + 3x_1 t(1-t)^2 + 3x_2 t^2(1-t) + x_3 t^3$$
$$y = y_0(1-t)^3 + 3y_1 t(1-t)^2 + 3y_2 t^2(1-t) + y_3 t^3$$

where $0 \leq t \leq 1$. Notice that when $t = 0$ we have $(x, y) = (x_0, y_0)$ and when $t = 1$ we have $(x, y) = (x_3, y_3)$, so the curve starts at P_0 and ends at P_3.

1. Graph the Bézier curve with control points $P_0(4, 1)$, $P_1(28, 48)$, $P_2(50, 42)$, and $P_3(40, 5)$. Then, on the same screen, graph the line segments $P_0 P_1$, $P_1 P_2$, and $P_2 P_3$. (Exercise 31 in Section 11.1 shows how to do this.) Notice that the middle control points P_1 and P_2 don't lie on the curve; the curve starts at P_0, heads toward P_1 and P_2 without reaching them, and ends at P_3.

2. From the graph in Problem 1 it appears that the tangent at P_0 passes through P_1 and the tangent at P_3 passes through P_2. Prove it.

3. Try to produce a Bézier curve with a loop by changing the second control point in Problem 1.

4. Some laser printers use Bézier curves to represent letters and other symbols. Experiment with control points until you find a Bézier curve that gives a reasonable representation of the letter C.

5. More complicated shapes can be represented by piecing together two or more Bézier curves. Suppose the first Bézier curve has control points P_0, P_1, P_2, P_3 and the second one has control points P_3, P_4, P_5, P_6. If we want these two pieces to join together smoothly, then the tangents at P_3 should match and so the points P_2, P_3, and P_4 all have to lie on this common tangent line. Using this principle, find control points for a pair of Bézier curves that represent the letter S.

11.3 Polar Coordinates

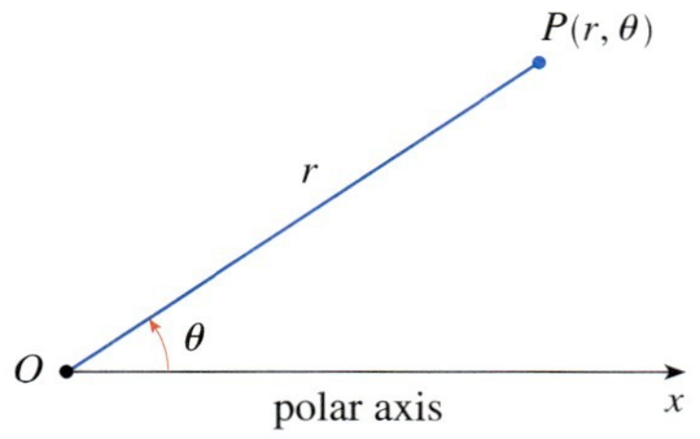

A coordinate system represents a point in the plane by an ordered pair of numbers called coordinates. Usually we use Cartesian coordinates, which are directed distances from two perpendicular axes. Here we describe a coordinate system introduced by Newton, called the **polar coordinate system**, which is more convenient for many purposes.

We choose a point in the plane that is called the **pole** (or origin) and is labeled O. Then we draw a ray (half-line) starting at O called the **polar axis**. This axis is usually drawn horizontally to the right and corresponds to the positive x-axis in Cartesian coordinates.

If P is any other point in the plane, let r be the distance from O to P and let θ be the angle (usually measured in radians) between the polar axis and the line OP as in Figure 1. Then the point P is represented by the ordered pair (r, θ) and r, θ are called **polar coordinates** of P. We use the convention that an angle is positive if measured in the counterclockwise direction from the polar axis and negative in the clockwise direction. If $P = O$, then $r = 0$ and we agree that $(0, \theta)$ represents the pole for any value of θ.

FIGURE 1

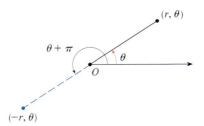

FIGURE 2

We extend the meaning of polar coordinates (r, θ) to the case in which r is negative by agreeing that, as in Figure 2, the points $(-r, \theta)$ and (r, θ) lie on the same line through O and at the same distance $|r|$ from O, but on opposite sides of O. If $r > 0$, the point (r, θ) lies in the same quadrant as θ; if $r < 0$, it lies in the quadrant on the opposite side of the pole. Notice that $(-r, \theta)$ represents the same point as $(r, \theta + \pi)$.

EXAMPLE 1 Plot the points whose polar coordinates are given.
(a) $(1, 5\pi/4)$ (b) $(2, 3\pi)$ (c) $(2, -2\pi/3)$ (d) $(-3, 3\pi/4)$

SOLUTION The points are plotted in Figure 3. In part (d) the point $(-3, 3\pi/4)$ is located three units from the pole in the fourth quadrant because the angle $3\pi/4$ is in the second quadrant and $r = -3$ is negative.

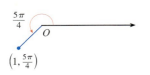
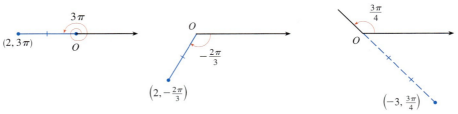

FIGURE 3

In the Cartesian coordinate system every point has only one representation, but in the polar coordinate system each point has many representations. For instance, the point $(1, 5\pi/4)$ in Example 1(a) could be written as $(1, -3\pi/4)$ or $(1, 13\pi/4)$ or $(-1, \pi/4)$. (See Figure 4.)

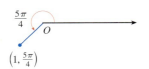
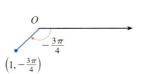
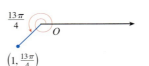
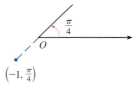

FIGURE 4

In fact, since a complete counterclockwise rotation is given by an angle 2π, the point represented by polar coordinates (r, θ) is also represented by

$$(r, \theta + 2n\pi) \quad \text{and} \quad (-r, \theta + (2n+1)\pi)$$

where n is any integer.

The connection between polar and Cartesian coordinates can be seen from Figure 5, in which the pole corresponds to the origin and the polar axis coincides with the positive x-axis. If the point P has Cartesian coordinates (x, y) and polar coordinates (r, θ), then, from the figure, we have

$$\cos\theta = \frac{x}{r} \qquad \sin\theta = \frac{y}{r}$$

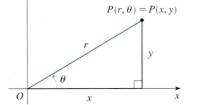

FIGURE 5

and so

 $$x = r\cos\theta \qquad y = r\sin\theta$$

Although Equations 1 were deduced from Figure 5, which illustrates the case where $r > 0$ and $0 < \theta < \pi/2$, these equations are valid for all values of r and θ. (See the general definition of $\sin\theta$ and $\cos\theta$ in Appendix D.)

Equations 1 allow us to find the Cartesian coordinates of a point when the polar coordinates are known. To find r and θ when x and y are known, we use the equations

$$\boxed{2} \qquad r^2 = x^2 + y^2 \qquad \tan\theta = \frac{y}{x}$$

which can be deduced from Equations 1 or simply read from Figure 5.

EXAMPLE 2 Convert the point $(2, \pi/3)$ from polar to Cartesian coordinates.

SOLUTION Since $r = 2$ and $\theta = \pi/3$, Equations 1 give

$$x = r\cos\theta = 2\cos\frac{\pi}{3} = 2 \cdot \frac{1}{2} = 1$$

$$y = r\sin\theta = 2\sin\frac{\pi}{3} = 2 \cdot \frac{\sqrt{3}}{2} = \sqrt{3}$$

Therefore, the point is $(1, \sqrt{3})$ in Cartesian coordinates.

EXAMPLE 3 Represent the point with Cartesian coordinates $(1, -1)$ in terms of polar coordinates.

SOLUTION If we choose r to be positive, then Equations 2 give

$$r = \sqrt{x^2 + y^2} = \sqrt{1^2 + (-1)^2} = \sqrt{2}$$

$$\tan\theta = \frac{y}{x} = -1$$

Since the point $(1, -1)$ lies in the fourth quadrant, we can choose $\theta = -\pi/4$ or $\theta = 7\pi/4$. Thus, one possible answer is $(\sqrt{2}, -\pi/4)$; another is $(\sqrt{2}, 7\pi/4)$.

NOTE □ Equations 2 do not uniquely determine θ when x and y are given because, as θ increases through the interval $0 \leq \theta < 2\pi$, each value of $\tan\theta$ occurs twice. Therefore, in converting from Cartesian to polar coordinates, it's not good enough just to find r and θ that satisfy Equations 2. As in Example 3, we must choose θ so that the point (r, θ) lies in the correct quadrant.

Polar Curves

The **graph of a polar equation** $r = f(\theta)$, or more generally $F(r, \theta) = 0$, consists of all points P that have at least one polar representation (r, θ) whose coordinates satisfy the equation.

EXAMPLE 4 What curve is represented by the polar equation $r = 2$?

SOLUTION The curve consists of all points (r, θ) with $r = 2$. Since r represents the distance from the point to the pole, the curve $r = 2$ represents the circle with center O and radius 2. In general, the equation $r = a$ represents a circle with center O and radius $|a|$. (See Figure 6.)

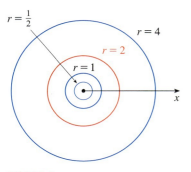

FIGURE 6

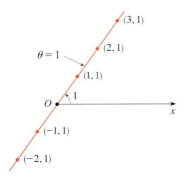

FIGURE 7

EXAMPLE 5 Sketch the polar curve $\theta = 1$.

SOLUTION This curve consists of all points (r, θ) such that the polar angle θ is 1 radian. It is the straight line that passes through O and makes an angle of 1 radian with the polar axis (see Figure 7). Notice that the points $(r, 1)$ on the line with $r > 0$ are in the first quadrant, whereas those with $r < 0$ are in the third quadrant.

EXAMPLE 6
(a) Sketch the curve with polar equation $r = 2 \cos \theta$.
(b) Find a Cartesian equation for this curve.

SOLUTION
(a) In Figure 8 we find the values of r for some convenient values of θ and plot the corresponding points (r, θ). Then we join these points to sketch the curve, which appears to be a circle. We have used only values of θ between 0 and π, since if we let θ increase beyond π, we obtain the same points again.

θ	$r = 2\cos\theta$
0	2
$\pi/6$	$\sqrt{3}$
$\pi/4$	$\sqrt{2}$
$\pi/3$	1
$\pi/2$	0
$2\pi/3$	-1
$3\pi/4$	$-\sqrt{2}$
$5\pi/6$	$-\sqrt{3}$
π	-2

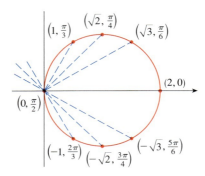

FIGURE 8
Table of values and graph of $r = 2 \cos \theta$

(b) To convert the given equation into a Cartesian equation we use Equations 1 and 2. From $x = r \cos \theta$ we have $\cos \theta = x/r$, so the equation $r = 2 \cos \theta$ becomes $r = 2x/r$, which gives

$$2x = r^2 = x^2 + y^2 \quad \text{or} \quad x^2 + y^2 - 2x = 0$$

Completing the square, we obtain

$$(x - 1)^2 + y^2 = 1$$

which is an equation of a circle with center $(1, 0)$ and radius 1.

|||| Figure 9 shows a geometrical illustration that the circle in Example 6 has the equation $r = 2 \cos \theta$. The angle OPQ is a right angle (Why?) and so $r/2 = \cos \theta$.

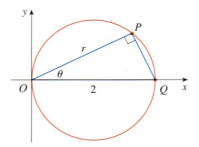

FIGURE 9

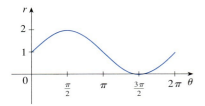

FIGURE 10
$r = 1 + \sin\theta$ in Cartesian coordinates, $0 \le \theta \le 2\pi$

EXAMPLE 7 Sketch the curve $r = 1 + \sin\theta$.

SOLUTION Instead of plotting points as in Example 6, we first sketch the graph of $r = 1 + \sin\theta$ in *Cartesian* coordinates in Figure 10 by shifting the sine curve up one unit. This enables us to read at a glance the values of r that correspond to increasing values of θ. For instance, we see that as θ increases from 0 to $\pi/2$, r (the distance from O) increases from 1 to 2, so we sketch the corresponding part of the polar curve in Figure 11(a). As θ increases from $\pi/2$ to π, Figure 10 shows that r decreases from 2 to 1, so we sketch the next part of the curve as in Figure 11(b). As θ increases from π to $3\pi/2$, r decreases from 1 to 0 as shown in part (c). Finally, as θ increases from $3\pi/2$ to 2π, r increases from 0 to 1 as shown in part (d). If we let θ increase beyond 2π or decrease beyond 0, we would simply retrace our path. Putting together the parts of the curve from Figure 11(a)–(d), we sketch the complete curve in part (e). It is called a **cardioid** because it's shaped like a heart.

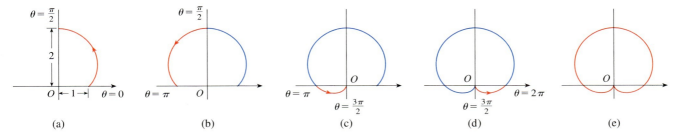

FIGURE 11
Stages in sketching the cardioid $r = 1 + \sin\theta$

Module 11.3 helps you see how polar curves are traced out by showing animations similar to Figures 10–13. Tangents to these polar curves can also be visualized as in Figure 15 (see page 712).

EXAMPLE 8 Sketch the curve $r = \cos 2\theta$.

SOLUTION As in Example 7, we first sketch $r = \cos 2\theta$, $0 \le \theta \le 2\pi$, in Cartesian coordinates in Figure 12. As θ increases from 0 to $\pi/4$, Figure 12 shows that r decreases from 1 to 0 and so we draw the corresponding portion of the polar curve in Figure 13 (indicated by ①). As θ increases from $\pi/4$ to $\pi/2$, r goes from 0 to -1. This means that the distance from O increases from 0 to 1, but instead of being in the first quadrant this portion of the polar curve (indicated by ②) lies on the opposite side of the pole in the third quadrant. The remainder of the curve is drawn in a similar fashion, with the arrows and numbers indicating the order in which the portions are traced out. The resulting curve has four loops and is called a **four-leaved rose**.

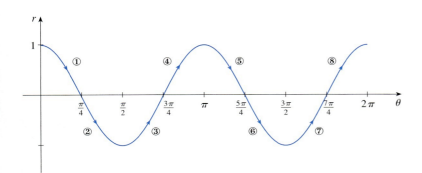

FIGURE 12
$r = \cos 2\theta$ in Cartesian coordinates

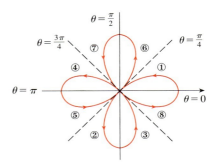

FIGURE 13
Four-leaved rose $r = \cos 2\theta$

Symmetry

When we sketch polar curves it is sometimes helpful to take advantage of symmetry. The following three rules are explained by Figure 14.

(a) If a polar equation is unchanged when θ is replaced by $-\theta$, the curve is symmetric about the polar axis.

(b) If the equation is unchanged when r is replaced by $-r$, or when θ is replaced by $\theta + \pi$, the curve is symmetric about the pole. (This means that the curve remains unchanged if we rotate it through 180° about the origin.)

(c) If the equation is unchanged when θ is replaced by $\pi - \theta$, the curve is symmetric about the vertical line $\theta = \pi/2$.

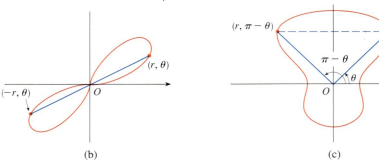

FIGURE 14

The curves sketched in Examples 6 and 8 are symmetric about the polar axis, since $\cos(-\theta) = \cos\theta$. The curves in Examples 7 and 8 are symmetric about $\theta = \pi/2$ because $\sin(\pi - \theta) = \sin\theta$ and $\cos 2(\pi - \theta) = \cos 2\theta$. The four-leaved rose is also symmetric about the pole. These symmetry properties could have been used in sketching the curves. For instance, in Example 6 we need only have plotted points for $0 \leq \theta \leq \pi/2$ and then reflected about the polar axis to obtain the complete circle.

Tangents to Polar Curves

To find a tangent line to a polar curve $r = f(\theta)$ we regard θ as a parameter and write its parametric equations as

$$x = r\cos\theta = f(\theta)\cos\theta \qquad y = r\sin\theta = f(\theta)\sin\theta$$

Then, using the method for finding slopes of parametric curves (Equation 11.2.2) and the Product Rule, we have

$$\frac{dy}{dx} = \frac{\dfrac{dy}{d\theta}}{\dfrac{dx}{d\theta}} = \frac{\dfrac{dr}{d\theta}\sin\theta + r\cos\theta}{\dfrac{dr}{d\theta}\cos\theta - r\sin\theta}$$

We locate horizontal tangents by finding the points where $dy/d\theta = 0$ (provided that $dx/d\theta \neq 0$). Likewise, we locate vertical tangents at the points where $dx/d\theta = 0$ (provided that $dy/d\theta \neq 0$).

Notice that if we are looking for tangent lines at the pole, then $r = 0$ and Equation 3 simplifies to

$$\frac{dy}{dx} = \tan\theta \qquad \text{if} \quad \frac{dr}{d\theta} \neq 0$$

For instance, in Example 8 we found that $r = \cos 2\theta = 0$ when $\theta = \pi/4$ or $3\pi/4$. This means that the lines $\theta = \pi/4$ and $\theta = 3\pi/4$ (or $y = x$ and $y = -x$) are tangent lines to $r = \cos 2\theta$ at the origin.

EXAMPLE 9
(a) For the cardioid $r = 1 + \sin \theta$ of Example 7, find the slope of the tangent line when $\theta = \pi/3$.
(b) Find the points on the cardioid where the tangent line is horizontal or vertical.

SOLUTION Using Equation 3 with $r = 1 + \sin \theta$, we have

$$\frac{dy}{dx} = \frac{\dfrac{dr}{d\theta} \sin \theta + r \cos \theta}{\dfrac{dr}{d\theta} \cos \theta - r \sin \theta} = \frac{\cos \theta \sin \theta + (1 + \sin \theta) \cos \theta}{\cos \theta \cos \theta - (1 + \sin \theta) \sin \theta}$$

$$= \frac{\cos \theta (1 + 2 \sin \theta)}{1 - 2 \sin^2 \theta - \sin \theta} = \frac{\cos \theta (1 + 2 \sin \theta)}{(1 + \sin \theta)(1 - 2 \sin \theta)}$$

(a) The slope of the tangent at the point where $\theta = \pi/3$ is

$$\left. \frac{dy}{dx} \right|_{\theta=\pi/3} = \frac{\cos(\pi/3)(1 + 2 \sin(\pi/3))}{(1 + \sin(\pi/3))(1 - 2 \sin(\pi/3))}$$

$$= \frac{\frac{1}{2}(1 + \sqrt{3})}{(1 + \sqrt{3}/2)(1 - \sqrt{3})} = \frac{1 + \sqrt{3}}{(2 + \sqrt{3})(1 - \sqrt{3})}$$

$$= \frac{1 + \sqrt{3}}{-1 - \sqrt{3}} = -1$$

(b) Observe that

$$\frac{dy}{d\theta} = \cos \theta (1 + 2 \sin \theta) = 0 \qquad \text{when } \theta = \frac{\pi}{2}, \frac{3\pi}{2}, \frac{7\pi}{6}, \frac{11\pi}{6}$$

$$\frac{dx}{d\theta} = (1 + \sin \theta)(1 - 2 \sin \theta) = 0 \qquad \text{when } \theta = \frac{3\pi}{2}, \frac{\pi}{6}, \frac{5\pi}{6}$$

Therefore, there are horizontal tangents at the points $(2, \pi/2)$, $(\frac{1}{2}, 7\pi/6)$, $(\frac{1}{2}, 11\pi/6)$ and vertical tangents at $(\frac{3}{2}, \pi/6)$ and $(\frac{3}{2}, 5\pi/6)$. When $\theta = 3\pi/2$, both $dy/d\theta$ and $dx/d\theta$ are 0, so we must be careful. Using l'Hospital's Rule, we have

$$\lim_{\theta \to (3\pi/2)^-} \frac{dy}{dx} = \left(\lim_{\theta \to (3\pi/2)^-} \frac{1 + 2 \sin \theta}{1 - 2 \sin \theta} \right) \left(\lim_{\theta \to (3\pi/2)^-} \frac{\cos \theta}{1 + \sin \theta} \right)$$

$$= -\frac{1}{3} \lim_{\theta \to (3\pi/2)^-} \frac{\cos \theta}{1 + \sin \theta}$$

$$= -\frac{1}{3} \lim_{\theta \to (3\pi/2)^-} \frac{-\sin \theta}{\cos \theta} = \infty$$

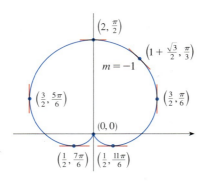

FIGURE 15
Tangent lines for $r = 1 + \sin\theta$

By symmetry,
$$\lim_{\theta \to (3\pi/2)^+} \frac{dy}{dx} = -\infty$$

Thus, there is a vertical tangent line at the pole (see Figure 15).

NOTE ○ Instead of having to remember Equation 3, we could employ the method used to derive it. For instance, in Example 9 we could have written

$$x = r\cos\theta = (1 + \sin\theta)\cos\theta = \cos\theta + \tfrac{1}{2}\sin 2\theta$$

$$y = r\sin\theta = (1 + \sin\theta)\sin\theta = \sin\theta + \sin^2\theta$$

Then we have

$$\frac{dy}{dx} = \frac{dy/d\theta}{dx/d\theta} = \frac{\cos\theta + 2\sin\theta\cos\theta}{-\sin\theta + \cos 2\theta} = \frac{\cos\theta + \sin 2\theta}{-\sin\theta + \cos 2\theta}$$

which is equivalent to our previous expression.

Graphing Polar Curves with Graphing Devices

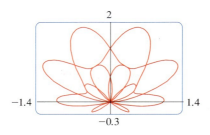

FIGURE 16
$r = \sin\theta + \sin^3(5\theta/2)$

Although it's useful to be able to sketch simple polar curves by hand, we need to use a graphing calculator or computer when we are faced with a curve as complicated as the one shown in Figure 16.

Some graphing devices have commands that enable us to graph polar curves directly. With other machines we need to convert to parametric equations first. In this case we take the polar equation $r = f(\theta)$ and write its parametric equations as

$$x = r\cos\theta = f(\theta)\cos\theta \qquad y = r\sin\theta = f(\theta)\sin\theta$$

Some machines require that the parameter be called t rather than θ.

EXAMPLE 10 Graph the curve $r = \sin(8\theta/5)$.

SOLUTION Let's assume that our graphing device doesn't have a built-in polar graphing command. In this case we need to work with the corresponding parametric equations, which are

$$x = r\cos\theta = \sin(8\theta/5)\cos\theta \qquad y = r\sin\theta = \sin(8\theta/5)\sin\theta$$

In any case we need to determine the domain for θ. So we ask ourselves: How many complete rotations are required until the curve starts to repeat itself? If the answer is n, then

$$\sin\frac{8(\theta + 2n\pi)}{5} = \sin\left(\frac{8\theta}{5} + \frac{16n\pi}{5}\right) = \sin\frac{8\theta}{5}$$

and so we require that $16n\pi/5$ be an even multiple of π. This will first occur when $n = 5$. Therefore, we will graph the entire curve if we specify that $0 \leq \theta \leq 10\pi$. Switching from θ to t, we have the equations

$$x = \sin(8t/5)\cos t \qquad y = \sin(8t/5)\sin t \qquad 0 \leq t \leq 10\pi$$

and Figure 17 shows the resulting curve. Notice that this rose has 16 loops.

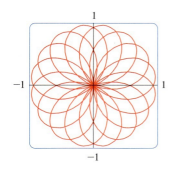

FIGURE 17
$r = \sin(8\theta/5)$

EXAMPLE 11 Investigate the family of polar curves given by $r = 1 + c \sin \theta$. How does the shape change as c changes? (These curves are called **limaçons**, after a French word for snail, because of the shape of the curves for certain values of c.)

SOLUTION Figure 18 shows computer-drawn graphs for various values of c. For $c > 1$ there is a loop that decreases in size as c decreases. When $c = 1$ the loop disappears and the curve becomes the cardioid that we sketched in Example 7. For c between 1 and $\frac{1}{2}$ the cardioid's cusp is smoothed out and becomes a "dimple." When c decreases from $\frac{1}{2}$ to 0, the limaçon is shaped like an oval. This oval becomes more circular as $c \to 0$, and when $c = 0$ the curve is just the circle $r = 1$.

|||| In Exercise 53 you are asked to prove analytically what we have discovered from the graphs in Figure 18.

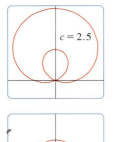
$c = 2.5$

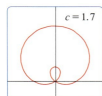
$c = 1.7$

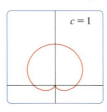
$c = 1$

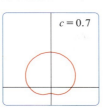
$c = 0.7$

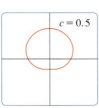
$c = 0.5$

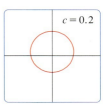
$c = 0.2$

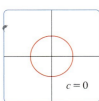
$c = 0$

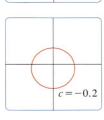
$c = -0.2$

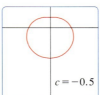
$c = -0.5$

$c = -0.8$

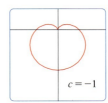
$c = -1$

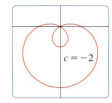
$c = -2$

FIGURE 18
Members of the family of limaçons $r = 1 + c \sin \theta$

The remaining parts of Figure 18 show that as c becomes negative, the shapes change in reverse order. In fact, these curves are reflections about the horizontal axis of the corresponding curves with positive c.

11.3 Exercises

1–2 |||| Plot the point whose polar coordinates are given. Then find two other pairs of polar coordinates of this point, one with $r > 0$ and one with $r < 0$.

1. (a) $(1, \pi/2)$ (b) $(-2, \pi/4)$ (c) $(3, 2)$

2. (a) $(3, 0)$ (b) $(2, -\pi/7)$ (c) $(-1, -\pi/2)$

3–4 |||| Plot the point whose polar coordinates are given. Then find the Cartesian coordinates of the point.

3. (a) $(3, \pi/2)$ (b) $(2\sqrt{2}, 3\pi/4)$ (c) $(-1, \pi/3)$

4. (a) $(2, 2\pi/3)$ (b) $(4, 3\pi)$ (c) $(-2, -5\pi/6)$

5–6 |||| The Cartesian coordinates of a point are given.
(i) Find polar coordinates (r, θ) of the point, where $r > 0$ and $0 \leq \theta < 2\pi$.
(ii) Find polar coordinates (r, θ) of the point, where $r < 0$ and $0 \leq \theta < 2\pi$.

5. (a) $(1, 1)$ (b) $(2\sqrt{3}, -2)$

6. (a) $(-1, -\sqrt{3})$ (b) $(-2, 3)$

7–12 |||| Sketch the region in the plane consisting of points whose polar coordinates satisfy the given conditions.

7. $1 \leq r \leq 2$

8. $r \geq 0$, $\pi/3 \leq \theta \leq 2\pi/3$

9. $0 \leq r < 4$, $-\pi/2 \leq \theta < \pi/6$

10. $2 < r \leq 5$, $3\pi/4 < \theta < 5\pi/4$

11. $2 < r < 3$, $5\pi/3 \leq \theta \leq 7\pi/3$

12. $-1 \leq r \leq 1$, $\pi/4 \leq \theta \leq 3\pi/4$

13. Find the distance between the points with polar coordinates $(1, \pi/6)$ and $(3, 3\pi/4)$.

14. Find a formula for the distance between the points with polar coordinates (r_1, θ_1) and (r_2, θ_2).

15–20 ▮ Identify the curve by finding a Cartesian equation for the curve.

15. $r = 2$

16. $r \cos \theta = 1$

17. $r = 3 \sin \theta$

18. $r = 2 \sin \theta + 2 \cos \theta$

19. $r = \csc \theta$

20. $r = \tan \theta \sec \theta$

21–26 ▮ Find a polar equation for the curve represented by the given Cartesian equation.

21. $x = 3$

22. $x^2 + y^2 = 9$

23. $x = -y^2$

24. $x + y = 9$

25. $x^2 + y^2 = 2cx$

26. $x^2 - y^2 = 1$

27–28 ▮ For each of the described curves, decide if the curve would be more easily given by a polar equation or a Cartesian equation. Then write an equation for the curve.

27. (a) A line through the origin that makes an angle of $\pi/6$ with the positive x-axis
(b) A vertical line through the point $(3, 3)$

28. (a) A circle with radius 5 and center $(2, 3)$
(b) A circle centered at the origin with radius 4

29–46 ▮ Sketch the curve with the given polar equation.

29. $\theta = -\pi/6$

30. $r^2 - 3r + 2 = 0$

31. $r = \sin \theta$

32. $r = -3 \cos \theta$

33. $r = 2(1 - \sin \theta)$

34. $r = 1 - 3 \cos \theta$

35. $r = \theta, \quad \theta \geq 0$

36. $r = \ln \theta, \quad \theta \geq 1$

37. $r = \sin 2\theta$

38. $r = 2 \cos 3\theta$

39. $r = 2 \cos 4\theta$

40. $r = \sin 5\theta$

41. $r^2 = 4 \cos 2\theta$

42. $r^2 = \sin 2\theta$

43. $r = 2 \cos(3\theta/2)$

44. $r^2 \theta = 1$

45. $r = 1 + 2 \cos 2\theta$

46. $r = 1 + 2 \cos(\theta/2)$

47–48 ▮ The figure shows the graph of r as a function of θ in Cartesian coordinates. Use it to sketch the corresponding polar curve.

47.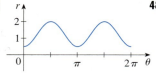

48.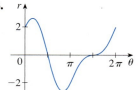

49. Show that the polar curve $r = 4 + 2 \sec \theta$ (called a **conchoid**) has the line $x = 2$ as a vertical asymptote by showing that $\lim_{r \to \pm \infty} x = 2$. Use this fact to help sketch the conchoid.

50. Show that the curve $r = 2 - \csc \theta$ (also a conchoid) has the line $y = -1$ as a horizontal asymptote by showing that $\lim_{r \to \pm \infty} y = -1$. Use this fact to help sketch the conchoid.

51. Show that the curve $r = \sin \theta \tan \theta$ (called a **cissoid of Diocles**) has the line $x = 1$ as a vertical asymptote. Show also that the curve lies entirely within the vertical strip $0 \leq x < 1$. Use these facts to help sketch the cissoid.

52. Sketch the curve $(x^2 + y^2)^3 = 4x^2 y^2$.

53. (a) In Example 11 the graphs suggest that the limaçon $r = 1 + c \sin \theta$ has an inner loop when $|c| > 1$. Prove that this is true, and find the values of θ that correspond to the inner loop.
(b) From Figure 18 it appears that the limaçon loses its dimple when $c = \frac{1}{2}$. Prove this.

54. Match the polar equations with the graphs labeled I–VI. Give reasons for your choices. (Don't use a graphing device.)
(a) $r = \sin(\theta/2)$
(b) $r = \sin(\theta/4)$
(c) $r = \sec(3\theta)$
(d) $r = \theta \sin \theta$
(e) $r = 1 + 4 \cos 5\theta$
(f) $r = 1/\sqrt{\theta}$

I  II 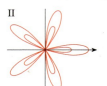 III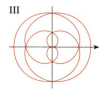

IV 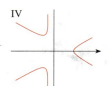 V 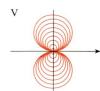 VI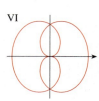

55–60 ▮ Find the slope of the tangent line to the given polar curve at the point specified by the value of θ.

55. $r = 2 \sin \theta, \quad \theta = \pi/6$

56. $r = 2 - \sin \theta, \quad \theta = \pi/3$

57. $r = 1/\theta, \quad \theta = \pi$

58. $r = \ln \theta, \quad \theta = e$

59. $r = 1 + \cos \theta, \quad \theta = \pi/6$

60. $r = \sin 3\theta, \quad \theta = \pi/6$

61–66 ▮ Find the points on the given curve where the tangent line is horizontal or vertical.

61. $r = 3 \cos \theta$

62. $r = \cos \theta + \sin \theta$

63. $r = 1 + \cos \theta$

64. $r = e^\theta$

65. $r = \cos 2\theta$

66. $r^2 = \sin 2\theta$

67. Show that the polar equation $r = a \sin \theta + b \cos \theta$, where $ab \neq 0$, represents a circle, and find its center and radius.

68. Show that the curves $r = a \sin \theta$ and $r = a \cos \theta$ intersect at right angles.

69–74 ▐▐▐ Use a graphing device to graph the polar curve. Choose the parameter interval to make sure that you produce the entire curve.

69. $r = 1 + 2 \sin(\theta/2)$ (nephroid of Freeth)

70. $r = \sqrt{1 - 0.8 \sin^2\theta}$ (hippopede)

71. $r = e^{\sin \theta} - 2 \cos(4\theta)$ (butterfly curve)

72. $r = \sin^2(4\theta) + \cos(4\theta)$

73. $r = 2 - 5 \sin(\theta/6)$

74. $r = \cos(\theta/2) + \cos(\theta/3)$

75. How are the graphs of $r = 1 + \sin(\theta - \pi/6)$ and $r = 1 + \sin(\theta - \pi/3)$ related to the graph of $r = 1 + \sin\theta$? In general, how is the graph of $r = f(\theta - \alpha)$ related to the graph of $r = f(\theta)$?

76. Use a graph to estimate the y-coordinate of the highest points on the curve $r = \sin 2\theta$. Then use calculus to find the exact value.

77. (a) Investigate the family of curves defined by the polar equations $r = \sin n\theta$, where n is a positive integer. How is the number of loops related to n?
(b) What happens if the equation in part (a) is replaced by $r = |\sin n\theta|$?

78. A family of curves is given by the equations $r = 1 + c \sin n\theta$, where c is a real number and n is a positive integer. How does the graph change as n increases? How does it change as c changes? Illustrate by graphing enough members of the family to support your conclusions.

79. A family of curves has polar equations

$$r = \frac{1 - a\cos\theta}{1 + a\cos\theta}$$

Investigate how the graph changes as the number a changes. In particular, you should identify the transitional values of a for which the basic shape of the curve changes.

80. The astronomer Giovanni Cassini (1625–1712) studied the family of curves with polar equations

$$r^4 - 2c^2 r^2 \cos 2\theta + c^4 - a^4 = 0$$

where a and c are positive real numbers. These curves are called the **ovals of Cassini** even though they are oval shaped only for certain values of a and c. (Cassini thought that these curves might represent planetary orbits better than Kepler's ellipses.) Investigate the variety of shapes that these curves may have. In particular, how are a and c related to each other when the curve splits into two parts?

81. Let P be any point (except the origin) on the curve $r = f(\theta)$. If ψ is the angle between the tangent line at P and the radial line OP, show that

$$\tan \psi = \frac{r}{dr/d\theta}$$

[*Hint:* Observe that $\psi = \phi - \theta$ in the figure.]

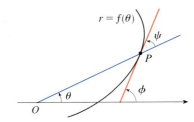

82. (a) Use Exercise 81 to show that the angle between the tangent line and the radial line is $\psi = \pi/4$ at every point on the curve $r = e^\theta$.
(b) Illustrate part (a) by graphing the curve and the tangent lines at the points where $\theta = 0$ and $\pi/2$.
(c) Prove that any polar curve $r = f(\theta)$ with the property that the angle ψ between the radial line and the tangent line is a constant must be of the form $r = Ce^{k\theta}$, where C and k are constants.

11.4 Areas and Lengths in Polar Coordinates

In this section we develop the formula for the area of a region whose boundary is given by a polar equation. We need to use the formula for the area of a sector of a circle

$$\boxed{1} \qquad A = \tfrac{1}{2} r^2 \theta$$

where, as in Figure 1, r is the radius and θ is the radian measure of the central angle. Formula 1 follows from the fact that the area of a sector is proportional to its central angle: $A = (\theta/2\pi)\pi r^2 = \tfrac{1}{2} r^2 \theta$. (See also Exercise 35 in Section 8.3.)

FIGURE 1

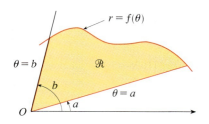

FIGURE 2

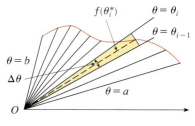

FIGURE 3

Let $\mathcal{R}$ be the region, illustrated in Figure 2, bounded by the polar curve $r = f(\theta)$ and by the rays $\theta = a$ and $\theta = b$, where f is a positive continuous function and where $0 < b - a \le 2\pi$. We divide the interval $[a, b]$ into subintervals with endpoints θ_0, θ_1, $\theta_2, \ldots, \theta_n$ and equal width $\Delta\theta$. The rays $\theta = \theta_i$ then divide $\mathcal{R}$ into n smaller regions with central angle $\Delta\theta = \theta_i - \theta_{i-1}$. If we choose θ_i^* in the ith subinterval $[\theta_{i-1}, \theta_i]$, then the area ΔA_i of the ith region is approximated by the area of the sector of a circle with central angle $\Delta\theta$ and radius $f(\theta_i^*)$. (See Figure 3.)

Thus, from Formula 1 we have

$$\Delta A_i \approx \tfrac{1}{2}[f(\theta_i^*)]^2 \Delta\theta$$

and so an approximation to the total area A of $\mathcal{R}$ is

$$\boxed{2} \qquad A \approx \sum_{i=1}^{n} \tfrac{1}{2}[f(\theta_i^*)]^2 \Delta\theta$$

It appears from Figure 3 that the approximation in (2) improves as $n \to \infty$. But the sums in (2) are Riemann sums for the function $g(\theta) = \tfrac{1}{2}[f(\theta)]^2$, so

$$\lim_{n\to\infty} \sum_{i=1}^{n} \tfrac{1}{2}[f(\theta_i^*)]^2 \Delta\theta = \int_a^b \tfrac{1}{2}[f(\theta)]^2\, d\theta$$

It therefore appears plausible (and can in fact be proved) that the formula for the area A of the polar region $\mathcal{R}$ is

$$\boxed{3} \qquad A = \int_a^b \tfrac{1}{2}[f(\theta)]^2\, d\theta$$

Formula 3 is often written as

$$\boxed{4} \qquad A = \int_a^b \tfrac{1}{2} r^2\, d\theta$$

with the understanding that $r = f(\theta)$. Note the similarity between Formulas 1 and 4.

When we apply Formula 3 or 4 it is helpful to think of the area as being swept out by a rotating ray through O that starts with angle a and ends with angle b.

EXAMPLE 1 Find the area enclosed by one loop of the four-leaved rose $r = \cos 2\theta$.

SOLUTION The curve $r = \cos 2\theta$ was sketched in Example 8 in Section 11.3. Notice from Figure 4 that the region enclosed by the right loop is swept out by a ray that rotates from $\theta = -\pi/4$ to $\theta = \pi/4$. Therefore, Formula 4 gives

$$A = \int_{-\pi/4}^{\pi/4} \tfrac{1}{2} r^2\, d\theta = \tfrac{1}{2} \int_{-\pi/4}^{\pi/4} \cos^2 2\theta\, d\theta$$

$$= \int_0^{\pi/4} \cos^2 2\theta\, d\theta = \int_0^{\pi/4} \tfrac{1}{2}(1 + \cos 4\theta)\, d\theta$$

$$= \tfrac{1}{2}\left[\theta + \tfrac{1}{4}\sin 4\theta\right]_0^{\pi/4} = \frac{\pi}{8}$$

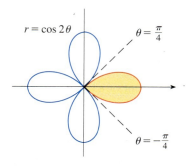

FIGURE 4

EXAMPLE 2 Find the area of the region that lies inside the circle $r = 3 \sin \theta$ and outside the cardioid $r = 1 + \sin \theta$.

SOLUTION The cardioid (see Example 7 in Section 11.3) and the circle are sketched in Figure 5 and the desired region is shaded. The values of a and b in Formula 4 are determined by finding the points of intersection of the two curves. They intersect when $3 \sin \theta = 1 + \sin \theta$, which gives $\sin \theta = \frac{1}{2}$, so $\theta = \pi/6, 5\pi/6$. The desired area can be found by subtracting the area inside the cardioid between $\theta = \pi/6$ and $\theta = 5\pi/6$ from the area inside the circle from $\pi/6$ to $5\pi/6$. Thus

$$A = \tfrac{1}{2} \int_{\pi/6}^{5\pi/6} (3 \sin \theta)^2 \, d\theta - \tfrac{1}{2} \int_{\pi/6}^{5\pi/6} (1 + \sin \theta)^2 \, d\theta$$

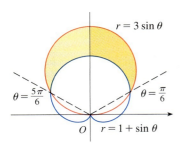

FIGURE 5

Since the region is symmetric about the vertical axis $\theta = \pi/2$, we can write

$$A = 2 \left[\tfrac{1}{2} \int_{\pi/6}^{\pi/2} 9 \sin^2 \theta \, d\theta - \tfrac{1}{2} \int_{\pi/6}^{\pi/2} (1 + 2 \sin \theta + \sin^2 \theta) \, d\theta \right]$$

$$= \int_{\pi/6}^{\pi/2} (8 \sin^2 \theta - 1 - 2 \sin \theta) \, d\theta$$

$$= \int_{\pi/6}^{\pi/2} (3 - 4 \cos 2\theta - 2 \sin \theta) \, d\theta \qquad [\text{because } \sin^2 \theta = \tfrac{1}{2}(1 - \cos 2\theta)]$$

$$= 3\theta - 2 \sin 2\theta + 2 \cos \theta \Big]_{\pi/6}^{\pi/2} = \pi$$

Example 2 illustrates the procedure for finding the area of the region bounded by two polar curves. In general, let $\mathcal{R}$ be a region, as illustrated in Figure 6, that is bounded by curves with polar equations $r = f(\theta)$, $r = g(\theta)$, $\theta = a$, and $\theta = b$, where $f(\theta) \geq g(\theta) \geq 0$ and $0 < b - a \leq 2\pi$. The area A of $\mathcal{R}$ is found by subtracting the area inside $r = g(\theta)$ from the area inside $r = f(\theta)$, so using Formula 3 we have

$$A = \int_a^b \tfrac{1}{2} [f(\theta)]^2 \, d\theta - \int_a^b \tfrac{1}{2} [g(\theta)]^2 \, d\theta$$

$$= \tfrac{1}{2} \int_a^b ([f(\theta)]^2 - [g(\theta)]^2) \, d\theta$$

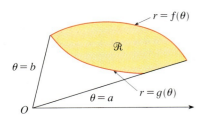

FIGURE 6

⊘ **CAUTION** ▫ The fact that a single point has many representations in polar coordinates sometimes makes it difficult to find all the points of intersection of two polar curves. For instance, it is obvious from Figure 5 that the circle and the cardioid have three points of intersection; however, in Example 2 we solved the equations $r = 3 \sin \theta$ and $r = 1 + \sin \theta$ and found only two such points, $(\tfrac{3}{2}, \pi/6)$ and $(\tfrac{3}{2}, 5\pi/6)$. The origin is also a point of intersection, but we can't find it by solving the equations of the curves because the origin has no single representation in polar coordinates that satisfies both equations. Notice that, when represented as $(0, 0)$ or $(0, \pi)$, the origin satisfies $r = 3 \sin \theta$ and so it lies on the circle; when represented as $(0, 3\pi/2)$, it satisfies $r = 1 + \sin \theta$ and so it lies on the cardioid. Think of two points moving along the curves as the parameter value θ increases from 0 to 2π. On one curve the origin is reached at $\theta = 0$ and $\theta = \pi$; on the other curve it is reached at $\theta = 3\pi/2$. The points don't collide at the origin because they reach the origin at different times, but the curves intersect there nonetheless.

Thus, to find *all* points of intersection of two polar curves, it is recommended that you draw the graphs of both curves. It is especially convenient to use a graphing calculator or computer to help with this task.

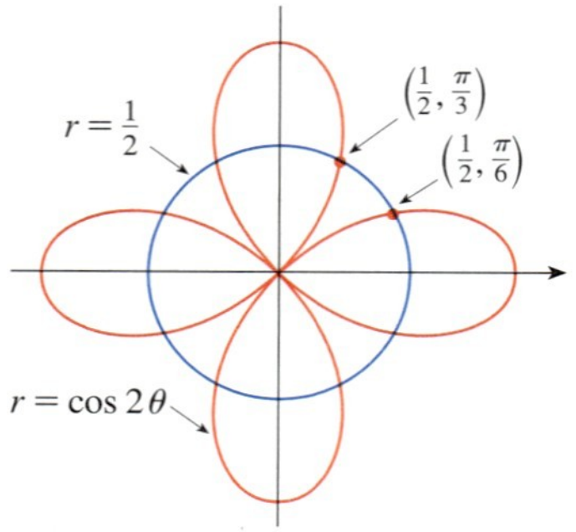

EXAMPLE 3 Find all points of intersection of the curves $r = \cos 2\theta$ and $r = \frac{1}{2}$.

SOLUTION If we solve the equations $r = \cos 2\theta$ and $r = \frac{1}{2}$, we get $\cos 2\theta = \frac{1}{2}$ and, therefore, $2\theta = \pi/3, 5\pi/3, 7\pi/3, 11\pi/3$. Thus, the values of θ between 0 and 2π that satisfy both equations are $\theta = \pi/6, 5\pi/6, 7\pi/6, 11\pi/6$. We have found four points of intersection: $(\frac{1}{2}, \pi/6), (\frac{1}{2}, 5\pi/6), (\frac{1}{2}, 7\pi/6)$, and $(\frac{1}{2}, 11\pi/6)$.

However, you can see from Figure 7 that the curves have four other points of intersection—namely, $(\frac{1}{2}, \pi/3), (\frac{1}{2}, 2\pi/3), (\frac{1}{2}, 4\pi/3)$, and $(\frac{1}{2}, 5\pi/3)$. These can be found using symmetry or by noticing that another equation of the circle is $r = -\frac{1}{2}$ and then solving the equations $r = \cos 2\theta$ and $r = -\frac{1}{2}$.

FIGURE 7

Arc Length

To find the length of a polar curve $r = f(\theta)$, $a \leq \theta \leq b$, we regard θ as a parameter and write the parametric equations of the curve as

$$x = r \cos \theta = f(\theta) \cos \theta \qquad y = r \sin \theta = f(\theta) \sin \theta$$

Using the Product Rule and differentiating with respect to θ, we obtain

$$\frac{dx}{d\theta} = \frac{dr}{d\theta} \cos \theta - r \sin \theta \qquad \frac{dy}{d\theta} = \frac{dr}{d\theta} \sin \theta + r \cos \theta$$

so, using $\cos^2 \theta + \sin^2 \theta = 1$, we have

$$\left(\frac{dx}{d\theta}\right)^2 + \left(\frac{dy}{d\theta}\right)^2 = \left(\frac{dr}{d\theta}\right)^2 \cos^2 \theta - 2r \frac{dr}{d\theta} \cos \theta \sin \theta + r^2 \sin^2 \theta$$
$$+ \left(\frac{dr}{d\theta}\right)^2 \sin^2 \theta + 2r \frac{dr}{d\theta} \sin \theta \cos \theta + r^2 \cos^2 \theta$$
$$= \left(\frac{dr}{d\theta}\right)^2 + r^2$$

Assuming that f' is continuous, we can use Theorem 11.2.6 to write the arc length as

$$L = \int_a^b \sqrt{\left(\frac{dx}{d\theta}\right)^2 + \left(\frac{dy}{d\theta}\right)^2} \, d\theta$$

Therefore, the length of a curve with polar equation $r = f(\theta)$, $a \leq \theta \leq b$, is

[5]
$$L = \int_a^b \sqrt{r^2 + \left(\frac{dr}{d\theta}\right)^2} \, d\theta$$

EXAMPLE 4 Find the length of the cardioid $r = 1 + \sin \theta$.

SOLUTION The cardioid is shown in Figure 8. (We sketched it in Example 7 in Section 11.3.) Its full length is given by the parameter interval $0 \leq \theta \leq 2\pi$, so Formula 5 gives

$$L = \int_0^{2\pi} \sqrt{r^2 + \left(\frac{dr}{d\theta}\right)^2} \, d\theta = \int_0^{2\pi} \sqrt{(1 + \sin \theta)^2 + \cos^2 \theta} \, d\theta$$
$$= \int_0^{2\pi} \sqrt{2 + 2 \sin \theta} \, d\theta$$

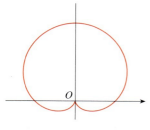

FIGURE 8
$r = 1 + \sin \theta$

11.4 Exercises

1–4 ⅢⅠ Find the area of the region that is bounded by the given curve and lies in the specified sector.

1. $r = \sqrt{\theta}$, $0 \leq \theta \leq \pi/4$
2. $r = e^{\theta/2}$, $\pi \leq \theta \leq 2\pi$
3. $r = \sin \theta$, $\pi/3 \leq \theta \leq 2\pi/3$
4. $r = \sqrt{\sin \theta}$, $0 \leq \theta \leq \pi$

5–8 ⅢⅠ Find the area of the shaded region.

5.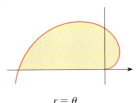
$r = \theta$

6.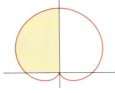
$r = 1 + \sin \theta$

7.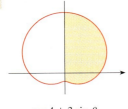
$r = 4 + 3 \sin \theta$

8.
$r = \sin 4\theta$

9–14 ⅢⅠ Sketch the curve and find the area that it encloses.

9. $r = 3 \cos \theta$
10. $r = 3(1 + \cos \theta)$
11. $r^2 = 4 \cos 2\theta$
12. $r^2 = \sin 2\theta$
13. $r = 2 \cos 3\theta$
14. $r = 2 + \cos 2\theta$

15–16 ⅢⅠ Graph the curve and find the area that it encloses.

15. $r = 1 + 2 \sin 6\theta$
16. $r = 2 \sin \theta + 3 \sin 9\theta$

17–21 ⅢⅠ Find the area of the region enclosed by one loop of the curve.

17. $r = \sin 2\theta$
18. $r = 4 \sin 3\theta$
19. $r = 3 \cos 5\theta$
20. $r = 2 \cos 4\theta$
21. $r = 1 + 2 \sin \theta$ (inner loop)

22. Find the area enclosed by the loop of the **strophoid** $r = 2 \cos \theta - \sec \theta$.

23–28 ⅢⅠ Find the area of the region that lies inside the first curve and outside the second curve.

23. $r = 4 \sin \theta$, $r = 2$
24. $r = 1 - \sin \theta$, $r = 1$
25. $r^2 = 8 \cos 2\theta$, $r = 2$
26. $r = 2 + \sin \theta$, $r = 3 \sin \theta$
27. $r = 3 \cos \theta$, $r = 1 + \cos \theta$
28. $r = 1 + \cos \theta$, $r = 3 \cos \theta$

29–34 ⅢⅠ Find the area of the region that lies inside both curves.

29. $r = \sin \theta$, $r = \cos \theta$
30. $r = \sin 2\theta$, $r = \sin \theta$
31. $r = \sin 2\theta$, $r = \cos 2\theta$
32. $r^2 = 2 \sin 2\theta$, $r = 1$
33. $r = 3 + 2 \sin \theta$, $r = 2$
34. $r = a \sin \theta$, $r = b \cos \theta$, $a > 0$, $b > 0$

35. Find the area inside the larger loop and outside the smaller loop of the limaçon $r = \tfrac{1}{2} + \cos \theta$.

36. Find the area between a large loop and the enclosed small loop of the curve $r = 1 + 2 \cos 3\theta$.

37–42 ⅢⅠ Find all points of intersection of the given curves.

37. $r = \sin \theta$, $r = \cos \theta$
38. $r = 2$, $r = 2 \cos 2\theta$
39. $r = \cos \theta$, $r = 1 - \cos \theta$
40. $r = \cos 3\theta$, $r = \sin 3\theta$
41. $r = \sin \theta$, $r = \sin 2\theta$
42. $r^2 = \sin 2\theta$, $r^2 = \cos 2\theta$

43. The points of intersection of the cardioid $r = 1 + \sin \theta$ and the spiral loop $r = 2\theta$, $-\pi/2 \leq \theta \leq \pi/2$, can't be found exactly. Use a graphing device to find the approximate values of θ at which they intersect. Then use these values to estimate the area that lies inside both curves.

44. Use a graph to estimate the values of θ for which the curves $r = 3 + \sin 5\theta$ and $r = 6 \sin \theta$ intersect. Then estimate the area that lies inside both curves.

45–48 ▮▮▮▮ Find the exact length of the polar curve.

45. $r = 3 \sin \theta$, $0 \leq \theta \leq \pi/3$ **46.** $r = e^{2\theta}$, $0 \leq \theta \leq 2\pi$

47. $r = \theta^2$, $0 \leq \theta \leq 2\pi$ **48.** $r = \theta$, $0 \leq \theta \leq 2\pi$

49–52 ▮▮▮▮ Use a calculator to find the length of the curve correct to four decimal places.

49. $r = 3 \sin 2\theta$ **50.** $r = 4 \sin 3\theta$

51. $r = \sin(\theta/2)$ **52.** $r = 1 + \cos(\theta/3)$

53–54 ▮▮▮▮ Graph the curve and find its length.

53. $r = \cos^4(\theta/4)$ **54.** $r = \cos^2(\theta/2)$

55. (a) Use Formula 11.2.7 to show that the area of the surface generated by rotating the polar curve
$$r = f(\theta) \qquad a \leq \theta \leq b$$
(where f' is continuous and $0 \leq a < b \leq \pi$) about the polar axis is
$$S = \int_a^b 2\pi r \sin \theta \sqrt{r^2 + \left(\frac{dr}{d\theta}\right)^2}\, d\theta$$
(b) Use the formula in part (a) to find the surface area generated by rotating the lemniscate $r^2 = \cos 2\theta$ about the polar axis.

56. (a) Find a formula for the area of the surface generated by rotating the polar curve $r = f(\theta)$, $a \leq \theta \leq b$ (where f' is continuous and $0 \leq a < b \leq \pi$), about the line $\theta = \pi/2$.
(b) Find the surface area generated by rotating the lemniscate $r^2 = \cos 2\theta$ about the line $\theta = \pi/2$.

11.5 Conic Sections

In this section we give geometric definitions of parabolas, ellipses, and hyperbolas and derive their standard equations. They are called **conic sections**, or **conics**, because they result from intersecting a cone with a plane as shown in Figure 1.

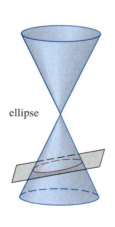
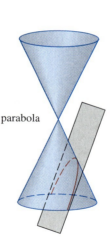
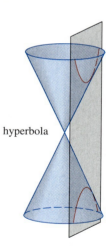

FIGURE 1
Conics

Parabolas

A **parabola** is the set of points in a plane that are equidistant from a fixed point F (called the **focus**) and a fixed line (called the **directrix**). This definition is illustrated by Figure 2. Notice that the point halfway between the focus and the directrix lies on the parabola; it is called the **vertex**. The line through the focus perpendicular to the directrix is called the **axis** of the parabola.

In the 16th century Galileo showed that the path of a projectile that is shot into the air at an angle to the ground is a parabola. Since then, parabolic shapes have been used in designing automobile headlights, reflecting telescopes, and suspension bridges.

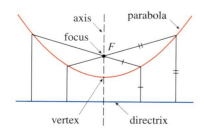

FIGURE 2

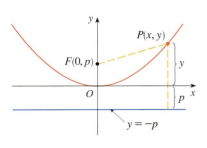

FIGURE 3

(See Problem 14 on page 220 for the reflection property of parabolas that makes them so useful.)

We obtain a particularly simple equation for a parabola if we place its vertex at the origin O and its directrix parallel to the x-axis as in Figure 3. If the focus is the point $(0, p)$, then the directrix has the equation $y = -p$. If $P(x, y)$ is any point on the parabola, then the distance from P to the focus is

$$|PF| = \sqrt{x^2 + (y - p)^2}$$

and the distance from P to the directrix is $|y + p|$. (Figure 3 illustrates the case where $p > 0$.) The defining property of a parabola is that these distances are equal:

$$\sqrt{x^2 + (y - p)^2} = |y + p|$$

We get an equivalent equation by squaring and simplifying:

$$x^2 + (y - p)^2 = |y + p|^2 = (y + p)^2$$
$$x^2 + y^2 - 2py + p^2 = y^2 + 2py + p^2$$
$$x^2 = 4py$$

1 An equation of the parabola with focus $(0, p)$ and directrix $y = -p$ is

$$x^2 = 4py$$

If we write $a = 1/(4p)$, then the standard equation of a parabola (1) becomes $y = ax^2$. It opens upward if $p > 0$ and downward if $p < 0$ [see Figure 4, parts (a) and (b)]. The graph is symmetric with respect to the y-axis because (1) is unchanged when x is replaced by $-x$.

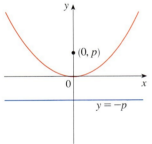
(a) $x^2 = 4py, p > 0$

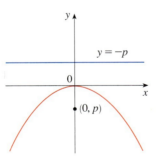
(b) $x^2 = 4py, p < 0$

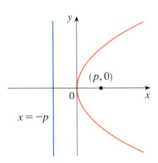
(c) $y^2 = 4px, p > 0$

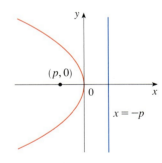
(d) $y^2 = 4px, p < 0$

FIGURE 4

If we interchange x and y in (1), we obtain

2
$$y^2 = 4px$$

which is an equation of the parabola with focus $(p, 0)$ and directrix $x = -p$. (Interchanging x and y amounts to reflecting about the diagonal line $y = x$.) The parabola opens to the right if $p > 0$ and to the left if $p < 0$ [see Figure 4, parts (c) and (d)]. In both cases the graph is symmetric with respect to the x-axis, which is the axis of the parabola.

EXAMPLE 1 Find the focus and directrix of the parabola $y^2 + 10x = 0$ and sketch the graph.

SOLUTION If we write the equation as $y^2 = -10x$ and compare it with Equation 2, we see that $4p = -10$, so $p = -\frac{5}{2}$. Thus, the focus is $(p, 0) = \left(-\frac{5}{2}, 0\right)$ and the directrix is $x = \frac{5}{2}$. The sketch is shown in Figure 5.

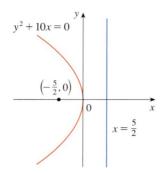

FIGURE 5

Ellipses

An **ellipse** is the set of points in a plane the sum of whose distances from two fixed points F_1 and F_2 is a constant (see Figure 6). These two fixed points are called the **foci** (plural of **focus**). One of Kepler's laws is that the orbits of the planets in the solar system are ellipses with the Sun at one focus.

In order to obtain the simplest equation for an ellipse, we place the foci on the x-axis at the points $(-c, 0)$ and $(c, 0)$ as in Figure 7 so that the origin is halfway between the foci. Let the sum of the distances from a point on the ellipse to the foci be $2a > 0$. Then $P(x, y)$ is a point on the ellipse when

$$|PF_1| + |PF_2| = 2a$$

that is,
$$\sqrt{(x+c)^2 + y^2} + \sqrt{(x-c)^2 + y^2} = 2a$$

or
$$\sqrt{(x-c)^2 + y^2} = 2a - \sqrt{(x+c)^2 + y^2}$$

FIGURE 6

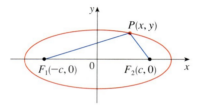

FIGURE 7

Squaring both sides, we have

$$x^2 - 2cx + c^2 + y^2 = 4a^2 - 4a\sqrt{(x+c)^2 + y^2} + x^2 + 2cx + c^2 + y^2$$

which simplifies to

$$a\sqrt{(x+c)^2 + y^2} = a^2 + cx$$

We square again:

$$a^2(x^2 + 2cx + c^2 + y^2) = a^4 + 2a^2cx + c^2x^2$$

which becomes
$$(a^2 - c^2)x^2 + a^2y^2 = a^2(a^2 - c^2)$$

From triangle F_1F_2P in Figure 7 we see that $2c < 2a$, so $c < a$ and, therefore, $a^2 - c^2 > 0$. For convenience, let $b^2 = a^2 - c^2$. Then the equation of the ellipse becomes $b^2x^2 + a^2y^2 = a^2b^2$ or, if both sides are divided by a^2b^2,

$$\boxed{3} \qquad \frac{x^2}{a^2} + \frac{y^2}{b^2} = 1$$

Since $b^2 = a^2 - c^2 < a^2$, it follows that $b < a$. The x-intercepts are found by setting $y = 0$. Then $x^2/a^2 = 1$, or $x^2 = a^2$, so $x = \pm a$. The corresponding points $(a, 0)$ and $(-a, 0)$ are called the **vertices** of the ellipse and the line segment joining the vertices is called the **major axis**. To find the y-intercepts we set $x = 0$ and obtain $y^2 = b^2$, so $y = \pm b$. Equation 3 is unchanged if x is replaced by $-x$ or y is replaced by $-y$, so the ellipse is symmetric about both axes. Notice that if the foci coincide, then $c = 0$, so $a = b$ and the ellipse becomes a circle with radius $r = a = b$.

We summarize this discussion as follows (see also Figure 8).

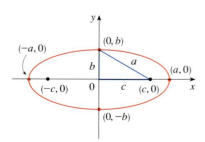

FIGURE 8
$\dfrac{x^2}{a^2} + \dfrac{y^2}{b^2} = 1$

4 The ellipse
$$\frac{x^2}{a^2} + \frac{y^2}{b^2} = 1 \qquad a \geq b > 0$$

has foci $(\pm c, 0)$, where $c^2 = a^2 - b^2$, and vertices $(\pm a, 0)$.

If the foci of an ellipse are located on the y-axis at $(0, \pm c)$, then we can find its equation by interchanging x and y in (4). (See Figure 9.)

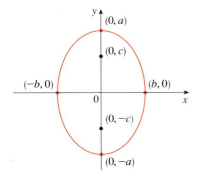

FIGURE 9
$\dfrac{x^2}{b^2} + \dfrac{y^2}{a^2} = 1$, $a \geq b$

5 The ellipse
$$\frac{x^2}{b^2} + \frac{y^2}{a^2} = 1 \qquad a \geq b > 0$$
has foci $(0, \pm c)$, where $c^2 = a^2 - b^2$, and vertices $(0, \pm a)$.

EXAMPLE 2 Sketch the graph of $9x^2 + 16y^2 = 144$ and locate the foci.

SOLUTION Divide both sides of the equation by 144:
$$\frac{x^2}{16} + \frac{y^2}{9} = 1$$

The equation is now in the standard form for an ellipse, so we have $a^2 = 16$, $b^2 = 9$, $a = 4$, and $b = 3$. The x-intercepts are ± 4 and the y-intercepts are ± 3. Also, $c^2 = a^2 - b^2 = 7$, so $c = \sqrt{7}$ and the foci are $(\pm\sqrt{7}, 0)$. The graph is sketched in Figure 10.

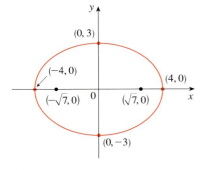

FIGURE 10
$9x^2 + 16y^2 = 144$

EXAMPLE 3 Find an equation of the ellipse with foci $(0, \pm 2)$ and vertices $(0, \pm 3)$.

SOLUTION Using the notation of (5), we have $c = 2$ and $a = 3$. Then we obtain $b^2 = a^2 - c^2 = 9 - 4 = 5$, so an equation of the ellipse is
$$\frac{x^2}{5} + \frac{y^2}{9} = 1$$

Another way of writing the equation is $9x^2 + 5y^2 = 45$.

Like parabolas, ellipses have an interesting reflection property that has practical consequences. If a source of light or sound is placed at one focus of a surface with elliptical cross-sections, then all the light or sound is reflected off the surface to the other focus (see Exercise 59). This principle is used in *lithotripsy*, a treatment for kidney stones. A reflector with elliptical cross-section is placed in such a way that the kidney stone is at one focus. High-intensity sound waves generated at the other focus are reflected to the stone and destroy it without damaging surrounding tissue. The patient is spared the trauma of surgery and recovers within a few days.

Hyperbolas

A **hyperbola** is the set of all points in a plane the difference of whose distances from two fixed points F_1 and F_2 (the foci) is a constant. This definition is illustrated in Figure 11.

Hyperbolas occur frequently as graphs of equations in chemistry, physics, biology, and economics (Boyle's Law, Ohm's Law, supply and demand curves). A particularly significant application of hyperbolas is found in the navigation systems developed in World Wars I and II (see Exercise 51).

Notice that the definition of a hyperbola is similar to that of an ellipse; the only change is that the sum of distances has become a difference of distances. In fact, the derivation of the equation of a hyperbola is also similar to the one given earlier for an ellipse. It is left

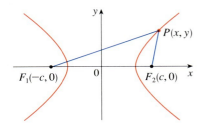

FIGURE 11
P is on the hyperbola when $|PF_1| - |PF_2| = \pm 2a$

as Exercise 52 to show that when the foci are on the x-axis at $(\pm c, 0)$ and the difference of distances is $|PF_1| - |PF_2| = \pm 2a$, then the equation of the hyperbola is

$$\boxed{6} \qquad \frac{x^2}{a^2} - \frac{y^2}{b^2} = 1$$

where $c^2 = a^2 + b^2$. Notice that the x-intercepts are again $\pm a$ and the points $(a, 0)$ and $(-a, 0)$ are the **vertices** of the hyperbola. But if we put $x = 0$ in Equation 6 we get $y^2 = -b^2$, which is impossible, so there is no y-intercept. The hyperbola is symmetric with respect to both axes.

To analyze the hyperbola further, we look at Equation 6 and obtain

$$\frac{x^2}{a^2} = 1 + \frac{y^2}{b^2} \geq 1$$

This shows that $x^2 \geq a^2$, so $|x| = \sqrt{x^2} \geq a$. Therefore, we have $x \geq a$ or $x \leq -a$. This means that the hyperbola consists of two parts, called its *branches*.

When we draw a hyperbola it is useful to first draw its **asymptotes**, which are the dashed lines $y = (b/a)x$ and $y = -(b/a)x$ shown in Figure 12. Both branches of the hyperbola approach the asymptotes; that is, they come arbitrarily close to the asymptotes. [See Exercise 55 in Section 4.5, where $y = (b/a)x$ is shown to be a slant asymptote.]

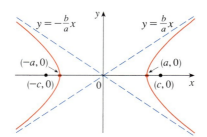

FIGURE 12
$\dfrac{x^2}{a^2} - \dfrac{y^2}{b^2} = 1$

$\boxed{7}$ The hyperbola

$$\frac{x^2}{a^2} - \frac{y^2}{b^2} = 1$$

has foci $(\pm c, 0)$, where $c^2 = a^2 + b^2$, vertices $(\pm a, 0)$, and asymptotes $y = \pm(b/a)x$.

If the foci of a hyperbola are on the y-axis, then by reversing the roles of x and y we obtain the following information, which is illustrated in Figure 13.

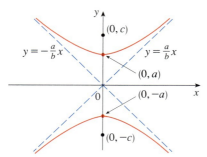

FIGURE 13
$\dfrac{y^2}{a^2} - \dfrac{x^2}{b^2} = 1$

$\boxed{8}$ The hyperbola

$$\frac{y^2}{a^2} - \frac{x^2}{b^2} = 1$$

has foci $(0, \pm c)$, where $c^2 = a^2 + b^2$, vertices $(0, \pm a)$, and asymptotes $y = \pm(a/b)x$.

EXAMPLE 4 Find the foci and asymptotes of the hyperbola $9x^2 - 16y^2 = 144$ and sketch its graph.

SOLUTION If we divide both sides of the equation by 144, it becomes

$$\frac{x^2}{16} - \frac{y^2}{9} = 1$$

which is of the form given in (7) with $a = 4$ and $b = 3$. Since $c^2 = 16 + 9 = 25$, the foci are $(\pm 5, 0)$. The asymptotes are the lines $y = \tfrac{3}{4}x$ and $y = -\tfrac{3}{4}x$. The graph is shown in Figure 14.

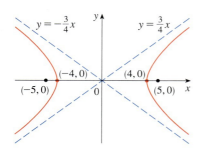

FIGURE 14
$9x^2 - 16y^2 = 144$

EXAMPLE 5 Find the foci and equation of the hyperbola with vertices $(0, \pm 1)$ and asymptote $y = 2x$.

SOLUTION From (8) and the given information, we see that $a = 1$ and $a/b = 2$. Thus, $b = a/2 = \frac{1}{2}$ and $c^2 = a^2 + b^2 = \frac{5}{4}$. The foci are $(0, \pm\sqrt{5}/2)$ and the equation of the hyperbola is

$$y^2 - 4x^2 = 1$$

Shifted Conics

As discussed in Appendix C, we shift conics by taking the standard equations (1), (2), (4), (5), (7), and (8) and replacing x and y by $x - h$ and $y - k$.

EXAMPLE 6 Find an equation of the ellipse with foci $(2, -2)$, $(4, -2)$ and vertices $(1, -2)$, $(5, -2)$.

SOLUTION The major axis is the line segment that joins the vertices $(1, -2)$, $(5, -2)$ and has length 4, so $a = 2$. The distance between the foci is 2, so $c = 1$. Thus, $b^2 = a^2 - c^2 = 3$. Since the center of the ellipse is $(3, -2)$, we replace x and y in (4) by $x - 3$ and $y + 2$ to obtain

$$\frac{(x-3)^2}{4} + \frac{(y+2)^2}{3} = 1$$

as the equation of the ellipse.

EXAMPLE 7 Sketch the conic

$$9x^2 - 4y^2 - 72x + 8y + 176 = 0$$

and find its foci.

SOLUTION We complete the squares as follows:

$$4(y^2 - 2y) - 9(x^2 - 8x) = 176$$

$$4(y^2 - 2y + 1) - 9(x^2 - 8x + 16) = 176 + 4 - 144$$

$$4(y - 1)^2 - 9(x - 4)^2 = 36$$

$$\frac{(y-1)^2}{9} - \frac{(x-4)^2}{4} = 1$$

This is in the form (8) except that x and y are replaced by $x - 4$ and $y - 1$. Thus, $a^2 = 9$, $b^2 = 4$, and $c^2 = 13$. The hyperbola is shifted four units to the right and one unit upward. The foci are $(4, 1 + \sqrt{13})$ and $(4, 1 - \sqrt{13})$ and the vertices are $(4, 4)$ and $(4, -2)$. The asymptotes are $y - 1 = \pm\frac{3}{2}(x - 4)$. The hyperbola is sketched in Figure 15.

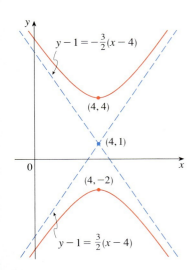

FIGURE 15
$9x^2 - 4y^2 - 72x + 8y + 176 = 0$

11.5 Exercises

1–8 Find the vertex, focus, and directrix of the parabola and sketch its graph.

1. $x = 2y^2$
2. $4y + x^2 = 0$
3. $4x^2 = -y$
4. $y^2 = 12x$
5. $(x + 2)^2 = 8(y - 3)$
6. $x - 1 = (y + 5)^2$
7. $y^2 + 2y + 12x + 25 = 0$
8. $y + 12x - 2x^2 = 16$

9–10 Find an equation of the parabola. Then find the focus and directrix.

9.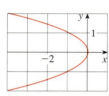
10.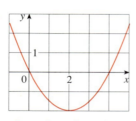

11–16 Find the vertices and foci of the ellipse and sketch its graph.

11. $\dfrac{x^2}{9} + \dfrac{y^2}{5} = 1$
12. $\dfrac{x^2}{64} + \dfrac{y^2}{100} = 1$
13. $4x^2 + y^2 = 16$
14. $4x^2 + 25y^2 = 25$
15. $9x^2 - 18x + 4y^2 = 27$
16. $x^2 + 2y^2 - 6x + 4y + 7 = 0$

17–18 Find an equation of the ellipse. Then find its foci.

17.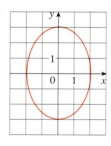
18.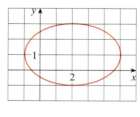

19–24 Find the vertices, foci, and asymptotes of the hyperbola and sketch its graph.

19. $\dfrac{x^2}{144} - \dfrac{y^2}{25} = 1$
20. $\dfrac{y^2}{16} - \dfrac{x^2}{36} = 1$
21. $y^2 - x^2 = 4$
22. $9x^2 - 4y^2 = 36$
23. $2y^2 - 3x^2 - 4y + 12x + 8 = 0$
24. $16x^2 - 9y^2 + 64x - 90y = 305$

25–30 Identify the type of conic section whose equation is given and find the vertices and foci.

25. $x^2 = y + 1$
26. $x^2 = y^2 + 1$
27. $x^2 = 4y - 2y^2$
28. $y^2 - 8y = 6x - 16$
29. $y^2 + 2y = 4x^2 + 3$
30. $4x^2 + 4x + y^2 = 0$

31–48 Find an equation for the conic that satisfies the given conditions.

31. Parabola, vertex $(0, 0)$, focus $(0, -2)$
32. Parabola, vertex $(1, 0)$, directrix $x = -5$
33. Parabola, focus $(-4, 0)$, directrix $x = 2$
34. Parabola, focus $(3, 6)$, vertex $(3, 2)$
35. Parabola, vertex $(0, 0)$, axis the x-axis, passing through $(1, -4)$
36. Parabola, vertical axis, passing through $(-2, 3)$, $(0, 3)$, and $(1, 9)$
37. Ellipse, foci $(\pm 2, 0)$, vertices $(\pm 5, 0)$
38. Ellipse, foci $(0, \pm 5)$, vertices $(0, \pm 13)$
39. Ellipse, foci $(0, 2), (0, 6)$ vertices $(0, 0), (0, 8)$
40. Ellipse, foci $(0, -1), (8, -1)$, vertex $(9, -1)$
41. Ellipse, center $(2, 2)$, focus $(0, 2)$, vertex $(5, 2)$
42. Ellipse, foci $(\pm 2, 0)$, passing through $(2, 1)$
43. Hyperbola, foci $(0, \pm 3)$, vertices $(0, \pm 1)$
44. Hyperbola, foci $(\pm 6, 0)$, vertices $(\pm 4, 0)$
45. Hyperbola, foci $(1, 3)$ and $(7, 3)$, vertices $(2, 3)$ and $(6, 3)$
46. Hyperbola, foci $(2, -2)$ and $(2, 8)$, vertices $(2, 0)$ and $(2, 6)$
47. Hyperbola, vertices $(\pm 3, 0)$, asymptotes $y = \pm 2x$
48. Hyperbola, foci $(2, 2)$ and $(6, 2)$, asymptotes $y = x - 2$ and $y = 6 - x$

49. The point in a lunar orbit nearest the surface of the moon is called *perilune* and the point farthest from the surface is called *apolune*. The Apollo 11 spacecraft was placed in an elliptical lunar orbit with perilune altitude 110 km and apolune altitude 314 km (above the moon). Find an equation of this ellipse if the radius of the moon is 1728 km and the center of the moon is at one focus.

50. A cross-section of a parabolic reflector is shown in the figure. The bulb is located at the focus and the opening at the focus is 10 cm.
(a) Find an equation of the parabola.
(b) Find the diameter of the opening $|CD|$, 11 cm from the vertex.

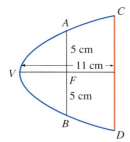

51. In the LORAN (LOng RAnge Navigation) radio navigation system, two radio stations located at A and B transmit simultaneous signals to a ship or an aircraft located at P. The onboard computer converts the time difference in receiving these signals into a distance difference $|PA| - |PB|$, and this, according to the definition of a hyperbola, locates the ship or aircraft on one branch of a hyperbola (see the figure). Suppose that station B is located 400 mi due east of station A on a coastline. A ship received the signal from B 1200 microseconds (μs) before it received the signal from A.
(a) Assuming that radio signals travel at a speed of 980 ft/μs, find an equation of the hyperbola on which the ship lies.
(b) If the ship is due north of B, how far off the coastline is the ship?

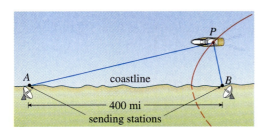

52. Use the definition of a hyperbola to derive Equation 6 for a hyperbola with foci $(\pm c, 0)$ and vertices $(\pm a, 0)$.

53. Show that the function defined by the upper branch of the hyperbola $y^2/a^2 - x^2/b^2 = 1$ is concave upward.

54. Find an equation for the ellipse with foci $(1, 1)$ and $(-1, -1)$ and major axis of length 4.

55. Determine the type of curve represented by the equation
$$\frac{x^2}{k} + \frac{y^2}{k-16} = 1$$
in each of the following cases: (a) $k > 16$, (b) $0 < k < 16$, and (c) $k < 0$.
(d) Show that all the curves in parts (a) and (b) have the same foci, no matter what the value of k is.

56. (a) Show that the equation of the tangent line to the parabola $y^2 = 4px$ at the point (x_0, y_0) can be written as
$$y_0 y = 2p(x + x_0)$$
(b) What is the x-intercept of this tangent line? Use this fact to draw the tangent line.

57. Use Simpson's Rule with $n = 10$ to estimate the length of the ellipse $x^2 + 4y^2 = 4$.

58. The planet Pluto travels in an elliptical orbit around the Sun (at one focus). The length of the major axis is 1.18×10^{10} km and the length of the minor axis is 1.14×10^{10} km. Use Simpson's Rule with $n = 10$ to estimate the distance traveled by the planet during one complete orbit around the Sun.

59. Let $P(x_1, y_1)$ be a point on the ellipse $x^2/a^2 + y^2/b^2 = 1$ with foci F_1 and F_2 and let α and β be the angles between the lines PF_1, PF_2 and the ellipse as in the figure. Prove that $\alpha = \beta$. This explains how whispering galleries and lithotripsy work. Sound coming from one focus is reflected and passes through the other focus. [*Hint:* Use the formula in Problem 13 on page 220 to show that $\tan \alpha = \tan \beta$.]

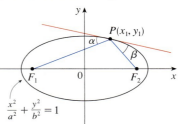

60. Let $P(x_1, y_1)$ be a point on the hyperbola $x^2/a^2 - y^2/b^2 = 1$ with foci F_1 and F_2 and let α and β be the angles between the lines PF_1, PF_2 and the hyperbola as shown in the figure. Prove that $\alpha = \beta$. (This is the reflection property of the hyperbola. It shows that light aimed at a focus F_2 of a hyperbolic mirror is reflected toward the other focus F_1.)

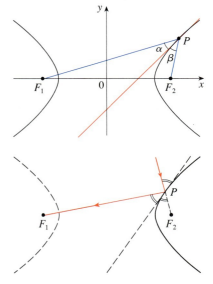

11.6 Conic Sections in Polar Coordinates

In the preceding section we defined the parabola in terms of a focus and directrix, but we defined the ellipse and hyperbola in terms of two foci. In this section we give a more unified treatment of all three types of conic sections in terms of a focus and directrix. Furthermore, if we place the focus at the origin, then a conic section has a simple polar equation. In Chapter 14 we will use the polar equation of an ellipse to derive Kepler's laws of planetary motion.

[1] Theorem Let F be a fixed point (called the **focus**) and l be a fixed line (called the **directrix**) in a plane. Let e be a fixed positive number (called the **eccentricity**). The set of all points P in the plane such that

$$\frac{|PF|}{|Pl|} = e$$

(The ratio of the distance from F to the distance from l is the constant e.)

is a conic section. The conic is

(a) an ellipse if $e < 1$

(b) a parabola if $e = 1$

(c) a hyperbola if $e > 1$

Proof Notice that if the eccentricity is $e = 1$, then $|PF| = |Pl|$ and so the given condition simply becomes the definition of a parabola as given in Section 11.5.

Let us place the focus F at the origin and the directrix parallel to the y-axis and d units to the right. Thus, the directrix has equation $x = d$ and is perpendicular to the polar axis. If the point P has polar coordinates (r, θ), we see from Figure 1 that

$$|PF| = r \qquad |Pl| = d - r\cos\theta$$

Thus, the condition $|PF|/|Pl| = e$, or $|PF| = e|Pl|$, becomes

$$[2] \qquad r = e(d - r\cos\theta)$$

If we square both sides of this polar equation and convert to rectangular coordinates, we get

$$x^2 + y^2 = e^2(d-x)^2 = e^2(d^2 - 2dx + x^2)$$

or

$$(1 - e^2)x^2 + 2de^2x + y^2 = e^2d^2$$

After completing the square, we have

$$[3] \qquad \left(x + \frac{e^2d}{1-e^2}\right)^2 + \frac{y^2}{1-e^2} = \frac{e^2d^2}{(1-e^2)^2}$$

If $e < 1$, we recognize Equation 3 as the equation of an ellipse. In fact, it is of the form

$$\frac{(x-h)^2}{a^2} + \frac{y^2}{b^2} = 1$$

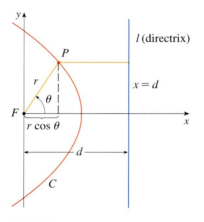

FIGURE 1

where

$$\boxed{4} \qquad h = -\frac{e^2 d}{1 - e^2} \qquad a^2 = \frac{e^2 d^2}{(1 - e^2)^2} \qquad b^2 = \frac{e^2 d^2}{1 - e^2}$$

In Section 11.5 we found that the foci of an ellipse are at a distance c from the center, where

$$\boxed{5} \qquad c^2 = a^2 - b^2 = \frac{e^4 d^2}{(1 - e^2)^2}$$

This shows that

$$c = \frac{e^2 d}{1 - e^2} = -h$$

and confirms that the focus as defined in Theorem 1 means the same as the focus defined in Section 11.5. It also follows from Equations 4 and 5 that the eccentricity is given by

$$e = \frac{c}{a}$$

If $e > 1$, then $1 - e^2 < 0$ and we see that Equation 3 represents a hyperbola. Just as we did before, we could rewrite Equation 3 in the form

$$\frac{(x - h)^2}{a^2} - \frac{y^2}{b^2} = 1$$

and see that

$$e = \frac{c}{a} \qquad \text{where} \qquad c^2 = a^2 + b^2$$

By solving Equation 2 for r, we see that the polar equation of the conic shown in Figure 1 can be written as

$$r = \frac{ed}{1 + e \cos \theta}$$

If the directrix is chosen to be to the left of the focus as $x = -d$, or if the directrix is chosen to be parallel to the polar axis as $y = \pm d$, then the polar equation of the conic is given by the following theorem, which is illustrated by Figure 2. (See Exercises 21–23.)

$\boxed{6}$ Theorem   A polar equation of the form

$$r = \frac{ed}{1 \pm e \cos \theta} \qquad \text{or} \qquad r = \frac{ed}{1 \pm e \sin \theta}$$

represents a conic section with eccentricity e. The conic is an ellipse if $e < 1$, a parabola if $e = 1$, or a hyperbola if $e > 1$.

EXAMPLE 1 Find a polar equation for a parabola that has its focus at the origin and whose directrix is the line $y = -6$.

SOLUTION Using Theorem 6 with $e = 1$ and $d = 6$, and using part (d) of Figure 2, we see that the equation of the parabola is

$$r = \frac{6}{1 - \sin \theta}$$

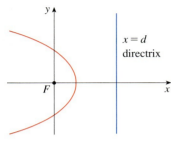

(a) $r = \dfrac{ed}{1 + e \cos \theta}$

(b) $r = \dfrac{ed}{1 - e \cos \theta}$

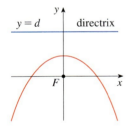

(c) $r = \dfrac{ed}{1 + e \sin \theta}$

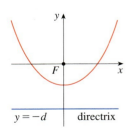

(d) $r = \dfrac{ed}{1 - e \sin \theta}$

FIGURE 2
Polar equations of conics

EXAMPLE 2 A conic is given by the polar equation

$$r = \frac{10}{3 - 2\cos\theta}$$

Find the eccentricity, identify the conic, locate the directrix, and sketch the conic.

SOLUTION Dividing numerator and denominator by 3, we write the equation as

$$r = \frac{\frac{10}{3}}{1 - \frac{2}{3}\cos\theta}$$

From Theorem 6 we see that this represents an ellipse with $e = \frac{2}{3}$. Since $ed = \frac{10}{3}$, we have

$$d = \frac{\frac{10}{3}}{e} = \frac{\frac{10}{3}}{\frac{2}{3}} = 5$$

so the directrix has Cartesian equation $x = -5$. When $\theta = 0$, $r = 10$; when $\theta = \pi$, $r = 2$. So the vertices have polar coordinates $(10, 0)$ and $(2, \pi)$. The ellipse is sketched in Figure 3.

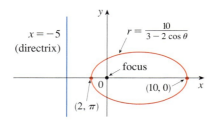

FIGURE 3

EXAMPLE 3 Sketch the conic $r = \dfrac{12}{2 + 4\sin\theta}$.

SOLUTION Writing the equation in the form

$$r = \frac{6}{1 + 2\sin\theta}$$

we see that the eccentricity is $e = 2$ and the equation therefore represents a hyperbola. Since $ed = 6$, $d = 3$ and the directrix has equation $y = 3$. The vertices occur when $\theta = \pi/2$ and $3\pi/2$, so they are $(2, \pi/2)$ and $(-6, 3\pi/2) = (6, \pi/2)$. It is also useful to plot the x-intercepts. These occur when $\theta = 0, \pi$; in both cases $r = 6$. For additional accuracy we could draw the asymptotes. Note that $r \to \pm\infty$ when $2 + 4\sin\theta \to 0^+$ or 0^- and $2 + 4\sin\theta = 0$ when $\sin\theta = -\frac{1}{2}$. Thus, the asymptotes are parallel to the rays $\theta = 7\pi/6$ and $\theta = 11\pi/6$. The hyperbola is sketched in Figure 4.

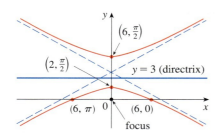

FIGURE 4
$r = \dfrac{12}{2 + 4\sin\theta}$

When rotating conic sections, we find it much more convenient to use polar equations than Cartesian equations. We just use the fact (see Exercise 75 in Section 11.3) that the graph of $r = f(\theta - \alpha)$ is the graph of $r = f(\theta)$ rotated counterclockwise about the origin through an angle α.

EXAMPLE 4 If the ellipse of Example 2 is rotated through an angle $\pi/4$ about the origin, find a polar equation and graph the resulting ellipse.

SOLUTION We get the equation of the rotated ellipse by replacing θ with $\theta - \pi/4$ in the equation given in Example 2. So the new equation is

$$r = \frac{10}{3 - 2\cos(\theta - \pi/4)}$$

We use this equation to graph the rotated ellipse in Figure 5. Notice that the ellipse has been rotated about its left focus.

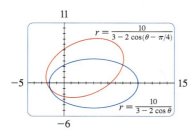

FIGURE 5

In Figure 6 we use a computer to sketch a number of conics to demonstrate the effect of varying the eccentricity e. Notice that when e is close to 0 the ellipse is nearly circular, whereas it becomes more elongated as $e \to 1^-$. When $e = 1$, of course, the conic is a parabola.

$e = 0.1$

$e = 0.5$

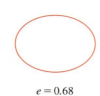
$e = 0.68$

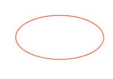
$e = 0.86$

$e = 0.96$

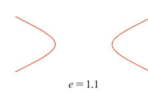

$e = 1$ $e = 1.1$

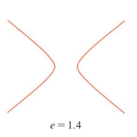

$e = 1.4$

$e = 4$

FIGURE 6

11.6 Exercises

1–8 Write a polar equation of a conic with the focus at the origin and the given data.

1. Hyperbola, eccentricity $\tfrac{7}{4}$, directrix $y = 6$
2. Parabola, directrix $x = 4$
3. Ellipse, eccentricity $\tfrac{3}{4}$, directrix $x = -5$
4. Hyperbola, eccentricity 2, directrix $y = -2$
5. Parabola, vertex $(4, 3\pi/2)$
6. Ellipse, eccentricity 0.8, vertex $(1, \pi/2)$
7. Ellipse, eccentricity $\tfrac{1}{2}$, directrix $r = 4 \sec\theta$
8. Hyperbola, eccentricity 3, directrix $r = -6 \csc\theta$

9–16 (a) Find the eccentricity, (b) identify the conic, (c) give an equation of the directrix, and (d) sketch the conic.

9. $r = \dfrac{1}{1 + \sin\theta}$
10. $r = \dfrac{6}{3 + 2\sin\theta}$
11. $r = \dfrac{12}{4 - \sin\theta}$
12. $r = \dfrac{4}{2 - 3\cos\theta}$
13. $r = \dfrac{9}{6 + 2\cos\theta}$
14. $r = \dfrac{5}{2 - 2\sin\theta}$
15. $r = \dfrac{3}{4 - 8\cos\theta}$
16. $r = \dfrac{4}{2 + \cos\theta}$

17. (a) Find the eccentricity and directrix of the conic $r = 1/(4 - 3\cos\theta)$ and graph the conic and its directrix.
    (b) If this conic is rotated counterclockwise about the origin through an angle $\pi/3$, write the resulting equation and graph its curve.

18. Graph the parabola $r = 5/(2 + 2\sin\theta)$ and its directrix. Also graph the curve obtained by rotating this parabola about its focus through an angle $\pi/6$.

19. Graph the conics $r = e/(1 - e\cos\theta)$ with $e = 0.4, 0.6, 0.8$, and 1.0 on a common screen. How does the value of e affect the shape of the curve?

20. (a) Graph the conics $r = ed/(1 + e\sin\theta)$ for $e = 1$ and various values of d. How does the value of d affect the shape of the conic?
    (b) Graph these conics for $d = 1$ and various values of e. How does the value of e affect the shape of the conic?

21. Show that a conic with focus at the origin, eccentricity e, and directrix $x = -d$ has polar equation
$$r = \frac{ed}{1 - e\cos\theta}$$

22. Show that a conic with focus at the origin, eccentricity e, and directrix $y = d$ has polar equation
$$r = \frac{ed}{1 + e\sin\theta}$$

23. Show that a conic with focus at the origin, eccentricity e, and directrix $y = -d$ has polar equation
$$r = \frac{ed}{1 - e \sin \theta}$$

24. Show that the parabolas $r = c/(1 + \cos \theta)$ and $r = d/(1 - \cos \theta)$ intersect at right angles.

25. (a) Show that the polar equation of an ellipse with directrix $x = -d$ can be written in the form
$$r = \frac{a(1 - e^2)}{1 - e \cos \theta}$$
(b) Find an approximate polar equation for the elliptical orbit of Earth around the Sun (at one focus) given that the eccentricity is about 0.017 and the length of the major axis is about 2.99×10^8 km.

26. (a) The planets move around the Sun in elliptical orbits with the Sun at one focus. The positions of a planet that are closest to and farthest from the Sun are called its *perihelion* and

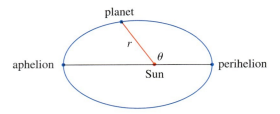

aphelion, respectively. Use Exercise 25(a) to show that the perihelion distance from a planet to the Sun is $a(1 - e)$ and the aphelion distance is $a(1 + e)$.
(b) Use the data of Exercise 25(b) to find the distances from Earth to the Sun at perihelion and at aphelion.

27. The orbit of Halley's comet, last seen in 1986 and due to return in 2062, is an ellipse with eccentricity 0.97 and one focus at the Sun. The length of its major axis is 36.18 AU. [An astronomical unit (AU) is the mean distance between Earth and the Sun, about 93 million miles.] Find a polar equation for the orbit of Halley's comet. What is the maximum distance from the comet to the Sun?

28. The Hale-Bopp comet, discovered in 1995, has an elliptical orbit with eccentricity 0.9951 and the length of the major axis is 356.5 AU. Find a polar equation for the orbit of this comet. How close to the Sun does it come?

29. The planet Mercury travels in an elliptical orbit with eccentricity 0.206. Its minimum distance from the Sun is 4.6×10^7 km. Use the results of Exercise 26(a) to find its maximum distance from the Sun.

30. The distance from the planet Pluto to the Sun is 4.43×10^9 km at perihelion and 7.37×10^9 km at aphelion. Use Exercise 26 to find the eccentricity of Pluto's orbit.

31. Using the data from Exercise 29, find the distance traveled by the planet Mercury during one complete orbit around the Sun. (If your calculator or computer algebra system evaluates definite integrals, use it. Otherwise, use Simpson's Rule.)

11 Review

CONCEPT CHECK

1. (a) What is a parametric curve?
(b) How do you sketch a parametric curve?

2. (a) How do you find the slope of a tangent to a parametric curve?
(b) How do you find the area under a parametric curve?

3. Write an expression for each of the following:
(a) The length of a parametric curve
(b) The area of the surface obtained by rotating a parametric curve about the x-axis

4. (a) Use a diagram to explain the meaning of the polar coordinates (r, θ) of a point.
(b) Write equations that express the Cartesian coordinates (x, y) of a point in terms of the polar coordinates.
(c) What equations would you use to find the polar coordinates of a point if you knew the Cartesian coordinates?

5. (a) How do you find the slope of a tangent line to a polar curve?
(b) How do you find the area of a region bounded by a polar curve?

(c) How do you find the length of a polar curve?

6. (a) Give a geometric definition of a parabola.
(b) Write an equation of a parabola with focus $(0, p)$ and directrix $y = -p$. What if the focus is $(p, 0)$ and the directrix is $x = -p$?

7. (a) Give a definition of an ellipse in terms of foci.
(b) Write an equation for the ellipse with foci $(\pm c, 0)$ and vertices $(\pm a, 0)$.

8. (a) Give a definition of a hyperbola in terms of foci.
(b) Write an equation for the hyperbola with foci $(\pm c, 0)$ and vertices $(\pm a, 0)$.
(c) Write equations for the asymptotes of the hyperbola in part (b).

9. (a) What is the eccentricity of a conic section?
(b) What can you say about the eccentricity if the conic section is an ellipse? A hyperbola? A parabola?
(c) Write a polar equation for a conic section with eccentricity e and directrix $x = d$. What if the directrix is $x = -d$? $y = d$? $y = -d$?

TRUE-FALSE QUIZ

Determine whether the statement is true or false. If it is true, explain why. If it is false, explain why or give an example that disproves the statement.

1. If the parametric curve $x = f(t)$, $y = g(t)$ satisfies $g'(1) = 0$, then it has a horizontal tangent when $t = 1$.

2. If $x = f(t)$ and $y = g(t)$ are twice differentiable, then $d^2y/dx^2 = (d^2y/dt^2)/(d^2x/dt^2)$.

3. The length of the curve $x = f(t)$, $y = g(t)$, $a \le t \le b$, is $\int_a^b \sqrt{[f'(t)]^2 + [g'(t)]^2}\, dt$.

4. If a point is represented by (x, y) in Cartesian coordinates (where $x \ne 0$) and (r, θ) in polar coordinates, then $\theta = \tan^{-1}(y/x)$.

5. The polar curves $r = 1 - \sin 2\theta$ and $r = \sin 2\theta - 1$ have the same graph.

6. The equations $r = 2$, $x^2 + y^2 = 4$, and $x = 2 \sin 3t$, $y = 2 \cos 3t$ $(0 \le t \le 2\pi)$ all have the same graph.

7. The parametric equations $x = t^2$, $y = t^4$ have the same graph as $x = t^3$, $y = t^6$.

8. The graph of $y^2 = 2y + 3x$ is a parabola.

9. A tangent line to a parabola intersects the parabola only once.

10. A hyperbola never intersects its directrix.

EXERCISES

1–4 Sketch the parametric curve and eliminate the parameter to find the Cartesian equation of the curve.

1. $x = t^2 + 4t$, $y = 2 - t$, $-4 \le t \le 1$
2. $x = 1 + e^{2t}$, $y = e^t$
3. $x = \tan\theta$, $y = \cot\theta$
4. $x = 2\cos\theta$, $y = 1 + \sin\theta$

5. Write three different sets of parametric equations for the curve $y = \sqrt{x}$.

6. Use the graphs of $x = f(t)$ and $y = g(t)$ to sketch the parametric curve $x = f(t)$, $y = g(t)$. Indicate with arrows the direction in which the curve is traced as t increases.

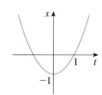
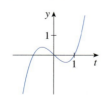

7–14 Sketch the polar curve.

7. $r = 1 - \cos\theta$
8. $r = \sin 4\theta$
9. $r = 1 + \cos 2\theta$
10. $r = 3 + \cos 3\theta$
11. $r^2 = \sec 2\theta$
12. $r = 2\cos(\theta/2)$
13. $r = \dfrac{1}{1 + \cos\theta}$
14. $r = \dfrac{8}{4 + 3\sin\theta}$

15–16 Find a polar equation for the curve represented by the given Cartesian equation.

15. $x + y = 2$
16. $x^2 + y^2 = 2$

17. The curve with polar equation $r = (\sin\theta)/\theta$ is called a **cochleoid**. Use a graph of r as a function of θ in Cartesian coordinates to sketch the cochleoid by hand. Then graph it with a machine to check your sketch.

18. Graph the ellipse $r = 2/(4 - 3\cos\theta)$ and its directrix. Also graph the ellipse obtained by rotation about the origin through an angle $2\pi/3$.

19–22 Find the slope of the tangent line to the given curve at the point corresponding to the specified value of the parameter.

19. $x = \ln t$, $y = 1 + t^2$; $t = 1$
20. $x = t^3 + 6t + 1$, $y = 2t - t^2$; $t = -1$
21. $r = e^{-\theta}$; $\theta = \pi$
22. $r = 3 + \cos 3\theta$; $\theta = \pi/2$

23–24 Find dy/dx and d^2y/dx^2.

23. $x = t\cos t$, $y = t\sin t$
24. $x = 1 + t^2$, $y = t - t^3$

25. Use a graph to estimate the coordinates of the lowest point on the curve $x = t^3 - 3t$, $y = t^2 + t + 1$. Then use calculus to find the exact coordinates.

26. Find the area enclosed by the loop of the curve in Exercise 25.

27. At what points does the curve

$$x = 2a \cos t - a \cos 2t \qquad y = 2a \sin t - a \sin 2t$$

have vertical or horizontal tangents? Use this information to help sketch the curve.

28. Find the area enclosed by the curve in Exercise 27.

29. Find the area enclosed by the curve $r^2 = 9 \cos 5\theta$.

30. Find the area enclosed by the inner loop of the curve $r = 1 - 3 \sin \theta$.

31. Find the points of intersection of the curves $r = 2$ and $r = 4 \cos \theta$.

32. Find the points of intersection of the curves $r = \cot \theta$ and $r = 2 \cos \theta$.

33. Find the area of the region that lies inside both of the circles $r = 2 \sin \theta$ and $r = \sin \theta + \cos \theta$.

34. Find the area of the region that lies inside the curve $r = 2 + \cos 2\theta$ but outside the curve $r = 2 + \sin \theta$.

35–38 |||| Find the length of the curve.

35. $x = 3t^2, \quad y = 2t^3, \quad 0 \le t \le 2$

36. $x = 2 + 3t, \quad y = \cosh 3t, \quad 0 \le t \le 1$

37. $r = 1/\theta, \quad \pi \le \theta \le 2\pi$

38. $r = \sin^3(\theta/3), \quad 0 \le \theta \le \pi$

▫ ▫ ▫ ▫ ▫ ▫ ▫ ▫ ▫ ▫ ▫

39–40 |||| Find the area of the surface obtained by rotating the given curve about the x-axis.

39. $x = 4\sqrt{t}, \quad y = \dfrac{t^3}{3} + \dfrac{1}{2t^2}, \quad 1 \le t \le 4$

40. $x = 2 + 3t, \quad y = \cosh 3t, \quad 0 \le t \le 1$

▫ ▫ ▫ ▫ ▫ ▫ ▫ ▫ ▫ ▫ ▫

41. The curves defined by the parametric equations

$$x = \dfrac{t^2 - c}{t^2 + 1} \qquad y = \dfrac{t(t^2 - c)}{t^2 + 1}$$

are called **strophoids** (from a Greek word meaning "to turn or twist"). Investigate how these curves vary as c varies.

42. A family of curves has polar equations $r^a = |\sin 2\theta|$ where a is a positive number. Investigate how the curves change as a changes.

43–46 |||| Find the foci and vertices and sketch the graph.

43. $\dfrac{x^2}{9} + \dfrac{y^2}{8} = 1$

44. $4x^2 - y^2 = 16$

45. $6y^2 + x - 36y + 55 = 0$

46. $25x^2 + 4y^2 + 50x - 16y = 59$

▫ ▫ ▫ ▫ ▫ ▫ ▫ ▫ ▫ ▫ ▫

47. Find an equation of the parabola with focus $(0, 6)$ and directrix $y = 2$.

48. Find an equation of the hyperbola with foci $(0, \pm 5)$ and vertices $(0, \pm 2)$.

49. Find an equation of the hyperbola with foci $(\pm 3, 0)$ and asymptotes $2y = \pm x$.

50. Find an equation of the ellipse with foci $(3, \pm 2)$ and major axis with length 8.

51. Find an equation for the ellipse that shares a vertex and a focus with the parabola $x^2 + y = 100$ and that has its other focus at the origin.

52. Show that if m is any real number, then there are exactly two lines of slope m that are tangent to the ellipse $x^2/a^2 + y^2/b^2 = 1$ and their equations are $y = mx \pm \sqrt{a^2 m^2 + b^2}$.

53. Find a polar equation for the ellipse with focus at the origin, eccentricity $\tfrac{1}{3}$, and directrix with equation $r = 4 \sec \theta$.

54. Show that the angles between the polar axis and the asymptotes of the hyperbola $r = ed/(1 - e \cos \theta), e > 1$, are given by $\cos^{-1}(\pm 1/e)$.

55. In the figure the circle of radius a is stationary, and for every θ, the point P is the midpoint of the segment QR. The curve traced out by P for $0 < \theta < \pi$ is called the **longbow curve**. Find parametric equations for this curve.

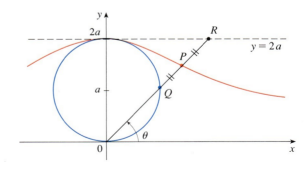

PROBLEMS PLUS

1. A curve is defined by the parametric equations

$$x = \int_1^t \frac{\cos u}{u}\, du \qquad y = \int_1^t \frac{\sin u}{u}\, du$$

   Find the length of the arc of the curve from the origin to the nearest point where there is a vertical tangent line.

2. (a) Find the highest and lowest points on the curve $x^4 + y^4 = x^2 + y^2$.
   (b) Sketch the curve. (Notice that it is symmetric with respect to both axes and both of the lines $y = \pm x$, so it suffices to consider $y \geq x \geq 0$ initially.)
   [CAS] (c) Use polar coordinates and a computer algebra system to find the area enclosed by the curve.

3. What is the smallest viewing rectangle that contains every member of the family of polar curves $r = 1 + c \sin \theta$, where $0 \leq c \leq 1$? Illustrate your answer by graphing several members of the family in this viewing rectangle.

4. Four bugs are placed at the four corners of a square with side length a. The bugs crawl counterclockwise at the same speed and each bug crawls directly toward the next bug at all times. They approach the center of the square along spiral paths.
   (a) Find the polar equation of a bug's path assuming the pole is at the center of the square. (Use the fact that the line joining one bug to the next is tangent to the bug's path.)
   (b) Find the distance traveled by a bug by the time it meets the other bugs at the center.

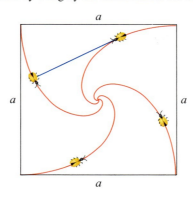

5. A curve called the **folium of Descartes** is defined by the parametric equations

$$x = \frac{3t}{1+t^3} \qquad y = \frac{3t^2}{1+t^3}$$

   (a) Show that if (a, b) lies on the curve, then so does (b, a); that is, the curve is symmetric with respect to the line $y = x$. Where does the curve intersect this line?
   (b) Find the points on the curve where the tangent lines are horizontal or vertical.
   (c) Show that the line $y = -x - 1$ is a slant asymptote.
   (d) Sketch the curve.
   (e) Show that a Cartesian equation of this curve is $x^3 + y^3 = 3xy$.
   (f) Show that the polar equation can be written in the form

$$r = \frac{3 \sec \theta \tan \theta}{1 + \tan^3 \theta}$$

   (g) Find the area enclosed by the loop of this curve.
   [CAS] (h) Show that the area of the loop is the same as the area that lies between the asymptote and the infinite branches of the curve. (Use a computer algebra system to evaluate the integral.)

Chapter 12

Bessel functions, which are used to model the vibrations of drumheads and cymbals, are defined as sums of infinite series in Section 12.8. Notice how closely the computer-generated models (which involve Bessel functions and cosine functions) match the photographs of a vibrating rubber membrane.

Infinite Sequences and Series

Infinite sequences and series were introduced briefly in *A Preview of Calculus* in connection with Zeno's paradoxes and the decimal representation of numbers. Their importance in calculus stems from Newton's idea of representing functions as sums of infinite series. For instance, in finding areas he often integrated a function by first expressing it as a series and then integrating each term of the series. We will pursue his idea in Section 12.10 in order to integrate such functions as e^{-x^2}. (Recall that we have previously been unable to do this.) Many of the functions that arise in mathematical physics and chemistry, such as Bessel functions, are defined as sums of series, so it is important to be familiar with the basic concepts of convergence of infinite sequences and series.

Physicists also use series in another way, as we will see in Section 11.12. In studying fields as diverse as optics, special relativity, and electromagnetism, they analyze phenomena by replacing a function with the first few terms in the series that represents it.

12.1 Sequences

A **sequence** can be thought of as a list of numbers written in a definite order:

$$a_1, a_2, a_3, a_4, \ldots, a_n, \ldots$$

The number a_1 is called the *first term*, a_2 is the *second term*, and in general a_n is the *nth term*. We will deal exclusively with infinite sequences and so each term a_n will have a successor a_{n+1}.

Notice that for every positive integer n there is a corresponding number a_n and so a sequence can be defined as a function whose domain is the set of positive integers. But we usually write a_n instead of the function notation $f(n)$ for the value of the function at the number n.

NOTATION ◦ The sequence $\{a_1, a_2, a_3, \ldots\}$ is also denoted by

$$\{a_n\} \quad \text{or} \quad \{a_n\}_{n=1}^{\infty}$$

EXAMPLE 1 Some sequences can be defined by giving a formula for the nth term. In the following examples we give three descriptions of the sequence: one by using the preceding notation, another by using the defining formula, and a third by writing out the terms of the sequence. Notice that n doesn't have to start at 1.

(a) $\left\{\dfrac{n}{n+1}\right\}_{n=1}^{\infty}$ $\quad a_n = \dfrac{n}{n+1} \quad$ $\left\{\dfrac{1}{2}, \dfrac{2}{3}, \dfrac{3}{4}, \dfrac{4}{5}, \ldots, \dfrac{n}{n+1}, \ldots\right\}$

(b) $\left\{\dfrac{(-1)^n(n+1)}{3^n}\right\}$ $\quad a_n = \dfrac{(-1)^n(n+1)}{3^n} \quad$ $\left\{-\dfrac{2}{3}, \dfrac{3}{9}, -\dfrac{4}{27}, \dfrac{5}{81}, \ldots, \dfrac{(-1)^n(n+1)}{3^n}, \ldots\right\}$

(c) $\{\sqrt{n-3}\}_{n=3}^{\infty}$ $\quad a_n = \sqrt{n-3},\ n \geq 3 \quad \{0, 1, \sqrt{2}, \sqrt{3}, \ldots, \sqrt{n-3}, \ldots\}$

(d) $\left\{\cos\dfrac{n\pi}{6}\right\}_{n=0}^{\infty}$ $\quad a_n = \cos\dfrac{n\pi}{6},\ n \geq 0 \quad \left\{1, \dfrac{\sqrt{3}}{2}, \dfrac{1}{2}, 0, \ldots, \cos\dfrac{n\pi}{6}, \ldots\right\}$

EXAMPLE 2 Find a formula for the general term a_n of the sequence

$$\left\{\dfrac{3}{5}, -\dfrac{4}{25}, \dfrac{5}{125}, -\dfrac{6}{625}, \dfrac{7}{3125}, \ldots\right\}$$

assuming that the pattern of the first few terms continues.

SOLUTION We are given that

$$a_1 = \dfrac{3}{5} \quad a_2 = -\dfrac{4}{25} \quad a_3 = \dfrac{5}{125} \quad a_4 = -\dfrac{6}{625} \quad a_5 = \dfrac{7}{3125}$$

Notice that the numerators of these fractions start with 3 and increase by 1 whenever we go to the next term. The second term has numerator 4, the third term has numerator 5; in general, the nth term will have numerator $n + 2$. The denominators are the powers of 5, so a_n has denominator 5^n. The signs of the terms are alternately positive and negative, so we need to multiply by a power of -1. In Example 1(b) the factor $(-1)^n$ meant we started with a negative term. Here we want to start with a positive term and so we use $(-1)^{n-1}$ or $(-1)^{n+1}$. Therefore,

$$a_n = (-1)^{n-1}\dfrac{n+2}{5^n}$$

EXAMPLE 3 Here are some sequences that don't have a simple defining equation.
(a) The sequence $\{p_n\}$, where p_n is the population of the world as of January 1 in the year n.
(b) If we let a_n be the digit in the nth decimal place of the number e, then $\{a_n\}$ is a well-defined sequence whose first few terms are

$$\{7, 1, 8, 2, 8, 1, 8, 2, 8, 4, 5, \ldots\}$$

(c) The **Fibonacci sequence** $\{f_n\}$ is defined recursively by the conditions

$$f_1 = 1 \quad f_2 = 1 \quad f_n = f_{n-1} + f_{n-2} \quad n \geq 3$$

Each term is the sum of the two preceding terms. The first few terms are

$$\{1, 1, 2, 3, 5, 8, 13, 21, \ldots\}$$

This sequence arose when the 13th-century Italian mathematician known as Fibonacci solved a problem concerning the breeding of rabbits (see Exercise 65).

A sequence such as the one in Example 1(a), $a_n = n/(n + 1)$, can be pictured either by plotting its terms on a number line as in Figure 1 or by plotting its graph as in Figure 2. Note that, since a sequence is a function whose domain is the set of positive integers, its graph consists of isolated points with coordinates

$$(1, a_1) \quad (2, a_2) \quad (3, a_3) \quad \ldots \quad (n, a_n) \quad \ldots$$

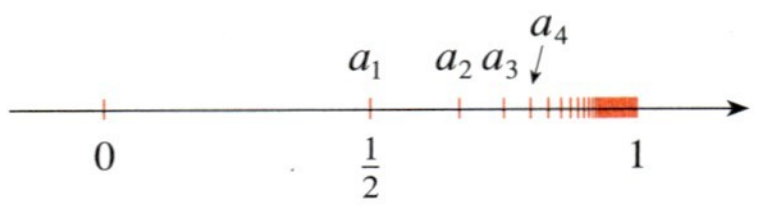

FIGURE 1

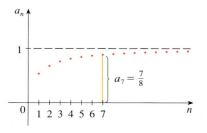

FIGURE 2

From Figure 1 or 2 it appears that the terms of the sequence $a_n = n/(n+1)$ are approaching 1 as n becomes large. In fact, the difference

$$1 - \frac{n}{n+1} = \frac{1}{n+1}$$

can be made as small as we like by taking n sufficiently large. We indicate this by writing

$$\lim_{n \to \infty} \frac{n}{n+1} = 1$$

In general, the notation

$$\lim_{n \to \infty} a_n = L$$

means that the terms of the sequence $\{a_n\}$ approach L as n becomes large. Notice that the following definition of the limit of a sequence is very similar to the definition of a limit of a function at infinity given in Section 4.4.

> **1 Definition** A sequence $\{a_n\}$ has the **limit** L and we write
>
> $$\lim_{n \to \infty} a_n = L \quad \text{or} \quad a_n \to L \text{ as } n \to \infty$$
>
> if we can make the terms a_n as close to L as we like by taking n sufficiently large. If $\lim_{n \to \infty} a_n$ exists, we say the sequence **converges** (or is **convergent**). Otherwise, we say the sequence **diverges** (or is **divergent**).

Figure 3 illustrates Definition 1 by showing the graphs of two sequences that have the limit L.

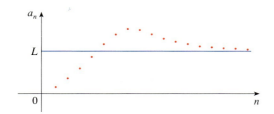 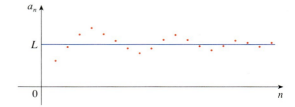

FIGURE 3
Graphs of two sequences with $\lim_{n \to \infty} a_n = L$

A more precise version of Definition 1 is as follows.

|||| Compare this definition with Definition 4.4.5.

> **2 Definition** A sequence $\{a_n\}$ has the **limit** L and we write
>
> $$\lim_{n \to \infty} a_n = L \quad \text{or} \quad a_n \to L \text{ as } n \to \infty$$
>
> if for every $\varepsilon > 0$ there is a corresponding integer N such that
>
> $$|a_n - L| < \varepsilon \quad \text{whenever } n > N$$

Definition 2 is illustrated by Figure 4, in which the terms $a_1, a_2, a_3, \ldots$ are plotted on a number line. No matter how small an interval $(L - \varepsilon, L + \varepsilon)$ is chosen, there exists an N such that all terms of the sequence from a_{N+1} onward must lie in that interval.

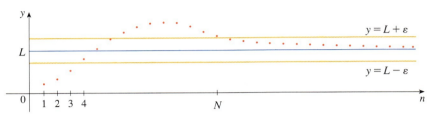

FIGURE 4

Another illustration of Definition 2 is given in Figure 5. The points on the graph of $\{a_n\}$ must lie between the horizontal lines $y = L + \varepsilon$ and $y = L - \varepsilon$ if $n > N$. This picture must be valid no matter how small ε is chosen, but usually a smaller ε requires a larger N.

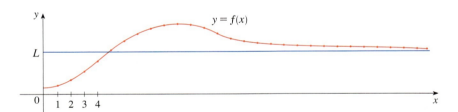

FIGURE 5

Comparison of Definition 2 and Definition 4.4.5 shows that the only difference between $\lim_{n \to \infty} a_n = L$ and $\lim_{x \to \infty} f(x) = L$ is that n is required to be an integer. Thus, we have the following theorem, which is illustrated by Figure 6.

3 Theorem If $\lim_{x \to \infty} f(x) = L$ and $f(n) = a_n$ when n is an integer, then $\lim_{n \to \infty} a_n = L$.

FIGURE 6

In particular, since we know that $\lim_{x \to \infty} (1/x^r) = 0$ when $r > 0$ (Theorem 4.4.4), we have

4
$$\lim_{n \to \infty} \frac{1}{n^r} = 0 \quad \text{if } r > 0$$

If a_n becomes large as n becomes large, we use the notation $\lim_{n \to \infty} a_n = \infty$. The following precise definition is similar to Definition 4.4.7.

5 Definition $\lim_{n \to \infty} a_n = \infty$ means that for every positive number M there is an integer N such that

$$a_n > M \quad \text{whenever } n > N$$

If $\lim_{n \to \infty} a_n = \infty$, then the sequence $\{a_n\}$ is divergent but in a special way. We say that $\{a_n\}$ diverges to ∞.

The Limit Laws given in Section 2.3 also hold for the limits of sequences and their proofs are similar.

Limit Laws for Sequences

If $\{a_n\}$ and $\{b_n\}$ are convergent sequences and c is a constant, then

$$\lim_{n \to \infty} (a_n + b_n) = \lim_{n \to \infty} a_n + \lim_{n \to \infty} b_n$$

$$\lim_{n \to \infty} (a_n - b_n) = \lim_{n \to \infty} a_n - \lim_{n \to \infty} b_n$$

$$\lim_{n \to \infty} c a_n = c \lim_{n \to \infty} a_n \qquad \lim_{n \to \infty} c = c$$

$$\lim_{n \to \infty} (a_n b_n) = \lim_{n \to \infty} a_n \cdot \lim_{n \to \infty} b_n$$

$$\lim_{n \to \infty} \frac{a_n}{b_n} = \frac{\lim_{n \to \infty} a_n}{\lim_{n \to \infty} b_n} \quad \text{if } \lim_{n \to \infty} b_n \neq 0$$

$$\lim_{n \to \infty} a_n^p = \left[\lim_{n \to \infty} a_n \right]^p \quad \text{if } p > 0 \text{ and } a_n > 0$$

The Squeeze Theorem can also be adapted for sequences as follows (see Figure 7).

Squeeze Theorem for Sequences

If $a_n \leq b_n \leq c_n$ for $n \geq n_0$ and $\lim_{n \to \infty} a_n = \lim_{n \to \infty} c_n = L$, then $\lim_{n \to \infty} b_n = L$.

Another useful fact about limits of sequences is given by the following theorem, whose proof is left as Exercise 69.

6 Theorem If $\lim_{n \to \infty} |a_n| = 0$, then $\lim_{n \to \infty} a_n = 0$.

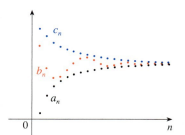

FIGURE 7
The sequence $\{b_n\}$ is squeezed between the sequences $\{a_n\}$ and $\{c_n\}$.

EXAMPLE 4 Find $\lim_{n \to \infty} \dfrac{n}{n+1}$.

SOLUTION The method is similar to the one we used in Section 4.4: Divide numerator and denominator by the highest power of n and then use the Limit Laws.

$$\lim_{n \to \infty} \frac{n}{n+1} = \lim_{n \to \infty} \frac{1}{1 + \dfrac{1}{n}} = \frac{\lim_{n \to \infty} 1}{\lim_{n \to \infty} 1 + \lim_{n \to \infty} \dfrac{1}{n}}$$

$$= \frac{1}{1 + 0} = 1$$

|||| This shows that the guess we made earlier from Figures 1 and 2 was correct.

Here we used Equation 4 with $r = 1$.

EXAMPLE 5 Calculate $\lim_{n \to \infty} \dfrac{\ln n}{n}$.

SOLUTION Notice that both numerator and denominator approach infinity as $n \to \infty$. We can't apply l'Hospital's Rule directly because it applies not to sequences but to functions of a real variable. However, we can apply l'Hospital's Rule to the related function $f(x) = (\ln x)/x$ and obtain

$$\lim_{x \to \infty} \frac{\ln x}{x} = \lim_{x \to \infty} \frac{1/x}{1} = 0$$

Therefore, by Theorem 3 we have

$$\lim_{n \to \infty} \frac{\ln n}{n} = 0$$

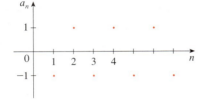

FIGURE 8

EXAMPLE 6 Determine whether the sequence $a_n = (-1)^n$ is convergent or divergent.

SOLUTION If we write out the terms of the sequence, we obtain

$$\{-1, 1, -1, 1, -1, 1, -1, \ldots\}$$

The graph of this sequence is shown in Figure 8. Since the terms oscillate between 1 and -1 infinitely often, a_n does not approach any number. Thus, $\lim_{n \to \infty} (-1)^n$ does not exist; that is, the sequence $\{(-1)^n\}$ is divergent.

|||| The graph of the sequence in Example 7 is shown in Figure 9 and supports the answer.

EXAMPLE 7 Evaluate $\lim_{n \to \infty} \dfrac{(-1)^n}{n}$ if it exists.

SOLUTION

$$\lim_{n \to \infty} \left| \frac{(-1)^n}{n} \right| = \lim_{n \to \infty} \frac{1}{n} = 0$$

Therefore, by Theorem 6,

$$\lim_{n \to \infty} \frac{(-1)^n}{n} = 0$$

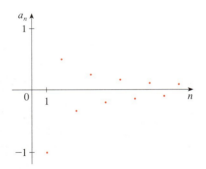

FIGURE 9

EXAMPLE 8 Discuss the convergence of the sequence $a_n = n!/n^n$, where $n! = 1 \cdot 2 \cdot 3 \cdot \cdots \cdot n$.

SOLUTION Both numerator and denominator approach infinity as $n \to \infty$ but here we have no corresponding function for use with l'Hospital's Rule ($x!$ is not defined when x is not an integer). Let's write out a few terms to get a feeling for what happens to a_n as n gets large:

$$a_1 = 1 \qquad a_2 = \frac{1 \cdot 2}{2 \cdot 2} \qquad a_3 = \frac{1 \cdot 2 \cdot 3}{3 \cdot 3 \cdot 3}$$

$$\boxed{7} \qquad a_n = \frac{1 \cdot 2 \cdot 3 \cdot \cdots \cdot n}{n \cdot n \cdot n \cdot \cdots \cdot n}$$

It appears from these expressions and the graph in Figure 10 that the terms are decreas-

CREATING GRAPHS OF SEQUENCES

Some computer algebra systems have special commands that enable us to create sequences and graph them directly. With most graphing calculators, however, sequences can be graphed by using parametric equations. For instance, the sequence in Example 8 can be graphed by entering the parametric equations

$$x = t \qquad y = t!/t^t$$

and graphing in dot mode starting with $t = 1$, setting the t-step equal to 1. The result is shown in Figure 10.

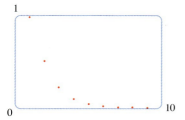

FIGURE 10

ing and perhaps approach 0. To confirm this, observe from Equation 7 that

$$a_n = \frac{1}{n} \left(\frac{2 \cdot 3 \cdot \cdots \cdot n}{n \cdot n \cdot \cdots \cdot n} \right)$$

Notice that the expression in parentheses is at most 1 because the numerator is less than (or equal to) the denominator. So

$$0 < a_n \le \frac{1}{n}$$

We know that $1/n \to 0$ as $n \to \infty$. Therefore, $a_n \to 0$ as $n \to \infty$ by the Squeeze Theorem.

EXAMPLE 9 For what values of r is the sequence $\{r^n\}$ convergent?

SOLUTION We know from Section 4.4 and the graphs of the exponential functions in Section 7.2 (or Section 7.4*) that $\lim_{x \to \infty} a^x = \infty$ for $a > 1$ and $\lim_{x \to \infty} a^x = 0$ for $0 < a < 1$. Therefore, putting $a = r$ and using Theorem 2, we have

$$\lim_{n \to \infty} r^n = \begin{cases} \infty & \text{if } r > 1 \\ 0 & \text{if } 0 < r < 1 \end{cases}$$

It is obvious that

$$\lim_{n \to \infty} 1^n = 1 \quad \text{and} \quad \lim_{n \to \infty} 0^n = 0$$

If $-1 < r < 0$, then $0 < |r| < 1$, so

$$\lim_{n \to \infty} |r^n| = \lim_{n \to \infty} |r|^n = 0$$

and therefore $\lim_{n \to \infty} r^n = 0$ by Theorem 6. If $r \le -1$, then $\{r^n\}$ diverges as in Example 6. Figure 11 shows the graphs for various values of r. (The case $r = -1$ is shown in Figure 8.)

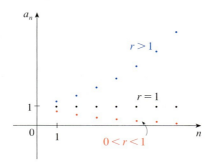
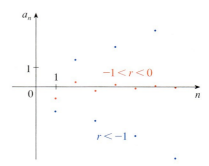

FIGURE 11
The sequence $a_n = r^n$

The results of Example 9 are summarized for future use as follows.

8 The sequence $\{r^n\}$ is convergent if $-1 < r \le 1$ and divergent for all other values of r.

$$\lim_{n \to \infty} r^n = \begin{cases} 0 & \text{if } -1 < r < 1 \\ 1 & \text{if } r = 1 \end{cases}$$

9 Definition A sequence $\{a_n\}$ is called **increasing** if $a_n < a_{n+1}$ for all $n \geq 1$, that is, $a_1 < a_2 < a_3 < \cdots$. It is called **decreasing** if $a_n > a_{n+1}$ for all $n \geq 1$. It is called **monotonic** if it is either increasing or decreasing.

EXAMPLE 10 The sequence $\left\{\dfrac{3}{n+5}\right\}$ is decreasing because

$$\frac{3}{n+5} > \frac{3}{(n+1)+5} = \frac{3}{n+6}$$

for all $n \geq 1$. (The right side is smaller because it has a larger denominator.)

EXAMPLE 11 Show that the sequence $a_n = \dfrac{n}{n^2+1}$ is decreasing.

SOLUTION 1 We must show that $a_{n+1} < a_n$, that is,

$$\frac{n+1}{(n+1)^2+1} < \frac{n}{n^2+1}$$

This inequality is equivalent to the one we get by cross-multiplication:

$$\frac{n+1}{(n+1)^2+1} < \frac{n}{n^2+1} \iff (n+1)(n^2+1) < n[(n+1)^2+1]$$

$$\iff n^3 + n^2 + n + 1 < n^3 + 2n^2 + 2n$$

$$\iff 1 < n^2 + n$$

Since $n \geq 1$, we know that the inequality $n^2 + n > 1$ is true. Therefore, $a_{n+1} < a_n$ and so $\{a_n\}$ is decreasing.

SOLUTION 2 Consider the function $f(x) = \dfrac{x}{x^2+1}$:

$$f'(x) = \frac{x^2+1-2x^2}{(x^2+1)^2} = \frac{1-x^2}{(x^2+1)^2} < 0 \qquad \text{whenever } x^2 > 1$$

Thus, f is decreasing on $(1, \infty)$ and so $f(n) > f(n+1)$. Therefore, $\{a_n\}$ is decreasing.

10 Definition A sequence $\{a_n\}$ is **bounded above** if there is a number M such that

$$a_n \leq M \qquad \text{for all } n \geq 1$$

It is **bounded below** if there is a number m such that

$$m \leq a_n \qquad \text{for all } n \geq 1$$

If it is bounded above and below, then $\{a_n\}$ is a **bounded sequence**.

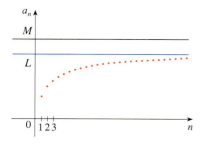

FIGURE 12

For instance, the sequence $a_n = n$ is bounded below ($a_n > 0$) but not above. The sequence $a_n = n/(n + 1)$ is bounded because $0 < a_n < 1$ for all n.

We know that not every bounded sequence is convergent [for instance, the sequence $a_n = (-1)^n$ satisfies $-1 \leq a_n \leq 1$ but is divergent from Example 6] and not every monotonic sequence is convergent ($a_n = n \to \infty$). But if a sequence is both bounded *and* monotonic, then it must be convergent. This fact is proved as Theorem 11, but intuitively you can understand why it is true by looking at Figure 12. If $\{a_n\}$ is increasing and $a_n \leq M$ for all n, then the terms are forced to crowd together and approach some number L.

The proof of Theorem 11 is based on the **Completeness Axiom** for the set $\mathbb{R}$ of real numbers, which says that if S is a nonempty set of real numbers that has an upper bound M ($x \leq M$ for all x in S), then S has a **least upper bound** b. (This means that b is an upper bound for S, but if M is any other upper bound, then $b \leq M$.) The Completeness Axiom is an expression of the fact that there is no gap or hole in the real number line.

11 **Monotonic Sequence Theorem** Every bounded, monotonic sequence is convergent.

Proof Suppose $\{a_n\}$ is an increasing sequence. Since $\{a_n\}$ is bounded, the set $S = \{a_n \mid n \geq 1\}$ has an upper bound. By the Completeness Axiom it has a least upper bound L. Given $\varepsilon > 0$, $L - \varepsilon$ is *not* an upper bound for S (since L is the *least* upper bound). Therefore

$$a_N > L - \varepsilon \qquad \text{for some integer } N$$

But the sequence is increasing so $a_n \geq a_N$ for every $n > N$. Thus, if $n > N$ we have

$$a_n > L - \varepsilon$$

so

$$0 \leq L - a_n < \varepsilon$$

since $a_n \leq L$. Thus

$$|L - a_n| < \varepsilon \qquad \text{whenever } n > N$$

so $\lim_{n \to \infty} a_n = L$.

A similar proof (using the greatest lower bound) works if $\{a_n\}$ is decreasing.

The proof of Theorem 11 shows that a sequence that is increasing and bounded above is convergent. (Likewise, a decreasing sequence that is bounded below is convergent.) This fact is used many times in dealing with infinite series.

EXAMPLE 12 Investigate the sequence $\{a_n\}$ defined by the recurrence relation

$$a_1 = 2 \qquad a_{n+1} = \tfrac{1}{2}(a_n + 6) \qquad \text{for } n = 1, 2, 3, \ldots$$

SOLUTION We begin by computing the first several terms:

$$a_1 = 2 \qquad a_2 = \tfrac{1}{2}(2 + 6) = 4 \qquad a_3 = \tfrac{1}{2}(4 + 6) = 5$$

$$a_4 = \tfrac{1}{2}(5 + 6) = 5.5 \qquad a_5 = 5.75 \qquad a_6 = 5.875$$

$$a_7 = 5.9375 \qquad a_8 = 5.96875 \qquad a_9 = 5.984375$$

> Mathematical induction is often used in dealing with recursive sequences. See page 59 for a discussion of the Principle of Mathematical Induction.

These initial terms suggest that the sequence is increasing and the terms are approaching 6. To confirm that the sequence is increasing we use mathematical induction to show that $a_{n+1} > a_n$ for all $n \geq 1$. This is true for $n = 1$ because $a_2 = 4 > a_1$. If we assume that it is true for $n = k$, then we have

$$a_{k+1} > a_k$$

so

$$a_{k+1} + 6 > a_k + 6$$

and

$$\tfrac{1}{2}(a_{k+1} + 6) > \tfrac{1}{2}(a_k + 6)$$

Thus

$$a_{k+2} > a_{k+1}$$

We have deduced that $a_{n+1} > a_n$ is true for $n = k + 1$. Therefore, the inequality is true for all n by induction.

Next we verify that $\{a_n\}$ is bounded by showing that $a_n < 6$ for all n. (Since the sequence is increasing, we already know that it has a lower bound: $a_n \geq a_1 = 2$ for all n.) We know that $a_1 < 6$, so the assertion is true for $n = 1$. Suppose it is true for $n = k$. Then

$$a_k < 6$$

so

$$a_k + 6 < 12$$

$$\tfrac{1}{2}(a_k + 6) < \tfrac{1}{2}(12) = 6$$

Thus

$$a_{k+1} < 6$$

This shows, by mathematical induction, that $a_n < 6$ for all n.

Since the sequence $\{a_n\}$ is increasing and bounded, Theorem 11 guarantees that it has a limit. The theorem doesn't tell us what the value of the limit is. But now that we know $L = \lim_{n \to \infty} a_n$ exists, we can use the recurrence relation to write

$$\lim_{n \to \infty} a_{n+1} = \lim_{n \to \infty} \tfrac{1}{2}(a_n + 6) = \tfrac{1}{2}\left(\lim_{n \to \infty} a_n + 6\right) = \tfrac{1}{2}(L + 6)$$

> A proof of this fact is requested in Exercise 52.

Since $a_n \to L$, it follows that $a_{n+1} \to L$, too (as $n \to \infty$, $n + 1 \to \infty$, too). So we have

$$L = \tfrac{1}{2}(L + 6)$$

Solving this equation for L, we get $L = 6$, as predicted.

12.1 Exercises

1. (a) What is a sequence?
   (b) What does it mean to say that $\lim_{n \to \infty} a_n = 8$?
   (c) What does it mean to say that $\lim_{n \to \infty} a_n = \infty$?

2. (a) What is a convergent sequence? Give two examples.
   (b) What is a divergent sequence? Give two examples.

3–8 List the first five terms of the sequence.

3. $a_n = 1 - (0.2)^n$

4. $a_n = \dfrac{n+1}{3n-1}$

5. $a_n = \dfrac{3(-1)^n}{n!}$

6. $\{2 \cdot 4 \cdot 6 \cdots (2n)\}$

7. $a_1 = 3$, $a_{n+1} = 2a_n - 1$ **8.** $a_1 = 4$, $a_{n+1} = \dfrac{a_n}{a_n - 1}$

9–14 Find a formula for the general term a_n of the sequence, assuming that the pattern of the first few terms continues.

9. $\left\{\dfrac{1}{2}, \dfrac{1}{4}, \dfrac{1}{8}, \dfrac{1}{16}, \ldots\right\}$ **10.** $\left\{\dfrac{1}{2}, \dfrac{1}{4}, \dfrac{1}{6}, \dfrac{1}{8}, \ldots\right\}$

11. $\{2, 7, 12, 17, \ldots\}$ **12.** $\left\{-\dfrac{1}{4}, \dfrac{2}{9}, -\dfrac{3}{16}, \dfrac{4}{25}, \ldots\right\}$

13. $\left\{1, -\dfrac{2}{3}, \dfrac{4}{9}, -\dfrac{8}{27}, \ldots\right\}$ **14.** $\{5, 1, 5, 1, 5, 1, \ldots\}$

15–40 Determine whether the sequence converges or diverges. If it converges, find the limit.

15. $a_n = n(n-1)$ **16.** $a_n = \dfrac{n+1}{3n-1}$

17. $a_n = \dfrac{3 + 5n^2}{n + n^2}$ **18.** $a_n = \dfrac{\sqrt{n}}{1 + \sqrt{n}}$

19. $a_n = \dfrac{2^n}{3^{n+1}}$ **20.** $a_n = \dfrac{n}{1 + \sqrt{n}}$

21. $a_n = \dfrac{(-1)^{n-1} n}{n^2 + 1}$ **22.** $a_n = \dfrac{(-1)^n n^3}{n^3 + 2n^2 + 1}$

23. $a_n = \cos(n/2)$ **24.** $a_n = \cos(2/n)$

25. $\left\{\dfrac{(2n-1)!}{(2n+1)!}\right\}$ **26.** $\{\arctan 2n\}$

27. $\left\{\dfrac{e^n + e^{-n}}{e^{2n} - 1}\right\}$ **28.** $\left\{\dfrac{\ln n}{\ln 2n}\right\}$

29. $\{n^2 e^{-n}\}$ **30.** $\{n \cos n\pi\}$

31. $a_n = \dfrac{\cos^2 n}{2^n}$ **32.** $a_n = \ln(n+1) - \ln n$

33. $a_n = n \sin(1/n)$ **34.** $a_n = \sqrt{n} - \sqrt{n^2 - 1}$

35. $a_n = \left(1 + \dfrac{2}{n}\right)^{1/n}$ **36.** $a_n = \dfrac{\sin 2n}{1 + \sqrt{n}}$

37. $\{0, 1, 0, 0, 1, 0, 0, 0, 1, \ldots\}$ **38.** $\left\{\dfrac{1}{1}, \dfrac{1}{3}, \dfrac{1}{2}, \dfrac{1}{4}, \dfrac{1}{3}, \dfrac{1}{5}, \dfrac{1}{4}, \dfrac{1}{6}, \ldots\right\}$

39. $a_n = \dfrac{n!}{2^n}$ **40.** $a_n = \dfrac{(-3)^n}{n!}$

41–48 Use a graph of the sequence to decide whether the sequence is convergent or divergent. If the sequence is convergent, guess the value of the limit from the graph and then prove your guess. (See the margin note on page 743 for advice on graphing sequences.)

41. $a_n = (-1)^n \dfrac{n+1}{n}$ **42.** $a_n = 2 + (-2/\pi)^n$

43. $\left\{\arctan\left(\dfrac{2n}{2n+1}\right)\right\}$ **44.** $\left\{\dfrac{\sin n}{\sqrt{n}}\right\}$

45. $a_n = \dfrac{n^3}{n!}$ **46.** $a_n = \sqrt[n]{3^n + 5^n}$

47. $a_n = \dfrac{1 \cdot 3 \cdot 5 \cdots (2n-1)}{(2n)^n}$

48. $a_n = \dfrac{1 \cdot 3 \cdot 5 \cdots (2n-1)}{n!}$

49. If \$1000 is invested at 6% interest, compounded annually, then after n years the investment is worth $a_n = 1000(1.06)^n$ dollars.
(a) Find the first five terms of the sequence $\{a_n\}$.
(b) Is the sequence convergent or divergent? Explain.

50. Find the first 40 terms of the sequence defined by
$$a_{n+1} = \begin{cases} \tfrac{1}{2} a_n & \text{if } a_n \text{ is an even number} \\ 3a_n + 1 & \text{if } a_n \text{ is an odd number} \end{cases}$$
and $a_1 = 11$. Do the same if $a_1 = 25$. Make a conjecture about this type of sequence.

51. For what values of r is the sequence $\{nr^n\}$ convergent?

52. (a) If $\{a_n\}$ is convergent, show that
$$\lim_{n \to \infty} a_{n+1} = \lim_{n \to \infty} a_n$$
(b) A sequence $\{a_n\}$ is defined by $a_1 = 1$ and $a_{n+1} = 1/(1 + a_n)$ for $n \geq 1$. Assuming that $\{a_n\}$ is convergent, find its limit.

53. Suppose you know that $\{a_n\}$ is a decreasing sequence and all its terms lie between the numbers 5 and 8. Explain why the sequence has a limit. What can you say about the value of the limit?

54–60 Determine whether the sequence is increasing, decreasing, or not monotonic. Is the sequence bounded?

54. $a_n = \dfrac{1}{5^n}$ **55.** $a_n = \dfrac{1}{2n+3}$

56. $a_n = \dfrac{2n-3}{3n+4}$ **57.** $a_n = \cos(n\pi/2)$

58. $a_n = ne^{-n}$ **59.** $a_n = \dfrac{n}{n^2+1}$

60. $a_n = n + \dfrac{1}{n}$

61. Find the limit of the sequence
$$\left\{\sqrt{2}, \sqrt{2\sqrt{2}}, \sqrt{2\sqrt{2\sqrt{2}}}, \ldots\right\}$$

62. A sequence $\{a_n\}$ is given by $a_1 = \sqrt{2}$, $a_{n+1} = \sqrt{2 + a_n}$.
(a) By induction or otherwise, show that $\{a_n\}$ is increasing and bounded above by 3. Apply Theorem 11 to show that $\lim_{n \to \infty} a_n$ exists.
(b) Find $\lim_{n \to \infty} a_n$.

63. Show that the sequence defined by $a_1 = 1$, $a_{n+1} = 3 - 1/a_n$ is increasing and $a_n < 3$ for all n. Deduce that $\{a_n\}$ is convergent and find its limit.

64. Show that the sequence defined by
$$a_1 = 2 \qquad a_{n+1} = \frac{1}{3 - a_n}$$
satisfies $0 < a_n \leq 2$ and is decreasing. Deduce that the sequence is convergent and find its limit.

65. (a) Fibonacci posed the following problem: Suppose that rabbits live forever and that every month each pair produces a new pair which becomes productive at age 2 months. If we start with one newborn pair, how many pairs of rabbits will we have in the nth month? Show that the answer is f_n, where $\{f_n\}$ is the Fibonacci sequence defined in Example 3(c).
(b) Let $a_n = f_{n+1}/f_n$ and show that $a_{n-1} = 1 + 1/a_{n-2}$. Assuming that $\{a_n\}$ is convergent, find its limit.

66. (a) Let $a_1 = a$, $a_2 = f(a)$, $a_3 = f(a_2) = f(f(a))$, ..., $a_{n+1} = f(a_n)$, where f is a continuous function. If $\lim_{n \to \infty} a_n = L$, show that $f(L) = L$.
(b) Illustrate part (a) by taking $f(x) = \cos x$, $a = 1$, and estimating the value of L to five decimal places.

67. (a) Use a graph to guess the value of the limit
$$\lim_{n \to \infty} \frac{n^5}{n!}$$
(b) Use a graph of the sequence in part (a) to find the smallest values of N that correspond to $\varepsilon = 0.1$ and $\varepsilon = 0.001$ in Definition 2.

68. Use Definition 2 directly to prove that $\lim_{n \to \infty} r^n = 0$ when $|r| < 1$.

69. Prove Theorem 6.
[*Hint:* Use either Definition 2 or the Squeeze Theorem.]

70. Let $a_n = \left(1 + \dfrac{1}{n}\right)^n$.
(a) Show that if $0 \leq a < b$, then
$$\frac{b^{n+1} - a^{n+1}}{b - a} < (n + 1)b^n$$
(b) Deduce that $b^n[(n + 1)a - nb] < a^{n+1}$.
(c) Use $a = 1 + 1/(n + 1)$ and $b = 1 + 1/n$ in part (b) to show that $\{a_n\}$ is increasing.
(d) Use $a = 1$ and $b = 1 + 1/(2n)$ in part (b) to show that $a_{2n} < 4$.
(e) Use parts (c) and (d) to show that $a_n < 4$ for all n.

(f) Use Theorem 11 to show that $\lim_{n \to \infty} (1 + 1/n)^n$ exists. (The limit is e. See Equation 7.4.9.)

71. Let a and b be positive numbers with $a > b$. Let a_1 be their arithmetic mean and b_1 their geometric mean:
$$a_1 = \frac{a + b}{2} \qquad b_1 = \sqrt{ab}$$
Repeat this process so that, in general,
$$a_{n+1} = \frac{a_n + b_n}{2} \qquad b_{n+1} = \sqrt{a_n b_n}$$
(a) Use mathematical induction to show that
$$a_n > a_{n+1} > b_{n+1} > b_n$$
(b) Deduce that both $\{a_n\}$ and $\{b_n\}$ are convergent.
(c) Show that $\lim_{n \to \infty} a_n = \lim_{n \to \infty} b_n$. Gauss called the common value of these limits the **arithmetic-geometric mean** of the numbers a and b.

72. (a) Show that if $\lim_{n \to \infty} a_{2n} = L$ and $\lim_{n \to \infty} a_{2n+1} = L$, then $\{a_n\}$ is convergent and $\lim_{n \to \infty} a_n = L$.
(b) If $a_1 = 1$ and
$$a_{n+1} = 1 + \frac{1}{1 + a_n}$$
find the first eight terms of the sequence $\{a_n\}$. Then use part (a) to show that $\lim_{n \to \infty} a_n = \sqrt{2}$. This gives the **continued fraction expansion**
$$\sqrt{2} = 1 + \cfrac{1}{2 + \cfrac{1}{2 + \cdots}}$$

73. The size of an undisturbed fish population has been modeled by the formula
$$p_{n+1} = \frac{bp_n}{a + p_n}$$
where p_n is the fish population after n years and a and b are positive constants that depend on the species and its environment. Suppose that the population in year 0 is $p_0 > 0$.
(a) Show that if $\{p_n\}$ is convergent, then the only possible values for its limit are 0 and $b - a$.
(b) Show that $p_{n+1} < (b/a)p_n$.
(c) Use part (b) to show that if $a > b$, then $\lim_{n \to \infty} p_n = 0$; in other words, the population dies out.
(d) Now assume that $a < b$. Show that if $p_0 < b - a$, then $\{p_n\}$ is increasing and $0 < p_n < b - a$. Show also that if $p_0 > b - a$, then $\{p_n\}$ is decreasing and $p_n > b - a$. Deduce that if $a < b$, then $\lim_{n \to \infty} p_n = b - a$.

LABORATORY PROJECT

⊡ Logistic Sequences

A sequence that arises in ecology as a model for population growth is defined by the **logistic difference equation**

$$p_{n+1} = kp_n(1 - p_n)$$

where p_n measures the size of the population of the nth generation of a single species. To keep the numbers manageable, p_n is a fraction of the maximal size of the population, so $0 \leq p_n \leq 1$. Notice that the form of this equation is similar to the logistic differential equation in Section 10.5. The discrete model—with sequences instead of continuous functions—is preferable for modeling insect populations, where mating and death occur in a periodic fashion.

An ecologist is interested in predicting the size of the population as time goes on, and asks these questions: Will it stabilize at a limiting value? Will it change in a cyclical fashion? Or will it exhibit random behavior?

Write a program to compute the first n terms of this sequence starting with an initial population p_0, where $0 < p_0 < 1$. Use this program to do the following.

1. Calculate 20 or 30 terms of the sequence for $p_0 = \frac{1}{2}$ and for two values of k such that $1 < k < 3$. Graph the sequences. Do they appear to converge? Repeat for a different value of p_0 between 0 and 1. Does the limit depend on the choice of p_0? Does it depend on the choice of k?

2. Calculate terms of the sequence for a value of k between 3 and 3.4 and plot them. What do you notice about the behavior of the terms?

3. Experiment with values of k between 3.4 and 3.5. What happens to the terms?

4. For values of k between 3.6 and 4, compute and plot at least 100 terms and comment on the behavior of the sequence. What happens if you change p_0 by 0.001? This type of behavior is called *chaotic* and is exhibited by insect populations under certain conditions.

12.2 Series

If we try to add the terms of an infinite sequence $\{a_n\}_{n=1}^{\infty}$ we get an expression of the form

$$\boxed{1} \qquad a_1 + a_2 + a_3 + \cdots + a_n + \cdots$$

which is called an **infinite series** (or just a **series**) and is denoted, for short, by the symbol

$$\sum_{n=1}^{\infty} a_n \quad \text{or} \quad \sum a_n$$

But does it make sense to talk about the sum of infinitely many terms?

It would be impossible to find a finite sum for the series

$$1 + 2 + 3 + 4 + 5 + \cdots + n + \cdots$$

because if we start adding the terms we get the cumulative sums 1, 3, 6, 10, 15, 21, . . . and, after the nth term, we get $n(n + 1)/2$, which becomes very large as n increases.

However, if we start to add the terms of the series

$$\frac{1}{2} + \frac{1}{4} + \frac{1}{8} + \frac{1}{16} + \frac{1}{32} + \frac{1}{64} + \cdots + \frac{1}{2^n} + \cdots$$

n	Sum of first n terms
1	0.50000000
2	0.75000000
3	0.87500000
4	0.93750000
5	0.96875000
6	0.98437500
7	0.99218750
10	0.99902344
15	0.99996948
20	0.99999905
25	0.99999997

we get $\frac{1}{2}, \frac{3}{4}, \frac{7}{8}, \frac{15}{16}, \frac{31}{32}, \frac{63}{64}, \ldots, 1 - 1/2^n, \ldots$. The table shows that as we add more and more terms, these *partial sums* become closer and closer to 1. (See also Figure 11 in *A Preview of Calculus,* page 7.) In fact, by adding sufficiently many terms of the series we can make the partial sums as close as we like to 1. So it seems reasonable to say that the sum of this infinite series is 1 and to write

$$\sum_{n=1}^{\infty} \frac{1}{2^n} = \frac{1}{2} + \frac{1}{4} + \frac{1}{8} + \frac{1}{16} + \cdots + \frac{1}{2^n} + \cdots = 1$$

We use a similar idea to determine whether or not a general series (1) has a sum. We consider the **partial sums**

$$s_1 = a_1$$
$$s_2 = a_1 + a_2$$
$$s_3 = a_1 + a_2 + a_3$$
$$s_4 = a_1 + a_2 + a_3 + a_4$$

and, in general,

$$s_n = a_1 + a_2 + a_3 + \cdots + a_n = \sum_{i=1}^{n} a_i$$

These partial sums form a new sequence $\{s_n\}$, which may or may not have a limit. If $\lim_{n \to \infty} s_n = s$ exists (as a finite number), then, as in the preceding example, we call it the sum of the infinite series $\Sigma\, a_n$.

2 Definition Given a series $\sum_{n=1}^{\infty} a_n = a_1 + a_2 + a_3 + \cdots$, let s_n denote its nth partial sum:

$$s_n = \sum_{i=1}^{n} a_i = a_1 + a_2 + \cdots + a_n$$

If the sequence $\{s_n\}$ is convergent and $\lim_{n \to \infty} s_n = s$ exists as a real number, then the series $\Sigma\, a_n$ is called **convergent** and we write

$$a_1 + a_2 + \cdots + a_n + \cdots = s \qquad \text{or} \qquad \sum_{n=1}^{\infty} a_n = s$$

The number s is called the **sum** of the series. Otherwise, the series is called **divergent**.

Thus, when we write $\sum_{n=1}^{\infty} a_n = s$ we mean that by adding sufficiently many terms of the series we can get as close as we like to the number s. Notice that

$$\sum_{n=1}^{\infty} a_n = \lim_{n \to \infty} \sum_{i=1}^{n} a_i$$

|||| Compare with the improper integral

$$\int_1^{\infty} f(x)\, dx = \lim_{t \to \infty} \int_1^{t} f(x)\, dx$$

To find this integral we integrate from 1 to t and then let $t \to \infty$. For a series, we sum from 1 to n and then let $n \to \infty$.

EXAMPLE 1 An important example of an infinite series is the **geometric series**

$$a + ar + ar^2 + ar^3 + \cdots + ar^{n-1} + \cdots = \sum_{n=1}^{\infty} ar^{n-1} \qquad a \neq 0$$

Each term is obtained from the preceding one by multiplying it by the common ratio r. (We have already considered the special case where $a = \frac{1}{2}$ and $r = \frac{1}{2}$ on page 749.)

If $r = 1$, then $s_n = a + a + \cdots + a = na \to \pm\infty$. Since $\lim_{n\to\infty} s_n$ doesn't exist, the geometric series diverges in this case.

If $r \neq 1$, we have

$$s_n = a + ar + ar^2 + \cdots + ar^{n-1}$$

and

$$rs_n = ar + ar^2 + \cdots + ar^{n-1} + ar^n$$

Subtracting these equations, we get

$$s_n - rs_n = a - ar^n$$

$$\boxed{3} \qquad s_n = \frac{a(1 - r^n)}{1 - r}$$

If $-1 < r < 1$, we know from (12.1.8) that $r^n \to 0$ as $n \to \infty$, so

$$\lim_{n\to\infty} s_n = \lim_{n\to\infty} \frac{a(1 - r^n)}{1 - r} = \frac{a}{1 - r} - \frac{a}{1 - r} \lim_{n\to\infty} r^n = \frac{a}{1 - r}$$

Thus, when $|r| < 1$ the geometric series is convergent and its sum is $a/(1 - r)$.

If $r \leq -1$ or $r > 1$, the sequence $\{r^n\}$ is divergent by (12.1.8) and so, by Equation 3, $\lim_{n\to\infty} s_n$ does not exist. Therefore, the geometric series diverges in those cases.

We summarize the results of Example 1 as follows.

> $\boxed{4}$ The geometric series
>
> $$\sum_{n=1}^{\infty} ar^{n-1} = a + ar + ar^2 + \cdots$$
>
> is convergent if $|r| < 1$ and its sum is
>
> $$\sum_{n=1}^{\infty} ar^{n-1} = \frac{a}{1 - r} \qquad |r| < 1$$
>
> If $|r| \geq 1$, the geometric series is divergent.

EXAMPLE 2 Find the sum of the geometric series

$$5 - \frac{10}{3} + \frac{20}{9} - \frac{40}{27} + \cdots$$

SOLUTION The first term is $a = 5$ and the common ratio is $r = -\frac{2}{3}$. Since $|r| = \frac{2}{3} < 1$, the series is convergent by (4) and its sum is

$$5 - \frac{10}{3} + \frac{20}{9} - \frac{40}{27} + \cdots = \frac{5}{1 - \left(-\frac{2}{3}\right)} = \frac{5}{\frac{5}{3}} = 3$$

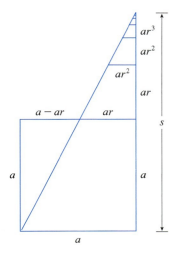

|||| Figure 1 provides a geometric demonstration of the result in Example 1. If the triangles are constructed as shown and s is the sum of the series, then, by similar triangles,

$$\frac{s}{a} = \frac{a}{a - ar} \qquad \text{so} \qquad s = \frac{a}{1 - r}$$

FIGURE 1

|||| In words: The sum of a convergent geometric series is

$$\frac{\text{first term}}{1 - \text{common ratio}}$$

What do we really mean when we say that the sum of the series in Example 2 is 3? Of course, we can't literally add an infinite number of terms, one by one. But, according to Definition 2, the total sum is the limit of the sequence of partial sums. So, by taking the sum of sufficiently many terms, we can get as close as we like to the number 3. The table shows the first ten partial sums s_n and the graph in Figure 2 shows how the sequence of partial sums approaches 3.

n	s_n
1	5.000000
2	1.666667
3	3.888889
4	2.407407
5	3.395062
6	2.736626
7	3.175583
8	2.882945
9	3.078037
10	2.947975

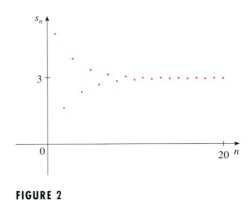

FIGURE 2

EXAMPLE 3 Is the series $\sum_{n=1}^{\infty} 2^{2n} 3^{1-n}$ convergent or divergent?

SOLUTION Let's rewrite the nth term of the series in the form ar^{n-1}:

$$\sum_{n=1}^{\infty} 2^{2n} 3^{1-n} = \sum_{n=1}^{\infty} \frac{4^n}{3^{n-1}} = \sum_{n=1}^{\infty} 4\left(\tfrac{4}{3}\right)^{n-1}$$

We recognize this series as a geometric series with $a = 4$ and $r = \tfrac{4}{3}$. Since $r > 1$, the series diverges by (4).

|||| Another way to identify a and r is to write out the first few terms:
$$4 + \tfrac{16}{3} + \tfrac{64}{9} + \cdots$$

EXAMPLE 4 Write the number $2.3\overline{17} = 2.3171717\ldots$ as a ratio of integers.

SOLUTION

$$2.3171717\ldots = 2.3 + \frac{17}{10^3} + \frac{17}{10^5} + \frac{17}{10^7} + \cdots$$

After the first term we have a geometric series with $a = 17/10^3$ and $r = 1/10^2$. Therefore

$$2.3\overline{17} = 2.3 + \frac{\dfrac{17}{10^3}}{1 - \dfrac{1}{10^2}} = 2.3 + \frac{\dfrac{17}{1000}}{\dfrac{99}{100}}$$

$$= \frac{23}{10} + \frac{17}{990} = \frac{1147}{495}$$

EXAMPLE 5 Find the sum of the series $\sum_{n=0}^{\infty} x^n$, where $|x| < 1$.

SOLUTION Notice that this series starts with $n = 0$ and so the first term is $x^0 = 1$. (With series, we adopt the convention that $x^0 = 1$ even when $x = 0$.) Thus

$$\sum_{n=0}^{\infty} x^n = 1 + x + x^2 + x^3 + x^4 + \cdots$$

Module 12.2 explores a series that depends on an angle θ in a triangle and enables you to see how rapidly the series converges when θ varies.

This is a geometric series with $a = 1$ and $r = x$. Since $|r| = |x| < 1$, it converges and (4) gives

$$\boxed{5} \qquad \sum_{n=0}^{\infty} x^n = \frac{1}{1-x}$$

EXAMPLE 6 Show that the series $\sum_{n=1}^{\infty} \dfrac{1}{n(n+1)}$ is convergent, and find its sum.

SOLUTION This is not a geometric series, so we go back to the definition of a convergent series and compute the partial sums.

$$s_n = \sum_{i=1}^{n} \frac{1}{i(i+1)} = \frac{1}{1 \cdot 2} + \frac{1}{2 \cdot 3} + \frac{1}{3 \cdot 4} + \cdots + \frac{1}{n(n+1)}$$

We can simplify this expression if we use the partial fraction decomposition

$$\frac{1}{i(i+1)} = \frac{1}{i} - \frac{1}{i+1}$$

(see Section 8.4). Thus, we have

$$s_n = \sum_{i=1}^{n} \frac{1}{i(i+1)} = \sum_{i=1}^{n} \left(\frac{1}{i} - \frac{1}{i+1} \right)$$

$$= \left(1 - \frac{1}{2}\right) + \left(\frac{1}{2} - \frac{1}{3}\right) + \left(\frac{1}{3} - \frac{1}{4}\right) + \cdots + \left(\frac{1}{n} - \frac{1}{n+1}\right)$$

$$= 1 - \frac{1}{n+1}$$

|||| Notice that the terms cancel in pairs. This is an example of a **telescoping sum**: Because of all the cancellations, the sum collapses (like an old-fashioned collapsing telescope) into just two terms.

and so

$$\lim_{n \to \infty} s_n = \lim_{n \to \infty} \left(1 - \frac{1}{n+1}\right) = 1 - 0 = 1$$

|||| Figure 3 illustrates Example 6 by showing the graphs of the sequence of terms $a_n = 1/[n(n+1)]$ and the sequence $\{s_n\}$ of partial sums. Notice that $a_n \to 0$ and $s_n \to 1$. See Exercises 54 and 55 for two geometric interpretations of Example 6.

Therefore, the given series is convergent and

$$\sum_{n=1}^{\infty} \frac{1}{n(n+1)} = 1$$

EXAMPLE 7 Show that the **harmonic series**

$$\sum_{n=1}^{\infty} \frac{1}{n} = 1 + \frac{1}{2} + \frac{1}{3} + \frac{1}{4} + \cdots$$

is divergent.

SOLUTION

$$s_1 = 1$$

$$s_2 = 1 + \tfrac{1}{2}$$

$$s_4 = 1 + \tfrac{1}{2} + \left(\tfrac{1}{3} + \tfrac{1}{4}\right) > 1 + \tfrac{1}{2} + \left(\tfrac{1}{4} + \tfrac{1}{4}\right) = 1 + \tfrac{2}{2}$$

$$s_8 = 1 + \tfrac{1}{2} + \left(\tfrac{1}{3} + \tfrac{1}{4}\right) + \left(\tfrac{1}{5} + \tfrac{1}{6} + \tfrac{1}{7} + \tfrac{1}{8}\right)$$
$$> 1 + \tfrac{1}{2} + \left(\tfrac{1}{4} + \tfrac{1}{4}\right) + \left(\tfrac{1}{8} + \tfrac{1}{8} + \tfrac{1}{8} + \tfrac{1}{8}\right)$$
$$= 1 + \tfrac{1}{2} + \tfrac{1}{2} + \tfrac{1}{2} = 1 + \tfrac{3}{2}$$

$$s_{16} = 1 + \tfrac{1}{2} + \left(\tfrac{1}{3} + \tfrac{1}{4}\right) + \left(\tfrac{1}{5} + \cdots + \tfrac{1}{8}\right) + \left(\tfrac{1}{9} + \cdots + \tfrac{1}{16}\right)$$
$$> 1 + \tfrac{1}{2} + \left(\tfrac{1}{4} + \tfrac{1}{4}\right) + \left(\tfrac{1}{8} + \cdots + \tfrac{1}{8}\right) + \left(\tfrac{1}{16} + \cdots + \tfrac{1}{16}\right)$$
$$= 1 + \tfrac{1}{2} + \tfrac{1}{2} + \tfrac{1}{2} + \tfrac{1}{2} = 1 + \tfrac{4}{2}$$

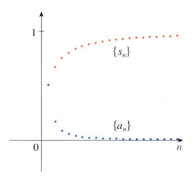

FIGURE 3

Similarly, $s_{32} > 1 + \frac{5}{2}$, $s_{64} > 1 + \frac{6}{2}$, and in general

$$s_{2^n} > 1 + \frac{n}{2}$$

|||| The method used in Example 7 for showing that the harmonic series diverges is due to the French scholar Nicole Oresme (1323–1382).

This shows that $s_{2^n} \to \infty$ as $n \to \infty$ and so $\{s_n\}$ is divergent. Therefore, the harmonic series diverges.

6 Theorem If the series $\sum_{n=1}^{\infty} a_n$ is convergent, then $\lim_{n \to \infty} a_n = 0$.

Proof Let $s_n = a_1 + a_2 + \cdots + a_n$. Then $a_n = s_n - s_{n-1}$. Since $\sum a_n$ is convergent, the sequence $\{s_n\}$ is convergent. Let $\lim_{n \to \infty} s_n = s$. Since $n - 1 \to \infty$ as $n \to \infty$, we also have $\lim_{n \to \infty} s_{n-1} = s$. Therefore

$$\lim_{n \to \infty} a_n = \lim_{n \to \infty} (s_n - s_{n-1}) = \lim_{n \to \infty} s_n - \lim_{n \to \infty} s_{n-1}$$
$$= s - s = 0$$

NOTE 1 ▫ With any *series* $\sum a_n$ we associate two *sequences:* the sequence $\{s_n\}$ of its partial sums and the sequence $\{a_n\}$ of its terms. If $\sum a_n$ is convergent, then the limit of the sequence $\{s_n\}$ is s (the sum of the series) and, as Theorem 6 asserts, the limit of the sequence $\{a_n\}$ is 0.

NOTE 2 ▫ The converse of Theorem 6 is not true in general. If $\lim_{n \to \infty} a_n = 0$, we cannot conclude that $\sum a_n$ is convergent. Observe that for the harmonic series $\sum 1/n$ we have $a_n = 1/n \to 0$ as $n \to \infty$, but we showed in Example 7 that $\sum 1/n$ is divergent.

7 The Test for Divergence If $\lim_{n \to \infty} a_n$ does not exist or if $\lim_{n \to \infty} a_n \neq 0$, then the series $\sum_{n=1}^{\infty} a_n$ is divergent.

The Test for Divergence follows from Theorem 6 because, if the series is not divergent, then it is convergent, and so $\lim_{n \to \infty} a_n = 0$.

EXAMPLE 8 Show that the series $\sum_{n=1}^{\infty} \frac{n^2}{5n^2 + 4}$ diverges.

SOLUTION

$$\lim_{n \to \infty} a_n = \lim_{n \to \infty} \frac{n^2}{5n^2 + 4} = \lim_{n \to \infty} \frac{1}{5 + 4/n^2} = \frac{1}{5} \neq 0$$

So the series diverges by the Test for Divergence.

NOTE 3 ▫ If we find that $\lim_{n \to \infty} a_n \neq 0$, we know that $\sum a_n$ is divergent. If we find that $\lim_{n \to \infty} a_n = 0$, we know *nothing* about the convergence or divergence of $\sum a_n$. Remember the warning in Note 2: If $\lim_{n \to \infty} a_n = 0$, the series $\sum a_n$ might converge or it might diverge.

8 Theorem If $\Sigma \, a_n$ and $\Sigma \, b_n$ are convergent series, then so are the series $\Sigma \, ca_n$ (where c is a constant), $\Sigma \, (a_n + b_n)$, and $\Sigma \, (a_n - b_n)$, and

(i) $\displaystyle\sum_{n=1}^{\infty} ca_n = c \sum_{n=1}^{\infty} a_n$ (ii) $\displaystyle\sum_{n=1}^{\infty} (a_n + b_n) = \sum_{n=1}^{\infty} a_n + \sum_{n=1}^{\infty} b_n$

(iii) $\displaystyle\sum_{n=1}^{\infty} (a_n - b_n) = \sum_{n=1}^{\infty} a_n - \sum_{n=1}^{\infty} b_n$

These properties of convergent series follow from the corresponding Limit Laws for Sequences in Section 11.1. For instance, here is how part (ii) of Theorem 8 is proved:
Let

$$s_n = \sum_{i=1}^{n} a_i \qquad s = \sum_{n=1}^{\infty} a_n \qquad t_n = \sum_{i=1}^{n} b_i \qquad t = \sum_{n=1}^{\infty} b_n$$

The nth partial sum for the series $\Sigma \, (a_n + b_n)$ is

$$u_n = \sum_{i=1}^{n} (a_i + b_i)$$

and, using Equation 5.2.9, we have

$$\lim_{n \to \infty} u_n = \lim_{n \to \infty} \sum_{i=1}^{n} (a_i + b_i) = \lim_{n \to \infty} \left(\sum_{i=1}^{n} a_i + \sum_{i=1}^{n} b_i \right)$$

$$= \lim_{n \to \infty} \sum_{i=1}^{n} a_i + \lim_{n \to \infty} \sum_{i=1}^{n} b_i$$

$$= \lim_{n \to \infty} s_n + \lim_{n \to \infty} t_n = s + t$$

Therefore, $\Sigma \, (a_n + b_n)$ is convergent and its sum is

$$\sum_{n=1}^{\infty} (a_n + b_n) = s + t = \sum_{n=1}^{\infty} a_n + \sum_{n=1}^{\infty} b_n$$

EXAMPLE 9 Find the sum of the series $\displaystyle\sum_{n=1}^{\infty} \left(\frac{3}{n(n+1)} + \frac{1}{2^n} \right)$.

SOLUTION The series $\Sigma \, 1/2^n$ is a geometric series with $a = \frac{1}{2}$ and $r = \frac{1}{2}$, so

$$\sum_{n=1}^{\infty} \frac{1}{2^n} = \frac{\frac{1}{2}}{1 - \frac{1}{2}} = 1$$

In Example 6 we found that

$$\sum_{n=1}^{\infty} \frac{1}{n(n+1)} = 1$$

So, by Theorem 8, the given series is convergent and

$$\sum_{n=1}^{\infty} \left(\frac{3}{n(n+1)} + \frac{1}{2^n} \right) = 3 \sum_{n=1}^{\infty} \frac{1}{n(n+1)} + \sum_{n=1}^{\infty} \frac{1}{2^n}$$

$$= 3 \cdot 1 + 1 = 4$$

NOTE 4 · A finite number of terms doesn't affect the convergence or divergence of a series. For instance, suppose that we were able to show that the series

$$\sum_{n=4}^{\infty} \frac{n}{n^3 + 1}$$

is convergent. Since

$$\sum_{n=1}^{\infty} \frac{n}{n^3 + 1} = \frac{1}{2} + \frac{2}{9} + \frac{3}{28} + \sum_{n=4}^{\infty} \frac{n}{n^3 + 1}$$

it follows that the entire series $\sum_{n=1}^{\infty} n/(n^3 + 1)$ is convergent. Similarly, if it is known that the series $\sum_{n=N+1}^{\infty} a_n$ converges, then the full series

$$\sum_{n=1}^{\infty} a_n = \sum_{n=1}^{N} a_n + \sum_{n=N+1}^{\infty} a_n$$

is also convergent.

12.2 Exercises

1. (a) What is the difference between a sequence and a series?
(b) What is a convergent series? What is a divergent series?

2. Explain what it means to say that $\sum_{n=1}^{\infty} a_n = 5$.

3–8 ▪ Find at least 10 partial sums of the series. Graph both the sequence of terms and the sequence of partial sums on the same screen. Does it appear that the series is convergent or divergent? If it is convergent, find the sum. If it is divergent, explain why.

3. $\sum_{n=1}^{\infty} \frac{12}{(-5)^n}$

4. $\sum_{n=1}^{\infty} \frac{2n^2 - 1}{n^2 + 1}$

5. $\sum_{n=1}^{\infty} \tan n$

6. $\sum_{n=1}^{\infty} (0.6)^{n-1}$

7. $\sum_{n=1}^{\infty} \left(\frac{1}{n^{1.5}} - \frac{1}{(n + 1)^{1.5}} \right)$

8. $\sum_{n=2}^{\infty} \frac{1}{n(n - 1)}$

9. Let $a_n = \frac{2n}{3n + 1}$.
(a) Determine whether $\{a_n\}$ is convergent.
(b) Determine whether $\sum_{n=1}^{\infty} a_n$ is convergent.

10. (a) Explain the difference between

$$\sum_{i=1}^{n} a_i \quad \text{and} \quad \sum_{j=1}^{n} a_j$$

(b) Explain the difference between

$$\sum_{i=1}^{n} a_i \quad \text{and} \quad \sum_{i=1}^{n} a_j$$

11–34 ▪ Determine whether the series is convergent or divergent. If it is convergent, find its sum.

11. $3 + 2 + \frac{4}{3} + \frac{8}{9} + \cdots$

12. $\frac{1}{8} - \frac{1}{4} + \frac{1}{2} - 1 + \cdots$

13. $-2 + \frac{5}{2} - \frac{25}{8} + \frac{125}{32} - \cdots$

14. $1 + 0.4 + 0.16 + 0.064 + \cdots$

15. $\sum_{n=1}^{\infty} 5 \left(\frac{2}{3} \right)^{n-1}$

16. $\sum_{n=1}^{\infty} \frac{(-6)^{n-1}}{5^{n-1}}$

17. $\sum_{n=1}^{\infty} \frac{(-3)^{n-1}}{4^n}$

18. $\sum_{n=0}^{\infty} \frac{1}{(\sqrt{2})^n}$

19. $\sum_{n=0}^{\infty} \frac{\pi^n}{3^{n+1}}$

20. $\sum_{n=1}^{\infty} \frac{e^n}{3^{n-1}}$

21. $\sum_{n=1}^{\infty} \frac{n}{n + 5}$

22. $\sum_{n=1}^{\infty} \frac{3}{n}$

23. $\sum_{n=2}^{\infty} \frac{2}{n^2 - 1}$

24. $\sum_{n=1}^{\infty} \frac{(n + 1)^2}{n(n + 2)}$

25. $\sum_{k=2}^{\infty} \frac{k^2}{k^2 - 1}$

26. $\sum_{n=1}^{\infty} \frac{2}{n^2 + 4n + 3}$

27. $\sum_{n=1}^{\infty} \frac{3^n + 2^n}{6^n}$

28. $\sum_{n=1}^{\infty} [(0.8)^{n-1} - (0.3)^n]$

29. $\sum_{n=1}^{\infty} \sqrt[n]{2}$

30. $\sum_{n=1}^{\infty} \ln \left(\frac{n}{2n + 5} \right)$

31. $\sum_{n=1}^{\infty} \arctan n$

32. $\sum_{k=1}^{\infty} (\cos 1)^k$

33. $\sum_{n=1}^{\infty} \left(\frac{3}{n(n + 3)} + \frac{5}{2^n} \right)$

34. $\sum_{n=1}^{\infty} \left(\frac{3}{5^n} + \frac{2}{n} \right)$

35–40 ▪ Express the number as a ratio of integers.

35. $0.\overline{2} = 0.2222\ldots$

36. $0.\overline{73} = 0.73737373\ldots$

37. $3.\overline{417} = 3.417417417\ldots$

38. $6.2\overline{54}$

39. $0.12\overline{3456}$

40. 5.6021

41–45 ▪ Find the values of x for which the series converges. Find the sum of the series for those values of x.

41. $\sum_{n=1}^{\infty} \dfrac{x^n}{3^n}$

42. $\sum_{n=1}^{\infty} (x-4)^n$

43. $\sum_{n=0}^{\infty} 4^n x^n$

44. $\sum_{n=0}^{\infty} \dfrac{(x+3)^n}{2^n}$

45. $\sum_{n=0}^{\infty} \dfrac{\cos^n x}{2^n}$

---

46. We have seen that the harmonic series is a divergent series whose terms approach 0. Show that

$$\sum_{n=1}^{\infty} \ln\!\left(1 + \dfrac{1}{n}\right)$$

is another series with this property.

CAS 47–48 ▪ Use the partial fraction command on your CAS to find a convenient expression for the partial sum, and then use this expression to find the sum of the series. Check your answer by using the CAS to sum the series directly.

47. $\sum_{n=1}^{\infty} \dfrac{1}{(4n+1)(4n-3)}$

48. $\sum_{n=1}^{\infty} \dfrac{n^2+3n+1}{(n^2+n)^2}$

---

49. If the nth partial sum of a series $\sum_{n=1}^{\infty} a_n$ is

$$s_n = \dfrac{n-1}{n+1}$$

find a_n and $\sum_{n=1}^{\infty} a_n$.

50. If the nth partial sum of a series $\sum_{n=1}^{\infty} a_n$ is $s_n = 3 - n2^{-n}$, find a_n and $\sum_{n=1}^{\infty} a_n$.

51. When money is spent on goods and services, those that receive the money also spend some of it. The people receiving some of the twice-spent money will spend some of that, and so on. Economists call this chain reaction the *multiplier effect*. In a hypothetical isolated community, the local government begins the process by spending D dollars. Suppose that each recipient of spent money spends $100c\%$ and saves $100s\%$ of the money that he or she receives. The values c and s are called the *marginal propensity to consume* and the *marginal propensity to save* and, of course, $c + s = 1$.
(a) Let S_n be the total spending that has been generated after n transactions. Find an equation for S_n.
(b) Show that $\lim_{n\to\infty} S_n = kD$, where $k = 1/s$. The number k is called the *multiplier*. What is the multiplier if the marginal propensity to consume is 80%?

Note: The federal government uses this principle to justify deficit spending. Banks use this principle to justify lending a large percentage of the money that they receive in deposits.

52. A certain ball has the property that each time it falls from a height h onto a hard, level surface, it rebounds to a height rh, where $0 < r < 1$. Suppose that the ball is dropped from an initial height of H meters.

(a) Assuming that the ball continues to bounce indefinitely, find the total distance that it travels. (Use the fact that the ball falls $\tfrac{1}{2}gt^2$ meters in t seconds.)
(b) Calculate the total time that the ball travels.
(c) Suppose that each time the ball strikes the surface with velocity v it rebounds with velocity $-kv$, where $0 < k < 1$. How long will it take for the ball to come to rest?

53. What is the value of c if $\sum_{n=2}^{\infty} (1+c)^{-n} = 2$?

54. Graph the curves $y = x^n$, $0 \leq x \leq 1$, for $n = 0, 1, 2, 3, 4, \ldots$ on a common screen. By finding the areas between successive curves, give a geometric demonstration of the fact, shown in Example 6, that

$$\sum_{n=1}^{\infty} \dfrac{1}{n(n+1)} = 1$$

55. The figure shows two circles C and D of radius 1 that touch at P. T is a common tangent line; C_1 is the circle that touches C, D, and T; C_2 is the circle that touches C, D, and C_1; C_3 is the circle that touches C, D, and C_2. This procedure can be continued indefinitely and produces an infinite sequence of circles $\{C_n\}$. Find an expression for the diameter of C_n and thus provide another geometric demonstration of Example 6.

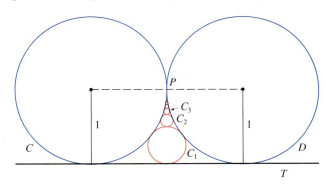

56. A right triangle ABC is given with $\angle A = \theta$ and $|AC| = b$. CD is drawn perpendicular to AB, DE is drawn perpendicular to BC, $EF \perp AB$, and this process is continued indefinitely as shown in the figure. Find the total length of all the perpendiculars

$$|CD| + |DE| + |EF| + |FG| + \cdots$$

in terms of b and θ.

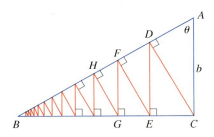

57. What is wrong with the following calculation?

$$0 = 0 + 0 + 0 + \cdots$$
$$= (1 - 1) + (1 - 1) + (1 - 1) + \cdots$$
$$= 1 - 1 + 1 - 1 + 1 - 1 + \cdots$$
$$= 1 + (-1 + 1) + (-1 + 1) + (-1 + 1) + \cdots$$
$$= 1 + 0 + 0 + 0 + \cdots = 1$$

(Guido Ubaldus thought that this proved the existence of God because "something has been created out of nothing.")

58. Suppose that $\sum_{n=1}^{\infty} a_n$ ($a_n \neq 0$) is known to be a convergent series. Prove that $\sum_{n=1}^{\infty} 1/a_n$ is a divergent series.

59. Prove part (i) of Theorem 8.

60. If Σa_n is divergent and $c \neq 0$, show that Σca_n is divergent.

61. If Σa_n is convergent and Σb_n is divergent, show that the series $\Sigma (a_n + b_n)$ is divergent. [*Hint:* Argue by contradiction.]

62. If Σa_n and Σb_n are both divergent, is $\Sigma (a_n + b_n)$ necessarily divergent?

63. Suppose that a series Σa_n has positive terms and its partial sums s_n satisfy the inequality $s_n \leq 1000$ for all n. Explain why Σa_n must be convergent.

64. The Fibonacci sequence was defined in Section 12.1 by the equations
$$f_1 = 1, \quad f_2 = 1, \quad f_n = f_{n-1} + f_{n-2} \quad n \geq 3$$
Show that each of the following statements is true.

(a) $\dfrac{1}{f_{n-1}f_{n+1}} = \dfrac{1}{f_{n-1}f_n} - \dfrac{1}{f_n f_{n+1}}$

(b) $\displaystyle\sum_{n=2}^{\infty} \dfrac{1}{f_{n-1}f_{n+1}} = 1$

(c) $\displaystyle\sum_{n=2}^{\infty} \dfrac{f_n}{f_{n-1}f_{n+1}} = 2$

65. The **Cantor set**, named after the German mathematician Georg Cantor (1845–1918), is constructed as follows. We start with the closed interval [0, 1] and remove the open interval $(\frac{1}{3}, \frac{2}{3})$. That leaves the two intervals $[0, \frac{1}{3}]$ and $[\frac{2}{3}, 1]$ and we remove the open middle third of each. Four intervals remain and again we remove the open middle third of each of them. We continue this procedure indefinitely, at each step removing the open middle third of every interval that remains from the preceding step. The Cantor set consists of the numbers that remain in [0, 1] after all those intervals have been removed.

(a) Show that the total length of all the intervals that are removed is 1. Despite that, the Cantor set contains infinitely many numbers. Give examples of some numbers in the Cantor set.

(b) The **Sierpinski carpet** is a two-dimensional counterpart of the Cantor set. It is constructed by removing the center one-ninth of a square of side 1, then removing the centers of the eight smaller remaining squares, and so on. (The figure shows the first three steps of the construction.) Show that the sum of the areas of the removed squares is 1. This implies that the Sierpinski carpet has area 0.

66. (a) A sequence $\{a_n\}$ is defined recursively by the equation $a_n = \frac{1}{2}(a_{n-1} + a_{n-2})$ for $n \geq 3$, where a_1 and a_2 can be any real numbers. Experiment with various values of a_1 and a_2 and use your calculator to guess the limit of the sequence.

(b) Find $\lim_{n\to\infty} a_n$ in terms of a_1 and a_2 by expressing $a_{n+1} - a_n$ in terms of $a_2 - a_1$ and summing a series.

67. Consider the series
$$\sum_{n=1}^{\infty} \frac{n}{(n+1)!}$$

(a) Find the partial sums s_1, s_2, s_3, and s_4. Do you recognize the denominators? Use the pattern to guess a formula for s_n.

(b) Use mathematical induction to prove your guess.

(c) Show that the given infinite series is convergent, and find its sum.

68. In the figure there are infinitely many circles approaching the vertices of an equilateral triangle, each circle touching other circles and sides of the triangle. If the triangle has sides of length 1, find the total area occupied by the circles.

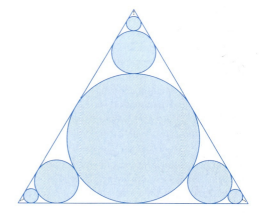

12.3 The Integral Test and Estimates of Sums

In general, it is difficult to find the exact sum of a series. We were able to accomplish this for geometric series and the series $\Sigma\ 1/[n(n+1)]$ because in each of those cases we could find a simple formula for the nth partial sum s_n. But usually it isn't easy to compute $\lim_{n\to\infty} s_n$. Therefore, in the next few sections we develop several tests that enable us to determine whether a series is convergent or divergent without explicitly finding its sum. (In some cases, however, our methods will enable us to find good estimates of the sum.) Our first test involves improper integrals.

We begin by investigating the series whose terms are the reciprocals of the squares of the positive integers:

$$\sum_{n=1}^{\infty} \frac{1}{n^2} = \frac{1}{1^2} + \frac{1}{2^2} + \frac{1}{3^2} + \frac{1}{4^2} + \frac{1}{5^2} + \cdots$$

n	$s_n = \sum_{i=1}^{n} \frac{1}{i^2}$
5	1.4636
10	1.5498
50	1.6251
100	1.6350
500	1.6429
1000	1.6439
5000	1.6447

There's no simple formula for the sum s_n of the first n terms, but the computer-generated table of values given in the margin suggests that the partial sums are approaching a number near 1.64 as $n \to \infty$ and so it looks as if the series is convergent.

We can confirm this impression with a geometric argument. Figure 1 shows the curve $y = 1/x^2$ and rectangles that lie below the curve. The base of each rectangle is an interval of length 1; the height is equal to the value of the function $y = 1/x^2$ at the right endpoint of the interval. So the sum of the areas of the rectangles is

$$\frac{1}{1^2} + \frac{1}{2^2} + \frac{1}{3^2} + \frac{1}{4^2} + \frac{1}{5^2} + \cdots = \sum_{n=1}^{\infty} \frac{1}{n^2}$$

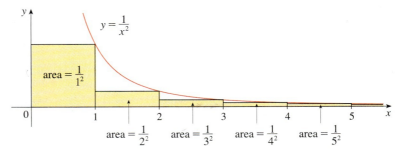

FIGURE 1

If we exclude the first rectangle, the total area of the remaining rectangles is smaller than the area under the curve $y = 1/x^2$ for $x \geq 1$, which is the value of the integral $\int_1^{\infty} (1/x^2)\, dx$. In Section 8.8 we discovered that this improper integral is convergent and has value 1. So the picture shows that all the partial sums are less than

$$\frac{1}{1^2} + \int_1^{\infty} \frac{1}{x^2}\, dx = 2$$

Thus, the partial sums are bounded. We also know that the partial sums are increasing (because all the terms are positive). Therefore, the partial sums converge (by the Monotonic Sequence Theorem) and so the series is convergent. The sum of the series (the limit of the partial sums) is also less than 2:

$$\sum_{n=1}^{\infty} \frac{1}{n^2} = \frac{1}{1^2} + \frac{1}{2^2} + \frac{1}{3^2} + \frac{1}{4^2} + \cdots < 2$$

[The exact sum of this series was found by the Swiss mathematician Leonhard Euler (1707–1783) to be $\pi^2/6$, but the proof of this fact is quite difficult. (See Problem 6 in the Problems Plus following Chapter 16.)]

Now let's look at the series

$$\sum_{n=1}^{\infty} \frac{1}{\sqrt{n}} = \frac{1}{\sqrt{1}} + \frac{1}{\sqrt{2}} + \frac{1}{\sqrt{3}} + \frac{1}{\sqrt{4}} + \frac{1}{\sqrt{5}} + \cdots$$

n	$s_n = \sum_{i=1}^{n} \frac{1}{\sqrt{i}}$
5	3.2317
10	5.0210
50	12.7524
100	18.5896
500	43.2834
1000	61.8010
5000	139.9681

The table of values of s_n suggests that the partial sums aren't approaching a finite number, so we suspect that the given series may be divergent. Again we use a picture for confirmation. Figure 2 shows the curve $y = 1/\sqrt{x}$, but this time we use rectangles whose tops lie *above* the curve.

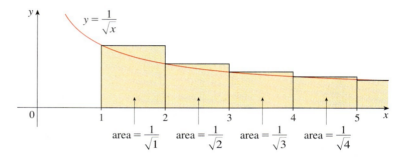

FIGURE 2

The base of each rectangle is an interval of length 1. The height is equal to the value of the function $y = 1/\sqrt{x}$ at the *left* endpoint of the interval. So the sum of the areas of all the rectangles is

$$\frac{1}{\sqrt{1}} + \frac{1}{\sqrt{2}} + \frac{1}{\sqrt{3}} + \frac{1}{\sqrt{4}} + \frac{1}{\sqrt{5}} + \cdots = \sum_{n=1}^{\infty} \frac{1}{\sqrt{n}}$$

This total area is greater than the area under the curve $y = 1/\sqrt{x}$ for $x \geq 1$, which is equal to the integral $\int_1^{\infty} (1/\sqrt{x}) \, dx$. But we know from Section 8.8 that this improper integral is divergent. In other words, the area under the curve is infinite. So the sum of the series must be infinite; that is, the series is divergent.

The same sort of geometric reasoning that we used for these two series can be used to prove the following test. (The proof is given at the end of this section.)

The Integral Test Suppose f is a continuous, positive, decreasing function on $[1, \infty)$ and let $a_n = f(n)$. Then the series $\sum_{n=1}^{\infty} a_n$ is convergent if and only if the improper integral $\int_1^{\infty} f(x) \, dx$ is convergent. In other words:

(i) If $\int_1^{\infty} f(x) \, dx$ is convergent, then $\sum_{n=1}^{\infty} a_n$ is convergent.

(ii) If $\int_1^{\infty} f(x) \, dx$ is divergent, then $\sum_{n=1}^{\infty} a_n$ is divergent.

NOTE ▫ When we use the Integral Test it is not necessary to start the series or the integral at $n = 1$. For instance, in testing the series

$$\sum_{n=4}^{\infty} \frac{1}{(n-3)^2} \quad \text{we use} \quad \int_4^{\infty} \frac{1}{(x-3)^2} \, dx$$

Also, it is not necessary that f be always decreasing. What is important is that f be *ultimately* decreasing, that is, decreasing for x larger than some number N. Then $\sum_{n=N}^{\infty} a_n$ is convergent, so $\sum_{n=1}^{\infty} a_n$ is convergent by Note 4 of Section 12.2.

EXAMPLE 1 Test the series $\displaystyle\sum_{n=1}^{\infty} \frac{1}{n^2 + 1}$ for convergence or divergence.

SOLUTION The function $f(x) = 1/(x^2 + 1)$ is continuous, positive, and decreasing on $[1, \infty)$ so we use the Integral Test:

$$\int_1^\infty \frac{1}{x^2 + 1}\, dx = \lim_{t \to \infty} \int_1^t \frac{1}{x^2 + 1}\, dx = \lim_{t \to \infty} \tan^{-1} x \Big]_1^t$$

$$= \lim_{t \to \infty} \left(\tan^{-1} t - \frac{\pi}{4} \right) = \frac{\pi}{2} - \frac{\pi}{4} = \frac{\pi}{4}$$

Thus, $\int_1^\infty 1/(x^2+1)\, dx$ is a convergent integral and so, by the Integral Test, the series $\sum 1/(n^2 + 1)$ is convergent.

EXAMPLE 2 For what values of p is the series $\displaystyle\sum_{n=1}^{\infty} \frac{1}{n^p}$ convergent?

SOLUTION If $p < 0$, then $\lim_{n \to \infty} (1/n^p) = \infty$. If $p = 0$, then $\lim_{n \to \infty} (1/n^p) = 1$. In either case $\lim_{n \to \infty} (1/n^p) \ne 0$, so the given series diverges by the Test for Divergence (12.2.7).

If $p > 0$, then the function $f(x) = 1/x^p$ is clearly continuous, positive, and decreasing on $[1, \infty)$. We found in Chapter 8 [see (8.8.2)] that

$$\int_1^\infty \frac{1}{x^p}\, dx \text{ converges if } p > 1 \text{ and diverges if } p \le 1$$

It follows from the Integral Test that the series $\sum 1/n^p$ converges if $p > 1$ and diverges if $0 < p \le 1$. (For $p = 1$, this series is the harmonic series discussed in Example 7 in Section 12.2.)

The series in Example 2 is called the *p-series*. It is important in the rest of this chapter, so we summarize the results of Example 2 for future reference as follows.

> **1** The *p*-series $\displaystyle\sum_{n=1}^{\infty} \frac{1}{n^p}$ is convergent if $p > 1$ and divergent if $p \le 1$.

EXAMPLE 3
(a) The series

$$\sum_{n=1}^{\infty} \frac{1}{n^3} = \frac{1}{1^3} + \frac{1}{2^3} + \frac{1}{3^3} + \frac{1}{4^3} + \cdots$$

is convergent because it is a *p*-series with $p = 3 > 1$.
(b) The series

$$\sum_{n=1}^{\infty} \frac{1}{n^{1/3}} = \sum_{n=1}^{\infty} \frac{1}{\sqrt[3]{n}} = 1 + \frac{1}{\sqrt[3]{2}} + \frac{1}{\sqrt[3]{3}} + \frac{1}{\sqrt[3]{4}} + \cdots$$

is divergent because it is a *p*-series with $p = \frac{1}{3} < 1$.

NOTE We should *not* infer from the Integral Test that the sum of the series is equal to the value of the integral. In fact,

$$\sum_{n=1}^{\infty} \frac{1}{n^2} = \frac{\pi^2}{6} \quad \text{whereas} \quad \int_1^{\infty} \frac{1}{x^2} dx = 1$$

Therefore, in general,

$$\sum_{n=1}^{\infty} a_n \neq \int_1^{\infty} f(x) \, dx$$

EXAMPLE 4 Determine whether the series $\sum_{n=1}^{\infty} \frac{\ln n}{n}$ converges or diverges.

SOLUTION The function $f(x) = (\ln x)/x$ is positive and continuous for $x > 1$ because the logarithm function is continuous. But it is not obvious whether or not f is decreasing, so we compute its derivative:

$$f'(x) = \frac{(1/x)x - \ln x}{x^2} = \frac{1 - \ln x}{x^2}$$

Thus, $f'(x) < 0$ when $\ln x > 1$, that is, $x > e$. It follows that f is decreasing when $x > e$ and so we can apply the Integral Test:

$$\int_1^{\infty} \frac{\ln x}{x} dx = \lim_{t \to \infty} \int_1^t \frac{\ln x}{x} dx = \lim_{t \to \infty} \frac{(\ln x)^2}{2} \bigg]_1^t$$

$$= \lim_{t \to \infty} \frac{(\ln t)^2}{2} = \infty$$

Since this improper integral is divergent, the series $\sum (\ln n)/n$ is also divergent by the Integral Test.

Estimating the Sum of a Series

Suppose we have been able to use the Integral Test to show that a series $\sum a_n$ is convergent and we now want to find an approximation to the sum s of the series. Of course, any partial sum s_n is an approximation to s because $\lim_{n \to \infty} s_n = s$. But how good is such an approximation? To find out, we need to estimate the size of the **remainder**

$$R_n = s - s_n = a_{n+1} + a_{n+2} + a_{n+3} + \cdots$$

The remainder R_n is the error made when s_n, the sum of the first n terms, is used as an approximation to the total sum.

We use the same notation and ideas as in the Integral Test, assuming that f is decreasing on $[n, \infty)$. Comparing the areas of the rectangles with the area under $y = f(x)$ for $x > n$ in Figure 3, we see that

$$R_n = a_{n+1} + a_{n+2} + \cdots \leq \int_n^{\infty} f(x) \, dx$$

Similarly, we see from Figure 4 that

$$R_n = a_{n+1} + a_{n+2} + \cdots \geq \int_{n+1}^{\infty} f(x) \, dx$$

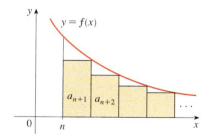

FIGURE 3

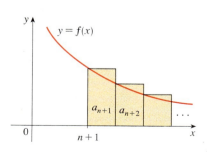

FIGURE 4

So we have proved the following error estimate.

> **2 Remainder Estimate for the Integral Test** Suppose $f(k) = a_k$, where f is a continuous, positive, decreasing function for $x \geq n$ and $\Sigma\, a_n$ is convergent. If $R_n = s - s_n$, then
> $$\int_{n+1}^{\infty} f(x)\, dx \leq R_n \leq \int_{n}^{\infty} f(x)\, dx$$

EXAMPLE 5
(a) Approximate the sum of the series $\Sigma\, 1/n^3$ by using the sum of the first 10 terms. Estimate the error involved in this approximation.
(b) How many terms are required to ensure that the sum is accurate to within 0.0005?

SOLUTION In both parts (a) and (b) we need to know $\int_n^{\infty} f(x)\, dx$. With $f(x) = 1/x^3$, we have

$$\int_n^{\infty} \frac{1}{x^3}\, dx = \lim_{t \to \infty} \left[-\frac{1}{2x^2} \right]_n^t = \lim_{t \to \infty} \left(-\frac{1}{2t^2} + \frac{1}{2n^2} \right) = \frac{1}{2n^2}$$

(a)
$$\sum_{n=1}^{\infty} \frac{1}{n^3} \approx s_{10} = \frac{1}{1^3} + \frac{1}{2^3} + \frac{1}{3^3} + \cdots + \frac{1}{10^3} \approx 1.1975$$

According to the remainder estimate in (2), we have

$$R_{10} \leq \int_{10}^{\infty} \frac{1}{x^3}\, dx = \frac{1}{2(10)^2} = \frac{1}{200}$$

So the size of the error is at most 0.005.

(b) Accuracy to within 0.0005 means that we have to find a value of n such that $R_n \leq 0.0005$. Since

$$R_n \leq \int_n^{\infty} \frac{1}{x^3}\, dx = \frac{1}{2n^2}$$

we want
$$\frac{1}{2n^2} < 0.0005$$

Solving this inequality, we get

$$n^2 > \frac{1}{0.001} = 1000 \quad \text{or} \quad n > \sqrt{1000} \approx 31.6$$

We need 32 terms to ensure accuracy to within 0.0005.

If we add s_n to each side of the inequalities in (2), we get

> **3**
> $$s_n + \int_{n+1}^{\infty} f(x)\, dx \leq s \leq s_n + \int_n^{\infty} f(x)\, dx$$

because $s_n + R_n = s$. The inequalities in (3) give a lower bound and an upper bound for s. They provide a more accurate approximation to the sum of the series than the partial sum s_n does.

EXAMPLE 6 Use (3) with $n = 10$ to estimate the sum of the series $\sum_{n=1}^{\infty} \dfrac{1}{n^3}$.

SOLUTION The inequalities in (3) become

$$s_{10} + \int_{11}^{\infty} \frac{1}{x^3}\,dx \leq s \leq s_{10} + \int_{10}^{\infty} \frac{1}{x^3}\,dx$$

From Example 5 we know that

$$\int_{n}^{\infty} \frac{1}{x^3}\,dx = \frac{1}{2n^2}$$

so

$$s_{10} + \frac{1}{2(11)^2} \leq s \leq s_{10} + \frac{1}{2(10)^2}$$

Using $s_{10} \approx 1.197532$, we get

$$1.201664 \leq s \leq 1.202532$$

If we approximate s by the midpoint of this interval, then the error is at most half the length of the interval. So

$$\sum_{n=1}^{\infty} \frac{1}{n^3} \approx 1.2021 \qquad \text{with error} < 0.0005$$

If we compare Example 6 with Example 5, we see that the improved estimate in (3) can be much better than the estimate $s \approx s_n$. To make the error smaller than 0.0005 we had to use 32 terms in Example 5 but only 10 terms in Example 6.

Proof of the Integral Test

We have already seen the basic idea behind the proof of the Integral Test in Figures 1 and 2 for the series $\sum 1/n^2$ and $\sum 1/\sqrt{n}$. For the general series $\sum a_n$ look at Figures 5 and 6. The area of the first shaded rectangle in Figure 5 is the value of f at the right endpoint of $[1, 2]$, that is, $f(2) = a_2$. So, comparing the areas of the shaded rectangles with the area under $y = f(x)$ from 1 to n, we see that

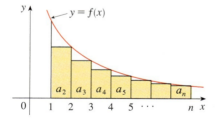

FIGURE 5

4
$$a_2 + a_3 + \cdots + a_n \leq \int_{1}^{n} f(x)\,dx$$

(Notice that this inequality depends on the fact that f is decreasing.) Likewise, Figure 6 shows that

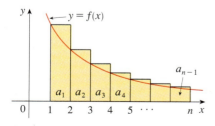

FIGURE 6

5
$$\int_{1}^{n} f(x)\,dx \leq a_1 + a_2 + \cdots + a_{n-1}$$

(i) If $\int_1^\infty f(x)\,dx$ is convergent, then (4) gives

$$\sum_{i=2}^n a_i \leq \int_1^n f(x)\,dx \leq \int_1^\infty f(x)\,dx$$

since $f(x) \geq 0$. Therefore

$$s_n = a_1 + \sum_{i=2}^n a_i \leq a_1 + \int_1^\infty f(x)\,dx = M, \text{ say}$$

Since $s_n \leq M$ for all n, the sequence $\{s_n\}$ is bounded above. Also

$$s_{n+1} = s_n + a_{n+1} \geq s_n$$

since $a_{n+1} = f(n+1) \geq 0$. Thus, $\{s_n\}$ is an increasing bounded sequence and so it is convergent by the Monotonic Sequence Theorem (12.1.11). This means that $\Sigma\, a_n$ is convergent.

(ii) If $\int_1^\infty f(x)\,dx$ is divergent, then $\int_1^n f(x)\,dx \to \infty$ as $n \to \infty$ because $f(x) \geq 0$. But (5) gives

$$\int_1^n f(x)\,dx \leq \sum_{i=1}^{n-1} a_i = s_{n-1}$$

and so $s_{n-1} \to \infty$. This implies that $s_n \to \infty$ and so $\Sigma\, a_n$ diverges.

12.3 Exercises

1. Draw a picture to show that

$$\sum_{n=2}^\infty \frac{1}{n^{1.3}} < \int_1^\infty \frac{1}{x^{1.3}}\,dx$$

What can you conclude about the series?

2. Suppose f is a continuous positive decreasing function for $x \geq 1$ and $a_n = f(n)$. By drawing a picture, rank the following three quantities in increasing order:

$$\int_1^6 f(x)\,dx \qquad \sum_{i=1}^5 a_i \qquad \sum_{i=2}^6 a_i$$

3–8 Use the Integral Test to determine whether the series is convergent or divergent.

3. $\sum_{n=1}^\infty \frac{1}{n^4}$

4. $\sum_{n=1}^\infty \frac{1}{\sqrt[4]{n}}$

5. $\sum_{n=1}^\infty \frac{1}{3n+1}$

6. $\sum_{n=1}^\infty e^{-n}$

7. $\sum_{n=1}^\infty ne^{-n}$

8. $\sum_{n=1}^\infty \frac{n+2}{n+1}$

9–24 Determine whether the series is convergent or divergent.

9. $\sum_{n=1}^\infty \frac{2}{n^{0.85}}$

10. $\sum_{n=1}^\infty (n^{-1.4} + 3n^{-1.2})$

11. $1 + \frac{1}{8} + \frac{1}{27} + \frac{1}{64} + \frac{1}{125} + \cdots$

12. $1 + \frac{1}{2\sqrt{2}} + \frac{1}{3\sqrt{3}} + \frac{1}{4\sqrt{4}} + \frac{1}{5\sqrt{5}} + \cdots$

13. $\sum_{n=1}^\infty \frac{5 - 2\sqrt{n}}{n^3}$

14. $\sum_{n=3}^\infty \frac{5}{n-2}$

15. $\sum_{n=1}^\infty \frac{1}{n^2+4}$

16. $\sum_{n=1}^\infty \frac{3n+2}{n(n+1)}$

17. $\sum_{n=1}^\infty \frac{n}{n^2+1}$

18. $\sum_{n=1}^\infty \frac{1}{n^2-4n+5}$

19. $\sum_{n=1}^\infty ne^{-n^2}$

20. $\sum_{n=1}^\infty \frac{\ln n}{n^2}$

21. $\sum_{n=2}^\infty \frac{1}{n \ln n}$

22. $\sum_{n=1}^\infty \frac{n}{n^4+1}$

23. $\sum_{n=1}^\infty \frac{1}{n^3+n}$

24. $\sum_{n=3}^\infty \frac{1}{n \ln n \ln(\ln n)}$

25–28 Find the values of p for which the series is convergent.

25. $\sum_{n=2}^\infty \frac{1}{n(\ln n)^p}$

26. $\sum_{n=3}^\infty \frac{1}{n \ln n\, [\ln(\ln n)]^p}$

27. $\sum_{n=1}^{\infty} n(1 + n^2)^p$

28. $\sum_{n=1}^{\infty} \dfrac{\ln n}{n^p}$

29. The Riemann zeta-function ζ is defined by
$$\zeta(x) = \sum_{n=1}^{\infty} \dfrac{1}{n^x}$$
and is used in number theory to study the distribution of prime numbers. What is the domain of ζ?

30. (a) Find the partial sum s_{10} of the series $\sum_{n=1}^{\infty} 1/n^4$. Estimate the error in using s_{10} as an approximation to the sum of the series.
(b) Use (3) with $n = 10$ to give an improved estimate of the sum.
(c) Find a value of n so that s_n is within 0.00001 of the sum.

31. (a) Use the sum of the first 10 terms to estimate the sum of the series $\sum_{n=1}^{\infty} 1/n^2$. How good is this estimate?
(b) Improve this estimate using (3) with $n = 10$.
(c) Find a value of n that will ensure that the error in the approximation $s \approx s_n$ is less than 0.001.

32. Find the sum of the series $\sum_{n=1}^{\infty} 1/n^5$ correct to three decimal places.

33. Estimate $\sum_{n=1}^{\infty} n^{-3/2}$ to within 0.01.

34. How many terms of the series $\sum_{n=2}^{\infty} 1/[n(\ln n)^2]$ would you need to add to find its sum to within 0.01?

35. Show that if we want to approximate the sum of the series $\sum_{n=1}^{\infty} n^{-1.001}$ so that the error is less than 5 in the ninth decimal place, then we need to add more than $10^{11,301}$ terms!

[CAS] **36.** (a) Show that the series $\sum_{n=1}^{\infty} (\ln n)^2/n^2$ is convergent.
(b) Find an upper bound for the error in the approximation $s \approx s_n$.
(c) What is the smallest value of n such that this upper bound is less than 0.05?
(d) Find s_n for this value of n.

37. (a) Use (4) to show that if s_n is the nth partial sum of the harmonic series, then
$$s_n \leq 1 + \ln n$$
(b) The harmonic series diverges, but very slowly. Use part (a) to show that the sum of the first million terms is less than 15 and the sum of the first billion terms is less than 22.

38. Use the following steps to show that the sequence
$$t_n = 1 + \dfrac{1}{2} + \dfrac{1}{3} + \cdots + \dfrac{1}{n} - \ln n$$
has a limit. (The value of the limit is denoted by γ and is called Euler's constant.)
(a) Draw a picture like Figure 6 with $f(x) = 1/x$ and interpret t_n as an area [or use (5)] to show that $t_n > 0$ for all n.
(b) Interpret
$$t_n - t_{n+1} = [\ln(n + 1) - \ln n] - \dfrac{1}{n + 1}$$
as a difference of areas to show that $t_n - t_{n+1} > 0$. Therefore, $\{t_n\}$ is a decreasing sequence.
(c) Use the Monotonic Sequence Theorem to show that $\{t_n\}$ is convergent.

39. Find all positive values of b for which the series $\sum_{n=1}^{\infty} b^{\ln n}$ converges.

12.4 The Comparison Tests

In the comparison tests the idea is to compare a given series with a series that is known to be convergent or divergent. For instance, the series

$$\sum_{n=1}^{\infty} \dfrac{1}{2^n + 1}$$

reminds us of the series $\sum_{n=1}^{\infty} 1/2^n$, which is a geometric series with $a = \tfrac{1}{2}$ and $r = \tfrac{1}{2}$ and is therefore convergent. Because the series (1) is so similar to a convergent series, we have the feeling that it too must be convergent. Indeed, it is. The inequality

$$\dfrac{1}{2^n + 1} < \dfrac{1}{2^n}$$

shows that our given series (1) has smaller terms than those of the geometric series and therefore all its partial sums are also smaller than 1 (the sum of the geometric series). This means that its partial sums form a bounded increasing sequence, which is convergent. It also follows that the sum of the series is less than the sum of the geometric series:

$$\sum_{n=1}^{\infty} \frac{1}{2^n + 1} < 1$$

Similar reasoning can be used to prove the following test, which applies only to series whose terms are positive. The first part says that if we have a series whose terms are *smaller* than those of a known *convergent* series, then our series is also convergent. The second part says that if we start with a series whose terms are *larger* than those of a known *divergent* series, then it too is divergent.

> **The Comparison Test** Suppose that $\Sigma\, a_n$ and $\Sigma\, b_n$ are series with positive terms.
> (i) If $\Sigma\, b_n$ is convergent and $a_n \le b_n$ for all n, then $\Sigma\, a_n$ is also convergent.
> (ii) If $\Sigma\, b_n$ is divergent and $a_n \ge b_n$ for all n, then $\Sigma\, a_n$ is also divergent.

|||| It is important to keep in mind the distinction between a sequence and a series. A sequence is a list of numbers, whereas a series is a sum. With every series $\Sigma\, a_n$ there are associated two sequences: the sequence $\{a_n\}$ of terms and the sequence $\{s_n\}$ of partial sums.

Proof
(i) Let

$$s_n = \sum_{i=1}^{n} a_i \qquad t_n = \sum_{i=1}^{n} b_i \qquad t = \sum_{n=1}^{\infty} b_n$$

Since both series have positive terms, the sequences $\{s_n\}$ and $\{t_n\}$ are increasing ($s_{n+1} = s_n + a_{n+1} \ge s_n$). Also $t_n \to t$, so $t_n \le t$ for all n. Since $a_i \le b_i$, we have $s_n \le t_n$. Thus, $s_n \le t$ for all n. This means that $\{s_n\}$ is increasing and bounded above and therefore converges by the Monotonic Sequence Theorem. Thus, $\Sigma\, a_n$ converges.

(ii) If $\Sigma\, b_n$ is divergent, then $t_n \to \infty$ (since $\{t_n\}$ is increasing). But $a_i \ge b_i$ so $s_n \ge t_n$. Thus, $s_n \to \infty$. Therefore, $\Sigma\, a_n$ diverges.

Standard Series for Use with the Comparison Test

In using the Comparison Test we must, of course, have some known series $\Sigma\, b_n$ for the purpose of comparison. Most of the time we use either a *p*-series $\left[\Sigma\, 1/n^p \text{ converges if } p > 1 \text{ and diverges if } p \le 1;\ \text{see (12.3.1)}\right]$ or a geometric series $\left[\Sigma\, ar^{n-1} \text{ converges if } |r| < 1 \text{ and diverges if } |r| \ge 1;\ \text{see (12.2.4)}\right]$.

EXAMPLE 1 Determine whether the series $\displaystyle\sum_{n=1}^{\infty} \frac{5}{2n^2 + 4n + 3}$ converges or diverges.

SOLUTION For large n the dominant term in the denominator is $2n^2$ so we compare the given series with the series $\Sigma\, 5/(2n^2)$. Observe that

$$\frac{5}{2n^2 + 4n + 3} < \frac{5}{2n^2}$$

because the left side has a bigger denominator. (In the notation of the Comparison Test, a_n is the left side and b_n is the right side.) We know that

$$\sum_{n=1}^{\infty} \frac{5}{2n^2} = \frac{5}{2} \sum_{n=1}^{\infty} \frac{1}{n^2}$$

is convergent because it's a constant times a *p*-series with $p = 2 > 1$. Therefore

$$\sum_{n=1}^{\infty} \frac{5}{2n^2 + 4n + 3}$$

is convergent by part (i) of the Comparison Test.

NOTE 1 ▫ Although the condition $a_n \leq b_n$ or $a_n \geq b_n$ in the Comparison Test is given for all n, we need verify only that it holds for $n \geq N$, where N is some fixed integer, because the convergence of a series is not affected by a finite number of terms. This is illustrated in the next example.

EXAMPLE 2 Test the series $\sum_{n=1}^{\infty} \dfrac{\ln n}{n}$ for convergence or divergence.

SOLUTION This series was tested (using the Integral Test) in Example 4 in Section 12.3, but it is also possible to test it by comparing it with the harmonic series. Observe that $\ln n > 1$ for $n \geq 3$ and so

$$\frac{\ln n}{n} > \frac{1}{n} \qquad n \geq 3$$

We know that $\Sigma\, 1/n$ is divergent (p-series with $p = 1$). Thus, the given series is divergent by the Comparison Test.

NOTE 2 ▫ The terms of the series being tested must be smaller than those of a convergent series or larger than those of a divergent series. If the terms are larger than the terms of a convergent series or smaller than those of a divergent series, then the Comparison Test doesn't apply. Consider, for instance, the series

$$\sum_{n=1}^{\infty} \frac{1}{2^n - 1}$$

The inequality

$$\frac{1}{2^n - 1} > \frac{1}{2^n}$$

is useless as far as the Comparison Test is concerned because $\Sigma\, b_n = \Sigma \left(\frac{1}{2}\right)^n$ is convergent and $a_n > b_n$. Nonetheless, we have the feeling that $\Sigma\, 1/(2^n - 1)$ ought to be convergent because it is very similar to the convergent geometric series $\Sigma \left(\frac{1}{2}\right)^n$. In such cases the following test can be used.

|||| Exercises 40 and 41 deal with the cases $c = 0$ and $c = \infty$.

The Limit Comparison Test Suppose that $\Sigma\, a_n$ and $\Sigma\, b_n$ are series with positive terms. If

$$\lim_{n \to \infty} \frac{a_n}{b_n} = c$$

where c is a finite number and $c > 0$, then either both series converge or both diverge.

Proof Let m and M be positive numbers such that $m < c < M$. Because a_n/b_n is close to c for large n, there is an integer N such that

$$m < \frac{a_n}{b_n} < M \qquad \text{when } n > N$$

and so

$$mb_n < a_n < Mb_n \qquad \text{when } n > N$$

If $\Sigma\, b_n$ converges, so does $\Sigma\, Mb_n$. Thus, $\Sigma\, a_n$ converges by part (i) of the Comparison Test. If $\Sigma\, b_n$ diverges, so does $\Sigma\, mb_n$ and part (ii) of the Comparison Test shows that $\Sigma\, a_n$ diverges.

EXAMPLE 3 Test the series $\displaystyle\sum_{n=1}^{\infty} \frac{1}{2^n - 1}$ for convergence or divergence.

SOLUTION We use the Limit Comparison Test with

$$a_n = \frac{1}{2^n - 1} \qquad b_n = \frac{1}{2^n}$$

and obtain

$$\lim_{n \to \infty} \frac{a_n}{b_n} = \lim_{n \to \infty} \frac{1/(2^n - 1)}{1/2^n} = \lim_{n \to \infty} \frac{2^n}{2^n - 1} = \lim_{n \to \infty} \frac{1}{1 - 1/2^n} = 1 > 0$$

Since this limit exists and $\Sigma\, 1/2^n$ is a convergent geometric series, the given series converges by the Limit Comparison Test.

EXAMPLE 4 Determine whether the series $\displaystyle\sum_{n=1}^{\infty} \frac{2n^2 + 3n}{\sqrt{5 + n^5}}$ converges or diverges.

SOLUTION The dominant part of the numerator is $2n^2$ and the dominant part of the denominator is $\sqrt{n^5} = n^{5/2}$. This suggests taking

$$a_n = \frac{2n^2 + 3n}{\sqrt{5 + n^5}} \qquad b_n = \frac{2n^2}{n^{5/2}} = \frac{2}{n^{1/2}}$$

$$\lim_{n \to \infty} \frac{a_n}{b_n} = \lim_{n \to \infty} \frac{2n^2 + 3n}{\sqrt{5 + n^5}} \cdot \frac{n^{1/2}}{2} = \lim_{n \to \infty} \frac{2n^{5/2} + 3n^{3/2}}{2\sqrt{5 + n^5}}$$

$$= \lim_{n \to \infty} \frac{2 + \dfrac{3}{n}}{2\sqrt{\dfrac{5}{n^5} + 1}} = \frac{2 + 0}{2\sqrt{0 + 1}} = 1$$

Since $\Sigma\, b_n = 2\, \Sigma\, 1/n^{1/2}$ is divergent (p-series with $p = \tfrac{1}{2} < 1$), the given series diverges by the Limit Comparison Test.

Notice that in testing many series we find a suitable comparison series $\Sigma\, b_n$ by keeping only the highest powers in the numerator and denominator.

Estimating Sums

If we have used the Comparison Test to show that a series $\Sigma\, a_n$ converges by comparison with a series $\Sigma\, b_n$, then we may be able to estimate the sum $\Sigma\, a_n$ by comparing remainders. As in Section 12.3, we consider the remainder

$$R_n = s - s_n = a_{n+1} + a_{n+2} + \cdots$$

For the comparison series $\Sigma\, b_n$ we consider the corresponding remainder

$$T_n = t - t_n = b_{n+1} + b_{n+2} + \cdots$$

Since $a_n \leq b_n$ for all n, we have $R_n \leq T_n$. If Σb_n is a p-series, we can estimate its remainder T_n as in Section 12.3. If Σb_n is a geometric series, then T_n is the sum of a geometric series and we can sum it exactly (see Exercises 35 and 36). In either case we know that R_n is smaller than T_n.

EXAMPLE 5 Use the sum of the first 100 terms to approximate the sum of the series $\Sigma 1/(n^3 + 1)$. Estimate the error involved in this approximation.

SOLUTION Since

$$\frac{1}{n^3 + 1} < \frac{1}{n^3}$$

the given series is convergent by the Comparison Test. The remainder T_n for the comparison series $\Sigma 1/n^3$ was estimated in Example 5 in Section 12.3 using the Remainder Estimate for the Integral Test. There we found that

$$T_n \leq \int_n^\infty \frac{1}{x^3} \, dx = \frac{1}{2n^2}$$

Therefore, the remainder R_n for the given series satisfies

$$R_n \leq T_n \leq \frac{1}{2n^2}$$

With $n = 100$ we have

$$R_{100} \leq \frac{1}{2(100)^2} = 0.00005$$

Using a programmable calculator or a computer, we find that

$$\sum_{n=1}^\infty \frac{1}{n^3 + 1} \approx \sum_{n=1}^{100} \frac{1}{n^3 + 1} \approx 0.6864538$$

with error less than 0.00005.

12.4 Exercises

1. Suppose Σa_n and Σb_n are series with positive terms and Σb_n is known to be convergent.
(a) If $a_n > b_n$ for all n, what can you say about Σa_n? Why?
(b) If $a_n < b_n$ for all n, what can you say about Σa_n? Why?

2. Suppose Σa_n and Σb_n are series with positive terms and Σb_n is known to be divergent.
(a) If $a_n > b_n$ for all n, what can you say about Σa_n? Why?
(b) If $a_n < b_n$ for all n, what can you say about Σa_n? Why?

3–32 ▮▮▮▮ Determine whether the series converges or diverges.

3. $\displaystyle\sum_{n=1}^\infty \frac{1}{n^2 + n + 1}$

4. $\displaystyle\sum_{n=1}^\infty \frac{2}{n^3 + 4}$

5. $\displaystyle\sum_{n=1}^\infty \frac{5}{2 + 3^n}$

6. $\displaystyle\sum_{n=2}^\infty \frac{1}{n - \sqrt{n}}$

7. $\displaystyle\sum_{n=1}^\infty \frac{n + 1}{n^2}$

8. $\displaystyle\sum_{n=1}^\infty \frac{4 + 3^n}{2^n}$

9. $\displaystyle\sum_{n=1}^\infty \frac{\cos^2 n}{n^2 + 1}$

10. $\displaystyle\sum_{n=1}^\infty \frac{n^2 - 1}{3n^4 + 1}$

11. $\displaystyle\sum_{n=2}^\infty \frac{n^2 + 1}{n^3 - 1}$

12. $\displaystyle\sum_{n=0}^\infty \frac{1 + \sin n}{10^n}$

13. $\displaystyle\sum_{n=1}^\infty \frac{n - 1}{n 4^n}$

14. $\displaystyle\sum_{n=2}^\infty \frac{\sqrt{n}}{n - 1}$

15. $\displaystyle\sum_{n=1}^\infty \frac{2 + (-1)^n}{n\sqrt{n}}$

16. $\displaystyle\sum_{n=1}^\infty \frac{1}{\sqrt{n^3 + 1}}$

17. $\displaystyle\sum_{n=1}^\infty \frac{1}{\sqrt{n^2 + 1}}$

18. $\displaystyle\sum_{n=1}^\infty \frac{1}{2n + 3}$

19. $\sum_{n=1}^{\infty} \dfrac{2^n}{1+3^n}$

20. $\sum_{n=1}^{\infty} \dfrac{1+2^n}{1+3^n}$

21. $\sum_{n=1}^{\infty} \dfrac{1}{1+\sqrt{n}}$

22. $\sum_{n=3}^{\infty} \dfrac{n+2}{(n+1)^3}$

23. $\sum_{n=1}^{\infty} \dfrac{5+2n}{(1+n^2)^2}$

24. $\sum_{n=1}^{\infty} \dfrac{n^2-5n}{n^3+n+1}$

25. $\sum_{n=1}^{\infty} \dfrac{1+n+n^2}{\sqrt{1+n^2+n^6}}$

26. $\sum_{n=1}^{\infty} \dfrac{n+5}{\sqrt[3]{n^7+n^2}}$

27. $\sum_{n=1}^{\infty} \left(1+\dfrac{1}{n}\right)^2 e^{-n}$

28. $\sum_{n=1}^{\infty} \dfrac{2n^2+7n}{3^n(n^2+5n-1)}$

29. $\sum_{n=1}^{\infty} \dfrac{1}{n!}$

30. $\sum_{n=1}^{\infty} \dfrac{n!}{n^n}$

31. $\sum_{n=1}^{\infty} \sin\left(\dfrac{1}{n}\right)$

32. $\sum_{n=1}^{\infty} \dfrac{1}{n^{1+1/n}}$

33–36 ▐▐▐▐ Use the sum of the first 10 terms to approximate the sum of the series. Estimate the error.

33. $\sum_{n=1}^{\infty} \dfrac{1}{n^4+n^2}$

34. $\sum_{n=1}^{\infty} \dfrac{1+\cos n}{n^5}$

35. $\sum_{n=1}^{\infty} \dfrac{1}{1+2^n}$

36. $\sum_{n=1}^{\infty} \dfrac{n}{(n+1)3^n}$

37. The meaning of the decimal representation of a number $0.d_1 d_2 d_3 \ldots$ (where the digit d_i is one of the numbers 0, 1, 2, ..., 9) is that

$$0.d_1 d_2 d_3 d_4 \ldots = \dfrac{d_1}{10} + \dfrac{d_2}{10^2} + \dfrac{d_3}{10^3} + \dfrac{d_4}{10^4} + \cdots$$

Show that this series always converges.

38. For what values of p does the series $\sum_{n=2}^{\infty} 1/(n^p \ln n)$ converge?

39. Prove that if $a_n \geq 0$ and $\sum a_n$ converges, then $\sum a_n^2$ also converges.

40. (a) Suppose that $\sum a_n$ and $\sum b_n$ are series with positive terms and $\sum b_n$ is convergent. Prove that if

$$\lim_{n \to \infty} \dfrac{a_n}{b_n} = 0$$

then $\sum a_n$ is also convergent.
(b) Use part (a) to show that the series converges.
(i) $\sum_{n=1}^{\infty} \dfrac{\ln n}{n^3}$ (ii) $\sum_{n=1}^{\infty} \dfrac{\ln n}{\sqrt{n}e^n}$

41. (a) Suppose that $\sum a_n$ and $\sum b_n$ are series with positive terms and $\sum b_n$ is divergent. Prove that if

$$\lim_{n \to \infty} \dfrac{a_n}{b_n} = \infty$$

then $\sum a_n$ is also divergent.
(b) Use part (a) to show that the series diverges.
(i) $\sum_{n=2}^{\infty} \dfrac{1}{\ln n}$ (ii) $\sum_{n=1}^{\infty} \dfrac{\ln n}{n}$

42. Give an example of a pair of series $\sum a_n$ and $\sum b_n$ with positive terms where $\lim_{n \to \infty} (a_n/b_n) = 0$ and $\sum b_n$ diverges, but $\sum a_n$ converges. [Compare with Exercise 40.]

43. Show that if $a_n > 0$ and $\lim_{n \to \infty} n a_n \neq 0$, then $\sum a_n$ is divergent.

44. Show that if $a_n > 0$ and $\sum a_n$ is convergent, then $\sum \ln(1+a_n)$ is convergent.

45. If $\sum a_n$ is a convergent series with positive terms, is it true that $\sum \sin(a_n)$ is also convergent?

46. If $\sum a_n$ and $\sum b_n$ are both convergent series with positive terms, is it true that $\sum a_n b_n$ is also convergent?

12.5 Alternating Series

The convergence tests that we have looked at so far apply only to series with positive terms. In this section and the next we learn how to deal with series whose terms are not necessarily positive. Of particular importance are *alternating series*, whose terms alternate in sign.

An **alternating series** is a series whose terms are alternately positive and negative. Here are two examples:

$$1 - \dfrac{1}{2} + \dfrac{1}{3} - \dfrac{1}{4} + \dfrac{1}{5} - \dfrac{1}{6} + \cdots = \sum_{n=1}^{\infty} \dfrac{(-1)^{n-1}}{n}$$

$$-\dfrac{1}{2} + \dfrac{2}{3} - \dfrac{3}{4} + \dfrac{4}{5} - \dfrac{5}{6} + \dfrac{6}{7} - \cdots = \sum_{n=1}^{\infty} (-1)^n \dfrac{n}{n+1}$$

We see from these examples that the nth term of an alternating series is of the form

$$a_n = (-1)^{n-1} b_n \quad \text{or} \quad a_n = (-1)^n b_n$$

where b_n is a positive number. (In fact, $b_n = |a_n|$.)

The following test says that if the terms of an alternating series decrease toward 0 in absolute value, then the series converges.

> **The Alternating Series Test** If the alternating series
>
> $$\sum_{n=1}^{\infty} (-1)^{n-1} b_n = b_1 - b_2 + b_3 - b_4 + b_5 - b_6 + \cdots \quad (b_n > 0)$$
>
> satisfies
>
> (i) $b_{n+1} \leq b_n$ for all n
>
> (ii) $\lim_{n \to \infty} b_n = 0$
>
> then the series is convergent.

Before giving the proof let's look at Figure 1, which gives a picture of the idea behind the proof. We first plot $s_1 = b_1$ on a number line. To find s_2 we subtract b_2, so s_2 is to the left of s_1. Then to find s_3 we add b_3, so s_3 is to the right of s_2. But, since $b_3 < b_2$, s_3 is to the left of s_1. Continuing in this manner, we see that the partial sums oscillate back and forth. Since $b_n \to 0$, the successive steps are becoming smaller and smaller. The even partial sums $s_2, s_4, s_6, \ldots$ are increasing and the odd partial sums $s_1, s_3, s_5, \ldots$ are decreasing. Thus, it seems plausible that both are converging to some number s, which is the sum of the series. Therefore, in the following proof we consider the even and odd partial sums separately.

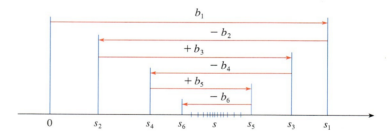

FIGURE 1

Proof of the Alternating Series Test We first consider the even partial sums:

$$s_2 = b_1 - b_2 \geq 0 \quad \text{since } b_2 \leq b_1$$

$$s_4 = s_2 + (b_3 - b_4) \geq s_2 \quad \text{since } b_4 \leq b_3$$

In general $\quad s_{2n} = s_{2n-2} + (b_{2n-1} - b_{2n}) \geq s_{2n-2} \quad \text{since } b_{2n} \leq b_{2n-1}$

Thus $\quad 0 \leq s_2 \leq s_4 \leq s_6 \leq \cdots \leq s_{2n} \leq \cdots$

But we can also write

$$s_{2n} = b_1 - (b_2 - b_3) - (b_4 - b_5) - \cdots - (b_{2n-2} - b_{2n-1}) - b_{2n}$$

Every term in brackets is positive, so $s_{2n} \leq b_1$ for all n. Therefore, the sequence $\{s_{2n}\}$ of even partial sums is increasing and bounded above. It is therefore convergent by the Monotonic Sequence Theorem. Let's call its limit s, that is,

$$\lim_{n \to \infty} s_{2n} = s$$

Now we compute the limit of the odd partial sums:

$$\lim_{n \to \infty} s_{2n+1} = \lim_{n \to \infty} (s_{2n} + b_{2n+1})$$
$$= \lim_{n \to \infty} s_{2n} + \lim_{n \to \infty} b_{2n+1}$$
$$= s + 0 \qquad \text{[by condition (ii)]}$$
$$= s$$

Since both the even and odd partial sums converge to s, we have $\lim_{n \to \infty} s_n = s$ (see Exercise 72 in Section 12.1) and so the series is convergent.

EXAMPLE 1 The alternating harmonic series

$$1 - \frac{1}{2} + \frac{1}{3} - \frac{1}{4} + \cdots = \sum_{n=1}^{\infty} \frac{(-1)^{n-1}}{n}$$

satisfies

(i) $b_{n+1} < b_n$ because $\dfrac{1}{n+1} < \dfrac{1}{n}$

(ii) $\lim_{n \to \infty} b_n = \lim_{n \to \infty} \dfrac{1}{n} = 0$

so the series is convergent by the Alternating Series Test.

EXAMPLE 2 The series $\displaystyle\sum_{n=1}^{\infty} \frac{(-1)^n 3n}{4n - 1}$ is alternating but

$$\lim_{n \to \infty} b_n = \lim_{n \to \infty} \frac{3n}{4n - 1} = \lim_{n \to \infty} \frac{3}{4 - \dfrac{1}{n}} = \frac{3}{4}$$

so condition (ii) is not satisfied. Instead, we look at the limit of the nth term of the series:

$$\lim_{n \to \infty} a_n = \lim_{n \to \infty} \frac{(-1)^n 3n}{4n - 1}$$

This limit does not exist, so the series diverges by the Test for Divergence.

EXAMPLE 3 Test the series $\displaystyle\sum_{n=1}^{\infty} (-1)^{n+1} \frac{n^2}{n^3 + 1}$ for convergence or divergence.

SOLUTION The given series is alternating so we try to verify conditions (i) and (ii) of the Alternating Series Test.

Unlike the situation in Example 1, it is not obvious that the sequence given by $b_n = n^2/(n^3 + 1)$ is decreasing. However, if we consider the related function

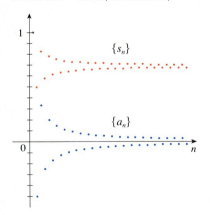

|||| Figure 2 illustrates Example 1 by showing the graphs of the terms $a_n = (-1)^{n-1}/n$ and the partial sums s_n. Notice how the values of s_n zigzag across the limiting value, which appears to be about 0.7. In fact, the exact sum of the series is $\ln 2 \approx 0.693$ (see Exercise 36).

FIGURE 2

$f(x) = x^2/(x^3 + 1)$, we find that

$$f'(x) = \frac{x(2 - x^3)}{(x^3 + 1)^2}$$

Since we are considering only positive x, we see that $f'(x) < 0$ if $2 - x^3 < 0$, that is, $x > \sqrt[3]{2}$. Thus, f is decreasing on the interval $(\sqrt[3]{2}, \infty)$. This means that $f(n + 1) < f(n)$ and therefore $b_{n+1} < b_n$ when $n \geq 2$. (The inequality $b_2 < b_1$ can be verified directly but all that really matters is that the sequence $\{b_n\}$ is eventually decreasing.)

Condition (ii) is readily verified:

$$\lim_{n \to \infty} b_n = \lim_{n \to \infty} \frac{n^2}{n^3 + 1} = \lim_{n \to \infty} \frac{\frac{1}{n}}{1 + \frac{1}{n^3}} = 0$$

Thus, the given series is convergent by the Alternating Series Test.

|||| Instead of verifying condition (i) of the Alternating Series Test by computing a derivative, we could verify that $b_{n+1} < b_n$ directly by using the technique of Solution 1 of Example 11 in Section 12.1.

Estimating Sums

A partial sum s_n of any convergent series can be used as an approximation to the total sum s, but this is not of much use unless we can estimate the accuracy of the approximation. The error involved in using $s \approx s_n$ is the remainder $R_n = s - s_n$. The next theorem says that for series that satisfy the conditions of the Alternating Series Test, the size of the error is smaller than b_{n+1}, which is the absolute value of the first neglected term.

> **Alternating Series Estimation Theorem** If $s = \Sigma (-1)^{n-1} b_n$ is the sum of an alternating series that satisfies
>
> (i) $0 \leq b_{n+1} \leq b_n$ and (ii) $\lim_{n \to \infty} b_n = 0$
>
> then
>
> $$|R_n| = |s - s_n| \leq b_{n+1}$$

|||| You can see geometrically why the Alternating Series Estimation Theorem is true by looking at Figure 1 (on page 772). Notice that $s - s_4 < b_5$, $|s - s_5| < b_6$, and so on. Notice also that s lies between any two consecutive partial sums.

Proof We know from the proof of the Alternating Series Test that s lies between any two consecutive partial sums s_n and s_{n+1}. It follows that

$$|s - s_n| \leq |s_{n+1} - s_n| = b_{n+1}$$

EXAMPLE 4 Find the sum of the series $\sum_{n=0}^{\infty} \frac{(-1)^n}{n!}$ correct to three decimal places. (By definition, $0! = 1$.)

SOLUTION We first observe that the series is convergent by the Alternating Series Test because

(i) $\dfrac{1}{(n + 1)!} = \dfrac{1}{n!(n + 1)} < \dfrac{1}{n!}$

(ii) $0 < \dfrac{1}{n!} < \dfrac{1}{n} \to 0$ so $\dfrac{1}{n!} \to 0$ as $n \to \infty$

To get a feel for how many terms we need to use in our approximation, let's write out the first few terms of the series:

$$s = \frac{1}{0!} - \frac{1}{1!} + \frac{1}{2!} - \frac{1}{3!} + \frac{1}{4!} - \frac{1}{5!} + \frac{1}{6!} - \frac{1}{7!} + \cdots$$

$$= 1 - 1 + \tfrac{1}{2} - \tfrac{1}{6} + \tfrac{1}{24} - \tfrac{1}{120} + \tfrac{1}{720} - \tfrac{1}{5040} + \cdots$$

Notice that
$$b_7 = \tfrac{1}{5040} < \tfrac{1}{5000} = 0.0002$$

and
$$s_6 = 1 - 1 + \tfrac{1}{2} - \tfrac{1}{6} + \tfrac{1}{24} - \tfrac{1}{120} + \tfrac{1}{720} \approx 0.368056$$

By the Alternating Series Estimation Theorem we know that

$$|s - s_6| \leq b_7 < 0.0002$$

This error of less than 0.0002 does not affect the third decimal place, so we have

$$s \approx 0.368$$

|||| In Section 12.10 we will prove that $e^x = \sum_{n=0}^{\infty} x^n/n!$ for all x, so what we have obtained in Example 4 is actually an approximation to the number e^{-1}.

correct to three decimal places.

NOTE ▫ The rule that the error (in using s_n to approximate s) is smaller than the first neglected term is, in general, valid only for alternating series that satisfy the conditions of the Alternating Series Estimation Theorem. The rule does not apply to other types of series.

12.5 Exercises

1. (a) What is an alternating series?
   (b) Under what conditions does an alternating series converge?
   (c) If these conditions are satisfied, what can you say about the remainder after n terms?

2–20 |||| Test the series for convergence or divergence.

2. $-\tfrac{1}{3} + \tfrac{2}{4} - \tfrac{3}{5} + \tfrac{4}{6} - \tfrac{5}{7} + \cdots$

3. $\tfrac{4}{7} - \tfrac{4}{8} + \tfrac{4}{9} - \tfrac{4}{10} + \tfrac{4}{11} - \cdots$

4. $\dfrac{1}{\ln 2} - \dfrac{1}{\ln 3} + \dfrac{1}{\ln 4} - \dfrac{1}{\ln 5} + \dfrac{1}{\ln 6} - \cdots$

5. $\displaystyle\sum_{n=1}^{\infty} \frac{(-1)^{n-1}}{\sqrt{n}}$

6. $\displaystyle\sum_{n=1}^{\infty} \frac{(-1)^{n-1}}{3n-1}$

7. $\displaystyle\sum_{n=1}^{\infty} (-1)^n \frac{3n-1}{2n+1}$

8. $\displaystyle\sum_{n=1}^{\infty} (-1)^n \frac{2n}{4n^2+1}$

9. $\displaystyle\sum_{n=1}^{\infty} \frac{(-1)^{n+1}}{4n^2+1}$

10. $\displaystyle\sum_{n=1}^{\infty} (-1)^n \frac{\sqrt{n}}{1+2\sqrt{n}}$

11. $\displaystyle\sum_{n=1}^{\infty} (-1)^{n+1} \frac{n^2}{n^3+4}$

12. $\displaystyle\sum_{n=1}^{\infty} (-1)^{n-1} \frac{e^{1/n}}{n}$

13. $\displaystyle\sum_{n=2}^{\infty} (-1)^n \frac{n}{\ln n}$

14. $\displaystyle\sum_{n=1}^{\infty} (-1)^{n-1} \frac{\ln n}{n}$

15. $\displaystyle\sum_{n=1}^{\infty} \frac{\cos n\pi}{n^{3/4}}$

16. $\displaystyle\sum_{n=1}^{\infty} \frac{\sin(n\pi/2)}{n!}$

17. $\displaystyle\sum_{n=1}^{\infty} (-1)^n \sin\left(\frac{\pi}{n}\right)$

18. $\displaystyle\sum_{n=1}^{\infty} (-1)^n \cos\left(\frac{\pi}{n}\right)$

19. $\displaystyle\sum_{n=1}^{\infty} (-1)^n \frac{n^n}{n!}$

20. $\displaystyle\sum_{n=1}^{\infty} \left(-\frac{n}{5}\right)^n$

21–22 |||| Calculate the first 10 partial sums of the series and graph both the sequence of terms and the sequence of partial sums on the same screen. Estimate the error in using the 10th partial sum to approximate the total sum.

21. $\displaystyle\sum_{n=1}^{\infty} \frac{(-1)^{n-1}}{n^{3/2}}$

22. $\displaystyle\sum_{n=1}^{\infty} \frac{(-1)^{n-1}}{n^3}$

23–26 |||| How many terms of the series do we need to add in order to find the sum to the indicated accuracy?

23. $\displaystyle\sum_{n=1}^{\infty} \frac{(-1)^{n-1}}{n^2}$ $(|\text{error}| < 0.01)$

24. $\displaystyle\sum_{n=1}^{\infty} \frac{(-1)^{n+1}}{n^4}$ $(|\text{error}| < 0.001)$

25. $\sum_{n=1}^{\infty} \dfrac{(-2)^n}{n!}$ ($|\text{error}| < 0.01$)

26. $\sum_{n=1}^{\infty} \dfrac{(-1)^n n}{4^n}$ ($|\text{error}| < 0.002$)

27–30 Approximate the sum of the series correct to four decimal places.

27. $\sum_{n=1}^{\infty} \dfrac{(-1)^{n+1}}{n^5}$

28. $\sum_{n=1}^{\infty} \dfrac{(-1)^n n}{8^n}$

29. $\sum_{n=1}^{\infty} \dfrac{(-1)^{n-1} n^2}{10^n}$

30. $\sum_{n=1}^{\infty} \dfrac{(-1)^n}{3^n n!}$

31. Is the 50th partial sum s_{50} of the alternating series $\sum_{n=1}^{\infty} (-1)^{n-1}/n$ an overestimate or an underestimate of the total sum? Explain.

32–34 For what values of p is each series convergent?

32. $\sum_{n=1}^{\infty} \dfrac{(-1)^{n-1}}{n^p}$

33. $\sum_{n=1}^{\infty} \dfrac{(-1)^n}{n+p}$

34. $\sum_{n=2}^{\infty} (-1)^{n-1} \dfrac{(\ln n)^p}{n}$

35. Show that the series $\sum (-1)^{n-1} b_n$, where $b_n = 1/n$ if n is odd and $b_n = 1/n^2$ if n is even, is divergent. Why does the Alternating Series Test not apply?

36. Use the following steps to show that
$$\sum_{n=1}^{\infty} \dfrac{(-1)^{n-1}}{n} = \ln 2$$
Let h_n and s_n be the partial sums of the harmonic and alternating harmonic series.
(a) Show that $s_{2n} = h_{2n} - h_n$.
(b) From Exercise 38 in Section 12.3 we have
$$h_n - \ln n \to \gamma \qquad \text{as } n \to \infty$$
and therefore
$$h_{2n} - \ln(2n) \to \gamma \qquad \text{as } n \to \infty$$
Use these facts together with part (a) to show that $s_{2n} \to \ln 2$ as $n \to \infty$.

12.6 Absolute Convergence and the Ratio and Root Tests

Given any series $\sum a_n$, we can consider the corresponding series
$$\sum_{n=1}^{\infty} |a_n| = |a_1| + |a_2| + |a_3| + \cdots$$
whose terms are the absolute values of the terms of the original series.

|||| We have convergence tests for series with positive terms and for alternating series. But what if the signs of the terms switch back and forth irregularly? We will see in Example 3 that the idea of absolute convergence sometimes helps in such cases.

1 Definition A series $\sum a_n$ is called **absolutely convergent** if the series of absolute values $\sum |a_n|$ is convergent.

Notice that if $\sum a_n$ is a series with positive terms, then $|a_n| = a_n$ and so absolute convergence is the same as convergence in this case.

EXAMPLE 1 The series
$$\sum_{n=1}^{\infty} \dfrac{(-1)^{n-1}}{n^2} = 1 - \dfrac{1}{2^2} + \dfrac{1}{3^2} - \dfrac{1}{4^2} + \cdots$$
is absolutely convergent because
$$\sum_{n=1}^{\infty} \left| \dfrac{(-1)^{n-1}}{n^2} \right| = \sum_{n=1}^{\infty} \dfrac{1}{n^2} = 1 + \dfrac{1}{2^2} + \dfrac{1}{3^2} + \dfrac{1}{4^2} + \cdots$$
is a convergent p-series ($p = 2$).

EXAMPLE 2 We know that the alternating harmonic series

$$\sum_{n=1}^{\infty} \frac{(-1)^{n-1}}{n} = 1 - \frac{1}{2} + \frac{1}{3} - \frac{1}{4} + \cdots$$

is convergent (see Example 1 in Section 12.5), but it is not absolutely convergent because the corresponding series of absolute values is

$$\sum_{n=1}^{\infty} \left| \frac{(-1)^{n-1}}{n} \right| = \sum_{n=1}^{\infty} \frac{1}{n} = 1 + \frac{1}{2} + \frac{1}{3} + \frac{1}{4} + \cdots$$

which is the harmonic series (p-series with $p = 1$) and is therefore divergent.

> **2 Definition** A series $\Sigma \, a_n$ is called **conditionally convergent** if it is convergent but not absolutely convergent.

Example 2 shows that the alternating harmonic series is conditionally convergent. Thus, it is possible for a series to be convergent but not absolutely convergent. However, the next theorem shows that absolute convergence implies convergence.

> **3 Theorem** If a series $\Sigma \, a_n$ is absolutely convergent, then it is convergent.

Proof Observe that the inequality

$$0 \leq a_n + |a_n| \leq 2|a_n|$$

is true because $|a_n|$ is either a_n or $-a_n$. If $\Sigma \, a_n$ is absolutely convergent, then $\Sigma \, |a_n|$ is convergent, so $\Sigma \, 2|a_n|$ is convergent. Therefore, by the Comparison Test, $\Sigma \, (a_n + |a_n|)$ is convergent. Then

$$\sum a_n = \sum (a_n + |a_n|) - \sum |a_n|$$

is the difference of two convergent series and is therefore convergent.

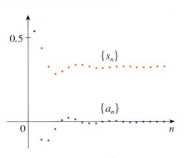

Figure 1 shows the graphs of the terms a_n and partial sums s_n of the series in Example 3. Notice that the series is not alternating but has positive and negative terms.

FIGURE 1

EXAMPLE 3 Determine whether the series

$$\sum_{n=1}^{\infty} \frac{\cos n}{n^2} = \frac{\cos 1}{1^2} + \frac{\cos 2}{2^2} + \frac{\cos 3}{3^2} + \cdots$$

is convergent or divergent.

SOLUTION This series has both positive and negative terms, but it is not alternating. (The first term is positive, the next three are negative, and the following three are positive. The signs change irregularly.) We can apply the Comparison Test to the series of absolute values

$$\sum_{n=1}^{\infty} \left| \frac{\cos n}{n^2} \right| = \sum_{n=1}^{\infty} \frac{|\cos n|}{n^2}$$

Since $|\cos n| \leq 1$ for all n, we have

$$\frac{|\cos n|}{n^2} \leq \frac{1}{n^2}$$

We know that $\Sigma\, 1/n^2$ is convergent (p-series with $p = 2$) and therefore $\Sigma\, |\cos n|/n^2$ is convergent by the Comparison Test. Thus, the given series $\Sigma\, (\cos n)/n^2$ is absolutely convergent and therefore convergent by Theorem 3.

The following test is very useful in determining whether a given series is absolutely convergent.

The Ratio Test

(i) If $\displaystyle\lim_{n\to\infty} \left|\frac{a_{n+1}}{a_n}\right| = L < 1$, then the series $\displaystyle\sum_{n=1}^{\infty} a_n$ is absolutely convergent (and therefore convergent).

(ii) If $\displaystyle\lim_{n\to\infty} \left|\frac{a_{n+1}}{a_n}\right| = L > 1$ or $\displaystyle\lim_{n\to\infty} \left|\frac{a_{n+1}}{a_n}\right| = \infty$, then the series $\displaystyle\sum_{n=1}^{\infty} a_n$ is divergent.

(iii) If $\displaystyle\lim_{n\to\infty} \left|\frac{a_{n+1}}{a_n}\right| = 1$, the Ratio Test is inconclusive; that is, no conclusion can be drawn about the convergence or divergence of $\Sigma\, a_n$.

Proof

(i) The idea is to compare the given series with a convergent geometric series. Since $L < 1$, we can choose a number r such that $L < r < 1$. Since

$$\lim_{n\to\infty} \left|\frac{a_{n+1}}{a_n}\right| = L \quad \text{and} \quad L < r$$

the ratio $|a_{n+1}/a_n|$ will eventually be less than r; that is, there exists an integer N such that

$$\left|\frac{a_{n+1}}{a_n}\right| < r \quad \text{whenever } n \geq N$$

or, equivalently,

$$\boxed{4} \qquad |a_{n+1}| < |a_n| r \quad \text{whenever } n \geq N$$

Putting n successively equal to $N, N+1, N+2, \ldots$ in (4), we obtain

$$|a_{N+1}| < |a_N| r$$
$$|a_{N+2}| < |a_{N+1}| r < |a_N| r^2$$
$$|a_{N+3}| < |a_{N+2}| r < |a_N| r^3$$

and, in general,

$$\boxed{5} \qquad |a_{N+k}| < |a_N| r^k \quad \text{for all } k \geq 1$$

Now the series

$$\sum_{k=1}^{\infty} |a_N| r^k = |a_N| r + |a_N| r^2 + |a_N| r^3 + \cdots$$

is convergent because it is a geometric series with $0 < r < 1$. So the inequality (5), together with the Comparison Test, shows that the series

$$\sum_{n=N+1}^{\infty} |a_n| = \sum_{k=1}^{\infty} |a_{N+k}| = |a_{N+1}| + |a_{N+2}| + |a_{N+3}| + \cdots$$

is also convergent. It follows that the series $\sum_{n=1}^{\infty} |a_n|$ is convergent. (Recall that a finite number of terms doesn't affect convergence.) Therefore, Σa_n is absolutely convergent.

(ii) If $|a_{n+1}/a_n| \to L > 1$ or $|a_{n+1}/a_n| \to \infty$, then the ratio $|a_{n+1}/a_n|$ will eventually be greater than 1; that is, there exists an integer N such that

$$\left| \frac{a_{n+1}}{a_n} \right| > 1 \quad \text{whenever } n \geq N$$

This means that $|a_{n+1}| > |a_n|$ whenever $n \geq N$ and so

$$\lim_{n \to \infty} a_n \neq 0$$

Therefore, Σa_n diverges by the Test for Divergence.

NOTE ▫ Part (iii) of the Ratio Test says that if $\lim_{n \to \infty} |a_{n+1}/a_n| = 1$, the test gives no information. For instance, for the convergent series $\Sigma 1/n^2$ we have

$$\left| \frac{a_{n+1}}{a_n} \right| = \frac{\dfrac{1}{(n+1)^2}}{\dfrac{1}{n^2}} = \frac{n^2}{(n+1)^2} = \frac{1}{\left(1 + \dfrac{1}{n}\right)^2} \to 1 \quad \text{as } n \to \infty$$

whereas for the divergent series $\Sigma 1/n$ we have

$$\left| \frac{a_{n+1}}{a_n} \right| = \frac{\dfrac{1}{n+1}}{\dfrac{1}{n}} = \frac{n}{n+1} = \frac{1}{1 + \dfrac{1}{n}} \to 1 \quad \text{as } n \to \infty$$

Therefore, if $\lim_{n \to \infty} |a_{n+1}/a_n| = 1$, the series Σa_n might converge or it might diverge. In this case the Ratio Test fails and we must use some other test.

EXAMPLE 4 Test the series $\displaystyle\sum_{n=1}^{\infty} (-1)^n \frac{n^3}{3^n}$ for absolute convergence.

SOLUTION We use the Ratio Test with $a_n = (-1)^n n^3/3^n$:

$$\left| \frac{a_{n+1}}{a_n} \right| = \left| \frac{\dfrac{(-1)^{n+1}(n+1)^3}{3^{n+1}}}{\dfrac{(-1)^n n^3}{3^n}} \right| = \frac{(n+1)^3}{3^{n+1}} \cdot \frac{3^n}{n^3}$$

$$= \frac{1}{3} \left(\frac{n+1}{n} \right)^3 = \frac{1}{3} \left(1 + \frac{1}{n} \right)^3 \to \frac{1}{3} < 1$$

Thus, by the Ratio Test, the given series is absolutely convergent and therefore convergent.

IIII ESTIMATING SUMS

In the last three sections we used various methods for estimating the sum of a series—the method depended on which test was used to prove convergence. What about series for which the Ratio Test works? There are two possibilities: If the series happens to be an alternating series, as in Example 4, then it is best to use the methods of Section 12.5. If the terms are all positive, then use the special methods explained in Exercise 34.

EXAMPLE 5 Test the convergence of the series $\sum_{n=1}^{\infty} \dfrac{n^n}{n!}$.

SOLUTION Since the terms $a_n = n^n/n!$ are positive, we don't need the absolute value signs.

$$\frac{a_{n+1}}{a_n} = \frac{(n+1)^{n+1}}{(n+1)!} \cdot \frac{n!}{n^n} = \frac{(n+1)(n+1)^n}{(n+1)n!} \cdot \frac{n!}{n^n}$$

$$= \left(\frac{n+1}{n}\right)^n = \left(1 + \frac{1}{n}\right)^n \to e \quad \text{as } n \to \infty$$

(See Equation 7.4.9 or 7.4*.9.) Since $e > 1$, the given series is divergent by the Ratio Test.

NOTE Although the Ratio Test works in Example 5, an easier method is to use the Test for Divergence. Since

$$a_n = \frac{n^n}{n!} = \frac{n \cdot n \cdot n \cdot \cdots \cdot n}{1 \cdot 2 \cdot 3 \cdot \cdots \cdot n} \geq n$$

it follows that a_n does not approach 0 as $n \to \infty$. Therefore, the given series is divergent by the Test for Divergence.

The following test is convenient to apply when nth powers occur. Its proof is similar to the proof of the Ratio Test and is left as Exercise 38.

The Root Test

(i) If $\lim_{n \to \infty} \sqrt[n]{|a_n|} = L < 1$, then the series $\sum_{n=1}^{\infty} a_n$ is absolutely convergent (and therefore convergent).

(ii) If $\lim_{n \to \infty} \sqrt[n]{|a_n|} = L > 1$ or $\lim_{n \to \infty} \sqrt[n]{|a_n|} = \infty$, then the series $\sum_{n=1}^{\infty} a_n$ is divergent.

(iii) If $\lim_{n \to \infty} \sqrt[n]{|a_n|} = 1$, the Root Test is inconclusive.

If $\lim_{n \to \infty} \sqrt[n]{|a_n|} = 1$, then part (iii) of the Root Test says that the test gives no information. The series $\sum a_n$ could converge or diverge. (If $L = 1$ in the Ratio Test, don't try the Root Test because L will again be 1.)

EXAMPLE 6 Test the convergence of the series $\sum_{n=1}^{\infty} \left(\dfrac{2n+3}{3n+2}\right)^n$.

SOLUTION

$$a_n = \left(\frac{2n+3}{3n+2}\right)^n$$

$$\sqrt[n]{|a_n|} = \frac{2n+3}{3n+2} = \frac{2 + \dfrac{3}{n}}{3 + \dfrac{2}{n}} \to \frac{2}{3} < 1$$

Thus, the given series converges by the Root Test.

Rearrangements

The question of whether a given convergent series is absolutely convergent or conditionally convergent has a bearing on the question of whether infinite sums behave like finite sums.

If we rearrange the order of the terms in a finite sum, then of course the value of the sum remains unchanged. But this is not always the case for an infinite series. By a **rearrangement** of an infinite series $\sum a_n$ we mean a series obtained by simply changing the order of the terms. For instance, a rearrangement of $\sum a_n$ could start as follows:

$$a_1 + a_2 + a_5 + a_3 + a_4 + a_{15} + a_6 + a_7 + a_{20} + \cdots$$

It turns out that

if $\sum a_n$ is an absolutely convergent series with sum s, then any rearrangement of $\sum a_n$ has the same sum s.

However, any conditionally convergent series can be rearranged to give a different sum. To illustrate this fact let's consider the alternating harmonic series

$$\boxed{6} \qquad 1 - \tfrac{1}{2} + \tfrac{1}{3} - \tfrac{1}{4} + \tfrac{1}{5} - \tfrac{1}{6} + \tfrac{1}{7} - \tfrac{1}{8} + \cdots = \ln 2$$

(See Exercise 36 in Section 12.5.) If we multiply this series by $\tfrac{1}{2}$, we get

$$\tfrac{1}{2} - \tfrac{1}{4} + \tfrac{1}{6} - \tfrac{1}{8} + \cdots = \tfrac{1}{2} \ln 2$$

Inserting zeros between the terms of this series, we have

|||| Adding these zeros does not affect the sum of the series; each term in the sequence of partial sums is repeated, but the limit is the same.

$$\boxed{7} \qquad 0 + \tfrac{1}{2} + 0 - \tfrac{1}{4} + 0 + \tfrac{1}{6} + 0 - \tfrac{1}{8} + \cdots = \tfrac{1}{2} \ln 2$$

Now we add the series in Equations 6 and 7 using Theorem 12.2.8:

$$\boxed{8} \qquad 1 + \tfrac{1}{3} - \tfrac{1}{2} + \tfrac{1}{5} + \tfrac{1}{7} - \tfrac{1}{4} + \cdots = \tfrac{3}{2} \ln 2$$

Notice that the series in (8) contains the same terms as in (6), but rearranged so that one negative term occurs after each pair of positive terms. The sums of these series, however, are different. In fact, Riemann proved that

if $\sum a_n$ is a conditionally convergent series and r is any real number whatsoever, then there is a rearrangement of $\sum a_n$ that has a sum equal to r.

A proof of this fact is outlined in Exercise 40.

12.6 Exercises

1. What can you say about the series $\sum a_n$ in each of the following cases?

   (a) $\lim\limits_{n \to \infty} \left| \dfrac{a_{n+1}}{a_n} \right| = 8$ (b) $\lim\limits_{n \to \infty} \left| \dfrac{a_{n+1}}{a_n} \right| = 0.8$

   (c) $\lim\limits_{n \to \infty} \left| \dfrac{a_{n+1}}{a_n} \right| = 1$

2–28 |||| Determine whether the series is absolutely convergent, conditionally convergent, or divergent.

2. $\sum\limits_{n=1}^{\infty} \dfrac{n^2}{2^n}$

3. $\sum\limits_{n=0}^{\infty} \dfrac{(-10)^n}{n!}$

4. $\sum\limits_{n=1}^{\infty} (-1)^{n-1} \dfrac{2^n}{n^4}$

5. $\sum\limits_{n=1}^{\infty} \dfrac{(-1)^{n+1}}{\sqrt[4]{n}}$

6. $\sum\limits_{n=1}^{\infty} \dfrac{(-1)^n}{n^4}$

7. $\sum\limits_{n=1}^{\infty} (-1)^n \dfrac{n}{5+n}$

8. $\sum\limits_{n=1}^{\infty} (-1)^{n-1} \dfrac{n}{n^2+1}$

9. $\sum\limits_{n=1}^{\infty} \dfrac{1}{(2n)!}$

10. $\sum\limits_{n=1}^{\infty} e^{-n} n!$

11. $\sum\limits_{n=1}^{\infty} \dfrac{(-1)^n e^{1/n}}{n^3}$

12. $\sum\limits_{n=1}^{\infty} \dfrac{\sin 4n}{4^n}$

13. $\sum_{n=1}^{\infty} \dfrac{n(-3)^n}{4^{n-1}}$

14. $\sum_{n=1}^{\infty} (-1)^{n+1} \dfrac{n^2 2^n}{n!}$

15. $\sum_{n=1}^{\infty} \dfrac{10^n}{(n+1)4^{2n+1}}$

16. $\sum_{n=1}^{\infty} \dfrac{3 - \cos n}{n^{2/3} - 2}$

17. $\sum_{n=2}^{\infty} \dfrac{(-1)^n}{\ln n}$

18. $\sum_{n=1}^{\infty} \dfrac{n!}{n^n}$

19. $\sum_{n=1}^{\infty} \dfrac{\cos(n\pi/3)}{n!}$

20. $\sum_{n=2}^{\infty} \dfrac{(-1)^n}{(\ln n)^n}$

21. $\sum_{n=1}^{\infty} \dfrac{n^n}{3^{1+3n}}$

22. $\sum_{n=2}^{\infty} \dfrac{(-1)^n}{n \ln n}$

23. $\sum_{n=1}^{\infty} \left(\dfrac{n^2+1}{2n^2+1} \right)^n$

24. $\sum_{n=1}^{\infty} \dfrac{(-1)^n}{(\arctan n)^n}$

25. $1 - \dfrac{1 \cdot 3}{3!} + \dfrac{1 \cdot 3 \cdot 5}{5!} - \dfrac{1 \cdot 3 \cdot 5 \cdot 7}{7!} + \cdots$
$+ (-1)^{n-1} \dfrac{1 \cdot 3 \cdot 5 \cdot \,\cdots\, \cdot (2n-1)}{(2n-1)!} + \cdots$

26. $\dfrac{2}{5} + \dfrac{2 \cdot 6}{5 \cdot 8} + \dfrac{2 \cdot 6 \cdot 10}{5 \cdot 8 \cdot 11} + \dfrac{2 \cdot 6 \cdot 10 \cdot 14}{5 \cdot 8 \cdot 11 \cdot 14} + \cdots$

27. $\sum_{n=1}^{\infty} \dfrac{2 \cdot 4 \cdot 6 \cdot \,\cdots\, \cdot (2n)}{n!}$

28. $\sum_{n=1}^{\infty} (-1)^n \dfrac{2^n n!}{5 \cdot 8 \cdot 11 \cdot \,\cdots\, \cdot (3n+2)}$

29. The terms of a series are defined recursively by the equations
$$a_1 = 2 \qquad a_{n+1} = \dfrac{5n+1}{4n+3} a_n$$
Determine whether $\Sigma\, a_n$ converges or diverges.

30. A series $\Sigma\, a_n$ is defined by the equations
$$a_1 = 1 \qquad a_{n+1} = \dfrac{2 + \cos n}{\sqrt{n}} a_n$$
Determine whether $\Sigma\, a_n$ converges or diverges.

31. For which of the following series is the Ratio Test inconclusive (that is, it fails to give a definite answer)?

(a) $\sum_{n=1}^{\infty} \dfrac{1}{n^3}$

(b) $\sum_{n=1}^{\infty} \dfrac{n}{2^n}$

(c) $\sum_{n=1}^{\infty} \dfrac{(-3)^{n-1}}{\sqrt{n}}$

(d) $\sum_{n=1}^{\infty} \dfrac{\sqrt{n}}{1+n^2}$

32. For which positive integers k is the following series convergent?
$$\sum_{n=1}^{\infty} \dfrac{(n!)^2}{(kn)!}$$

33. (a) Show that $\sum_{n=0}^{\infty} x^n/n!$ converges for all x.
    (b) Deduce that $\lim_{n \to \infty} x^n/n! = 0$ for all x.

34. Let $\Sigma\, a_n$ be a series with positive terms and let $r_n = a_{n+1}/a_n$. Suppose that $\lim_{n \to \infty} r_n = L < 1$, so $\Sigma\, a_n$ converges by the Ratio Test. As usual, we let R_n be the remainder after n terms, that is,
$$R_n = a_{n+1} + a_{n+2} + a_{n+3} + \cdots$$

(a) If $\{r_n\}$ is a decreasing sequence and $r_{n+1} < 1$, show, by summing a geometric series, that
$$R_n \leq \dfrac{a_{n+1}}{1 - r_{n+1}}$$

(b) If $\{r_n\}$ is an increasing sequence, show that
$$R_n \leq \dfrac{a_{n+1}}{1 - L}$$

35. (a) Find the partial sum s_5 of the series $\sum_{n=1}^{\infty} 1/n2^n$. Use Exercise 34 to estimate the error in using s_5 as an approximation to the sum of the series.
    (b) Find a value of n so that s_n is within 0.00005 of the sum. Use this value of n to approximate the sum of the series.

36. Use the sum of the first 10 terms to approximate the sum of the series
$$\sum_{n=1}^{\infty} \dfrac{n}{2^n}$$
Use Exercise 34 to estimate the error.

37. Prove that if $\Sigma\, a_n$ is absolutely convergent, then
$$\left| \sum_{n=1}^{\infty} a_n \right| \leq \sum_{n=1}^{\infty} |a_n|$$

38. Prove the Root Test. [Hint for part (i): Take any number r such that $L < r < 1$ and use the fact that there is an integer N such that $\sqrt[n]{|a_n|} < r$ whenever $n \geq N$.]

39. Given any series $\Sigma\, a_n$ we define a series $\Sigma\, a_n^+$ whose terms are all the positive terms of $\Sigma\, a_n$ and a series $\Sigma\, a_n^-$ whose terms are all the negative terms of $\Sigma\, a_n$. To be specific, we let
$$a_n^+ = \dfrac{a_n + |a_n|}{2} \qquad a_n^- = \dfrac{a_n - |a_n|}{2}$$
Notice that if $a_n > 0$, then $a_n^+ = a_n$ and $a_n^- = 0$, whereas if $a_n < 0$, then $a_n^- = a_n$ and $a_n^+ = 0$.
(a) If $\Sigma\, a_n$ is absolutely convergent, show that both of the series $\Sigma\, a_n^+$ and $\Sigma\, a_n^-$ are convergent.
(b) If $\Sigma\, a_n$ is conditionally convergent, show that both of the series $\Sigma\, a_n^+$ and $\Sigma\, a_n^-$ are divergent.

40. Prove that if $\Sigma\, a_n$ is a conditionally convergent series and r is any real number, then there is a rearrangement of $\Sigma\, a_n$ whose sum is r. [Hints: Use the notation of Exercise 39. Take just enough positive terms a_n^+ so that their sum is greater than r. Then add just enough negative terms a_n^- so that the cumulative sum is less than r. Continue in this manner and use Theorem 12.2.6.]

12.7 Strategy for Testing Series

We now have several ways of testing a series for convergence or divergence; the problem is to decide which test to use on which series. In this respect testing series is similar to integrating functions. Again there are no hard and fast rules about which test to apply to a given series, but you may find the following advice of some use.

It is not wise to apply a list of the tests in a specific order until one finally works. That would be a waste of time and effort. Instead, as with integration, the main strategy is to classify the series according to its *form*.

1. If the series is of the form $\Sigma\, 1/n^p$, it is a *p*-series, which we know to be convergent if $p > 1$ and divergent if $p \leq 1$.

2. If the series has the form $\Sigma\, ar^{n-1}$ or $\Sigma\, ar^n$, it is a geometric series, which converges if $|r| < 1$ and diverges if $|r| \geq 1$. Some preliminary algebraic manipulation may be required to bring the series into this form.

3. If the series has a form that is similar to a *p*-series or a geometric series, then one of the comparison tests should be considered. In particular, if a_n is a rational function or algebraic function of n (involving roots of polynomials), then the series should be compared with a *p*-series. Notice that most of the series in Exercises 12.4 have this form. (The value of p should be chosen as in Section 12.4 by keeping only the highest powers of n in the numerator and denominator.) The comparison tests apply only to series with positive terms, but if $\Sigma\, a_n$ has some negative terms, then we can apply the Comparison Test to $\Sigma\, |a_n|$ and test for absolute convergence.

4. If you can see at a glance that $\lim_{n \to \infty} a_n \neq 0$, then the Test for Divergence should be used.

5. If the series is of the form $\Sigma\, (-1)^{n-1} b_n$ or $\Sigma\, (-1)^n b_n$, then the Alternating Series Test is an obvious possibility.

6. Series that involve factorials or other products (including a constant raised to the *n*th power) are often conveniently tested using the Ratio Test. Bear in mind that $|a_{n+1}/a_n| \to 1$ as $n \to \infty$ for all *p*-series and therefore all rational or algebraic functions of n. Thus, the Ratio Test should not be used for such series.

7. If a_n is of the form $(b_n)^n$, then the Root Test may be useful.

8. If $a_n = f(n)$, where $\int_1^\infty f(x)\, dx$ is easily evaluated, then the Integral Test is effective (assuming the hypotheses of this test are satisfied).

In the following examples we don't work out all the details but simply indicate which tests should be used.

EXAMPLE 1 $\sum_{n=1}^{\infty} \dfrac{n-1}{2n+1}$

Since $a_n \to \frac{1}{2} \neq 0$ as $n \to \infty$, we should use the Test for Divergence.

EXAMPLE 2 $\sum_{n=1}^{\infty} \dfrac{\sqrt{n^3+1}}{3n^3+4n^2+2}$

Since a_n is an algebraic function of n, we compare the given series with a *p*-series.

The comparison series for the Limit Comparison Test is $\Sigma\, b_n$, where

$$b_n = \frac{\sqrt{n^3}}{3n^3} = \frac{n^{3/2}}{3n^3} = \frac{1}{3n^{3/2}}$$

EXAMPLE 3 $\displaystyle\sum_{n=1}^{\infty} ne^{-n^2}$

Since the integral $\int_1^{\infty} xe^{-x^2}\,dx$ is easily evaluated, we use the Integral Test. The Ratio Test also works.

EXAMPLE 4 $\displaystyle\sum_{n=1}^{\infty} (-1)^n \frac{n^3}{n^4+1}$

Since the series is alternating, we use the Alternating Series Test.

EXAMPLE 5 $\displaystyle\sum_{k=1}^{\infty} \frac{2^k}{k!}$

Since the series involves $k!$, we use the Ratio Test.

EXAMPLE 6 $\displaystyle\sum_{n=1}^{\infty} \frac{1}{2+3^n}$

Since the series is closely related to the geometric series $\Sigma\, 1/3^n$, we use the Comparison Test.

12.7 Exercises

1–38 Test the series for convergence or divergence.

1. $\displaystyle\sum_{n=1}^{\infty} \frac{n^2-1}{n^2+n}$

2. $\displaystyle\sum_{n=1}^{\infty} \frac{n-1}{n^2+n}$

3. $\displaystyle\sum_{n=1}^{\infty} \frac{1}{n^2+n}$

4. $\displaystyle\sum_{n=1}^{\infty} (-1)^{n-1} \frac{n-1}{n^2+n}$

5. $\displaystyle\sum_{n=1}^{\infty} \frac{(-3)^{n+1}}{2^{3n}}$

6. $\displaystyle\sum_{n=1}^{\infty} \left(\frac{3n}{1+8n}\right)^n$

7. $\displaystyle\sum_{n=2}^{\infty} \frac{1}{n\sqrt{\ln n}}$

8. $\displaystyle\sum_{k=1}^{\infty} \frac{2^k k!}{(k+2)!}$

9. $\displaystyle\sum_{k=1}^{\infty} k^2 e^{-k}$

10. $\displaystyle\sum_{n=1}^{\infty} n^2 e^{-n^3}$

11. $\displaystyle\sum_{n=2}^{\infty} \frac{(-1)^{n+1}}{n \ln n}$

12. $\displaystyle\sum_{n=1}^{\infty} (-1)^n \frac{n}{n^2+25}$

13. $\displaystyle\sum_{n=1}^{\infty} \frac{3^n n^2}{n!}$

14. $\displaystyle\sum_{n=1}^{\infty} \sin n$

15. $\displaystyle\sum_{n=0}^{\infty} \frac{n!}{2\cdot 5\cdot 8\cdot\,\cdots\,\cdot(3n+2)}$

16. $\displaystyle\sum_{n=1}^{\infty} \frac{n^2+1}{n^3+1}$

17. $\displaystyle\sum_{n=1}^{\infty} (-1)^n 2^{1/n}$

18. $\displaystyle\sum_{n=2}^{\infty} \frac{(-1)^{n-1}}{\sqrt{n-1}}$

19. $\displaystyle\sum_{n=1}^{\infty} (-1)^n \frac{\ln n}{\sqrt{n}}$

20. $\displaystyle\sum_{k=1}^{\infty} \frac{k+5}{5^k}$

21. $\displaystyle\sum_{n=1}^{\infty} \frac{(-2)^{2n}}{n^n}$

22. $\displaystyle\sum_{n=1}^{\infty} \frac{\sqrt{n^2-1}}{n^3+2n^2+5}$

23. $\displaystyle\sum_{n=1}^{\infty} \tan(1/n)$

24. $\displaystyle\sum_{n=1}^{\infty} \frac{\cos(n/2)}{n^2+4n}$

25. $\displaystyle\sum_{n=1}^{\infty} \frac{n!}{e^{n^2}}$

26. $\displaystyle\sum_{n=1}^{\infty} \frac{n^2+1}{5^n}$

27. $\displaystyle\sum_{k=1}^{\infty} \frac{k \ln k}{(k+1)^3}$

28. $\displaystyle\sum_{n=1}^{\infty} \frac{e^{1/n}}{n^2}$

29. $\displaystyle\sum_{n=1}^{\infty} \frac{\tan^{-1} n}{n\sqrt{n}}$

30. $\displaystyle\sum_{j=1}^{\infty} (-1)^j \frac{\sqrt{j}}{j+5}$

31. $\displaystyle\sum_{k=1}^{\infty} \frac{5^k}{3^k+4^k}$

32. $\displaystyle\sum_{n=1}^{\infty} \frac{(2n)^n}{n^{2n}}$

33. $\displaystyle\sum_{n=1}^{\infty} \frac{\sin(1/n)}{\sqrt{n}}$

34. $\displaystyle\sum_{n=1}^{\infty} \frac{1}{n+n\cos^2 n}$

35. $\displaystyle\sum_{n=1}^{\infty} \left(\frac{n}{n+1}\right)^{n^2}$

36. $\displaystyle\sum_{n=2}^{\infty} \frac{1}{(\ln n)^{\ln n}}$

37. $\displaystyle\sum_{n=1}^{\infty} \left(\sqrt[n]{2}-1\right)^n$

38. $\displaystyle\sum_{n=1}^{\infty} \left(\sqrt[n]{2}-1\right)$

12.8 Power Series

A **power series** is a series of the form

$$\boxed{1} \quad \sum_{n=0}^{\infty} c_n x^n = c_0 + c_1 x + c_2 x^2 + c_3 x^3 + \cdots$$

where x is a variable and the c_n's are constants called the **coefficients** of the series. For each fixed x, the series (1) is a series of constants that we can test for convergence or divergence. A power series may converge for some values of x and diverge for other values of x. The sum of the series is a function

$$f(x) = c_0 + c_1 x + c_2 x^2 + \cdots + c_n x^n + \cdots$$

whose domain is the set of all x for which the series converges. Notice that f resembles a polynomial. The only difference is that f has infinitely many terms.

For instance, if we take $c_n = 1$ for all n, the power series becomes the geometric series

$$\sum_{n=0}^{\infty} x^n = 1 + x + x^2 + \cdots + x^n + \cdots$$

which converges when $-1 < x < 1$ and diverges when $|x| \geq 1$ (see Equation 12.2.5).

More generally, a series of the form

$$\boxed{2} \quad \sum_{n=0}^{\infty} c_n (x - a)^n = c_0 + c_1 (x - a) + c_2 (x - a)^2 + \cdots$$

is called a **power series in** $(x - a)$ or a **power series centered at** a or a **power series about** a. Notice that in writing out the term corresponding to $n = 0$ in Equations 1 and 2 we have adopted the convention that $(x - a)^0 = 1$ even when $x = a$. Notice also that when $x = a$ all of the terms are 0 for $n \geq 1$ and so the power series (2) always converges when $x = a$.

EXAMPLE 1 For what values of x is the series $\sum_{n=0}^{\infty} n! x^n$ convergent?

SOLUTION We use the Ratio Test. If we let a_n, as usual, denote the nth term of the series, then $a_n = n! x^n$. If $x \neq 0$, we have

$$\lim_{n \to \infty} \left| \frac{a_{n+1}}{a_n} \right| = \lim_{n \to \infty} \left| \frac{(n+1)! x^{n+1}}{n! x^n} \right| = \lim_{n \to \infty} (n+1)|x| = \infty$$

By the Ratio Test, the series diverges when $x \neq 0$. Thus, the given series converges only when $x = 0$.

EXAMPLE 2 For what values of x does the series $\sum_{n=1}^{\infty} \frac{(x-3)^n}{n}$ converge?

SOLUTION Let $a_n = (x-3)^n/n$. Then

$$\left| \frac{a_{n+1}}{a_n} \right| = \left| \frac{(x-3)^{n+1}}{n+1} \cdot \frac{n}{(x-3)^n} \right|$$

$$= \frac{1}{1 + \dfrac{1}{n}} |x - 3| \to |x - 3| \quad \text{as } n \to \infty$$

|||| TRIGONOMETRIC SERIES

A power series is a series in which each term is a power function. A **trigonometric series**

$$\sum_{n=0}^{\infty} (a_n \cos nx + b_n \sin nx)$$

is a series whose terms are trigonometric functions. This type of series is discussed on the web site

www.stewartcalculus.com

Click on *Additional Topics* and then on *Fourier Series*.

By the Ratio Test, the given series is absolutely convergent, and therefore convergent, when $|x - 3| < 1$ and divergent when $|x - 3| > 1$. Now

$$|x - 3| < 1 \iff -1 < x - 3 < 1 \iff 2 < x < 4$$

so the series converges when $2 < x < 4$ and diverges when $x < 2$ or $x > 4$.

The Ratio Test gives no information when $|x - 3| = 1$ so we must consider $x = 2$ and $x = 4$ separately. If we put $x = 4$ in the series, it becomes $\Sigma \, 1/n$, the harmonic series, which is divergent. If $x = 2$, the series is $\Sigma \, (-1)^n/n$, which converges by the Alternating Series Test. Thus, the given power series converges for $2 \leq x < 4$.

We will see that the main use of a power series is that it provides a way to represent some of the most important functions that arise in mathematics, physics, and chemistry. In particular, the sum of the power series in the next example is called a **Bessel function**, after the German astronomer Friedrich Bessel (1784–1846), and the function given in Exercise 33 is another example of a Bessel function. In fact, these functions first arose when Bessel solved Kepler's equation for describing planetary motion. Since that time, these functions have been applied in many different physical situations, including the temperature distribution in a circular plate and the shape of a vibrating drumhead (see the photographs on page 736).

EXAMPLE 3 Find the domain of the Bessel function of order 0 defined by

$$J_0(x) = \sum_{n=0}^{\infty} \frac{(-1)^n x^{2n}}{2^{2n}(n!)^2}$$

SOLUTION Let $a_n = (-1)^n x^{2n}/[2^{2n}(n!)^2]$. Then

$$\left| \frac{a_{n+1}}{a_n} \right| = \left| \frac{(-1)^{n+1} x^{2(n+1)}}{2^{2(n+1)}[(n+1)!]^2} \cdot \frac{2^{2n}(n!)^2}{(-1)^n x^{2n}} \right|$$

$$= \frac{x^{2n+2}}{2^{2n+2}(n+1)^2(n!)^2} \cdot \frac{2^{2n}(n!)^2}{x^{2n}}$$

$$= \frac{x^2}{4(n+1)^2} \to 0 < 1 \quad \text{for all } x$$

Thus, by the Ratio Test, the given series converges for all values of x. In other words, the domain of the Bessel function J_0 is $(-\infty, \infty) = \mathbb{R}$.

Recall that the sum of a series is equal to the limit of the sequence of partial sums. So when we define the Bessel function in Example 3 as the sum of a series we mean that, for every real number x,

$$J_0(x) = \lim_{n \to \infty} s_n(x) \quad \text{where} \quad s_n(x) = \sum_{i=0}^{n} \frac{(-1)^i x^{2i}}{2^{2i}(i!)^2}$$

The first few partial sums are

$$s_0(x) = 1 \qquad s_1(x) = 1 - \frac{x^2}{4} \qquad s_2(x) = 1 - \frac{x^2}{4} + \frac{x^4}{64}$$

$$s_3(x) = 1 - \frac{x^2}{4} + \frac{x^4}{64} - \frac{x^6}{2304} \qquad s_4(x) = 1 - \frac{x^2}{4} + \frac{x^4}{64} - \frac{x^6}{2304} + \frac{x^8}{147{,}456}$$

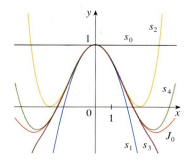

FIGURE 1
Partial sums of the Bessel function J_0

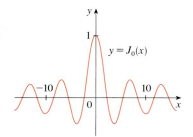

FIGURE 2

Figure 1 shows the graphs of these partial sums, which are polynomials. They are all approximations to the function J_0, but notice that the approximations become better when more terms are included. Figure 2 shows a more complete graph of the Bessel function.

For the power series that we have looked at so far, the set of values of x for which the series is convergent has always turned out to be an interval [a finite interval for the geometric series and the series in Example 2, the infinite interval $(-\infty, \infty)$ in Example 3, and a collapsed interval $[0, 0] = \{0\}$ in Example 1]. The following theorem, proved in Appendix F, says that this is true in general.

---

3 Theorem For a given power series $\sum_{n=0}^{\infty} c_n(x - a)^n$ there are only three possibilities:

(i) The series converges only when $x = a$.

(ii) The series converges for all x.

(iii) There is a positive number R such that the series converges if $|x - a| < R$ and diverges if $|x - a| > R$.

The number R in case (iii) is called the **radius of convergence** of the power series. By convention, the radius of convergence is $R = 0$ in case (i) and $R = \infty$ in case (ii). The **interval of convergence** of a power series is the interval that consists of all values of x for which the series converges. In case (i) the interval consists of just a single point a. In case (ii) the interval is $(-\infty, \infty)$. In case (iii) note that the inequality $|x - a| < R$ can be rewritten as $a - R < x < a + R$. When x is an *endpoint* of the interval, that is, $x = a \pm R$, anything can happen—the series might converge at one or both endpoints or it might diverge at both endpoints. Thus, in case (iii) there are four possibilities for the interval of convergence:

$$(a - R, a + R) \qquad (a - R, a + R] \qquad [a - R, a + R) \qquad [a - R, a + R]$$

The situation is illustrated in Figure 3.

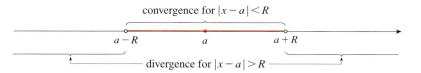

FIGURE 3

We summarize here the radius and interval of convergence for each of the examples already considered in this section.

	Series	Radius of convergence	Interval of convergence
Geometric series	$\sum_{n=0}^{\infty} x^n$	$R = 1$	$(-1, 1)$
Example 1	$\sum_{n=0}^{\infty} n!\, x^n$	$R = 0$	$\{0\}$
Example 2	$\sum_{n=1}^{\infty} \dfrac{(x - 3)^n}{n}$	$R = 1$	$[2, 4)$
Example 3	$\sum_{n=0}^{\infty} \dfrac{(-1)^n x^{2n}}{2^{2n}(n!)^2}$	$R = \infty$	$(-\infty, \infty)$

In general, the Ratio Test (or sometimes the Root Test) should be used to determine the radius of convergence R. The Ratio and Root Tests always fail when x is an endpoint of the interval of convergence, so the endpoints must be checked with some other test.

EXAMPLE 4 Find the radius of convergence and interval of convergence of the series

$$\sum_{n=0}^{\infty} \frac{(-3)^n x^n}{\sqrt{n+1}}$$

SOLUTION Let $a_n = (-3)^n x^n / \sqrt{n+1}$. Then

$$\left| \frac{a_{n+1}}{a_n} \right| = \left| \frac{(-3)^{n+1} x^{n+1}}{\sqrt{n+2}} \cdot \frac{\sqrt{n+1}}{(-3)^n x^n} \right| = \left| -3x \sqrt{\frac{n+1}{n+2}} \right|$$

$$= 3 \sqrt{\frac{1 + (1/n)}{1 + (2/n)}} |x| \to 3|x| \quad \text{as } n \to \infty$$

By the Ratio Test, the given series converges if $3|x| < 1$ and diverges if $3|x| > 1$. Thus, it converges if $|x| < \tfrac{1}{3}$ and diverges if $|x| > \tfrac{1}{3}$. This means that the radius of convergence is $R = \tfrac{1}{3}$.

We know the series converges in the interval $(-\tfrac{1}{3}, \tfrac{1}{3})$, but we must now test for convergence at the endpoints of this interval. If $x = -\tfrac{1}{3}$, the series becomes

$$\sum_{n=0}^{\infty} \frac{(-3)^n \left(-\tfrac{1}{3}\right)^n}{\sqrt{n+1}} = \sum_{n=0}^{\infty} \frac{1}{\sqrt{n+1}} = \frac{1}{\sqrt{1}} + \frac{1}{\sqrt{2}} + \frac{1}{\sqrt{3}} + \frac{1}{\sqrt{4}} + \cdots$$

which diverges. (Use the Integral Test or simply observe that it is a p-series with $p = \tfrac{1}{2} < 1$.) If $x = \tfrac{1}{3}$, the series is

$$\sum_{n=0}^{\infty} \frac{(-3)^n \left(\tfrac{1}{3}\right)^n}{\sqrt{n+1}} = \sum_{n=0}^{\infty} \frac{(-1)^n}{\sqrt{n+1}}$$

which converges by the Alternating Series Test. Therefore, the given power series converges when $-\tfrac{1}{3} < x \leq \tfrac{1}{3}$, so the interval of convergence is $\left(-\tfrac{1}{3}, \tfrac{1}{3}\right]$.

EXAMPLE 5 Find the radius of convergence and interval of convergence of the series

$$\sum_{n=0}^{\infty} \frac{n(x+2)^n}{3^{n+1}}$$

SOLUTION If $a_n = n(x+2)^n / 3^{n+1}$, then

$$\left| \frac{a_{n+1}}{a_n} \right| = \left| \frac{(n+1)(x+2)^{n+1}}{3^{n+2}} \cdot \frac{3^{n+1}}{n(x+2)^n} \right|$$

$$= \left(1 + \frac{1}{n}\right) \frac{|x+2|}{3} \to \frac{|x+2|}{3} \quad \text{as } n \to \infty$$

Using the Ratio Test, we see that the series converges if $|x+2|/3 < 1$ and it diverges if $|x+2|/3 > 1$. So it converges if $|x+2| < 3$ and diverges if $|x+2| > 3$. Thus, the radius of convergence is $R = 3$.

The inequality $|x + 2| < 3$ can be written as $-5 < x < 1$, so we test the series at the endpoints -5 and 1. When $x = -5$, the series is

$$\sum_{n=0}^{\infty} \frac{n(-3)^n}{3^{n+1}} = \frac{1}{3} \sum_{n=0}^{\infty} (-1)^n n$$

which diverges by the Test for Divergence [$(-1)^n n$ doesn't converge to 0]. When $x = 1$, the series is

$$\sum_{n=0}^{\infty} \frac{n(3)^n}{3^{n+1}} = \frac{1}{3} \sum_{n=0}^{\infty} n$$

which also diverges by the Test for Divergence. Thus, the series converges only when $-5 < x < 1$, so the interval of convergence is $(-5, 1)$.

12.8 Exercises

1. What is a power series?

2. (a) What is the radius of convergence of a power series? How do you find it?
   (b) What is the interval of convergence of a power series? How do you find it?

3–28 ▪▪▪ Find the radius of convergence and interval of convergence of the series.

3. $\sum_{n=1}^{\infty} \frac{x^n}{\sqrt{n}}$

4. $\sum_{n=0}^{\infty} \frac{(-1)^n x^n}{n+1}$

5. $\sum_{n=1}^{\infty} \frac{(-1)^{n-1} x^n}{n^3}$

6. $\sum_{n=1}^{\infty} \sqrt{n} x^n$

7. $\sum_{n=0}^{\infty} \frac{x^n}{n!}$

8. $\sum_{n=1}^{\infty} n^n x^n$

9. $\sum_{n=1}^{\infty} (-1)^n 4^n x^n$

10. $\sum_{n=1}^{\infty} \frac{x^n}{n3^n}$

11. $\sum_{n=1}^{\infty} \frac{(-2)^n x^n}{\sqrt[4]{n}}$

12. $\sum_{n=1}^{\infty} \frac{x^n}{5^n n^5}$

13. $\sum_{n=2}^{\infty} (-1)^n \frac{x^n}{4^n \ln n}$

14. $\sum_{n=0}^{\infty} (-1)^n \frac{x^{2n}}{(2n)!}$

15. $\sum_{n=0}^{\infty} \sqrt{n}\,(x-1)^n$

16. $\sum_{n=0}^{\infty} n^3(x-5)^n$

17. $\sum_{n=1}^{\infty} (-1)^n \frac{(x+2)^n}{n 2^n}$

18. $\sum_{n=1}^{\infty} \frac{(-2)^n}{\sqrt{n}} (x+3)^n$

19. $\sum_{n=1}^{\infty} \frac{(x-2)^n}{n^n}$

20. $\sum_{n=1}^{\infty} \frac{(3x-2)^n}{n 3^n}$

21. $\sum_{n=1}^{\infty} \frac{n}{b^n}(x-a)^n, \quad b > 0$

22. $\sum_{n=1}^{\infty} \frac{n(x-4)^n}{n^3+1}$

23. $\sum_{n=1}^{\infty} n!(2x-1)^n$

24. $\sum_{n=1}^{\infty} \frac{n^2 x^n}{2 \cdot 4 \cdot 6 \cdots (2n)}$

25. $\sum_{n=1}^{\infty} \frac{(4x+1)^n}{n^2}$

26. $\sum_{n=2}^{\infty} (-1)^n \frac{(2x+3)^n}{n \ln n}$

27. $\sum_{n=2}^{\infty} \frac{x^n}{(\ln n)^n}$

28. $\sum_{n=1}^{\infty} \frac{2 \cdot 4 \cdot 6 \cdots (2n)}{1 \cdot 3 \cdot 5 \cdots (2n-1)} x^n$

29. If $\sum_{n=0}^{\infty} c_n 4^n$ is convergent, does it follow that the following series are convergent?

   (a) $\sum_{n=0}^{\infty} c_n (-2)^n$
   (b) $\sum_{n=0}^{\infty} c_n (-4)^n$

30. Suppose that $\sum_{n=0}^{\infty} c_n x^n$ converges when $x = -4$ and diverges when $x = 6$. What can be said about the convergence or divergence of the following series?

   (a) $\sum_{n=0}^{\infty} c_n$
   (b) $\sum_{n=0}^{\infty} c_n 8^n$
   (c) $\sum_{n=0}^{\infty} c_n (-3)^n$
   (d) $\sum_{n=0}^{\infty} (-1)^n c_n 9^n$

31. If k is a positive integer, find the radius of convergence of the series

$$\sum_{n=0}^{\infty} \frac{(n!)^k}{(kn)!} x^n$$

32. Graph the first several partial sums $s_n(x)$ of the series $\sum_{n=0}^{\infty} x^n$, together with the sum function $f(x) = 1/(1-x)$, on a common screen. On what interval do these partial sums appear to be converging to $f(x)$?

33. The function J_1 defined by

$$J_1(x) = \sum_{n=0}^{\infty} \frac{(-1)^n x^{2n+1}}{n!(n+1)! 2^{2n+1}}$$

is called the *Bessel function of order 1*.
   (a) Find its domain.
   (b) Graph the first several partial sums on a common screen.

790 ||| CHAPTER 12 INFINITE SEQUENCES AND SERIES

[CAS] (c) If your CAS has built-in Bessel functions, graph J_1 on the same screen as the partial sums in part (b) and observe how the partial sums approximate J_1.

34. The function A defined by

$$A(x) = 1 + \frac{x^3}{2 \cdot 3} + \frac{x^6}{2 \cdot 3 \cdot 5 \cdot 6} + \frac{x^9}{2 \cdot 3 \cdot 5 \cdot 6 \cdot 8 \cdot 9} + \cdots$$

is called the *Airy function* after the English mathematician and astronomer Sir George Airy (1801–1892).
(a) Find the domain of the Airy function.
(b) Graph the first several partial sums $s_n(x)$ on a common screen.
[CAS] (c) If your CAS has built-in Airy functions, graph A on the same screen as the partial sums in part (b) and observe how the partial sums approximate A.

35. A function f is defined by

$$f(x) = 1 + 2x + x^2 + 2x^3 + x^4 + \cdots$$

that is, its coefficients are $c_{2n} = 1$ and $c_{2n+1} = 2$ for all $n \geq 0$. Find the interval of convergence of the series and find an explicit formula for $f(x)$.

36. If $f(x) = \sum_{n=0}^{\infty} c_n x^n$, where $c_{n+4} = c_n$ for all $n \geq 0$, find the interval of convergence of the series and a formula for $f(x)$.

37. Show that if $\lim_{n \to \infty} \sqrt[n]{|c_n|} = c$, where $c \neq 0$, then the radius of convergence of the power series $\sum c_n x^n$ is $R = 1/c$.

38. Suppose that the power series $\sum c_n(x - a)^n$ satisfies $c_n \neq 0$ for all n. Show that if $\lim_{n \to \infty} |c_n/c_{n+1}|$ exists, then it is equal to the radius of convergence of the power series.

39. Suppose the series $\sum c_n x^n$ has radius of convergence 2 and the series $\sum d_n x^n$ has radius of convergence 3. What is the radius of convergence of the series $\sum (c_n + d_n) x^n$?

40. Suppose that the radius of convergence of the power series $\sum c_n x^n$ is R. What is the radius of convergence of the power series $\sum c_n x^{2n}$?

12.9 Representations of Functions as Power Series

In this section we learn how to represent certain types of functions as sums of power series by manipulating geometric series or by differentiating or integrating such a series. You might wonder why we would ever want to express a known function as a sum of infinitely many terms. We will see later that this strategy is useful for integrating functions that don't have elementary antiderivatives, for solving differential equations, and for approximating functions by polynomials. (Scientists do this to simplify the expressions they deal with; computer scientists do this to represent functions on calculators and computers.)

We start with an equation that we have seen before:

$$\boxed{1} \quad \frac{1}{1-x} = 1 + x + x^2 + x^3 + \cdots = \sum_{n=0}^{\infty} x^n \qquad |x| < 1$$

We first encountered this equation in Example 5 in Section 12.2, where we obtained it by observing that it is a geometric series with $a = 1$ and $r = x$. But here our point of view is different. We now regard Equation 1 as expressing the function $f(x) = 1/(1 - x)$ as a sum of a power series.

|||| A geometric illustration of Equation 1 is shown in Figure 1. Because the sum of a series is the limit of the sequence of partial sums, we have

$$\frac{1}{1-x} = \lim_{n \to \infty} s_n(x)$$

where

$$s_n(x) = 1 + x + x^2 + \cdots + x^n$$

is the nth partial sum. Notice that as n increases, $s_n(x)$ becomes a better approximation to $f(x)$ for $-1 < x < 1$.

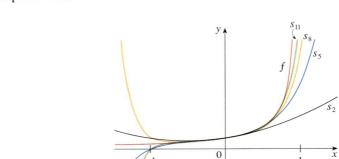

FIGURE 1
$f(x) = \dfrac{1}{1-x}$ and some partial sums

EXAMPLE 1 Express $1/(1 + x^2)$ as the sum of a power series and find the interval of convergence.

SOLUTION Replacing x by $-x^2$ in Equation 1, we have

$$\frac{1}{1 + x^2} = \frac{1}{1 - (-x^2)} = \sum_{n=0}^{\infty} (-x^2)^n$$

$$= \sum_{n=0}^{\infty} (-1)^n x^{2n} = 1 - x^2 + x^4 - x^6 + x^8 - \cdots$$

Because this is a geometric series, it converges when $|-x^2| < 1$, that is, $x^2 < 1$, or $|x| < 1$. Therefore, the interval of convergence is $(-1, 1)$. (Of course, we could have determined the radius of convergence by applying the Ratio Test, but that much work is unnecessary here.)

EXAMPLE 2 Find a power series representation for $1/(x + 2)$.

SOLUTION In order to put this function in the form of the left side of Equation 1 we first factor a 2 from the denominator:

$$\frac{1}{2 + x} = \frac{1}{2\left(1 + \frac{x}{2}\right)} = \frac{1}{2\left[1 - \left(-\frac{x}{2}\right)\right]}$$

$$= \frac{1}{2} \sum_{n=0}^{\infty} \left(-\frac{x}{2}\right)^n = \sum_{n=0}^{\infty} \frac{(-1)^n}{2^{n+1}} x^n$$

This series converges when $|-x/2| < 1$, that is, $|x| < 2$. So the interval of convergence is $(-2, 2)$.

EXAMPLE 3 Find a power series representation of $x^3/(x + 2)$.

SOLUTION Since this function is just x^3 times the function in Example 2, all we have to do is to multiply that series by x^3:

|||| It's legitimate to move x^3 across the sigma sign because it doesn't depend on n. [Use Theorem 12.2.8(i) with $c = x^3$.]

$$\frac{x^3}{x + 2} = x^3 \cdot \frac{1}{x + 2} = x^3 \sum_{n=0}^{\infty} \frac{(-1)^n}{2^{n+1}} x^n = \sum_{n=0}^{\infty} \frac{(-1)^n}{2^{n+1}} x^{n+3}$$

$$= \tfrac{1}{2}x^3 - \tfrac{1}{4}x^4 + \tfrac{1}{8}x^5 - \tfrac{1}{16}x^6 + \cdots$$

Another way of writing this series is as follows:

$$\frac{x^3}{x + 2} = \sum_{n=3}^{\infty} \frac{(-1)^{n-1}}{2^{n-2}} x^n$$

As in Example 2, the interval of convergence is $(-2, 2)$.

Differentiation and Integration of Power Series

The sum of a power series is a function $f(x) = \sum_{n=0}^{\infty} c_n(x - a)^n$ whose domain is the interval of convergence of the series. We would like to be able to differentiate and integrate such functions, and the following theorem (which we won't prove) says that we can do so by differentiating or integrating each individual term in the series, just as we would for a polynomial. This is called **term-by-term differentiation and integration**.

2 Theorem If the power series $\sum c_n(x - a)^n$ has radius of convergence $R > 0$, then the function f defined by

$$f(x) = c_0 + c_1(x - a) + c_2(x - a)^2 + \cdots = \sum_{n=0}^{\infty} c_n(x - a)^n$$

is differentiable (and therefore continuous) on the interval $(a - R, a + R)$ and

(i) $f'(x) = c_1 + 2c_2(x - a) + 3c_3(x - a)^2 + \cdots = \sum_{n=1}^{\infty} nc_n(x - a)^{n-1}$

(ii) $\int f(x)\, dx = C + c_0(x - a) + c_1 \dfrac{(x - a)^2}{2} + c_2 \dfrac{(x - a)^3}{3} + \cdots$

$= C + \sum_{n=0}^{\infty} c_n \dfrac{(x - a)^{n+1}}{n + 1}$

The radii of convergence of the power series in Equations (i) and (ii) are both R.

|||| In part (ii), $\int c_0\, dx = c_0 x + C_1$ is written as $c_0(x - a) + C$, where $C = C_1 + ac_0$, so all the terms of the series have the same form.

NOTE 1 ▫ Equations (i) and (ii) in Theorem 2 can be rewritten in the form

(iii) $\dfrac{d}{dx}\left[\sum_{n=0}^{\infty} c_n(x - a)^n\right] = \sum_{n=0}^{\infty} \dfrac{d}{dx}[c_n(x - a)^n]$

(iv) $\int \left[\sum_{n=0}^{\infty} c_n(x - a)^n\right] dx = \sum_{n=0}^{\infty} \int c_n(x - a)^n\, dx$

We know that, for finite sums, the derivative of a sum is the sum of the derivatives and the integral of a sum is the sum of the integrals. Equations (iii) and (iv) assert that the same is true for infinite sums, provided we are dealing with *power series*. (For other types of series of functions the situation is not as simple; see Exercise 36.)

NOTE 2 ▫ Although Theorem 2 says that the radius of convergence remains the same when a power series is differentiated or integrated, this does not mean that the *interval* of convergence remains the same. It may happen that the original series converges at an endpoint, whereas the differentiated series diverges there. (See Exercise 37.)

NOTE 3 ▫ The idea of differentiating a power series term by term is the basis for a powerful method for solving differential equations. We will discuss this method in Chapter 18.

EXAMPLE 4 In Example 3 in Section 12.8 we saw that the Bessel function

$$J_0(x) = \sum_{n=0}^{\infty} \dfrac{(-1)^n x^{2n}}{2^{2n}(n!)^2}$$

is defined for all x. Thus, by Theorem 2, J_0 is differentiable for all x and its derivative is found by term-by-term differentiation as follows:

$$J_0'(x) = \sum_{n=0}^{\infty} \dfrac{d}{dx} \dfrac{(-1)^n x^{2n}}{2^{2n}(n!)^2} = \sum_{n=1}^{\infty} \dfrac{(-1)^n 2n x^{2n-1}}{2^{2n}(n!)^2}$$

EXAMPLE 5 Express $1/(1 - x)^2$ as a power series by differentiating Equation 1. What is the radius of convergence?

SOLUTION Differentiating each side of the equation

$$\frac{1}{1-x} = 1 + x + x^2 + x^3 + \cdots = \sum_{n=0}^{\infty} x^n$$

we get

$$\frac{1}{(1-x)^2} = 1 + 2x + 3x^2 + \cdots = \sum_{n=1}^{\infty} nx^{n-1}$$

If we wish, we can replace n by $n+1$ and write the answer as

$$\frac{1}{(1-x)^2} = \sum_{n=0}^{\infty} (n+1)x^n$$

According to Theorem 2, the radius of convergence of the differentiated series is the same as the radius of convergence of the original series, namely, $R = 1$.

EXAMPLE 6 Find a power series representation for $\ln(1-x)$ and its radius of convergence.

SOLUTION We notice that, except for a factor of -1, the derivative of this function is $1/(1-x)$. So we integrate both sides of Equation 1:

$$-\ln(1-x) = \int \frac{1}{1-x} dx = \int (1 + x + x^2 + \cdots) dx$$

$$= x + \frac{x^2}{2} + \frac{x^3}{3} + \cdots + C = \sum_{n=0}^{\infty} \frac{x^{n+1}}{n+1} + C$$

$$= \sum_{n=1}^{\infty} \frac{x^n}{n} + C \qquad |x| < 1$$

To determine the value of C we put $x = 0$ in this equation and obtain $-\ln(1-0) = C$. Thus, $C = 0$ and

$$\ln(1-x) = -x - \frac{x^2}{2} - \frac{x^3}{3} - \cdots = -\sum_{n=1}^{\infty} \frac{x^n}{n} \qquad |x| < 1$$

The radius of convergence is the same as for the original series: $R = 1$.

Notice what happens if we put $x = \frac{1}{2}$ in the result of Example 6. Since $\ln \frac{1}{2} = -\ln 2$, we see that

$$\ln 2 = \frac{1}{2} + \frac{1}{8} + \frac{1}{24} + \frac{1}{64} + \cdots = \sum_{n=1}^{\infty} \frac{1}{n2^n}$$

EXAMPLE 7 Find a power series representation for $f(x) = \tan^{-1}x$.

SOLUTION We observe that $f'(x) = 1/(1 + x^2)$ and find the required series by integrating the power series for $1/(1 + x^2)$ found in Example 1.

$$\tan^{-1}x = \int \frac{1}{1+x^2} dx = \int (1 - x^2 + x^4 - x^6 + \cdots) dx$$

$$= C + x - \frac{x^3}{3} + \frac{x^5}{5} - \frac{x^7}{7} + \cdots$$

▐▐▐▐ The power series for $\tan^{-1}x$ obtained in Example 7 is called *Gregory's series* after the Scottish mathematician James Gregory (1638–1675), who had anticipated some of Newton's discoveries. We have shown that Gregory's series is valid when $-1 < x < 1$, but it turns out (although it isn't easy to prove) that it is also valid when $x = \pm 1$. Notice that when $x = 1$ the series becomes

$$\frac{\pi}{4} = 1 - \frac{1}{3} + \frac{1}{5} - \frac{1}{7} + \cdots$$

This beautiful result is known as the Leibniz formula for π.

To find C we put $x = 0$ and obtain $C = \tan^{-1} 0 = 0$. Therefore

$$\tan^{-1} x = x - \frac{x^3}{3} + \frac{x^5}{5} - \frac{x^7}{7} + \cdots = \sum_{n=0}^{\infty} (-1)^n \frac{x^{2n+1}}{2n+1}$$

Since the radius of convergence of the series for $1/(1 + x^2)$ is 1, the radius of convergence of this series for $\tan^{-1} x$ is also 1.

EXAMPLE 8
(a) Evaluate $\int [1/(1 + x^7)] \, dx$ as a power series.
(b) Use part (a) to approximate $\int_0^{0.5} [1/(1 + x^7)] \, dx$ correct to within 10^{-7}.

SOLUTION
(a) The first step is to express the integrand, $1/(1 + x^7)$, as the sum of a power series. As in Example 1, we start with Equation 1 and replace x by $-x^7$:

$$\frac{1}{1 + x^7} = \frac{1}{1 - (-x^7)} = \sum_{n=0}^{\infty} (-x^7)^n$$

$$= \sum_{n=0}^{\infty} (-1)^n x^{7n} = 1 - x^7 + x^{14} - \cdots$$

▐▐▐▐ This example demonstrates one way in which power series representations are useful. Integrating $1/(1 + x^7)$ by hand is incredibly difficult. Different computer algebra systems return different forms of the answer, but they are all extremely complicated. (If you have a CAS, try it yourself.) The infinite series answer that we obtain in Example 8(a) is actually much easier to deal with than the finite answer provided by a CAS.

Now we integrate term by term:

$$\int \frac{1}{1 + x^7} \, dx = \int \sum_{n=0}^{\infty} (-1)^n x^{7n} \, dx = C + \sum_{n=0}^{\infty} (-1)^n \frac{x^{7n+1}}{7n+1}$$

$$= C + x - \frac{x^8}{8} + \frac{x^{15}}{15} - \frac{x^{22}}{22} + \cdots$$

This series converges for $|-x^7| < 1$, that is, for $|x| < 1$.

(b) In applying the Fundamental Theorem of Calculus it doesn't matter which antiderivative we use, so let's use the antiderivative from part (a) with $C = 0$:

$$\int_0^{0.5} \frac{1}{1 + x^7} \, dx = \left[x - \frac{x^8}{8} + \frac{x^{15}}{15} - \frac{x^{22}}{22} + \cdots \right]_0^{1/2}$$

$$= \frac{1}{2} - \frac{1}{8 \cdot 2^8} + \frac{1}{15 \cdot 2^{15}} - \frac{1}{22 \cdot 2^{22}} + \cdots + \frac{(-1)^n}{(7n+1)2^{7n+1}} + \cdots$$

This infinite series is the exact value of the definite integral, but since it is an alternating series, we can approximate the sum using the Alternating Series Estimation Theorem. If we stop adding after the term with $n = 3$, the error is smaller than the term with $n = 4$:

$$\frac{1}{29 \cdot 2^{29}} \approx 6.4 \times 10^{-11}$$

So we have

$$\int_0^{0.5} \frac{1}{1 + x^7} \, dx \approx \frac{1}{2} - \frac{1}{8 \cdot 2^8} + \frac{1}{15 \cdot 2^{15}} - \frac{1}{22 \cdot 2^{22}} \approx 0.49951374$$

12.9 Exercises

1. If the radius of convergence of the power series $\sum_{n=0}^{\infty} c_n x^n$ is 10, what is the radius of convergence of the series $\sum_{n=1}^{\infty} n c_n x^{n-1}$? Why?

2. Suppose you know that the series $\sum_{n=0}^{\infty} b_n x^n$ converges for $|x| < 2$. What can you say about the following series? Why?
$$\sum_{n=0}^{\infty} \frac{b_n}{n+1} x^{n+1}$$

3–10 Find a power series representation for the function and determine the interval of convergence.

3. $f(x) = \dfrac{1}{1+x}$

4. $f(x) = \dfrac{3}{1-x^4}$

5. $f(x) = \dfrac{1}{1-x^3}$

6. $f(x) = \dfrac{1}{1+9x^2}$

7. $f(x) = \dfrac{1}{x-5}$

8. $f(x) = \dfrac{x}{4x+1}$

9. $f(x) = \dfrac{x}{9+x^2}$

10. $f(x) = \dfrac{x^2}{a^3-x^3}$

11–12 Express the function as the sum of a power series by first using partial fractions. Find the interval of convergence.

11. $f(x) = \dfrac{3}{x^2+x-2}$

12. $f(x) = \dfrac{7x-1}{3x^2+2x-1}$

13. (a) Use differentiation to find a power series representation for
$$f(x) = \frac{1}{(1+x)^2}$$
What is the radius of convergence?
(b) Use part (a) to find a power series for
$$f(x) = \frac{1}{(1+x)^3}$$
(c) Use part (b) to find a power series for
$$f(x) = \frac{x^2}{(1+x)^3}$$

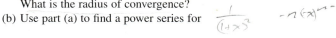

14. (a) Find a power series representation for $f(x) = \ln(1+x)$. What is the radius of convergence?
(b) Use part (a) to find a power series for $f(x) = x \ln(1+x)$.
(c) Use part (a) to find a power series for $f(x) = \ln(x^2+1)$.

15–18 Find a power series representation for the function and determine the radius of convergence.

15. $f(x) = \ln(5-x)$

16. $f(x) = \dfrac{x^2}{(1-2x)^2}$

17. $f(x) = \dfrac{x^3}{(x-2)^2}$

18. $f(x) = \arctan(x/3)$

19–22 Find a power series representation for f, and graph f and several partial sums $s_n(x)$ on the same screen. What happens as n increases?

19. $f(x) = \ln(3+x)$

20. $f(x) = \dfrac{1}{x^2+25}$

21. $f(x) = \ln\left(\dfrac{1+x}{1-x}\right)$

22. $f(x) = \tan^{-1}(2x)$

23–26 Evaluate the indefinite integral as a power series. What is the radius of convergence?

23. $\displaystyle\int \dfrac{t}{1-t^8} \, dt$

24. $\displaystyle\int \dfrac{\ln(1-t)}{t} \, dt$

25. $\displaystyle\int \dfrac{x - \tan^{-1} x}{x^3} \, dx$

26. $\displaystyle\int \tan^{-1}(x^2) \, dx$

27–30 Use a power series to approximate the definite integral to six decimal places.

27. $\displaystyle\int_0^{0.2} \dfrac{1}{1+x^5} \, dx$

28. $\displaystyle\int_0^{0.4} \ln(1+x^4) \, dx$

29. $\displaystyle\int_0^{1/3} x^2 \tan^{-1}(x^4) \, dx$

30. $\displaystyle\int_0^{0.5} \dfrac{dx}{1+x^6}$

31. Use the result of Example 6 to compute $\ln 1.1$ correct to five decimal places.

32. Show that the function
$$f(x) = \sum_{n=0}^{\infty} \frac{(-1)^n x^{2n}}{(2n)!}$$
is a solution of the differential equation
$$f''(x) + f(x) = 0$$

33. (a) Show that J_0 (the Bessel function of order 0 given in Example 4) satisfies the differential equation
$$x^2 J_0''(x) + x J_0'(x) + x^2 J_0(x) = 0$$
(b) Evaluate $\int_0^1 J_0(x) \, dx$ correct to three decimal places.

34. The Bessel function of order 1 is defined by

$$J_1(x) = \sum_{n=0}^{\infty} \frac{(-1)^n x^{2n+1}}{n!(n+1)!2^{2n+1}}$$

(a) Show that J_1 satisfies the differential equation

$$x^2 J_1''(x) + x J_1'(x) + (x^2 - 1) J_1(x) = 0$$

(b) Show that $J_0'(x) = -J_1(x)$.

35. (a) Show that the function

$$f(x) = \sum_{n=0}^{\infty} \frac{x^n}{n!}$$

is a solution of the differential equation

$$f'(x) = f(x)$$

(b) Show that $f(x) = e^x$.

36. Let $f_n(x) = (\sin nx)/n^2$. Show that the series $\Sigma f_n(x)$ converges for all values of x but the series of derivatives $\Sigma f_n'(x)$ diverges when $x = 2n\pi$, n an integer. For what values of x does the series $\Sigma f_n''(x)$ converge?

37. Let

$$f(x) = \sum_{n=1}^{\infty} \frac{x^n}{n^2}$$

Find the intervals of convergence for f, f', and f''.

38. (a) Starting with the geometric series $\Sigma_{n=0}^{\infty} x^n$, find the sum of the series

$$\sum_{n=1}^{\infty} n x^{n-1} \qquad |x| < 1$$

(b) Find the sum of each of the following series.

 (i) $\sum_{n=1}^{\infty} n x^n$, $|x| < 1$ (ii) $\sum_{n=1}^{\infty} \frac{n}{2^n}$

(c) Find the sum of each of the following series.

 (i) $\sum_{n=2}^{\infty} n(n-1) x^n$, $|x| < 1$

 (ii) $\sum_{n=2}^{\infty} \frac{n^2 - n}{2^n}$ (iii) $\sum_{n=1}^{\infty} \frac{n^2}{2^n}$

39. Use the power series for $\tan^{-1} x$ to prove the following expression for π as the sum of an infinite series:

$$\pi = 2\sqrt{3} \sum_{n=0}^{\infty} \frac{(-1)^n}{(2n+1)3^n}$$

40. (a) By completing the square, show that

$$\int_0^{1/2} \frac{dx}{x^2 - x + 1} = \frac{\pi}{3\sqrt{3}}$$

(b) By factoring $x^3 + 1$ as a sum of cubes, rewrite the integral in part (a). Then express $1/(x^3 + 1)$ as the sum of a power series and use it to prove the following formula for π:

$$\pi = \frac{3\sqrt{3}}{4} \sum_{n=0}^{\infty} \frac{(-1)^n}{8^n} \left(\frac{2}{3n+1} + \frac{1}{3n+2} \right)$$

12.10 Taylor and Maclaurin Series

In the preceding section we were able to find power series representations for a certain restricted class of functions. Here we investigate more general problems: Which functions have power series representations? How can we find such representations?

We start by supposing that f is any function that can be represented by a power series

$$\boxed{1} \quad f(x) = c_0 + c_1(x - a) + c_2(x - a)^2 + c_3(x - a)^3 + c_4(x - a)^4 + \cdots \qquad |x - a| < R$$

Let's try to determine what the coefficients c_n must be in terms of f. To begin, notice that if we put $x = a$ in Equation 1, then all terms after the first one are 0 and we get

$$f(a) = c_0$$

By Theorem 12.9.2, we can differentiate the series in Equation 1 term by term:

$$\boxed{2} \quad f'(x) = c_1 + 2c_2(x - a) + 3c_3(x - a)^2 + 4c_4(x - a)^3 + \cdots \qquad |x - a| < R$$

and substitution of $x = a$ in Equation 2 gives

$$f'(a) = c_1$$

Now we differentiate both sides of Equation 2 and obtain

$$\boxed{3} \quad f''(x) = 2c_2 + 2 \cdot 3c_3(x - a) + 3 \cdot 4c_4(x - a)^2 + \cdots \qquad |x - a| < R$$

Again we put $x = a$ in Equation 3. The result is

$$f''(a) = 2c_2$$

Let's apply the procedure one more time. Differentiation of the series in Equation 3 gives

$$\boxed{4} \quad f'''(x) = 2 \cdot 3c_3 + 2 \cdot 3 \cdot 4c_4(x - a) + 3 \cdot 4 \cdot 5c_5(x - a)^2 + \cdots \qquad |x - a| < R$$

and substitution of $x = a$ in Equation 4 gives

$$f'''(a) = 2 \cdot 3c_3 = 3!c_3$$

By now you can see the pattern. If we continue to differentiate and substitute $x = a$, we obtain

$$f^{(n)}(a) = 2 \cdot 3 \cdot 4 \cdot \cdots \cdot nc_n = n!c_n$$

Solving this equation for the nth coefficient c_n, we get

$$c_n = \frac{f^{(n)}(a)}{n!}$$

This formula remains valid even for $n = 0$ if we adopt the conventions that $0! = 1$ and $f^{(0)} = f$. Thus, we have proved the following theorem.

> $\boxed{5}$ **Theorem** If f has a power series representation (expansion) at a, that is, if
>
> $$f(x) = \sum_{n=0}^{\infty} c_n(x - a)^n \qquad |x - a| < R$$
>
> then its coefficients are given by the formula
>
> $$c_n = \frac{f^{(n)}(a)}{n!}$$

Substituting this formula for c_n back into the series, we see that *if f has a power series expansion at a, then it must be of the following form.*

> $$\boxed{6} \quad f(x) = \sum_{n=0}^{\infty} \frac{f^{(n)}(a)}{n!} (x - a)^n$$
>
> $$= f(a) + \frac{f'(a)}{1!}(x - a) + \frac{f''(a)}{2!}(x - a)^2 + \frac{f'''(a)}{3!}(x - a)^3 + \cdots$$

> The Taylor series is named after the English mathematician Brook Taylor (1685–1731) and the Maclaurin series is named in honor of the Scottish mathematician Colin Maclaurin (1698–1746) despite the fact that the Maclaurin series is really just a special case of the Taylor series. But the idea of representing particular functions as sums of power series goes back to Newton, and the general Taylor series was known to the Scottish mathematician James Gregory in 1668 and to the Swiss mathematician John Bernoulli in the 1690s. Taylor was apparently unaware of the work of Gregory and Bernoulli when he published his discoveries on series in 1715 in his book *Methodus incrementorum directa et inversa*. Maclaurin series are named after Colin Maclaurin because he popularized them in his calculus textbook *Treatise of Fluxions* published in 1742.

The series in Equation 6 is called the **Taylor series of the function f at a** (or **about a** or **centered at a**). For the special case $a = 0$ the Taylor series becomes

$$\boxed{7 \qquad f(x) = \sum_{n=0}^{\infty} \frac{f^{(n)}(0)}{n!} x^n = f(0) + \frac{f'(0)}{1!} x + \frac{f''(0)}{2!} x^2 + \cdots}$$

This case arises frequently enough that it is given the special name **Maclaurin series**.

NOTE ○ We have shown that if f can be represented as a power series about a, then f is equal to the sum of its Taylor series. But there exist functions that are not equal to the sum of their Taylor series. An example of such a function is given in Exercise 62.

EXAMPLE 1 Find the Maclaurin series of the function $f(x) = e^x$ and its radius of convergence.

SOLUTION If $f(x) = e^x$, then $f^{(n)}(x) = e^x$, so $f^{(n)}(0) = e^0 = 1$ for all n. Therefore, the Taylor series for f at 0 (that is, the Maclaurin series) is

$$\sum_{n=0}^{\infty} \frac{f^{(n)}(0)}{n!} x^n = \sum_{n=0}^{\infty} \frac{x^n}{n!} = 1 + \frac{x}{1!} + \frac{x^2}{2!} + \frac{x^3}{3!} + \cdots$$

To find the radius of convergence we let $a_n = x^n/n!$. Then

$$\left| \frac{a_{n+1}}{a_n} \right| = \left| \frac{x^{n+1}}{(n+1)!} \cdot \frac{n!}{x^n} \right| = \frac{|x|}{n+1} \to 0 < 1$$

so, by the Ratio Test, the series converges for all x and the radius of convergence is $R = \infty$.

The conclusion we can draw from Theorem 5 and Example 1 is that *if e^x has a power series expansion at 0*, then

$$e^x = \sum_{n=0}^{\infty} \frac{x^n}{n!}$$

So how can we determine whether e^x *does* have a power series representation?

Let's investigate the more general question: Under what circumstances is a function equal to the sum of its Taylor series? In other words, if f has derivatives of all orders, when is it true that

$$f(x) = \sum_{n=0}^{\infty} \frac{f^{(n)}(a)}{n!} (x - a)^n$$

As with any convergent series, this means that $f(x)$ is the limit of the sequence of partial sums. In the case of the Taylor series, the partial sums are

$$T_n(x) = \sum_{i=0}^{n} \frac{f^{(i)}(a)}{i!} (x - a)^i$$

$$= f(a) + \frac{f'(a)}{1!} (x - a) + \frac{f''(a)}{2!} (x - a)^2 + \cdots + \frac{f^{(n)}(a)}{n!} (x - a)^n$$

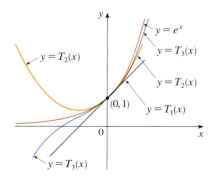

FIGURE 1

|||| As n increases, $T_n(x)$ appears to approach e^x in Figure 1. This suggests that e^x is equal to the sum of its Taylor series.

Notice that T_n is a polynomial of degree n called the **nth-degree Taylor polynomial of f at a**. For instance, for the exponential function $f(x) = e^x$, the result of Example 1 shows that the Taylor polynomials at 0 (or Maclaurin polynomials) with $n = 1, 2$, and 3 are

$$T_1(x) = 1 + x \qquad T_2(x) = 1 + x + \frac{x^2}{2!} \qquad T_3(x) = 1 + x + \frac{x^2}{2!} + \frac{x^3}{3!}$$

The graphs of the exponential function and these three Taylor polynomials are drawn in Figure 1.

In general, $f(x)$ is the sum of its Taylor series if

$$f(x) = \lim_{n \to \infty} T_n(x)$$

If we let

$$R_n(x) = f(x) - T_n(x) \qquad \text{so that} \qquad f(x) = T_n(x) + R_n(x)$$

then $R_n(x)$ is called the **remainder** of the Taylor series. If we can somehow show that $\lim_{n \to \infty} R_n(x) = 0$, then it follows that

$$\lim_{n \to \infty} T_n(x) = \lim_{n \to \infty} [f(x) - R_n(x)] = f(x) - \lim_{n \to \infty} R_n(x) = f(x)$$

We have therefore proved the following.

> **8 Theorem** If $f(x) = T_n(x) + R_n(x)$, where T_n is the nth-degree Taylor polynomial of f at a and
>
> $$\lim_{n \to \infty} R_n(x) = 0$$
>
> for $|x - a| < R$, then f is equal to the sum of its Taylor series on the interval $|x - a| < R$.

In trying to show that $\lim_{n \to \infty} R_n(x) = 0$ for a specific function f, we usually use the following fact.

> **9 Taylor's Inequality** If $|f^{(n+1)}(x)| \leq M$ for $|x - a| \leq d$, then the remainder $R_n(x)$ of the Taylor series satisfies the inequality
>
> $$|R_n(x)| \leq \frac{M}{(n+1)!} |x - a|^{n+1} \qquad \text{for } |x - a| \leq d$$

To see why this is true for $n = 1$, we assume that $|f''(x)| \leq M$. In particular, we have $f''(x) \leq M$, so for $a \leq x \leq a + d$ we have

$$\int_a^x f''(t)\, dt \leq \int_a^x M\, dt$$

An antiderivative of f'' is f', so by Part 2 of the Fundamental Theorem of Calculus, we have

$$f'(x) - f'(a) \leq M(x - a) \qquad \text{or} \qquad f'(x) \leq f'(a) + M(x - a)$$

|||| As alternatives to Taylor's Inequality, we have the following formulas for the remainder term. If $f^{(n+1)}$ is continuous on an interval I and $x \in I$, then

$$R_n(x) = \frac{1}{n!} \int_a^x (x-a)^n f^{(n+1)}(x) \, dx$$

This is called the *integral form of the remainder term*. Another formula, called *Lagrange's form of the remainder term*, states that there is a number z between x and a such that

$$R_n(x) = \frac{f^{(n+1)}(z)}{(n+1)!} (x-a)^{n+1}$$

This version is an extension of the Mean Value Theorem (which is the case $n = 0$).

Proofs of these formulas, together with discussions of how to use them to solve the examples of Sections 12.10 and 12.12, are given on the web site

www.stewartcalculus.com

Click on *Additional Topics* and then on *Formulas for the Remainder Term*.

Thus

$$\int_a^x f'(t) \, dt \leq \int_a^x [f'(a) + M(t-a)] \, dt$$

$$f(x) - f(a) \leq f'(a)(x-a) + M \frac{(x-a)^2}{2}$$

$$f(x) - f(a) - f'(a)(x-a) \leq \frac{M}{2}(x-a)^2$$

But $R_1(x) = f(x) - T_1(x) = f(x) - f(a) - f'(a)(x-a)$. So

$$R_1(x) \leq \frac{M}{2}(x-a)^2$$

A similar argument, using $f''(x) \geq -M$, shows that

$$R_1(x) \geq -\frac{M}{2}(x-a)^2$$

So

$$|R_1(x)| \leq \frac{M}{2}|x-a|^2$$

Although we have assumed that $x > a$, similar calculations show that this inequality is also true for $x < a$.

This proves Taylor's Inequality for the case where $n = 1$. The result for any n is proved in a similar way by integrating $n + 1$ times. (See Exercise 61 for the case $n = 2$.)

NOTE ▫ In Section 12.12 we will explore the use of Taylor's Inequality in approximating functions. Our immediate use of it is in conjunction with Theorem 8.

In applying Theorems 8 and 9 it is often helpful to make use of the following fact.

$$\boxed{10} \qquad \lim_{n \to \infty} \frac{x^n}{n!} = 0 \qquad \text{for every real number } x$$

This is true because we know from Example 1 that the series $\Sigma \, x^n/n!$ converges for all x and so its nth term approaches 0.

EXAMPLE 2 Prove that e^x is equal to the sum of its Maclaurin series.

SOLUTION If $f(x) = e^x$, then $f^{(n+1)}(x) = e^x$ for all n. If d is any positive number and $|x| \leq d$, then $|f^{(n+1)}(x)| = e^x \leq e^d$. So Taylor's Inequality, with $a = 0$ and $M = e^d$, says that

$$|R_n(x)| \leq \frac{e^d}{(n+1)!} |x|^{n+1} \qquad \text{for } |x| \leq d$$

Notice that the same constant $M = e^d$ works for every value of n. But, from Equation 10, we have

$$\lim_{n \to \infty} \frac{e^d}{(n+1)!} |x|^{n+1} = e^d \lim_{n \to \infty} \frac{|x|^{n+1}}{(n+1)!} = 0$$

It follows from the Squeeze Theorem that $\lim_{n\to\infty} |R_n(x)| = 0$ and therefore $\lim_{n\to\infty} R_n(x) = 0$ for all values of x. By Theorem 8, e^x is equal to the sum of its Maclaurin series, that is,

$$\boxed{11 \quad e^x = \sum_{n=0}^{\infty} \frac{x^n}{n!} \quad \text{for all } x}$$

In particular, if we put $x = 1$ in Equation 11, we obtain the following expression for the number e as a sum of an infinite series:

$$\boxed{12 \quad e = \sum_{n=0}^{\infty} \frac{1}{n!} = 1 + \frac{1}{1!} + \frac{1}{2!} + \frac{1}{3!} + \cdots}$$

|||| In 1748 Leonard Euler used Equation 12 to find the value of e correct to 23 digits. In 2000 Xavier Gourdon and S. Kondo, again using the series in (12), computed e to more than twelve billion decimal places. The special techniques they employed to speed up the computation are explained on the web page

www.numbers.computation.free.fr

EXAMPLE 3 Find the Taylor series for $f(x) = e^x$ at $a = 2$.

SOLUTION We have $f^{(n)}(2) = e^2$ and so, putting $a = 2$ in the definition of a Taylor series (6), we get

$$\sum_{n=0}^{\infty} \frac{f^{(n)}(2)}{n!} (x - 2)^n = \sum_{n=0}^{\infty} \frac{e^2}{n!} (x - 2)^n$$

Again it can be verified, as in Example 1, that the radius of convergence is $R = \infty$. As in Example 2 we can verify that $\lim_{n\to\infty} R_n(x) = 0$, so

$$\boxed{13 \quad e^x = \sum_{n=0}^{\infty} \frac{e^2}{n!} (x - 2)^n \quad \text{for all } x}$$

We have two power series expansions for e^x, the Maclaurin series in Equation 11 and the Taylor series in Equation 13. The first is better if we are interested in values of x near 0 and the second is better if x is near 2.

EXAMPLE 4 Find the Maclaurin series for $\sin x$ and prove that it represents $\sin x$ for all x.

SOLUTION We arrange our computation in two columns as follows:

$$f(x) = \sin x \qquad f(0) = 0$$
$$f'(x) = \cos x \qquad f'(0) = 1$$
$$f''(x) = -\sin x \qquad f''(0) = 0$$
$$f'''(x) = -\cos x \qquad f'''(0) = -1$$
$$f^{(4)}(x) = \sin x \qquad f^{(4)}(0) = 0$$

Since the derivatives repeat in a cycle of four, we can write the Maclaurin series as follows:

$$f(0) + \frac{f'(0)}{1!} x + \frac{f''(0)}{2!} x^2 + \frac{f'''(0)}{3!} x^3 + \cdots$$

$$= x - \frac{x^3}{3!} + \frac{x^5}{5!} - \frac{x^7}{7!} + \cdots = \sum_{n=0}^{\infty} (-1)^n \frac{x^{2n+1}}{(2n + 1)!}$$

Figure 2 shows the graph of sin x together with its Taylor (or Maclaurin) polynomials

$$T_1(x) = x$$
$$T_3(x) = x - \frac{x^3}{3!}$$
$$T_5(x) = x - \frac{x^3}{3!} + \frac{x^5}{5!}$$

Notice that, as n increases, $T_n(x)$ becomes a better approximation to sin x.

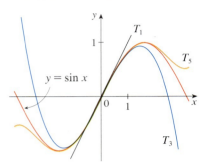

FIGURE 2

The Maclaurin series for e^x, sin x, and cos x that we found in Examples 2, 4, and 5 were discovered, using different methods, by Newton. (There is evidence that the series for sin x and cos x were known to Indian astronomers more than a century before Newton, but this knowledge didn't spread to the rest of the world.)

These equations are remarkable because they say we know everything about each of these functions if we know all its derivatives at the single number 0.

Since $f^{(n+1)}(x)$ is $\pm \sin x$ or $\pm \cos x$, we know that $|f^{(n+1)}(x)| \leq 1$ for all x. So we can take $M = 1$ in Taylor's Inequality:

$$\boxed{14} \quad |R_n(x)| \leq \frac{M}{(n+1)!}|x^{n+1}| = \frac{|x|^{n+1}}{(n+1)!}$$

By Equation 10 the right side of this inequality approaches 0 as $n \to \infty$, so $|R_n(x)| \to 0$ by the Squeeze Theorem. It follows that $R_n(x) \to 0$ as $n \to \infty$, so sin x is equal to the sum of its Maclaurin series by Theorem 8.

We state the result of Example 4 for future reference.

$$\boxed{15} \quad \sin x = x - \frac{x^3}{3!} + \frac{x^5}{5!} - \frac{x^7}{7!} + \cdots$$
$$= \sum_{n=0}^{\infty} (-1)^n \frac{x^{2n+1}}{(2n+1)!} \quad \text{for all } x$$

EXAMPLE 5 Find the Maclaurin series for cos x.

SOLUTION We could proceed directly as in Example 4 but it's easier to differentiate the Maclaurin series for sin x given by Equation 15:

$$\cos x = \frac{d}{dx}(\sin x) = \frac{d}{dx}\left(x - \frac{x^3}{3!} + \frac{x^5}{5!} - \frac{x^7}{7!} + \cdots\right)$$
$$= 1 - \frac{3x^2}{3!} + \frac{5x^4}{5!} - \frac{7x^6}{7!} + \cdots = 1 - \frac{x^2}{2!} + \frac{x^4}{4!} - \frac{x^6}{6!} + \cdots$$

Since the Maclaurin series for sin x converges for all x, Theorem 2 in Section 12.9 tells us that the differentiated series for cos x also converges for all x. Thus

$$\boxed{16} \quad \cos x = 1 - \frac{x^2}{2!} + \frac{x^4}{4!} - \frac{x^6}{6!} + \cdots$$
$$= \sum_{n=0}^{\infty} (-1)^n \frac{x^{2n}}{(2n)!} \quad \text{for all } x$$

EXAMPLE 6 Find the Maclaurin series for the function $f(x) = x \cos x$.

SOLUTION Instead of computing derivatives and substituting in Equation 7, it's easier to multiply the series for cos x (Equation 16) by x:

$$x \cos x = x \sum_{n=0}^{\infty} (-1)^n \frac{x^{2n}}{(2n)!} = \sum_{n=0}^{\infty} (-1)^n \frac{x^{2n+1}}{(2n)!}$$

EXAMPLE 7 Represent $f(x) = \sin x$ as the sum of its Taylor series centered at $\pi/3$.

SOLUTION Arranging our work in columns, we have

$$f(x) = \sin x \qquad f\left(\frac{\pi}{3}\right) = \frac{\sqrt{3}}{2}$$

$$f'(x) = \cos x \qquad f'\left(\frac{\pi}{3}\right) = \frac{1}{2}$$

$$f''(x) = -\sin x \qquad f''\left(\frac{\pi}{3}\right) = -\frac{\sqrt{3}}{2}$$

$$f'''(x) = -\cos x \qquad f'''\left(\frac{\pi}{3}\right) = -\frac{1}{2}$$

and this pattern repeats indefinitely. Therefore, the Taylor series at $\pi/3$ is

$$f\left(\frac{\pi}{3}\right) + \frac{f'\left(\frac{\pi}{3}\right)}{1!}\left(x - \frac{\pi}{3}\right) + \frac{f''\left(\frac{\pi}{3}\right)}{2!}\left(x - \frac{\pi}{3}\right)^2 + \frac{f'''\left(\frac{\pi}{3}\right)}{3!}\left(x - \frac{\pi}{3}\right)^3 + \cdots$$

$$= \frac{\sqrt{3}}{2} + \frac{1}{2 \cdot 1!}\left(x - \frac{\pi}{3}\right) - \frac{\sqrt{3}}{2 \cdot 2!}\left(x - \frac{\pi}{3}\right)^2 - \frac{1}{2 \cdot 3!}\left(x - \frac{\pi}{3}\right)^3 + \cdots$$

The proof that this series represents $\sin x$ for all x is very similar to that in Example 4. [Just replace x by $x - \pi/3$ in (14).] We can write the series in sigma notation if we separate the terms that contain $\sqrt{3}$:

$$\sin x = \sum_{n=0}^{\infty} \frac{(-1)^n \sqrt{3}}{2(2n)!}\left(x - \frac{\pi}{3}\right)^{2n} + \sum_{n=0}^{\infty} \frac{(-1)^n}{2(2n+1)!}\left(x - \frac{\pi}{3}\right)^{2n+1}$$

The power series that we obtained by indirect methods in Examples 5 and 6 and in Section 12.9 are indeed the Taylor or Maclaurin series of the given functions because Theorem 5 asserts that, no matter how a power series representation $f(x) = \Sigma\, c_n(x - a)^n$ is obtained, it is always true that $c_n = f^{(n)}(a)/n!$. In other words, the coefficients are uniquely determined.

We collect in the following table, for future reference, some important Maclaurin series that we have derived in this section and the preceding one.

|||| We have obtained two different series representations for $\sin x$, the Maclaurin series in Example 4 and the Taylor series in Example 7. It is best to use the Maclaurin series for values of x near 0 and the Taylor series for x near $\pi/3$. Notice that the third Taylor polynomial T_3 in Figure 3 is a good approximation to $\sin x$ near $\pi/3$ but not as good near 0. Compare it with the third Maclaurin polynomial T_3 in Figure 2, where the opposite is true.

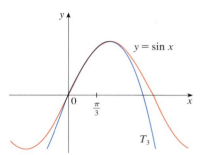

FIGURE 3

Important Maclaurin series and their intervals of convergence

$$\frac{1}{1-x} = \sum_{n=0}^{\infty} x^n = 1 + x + x^2 + x^3 + \cdots \qquad (-1, 1)$$

$$e^x = \sum_{n=0}^{\infty} \frac{x^n}{n!} = 1 + \frac{x}{1!} + \frac{x^2}{2!} + \frac{x^3}{3!} + \cdots \qquad (-\infty, \infty)$$

$$\sin x = \sum_{n=0}^{\infty} (-1)^n \frac{x^{2n+1}}{(2n+1)!} = x - \frac{x^3}{3!} + \frac{x^5}{5!} - \frac{x^7}{7!} + \cdots \qquad (-\infty, \infty)$$

$$\cos x = \sum_{n=0}^{\infty} (-1)^n \frac{x^{2n}}{(2n)!} = 1 - \frac{x^2}{2!} + \frac{x^4}{4!} - \frac{x^6}{6!} + \cdots \qquad (-\infty, \infty)$$

$$\tan^{-1} x = \sum_{n=0}^{\infty} (-1)^n \frac{x^{2n+1}}{2n+1} = x - \frac{x^3}{3} + \frac{x^5}{5} - \frac{x^7}{7} + \cdots \qquad [-1, 1]$$

Module 12.10/12.12 enables you to see how successive Taylor polynomials approach the original function.

One reason that Taylor series are important is that they enable us to integrate functions that we couldn't previously handle. In fact, in the introduction to this chapter we mentioned that Newton often integrated functions by first expressing them as power series and then integrating the series term by term. The function $f(x) = e^{-x^2}$ can't be integrated by techniques discussed so far because its antiderivative is not an elementary function (see Section 8.5). In the following example we use Newton's idea to integrate this function.

EXAMPLE 8
(a) Evaluate $\int e^{-x^2}\, dx$ as an infinite series.
(b) Evaluate $\int_0^1 e^{-x^2}\, dx$ correct to within an error of 0.001.

SOLUTION
(a) First we find the Maclaurin series for $f(x) = e^{-x^2}$. Although it's possible to use the direct method, let's find it simply by replacing x with $-x^2$ in the series for e^x given in the table of Maclaurin series. Thus, for all values of x,

$$e^{-x^2} = \sum_{n=0}^{\infty} \frac{(-x^2)^n}{n!} = \sum_{n=0}^{\infty} (-1)^n \frac{x^{2n}}{n!} = 1 - \frac{x^2}{1!} + \frac{x^4}{2!} - \frac{x^6}{3!} + \cdots$$

Now we integrate term by term:

$$\int e^{-x^2}\, dx = \int \left(1 - \frac{x^2}{1!} + \frac{x^4}{2!} - \frac{x^6}{3!} + \cdots + (-1)^n \frac{x^{2n}}{n!} + \cdots\right) dx$$

$$= C + x - \frac{x^3}{3 \cdot 1!} + \frac{x^5}{5 \cdot 2!} - \frac{x^7}{7 \cdot 3!} + \cdots + (-1)^n \frac{x^{2n+1}}{(2n+1)n!} + \cdots$$

This series converges for all x because the original series for e^{-x^2} converges for all x.

(b) The Evaluation Theorem gives

$$\int_0^1 e^{-x^2}\, dx = \left[x - \frac{x^3}{3 \cdot 1!} + \frac{x^5}{5 \cdot 2!} - \frac{x^7}{7 \cdot 3!} + \frac{x^9}{9 \cdot 4!} - \cdots\right]_0^1$$

$$= 1 - \tfrac{1}{3} + \tfrac{1}{10} - \tfrac{1}{42} + \tfrac{1}{216} - \cdots$$

$$\approx 1 - \tfrac{1}{3} + \tfrac{1}{10} - \tfrac{1}{42} + \tfrac{1}{216} \approx 0.7475$$

|||| We can take $C = 0$ in the antiderivative in part (a).

The Alternating Series Estimation Theorem shows that the error involved in this approximation is less than

$$\frac{1}{11 \cdot 5!} = \frac{1}{1320} < 0.001$$

Another use of Taylor series is illustrated in the next example. The limit could be found with l'Hospital's Rule, but instead we use a series.

EXAMPLE 9 Evaluate $\displaystyle\lim_{x \to 0} \frac{e^x - 1 - x}{x^2}$.

SOLUTION Using the Maclaurin series for e^x, we have

$$\lim_{x \to 0} \frac{e^x - 1 - x}{x^2} = \lim_{x \to 0} \frac{\left(1 + \dfrac{x}{1!} + \dfrac{x^2}{2!} + \dfrac{x^3}{3!} + \cdots\right) - 1 - x}{x^2}$$

$$= \lim_{x \to 0} \frac{\dfrac{x^2}{2!} + \dfrac{x^3}{3!} + \dfrac{x^4}{4!} + \cdots}{x^2}$$

$$= \lim_{x \to 0} \left(\frac{1}{2} + \frac{x}{3!} + \frac{x^2}{4!} + \frac{x^3}{5!} + \cdots\right)$$

$$= \frac{1}{2}$$

▕▎ Some computer algebra systems compute limits in this way.

because power series are continuous functions.

Multiplication and Division of Power Series

If power series are added or subtracted, they behave like polynomials (Theorem 12.2.8 shows this). In fact, as the following example illustrates, they can also be multiplied and divided like polynomials. We find only the first few terms because the calculations for the later terms become tedious and the initial terms are the most important ones.

EXAMPLE 10 Find the first three nonzero terms in the Maclaurin series for (a) $e^x \sin x$ and (b) $\tan x$.

SOLUTION
(a) Using the Maclaurin series for e^x and $\sin x$ in the table on page 803, we have

$$e^x \sin x = \left(1 + \frac{x}{1!} + \frac{x^2}{2!} + \frac{x^3}{3!} + \cdots\right)\left(x - \frac{x^3}{3!} + \cdots\right)$$

We multiply these expressions, collecting like terms just as for polynomials:

$$\begin{array}{r} 1 + x + \tfrac{1}{2}x^2 + \tfrac{1}{6}x^3 + \cdots \\ x \quad\quad\quad -\tfrac{1}{6}x^3 + \cdots \\ \hline x + x^2 + \tfrac{1}{2}x^3 + \tfrac{1}{6}x^4 + \cdots \\ -\tfrac{1}{6}x^3 - \tfrac{1}{6}x^4 - \cdots \\ \hline x + x^2 + \tfrac{1}{3}x^3 + \cdots \end{array}$$

Thus
$$e^x \sin x = x + x^2 + \tfrac{1}{3}x^3 + \cdots$$

(b) Using the Maclaurin series in the table, we have

$$\tan x = \frac{\sin x}{\cos x} = \frac{x - \dfrac{x^3}{3!} + \dfrac{x^5}{5!} - \cdots}{1 - \dfrac{x^2}{2!} + \dfrac{x^4}{4!} - \cdots}$$

We use a procedure like long division:

$$
\begin{array}{r}
x + \tfrac{1}{3}x^3 + \tfrac{2}{15}x^5 + \cdots \\
1 - \tfrac{1}{2}x^2 + \tfrac{1}{24}x^4 - \cdots \overline{\smash{\big)} x - \tfrac{1}{6}x^3 + \tfrac{1}{120}x^5 - \cdots} \\
\underline{x - \tfrac{1}{2}x^3 + \tfrac{1}{24}x^5 - \cdots} \\
\tfrac{1}{3}x^3 - \tfrac{1}{30}x^5 + \cdots \\
\underline{\tfrac{1}{3}x^3 - \tfrac{1}{6}x^5 + \cdots} \\
\tfrac{2}{15}x^5 + \cdots
\end{array}
$$

Thus $\quad \tan x = x + \tfrac{1}{3}x^3 + \tfrac{2}{15}x^5 + \cdots$

Although we have not attempted to justify the formal manipulations used in Example 10, they are legitimate. There is a theorem which states that if both $f(x) = \Sigma\, c_n x^n$ and $g(x) = \Sigma\, b_n x^n$ converge for $|x| < R$ and the series are multiplied as if they were polynomials, then the resulting series also converges for $|x| < R$ and represents $f(x)g(x)$. For division we require $b_0 \neq 0$; the resulting series converges for sufficiently small $|x|$.

12.10 Exercises

1. If $f(x) = \sum_{n=0}^{\infty} b_n(x-5)^n$ for all x, write a formula for b_8.

2. (a) The graph of f is shown. Explain why the series

$$1.6 - 0.8(x-1) + 0.4(x-1)^2 - 0.1(x-1)^3 + \cdots$$

is *not* the Taylor series of f centered at 1.

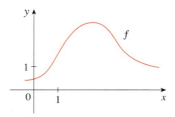

(b) Explain why the series

$$2.8 + 0.5(x-2) + 1.5(x-2)^2 - 0.1(x-2)^3 + \cdots$$

is *not* the Taylor series of f centered at 2.

3–10 Find the Maclaurin series for $f(x)$ using the definition of a Maclaurin series. [Assume that f has a power series expansion. Do not show that $R_n(x) \to 0$.] Also find the associated radius of convergence.

3. $f(x) = \cos x$

4. $f(x) = \sin 2x$

5. $f(x) = (1+x)^{-3}$

6. $f(x) = \ln(1+x)$

7. $f(x) = e^{5x}$

8. $f(x) = xe^x$

9. $f(x) = \sinh x$

10. $f(x) = \cosh x$

11–18 Find the Taylor series for $f(x)$ centered at the given value of a. [Assume that f has a power series expansion. Do not show that $R_n(x) \to 0$.]

11. $f(x) = 1 + x + x^2, \quad a = 2$

12. $f(x) = x^3, \quad a = -1$

13. $f(x) = e^x, \quad a = 3$

14. $f(x) = \ln x, \quad a = 2$

15. $f(x) = \cos x, \quad a = \pi$

16. $f(x) = \sin x, \quad a = \pi/2$

17. $f(x) = 1/\sqrt{x}, \quad a = 9$

18. $f(x) = x^{-2}, \quad a = 1$

19. Prove that the series obtained in Exercise 3 represents $\cos x$ for all x.

20. Prove that the series obtained in Exercise 16 represents $\sin x$ for all x.

21. Prove that the series obtained in Exercise 9 represents $\sinh x$ for all x.

22. Prove that the series obtained in Exercise 10 represents $\cosh x$ for all x.

23–32 Use a Maclaurin series derived in this section to obtain the Maclaurin series for the given function.

23. $f(x) = \cos \pi x$

24. $f(x) = e^{-x/2}$

25. $f(x) = x \tan^{-1} x$

26. $f(x) = \sin(x^4)$

27. $f(x) = x^2 e^{-x}$

28. $f(x) = x \cos 2x$

29. $f(x) = \sin^2 x$ [*Hint:* Use $\sin^2 x = \frac{1}{2}(1 - \cos 2x)$.]

30. $f(x) = \cos^2 x$

31. $f(x) = \begin{cases} \dfrac{\sin x}{x} & \text{if } x \neq 0 \\ 1 & \text{if } x = 0 \end{cases}$

32. $f(x) = \begin{cases} \dfrac{x - \sin x}{x^3} & \text{if } x \neq 0 \\ \dfrac{1}{6} & \text{if } x = 0 \end{cases}$

33–36 |||| Find the Maclaurin series of f (by any method) and its radius of convergence. Graph f and its first few Taylor polynomials on the same screen. What do you notice about the relationship between these polynomials and f?

33. $f(x) = \sqrt{1 + x}$

34. $f(x) = e^{-x^2} + \cos x$

35. $f(x) = \cos(x^2)$

36. $f(x) = 2^x$

37. Use the Maclaurin series for e^x to calculate $e^{-0.2}$ correct to five decimal places.

38. Use the Maclaurin series for $\sin x$ to compute $\sin 3°$ correct to five decimal places.

39–42 |||| Evaluate the indefinite integral as an infinite series.

39. $\displaystyle\int x \cos(x^3)\, dx$

40. $\displaystyle\int \frac{\sin x}{x}\, dx$

41. $\displaystyle\int \sqrt{x^3 + 1}\, dx$

42. $\displaystyle\int \frac{e^x - 1}{x}\, dx$

43–46 |||| Use series to approximate the definite integral to within the indicated accuracy.

43. $\displaystyle\int_0^1 x \cos(x^3)\, dx$ (three decimal places)

44. $\displaystyle\int_0^{0.2} [\tan^{-1}(x^3) + \sin(x^3)]\, dx$ (five decimal places)

45. $\displaystyle\int_0^{0.1} \frac{dx}{\sqrt{1 + x^3}}$ ($|\text{error}| < 10^{-8}$)

46. $\displaystyle\int_0^{0.5} x^2 e^{-x^2}\, dx$ ($|\text{error}| < 0.001$)

47–49 |||| Use series to evaluate the limit.

47. $\displaystyle\lim_{x \to 0} \frac{x - \tan^{-1} x}{x^3}$

48. $\displaystyle\lim_{x \to 0} \frac{1 - \cos x}{1 + x - e^x}$

49. $\displaystyle\lim_{x \to 0} \frac{\sin x - x + \frac{1}{6}x^3}{x^5}$

50. Use the series in Example 10(b) to evaluate
$$\lim_{x \to 0} \frac{\tan x - x}{x^3}$$
We found this limit in Example 4 in Section 7.7 using l'Hospital's Rule three times. Which method do you prefer?

51–54 |||| Use multiplication or division of power series to find the first three nonzero terms in the Maclaurin series for each function.

51. $y = e^{-x^2} \cos x$

52. $y = \sec x$

53. $y = \dfrac{x}{\sin x}$

54. $y = e^x \ln(1 - x)$

55–60 |||| Find the sum of the series.

55. $\displaystyle\sum_{n=0}^{\infty} (-1)^n \frac{x^{4n}}{n!}$

56. $\displaystyle\sum_{n=0}^{\infty} \frac{(-1)^n \pi^{2n}}{6^{2n}(2n)!}$

57. $\displaystyle\sum_{n=0}^{\infty} \frac{(-1)^n \pi^{2n+1}}{4^{2n+1}(2n + 1)!}$

58. $\displaystyle\sum_{n=0}^{\infty} \frac{3^n}{5^n n!}$

59. $3 + \dfrac{9}{2!} + \dfrac{27}{3!} + \dfrac{81}{4!} + \cdots$

60. $1 - \ln 2 + \dfrac{(\ln 2)^2}{2!} - \dfrac{(\ln 2)^3}{3!} + \cdots$

61. Prove Taylor's Inequality for $n = 2$, that is, prove that if $|f'''(x)| \leq M$ for $|x - a| \leq d$, then
$$|R_2(x)| \leq \frac{M}{6}|x - a|^3 \qquad \text{for } |x - a| \leq d$$

62. (a) Show that the function defined by
$$f(x) = \begin{cases} e^{-1/x^2} & \text{if } x \neq 0 \\ 0 & \text{if } x = 0 \end{cases}$$
is not equal to its Maclaurin series.
(b) Graph the function in part (a) and comment on its behavior near the origin.

LABORATORY PROJECT

CAS An Elusive Limit

This project deals with the function

$$f(x) = \frac{\sin(\tan x) - \tan(\sin x)}{\arcsin(\arctan x) - \arctan(\arcsin x)}$$

1. Use your computer algebra system to evaluate $f(x)$ for $x = 1, 0.1, 0.01, 0.001,$ and 0.0001. Does it appear that f has a limit as $x \to 0$?

2. Use the CAS to graph f near $x = 0$. Does it appear that f has a limit as $x \to 0$?

3. Try to evaluate $\lim_{x \to 0} f(x)$ with l'Hospital's Rule, using the CAS to find derivatives of the numerator and denominator. What do you discover? How many applications of l'Hospital's Rule are required?

4. Evaluate $\lim_{x \to 0} f(x)$ by using the CAS to find sufficiently many terms in the Taylor series of the numerator and denominator. (Use the command `taylor` in Maple or `Series` in Mathematica.)

5. Use the limit command on your CAS to find $\lim_{x \to 0} f(x)$ directly. (Most computer algebra systems use the method of Problem 4 to compute limits.)

6. In view of the answers to Problems 4 and 5, how do you explain the results of Problems 1 and 2?

12.11 The Binomial Series

You may be acquainted with the Binomial Theorem, which states that if a and b are any real numbers and k is a positive integer, then

$$(a + b)^k = a^k + ka^{k-1}b + \frac{k(k-1)}{2!}a^{k-2}b^2 + \frac{k(k-1)(k-2)}{3!}a^{k-3}b^3$$

$$+ \cdots + \frac{k(k-1)(k-2)\cdots(k-n+1)}{n!}a^{k-n}b^n$$

$$+ \cdots + kab^{k-1} + b^k$$

The traditional notation for the binomial coefficients is

$$\binom{k}{0} = 1 \qquad \binom{k}{n} = \frac{k(k-1)(k-2)\cdots(k-n+1)}{n!} \qquad n = 1, 2, \ldots, k$$

which enables us to write the Binomial Theorem in the abbreviated form

$$(a + b)^k = \sum_{n=0}^{k} \binom{k}{n} a^{k-n} b^n$$

In particular, if we put $a = 1$ and $b = x$, we get

$$\boxed{1} \qquad (1 + x)^k = \sum_{n=0}^{k} \binom{k}{n} x^n$$

One of Newton's accomplishments was to extend the Binomial Theorem (Equation 1) to the case in which k is no longer a positive integer. (See the Writing Project on page 812.) In this case the expression for $(1 + x)^k$ is no longer a finite sum; it becomes an infinite series. To find this series we compute the Maclaurin series of $(1 + x)^k$ in the usual way:

$$f(x) = (1 + x)^k \qquad\qquad f(0) = 1$$

$$f'(x) = k(1 + x)^{k-1} \qquad\qquad f'(0) = k$$

$$f''(x) = k(k-1)(1 + x)^{k-2} \qquad\qquad f''(0) = k(k-1)$$

$$f'''(x) = k(k-1)(k-2)(1 + x)^{k-3} \qquad\qquad f'''(0) = k(k-1)(k-2)$$

$$\vdots \qquad\qquad \vdots$$

$$f^{(n)}(x) = k(k-1)\cdots(k-n+1)(1+x)^{k-n} \qquad f^{(n)}(0) = k(k-1)\cdots(k-n+1)$$

Therefore, the Maclaurin series of $f(x) = (1 + x)^k$ is

$$\sum_{n=0}^{\infty} \frac{f^{(n)}(0)}{n!} x^n = \sum_{n=0}^{\infty} \frac{k(k-1)\cdots(k-n+1)}{n!} x^n$$

This series is called the **binomial series**. If its nth term is a_n, then

$$\left|\frac{a_{n+1}}{a_n}\right| = \left|\frac{k(k-1)\cdots(k-n+1)(k-n)x^{n+1}}{(n+1)!} \cdot \frac{n!}{k(k-1)\cdots(k-n+1)x^n}\right|$$

$$= \frac{|k-n|}{n+1}|x| = \frac{\left|1 - \dfrac{k}{n}\right|}{1 + \dfrac{1}{n}}|x| \to |x| \qquad \text{as } n \to \infty$$

Thus, by the Ratio Test, the binomial series converges if $|x| < 1$ and diverges if $|x| > 1$.

The following theorem states that $(1 + x)^k$ is equal to the sum of its Maclaurin series. It is possible to prove this by showing that the remainder term $R_n(x)$ approaches 0, but that turns out to be quite difficult. The proof outlined in Exercise 19 is much easier.

$\boxed{2}$ **The Binomial Series** If k is any real number and $|x| < 1$, then

$$(1 + x)^k = 1 + kx + \frac{k(k-1)}{2!}x^2 + \frac{k(k-1)(k-2)}{3!}x^3 + \cdots$$

$$= \sum_{n=0}^{\infty} \binom{k}{n} x^n$$

where $\binom{k}{n} = \dfrac{k(k-1)\cdots(k-n+1)}{n!} \quad (n \geq 1) \quad$ and $\quad \binom{k}{0} = 1$

Although the binomial series always converges when $|x| < 1$, the question of whether or not it converges at the endpoints, ± 1, depends on the value of k. It turns out that the series converges at 1 if $-1 < k \leq 0$ and at both endpoints if $k \geq 0$. Notice that if k is a positive integer and $n > k$, then the expression for $\binom{k}{n}$ contains a factor $(k - k)$, so $\binom{k}{n} = 0$ for $n > k$. This means that the series terminates and reduces to the ordinary Binomial Theorem (Equation 1) when k is a positive integer.

As we have seen, the binomial series is just a special case of the Maclaurin series; it occurs so frequently that it is worth remembering.

EXAMPLE 1 Expand $\dfrac{1}{(1 + x)^2}$ as a power series.

SOLUTION We use the binomial series with $k = -2$. The binomial coefficient is

$$\binom{-2}{n} = \frac{(-2)(-3)(-4) \cdots (-2 - n + 1)}{n!}$$

$$= \frac{(-1)^n 2 \cdot 3 \cdot 4 \cdot \cdots \cdot n(n + 1)}{n!} = (-1)^n(n + 1)$$

and so, when $|x| < 1$,

$$\frac{1}{(1 + x)^2} = (1 + x)^{-2} = \sum_{n=0}^{\infty} \binom{-2}{n} x^n$$

$$= \sum_{n=0}^{\infty} (-1)^n(n + 1)x^n = 1 - 2x + 3x^2 - 4x^3 + \cdots$$

EXAMPLE 2 Find the Maclaurin series for the function $f(x) = \dfrac{1}{\sqrt{4 - x}}$ and its radius of convergence.

SOLUTION As given, $f(x)$ is not quite of the form $(1 + x)^k$ so we rewrite it as follows:

$$\frac{1}{\sqrt{4 - x}} = \frac{1}{\sqrt{4\left(1 - \dfrac{x}{4}\right)}} = \frac{1}{2\sqrt{1 - \dfrac{x}{4}}} = \frac{1}{2}\left(1 - \frac{x}{4}\right)^{-1/2}$$

Using the binomial series with $k = -\tfrac{1}{2}$ and with x replaced by $-x/4$, we have

$$\frac{1}{\sqrt{4 - x}} = \frac{1}{2}\left(1 - \frac{x}{4}\right)^{-1/2} = \frac{1}{2}\sum_{n=0}^{\infty} \binom{-\tfrac{1}{2}}{n}\left(-\frac{x}{4}\right)^n$$

$$= \frac{1}{2}\left[1 + \left(-\frac{1}{2}\right)\left(-\frac{x}{4}\right) + \frac{\left(-\tfrac{1}{2}\right)\left(-\tfrac{3}{2}\right)}{2!}\left(-\frac{x}{4}\right)^2 + \frac{\left(-\tfrac{1}{2}\right)\left(-\tfrac{3}{2}\right)\left(-\tfrac{5}{2}\right)}{3!}\left(-\frac{x}{4}\right)^3\right.$$

$$\left. + \cdots + \frac{\left(-\tfrac{1}{2}\right)\left(-\tfrac{3}{2}\right)\left(-\tfrac{5}{2}\right)\cdots\left(-\tfrac{1}{2} - n + 1\right)}{n!}\left(-\frac{x}{4}\right)^n + \cdots\right]$$

$$= \frac{1}{2}\left[1 + \frac{1}{8}x + \frac{1 \cdot 3}{2!8^2}x^2 + \frac{1 \cdot 3 \cdot 5}{3!8^3}x^3 + \cdots + \frac{1 \cdot 3 \cdot 5 \cdot \cdots \cdot (2n - 1)}{n!8^n}x^n + \cdots\right]$$

We know from (2) that this series converges when $|-x/4| < 1$, that is, $|x| < 4$, so the radius of convergence is $R = 4$.

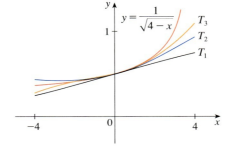

|||| A binomial series is a special case of a Taylor series. Figure 1 shows the graphs of the first three Taylor polynomials computed from the answer to Example 2.

FIGURE 1

12.11 Exercises

1–8 |||| Use the binomial series to expand the function as a power series. State the radius of convergence.

1. $\sqrt{1 + x}$
2. $\dfrac{1}{(1 + x)^4}$
3. $\dfrac{1}{(2 + x)^3}$
4. $(1 - x)^{2/3}$
5. $\sqrt[4]{1 - 8x}$
6. $\dfrac{1}{\sqrt[5]{32 - x}}$
7. $\dfrac{x}{\sqrt{4 + x^2}}$
8. $\dfrac{x^2}{\sqrt{2 + x}}$

9–10 |||| Use the binomial series to expand the function as a Maclaurin series and to find the first three Taylor polynomials T_1, T_2, and T_3. Graph the function and these Taylor polynomials in the interval of convergence.

9. $(1 + 2x)^{3/4}$
10. $\sqrt[3]{1 + 4x}$

11. (a) Use the binomial series to expand $1/\sqrt{1 - x^2}$.
    (b) Use part (a) to find the Maclaurin series for $\sin^{-1} x$.

12. (a) Use the binomial series to expand $1/\sqrt{1 + x^2}$.
    (b) Use part (a) to find the Maclaurin series for $\sinh^{-1} x$.

13. (a) Expand $\sqrt[3]{1 + x}$ as a power series.
    (b) Use part (a) to estimate $\sqrt[3]{1.01}$ correct to four decimal places.

14. (a) Expand $1/\sqrt[4]{1 + x}$ as a power series.
    (b) Use part (a) to estimate $1/\sqrt[4]{1.1}$ correct to three decimal places.

15. (a) Expand $f(x) = x/(1 - x)^2$ as a power series.
    (b) Use part (a) to find the sum of the series
    $$\sum_{n=1}^{\infty} \frac{n}{2^n}$$

16. (a) Expand $f(x) = (x + x^2)/(1 - x)^3$ as a power series.
    (b) Use part (a) to find the sum of the series
    $$\sum_{n=1}^{\infty} \frac{n^2}{2^n}$$

17. (a) Use the binomial series to find the Maclaurin series of $f(x) = \sqrt{1 + x^2}$.
    (b) Use part (a) to evaluate $f^{(10)}(0)$.

18. (a) Use the binomial series to find the Maclaurin series of $f(x) = 1/\sqrt{1 + x^3}$.
    (b) Use part (a) to evaluate $f^{(9)}(0)$.

19. Use the following steps to prove (2).
    (a) Let $g(x) = \sum_{n=0}^{\infty} \binom{k}{n} x^n$. Differentiate this series to show that
    $$g'(x) = \frac{kg(x)}{1 + x} \quad -1 < x < 1$$
    (b) Let $h(x) = (1 + x)^{-k} g(x)$ and show that $h'(x) = 0$.
    (c) Deduce that $g(x) = (1 + x)^k$.

20. In Exercise 53 in Section 11.2 it was shown that the length of the ellipse $x = a \sin \theta$, $y = a \cos \theta$, where $a > b > 0$, is
    $$L = 4a \int_0^{\pi/2} \sqrt{1 - e^2 \sin^2 \theta}\, d\theta$$
    where $e = \sqrt{a^2 - b^2}/a$ is the eccentricity of the ellipse. Expand the integrand as a binomial series and use the result of Exercise 44 in Section 8.1 to express L as a series in powers of the eccentricity up to the term in e^6.

WRITING PROJECT

How Newton Discovered the Binomial Series

The Binomial Theorem, which gives the expansion of $(a + b)^k$, was known to Chinese mathematicians many centuries before the time of Newton for the case where the exponent k is a positive integer. In 1665, when he was 22, Newton was the first to discover the infinite series expansion of $(a + b)^k$ when k is a fractional exponent (positive or negative). He didn't publish his discovery, but he stated it and gave examples of how to use it in a letter (now called the *epistola prior*) dated June 13, 1676, that he sent to Henry Oldenburg, secretary of the Royal Society of London, to transmit to Leibniz. When Leibniz replied, he asked how Newton had discovered the binomial series. Newton wrote a second letter, the *epistola posterior* of October 24, 1676, in which he explained in great detail how he arrived at his discovery by a very indirect route. He was investigating the areas under the curves $y = (1 - x^2)^{n/2}$ from 0 to x for $n = 0, 1, 2, 3, 4, \ldots$. These are easy to calculate if n is even. By observing patterns and interpolating, Newton was able to guess the answers for odd values of n. Then he realized he could get the same answers by expressing $(1 - x^2)^{n/2}$ as an infinite series.

Write a report on Newton's discovery of the binomial series. Start by giving the statement of the binomial series in Newton's notation (see the *epistola prior* on page 285 of [4] or page 402 of [2]). Explain why Newton's version is equivalent to Theorem 2 on page 809. Then read Newton's *epistola posterior* (page 287 in [4] or page 404 in [2]) and explain the patterns that Newton discovered in the areas under the curves $y = (1 - x^2)^{n/2}$. Show how he was able to guess the areas under the remaining curves and how he verified his answers. Finally, explain how these discoveries led to the binomial series. The books by Edwards [1] and Katz [3] contain commentaries on Newton's letters.

1. C. H. Edwards, *The Historical Development of the Calculus* (New York: Springer-Verlag, 1979), pp. 178–187.

2. John Fauvel and Jeremy Gray, eds., *The History of Mathematics: A Reader* (London: MacMillan Press, 1987).

3. Victor Katz, *A History of Mathematics: An Introduction* (New York: HarperCollins, 1993), pp. 463–466.

4. D. J. Struik, ed., *A Sourcebook in Mathematics, 1200–1800* (Princeton, N.J.: Princeton University Press, 1969).

12.12 Applications of Taylor Polynomials

In this section we explore two types of applications of Taylor polynomials. First we look at how they are used to approximate functions—computer scientists like them because polynomials are the simplest of functions. Then we investigate how physicists and engineers use them in such fields as relativity, optics, blackbody radiation, electric dipoles, the velocity of water waves, and building highways across a desert.

Approximating Functions by Polynomials

Suppose that $f(x)$ is equal to the sum of its Taylor series at a:

$$f(x) = \sum_{n=0}^{\infty} \frac{f^{(n)}(a)}{n!} (x - a)^n$$

In Section 12.10 we introduced the notation $T_n(x)$ for the nth partial sum of this series and called it the nth-degree Taylor polynomial of f at a. Thus

$$T_n(x) = \sum_{i=0}^{n} \frac{f^{(i)}(a)}{i!} (x-a)^i$$

$$= f(a) + \frac{f'(a)}{1!}(x-a) + \frac{f''(a)}{2!}(x-a)^2 + \cdots + \frac{f^{(n)}(a)}{n!}(x-a)^n$$

Since f is the sum of its Taylor series, we know that $T_n(x) \to f(x)$ as $n \to \infty$ and so T_n can be used as an approximation to f: $f(x) \approx T_n(x)$.

Notice that the first-degree Taylor polynomial

$$T_1(x) = f(a) + f'(a)(x-a)$$

is the same as the linearization of f at a that we discussed in Section 3.10. Notice also that T_1 and its derivative have the same values at a that f and f' have. In general, it can be shown that the derivatives of T_n at a agree with those of f up to and including derivatives of order n (see Exercise 36).

To illustrate these ideas let's take another look at the graphs of $y = e^x$ and its first few Taylor polynomials, as shown in Figure 1. The graph of T_1 is the tangent line to $y = e^x$ at $(0, 1)$; this tangent line is the best linear approximation to e^x near $(0, 1)$. The graph of T_2 is the parabola $y = 1 + x + x^2/2$, and the graph of T_3 is the cubic curve $y = 1 + x + x^2/2 + x^3/6$, which is a closer fit to the exponential curve $y = e^x$ than T_2. The next Taylor polynomial T_4 would be an even better approximation, and so on.

The values in the table give a numerical demonstration of the convergence of the Taylor polynomials $T_n(x)$ to the function $y = e^x$. We see that when $x = 0.2$ the convergence is very rapid, but when $x = 3$ it is somewhat slower. In fact, the farther x is from 0, the more slowly $T_n(x)$ converges to e^x.

When using a Taylor polynomial T_n to approximate a function f, we have to ask the questions: How good an approximation is it? How large should we take n to be in order to achieve a desired accuracy? To answer these questions we need to look at the absolute value of the remainder:

$$|R_n(x)| = |f(x) - T_n(x)|$$

There are three possible methods for estimating the size of the error:

1. If a graphing device is available, we can use it to graph $|R_n(x)|$ and thereby estimate the error.

2. If the series happens to be an alternating series, we can use the Alternating Series Estimation Theorem.

3. In all cases we can use Taylor's Inequality (Theorem 12.10.9), which says that if $|f^{(n+1)}(x)| \le M$, then

$$|R_n(x)| \le \frac{M}{(n+1)!} |x-a|^{n+1}$$

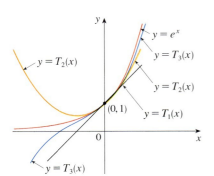

FIGURE 1

	$x = 0.2$	$x = 3.0$
$T_2(x)$	1.220000	8.500000
$T_4(x)$	1.221400	16.375000
$T_6(x)$	1.221403	19.412500
$T_8(x)$	1.221403	20.009152
$T_{10}(x)$	1.221403	20.079665
e^x	1.221403	20.085537

EXAMPLE 1
(a) Approximate the function $f(x) = \sqrt[3]{x}$ by a Taylor polynomial of degree 2 at $a = 8$.
(b) How accurate is this approximation when $7 \le x \le 9$?

SOLUTION
(a)
$$f(x) = \sqrt[3]{x} = x^{1/3} \qquad f(8) = 2$$
$$f'(x) = \tfrac{1}{3}x^{-2/3} \qquad f'(8) = \tfrac{1}{12}$$
$$f''(x) = -\tfrac{2}{9}x^{-5/3} \qquad f''(8) = -\tfrac{1}{144}$$
$$f'''(x) = \tfrac{10}{27}x^{-8/3}$$

Thus, the second-degree Taylor polynomial is

$$T_2(x) = f(8) + \frac{f'(8)}{1!}(x-8) + \frac{f''(8)}{2!}(x-8)^2$$
$$= 2 + \tfrac{1}{12}(x-8) - \tfrac{1}{288}(x-8)^2$$

The desired approximation is

$$\sqrt[3]{x} \approx T_2(x) = 2 + \tfrac{1}{12}(x-8) - \tfrac{1}{288}(x-8)^2$$

(b) The Taylor series is not alternating when $x < 8$, so we can't use the Alternating Series Estimation Theorem in this example. But we can use Taylor's Inequality with $n = 2$ and $a = 8$:

$$|R_2(x)| \leq \frac{M}{3!}|x-8|^3$$

where $|f'''(x)| \leq M$. Because $x \geq 7$, we have $x^{8/3} \geq 7^{8/3}$ and so

$$f'''(x) = \frac{10}{27} \cdot \frac{1}{x^{8/3}} \leq \frac{10}{27} \cdot \frac{1}{7^{8/3}} < 0.0021$$

Therefore, we can take $M = 0.0021$. Also $7 \leq x \leq 9$, so $-1 \leq x - 8 \leq 1$ and $|x - 8| \leq 1$. Then Taylor's Inequality gives

$$|R_2(x)| \leq \frac{0.0021}{3!} \cdot 1^3 = \frac{0.0021}{6} < 0.0004$$

Thus, if $7 \leq x \leq 9$, the approximation in part (a) is accurate to within 0.0004.

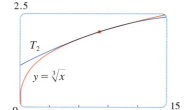

FIGURE 2

Let's use a graphing device to check the calculation in Example 1. Figure 2 shows that the graphs of $y = \sqrt[3]{x}$ and $y = T_2(x)$ are very close to each other when x is near 8. Figure 3 shows the graph of $|R_2(x)|$ computed from the expression

$$|R_2(x)| = |\sqrt[3]{x} - T_2(x)|$$

We see from the graph that

$$|R_2(x)| < 0.0003$$

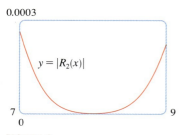

FIGURE 3

when $7 \leq x \leq 9$. Thus, the error estimate from graphical methods is slightly better than the error estimate from Taylor's Inequality in this case.

SECTION 12.12 APPLICATIONS OF TAYLOR POLYNOMIALS   815

EXAMPLE 2
(a) What is the maximum error possible in using the approximation

$$\sin x \approx x - \frac{x^3}{3!} + \frac{x^5}{5!}$$

when $-0.3 \leq x \leq 0.3$? Use this approximation to find $\sin 12°$ correct to six decimal places.
(b) For what values of x is this approximation accurate to within 0.00005?

SOLUTION
(a) Notice that the Maclaurin series

$$\sin x = x - \frac{x^3}{3!} + \frac{x^5}{5!} - \frac{x^7}{7!} + \cdots$$

is alternating for all nonzero values of x, and the successive terms decrease in size because $|x| < 1$, so we can use the Alternating Series Estimation Theorem. The error in approximating $\sin x$ by the first three terms of its Maclaurin series is at most

$$\left| \frac{x^7}{7!} \right| = \frac{|x|^7}{5040}$$

If $-0.3 \leq x \leq 0.3$, then $|x| \leq 0.3$, so the error is smaller than

$$\frac{(0.3)^7}{5040} \approx 4.3 \times 10^{-8}$$

To find $\sin 12°$ we first convert to radian measure.

$$\sin 12° = \sin\left(\frac{12\pi}{180}\right) = \sin\left(\frac{\pi}{15}\right)$$

$$\approx \frac{\pi}{15} - \left(\frac{\pi}{15}\right)^3 \frac{1}{3!} + \left(\frac{\pi}{15}\right)^5 \frac{1}{5!}$$

$$\approx 0.20791169$$

Thus, correct to six decimal places, $\sin 12° \approx 0.207912$.
(b) The error will be smaller than 0.00005 if

$$\frac{|x|^7}{5040} < 0.00005$$

Solving this inequality for x, we get

$$|x|^7 < 0.252 \quad \text{or} \quad |x| < (0.252)^{1/7} \approx 0.821$$

So the given approximation is accurate to within 0.00005 when $|x| < 0.82$.

Module 12.10/12.12 graphically shows the remainders in Taylor polynomial approximations.

What if we use Taylor's Inequality to solve Example 2? Since $f^{(7)}(x) = -\cos x$, we have $|f^{(7)}(x)| \leq 1$ and so

$$|R_6(x)| \leq \frac{1}{7!} |x|^7$$

So we get the same estimates as with the Alternating Series Estimation Theorem.

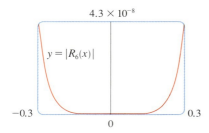

FIGURE 4

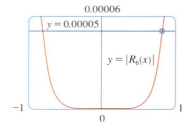

FIGURE 5

What about graphical methods? Figure 4 shows the graph of

$$|R_6(x)| = |\sin x - (x - \tfrac{1}{6}x^3 + \tfrac{1}{120}x^5)|$$

and we see from it that $|R_6(x)| < 4.3 \times 10^{-8}$ when $|x| \leq 0.3$. This is the same estimate that we obtained in Example 2. For part (b) we want $|R_6(x)| < 0.00005$, so we graph both $y = |R_6(x)|$ and $y = 0.00005$ in Figure 5. By placing the cursor on the right intersection point we find that the inequality is satisfied when $|x| < 0.82$. Again this is the same estimate that we obtained in the solution to Example 2.

If we had been asked to approximate $\sin 72°$ instead of $\sin 12°$ in Example 2, it would have been wise to use the Taylor polynomials at $a = \pi/3$ (instead of $a = 0$) because they are better approximations to $\sin x$ for values of x close to $\pi/3$. Notice that $72°$ is close to $60°$ (or $\pi/3$ radians) and the derivatives of $\sin x$ are easy to compute at $\pi/3$.

Figure 6 shows the graphs of the Maclaurin polynomial approximations

$$T_1(x) = x \qquad\qquad T_3(x) = x - \frac{x^3}{3!}$$

$$T_5(x) = x - \frac{x^3}{3!} + \frac{x^5}{5!} \qquad T_7(x) = x - \frac{x^3}{3!} + \frac{x^5}{5!} - \frac{x^7}{7!}$$

to the sine curve. You can see that as n increases, $T_n(x)$ is a good approximation to $\sin x$ on a larger and larger interval.

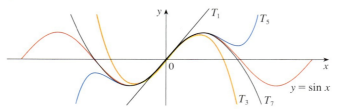

FIGURE 6

One use of the type of calculation done in Examples 1 and 2 occurs in calculators and computers. For instance, when you press the sin or e^x key on your calculator, or when a computer programmer uses a subroutine for a trigonometric or exponential or Bessel function, in many machines a polynomial approximation is calculated. The polynomial is often a Taylor polynomial that has been modified so that the error is spread more evenly throughout an interval.

Applications to Physics

Taylor polynomials are also used frequently in physics. In order to gain insight into an equation, a physicist often simplifies a function by considering only the first two or three terms in its Taylor series. In other words, the physicist uses a Taylor polynomial as an approximation to the function. Taylor's Inequality can then be used to gauge the accuracy of the approximation. The following example shows one way in which this idea is used in special relativity.

EXAMPLE 3 In Einstein's theory of special relativity the mass of an object moving with velocity v is

$$m = \frac{m_0}{\sqrt{1 - v^2/c^2}}$$

where m_0 is the mass of the object when at rest and c is the speed of light. The kinetic energy of the object is the difference between its total energy and its energy at rest:

$$K = mc^2 - m_0c^2$$

(a) Show that when v is very small compared with c, this expression for K agrees with classical Newtonian physics: $K = \frac{1}{2}m_0v^2$.

(b) Use Taylor's Inequality to estimate the difference in these expressions for K when $|v| \leq 100$ m/s.

SOLUTION

(a) Using the expressions given for K and m, we get

$$K = mc^2 - m_0c^2 = \frac{m_0c^2}{\sqrt{1 - v^2/c^2}} - m_0c^2$$

$$= m_0c^2\left[\left(1 - \frac{v^2}{c^2}\right)^{-1/2} - 1\right]$$

With $x = -v^2/c^2$, the Maclaurin series for $(1 + x)^{-1/2}$ is most easily computed as a binomial series with $k = -\frac{1}{2}$. (Notice that $|x| < 1$ because $v < c$.) Therefore, we have

$$(1 + x)^{-1/2} = 1 - \frac{1}{2}x + \frac{(-\frac{1}{2})(-\frac{3}{2})}{2!}x^2 + \frac{(-\frac{1}{2})(-\frac{3}{2})(-\frac{5}{2})}{3!}x^3 + \cdots$$

$$= 1 - \frac{1}{2}x + \frac{3}{8}x^2 - \frac{5}{16}x^3 + \cdots$$

and

$$K = m_0c^2\left[\left(1 + \frac{1}{2}\frac{v^2}{c^2} + \frac{3}{8}\frac{v^4}{c^4} + \frac{5}{16}\frac{v^6}{c^6} + \cdots\right) - 1\right]$$

$$= m_0c^2\left(\frac{1}{2}\frac{v^2}{c^2} + \frac{3}{8}\frac{v^4}{c^4} + \frac{5}{16}\frac{v^6}{c^6} + \cdots\right)$$

If v is much smaller than c, then all terms after the first are very small when compared with the first term. If we omit them, we get

$$K \approx m_0c^2\left(\frac{1}{2}\frac{v^2}{c^2}\right) = \frac{1}{2}m_0v^2$$

(b) If $x = -v^2/c^2$, $f(x) = m_0c^2[(1 + x)^{-1/2} - 1]$, and M is a number such that $|f''(x)| \leq M$, then we can use Taylor's Inequality to write

$$|R_1(x)| \leq \frac{M}{2!}x^2$$

We have $f''(x) = \frac{3}{4}m_0c^2(1 + x)^{-5/2}$ and we are given that $|v| \leq 100$ m/s, so

$$|f''(x)| = \frac{3m_0c^2}{4(1 - v^2/c^2)^{5/2}} \leq \frac{3m_0c^2}{4(1 - 100^2/c^2)^{5/2}} \quad (= M)$$

Thus, with $c = 3 \times 10^8$ m/s,

$$|R_1(x)| \leq \frac{1}{2} \cdot \frac{3m_0c^2}{4(1 - 100^2/c^2)^{5/2}} \cdot \frac{100^4}{c^4} < (4.17 \times 10^{-10})m_0$$

So when $|v| \leq 100$ m/s, the magnitude of the error in using the Newtonian expression for kinetic energy is at most $(4.2 \times 10^{-10})m_0$.

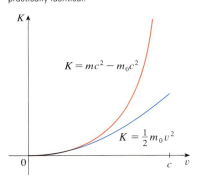

The upper curve in Figure 7 is the graph of the expression for the kinetic energy K of an object with velocity v in special relativity. The lower curve shows the function used for K in classical Newtonian physics. When v is much smaller than the speed of light, the curves are practically identical.

FIGURE 7

Another application to physics occurs in optics. Figure 8 is adapted from *Optics,* 4th ed., by Eugene Hecht (Reading, MA: Addison-Wesley, 2002), page 153. It depicts a wave from the point source S meeting a spherical interface of radius R centered at C. The ray SA is refracted toward P.

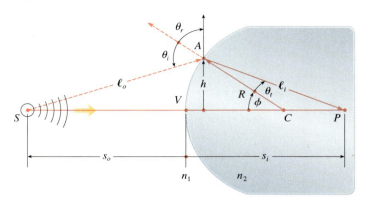

FIGURE 8
Refraction at a spherical interface

Using Fermat's principle that light travels so as to minimize the time taken, Hecht derives the equation

$$\boxed{1} \qquad \frac{n_1}{\ell_o} + \frac{n_2}{\ell_i} = \frac{1}{R}\left(\frac{n_2 s_i}{\ell_i} - \frac{n_1 s_o}{\ell_o}\right)$$

where n_1 and n_2 are indexes of refraction and ℓ_o, ℓ_i, s_o, and s_i are the distances indicated in Figure 8. By the Law of Cosines, applied to triangles ACS and ACP, we have

$$\boxed{2} \qquad \ell_o = \sqrt{R^2 + (s_o + R)^2 - 2R(s_o + R)\cos\phi}$$

$$\ell_i = \sqrt{R^2 + (s_i - R)^2 + 2R(s_i - R)\cos\phi}$$

▐▐▐▐ Here we use the identity
$$\cos(\pi - \phi) = -\cos\phi$$

Because Equation 1 is cumbersome to work with, Gauss, in 1841, simplified it by using the linear approximation $\cos\phi \approx 1$ for small values of ϕ. (This amounts to using the Taylor polynomial of degree 1.) Then Equation 1 becomes the following simpler equation [as you are asked to show in Exercise 32(a)]:

$$\boxed{3} \qquad \frac{n_1}{s_o} + \frac{n_2}{s_i} = \frac{n_2 - n_1}{R}$$

The resulting optical theory is known as *Gaussian optics,* or *first-order optics,* and has become the basic theoretical tool used to design lenses.

A more accurate theory is obtained by approximating $\cos\phi$ by its Taylor polynomial of degree 3 (which is the same as the Taylor polynomial of degree 2). This takes into account rays for which ϕ is not so small, that is, rays that strike the surface at greater distances h above the axis. In Exercise 32(b) you are asked to use this approximation to derive the more accurate equation

$$\boxed{4} \qquad \frac{n_1}{s_o} + \frac{n_2}{s_i} = \frac{n_2 - n_1}{R} + h^2\left[\frac{n_1}{2s_o}\left(\frac{1}{s_o} + \frac{1}{R}\right)^2 + \frac{n_2}{2s_i}\left(\frac{1}{R} - \frac{1}{s_i}\right)^2\right]$$

The resulting optical theory is known as *third-order optics.*

Other applications of Taylor polynomials to physics and engineering are explored in Exercises 30, 31, 33, 34, and 35 and in the Applied Project on page 821.

12.12 Exercises

1. (a) Find the Taylor polynomials up to degree 6 for $f(x) = \cos x$ centered at $a = 0$. Graph f and these polynomials on a common screen.
    (b) Evaluate f and these polynomials at $x = \pi/4$, $\pi/2$, and π.
    (c) Comment on how the Taylor polynomials converge to $f(x)$.

2. (a) Find the Taylor polynomials up to degree 3 for $f(x) = 1/x$ centered at $a = 1$. Graph f and these polynomials on a common screen.
    (b) Evaluate f and these polynomials at $x = 0.9$ and 1.3.
    (c) Comment on how the Taylor polynomials converge to $f(x)$.

3–10 Find the Taylor polynomial $T_n(x)$ for the function f at the number a. Graph f and T_n on the same screen.

3. $f(x) = \ln x$, $a = 1$, $n = 4$
4. $f(x) = e^x$, $a = 2$, $n = 3$
5. $f(x) = \sin x$, $a = \pi/6$, $n = 3$
6. $f(x) = \cos x$, $a = 2\pi/3$, $n = 4$
7. $f(x) = \arcsin x$, $a = 0$, $n = 3$
8. $f(x) = \dfrac{\ln x}{x}$, $a = 1$, $n = 3$
9. $f(x) = xe^{-2x}$, $a = 0$, $n = 3$
10. $f(x) = \sqrt{3 + x^2}$, $a = 1$, $n = 2$

CAS 11–12 Use a computer algebra system to find the Taylor polynomials T_n at $a = 0$ for the given values of n. Then graph these polynomials and f on the same screen.

11. $f(x) = \sec x$, $n = 2, 4, 6, 8$
12. $f(x) = \tan x$, $n = 1, 3, 5, 7, 9$

13–22
(a) Approximate f by a Taylor polynomial with degree n at the number a.
(b) Use Taylor's Inequality to estimate the accuracy of the approximation $f(x) \approx T_n(x)$ when x lies in the given interval.
(c) Check your result in part (b) by graphing $|R_n(x)|$.

13. $f(x) = \sqrt{x}$, $a = 4$, $n = 2$, $4 \le x \le 4.2$
14. $f(x) = x^{-2}$, $a = 1$, $n = 2$, $0.9 \le x \le 1.1$
15. $f(x) = x^{2/3}$, $a = 1$, $n = 3$, $0.8 \le x \le 1.2$
16. $f(x) = \cos x$, $a = \pi/3$, $n = 4$, $0 \le x \le 2\pi/3$
17. $f(x) = \tan x$, $a = 0$, $n = 3$, $0 \le x \le \pi/6$
18. $f(x) = \ln(1 + 2x)$, $a = 1$, $n = 3$, $0.5 \le x \le 1.5$
19. $f(x) = e^{x^2}$, $a = 0$, $n = 3$, $0 \le x \le 0.1$
20. $f(x) = x \ln x$, $a = 1$, $n = 3$, $0.5 \le x \le 1.5$
21. $f(x) = x \sin x$, $a = 0$, $n = 4$, $-1 \le x \le 1$
22. $f(x) = \sinh 2x$, $a = 0$, $n = 5$, $-1 \le x \le 1$

23. Use the information from Exercise 5 to estimate $\sin 35°$ correct to five decimal places.

24. Use the information from Exercise 16 to estimate $\cos 69°$ correct to five decimal places.

25. Use Taylor's Inequality to determine the number of terms of the Maclaurin series for e^x that should be used to estimate $e^{0.1}$ to within 0.00001.

26. How many terms of the Maclaurin series for $\ln(1 + x)$ do you need to use to estimate $\ln 1.4$ to within 0.001?

27–28 Use the Alternating Series Estimation Theorem or Taylor's Inequality to estimate the range of values of x for which the given approximation is accurate to within the stated error. Check your answer graphically.

27. $\sin x \approx x - \dfrac{x^3}{6}$ ($|\text{error}| < 0.01$)

28. $\cos x \approx 1 - \dfrac{x^2}{2} + \dfrac{x^4}{24}$ ($|\text{error}| < 0.005$)

29. A car is moving with speed 20 m/s and acceleration 2 m/s² at a given instant. Using a second-degree Taylor polynomial, estimate how far the car moves in the next second. Would it be reasonable to use this polynomial to estimate the distance traveled during the next minute?

30. The resistivity ρ of a conducting wire is the reciprocal of the conductivity and is measured in units of ohm-meters (Ω-m). The resistivity of a given metal depends on the temperature according to the equation

    $$\rho(t) = \rho_{20} e^{\alpha(t - 20)}$$

    where t is the temperature in °C. There are tables that list the values of α (called the temperature coefficient) and ρ_{20} (the resistivity at 20°C) for various metals. Except at very low temperatures, the resistivity varies almost linearly with temperature and so it is common to approximate the expression for $\rho(t)$ by its first- or second-degree Taylor polynomial at $t = 20$.
    (a) Find expressions for these linear and quadratic approximations.
    (b) For copper, the tables give $\alpha = 0.0039/\text{°C}$ and $\rho_{20} = 1.7 \times 10^{-8}$ Ω-m. Graph the resistivity of copper and the linear and quadratic approximations for $-250°C \le t \le 1000°C$.

(c) For what values of t does the linear approximation agree with the exponential expression to within one percent?

31. An electric dipole consists of two electric charges of equal magnitude and opposite signs. If the charges are q and $-q$ and are located at a distance d from each other, then the electric field E at the point P in the figure is

$$E = \frac{q}{D^2} - \frac{q}{(D+d)^2}$$

By expanding this expression for E as a series in powers of d/D, show that E is approximately proportional to $1/D^3$ when P is far away from the dipole.

32. (a) Derive Equation 3 for Gaussian optics from Equation 1 by approximating $\cos \phi$ in Equation 2 by its first-degree Taylor polynomial.
(b) Show that if $\cos \phi$ is replaced by its third-degree Taylor polynomial in Equation 2, then Equation 1 becomes Equation 4 for third-order optics. [*Hint:* Use the first two terms in the binomial series for ℓ_o^{-1} and ℓ_i^{-1}. Also, use $\phi \approx \sin \phi$.]

33. If a water wave with length L moves with velocity v across a body of water with depth d, as in the figure, then

$$v^2 = \frac{gL}{2\pi} \tanh \frac{2\pi d}{L}$$

(a) If the water is deep, show that $v \approx \sqrt{gL/(2\pi)}$.
(b) If the water is shallow, use the Maclaurin series for tanh to show that $v \approx \sqrt{gd}$. (Thus, in shallow water the velocity of a wave tends to be independent of the length of the wave.)
(c) Use the Alternating Series Estimation Theorem to show that if $L > 10d$, then the estimate $v^2 \approx gd$ is accurate to within $0.014gL$.

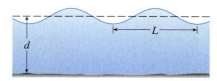

34. The period of a pendulum with length L that makes a maximum angle θ_0 with the vertical is

$$T = 4\sqrt{\frac{L}{g}} \int_0^{\pi/2} \frac{dx}{\sqrt{1 - k^2 \sin^2 x}}$$

where $k = \sin\left(\frac{1}{2}\theta_0\right)$ and g is the acceleration due to gravity. (In Exercise 40 in Section 8.7 we approximated this integral using Simpson's Rule.)
(a) Expand the integrand as a binomial series and use the result of Exercise 44 in Section 8.1 to show that

$$T = 2\pi\sqrt{\frac{L}{g}}\left[1 + \frac{1^2}{2^2}k^2 + \frac{1^2 3^2}{2^2 4^2}k^4 + \frac{1^2 3^2 5^2}{2^2 4^2 6^2}k^6 + \cdots\right]$$

If θ_0 is not too large, the approximation $T \approx 2\pi\sqrt{L/g}$, obtained by using only the first term in the series, is often used. A better approximation is obtained by using two terms:

$$T \approx 2\pi\sqrt{\frac{L}{g}}\left(1 + \tfrac{1}{4}k^2\right)$$

(b) Notice that all the terms in the series after the first one have coefficients that are at most $\frac{1}{4}$. Use this fact to compare this series with a geometric series and show that

$$2\pi\sqrt{\frac{L}{g}}\left(1 + \tfrac{1}{4}k^2\right) \leq T \leq 2\pi\sqrt{\frac{L}{g}}\frac{4 - 3k^2}{4 - 4k^2}$$

(c) Use the inequalities in part (b) to estimate the period of a pendulum with $L = 1$ meter and $\theta_0 = 10°$. How does it compare with the estimate $T \approx 2\pi\sqrt{L/g}$? What if $\theta_0 = 42°$?

35. If a surveyor measures differences in elevation when making plans for a highway across a desert, corrections must be made for the curvature of the Earth.
(a) If R is the radius of the Earth and L is the length of the highway, show that the correction is

$$C = R \sec(L/R) - R$$

(b) Use a Taylor polynomial to show that

$$C \approx \frac{L^2}{2R} + \frac{5L^4}{24R^3}$$

(c) Compare the corrections given by the formulas in parts (a) and (b) for a highway that is 100 km long. (Take the radius of Earth to be 6370 km.)

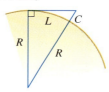

36. Show that T_n and f have the same derivatives at a up to order n.

37. In Section 4.9 we considered Newton's method for approximating a root r of the equation $f(x) = 0$, and from an initial approximation x_1 we obtained successive approximations x_2, $x_3, \ldots$, where

$$x_{n+1} = x_n - \frac{f(x_n)}{f'(x_n)}$$

Use Taylor's Inequality with $n = 1$, $a = x_n$, and $x = r$ to show that if $f''(x)$ exists on an interval I containing r, x_n, and x_{n+1}, and $|f''(x)| \leq M$, $|f'(x)| \geq K$ for all $x \in I$, then

$$|x_{n+1} - r| \leq \frac{M}{2K}|x_n - r|^2$$

[This means that if x_n is accurate to d decimal places, then x_{n+1} is accurate to about $2d$ decimal places. More precisely, if the error at stage n is at most 10^{-m}, then the error at stage $n + 1$ is at most $(M/2K)10^{-2m}$.]

APPLIED PROJECT

Radiation from the Stars

Any object emits radiation when heated. A *blackbody* is a system that absorbs all the radiation that falls on it. For instance, a matte black surface or a large cavity with a small hole in its wall (like a blastfurnace) is a blackbody and emits blackbody radiation. Even the radiation from the Sun is close to being blackbody radiation.

Proposed in the late 19th century, the Rayleigh-Jeans Law expresses the energy density of blackbody radiation of wavelength λ as

$$f(\lambda) = \frac{8\pi kT}{\lambda^4}$$

where λ is measured in meters, T is the temperature in kelvins (K), and k is Boltzmann's constant. The Rayleigh-Jeans Law agrees with experimental measurements for long wavelengths but disagrees drastically for short wavelengths. [The law predicts that $f(\lambda) \to \infty$ as $\lambda \to 0^+$ but experiments have shown that $f(\lambda) \to 0$.] This fact is known as the *ultraviolet catastrophe*.

In 1900 Max Planck found a better model (known now as Planck's Law) for blackbody radiation:

$$f(\lambda) = \frac{8\pi hc \lambda^{-5}}{e^{hc/(\lambda kT)} - 1}$$

where λ is measured in meters, T is the temperature (in kelvins), and

h = Planck's constant = 6.6262×10^{-34} J·s

c = speed of light = 2.997925×10^8 m/s

k = Boltzmann's constant = 1.3807×10^{-23} J/K

1. Use l'Hospital's Rule to show that

   $$\lim_{\lambda \to 0^+} f(\lambda) = 0 \quad \text{and} \quad \lim_{\lambda \to \infty} f(\lambda) = 0$$

   for Planck's Law. So this law models blackbody radiation better than the Rayleigh-Jeans Law for short wavelengths.

2. Use a Taylor polynomial to show that, for large wavelengths, Planck's Law gives approximately the same values as the Rayleigh-Jeans Law.

3. Graph f as given by both laws on the same screen and comment on the similarities and differences. Use $T = 5700$ K (the temperature of the Sun). (You may want to change from meters to the more convenient unit of micrometers: 1 µm = 10^{-6} m.)

4. Use your graph in Problem 3 to estimate the value of λ for which $f(\lambda)$ is a maximum under Planck's Law.

5. Investigate how the graph of f changes as T varies. (Use Planck's Law.) In particular, graph f for the stars Betelgeuse ($T = 3400$ K), Procyon ($T = 6400$ K), and Sirius ($T = 9200$ K) as well as the Sun. How does the total radiation emitted (the area under the curve) vary with T? Use the graph to comment on why Sirius is known as a blue star and Betelgeuse as a red star.

12 Review

CONCEPT CHECK

1. (a) What is a convergent sequence?
   (b) What is a convergent series?
   (c) What does $\lim_{n\to\infty} a_n = 3$ mean?
   (d) What does $\sum_{n=1}^{\infty} a_n = 3$ mean?

2. (a) What is a bounded sequence?
   (b) What is a monotonic sequence?
   (c) What can you say about a bounded monotonic sequence?

3. (a) What is a geometric series? Under what circumstances is it convergent? What is its sum?
   (b) What is a p-series? Under what circumstances is it convergent?

4. Suppose $\sum a_n = 3$ and s_n is the nth partial sum of the series. What is $\lim_{n\to\infty} a_n$? What is $\lim_{n\to\infty} s_n$?

5. State the following.
   (a) The Test for Divergence
   (b) The Integral Test
   (c) The Comparison Test
   (d) The Limit Comparison Test
   (e) The Alternating Series Test
   (f) The Ratio Test
   (g) The Root Test

6. (a) What is an absolutely convergent series?
   (b) What can you say about such a series?
   (c) What is a conditionally convergent series?

7. (a) If a series is convergent by the Integral Test, how do you estimate its sum?
   (b) If a series is convergent by the Comparison Test, how do you estimate its sum?
   (c) If a series is convergent by the Alternating Series Test, how do you estimate its sum?

8. (a) Write the general form of a power series.
   (b) What is the radius of convergence of a power series?
   (c) What is the interval of convergence of a power series?

9. Suppose $f(x)$ is the sum of a power series with radius of convergence R.
   (a) How do you differentiate f? What is the radius of convergence of the series for f'?
   (b) How do you integrate f? What is the radius of convergence of the series for $\int f(x)\,dx$?

10. (a) Write an expression for the nth-degree Taylor polynomial of f centered at a.
    (b) Write an expression for the Taylor series of f centered at a.
    (c) Write an expression for the Maclaurin series of f.
    (d) How do you show that $f(x)$ is equal to the sum of its Taylor series?
    (e) State Taylor's Inequality.

11. Write the Maclaurin series and the interval of convergence for each of the following functions.
    (a) $1/(1-x)$ (b) e^x
    (c) $\sin x$ (d) $\cos x$
    (e) $\tan^{-1} x$

12. Write the binomial series expansion of $(1+x)^k$. What is the radius of convergence of this series?

TRUE-FALSE QUIZ

Determine whether the statement is true or false. If it is true, explain why. If it is false, explain why or give an example that disproves the statement.

1. If $\lim_{n\to\infty} a_n = 0$, then $\sum a_n$ is convergent.

2. The series $\sum_{n=1}^{\infty} n^{-\sin 1}$ is convergent.

3. If $\lim_{n\to\infty} a_n = L$, then $\lim_{n\to\infty} a_{2n+1} = L$.

4. If $\sum c_n 6^n$ is convergent, then $\sum c_n(-2)^n$ is convergent.

5. If $\sum c_n 6^n$ is convergent, then $\sum c_n(-6)^n$ is convergent.

6. If $\sum c_n x^n$ diverges when $x = 6$, then it diverges when $x = 10$.

7. The Ratio Test can be used to determine whether $\sum 1/n^3$ converges.

8. The Ratio Test can be used to determine whether $\sum 1/n!$ converges.

9. If $0 \leq a_n \leq b_n$ and $\sum b_n$ diverges, then $\sum a_n$ diverges.

10. $\sum_{n=0}^{\infty} \dfrac{(-1)^n}{n!} = \dfrac{1}{e}$

11. If $-1 < \alpha < 1$, then $\lim_{n\to\infty} \alpha^n = 0$.

12. If $\sum a_n$ is divergent, then $\sum |a_n|$ is divergent.

13. If $f(x) = 2x - x^2 + \frac{1}{3}x^3 - \cdots$ converges for all x, then $f'''(0) = 2$.

14. If $\{a_n\}$ and $\{b_n\}$ are divergent, then $\{a_n + b_n\}$ is divergent.

15. If $\{a_n\}$ and $\{b_n\}$ are divergent, then $\{a_n b_n\}$ is divergent.

16. If $\{a_n\}$ is decreasing and $a_n > 0$ for all n, then $\{a_n\}$ is convergent.

17. If $a_n > 0$ and $\sum a_n$ converges, then $\sum (-1)^n a_n$ converges.

18. If $a_n > 0$ and $\lim_{n\to\infty}(a_{n+1}/a_n) < 1$, then $\lim_{n\to\infty} a_n = 0$.

EXERCISES

1–8 Determine whether the sequence is convergent or divergent. If it is convergent, find its limit.

1. $a_n = \dfrac{2 + n^3}{1 + 2n^3}$

2. $a_n = \dfrac{9^{n+1}}{10^n}$

3. $a_n = \dfrac{n^3}{1 + n^2}$

4. $a_n = \cos(n\pi/2)$

5. $a_n = \dfrac{n \sin n}{n^2 + 1}$

6. $a_n = \dfrac{\ln n}{\sqrt{n}}$

7. $\{(1 + 3/n)^{4n}\}$

8. $\{(-10)^n/n!\}$

9. A sequence is defined recursively by the equations $a_1 = 1$, $a_{n+1} = \tfrac{1}{3}(a_n + 4)$. Show that $\{a_n\}$ is increasing and $a_n < 2$ for all n. Deduce that $\{a_n\}$ is convergent and find its limit.

10. Show that $\lim_{n\to\infty} n^4 e^{-n} = 0$ and use a graph to find the smallest value of N that corresponds to $\varepsilon = 0.1$ in the precise definition of a limit.

11–22 Determine whether the series is convergent or divergent.

11. $\sum_{n=1}^{\infty} \dfrac{n}{n^3 + 1}$

12. $\sum_{n=1}^{\infty} \dfrac{n^2 + 1}{n^3 + 1}$

13. $\sum_{n=1}^{\infty} \dfrac{n^3}{5^n}$

14. $\sum_{n=1}^{\infty} \dfrac{(-1)^n}{\sqrt{n+1}}$

15. $\sum_{n=2}^{\infty} \dfrac{1}{n\sqrt{\ln n}}$

16. $\sum_{n=1}^{\infty} \ln\left(\dfrac{n}{3n + 1}\right)$

17. $\sum_{n=1}^{\infty} \dfrac{\cos 3n}{1 + (1.2)^n}$

18. $\sum_{n=1}^{\infty} \dfrac{n^{2n}}{(1 + 2n^2)^n}$

19. $\sum_{n=1}^{\infty} \dfrac{1 \cdot 3 \cdot 5 \cdots (2n - 1)}{5^n n!}$

20. $\sum_{n=1}^{\infty} \dfrac{(-5)^{2n}}{n^2 9^n}$

21. $\sum_{n=1}^{\infty} (-1)^{n-1} \dfrac{\sqrt{n}}{n + 1}$

22. $\sum_{n=1}^{\infty} \dfrac{\sqrt{n+1} - \sqrt{n-1}}{n}$

23–26 Determine whether the series is conditionally convergent, absolutely convergent, or divergent.

23. $\sum_{n=1}^{\infty} (-1)^{n-1} n^{-1/3}$

24. $\sum_{n=1}^{\infty} (-1)^{n-1} n^{-3}$

25. $\sum_{n=1}^{\infty} \dfrac{(-1)^n (n + 1) 3^n}{2^{2n+1}}$

26. $\sum_{n=2}^{\infty} \dfrac{(-1)^n \sqrt{n}}{\ln n}$

27–31 Find the sum of the series.

27. $\sum_{n=1}^{\infty} \dfrac{2^{2n+1}}{5^n}$

28. $\sum_{n=1}^{\infty} \dfrac{1}{n(n + 3)}$

29. $\sum_{n=1}^{\infty} [\tan^{-1}(n + 1) - \tan^{-1} n]$

30. $\sum_{n=0}^{\infty} \dfrac{(-1)^n \pi^n}{3^{2n}(2n)!}$

31. $1 - e + \dfrac{e^2}{2!} - \dfrac{e^3}{3!} + \dfrac{e^4}{4!} - \cdots$

32. Express the repeating decimal $4.17326326326\ldots$ as a fraction.

33. Show that $\cosh x \geq 1 + \tfrac{1}{2}x^2$ for all x.

34. For what values of x does the series $\sum_{n=1}^{\infty} (\ln x)^n$ converge?

35. Find the sum of the series $\sum_{n=1}^{\infty} \dfrac{(-1)^{n+1}}{n^5}$ correct to four decimal places.

36. (a) Find the partial sum s_5 of the series $\sum_{n=1}^{\infty} 1/n^6$ and estimate the error in using it as an approximation to the sum of the series.
    (b) Find the sum of this series correct to five decimal places.

37. Use the sum of the first eight terms to approximate the sum of the series $\sum_{n=1}^{\infty} (2 + 5^n)^{-1}$. Estimate the error involved in this approximation.

38. (a) Show that the series $\sum_{n=1}^{\infty} \dfrac{n^n}{(2n)!}$ is convergent.
    (b) Deduce that $\lim_{n \to \infty} \dfrac{n^n}{(2n)!} = 0$.

39. Prove that if the series $\sum_{n=1}^{\infty} a_n$ is absolutely convergent, then the series
$$\sum_{n=1}^{\infty} \left(\dfrac{n + 1}{n}\right) a_n$$
is also absolutely convergent.

40–43 Find the radius of convergence and interval of convergence of the series.

40. $\sum_{n=1}^{\infty} (-1)^n \dfrac{x^n}{n^2 5^n}$

41. $\sum_{n=1}^{\infty} \dfrac{(x + 2)^n}{n 4^n}$

42. $\sum_{n=1}^{\infty} \dfrac{2^n (x - 2)^n}{(n + 2)!}$

43. $\sum_{n=0}^{\infty} \dfrac{2^n (x - 3)^n}{\sqrt{n + 3}}$

44. Find the radius of convergence of the series
$$\sum_{n=1}^{\infty} \frac{(2n)!}{(n!)^2} x^n$$

45. Find the Taylor series of $f(x) = \sin x$ at $a = \pi/6$.

46. Find the Taylor series of $f(x) = \cos x$ at $a = \pi/3$.

47–54 Find the Maclaurin series for f and its radius of convergence. You may use either the direct method (definition of a Maclaurin series) or known series such as geometric series, binomial series, or the Maclaurin series for e^x, $\sin x$, and $\tan^{-1} x$.

47. $f(x) = \dfrac{x^2}{1+x}$ **48.** $f(x) = \tan^{-1}(x^2)$

49. $f(x) = \ln(1-x)$ **50.** $f(x) = xe^{2x}$

51. $f(x) = \sin(x^4)$ **52.** $f(x) = 10^x$

53. $f(x) = 1/\sqrt[4]{16-x}$ **54.** $f(x) = (1-3x)^{-5}$

55. Evaluate $\int \dfrac{e^x}{x}\,dx$ as an infinite series.

56. Use series to approximate $\int_0^1 \sqrt{1+x^4}\,dx$ correct to two decimal places.

57–58
(a) Approximate f by a Taylor polynomial with degree n at the number a.
(b) Graph f and T_n on a common screen.
(c) Use Taylor's Inequality to estimate the accuracy of the approximation $f(x) \approx T_n(x)$ when x lies in the given interval.
(d) Check your result in part (c) by graphing $|R_n(x)|$.

57. $f(x) = \sqrt{x}$, $a = 1$, $n = 3$, $0.9 \leq x \leq 1.1$

58. $f(x) = \sec x$, $a = 0$, $n = 2$, $0 \leq x \leq \pi/6$

59. Use series to evaluate the following limit.
$$\lim_{x \to 0} \frac{\sin x - x}{x^3}$$

60. The force due to gravity on an object with mass m at a height h above the surface of the Earth is
$$F = \frac{mgR^2}{(R+h)^2}$$
where R is the radius of the Earth and g is the acceleration due to gravity.
(a) Express F as a series in powers of h/R.
(b) Observe that if we approximate F by the first term in the series, we get the expression $F \approx mg$ that is usually used when h is much smaller than R. Use the Alternating Series Estimation Theorem to estimate the range of values of h for which the approximation $F \approx mg$ is accurate to within 1%. (Use $R = 6400$ km.)

61. Suppose that $f(x) = \sum_{n=0}^{\infty} c_n x^n$ for all x.
(a) If f is an odd function, show that
$$c_0 = c_2 = c_4 = \cdots = 0$$
(b) If f is an even function, show that
$$c_1 = c_3 = c_5 = \cdots = 0$$

62. If $f(x) = e^{x^2}$, show that $f^{(2n)}(0) = \dfrac{(2n)!}{n!}$.

PROBLEMS PLUS

1. If $f(x) = \sin(x^3)$, find $f^{(15)}(0)$.

2. A function f is defined by
$$f(x) = \lim_{n \to \infty} \frac{x^{2n} - 1}{x^{2n} + 1}$$
Where is f continuous?

3. (a) Show that $\tan \frac{1}{2}x = \cot \frac{1}{2}x - 2 \cot x$.

   (b) Find the sum of the series
$$\sum_{n=1}^{\infty} \frac{1}{2^n} \tan \frac{x}{2^n}$$

4. Let $\{P_n\}$ be a sequence of points determined as in the figure. Thus $|AP_1| = 1$, $|P_n P_{n+1}| = 2^{n-1}$, and angle $AP_n P_{n+1}$ is a right angle. Find $\lim_{n \to \infty} \angle P_n A P_{n+1}$.

5. To construct the **snowflake curve**, start with an equilateral triangle with sides of length 1. Step 1 in the construction is to divide each side into three equal parts, construct an equilateral triangle on the middle part, and then delete the middle part (see the figure). Step 2 is to repeat Step 1 for each side of the resulting polygon. This process is repeated at each succeeding step. The snowflake curve is the curve that results from repeating this process indefinitely.

   (a) Let s_n, l_n, and p_n represent the number of sides, the length of a side, and the total length of the nth approximating curve (the curve obtained after Step n of the construction), respectively. Find formulas for s_n, l_n, and p_n.

   (b) Show that $p_n \to \infty$ as $n \to \infty$.

   (c) Sum an infinite series to find the area enclosed by the snowflake curve.

   Parts (b) and (c) show that the snowflake curve is infinitely long but encloses only a finite area.

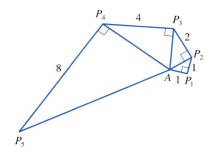

FIGURE FOR PROBLEM 4

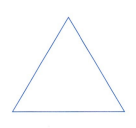 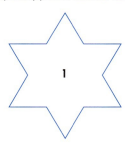 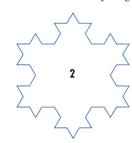 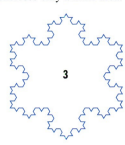

6. Find the sum of the series
$$1 + \frac{1}{2} + \frac{1}{3} + \frac{1}{4} + \frac{1}{6} + \frac{1}{8} + \frac{1}{9} + \frac{1}{12} + \cdots$$
where the terms are the reciprocals of the positive integers whose only prime factors are 2s and 3s.

7. (a) Show that for $xy \neq -1$,
$$\arctan x - \arctan y = \arctan \frac{x - y}{1 + xy}$$
if the left side lies between $-\pi/2$ and $\pi/2$.

   (b) Show that
$$\arctan \tfrac{120}{119} - \arctan \tfrac{1}{239} = \frac{\pi}{4}$$

   (c) Deduce the following formula of John Machin (1680–1751):
$$4 \arctan \tfrac{1}{5} - \arctan \tfrac{1}{239} = \frac{\pi}{4}$$

(d) Use the Maclaurin series for arctan to show that
$$0.197395560 < \arctan \tfrac{1}{5} < 0.197395562$$

(e) Show that
$$0.004184075 < \arctan \tfrac{1}{239} < 0.004184077$$

(f) Deduce that, correct to seven decimal places,
$$\pi \approx 3.1415927$$

Machin used this method in 1706 to find π correct to 100 decimal places. Recently, with the aid of computers, the value of π has been computed to increasingly greater accuracy. In 1999, Takahashi and Kanada, using methods of Borwein and Brent/Salamin, calculated the value of π to 206,158,430,000 decimal places!

8. (a) Prove a formula similar to the one in Problem 7(a) but involving arccot instead of arctan.
(b) Find the sum of the series
$$\sum_{n=0}^{\infty} \text{arccot}(n^2 + n + 1)$$

9. Find the interval of convergence of $\sum_{n=1}^{\infty} n^3 x^n$ and find its sum.

10. If $a_0 + a_1 + a_2 + \cdots + a_k = 0$, show that
$$\lim_{n \to \infty} \left(a_0 \sqrt{n} + a_1 \sqrt{n+1} + a_2 \sqrt{n+2} + \cdots + a_k \sqrt{n+k} \right) = 0$$

If you don't see how to prove this, try the problem-solving strategy of *using analogy* (see page 58). Try the special cases $k = 1$ and $k = 2$ first. If you can see how to prove the assertion for these cases, then you will probably see how to prove it in general.

11. Find the sum of the series $\displaystyle\sum_{n=2}^{\infty} \ln\left(1 - \frac{1}{n^2}\right)$.

12. Suppose you have a large supply of books, all the same size, and you stack them at the edge of a table, with each book extending farther beyond the edge of the table than the one beneath it. Show that it is possible to do this so that the top book extends entirely beyond the table. In fact, show that the top book can extend any distance at all beyond the edge of the table if the stack is high enough. Use the following method of stacking: The top book extends half its length beyond the second book. The second book extends a quarter of its length beyond the third. The third extends one-sixth of its length beyond the fourth, and so on. (Try it yourself with a deck of cards.) Consider centers of mass.

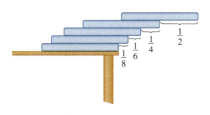

FIGURE FOR PROBLEM 12

13. Let
$$u = 1 + \frac{x^3}{3!} + \frac{x^6}{6!} + \frac{x^9}{9!} + \cdots$$
$$v = x + \frac{x^4}{4!} + \frac{x^7}{7!} + \frac{x^{10}}{10!} + \cdots$$
$$w = \frac{x^2}{2!} + \frac{x^5}{5!} + \frac{x^8}{8!} + \cdots$$

Show that $u^3 + v^3 + w^3 - 3uvw = 1$.

14. If $p > 1$, evaluate the expression
$$\frac{1 + \dfrac{1}{2^p} + \dfrac{1}{3^p} + \dfrac{1}{4^p} + \cdots}{1 - \dfrac{1}{2^p} + \dfrac{1}{3^p} - \dfrac{1}{4^p} + \cdots}$$

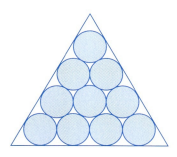

FIGURE FOR PROBLEM 15

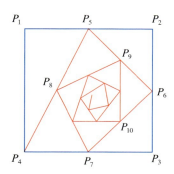

FIGURE FOR PROBLEM 18

15. Suppose that circles of equal diameter are packed tightly in n rows inside an equilateral triangle. (The figure illustrates the case $n = 4$.) If A is the area of the triangle and A_n is the total area occupied by the n rows of circles, show that

$$\lim_{n \to \infty} \frac{A_n}{A} = \frac{\pi}{2\sqrt{3}}$$

16. A sequence $\{a_n\}$ is defined recursively by the equations

$$a_0 = a_1 = 1 \qquad n(n-1)a_n = (n-1)(n-2)a_{n-1} - (n-3)a_{n-2}$$

Find the sum of the series $\sum_{n=0}^{\infty} a_n$.

17. Taking the value of x^x at 0 to be 1 and integrating a series term-by-term, show that

$$\int_0^1 x^x \, dx = \sum_{n=1}^{\infty} \frac{(-1)^{n-1}}{n^n}$$

18. Starting with the vertices $P_1(0, 1)$, $P_2(1, 1)$, $P_3(1, 0)$, $P_4(0, 0)$ of a square, we construct further points as shown in the figure: P_5 is the midpoint of P_1P_2, P_6 is the midpoint of P_2P_3, P_7 is the midpoint of P_3P_4, and so on. The polygonal spiral path $P_1P_2P_3P_4P_5P_6P_7 \ldots$ approaches a point P inside the square.
  (a) If the coordinates of P_n are (x_n, y_n), show that $\frac{1}{2}x_n + x_{n+1} + x_{n+2} + x_{n+3} = 2$ and find a similar equation for the y-coordinates.
  (b) Find the coordinates of P.

19. If $f(x) = \sum_{m=0}^{\infty} c_m x^m$ has positive radius of convergence and $e^{f(x)} = \sum_{n=0}^{\infty} d_n x^n$, show that

$$n d_n = \sum_{i=1}^{n} i c_i d_{n-i} \qquad n \geq 1$$

20. Right-angled triangles are constructed as in the figure. Each triangle has height 1 and its base is the hypotenuse of the preceding triangle. Show that this sequence of triangles makes indefinitely many turns around P by showing that $\sum \theta_n$ is a divergent series.

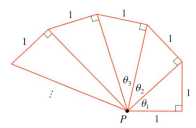

21. Consider the series whose terms are the reciprocals of the positive integers that can be written in base 10 notation without using the digit 0. Show that this series is convergent and the sum is less than 90.

22. (a) Show that the Maclaurin series of the function

$$f(x) = \frac{x}{1 - x - x^2} \qquad \text{is} \qquad \sum_{n=1}^{\infty} f_n x^n$$

where f_n is the nth Fibonacci number, that is, $f_1 = 1$, $f_2 = 1$, and $f_n = f_{n-1} + f_{n-2}$ for $n \geq 3$. [*Hint:* Write $x/(1 - x - x^2) = c_0 + c_1 x + c_2 x^2 + \cdots$ and multiply both sides of this equation by $1 - x - x^2$.]
  (b) By writing $f(x)$ as a sum of partial fractions and thereby obtaining the Maclaurin series in a different way, find an explicit formula for the nth Fibonacci number.

Appendixes

A Numbers, Inequalities, and Absolute Values   A2
B Coordinate Geometry and Lines   A10
C Graphs of Second-Degree Equations   A16
D Trigonometry   A24
E Sigma Notation   A34
F Proofs of Theorems   A39
G Complex Numbers   A41
H Answers to Odd-Numbered Exercises   A49

A Numbers, Inequalities, and Absolute Values

Calculus is based on the real number system. We start with the **integers**:

$$\ldots, \quad -3, \quad -2, \quad -1, \quad 0, \quad 1, \quad 2, \quad 3, \quad 4, \quad \ldots$$

Then we construct the **rational numbers**, which are ratios of integers. Thus, any rational number r can be expressed as

$$r = \frac{m}{n} \quad \text{where } m \text{ and } n \text{ are integers and } n \neq 0$$

Examples are

$$\tfrac{1}{2} \qquad -\tfrac{3}{7} \qquad 46 = \tfrac{46}{1} \qquad 0.17 = \tfrac{17}{100}$$

(Recall that division by 0 is always ruled out, so expressions like $\tfrac{3}{0}$ and $\tfrac{0}{0}$ are undefined.) Some real numbers, such as $\sqrt{2}$, can't be expressed as a ratio of integers and are therefore called **irrational numbers**. It can be shown, with varying degrees of difficulty, that the following are also irrational numbers:

$$\sqrt{3} \qquad \sqrt{5} \qquad \sqrt[3]{2} \qquad \pi \qquad \sin 1° \qquad \log_{10} 2$$

The set of all real numbers is usually denoted by the symbol $\mathbb{R}$. When we use the word *number* without qualification, we mean "real number."

Every number has a decimal representation. If the number is rational, then the corresponding decimal is repeating. For example,

$$\tfrac{1}{2} = 0.5000\ldots = 0.5\overline{0} \qquad \tfrac{2}{3} = 0.66666\ldots = 0.\overline{6}$$

$$\tfrac{157}{495} = 0.317171717\ldots = 0.3\overline{17} \qquad \tfrac{9}{7} = 1.285714285714\ldots = 1.\overline{285714}$$

(The bar indicates that the sequence of digits repeats forever.) On the other hand, if the number is irrational, the decimal is nonrepeating:

$$\sqrt{2} = 1.414213562373095\ldots \qquad \pi = 3.141592653589793\ldots$$

If we stop the decimal expansion of any number at a certain place, we get an approximation to the number. For instance, we can write

$$\pi \approx 3.14159265$$

where the symbol $\approx$ is read "is approximately equal to." The more decimal places we retain, the better the approximation we get.

The real numbers can be represented by points on a line as in Figure 1. The positive direction (to the right) is indicated by an arrow. We choose an arbitrary reference point O, called the **origin**, which corresponds to the real number 0. Given any convenient unit of measurement, each positive number x is represented by the point on the line a distance of x units to the right of the origin, and each negative number $-x$ is represented by the point x units to the left of the origin. Thus, every real number is represented by a point on the line, and every point P on the line corresponds to exactly one real number. The number associated with the point P is called the **coordinate** of P and the line is then called a **coor-**

dinate line, or a **real number line**, or simply a **real line**. Often we identify the point with its coordinate and think of a number as being a point on the real line.

FIGURE 1

The real numbers are ordered. We say *a is less than b* and write $a < b$ if $b - a$ is a positive number. Geometrically, this means that a lies to the left of b on the number line. (Equivalently, we say *b is greater than a* and write $b > a$.) The symbol $a \leq b$ (or $b \geq a$) means that either $a < b$ or $a = b$ and is read "a is less than or equal to b." For instance, the following are true inequalities:

$$7 < 7.4 < 7.5 \qquad -3 > -\pi \qquad \sqrt{2} < 2 \qquad \sqrt{2} \leq 2 \qquad 2 \leq 2$$

In what follows we need to use *set notation*. A **set** is a collection of objects, and these objects are called the **elements** of the set. If S is a set, the notation $a \in S$ means that a is an element of S, and $a \notin S$ means that a is not an element of S. For example, if Z represents the set of integers, then $-3 \in Z$ but $\pi \notin Z$. If S and T are sets, then their **union** $S \cup T$ is the set consisting of all elements that are in S or T (or in both S and T). The **intersection** of S and T is the set $S \cap T$ consisting of all elements that are in both S and T. In other words, $S \cap T$ is the common part of S and T. The empty set, denoted by $\emptyset$, is the set that contains no element.

Some sets can be described by listing their elements between braces. For instance, the set A consisting of all positive integers less than 7 can be written as

$$A = \{1, 2, 3, 4, 5, 6\}$$

We could also write A in *set-builder notation* as

$$A = \{x \mid x \text{ is an integer and } 0 < x < 7\}$$

which is read "A is the set of x such that x is an integer and $0 < x < 7$."

Intervals

Certain sets of real numbers, called **intervals,** occur frequently in calculus and correspond geometrically to line segments. For example, if $a < b$, the **open interval** from a to b consists of all numbers between a and b and is denoted by the symbol (a, b). Using set-builder notation, we can write

$$(a, b) = \{x \mid a < x < b\}$$

FIGURE 2
Open interval (a, b)

Notice that the endpoints of the interval—namely, a and b—are excluded. This is indicated by the round brackets ( ) and by the open dots in Figure 2. The **closed interval** from a to b is the set

$$[a, b] = \{x \mid a \leq x \leq b\}$$

FIGURE 3
Closed interval $[a, b]$

Here the endpoints of the interval are included. This is indicated by the square brackets [ ] and by the solid dots in Figure 3. It is also possible to include only one endpoint in an interval, as shown in Table 1.

1 **Table of Intervals**

Notation	Set description	Picture
(a, b)	$\{x \mid a < x < b\}$	
$[a, b]$	$\{x \mid a \leq x \leq b\}$	
$[a, b)$	$\{x \mid a \leq x < b\}$	
$(a, b]$	$\{x \mid a < x \leq b\}$	
(a, ∞)	$\{x \mid x > a\}$	
$[a, \infty)$	$\{x \mid x \geq a\}$	
$(-\infty, b)$	$\{x \mid x < b\}$	
$(-\infty, b]$	$\{x \mid x \leq b\}$	
$(-\infty, \infty)$	$\mathbb{R}$ (set of all real numbers)	

|||| Table 1 lists the nine possible types of intervals. When these intervals are discussed, it is always assumed that $a < b$.

We also need to consider infinite intervals such as

$$(a, \infty) = \{x \mid x > a\}$$

This does not mean that ∞ ("infinity") is a number. The notation (a, ∞) stands for the set of all numbers that are greater than a, so the symbol ∞ simply indicates that the interval extends indefinitely far in the positive direction.

Inequalities

When working with inequalities, note the following rules.

2 **Rules for Inequalities**

1. If $a < b$, then $a + c < b + c$.
2. If $a < b$ and $c < d$, then $a + c < b + d$.
3. If $a < b$ and $c > 0$, then $ac < bc$.
4. If $a < b$ and $c < 0$, then $ac > bc$.
5. If $0 < a < b$, then $1/a > 1/b$.

Rule 1 says that we can add any number to both sides of an inequality, and Rule 2 says that two inequalities can be added. However, we have to be careful with multiplication. Rule 3 says that we can multiply both sides of an inequality by a *positive* number, but Rule 4 says that if we multiply both sides of an inequality by a negative number, then we reverse the direction of the inequality. For example, if we take the inequality $3 < 5$ and multiply by 2, we get $6 < 10$, but if we multiply by -2, we get $-6 > -10$. Finally, Rule 5 says that if we take reciprocals, then we reverse the direction of an inequality (provided the numbers are positive).

EXAMPLE 1 Solve the inequality $1 + x < 7x + 5$.

SOLUTION The given inequality is satisfied by some values of x but not by others. To *solve* an inequality means to determine the set of numbers x for which the inequality is true. This is called the *solution set*.

First we subtract 1 from each side of the inequality (using Rule 1 with $c = -1$):

$$x < 7x + 4$$

Then we subtract $7x$ from both sides (Rule 1 with $c = -7x$):

$$-6x < 4$$

Now we divide both sides by -6 (Rule 4 with $c = -\tfrac{1}{6}$):

$$x > -\tfrac{4}{6} = -\tfrac{2}{3}$$

These steps can all be reversed, so the solution set consists of all numbers greater than $-\tfrac{2}{3}$. In other words, the solution of the inequality is the interval $\left(-\tfrac{2}{3}, \infty\right)$.

EXAMPLE 2 Solve the inequalities $4 \leq 3x - 2 < 13$.

SOLUTION Here the solution set consists of all values of x that satisfy both inequalities. Using the rules given in (2), we see that the following inequalities are equivalent:

$$4 \leq 3x - 2 < 13$$

$$6 \leq 3x < 15 \qquad \text{(add 2)}$$

$$2 \leq x < 5 \qquad \text{(divide by 3)}$$

Therefore, the solution set is $[2, 5)$.

EXAMPLE 3 Solve the inequality $x^2 - 5x + 6 \leq 0$.

SOLUTION First we factor the left side:

$$(x - 2)(x - 3) \leq 0$$

We know that the corresponding equation $(x - 2)(x - 3) = 0$ has the solutions 2 and 3. The numbers 2 and 3 divide the real line into three intervals:

$$(-\infty, 2) \qquad (2, 3) \qquad (3, \infty)$$

On each of these intervals we determine the signs of the factors. For instance,

$$x \in (-\infty, 2) \quad \Rightarrow \quad x < 2 \quad \Rightarrow \quad x - 2 < 0$$

Then we record these signs in the following chart:

Interval	$x - 2$	$x - 3$	$(x - 2)(x - 3)$
$x < 2$	−	−	+
$2 < x < 3$	+	−	−
$x > 3$	+	+	+

Another method for obtaining the information in the chart is to use *test values*. For instance, if we use the test value $x = 1$ for the interval $(-\infty, 2)$, then substitution in $x^2 - 5x + 6$ gives

$$1^2 - 5(1) + 6 = 2$$

|||| A visual method for solving Example 3 is to use a graphing device to graph the parabola $y = x^2 - 5x + 6$ (as in Figure 4) and observe that the curve lies on or below the x-axis when $2 \leq x \leq 3$.

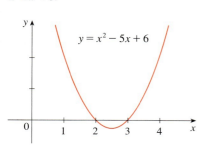

FIGURE 4

The polynomial $x^2 - 5x + 6$ doesn't change sign inside any of the three intervals, so we conclude that it is positive on $(-\infty, 2)$.

Then we read from the chart that $(x - 2)(x - 3)$ is negative when $2 < x < 3$. Thus, the solution of the inequality $(x - 2)(x - 3) \leq 0$ is

$$\{x \mid 2 \leq x \leq 3\} = [2, 3]$$

FIGURE 5

Notice that we have included the endpoints 2 and 3 because we are looking for values of x such that the product is either negative or zero. The solution is illustrated in Figure 5.

EXAMPLE 4 Solve $x^3 + 3x^2 > 4x$.

SOLUTION First we take all nonzero terms to one side of the inequality sign and factor the resulting expression:

$$x^3 + 3x^2 - 4x > 0 \quad \text{or} \quad x(x - 1)(x + 4) > 0$$

As in Example 3 we solve the corresponding equation $x(x - 1)(x + 4) = 0$ and use the solutions $x = -4$, $x = 0$, and $x = 1$ to divide the real line into four intervals $(-\infty, -4)$, $(-4, 0)$, $(0, 1)$, and $(1, \infty)$. On each interval the product keeps a constant sign as shown in the following chart:

Interval	x	$x - 1$	$x + 4$	$x(x - 1)(x + 4)$
$x < -4$	−	−	−	−
$-4 < x < 0$	−	−	+	+
$0 < x < 1$	+	−	+	−
$x > 1$	+	+	+	+

Then we read from the chart that the solution set is

$$\{x \mid -4 < x < 0 \text{ or } x > 1\} = (-4, 0) \cup (1, \infty)$$

The solution is illustrated in Figure 6.

FIGURE 6

Absolute Value

The **absolute value** of a number a, denoted by $|a|$, is the distance from a to 0 on the real number line. Distances are always positive or 0, so we have

$$|a| \geq 0 \quad \text{for every number } a$$

For example,

$$|3| = 3 \quad |-3| = 3 \quad |0| = 0 \quad |\sqrt{2} - 1| = \sqrt{2} - 1 \quad |3 - \pi| = \pi - 3$$

In general, we have

|||| Remember that if a is negative, then $-a$ is positive.

$$|a| = a \quad \text{if } a \geq 0$$
$$|a| = -a \quad \text{if } a < 0$$

EXAMPLE 5 Express $|3x - 2|$ without using the absolute-value symbol.

SOLUTION

$$|3x - 2| = \begin{cases} 3x - 2 & \text{if } 3x - 2 \geq 0 \\ -(3x - 2) & \text{if } 3x - 2 < 0 \end{cases}$$

$$= \begin{cases} 3x - 2 & \text{if } x \geq \frac{2}{3} \\ 2 - 3x & \text{if } x < \frac{2}{3} \end{cases}$$

Recall that the symbol $\sqrt{}$ means "the positive square root of." Thus, $\sqrt{r} = s$ means $s^2 = r$ and $s \geq 0$. Therefore, the equation $\sqrt{a^2} = a$ is not always true. It is true only when $a \geq 0$. If $a < 0$, then $-a > 0$, so we have $\sqrt{a^2} = -a$. In view of (3), we then have the equation

4 $$\sqrt{a^2} = |a|$$

which is true for all values of a.

Hints for the proofs of the following properties are given in the exercises.

5 **Properties of Absolute Values** Suppose a and b are any real numbers and n is an integer. Then

1. $|ab| = |a||b|$
2. $\left|\dfrac{a}{b}\right| = \dfrac{|a|}{|b|}$ $(b \neq 0)$
3. $|a^n| = |a|^n$

For solving equations or inequalities involving absolute values, it's often very helpful to use the following statements.

6 Suppose $a > 0$. Then

4. $|x| = a$ if and only if $x = \pm a$
5. $|x| < a$ if and only if $-a < x < a$
6. $|x| > a$ if and only if $x > a$ or $x < -a$

For instance, the inequality $|x| < a$ says that the distance from x to the origin is less than a, and you can see from Figure 7 that this is true if and only if x lies between $-a$ and a.

If a and b are any real numbers, then the distance between a and b is the absolute value of the difference, namely, $|a - b|$, which is also equal to $|b - a|$. (See Figure 8.)

EXAMPLE 6 Solve $|2x - 5| = 3$.

SOLUTION By Property 4 of (6), $|2x - 5| = 3$ is equivalent to

$$2x - 5 = 3 \quad \text{or} \quad 2x - 5 = -3$$

So $2x = 8$ or $2x = 2$. Thus, $x = 4$ or $x = 1$.

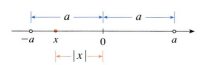

FIGURE 7

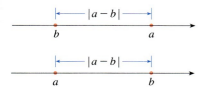

FIGURE 8
Length of a line segment $= |a - b|$

EXAMPLE 7 Solve $|x - 5| < 2$.

SOLUTION 1 By Property 5 of (6), $|x - 5| < 2$ is equivalent to

$$-2 < x - 5 < 2$$

Therefore, adding 5 to each side, we have

$$3 < x < 7$$

and the solution set is the open interval $(3, 7)$.

SOLUTION 2 Geometrically, the solution set consists of all numbers x whose distance from 5 is less than 2. From Figure 9 we see that this is the interval $(3, 7)$.

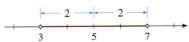

FIGURE 9

EXAMPLE 8 Solve $|3x + 2| \geq 4$.

SOLUTION By Properties 4 and 6 of (6), $|3x + 2| \geq 4$ is equivalent to

$$3x + 2 \geq 4 \quad \text{or} \quad 3x + 2 \leq -4$$

In the first case $3x \geq 2$, which gives $x \geq \frac{2}{3}$. In the second case $3x \leq -6$, which gives $x \leq -2$. So the solution set is

$$\{x \mid x \leq -2 \text{ or } x \geq \tfrac{2}{3}\} = (-\infty, -2] \cup [\tfrac{2}{3}, \infty)$$

Another important property of absolute value, called the Triangle Inequality, is used frequently not only in calculus but throughout mathematics in general.

> **7 The Triangle Inequality** If a and b are any real numbers, then
>
> $$|a + b| \leq |a| + |b|$$

Observe that if the numbers a and b are both positive or both negative, then the two sides in the Triangle Inequality are actually equal. But if a and b have opposite signs, the left side involves a subtraction and the right side does not. This makes the Triangle Inequality seem reasonable, but we can prove it as follows.

Notice that

$$-|a| \leq a \leq |a|$$

is always true because a equals either $|a|$ or $-|a|$. The corresponding statement for b is

$$-|b| \leq b \leq |b|$$

Adding these inequalities, we get

$$-(|a| + |b|) \leq a + b \leq |a| + |b|$$

If we now apply Properties 4 and 5 (with x replaced by $a + b$ and a by $|a| + |b|$), we obtain

$$|a + b| \leq |a| + |b|$$

which is what we wanted to show.

EXAMPLE 9 If $|x - 4| < 0.1$ and $|y - 7| < 0.2$, use the Triangle Inequality to estimate $|(x + y) - 11|$.

SOLUTION In order to use the given information, we use the Triangle Inequality with $a = x - 4$ and $b = y - 7$:

$$|(x + y) - 11| = |(x - 4) + (y - 7)|$$
$$\leq |x - 4| + |y - 7|$$
$$< 0.1 + 0.2 = 0.3$$

Thus $\qquad |(x + y) - 11| < 0.3$

A Exercises

1–12 ▪ Rewrite the expression without using the absolute value symbol.

1. $|5 - 23|$
2. $|5| - |-23|$
3. $|-\pi|$
4. $|\pi - 2|$
5. $|\sqrt{5} - 5|$
6. $||-2| - |-3||$
7. $|x - 2|$ if $x < 2$
8. $|x - 2|$ if $x > 2$
9. $|x + 1|$
10. $|2x - 1|$
11. $|x^2 + 1|$
12. $|1 - 2x^2|$

13–38 ▪ Solve the inequality in terms of intervals and illustrate the solution set on the real number line.

13. $2x + 7 > 3$
14. $3x - 11 < 4$
15. $1 - x \leq 2$
16. $4 - 3x \geq 6$
17. $2x + 1 < 5x - 8$
18. $1 + 5x > 5 - 3x$
19. $-1 < 2x - 5 < 7$
20. $1 < 3x + 4 \leq 16$
21. $0 \leq 1 - x < 1$
22. $-5 \leq 3 - 2x \leq 9$
23. $4x < 2x + 1 \leq 3x + 2$
24. $2x - 3 < x + 4 < 3x - 2$
25. $(x - 1)(x - 2) > 0$
26. $(2x + 3)(x - 1) \geq 0$
27. $2x^2 + x \leq 1$
28. $x^2 < 2x + 8$
29. $x^2 + x + 1 > 0$
30. $x^2 + x > 1$
31. $x^2 < 3$
32. $x^2 \geq 5$
33. $x^3 - x^2 \leq 0$
34. $(x + 1)(x - 2)(x + 3) \geq 0$
35. $x^3 > x$
36. $x^3 + 3x < 4x^2$
37. $\dfrac{1}{x} < 4$
38. $-3 < \dfrac{1}{x} \leq 1$

39. The relationship between the Celsius and Fahrenheit temperature scales is given by $C = \frac{5}{9}(F - 32)$, where C is the temperature in degrees Celsius and F is the temperature in degrees Fahrenheit. What interval on the Celsius scale corresponds to the temperature range $50 \leq F \leq 95$?

40. Use the relationship between C and F given in Exercise 39 to find the interval on the Fahrenheit scale corresponding to the temperature range $20 \leq C \leq 30$.

41. As dry air moves upward, it expands and in so doing cools at a rate of about 1°C for each 100-m rise, up to about 12 km.
    (a) If the ground temperature is 20°C, write a formula for the temperature at height h.
    (b) What range of temperature can be expected if a plane takes off and reaches a maximum height of 5 km?

42. If a ball is thrown upward from the top of a building 128 ft high with an initial velocity of 16 ft/s, then the height h above the ground t seconds later will be
    $$h = 128 + 16t - 16t^2$$
    During what time interval will the ball be at least 32 ft above the ground?

43–46 ▪ Solve the equation for x.

43. $|2x| = 3$
44. $|3x + 5| = 1$
45. $|x + 3| = |2x + 1|$
46. $\left|\dfrac{2x - 1}{x + 1}\right| = 3$

47–56 ▪ Solve the inequality.

47. $|x| < 3$
48. $|x| \geq 3$
49. $|x - 4| < 1$
50. $|x - 6| < 0.1$
51. $|x + 5| \geq 2$
52. $|x + 1| \geq 3$
53. $|2x - 3| \leq 0.4$
54. $|5x - 2| < 6$
55. $1 \leq |x| \leq 4$
56. $0 < |x - 5| < \frac{1}{2}$

57–58 ||| Solve for x, assuming a, b, and c are positive constants.

57. $a(bx - c) \geq bc$

58. $a \leq bx + c < 2a$

59–60 ||| Solve for x, assuming a, b, and c are negative constants.

59. $ax + b < c$

60. $\dfrac{ax + b}{c} \leq b$

61. Suppose that $|x - 2| < 0.01$ and $|y - 3| < 0.04$. Use the Triangle Inequality to show that $|(x + y) - 5| < 0.05$.

62. Show that if $|x + 3| < \frac{1}{2}$, then $|4x + 13| < 3$.

63. Show that if $a < b$, then $a < \dfrac{a + b}{2} < b$.

64. Use Rule 3 to prove Rule 5 of (2).

65. Prove that $|ab| = |a||b|$. [*Hint:* Use Equation 4.]

66. Prove that $\left|\dfrac{a}{b}\right| = \dfrac{|a|}{|b|}$.

67. Show that if $0 < a < b$, then $a^2 < b^2$.

68. Prove that $|x - y| \geq |x| - |y|$. [*Hint:* Use the Triangle Inequality with $a = x - y$ and $b = y$.]

69. Show that the sum, difference, and product of rational numbers are rational numbers.

70. (a) Is the sum of two irrational numbers always an irrational number?
(b) Is the product of two irrational numbers always an irrational number?

B Coordinate Geometry and Lines

Just as the points on a line can be identified with real numbers by assigning them coordinates, as described in Appendix A, so the points in a plane can be identified with ordered pairs of real numbers. We start by drawing two perpendicular coordinate lines that intersect at the origin O on each line. Usually one line is horizontal with positive direction to the right and is called the **x-axis**; the other line is vertical with positive direction upward and is called the **y-axis**.

Any point P in the plane can be located by a unique ordered pair of numbers as follows. Draw lines through P perpendicular to the x- and y-axes. These lines intersect the axes in points with coordinates a and b as shown in Figure 1. Then the point P is assigned the ordered pair (a, b). The first number a is called the **x-coordinate** of P; the second number b is called the **y-coordinate** of P. We say that P is the point with coordinates (a, b), and we denote the point by the symbol $P(a, b)$. Several points are labeled with their coordinates in Figure 2.

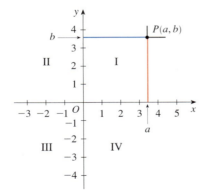

FIGURE 1

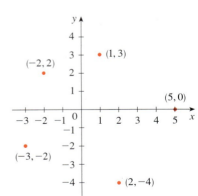

FIGURE 2

By reversing the preceding process we can start with an ordered pair (a, b) and arrive at the corresponding point P. Often we identify the point P with the ordered pair (a, b) and refer to "the point (a, b)." [Although the notation used for an open interval (a, b) is the same as the notation used for a point (a, b), you will be able to tell from the context which meaning is intended.]

This coordinate system is called the **rectangular coordinate system** or the **Cartesian coordinate system** in honor of the French mathematician René Descartes (1596–1650), even though another Frenchman, Pierre Fermat (1601–1665), invented the principles of analytic geometry at about the same time as Descartes. The plane supplied with this coordinate system is called the **coordinate plane** or the **Cartesian plane** and is denoted by $\mathbb{R}^2$.

The x- and y-axes are called the **coordinate axes** and divide the Cartesian plane into four quadrants, which are labeled I, II, III, and IV in Figure 1. Notice that the first quadrant consists of those points whose x- and y-coordinates are both positive.

EXAMPLE 1 Describe and sketch the regions given by the following sets.
(a) $\{(x, y) \mid x \geq 0\}$
(b) $\{(x, y) \mid y = 1\}$
(c) $\{(x, y) \mid |y| < 1\}$

SOLUTION
(a) The points whose x-coordinates are 0 or positive lie on the y-axis or to the right of it as indicated by the shaded region in Figure 3(a).

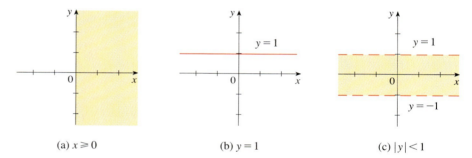

FIGURE 3

(a) $x \geq 0$
(b) $y = 1$
(c) $|y| < 1$

(b) The set of all points with y-coordinate 1 is a horizontal line one unit above the x-axis [see Figure 3(b)].

(c) Recall from Appendix A that

$$|y| < 1 \quad \text{if and only if} \quad -1 < y < 1$$

The given region consists of those points in the plane whose y-coordinates lie between -1 and 1. Thus, the region consists of all points that lie between (but not on) the horizontal lines $y = 1$ and $y = -1$. [These lines are shown as dashed lines in Figure 3(c) to indicate that the points on these lines don't lie in the set.]

Recall from Appendix A that the distance between points a and b on a number line is $|a - b| = |b - a|$. Thus, the distance between points $P_1(x_1, y_1)$ and $P_3(x_2, y_1)$ on a horizontal line must be $|x_2 - x_1|$ and the distance between $P_2(x_2, y_2)$ and $P_3(x_2, y_1)$ on a vertical line must be $|y_2 - y_1|$. (See Figure 4.)

To find the distance $|P_1 P_2|$ between any two points $P_1(x_1, y_1)$ and $P_2(x_2, y_2)$, we note that triangle $P_1 P_2 P_3$ in Figure 4 is a right triangle, and so by the Pythagorean Theorem we have

$$|P_1 P_2| = \sqrt{|P_1 P_3|^2 + |P_2 P_3|^2} = \sqrt{|x_2 - x_1|^2 + |y_2 - y_1|^2}$$
$$= \sqrt{(x_2 - x_1)^2 + (y_2 - y_1)^2}$$

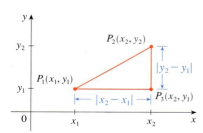

FIGURE 4

> **1 Distance Formula** The distance between the points $P_1(x_1, y_1)$ and $P_2(x_2, y_2)$ is
>
> $$|P_1P_2| = \sqrt{(x_2 - x_1)^2 + (y_2 - y_1)^2}$$

EXAMPLE 2 The distance between $(1, -2)$ and $(5, 3)$ is

$$\sqrt{(5-1)^2 + [3-(-2)]^2} = \sqrt{4^2 + 5^2} = \sqrt{41}$$

Lines

We want to find an equation of a given line L; such an equation is satisfied by the coordinates of the points on L and by no other point. To find the equation of L we use its *slope*, which is a measure of the steepness of the line.

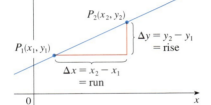

FIGURE 5

> **2 Definition** The **slope** of a nonvertical line that passes through the points $P_1(x_1, y_1)$ and $P_2(x_2, y_2)$ is
>
> $$m = \frac{\Delta y}{\Delta x} = \frac{y_2 - y_1}{x_2 - x_1}$$
>
> The slope of a vertical line is not defined.

Thus, the slope of a line is the ratio of the change in y, Δy, to the change in x, Δx. (See Figure 5.) The slope is therefore the rate of change of y with respect to x. The fact that the line is straight means that the rate of change is constant.

Figure 6 shows several lines labeled with their slopes. Notice that lines with positive slope slant upward to the right, whereas lines with negative slope slant downward to the right. Notice also that the steepest lines are the ones for which the absolute value of the slope is largest, and a horizontal line has slope 0.

Now let's find an equation of the line that passes through a given point $P_1(x_1, y_1)$ and has slope m. A point $P(x, y)$ with $x \neq x_1$ lies on this line if and only if the slope of the line through P_1 and P is equal to m; that is,

$$\frac{y - y_1}{x - x_1} = m$$

This equation can be rewritten in the form

$$y - y_1 = m(x - x_1)$$

and we observe that this equation is also satisfied when $x = x_1$ and $y = y_1$. Therefore, it is an equation of the given line.

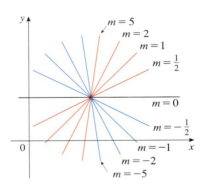

FIGURE 6

> **3 Point-Slope Form of the Equation of a Line** An equation of the line passing through the point $P_1(x_1, y_1)$ and having slope m is
>
> $$y - y_1 = m(x - x_1)$$

EXAMPLE 3 Find an equation of the line through $(1, -7)$ with slope $-\frac{1}{2}$.

SOLUTION Using (3) with $m = -\frac{1}{2}$, $x_1 = 1$, and $y_1 = -7$, we obtain an equation of the line as
$$y + 7 = -\tfrac{1}{2}(x - 1)$$
which we can rewrite as
$$2y + 14 = -x + 1 \quad \text{or} \quad x + 2y + 13 = 0$$

EXAMPLE 4 Find an equation of the line through the points $(-1, 2)$ and $(3, -4)$.

SOLUTION By Definition 2 the slope of the line is
$$m = \frac{-4 - 2}{3 - (-1)} = -\frac{3}{2}$$

Using the point-slope form with $x_1 = -1$ and $y_1 = 2$, we obtain
$$y - 2 = -\tfrac{3}{2}(x + 1)$$
which simplifies to
$$3x + 2y = 1$$

Suppose a nonvertical line has slope m and y-intercept b. (See Figure 7.) This means it intersects the y-axis at the point $(0, b)$, so the point-slope form of the equation of the line, with $x_1 = 0$ and $y_1 = b$, becomes
$$y - b = m(x - 0)$$
This simplifies as follows.

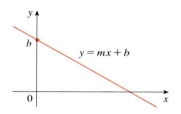

FIGURE 7

> **4 Slope-Intercept Form of the Equation of a Line** An equation of the line with slope m and y-intercept b is
> $$y = mx + b$$

In particular, if a line is horizontal, its slope is $m = 0$, so its equation is $y = b$, where b is the y-intercept (see Figure 8). A vertical line does not have a slope, but we can write its equation as $x = a$, where a is the x-intercept, because the x-coordinate of every point on the line is a.

Observe that the equation of every line can be written in the form

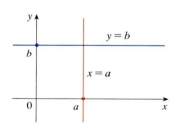

FIGURE 8

$$\boxed{5} \qquad Ax + By + C = 0$$

because a vertical line has the equation $x = a$ or $x - a = 0$ ($A = 1$, $B = 0$, $C = -a$) and a nonvertical line has the equation $y = mx + b$ or $-mx + y - b = 0$ ($A = -m$, $B = 1$, $C = -b$). Conversely, if we start with a general first-degree equation, that is, an equation of the form (5), where A, B, and C are constants and A and B are not both 0, then we can show that it is the equation of a line. If $B = 0$, the equation becomes $Ax + C = 0$ or $x = -C/A$, which represents a vertical line with x-intercept $-C/A$. If $B \neq 0$, the equation can be rewritten by solving for y:

$$y = -\frac{A}{B}x - \frac{C}{B}$$

and we recognize this as being the slope-intercept form of the equation of a line ($m = -A/B$, $b = -C/B$). Therefore, an equation of the form (5) is called a **linear equation** or the **general equation of a line**. For brevity, we often refer to "the line $Ax + By + C = 0$" instead of "the line whose equation is $Ax + By + C = 0$."

EXAMPLE 5 Sketch the graph of the equation $3x - 5y = 15$.

SOLUTION Since the equation is linear, its graph is a line. To draw the graph, we can simply find two points on the line. It's easiest to find the intercepts. Substituting $y = 0$ (the equation of the x-axis) in the given equation, we get $3x = 15$, so $x = 5$ is the x-intercept. Substituting $x = 0$ in the equation, we see that the y-intercept is -3. This allows us to sketch the graph as in Figure 9.

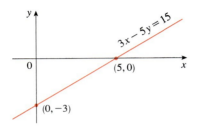

FIGURE 9

EXAMPLE 6 Graph the inequality $x + 2y > 5$.

SOLUTION We are asked to sketch the graph of the set $\{(x, y) \mid x + 2y > 5\}$ and we do so by solving the inequality for y:

$$x + 2y > 5$$

$$2y > -x + 5$$

$$y > -\tfrac{1}{2}x + \tfrac{5}{2}$$

Compare this inequality with the equation $y = -\tfrac{1}{2}x + \tfrac{5}{2}$, which represents a line with slope $-\tfrac{1}{2}$ and y-intercept $\tfrac{5}{2}$. We see that the given graph consists of points whose y-coordinates are *larger* than those on the line $y = -\tfrac{1}{2}x + \tfrac{5}{2}$. Thus, the graph is the region that lies *above* the line, as illustrated in Figure 10.

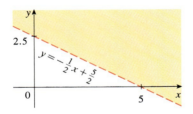

FIGURE 10

Parallel and Perpendicular Lines

Slopes can be used to show that lines are parallel or perpendicular. The following facts are proved, for instance, in *Precalculus: Mathematics for Calculus, Fourth Edition* by Stewart, Redlin, and Watson (Brooks/Cole Publishing Co., Pacific Grove, CA, 2002).

> **6 Parallel and Perpendicular Lines**
> 1. Two nonvertical lines are parallel if and only if they have the same slope.
> 2. Two lines with slopes m_1 and m_2 are perpendicular if and only if $m_1 m_2 = -1$; that is, their slopes are negative reciprocals:
>
> $$m_2 = -\frac{1}{m_1}$$

EXAMPLE 7 Find an equation of the line through the point $(5, 2)$ that is parallel to the line $4x + 6y + 5 = 0$.

SOLUTION The given line can be written in the form

$$y = -\tfrac{2}{3}x - \tfrac{5}{6}$$

which is in slope-intercept form with $m = -\frac{2}{3}$. Parallel lines have the same slope, so the required line has slope $-\frac{2}{3}$ and its equation in point-slope form is

$$y - 2 = -\tfrac{2}{3}(x - 5)$$

We can write this equation as $2x + 3y = 16$.

EXAMPLE 8 Show that the lines $2x + 3y = 1$ and $6x - 4y - 1 = 0$ are perpendicular.

SOLUTION The equations can be written as

$$y = -\tfrac{2}{3}x + \tfrac{1}{3} \quad \text{and} \quad y = \tfrac{3}{2}x - \tfrac{1}{4}$$

from which we see that the slopes are

$$m_1 = -\tfrac{2}{3} \quad \text{and} \quad m_2 = \tfrac{3}{2}$$

Since $m_1 m_2 = -1$, the lines are perpendicular.

B Exercises

1–6 ▮▮▮ Find the distance between the points.

1. $(1, 1)$, $(4, 5)$
2. $(1, -3)$, $(5, 7)$
3. $(6, -2)$, $(-1, 3)$
4. $(1, -6)$, $(-1, -3)$
5. $(2, 5)$, $(4, -7)$
6. (a, b), (b, a)

7–10 ▮▮▮ Find the slope of the line through P and Q.

7. $P(1, 5)$, $Q(4, 11)$
8. $P(-1, 6)$, $Q(4, -3)$
9. $P(-3, 3)$, $Q(-1, -6)$
10. $P(-1, -4)$, $Q(6, 0)$

11. Show that the triangle with vertices $A(0, 2)$, $B(-3, -1)$, and $C(-4, 3)$ is isosceles.

12. (a) Show that the triangle with vertices $A(6, -7)$, $B(11, -3)$, and $C(2, -2)$ is a right triangle using the converse of the Pythagorean Theorem.
(b) Use slopes to show that ABC is a right triangle.
(c) Find the area of the triangle.

13. Show that the points $(-2, 9)$, $(4, 6)$, $(1, 0)$, and $(-5, 3)$ are the vertices of a square.

14. (a) Show that the points $A(-1, 3)$, $B(3, 11)$, and $C(5, 15)$ are collinear (lie on the same line) by showing that $|AB| + |BC| = |AC|$.
(b) Use slopes to show that A, B, and C are collinear.

15. Show that $A(1, 1)$, $B(7, 4)$, $C(5, 10)$, and $D(-1, 7)$ are vertices of a parallelogram.

16. Show that $A(1, 1)$, $B(11, 3)$, $C(10, 8)$, and $D(0, 6)$ are vertices of a rectangle.

17–20 ▮▮▮ Sketch the graph of the equation.

17. $x = 3$
18. $y = -2$
19. $xy = 0$
20. $|y| = 1$

21–36 ▮▮▮ Find an equation of the line that satisfies the given conditions.

21. Through $(2, -3)$, slope 6
22. Through $(-1, 4)$, slope -3
23. Through $(1, 7)$, slope $\frac{2}{3}$
24. Through $(-3, -5)$, slope $-\frac{7}{2}$
25. Through $(2, 1)$ and $(1, 6)$
26. Through $(-1, -2)$ and $(4, 3)$
27. Slope 3, y-intercept -2
28. Slope $\frac{2}{5}$, y-intercept 4
29. x-intercept 1, y-intercept -3
30. x-intercept -8, y-intercept 6
31. Through $(4, 5)$, parallel to the x-axis
32. Through $(4, 5)$, parallel to the y-axis
33. Through $(1, -6)$, parallel to the line $x + 2y = 6$
34. y-intercept 6, parallel to the line $2x + 3y + 4 = 0$
35. Through $(-1, -2)$, perpendicular to the line $2x + 5y + 8 = 0$
36. Through $(\tfrac{1}{2}, -\tfrac{2}{3})$, perpendicular to the line $4x - 8y = 1$

37–42 ▮▮▮ Find the slope and y-intercept of the line and draw its graph.

37. $x + 3y = 0$
38. $2x - 5y = 0$

39. $y = -2$

40. $2x - 3y + 6 = 0$

41. $3x - 4y = 12$

42. $4x + 5y = 10$

43–52 ▪ Sketch the region in the xy-plane.

43. $\{(x, y) \mid x < 0\}$

44. $\{(x, y) \mid y > 0\}$

45. $\{(x, y) \mid xy < 0\}$

46. $\{(x, y) \mid x \geq 1 \text{ and } y < 3\}$

47. $\{(x, y) \mid |x| \leq 2\}$

48. $\{(x, y) \mid |x| < 3 \text{ and } |y| < 2\}$

49. $\{(x, y) \mid 0 \leq y \leq 4 \text{ and } x \leq 2\}$

50. $\{(x, y) \mid y > 2x - 1\}$

51. $\{(x, y) \mid 1 + x \leq y \leq 1 - 2x\}$

52. $\{(x, y) \mid -x \leq y < \frac{1}{2}(x + 3)\}$

---

53. Find a point on the y-axis that is equidistant from $(5, -5)$ and $(1, 1)$.

54. Show that the midpoint of the line segment from $P_1(x_1, y_1)$ to $P_2(x_2, y_2)$ is

$$\left(\frac{x_1 + x_2}{2}, \frac{y_1 + y_2}{2}\right)$$

55. Find the midpoint of the line segment joining the given points.
(a) $(1, 3)$ and $(7, 15)$
(b) $(-1, 6)$ and $(8, -12)$

56. Find the lengths of the medians of the triangle with vertices $A(1, 0)$, $B(3, 6)$, and $C(8, 2)$. (A median is a line segment from a vertex to the midpoint of the opposite side.)

57. Show that the lines $2x - y = 4$ and $6x - 2y = 10$ are not parallel and find their point of intersection.

58. Show that the lines $3x - 5y + 19 = 0$ and $10x + 6y - 50 = 0$ are perpendicular and find their point of intersection.

59. Find an equation of the perpendicular bisector of the line segment joining the points $A(1, 4)$ and $B(7, -2)$.

60. (a) Find equations for the sides of the triangle with vertices $P(1, 0)$, $Q(3, 4)$, and $R(-1, 6)$.
(b) Find equations for the medians of this triangle. Where do they intersect?

61. (a) Show that if the x- and y-intercepts of a line are nonzero numbers a and b, then the equation of the line can be put in the form

$$\frac{x}{a} + \frac{y}{b} = 1$$

This equation is called the **two-intercept form** of an equation of a line.
(b) Use part (a) to find an equation of the line whose x-intercept is 6 and whose y-intercept is -8.

62. A car leaves Detroit at 2:00 P.M., traveling at a constant speed west along I-96. It passes Ann Arbor, 40 mi from Detroit, at 2:50 P.M.
(a) Express the distance traveled in terms of the time elapsed.
(b) Draw the graph of the equation in part (a).
(c) What is the slope of this line? What does it represent?

C Graphs of Second-Degree Equations

In Appendix B we saw that a first-degree, or linear, equation $Ax + By + C = 0$ represents a line. In this section we discuss second-degree equations such as

$$x^2 + y^2 = 1 \qquad y = x^2 + 1 \qquad \frac{x^2}{9} + \frac{y^2}{4} = 1 \qquad x^2 - y^2 = 1$$

which represent a circle, a parabola, an ellipse, and a hyperbola, respectively.

The graph of such an equation in x and y is the set of all points (x, y) that satisfy the equation; it gives a visual representation of the equation. Conversely, given a curve in the xy-plane, we may have to find an equation that represents it, that is, an equation satisfied by the coordinates of the points on the curve and by no other point. This is the other half of the basic principle of analytic geometry as formulated by Descartes and Fermat. The idea is that if a geometric curve can be represented by an algebraic equation, then the rules of algebra can be used to analyze the geometric problem.

Circles

As an example of this type of problem, let's find an equation of the circle with radius r and center (h, k). By definition, the circle is the set of all points $P(x, y)$ whose distance from

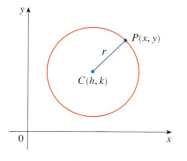

FIGURE 1

the center $C(h, k)$ is r. (See Figure 1.) Thus, P is on the circle if and only if $|PC| = r$. From the distance formula, we have

$$\sqrt{(x - h)^2 + (y - k)^2} = r$$

or equivalently, squaring both sides, we get

$$(x - h)^2 + (y - k)^2 = r^2$$

This is the desired equation.

> **1 Equation of a Circle** An equation of the circle with center (h, k) and radius r is
>
> $$(x - h)^2 + (y - k)^2 = r^2$$
>
> In particular, if the center is the origin $(0, 0)$, the equation is
>
> $$x^2 + y^2 = r^2$$

EXAMPLE 1 Find an equation of the circle with radius 3 and center $(2, -5)$.

SOLUTION From Equation 1 with $r = 3$, $h = 2$, and $k = -5$, we obtain

$$(x - 2)^2 + (y + 5)^2 = 9$$

EXAMPLE 2 Sketch the graph of the equation $x^2 + y^2 + 2x - 6y + 7 = 0$ by first showing that it represents a circle and then finding its center and radius.

SOLUTION We first group the x-terms and y-terms as follows:

$$(x^2 + 2x) + (y^2 - 6y) = -7$$

Then we complete the square within each grouping, adding the appropriate constants to both sides of the equation:

$$(x^2 + 2x + 1) + (y^2 - 6y + 9) = -7 + 1 + 9$$

or

$$(x + 1)^2 + (y - 3)^2 = 3$$

Comparing this equation with the standard equation of a circle (1), we see that $h = -1$, $k = 3$, and $r = \sqrt{3}$, so the given equation represents a circle with center $(-1, 3)$ and radius $\sqrt{3}$. It is sketched in Figure 2.

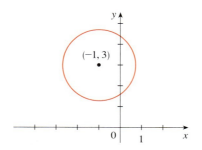

FIGURE 2
$x^2 + y^2 + 2x - 6y + 7 = 0$

Parabolas

The geometric properties of parabolas are reviewed in Section 11.5. Here we regard a parabola as a graph of an equation of the form $y = ax^2 + bx + c$.

EXAMPLE 3 Draw the graph of the parabola $y = x^2$.

SOLUTION We set up a table of values, plot points, and join them by a smooth curve to obtain the graph in Figure 3.

x	$y = x^2$
0	0
$\pm\frac{1}{2}$	$\frac{1}{4}$
± 1	1
± 2	4
± 3	9

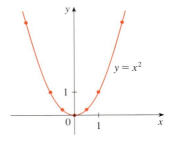

FIGURE 3

Figure 4 shows the graphs of several parabolas with equations of the form $y = ax^2$ for various values of the number a. In each case the *vertex*, the point where the parabola changes direction, is the origin. We see that the parabola $y = ax^2$ opens upward if $a > 0$ and downward if $a < 0$ (as in Figure 5).

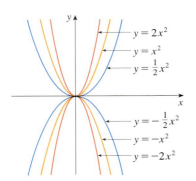

FIGURE 4

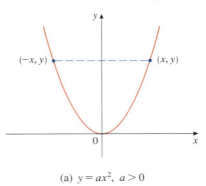

(a) $y = ax^2$, $a > 0$

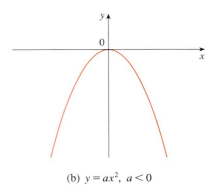

(b) $y = ax^2$, $a < 0$

FIGURE 5

Notice that if (x, y) satisfies $y = ax^2$, then so does $(-x, y)$. This corresponds to the geometric fact that if the right half of the graph is reflected about the y-axis, then the left half of the graph is obtained. We say that the graph is **symmetric with respect to the y-axis**.

> The graph of an equation is symmetric with respect to the y-axis if the equation is unchanged when x is replaced by $-x$.

If we interchange x and y in the equation $y = ax^2$, the result is $x = ay^2$, which also represents a parabola. (Interchanging x and y amounts to reflecting about the diagonal line $y = x$.) The parabola $x = ay^2$ opens to the right if $a > 0$ and to the left if $a < 0$. (See

Figure 6.) This time the parabola is symmetric with respect to the x-axis because if (x, y) satisfies $x = ay^2$, then so does $(x, -y)$.

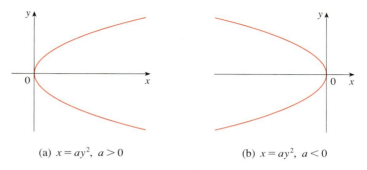

FIGURE 6

(a) $x = ay^2$, $a > 0$

(b) $x = ay^2$, $a < 0$

The graph of an equation is symmetric with respect to the x-axis if the equation is unchanged when y is replaced by $-y$.

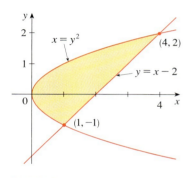

FIGURE 7

EXAMPLE 4 Sketch the region bounded by the parabola $x = y^2$ and the line $y = x - 2$.

SOLUTION First we find the points of intersection by solving the two equations. Substituting $x = y + 2$ into the equation $x = y^2$, we get $y + 2 = y^2$, which gives

$$0 = y^2 - y - 2 = (y - 2)(y + 1)$$

so $y = 2$ or -1. Thus, the points of intersection are $(4, 2)$ and $(1, -1)$, and we draw the line $y = x - 2$ passing through these points. We then sketch the parabola $x = y^2$ by referring to Figure 6(a) and having the parabola pass through $(4, 2)$ and $(1, -1)$. The region bounded by $x = y^2$ and $y = x - 2$ means the finite region whose boundaries are these curves. It is sketched in Figure 7.

Ellipses

The curve with equation

2
$$\frac{x^2}{a^2} + \frac{y^2}{b^2} = 1$$

where a and b are positive numbers, is called an **ellipse** in standard position. (Geometric properties of ellipses are discussed in Section 11.5.) Observe that Equation 2 is unchanged if x is replaced by $-x$ or y is replaced by $-y$, so the ellipse is symmetric with respect to both axes. As a further aid to sketching the ellipse, we find its intercepts.

The **x-intercepts** of a graph are the x-coordinates of the points where the graph intersects the x-axis. They are found by setting $y = 0$ in the equation of the graph.

The **y-intercepts** are the y-coordinates of the points where the graph intersects the y-axis. They are found by setting $x = 0$ in its equation.

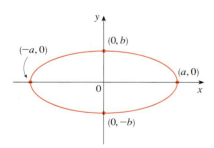

FIGURE 8
$\frac{x^2}{a^2} + \frac{y^2}{b^2} = 1$

If we set $y = 0$ in Equation 2, we get $x^2 = a^2$ and so the x-intercepts are $\pm a$. Setting $x = 0$, we get $y^2 = b^2$, so the y-intercepts are $\pm b$. Using this information, together with symmetry, we sketch the ellipse in Figure 8. If $a = b$, the ellipse is a circle with radius a.

EXAMPLE 5 Sketch the graph of $9x^2 + 16y^2 = 144$.

SOLUTION We divide both sides of the equation by 144:

$$\frac{x^2}{16} + \frac{y^2}{9} = 1$$

The equation is now in the standard form for an ellipse (2), so we have $a^2 = 16$, $b^2 = 9$, $a = 4$, and $b = 3$. The x-intercepts are ± 4; the y-intercepts are ± 3. The graph is sketched in Figure 9.

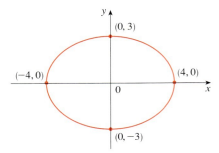

FIGURE 9
$9x^2 + 16y^2 = 144$

Hyperbolas

The curve with equation

$$\boxed{3} \qquad \frac{x^2}{a^2} - \frac{y^2}{b^2} = 1$$

is called a **hyperbola** in standard position. Again, Equation 3 is unchanged when x is replaced by $-x$ or y is replaced by $-y$, so the hyperbola is symmetric with respect to both axes. To find the x-intercepts we set $y = 0$ and obtain $x^2 = a^2$ and $x = \pm a$. However, if we put $x = 0$ in Equation 3, we get $y^2 = -b^2$, which is impossible, so there is no y-intercept. In fact, from Equation 3 we obtain

$$\frac{x^2}{a^2} = 1 + \frac{y^2}{b^2} \geq 1$$

which shows that $x^2 \geq a^2$ and so $|x| = \sqrt{x^2} \geq a$. Therefore, we have $x \geq a$ or $x \leq -a$. This means that the hyperbola consists of two parts, called its *branches*. It is sketched in Figure 10.

In drawing a hyperbola it is useful to draw first its *asymptotes*, which are the lines $y = (b/a)x$ and $y = -(b/a)x$ shown in Figure 10. Both branches of the hyperbola approach the asymptotes; that is, they come arbitrarily close to the asymptotes. This involves the idea of a limit, which is discussed in Chapter 2. (See also Exercise 55 in Section 4.5.)

By interchanging the roles of x and y we get an equation of the form

$$\frac{y^2}{a^2} - \frac{x^2}{b^2} = 1$$

which also represents a hyperbola and is sketched in Figure 11.

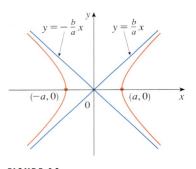

FIGURE 10
The hyperbola $\dfrac{x^2}{a^2} - \dfrac{y^2}{b^2} = 1$

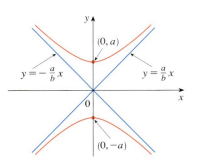

FIGURE 11
The hyperbola $\dfrac{y^2}{a^2} - \dfrac{x^2}{b^2} = 1$

EXAMPLE 6 Sketch the curve $9x^2 - 4y^2 = 36$.

SOLUTION Dividing both sides by 36, we obtain

$$\frac{x^2}{4} - \frac{y^2}{9} = 1$$

which is the standard form of the equation of a hyperbola (Equation 3). Since $a^2 = 4$, the x-intercepts are ± 2. Since $b^2 = 9$, we have $b = 3$ and the asymptotes are $y = \pm(\frac{3}{2})x$. The hyperbola is sketched in Figure 12.

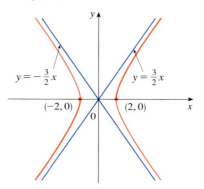

FIGURE 12
The hyperbola $9x^2 - 4y^2 = 36$

If $b = a$, a hyperbola has the equation $x^2 - y^2 = a^2$ (or $y^2 - x^2 = a^2$) and is called an *equilateral hyperbola* [see Figure 13(a)]. Its asymptotes are $y = \pm x$, which are perpendicular. If an equilateral hyperbola is rotated by 45°, the asymptotes become the x- and y-axes, and it can be shown that the new equation of the hyperbola is $xy = k$, where k is a constant [see Figure 13(b)].

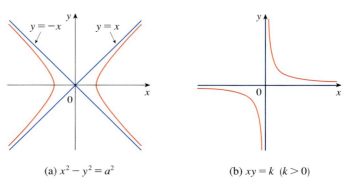

FIGURE 13
Equilateral hyperbolas

(a) $x^2 - y^2 = a^2$

(b) $xy = k$ $(k > 0)$

Shifted Conics

Recall that an equation of the circle with center the origin and radius r is $x^2 + y^2 = r^2$, but if the center is the point (h, k), then the equation of the circle becomes

$$(x - h)^2 + (y - k)^2 = r^2$$

Similarly, if we take the ellipse with equation

4
$$\frac{x^2}{a^2} + \frac{y^2}{b^2} = 1$$

and translate it (shift it) so that its center is the point (h, k), then its equation becomes

$$\boxed{\frac{(x - h)^2}{a^2} + \frac{(y - k)^2}{b^2} = 1}$$ ⑤

(See Figure 14.)

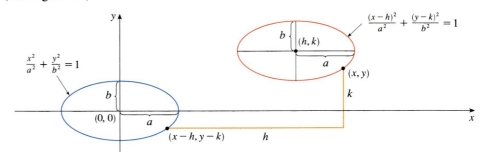

FIGURE 14

Notice that in shifting the ellipse, we replaced x by $x - h$ and y by $y - k$ in Equation 4 to obtain Equation 5. We use the same procedure to shift the parabola $y = ax^2$ so that its vertex (the origin) becomes the point (h, k) as in Figure 15. Replacing x by $x - h$ and y by $y - k$, we see that the new equation is

$$y - k = a(x - h)^2 \quad \text{or} \quad y = a(x - h)^2 + k$$

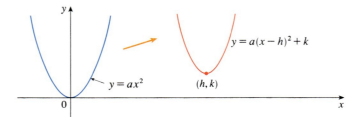

FIGURE 15

EXAMPLE 7 Sketch the graph of the equation $y = 2x^2 - 4x + 1$.

SOLUTION First we complete the square:

$$y = 2(x^2 - 2x) + 1 = 2(x - 1)^2 - 1$$

In this form we see that the equation represents the parabola obtained by shifting $y = 2x^2$ so that its vertex is at the point $(1, -1)$. The graph is sketched in Figure 16.

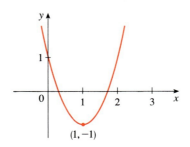

FIGURE 16
$y = 2x^2 - 4x + 1$

EXAMPLE 8 Sketch the curve $x = 1 - y^2$.

SOLUTION This time we start with the parabola $x = -y^2$ (as in Figure 6 with $a = -1$) and shift one unit to the right to get the graph of $x = 1 - y^2$. (See Figure 17.)

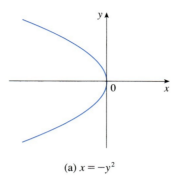

FIGURE 17

(a) $x = -y^2$

(b) $x = 1 - y^2$

C Exercises

1–4 Find an equation of a circle that satisfies the given conditions.

1. Center $(3, -1)$, radius 5

2. Center $(-2, -8)$, radius 10

3. Center at the origin, passes through $(4, 7)$

4. Center $(-1, 5)$, passes through $(-4, -6)$

5–9 Show that the equation represents a circle and find the center and radius.

5. $x^2 + y^2 - 4x + 10y + 13 = 0$

6. $x^2 + y^2 + 6y + 2 = 0$

7. $x^2 + y^2 + x = 0$

8. $16x^2 + 16y^2 + 8x + 32y + 1 = 0$

9. $2x^2 + 2y^2 - x + y = 1$

10. Under what condition on the coefficients a, b, and c does the equation $x^2 + y^2 + ax + by + c = 0$ represent a circle? When that condition is satisfied, find the center and radius of the circle.

11–32 Identify the type of curve and sketch the graph. Do not plot points. Just use the standard graphs given in Figures 5, 6, 8, 10, and 11 and shift if necessary.

11. $y = -x^2$

12. $y^2 - x^2 = 1$

13. $x^2 + 4y^2 = 16$

14. $x = -2y^2$

15. $16x^2 - 25y^2 = 400$

16. $25x^2 + 4y^2 = 100$

17. $4x^2 + y^2 = 1$

18. $y = x^2 + 2$

19. $x = y^2 - 1$

20. $9x^2 - 25y^2 = 225$

21. $9y^2 - x^2 = 9$

22. $2x^2 + 5y^2 = 10$

23. $xy = 4$

24. $y = x^2 + 2x$

25. $9(x - 1)^2 + 4(y - 2)^2 = 36$

26. $16x^2 + 9y^2 - 36y = 108$

27. $y = x^2 - 6x + 13$

28. $x^2 - y^2 - 4x + 3 = 0$

29. $x = 4 - y^2$

30. $y^2 - 2x + 6y + 5 = 0$

31. $x^2 + 4y^2 - 6x + 5 = 0$

32. $4x^2 + 9y^2 - 16x + 54y + 61 = 0$

33–34 Sketch the region bounded by the curves.

33. $y = 3x$, $y = x^2$

34. $y = 4 - x^2$, $x - 2y = 2$

35. Find an equation of the parabola with vertex $(1, -1)$ that passes through the points $(-1, 3)$ and $(3, 3)$.

36. Find an equation of the ellipse with center at the origin that passes through the points $\left(1, -10\sqrt{2}/3\right)$ and $\left(-2, 5\sqrt{5}/3\right)$.

37–40 Sketch the graph of the set.

37. $\{(x, y) \mid x^2 + y^2 \leq 1\}$

38. $\{(x, y) \mid x^2 + y^2 > 4\}$

39. $\{(x, y) \mid y \geq x^2 - 1\}$

40. $\{(x, y) \mid x^2 + 4y^2 \leq 4\}$

D Trigonometry

Angles

Angles can be measured in degrees or in radians (abbreviated as rad). The angle given by a complete revolution contains 360°, which is the same as 2π rad. Therefore

$$\boxed{\pi \text{ rad} = 180°} \quad \text{[1]}$$

and

$$\boxed{1 \text{ rad} = \left(\frac{180}{\pi}\right)° \approx 57.3° \qquad 1° = \frac{\pi}{180} \text{ rad} \approx 0.017 \text{ rad}} \quad \text{[2]}$$

EXAMPLE 1
(a) Find the radian measure of 60°. (b) Express $5\pi/4$ rad in degrees.

SOLUTION
(a) From Equation 1 or 2 we see that to convert from degrees to radians we multiply by $\pi/180$. Therefore

$$60° = 60\left(\frac{\pi}{180}\right) = \frac{\pi}{3} \text{ rad}$$

(b) To convert from radians to degrees we multiply by $180/\pi$. Thus

$$\frac{5\pi}{4} \text{ rad} = \frac{5\pi}{4}\left(\frac{180}{\pi}\right) = 225°$$

In calculus we use radians to measure angles except when otherwise indicated. The following table gives the correspondence between degree and radian measures of some common angles.

Degrees	0°	30°	45°	60°	90°	120°	135°	150°	180°	270°	360°
Radians	0	$\frac{\pi}{6}$	$\frac{\pi}{4}$	$\frac{\pi}{3}$	$\frac{\pi}{2}$	$\frac{2\pi}{3}$	$\frac{3\pi}{4}$	$\frac{5\pi}{6}$	π	$\frac{3\pi}{2}$	2π

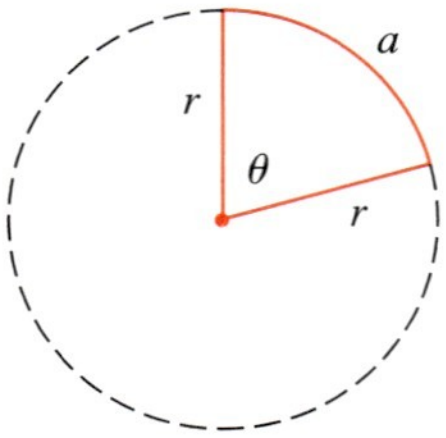

Figure 1 shows a sector of a circle with central angle θ and radius r subtending an arc with length a. Since the length of the arc is proportional to the size of the angle, and since the entire circle has circumference $2\pi r$ and central angle 2π, we have

$$\frac{\theta}{2\pi} = \frac{a}{2\pi r}$$

Solving this equation for θ and for a, we obtain

$$\boxed{\theta = \frac{a}{r}} \qquad \boxed{a = r\theta} \quad \text{[3]}$$

FIGURE 1

Remember that Equations 3 are valid only when θ is measured in radians.

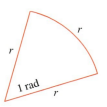

FIGURE 2

In particular, putting $a = r$ in Equation 3, we see that an angle of 1 rad is the angle subtended at the center of a circle by an arc equal in length to the radius of the circle (see Figure 2).

EXAMPLE 2
(a) If the radius of a circle is 5 cm, what angle is subtended by an arc of 6 cm?
(b) If a circle has radius 3 cm, what is the length of an arc subtended by a central angle of $3\pi/8$ rad?

SOLUTION
(a) Using Equation 3 with $a = 6$ and $r = 5$, we see that the angle is

$$\theta = \tfrac{6}{5} = 1.2 \text{ rad}$$

(b) With $r = 3$ cm and $\theta = 3\pi/8$ rad, the arc length is

$$a = r\theta = 3\left(\frac{3\pi}{8}\right) = \frac{9\pi}{8} \text{ cm}$$

The **standard position** of an angle occurs when we place its vertex at the origin of a coordinate system and its initial side on the positive x-axis as in Figure 3. A **positive** angle is obtained by rotating the initial side counterclockwise until it coincides with the terminal side. Likewise, **negative** angles are obtained by clockwise rotation as in Figure 4.

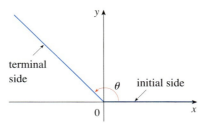

FIGURE 3 $\theta \geq 0$

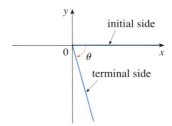

FIGURE 4 $\theta < 0$

Figure 5 shows several examples of angles in standard position. Notice that different angles can have the same terminal side. For instance, the angles $3\pi/4$, $-5\pi/4$, and $11\pi/4$ have the same initial and terminal sides because

$$\frac{3\pi}{4} - 2\pi = -\frac{5\pi}{4} \qquad \frac{3\pi}{4} + 2\pi = \frac{11\pi}{4}$$

and 2π rad represents a complete revolution.

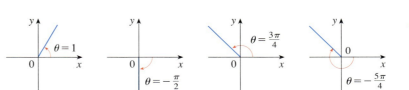

FIGURE 5
Angles in standard position

The Trigonometric Functions

For an acute angle θ the six trigonometric functions are defined as ratios of lengths of sides of a right triangle as follows (see Figure 6).

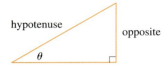

FIGURE 6

$$\boxed{4 \quad \begin{aligned} \sin\theta &= \frac{\text{opp}}{\text{hyp}} & \csc\theta &= \frac{\text{hyp}}{\text{opp}} \\ \cos\theta &= \frac{\text{adj}}{\text{hyp}} & \sec\theta &= \frac{\text{hyp}}{\text{adj}} \\ \tan\theta &= \frac{\text{opp}}{\text{adj}} & \cot\theta &= \frac{\text{adj}}{\text{opp}} \end{aligned}}$$

This definition doesn't apply to obtuse or negative angles, so for a general angle θ in standard position we let $P(x, y)$ be any point on the terminal side of θ and we let r be the distance $|OP|$ as in Figure 7. Then we define

FIGURE 7

$$\boxed{5 \quad \begin{aligned} \sin\theta &= \frac{y}{r} & \csc\theta &= \frac{r}{y} \\ \cos\theta &= \frac{x}{r} & \sec\theta &= \frac{r}{x} \\ \tan\theta &= \frac{y}{x} & \cot\theta &= \frac{x}{y} \end{aligned}}$$

Since division by 0 is not defined, $\tan\theta$ and $\sec\theta$ are undefined when $x = 0$ and $\csc\theta$ and $\cot\theta$ are undefined when $y = 0$. Notice that the definitions in (4) and (5) are consistent when θ is an acute angle.

If θ is a number, the convention is that $\sin\theta$ means the sine of the angle whose *radian* measure is θ. For example, the expression sin 3 implies that we are dealing with an angle of 3 rad. When finding a calculator approximation to this number we must remember to set our calculator in radian mode, and then we obtain

$$\sin 3 \approx 0.14112$$

If we want to know the sine of the angle 3° we would write sin 3° and, with our calculator in degree mode, we find that

$$\sin 3° \approx 0.05234$$

The exact trigonometric ratios for certain angles can be read from the triangles in Figure 8. For instance,

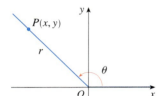

FIGURE 8

$$\sin\frac{\pi}{4} = \frac{1}{\sqrt{2}} \qquad \sin\frac{\pi}{6} = \frac{1}{2} \qquad \sin\frac{\pi}{3} = \frac{\sqrt{3}}{2}$$

$$\cos\frac{\pi}{4} = \frac{1}{\sqrt{2}} \qquad \cos\frac{\pi}{6} = \frac{\sqrt{3}}{2} \qquad \cos\frac{\pi}{3} = \frac{1}{2}$$

$$\tan\frac{\pi}{4} = 1 \qquad \tan\frac{\pi}{6} = \frac{1}{\sqrt{3}} \qquad \tan\frac{\pi}{3} = \sqrt{3}$$

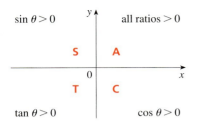

FIGURE 9

The signs of the trigonometric functions for angles in each of the four quadrants can be remembered by means of the rule "**A**ll **S**tudents **T**ake **C**alculus" shown in Figure 9.

EXAMPLE 3 Find the exact trigonometric ratios for $\theta = 2\pi/3$.

SOLUTION From Figure 10 we see that a point on the terminal line for $\theta = 2\pi/3$ is $P(-1, \sqrt{3})$. Therefore, taking

$$x = -1 \qquad y = \sqrt{3} \qquad r = 2$$

in the definitions of the trigonometric ratios, we have

$$\sin \frac{2\pi}{3} = \frac{\sqrt{3}}{2} \qquad \cos \frac{2\pi}{3} = -\frac{1}{2} \qquad \tan \frac{2\pi}{3} = -\sqrt{3}$$

$$\csc \frac{2\pi}{3} = \frac{2}{\sqrt{3}} \qquad \sec \frac{2\pi}{3} = -2 \qquad \cot \frac{2\pi}{3} = -\frac{1}{\sqrt{3}}$$

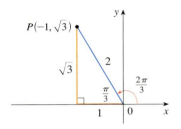

FIGURE 10

The following table gives some values of $\sin \theta$ and $\cos \theta$ found by the method of Example 3.

θ	0	$\frac{\pi}{6}$	$\frac{\pi}{4}$	$\frac{\pi}{3}$	$\frac{\pi}{2}$	$\frac{2\pi}{3}$	$\frac{3\pi}{4}$	$\frac{5\pi}{6}$	π	$\frac{3\pi}{2}$	2π
$\sin \theta$	0	$\frac{1}{2}$	$\frac{1}{\sqrt{2}}$	$\frac{\sqrt{3}}{2}$	1	$\frac{\sqrt{3}}{2}$	$\frac{1}{\sqrt{2}}$	$\frac{1}{2}$	0	-1	0
$\cos \theta$	1	$\frac{\sqrt{3}}{2}$	$\frac{1}{\sqrt{2}}$	$\frac{1}{2}$	0	$-\frac{1}{2}$	$-\frac{1}{\sqrt{2}}$	$-\frac{\sqrt{3}}{2}$	-1	0	1

EXAMPLE 4 If $\cos \theta = \frac{2}{5}$ and $0 < \theta < \pi/2$, find the other five trigonometric functions of θ.

SOLUTION Since $\cos \theta = \frac{2}{5}$, we can label the hypotenuse as having length 5 and the adjacent side as having length 2 in Figure 11. If the opposite side has length x, then the Pythagorean Theorem gives $x^2 + 4 = 25$ and so $x^2 = 21$, $x = \sqrt{21}$. We can now use the diagram to write the other five trigonometric functions:

$$\sin \theta = \frac{\sqrt{21}}{5} \qquad \tan \theta = \frac{\sqrt{21}}{2}$$

$$\csc \theta = \frac{5}{\sqrt{21}} \qquad \sec \theta = \frac{5}{2} \qquad \cot \theta = \frac{2}{\sqrt{21}}$$

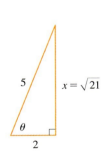

FIGURE 11

EXAMPLE 5 Use a calculator to approximate the value of x in Figure 12.

SOLUTION From the diagram we see that

$$\tan 40° = \frac{16}{x}$$

Therefore
$$x = \frac{16}{\tan 40°} \approx 19.07$$

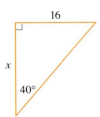

FIGURE 12

Trigonometric Identities

A trigonometric identity is a relationship among the trigonometric functions. The most elementary are the following, which are immediate consequences of the definitions of the trigonometric functions.

$$\boxed{6 \qquad \csc\theta = \frac{1}{\sin\theta} \qquad \sec\theta = \frac{1}{\cos\theta} \qquad \cot\theta = \frac{1}{\tan\theta}}$$

$$\tan\theta = \frac{\sin\theta}{\cos\theta} \qquad \cot\theta = \frac{\cos\theta}{\sin\theta}$$

For the next identity we refer back to Figure 7. The distance formula (or, equivalently, the Pythagorean Theorem) tells us that $x^2 + y^2 = r^2$. Therefore

$$\sin^2\theta + \cos^2\theta = \frac{y^2}{r^2} + \frac{x^2}{r^2} = \frac{x^2 + y^2}{r^2} = \frac{r^2}{r^2} = 1$$

We have therefore proved one of the most useful of all trigonometric identities:

$$\boxed{7 \qquad \sin^2\theta + \cos^2\theta = 1}$$

If we now divide both sides of Equation 7 by $\cos^2\theta$ and use Equations 6, we get

$$\boxed{8 \qquad \tan^2\theta + 1 = \sec^2\theta}$$

Similarly, if we divide both sides of Equation 7 by $\sin^2\theta$, we get

$$\boxed{9 \qquad 1 + \cot^2\theta = \csc^2\theta}$$

The identities

$$\boxed{\begin{aligned}10a \qquad & \sin(-\theta) = -\sin\theta \\ 10b \qquad & \cos(-\theta) = \cos\theta\end{aligned}}$$

|||| Odd functions and even functions are discussed in Section 1.1.

show that sin is an odd function and cos is an even function. They are easily proved by drawing a diagram showing θ and $-\theta$ in standard position (see Exercise 39).

Since the angles θ and $\theta + 2\pi$ have the same terminal side, we have

$$\boxed{11 \qquad \sin(\theta + 2\pi) = \sin\theta \qquad \cos(\theta + 2\pi) = \cos\theta}$$

These identities show that the sine and cosine functions are periodic with period 2π.

The remaining trigonometric identities are all consequences of two basic identities called the **addition formulas**:

12a
$$\sin(x + y) = \sin x \cos y + \cos x \sin y$$

12b
$$\cos(x + y) = \cos x \cos y - \sin x \sin y$$

The proofs of these addition formulas are outlined in Exercises 85, 86, and 87.

By substituting $-y$ for y in Equations 12a and 12b and using Equations 10a and 10b, we obtain the following **subtraction formulas**:

13a
$$\sin(x - y) = \sin x \cos y - \cos x \sin y$$

13b
$$\cos(x - y) = \cos x \cos y + \sin x \sin y$$

Then, by dividing the formulas in Equations 12 or Equations 13, we obtain the corresponding formulas for $\tan(x \pm y)$:

14a
$$\tan(x + y) = \frac{\tan x + \tan y}{1 - \tan x \tan y}$$

14b
$$\tan(x - y) = \frac{\tan x - \tan y}{1 + \tan x \tan y}$$

If we put $y = x$ in the addition formulas (12), we get the **double-angle formulas**:

15a
$$\sin 2x = 2 \sin x \cos x$$

15b
$$\cos 2x = \cos^2 x - \sin^2 x$$

Then, by using the identity $\sin^2 x + \cos^2 x = 1$, we obtain the following alternate forms of the double-angle formulas for $\cos 2x$:

16a
$$\cos 2x = 2 \cos^2 x - 1$$

16b
$$\cos 2x = 1 - 2 \sin^2 x$$

If we now solve these equations for $\cos^2 x$ and $\sin^2 x$, we get the following **half-angle formulas**, which are useful in integral calculus:

17a
$$\cos^2 x = \frac{1 + \cos 2x}{2}$$

17b
$$\sin^2 x = \frac{1 - \cos 2x}{2}$$

Finally, we state the **product formulas**, which can be deduced from Equations 12 and 13:

18a	$\sin x \cos y = \frac{1}{2}[\sin(x + y) + \sin(x - y)]$
18b	$\cos x \cos y = \frac{1}{2}[\cos(x + y) + \cos(x - y)]$
18c	$\sin x \sin y = \frac{1}{2}[\cos(x - y) - \cos(x + y)]$

There are many other trigonometric identities, but those we have stated are the ones used most often in calculus. If you forget any of them, remember that they can all be deduced from Equations 12a and 12b.

EXAMPLE 6 Find all values of x in the interval $[0, 2\pi]$ such that $\sin x = \sin 2x$.

SOLUTION Using the double-angle formula (15a), we rewrite the given equation as

$$\sin x = 2 \sin x \cos x \quad \text{or} \quad \sin x(1 - 2\cos x) = 0$$

Therefore, there are two possibilities:

$$\sin x = 0 \quad \text{or} \quad 1 - 2\cos x = 0$$
$$x = 0, \pi, 2\pi \quad\quad\quad \cos x = \tfrac{1}{2}$$
$$\quad\quad\quad\quad\quad\quad\quad\quad x = \frac{\pi}{3}, \frac{5\pi}{3}$$

The given equation has five solutions: $0, \pi/3, \pi, 5\pi/3$, and 2π.

Graphs of Trigonometric Functions

The graph of the function $f(x) = \sin x$, shown in Figure 13(a), is obtained by plotting points for $0 \le x \le 2\pi$ and then using the periodic nature of the function (from Equation 11) to complete the graph. Notice that the zeros of the sine function occur at the

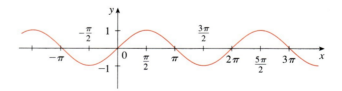

(a) $f(x) = \sin x$

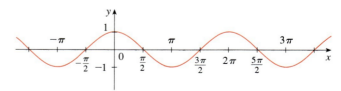

(b) $g(x) = \cos x$

FIGURE 13

integer multiples of π, that is,

$$\sin x = 0 \quad \text{whenever } x = n\pi, \quad n \text{ an integer}$$

Because of the identity

$$\cos x = \sin\left(x + \frac{\pi}{2}\right)$$

(which can be verified using Equation 12a), the graph of cosine is obtained by shifting the graph of sine by an amount $\pi/2$ to the left [see Figure 13(b)]. Note that for both the sine and cosine functions the domain is $(-\infty, \infty)$ and the range is the closed interval $[-1, 1]$. Thus, for all values of x, we have

$$-1 \leq \sin x \leq 1 \qquad -1 \leq \cos x \leq 1$$

The graphs of the remaining four trigonometric functions are shown in Figure 14 and their domains are indicated there. Notice that tangent and cotangent have range $(-\infty, \infty)$, whereas cosecant and secant have range $(-\infty, -1] \cup [1, \infty)$. All four functions are periodic: tangent and cotangent have period π, whereas cosecant and secant have period 2π.

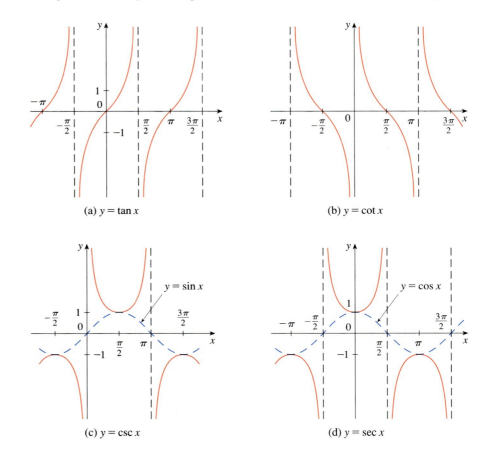

FIGURE 14

(a) $y = \tan x$

(b) $y = \cot x$

(c) $y = \csc x$

(d) $y = \sec x$

D Exercises

1–6 Convert from degrees to radians.

1. 210°
2. 300°
3. 9°
4. −315°
5. 900°
6. 36°

7–12 Convert from radians to degrees.

7. 4π
8. $-\dfrac{7\pi}{2}$
9. $\dfrac{5\pi}{12}$
10. $\dfrac{8\pi}{3}$
11. $-\dfrac{3\pi}{8}$
12. 5

13. Find the length of a circular arc subtended by an angle of $\pi/12$ rad if the radius of the circle is 36 cm.

14. If a circle has radius 10 cm, find the length of the arc subtended by a central angle of 72°.

15. A circle has radius 1.5 m. What angle is subtended at the center of the circle by an arc 1 m long?

16. Find the radius of a circular sector with angle $3\pi/4$ and arc length 6 cm.

17–22 Draw, in standard position, the angle whose measure is given.

17. 315°
18. −150°
19. $-\dfrac{3\pi}{4}$ rad
20. $\dfrac{7\pi}{3}$ rad
21. 2 rad
22. −3 rad

23–28 Find the exact trigonometric ratios for the angle whose radian measure is given.

23. $\dfrac{3\pi}{4}$
24. $\dfrac{4\pi}{3}$
25. $\dfrac{9\pi}{2}$
26. -5π
27. $\dfrac{5\pi}{6}$
28. $\dfrac{11\pi}{4}$

29–34 Find the remaining trigonometric ratios.

29. $\sin\theta = \dfrac{3}{5}$, $0 < \theta < \dfrac{\pi}{2}$
30. $\tan\alpha = 2$, $0 < \alpha < \dfrac{\pi}{2}$
31. $\sec\phi = -1.5$, $\dfrac{\pi}{2} < \phi < \pi$
32. $\cos x = -\dfrac{1}{3}$, $\pi < x < \dfrac{3\pi}{2}$
33. $\cot\beta = 3$, $\pi < \beta < 2\pi$
34. $\csc\theta = -\dfrac{4}{3}$, $\dfrac{3\pi}{2} < \theta < 2\pi$

35–38 Find, correct to five decimal places, the length of the side labeled x.

35.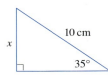
36.
37.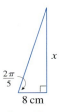
38.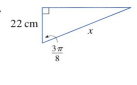

39–41 Prove each equation.

39. (a) Equation 10a (b) Equation 10b
40. (a) Equation 14a (b) Equation 14b
41. (a) Equation 18a (b) Equation 18b
    (c) Equation 18c

42–58 Prove the identity.

42. $\cos\left(\dfrac{\pi}{2} - x\right) = \sin x$
43. $\sin\left(\dfrac{\pi}{2} + x\right) = \cos x$
44. $\sin(\pi - x) = \sin x$
45. $\sin\theta\cot\theta = \cos\theta$
46. $(\sin x + \cos x)^2 = 1 + \sin 2x$
47. $\sec y - \cos y = \tan y\sin y$
48. $\tan^2\alpha - \sin^2\alpha = \tan^2\alpha\sin^2\alpha$
49. $\cot^2\theta + \sec^2\theta = \tan^2\theta + \csc^2\theta$
50. $2\csc 2t = \sec t\csc t$
51. $\tan 2\theta = \dfrac{2\tan\theta}{1 - \tan^2\theta}$
52. $\dfrac{1}{1 - \sin\theta} + \dfrac{1}{1 + \sin\theta} = 2\sec^2\theta$
53. $\sin x\sin 2x + \cos x\cos 2x = \cos x$
54. $\sin^2 x - \sin^2 y = \sin(x + y)\sin(x - y)$
55. $\dfrac{\sin\phi}{1 - \cos\phi} = \csc\phi + \cot\phi$

56. $\tan x + \tan y = \dfrac{\sin(x+y)}{\cos x \cos y}$

57. $\sin 3\theta + \sin \theta = 2 \sin 2\theta \cos \theta$

58. $\cos 3\theta = 4\cos^3\theta - 3\cos\theta$

59–64 ▪ If $\sin x = \tfrac{1}{3}$ and $\sec y = \tfrac{5}{4}$, where x and y lie between 0 and $\pi/2$, evaluate the expression.

59. $\sin(x+y)$ **60.** $\cos(x+y)$

61. $\cos(x-y)$ **62.** $\sin(x-y)$

63. $\sin 2y$ **64.** $\cos 2y$

65–72 ▪ Find all values of x in the interval $[0, 2\pi]$ that satisfy the equation.

65. $2\cos x - 1 = 0$ **66.** $3\cot^2 x = 1$

67. $2\sin^2 x = 1$ **68.** $|\tan x| = 1$

69. $\sin 2x = \cos x$ **70.** $2\cos x + \sin 2x = 0$

71. $\sin x = \tan x$ **72.** $2 + \cos 2x = 3\cos x$

73–76 ▪ Find all values of x in the interval $[0, 2\pi]$ that satisfy the inequality.

73. $\sin x \leq \tfrac{1}{2}$ **74.** $2\cos x + 1 > 0$

75. $-1 < \tan x < 1$ **76.** $\sin x > \cos x$

77–82 ▪ Graph the function by starting with the graphs in Figures 13 and 14 and applying the transformations of Section 1.3 where appropriate.

77. $y = \cos\left(x - \dfrac{\pi}{3}\right)$ **78.** $y = \tan 2x$

79. $y = \dfrac{1}{3}\tan\left(x - \dfrac{\pi}{2}\right)$ **80.** $y = 1 + \sec x$

81. $y = |\sin x|$ **82.** $y = 2 + \sin\left(x + \dfrac{\pi}{4}\right)$

83. Prove the **Law of Cosines**: If a triangle has sides with lengths a, b, and c, and θ is the angle between the sides with lengths a and b, then
$$c^2 = a^2 + b^2 - 2ab\cos\theta$$

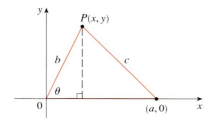

[*Hint:* Introduce a coordinate system so that θ is in standard position as in the figure. Express x and y in terms of θ and then use the distance formula to compute c.]

84. In order to find the distance $|AB|$ across a small inlet, a point C is located as in the figure and the following measurements were recorded:
$$\angle C = 103° \qquad |AC| = 820 \text{ m} \qquad |BC| = 910 \text{ m}$$

Use the Law of Cosines from Exercise 83 to find the required distance.

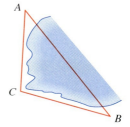

85. Use the figure to prove the subtraction formula
$$\cos(\alpha - \beta) = \cos\alpha\cos\beta + \sin\alpha\sin\beta$$

[*Hint:* Compute c^2 in two ways (using the Law of Cosines from Exercise 83 and also using the distance formula) and compare the two expressions.]

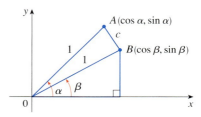

86. Use the formula in Exercise 85 to prove the addition formula for cosine (12b).

87. Use the addition formula for cosine and the identities
$$\cos\left(\dfrac{\pi}{2} - \theta\right) = \sin\theta \qquad \sin\left(\dfrac{\pi}{2} - \theta\right) = \cos\theta$$
to prove the subtraction formula for the sine function.

88. Show that the area of a triangle with sides of lengths a and b and with included angle θ is
$$A = \tfrac{1}{2}ab\sin\theta$$

89. Find the area of triangle ABC, correct to five decimal places, if
$$|AB| = 10 \text{ cm} \qquad |BC| = 3 \text{ cm} \qquad \angle ABC = 107°$$

E Sigma Notation

A convenient way of writing sums uses the Greek letter Σ (capital sigma, corresponding to our letter S) and is called **sigma notation**.

> **1** **Definition** If $a_m, a_{m+1}, \ldots, a_n$ are real numbers and m and n are integers such that $m \leq n$, then
> $$\sum_{i=m}^{n} a_i = a_m + a_{m+1} + a_{m+2} + \cdots + a_{n-1} + a_n$$

This tells us to end with $i = n$.
This tells us to add. $\sum_{i=m}^{n} a_i$
This tells us to start with $i = m$.

With function notation, Definition 1 can be written as

$$\sum_{i=m}^{n} f(i) = f(m) + f(m+1) + f(m+2) + \cdots + f(n-1) + f(n)$$

Thus, the symbol $\sum_{i=m}^{n}$ indicates a summation in which the letter i (called the **index of summation**) takes on consecutive integer values beginning with m and ending with n, that is, $m, m+1, \ldots, n$. Other letters can also be used as the index of summation.

EXAMPLE 1

(a) $\sum_{i=1}^{4} i^2 = 1^2 + 2^2 + 3^2 + 4^2 = 30$

(b) $\sum_{i=3}^{n} i = 3 + 4 + 5 + \cdots + (n-1) + n$

(c) $\sum_{j=0}^{5} 2^j = 2^0 + 2^1 + 2^2 + 2^3 + 2^4 + 2^5 = 63$

(d) $\sum_{k=1}^{n} \frac{1}{k} = 1 + \frac{1}{2} + \frac{1}{3} + \cdots + \frac{1}{n}$

(e) $\sum_{i=1}^{3} \frac{i-1}{i^2+3} = \frac{1-1}{1^2+3} + \frac{2-1}{2^2+3} + \frac{3-1}{3^2+3} = 0 + \frac{1}{7} + \frac{1}{6} = \frac{13}{42}$

(f) $\sum_{i=1}^{4} 2 = 2 + 2 + 2 + 2 = 8$

EXAMPLE 2 Write the sum $2^3 + 3^3 + \cdots + n^3$ in sigma notation.

SOLUTION There is no unique way of writing a sum in sigma notation. We could write

$$2^3 + 3^3 + \cdots + n^3 = \sum_{i=2}^{n} i^3$$

or

$$2^3 + 3^3 + \cdots + n^3 = \sum_{j=1}^{n-1} (j+1)^3$$

or

$$2^3 + 3^3 + \cdots + n^3 = \sum_{k=0}^{n-2} (k+2)^3$$

The following theorem gives three simple rules for working with sigma notation.

2 Theorem If c is any constant (that is, it does not depend on i), then

(a) $\displaystyle\sum_{i=m}^{n} ca_i = c \sum_{i=m}^{n} a_i$ 　　(b) $\displaystyle\sum_{i=m}^{n} (a_i + b_i) = \sum_{i=m}^{n} a_i + \sum_{i=m}^{n} b_i$

(c) $\displaystyle\sum_{i=m}^{n} (a_i - b_i) = \sum_{i=m}^{n} a_i - \sum_{i=m}^{n} b_i$

Proof To see why these rules are true, all we have to do is write both sides in expanded form. Rule (a) is just the distributive property of real numbers:

$$ca_m + ca_{m+1} + \cdots + ca_n = c(a_m + a_{m+1} + \cdots + a_n)$$

Rule (b) follows from the associative and commutative properties:

$$(a_m + b_m) + (a_{m+1} + b_{m+1}) + \cdots + (a_n + b_n)$$
$$= (a_m + a_{m+1} + \cdots + a_n) + (b_m + b_{m+1} + \cdots + b_n)$$

Rule (c) is proved similarly.

EXAMPLE 3 Find $\displaystyle\sum_{i=1}^{n} 1$.

SOLUTION
$$\sum_{i=1}^{n} 1 = \underbrace{1 + 1 + \cdots + 1}_{n \text{ terms}} = n$$

EXAMPLE 4 Prove the formula for the sum of the first n positive integers:

$$\sum_{i=1}^{n} i = 1 + 2 + 3 + \cdots + n = \frac{n(n+1)}{2}$$

SOLUTION This formula can be proved by mathematical induction (see page 59) or by the following method used by the German mathematician Karl Friedrich Gauss (1777–1855) when he was ten years old.

Write the sum S twice, once in the usual order and once in reverse order:

$$S = 1 + 2 + 3 + \cdots + (n-1) + n$$
$$S = n + (n-1) + (n-2) + \cdots + 2 + 1$$

Adding all columns vertically, we get

$$2S = (n+1) + (n+1) + (n+1) + \cdots + (n+1) + (n+1)$$

On the right side there are n terms, each of which is $n+1$, so

$$2S = n(n+1) \quad \text{or} \quad S = \frac{n(n+1)}{2}$$

EXAMPLE 5 Prove the formula for the sum of the squares of the first n positive integers:

$$\sum_{i=1}^{n} i^2 = 1^2 + 2^2 + 3^2 + \cdots + n^2 = \frac{n(n+1)(2n+1)}{6}$$

SOLUTION 1 Let S be the desired sum. We start with the *telescoping sum* (or collapsing sum):

Most terms cancel in pairs.

$$\sum_{i=1}^{n} [(1+i)^3 - i^3] = (2^3 - 1^3) + (3^3 - 2^3) + (4^3 - 3^3) + \cdots + [(n+1)^3 - n^3]$$

$$= (n+1)^3 - 1^3 = n^3 + 3n^2 + 3n$$

On the other hand, using Theorem 2 and Examples 3 and 4, we have

$$\sum_{i=1}^{n} [(1+i)^3 - i^3] = \sum_{i=1}^{n} [3i^2 + 3i + 1] = 3\sum_{i=1}^{n} i^2 + 3\sum_{i=1}^{n} i + \sum_{i=1}^{n} 1$$

$$= 3S + 3\frac{n(n+1)}{2} + n = 3S + \tfrac{3}{2}n^2 + \tfrac{3}{2}n$$

Thus, we have

$$n^3 + 3n^2 + 3n = 3S + \tfrac{3}{2}n^2 + \tfrac{5}{2}n$$

Solving this equation for S, we obtain

$$3S = n^3 + \tfrac{3}{2}n^2 + \tfrac{1}{2}n$$

or

$$S = \frac{2n^3 + 3n^2 + n}{6} = \frac{n(n+1)(2n+1)}{6}$$

IIII PRINCIPLE OF MATHEMATICAL INDUCTION
Let S_n be a statement involving the positive integer n. Suppose that
1. S_1 is true.
2. If S_k is true, then S_{k+1} is true.

Then S_n is true for all positive integers n.

IIII See pages 59 and 61 for a more thorough discussion of mathematical induction.

SOLUTION 2 Let S_n be the given formula.

1. S_1 is true because
$$1^2 = \frac{1(1+1)(2 \cdot 1 + 1)}{6}$$

2. Assume that S_k is true; that is,
$$1^2 + 2^2 + 3^2 + \cdots + k^2 = \frac{k(k+1)(2k+1)}{6}$$

Then
$$1^2 + 2^2 + 3^2 + \cdots + (k+1)^2 = (1^2 + 2^2 + 3^2 + \cdots + k^2) + (k+1)^2$$

$$= \frac{k(k+1)(2k+1)}{6} + (k+1)^2$$

$$= (k+1)\frac{k(2k+1) + 6(k+1)}{6}$$

$$= (k+1)\frac{2k^2 + 7k + 6}{6}$$

$$= \frac{(k+1)(k+2)(2k+3)}{6}$$

$$= \frac{(k+1)[(k+1)+1][2(k+1)+1]}{6}$$

So S_{k+1} is true.

By the Principle of Mathematical Induction, S_n is true for all n.

We list the results of Examples 3, 4, and 5 together with a similar result for cubes (see Exercises 37–40) as Theorem 3. These formulas are needed for finding areas and evaluating integrals in Chapter 5.

3 Theorem Let c be a constant and n a positive integer. Then

(a) $\sum_{i=1}^{n} 1 = n$

(b) $\sum_{i=1}^{n} c = nc$

(c) $\sum_{i=1}^{n} i = \dfrac{n(n+1)}{2}$

(d) $\sum_{i=1}^{n} i^2 = \dfrac{n(n+1)(2n+1)}{6}$

(e) $\sum_{i=1}^{n} i^3 = \left[\dfrac{n(n+1)}{2}\right]^2$

EXAMPLE 6 Evaluate $\sum_{i=1}^{n} i(4i^2 - 3)$.

SOLUTION Using Theorems 2 and 3, we have

$$\sum_{i=1}^{n} i(4i^2 - 3) = \sum_{i=1}^{n} (4i^3 - 3i) = 4\sum_{i=1}^{n} i^3 - 3\sum_{i=1}^{n} i$$

$$= 4\left[\frac{n(n+1)}{2}\right]^2 - 3\frac{n(n+1)}{2}$$

$$= \frac{n(n+1)[2n(n+1) - 3]}{2}$$

$$= \frac{n(n+1)(2n^2 + 2n - 3)}{2}$$

|||| The type of calculation in Example 7 arises in Chapter 5 when we compute areas.

EXAMPLE 7 Find $\displaystyle\lim_{n\to\infty} \sum_{i=1}^{n} \frac{3}{n}\left[\left(\frac{i}{n}\right)^2 + 1\right]$.

SOLUTION

$$\lim_{n\to\infty} \sum_{i=1}^{n} \frac{3}{n}\left[\left(\frac{i}{n}\right)^2 + 1\right] = \lim_{n\to\infty} \sum_{i=1}^{n} \left[\frac{3}{n^3} i^2 + \frac{3}{n}\right]$$

$$= \lim_{n\to\infty} \left[\frac{3}{n^3} \sum_{i=1}^{n} i^2 + \frac{3}{n} \sum_{i=1}^{n} 1\right]$$

$$= \lim_{n\to\infty} \left[\frac{3}{n^3} \frac{n(n+1)(2n+1)}{6} + \frac{3}{n} \cdot n\right]$$

$$= \lim_{n\to\infty} \left[\frac{1}{2} \cdot \frac{n}{n} \cdot \left(\frac{n+1}{n}\right)\left(\frac{2n+1}{n}\right) + 3\right]$$

$$= \lim_{n\to\infty} \left[\frac{1}{2} \cdot 1\left(1 + \frac{1}{n}\right)\left(2 + \frac{1}{n}\right) + 3\right]$$

$$= \tfrac{1}{2} \cdot 1 \cdot 1 \cdot 2 + 3 = 4$$

E Exercises

1–10 ■ Write the sum in expanded form.

1. $\sum_{i=1}^{5} \sqrt{i}$

2. $\sum_{i=1}^{6} \frac{1}{i+1}$

3. $\sum_{i=4}^{6} 3^i$

4. $\sum_{i=4}^{6} i^3$

5. $\sum_{k=0}^{4} \frac{2k-1}{2k+1}$

6. $\sum_{k=5}^{8} x^k$

7. $\sum_{i=1}^{n} i^{10}$

8. $\sum_{j=n}^{n+3} j^2$

9. $\sum_{j=0}^{n-1} (-1)^j$

10. $\sum_{i=1}^{n} f(x_i)\, \Delta x_i$

11–20 ■ Write the sum in sigma notation.

11. $1 + 2 + 3 + 4 + \cdots + 10$

12. $\sqrt{3} + \sqrt{4} + \sqrt{5} + \sqrt{6} + \sqrt{7}$

13. $\frac{1}{2} + \frac{2}{3} + \frac{3}{4} + \frac{4}{5} + \cdots + \frac{19}{20}$

14. $\frac{3}{7} + \frac{4}{8} + \frac{5}{9} + \frac{6}{10} + \cdots + \frac{23}{27}$

15. $2 + 4 + 6 + 8 + \cdots + 2n$

16. $1 + 3 + 5 + 7 + \cdots + (2n - 1)$

17. $1 + 2 + 4 + 8 + 16 + 32$

18. $\frac{1}{1} + \frac{1}{4} + \frac{1}{9} + \frac{1}{16} + \frac{1}{25} + \frac{1}{36}$

19. $x + x^2 + x^3 + \cdots + x^n$

20. $1 - x + x^2 - x^3 + \cdots + (-1)^n x^n$

21–35 ■ Find the value of the sum.

21. $\sum_{i=4}^{8} (3i - 2)$

22. $\sum_{i=3}^{6} i(i + 2)$

23. $\sum_{j=1}^{6} 3^{j+1}$

24. $\sum_{k=0}^{8} \cos k\pi$

25. $\sum_{n=1}^{20} (-1)^n$

26. $\sum_{i=1}^{100} 4$

27. $\sum_{i=0}^{4} (2^i + i^2)$

28. $\sum_{i=-2}^{4} 2^{3-i}$

29. $\sum_{i=1}^{n} 2i$

30. $\sum_{i=1}^{n} (2 - 5i)$

31. $\sum_{i=1}^{n} (i^2 + 3i + 4)$

32. $\sum_{i=1}^{n} (3 + 2i)^2$

33. $\sum_{i=1}^{n} (i + 1)(i + 2)$

34. $\sum_{i=1}^{n} i(i + 1)(i + 2)$

35. $\sum_{i=1}^{n} (i^3 - i - 2)$

36. Find the number n such that $\sum_{i=1}^{n} i = 78$.

37. Prove formula (b) of Theorem 3.

38. Prove formula (e) of Theorem 3 using mathematical induction.

39. Prove formula (e) of Theorem 3 using a method similar to that of Example 5, Solution 1 [start with $(1 + i)^4 - i^4$].

40. Prove formula (e) of Theorem 3 using the following method published by Abu Bekr Mohammed ibn Alhusain Alkarchi in about A.D. 1010. The figure shows a square $ABCD$ in which sides AB and AD have been divided into segments of lengths 1, 2, 3, $\ldots$, n. Thus, the side of the square has length $n(n + 1)/2$ so the area is $[n(n + 1)/2]^2$. But the area is also the sum of the areas of the n "gnomons" $G_1, G_2, \ldots, G_n$ shown in the figure. Show that the area of G_i is i^3 and conclude that formula (e) is true.

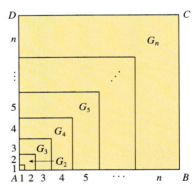

41. Evaluate each telescoping sum.

(a) $\sum_{i=1}^{n} [i^4 - (i - 1)^4]$

(b) $\sum_{i=1}^{100} (5^i - 5^{i-1})$

(c) $\sum_{i=3}^{99} \left(\frac{1}{i} - \frac{1}{i+1} \right)$

(d) $\sum_{i=1}^{n} (a_i - a_{i-1})$

42. Prove the generalized triangle inequality

$$\left| \sum_{i=1}^{n} a_i \right| \leq \sum_{i=1}^{n} |a_i|$$

43–46 ■ Find each limit.

43. $\lim_{n \to \infty} \sum_{i=1}^{n} \frac{1}{n} \left(\frac{i}{n} \right)^2$

44. $\lim_{n \to \infty} \sum_{i=1}^{n} \frac{1}{n} \left[\left(\frac{i}{n} \right)^3 + 1 \right]$

45. $\lim_{n \to \infty} \sum_{i=1}^{n} \frac{2}{n} \left[\left(\frac{2i}{n} \right)^3 + 5 \left(\frac{2i}{n} \right) \right]$

46. $\lim_{n \to \infty} \sum_{i=1}^{n} \frac{3}{n} \left[\left(1 + \frac{3i}{n}\right)^3 - 2\left(1 + \frac{3i}{n}\right) \right]$

47. Prove the formula for the sum of a finite geometric series with first term a and common ratio $r \neq 1$:

$$\sum_{i=1}^{n} ar^{i-1} = a + ar + ar^2 + \cdots + ar^{n-1} = \frac{a(r^n - 1)}{r - 1}$$

48. Evaluate $\sum_{i=1}^{n} \frac{3}{2^{i-1}}$.

49. Evaluate $\sum_{i=1}^{n} (2i + 2^i)$.

50. Evaluate $\sum_{i=1}^{m} \left[\sum_{j=1}^{n} (i + j) \right]$.

F Proofs of Theorems

In this appendix we present proofs of two theorems that are stated in the main body of the text. The sections in which they occur are indicated in the margin.

SECTION 7.1

6 Theorem If f is a one-to-one continuous function defined on an interval (a, b), then its inverse function f^{-1} is also continuous.

Proof First we show that if f is both one-to-one and continuous on (a, b), then it must be either increasing or decreasing on (a, b). If it were neither increasing nor decreasing, then there would exist numbers x_1, x_2, and x_3 in (a, b) with $x_1 < x_2 < x_3$ such that $f(x_2)$ does not lie between $f(x_1)$ and $f(x_3)$. There are two possibilities: either (1) $f(x_3)$ lies between $f(x_1)$ and $f(x_2)$ or (2) $f(x_1)$ lies between $f(x_2)$ and $f(x_3)$. (Draw a picture.) In case (1) we apply the Intermediate Value Theorem to the continuous function f to get a number c between x_1 and x_2 such that $f(c) = f(x_3)$. In case (2) the Intermediate Value Theorem gives a number c between x_2 and x_3 such that $f(c) = f(x_1)$. In either case we have contradicted the fact that f is one-to-one.

Let us assume, for the sake of definiteness, that f is increasing on (a, b). We take any number y_0 in the domain of f^{-1} and we let $f^{-1}(y_0) = x_0$; that is, x_0 is the number in (a, b) such that $f(x_0) = y_0$. To show that f^{-1} is continuous at y_0 we take any $\varepsilon > 0$ such that the interval $(x_0 - \varepsilon, x_0 + \varepsilon)$ is contained in the interval (a, b). Since f is increasing, it maps the numbers in the interval $(x_0 - \varepsilon, x_0 + \varepsilon)$ onto the numbers in the interval $(f(x_0 - \varepsilon), f(x_0 + \varepsilon))$ and f^{-1} reverses the correspondence. If we let δ denote the smaller of the numbers $\delta_1 = y_0 - f(x_0 - \varepsilon)$ and $\delta_2 = f(x_0 + \varepsilon) - y_0$, then the interval $(y_0 - \delta, y_0 + \delta)$ is contained in the interval $(f(x_0 - \varepsilon), f(x_0 + \varepsilon))$ and so is mapped into the interval $(x_0 - \varepsilon, x_0 + \varepsilon)$ by f^{-1}. (See the arrow diagram in Figure 3.) We have therefore found a number $\delta > 0$ such that

$$|f^{-1}(y) - f^{-1}(y_0)| < \varepsilon \quad \text{whenever} \quad |y - y_0| < \delta$$

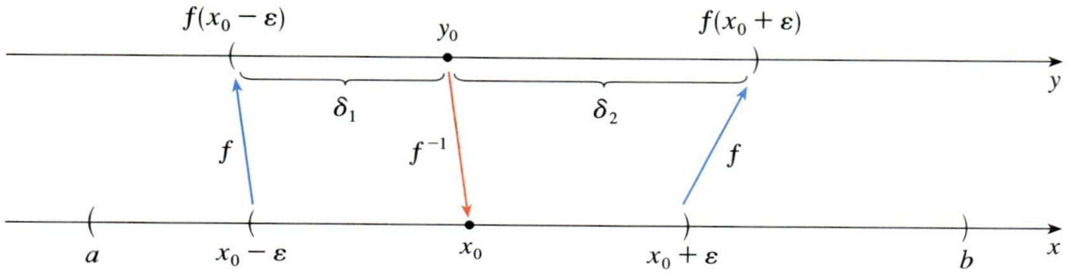

FIGURE 3

This shows that $\lim_{y \to y_0} f^{-1}(y) = f^{-1}(y_0)$ and so f^{-1} is continuous at any number y_0 in its domain.

APPENDIX F PROOFS OF THEOREMS

SECTION 12.8

In order to prove Theorem 12.8.3 we first need the following results.

> **Theorem**
> 1. If a power series $\sum c_n x^n$ converges when $x = b$ (where $b \neq 0$), then it converges whenever $|x| < |b|$.
> 2. If a power series $\sum c_n x^n$ diverges when $x = d$ (where $d \neq 0$), then it diverges whenever $|x| > |d|$.

Proof of 1 Suppose that $\sum c_n b^n$ converges. Then, by Theorem 12.2.6, we have $\lim_{n \to \infty} c_n b^n = 0$. According to Definition 12.1.2 with $\varepsilon = 1$, there is a positive integer N such that $|c_n b^n| < 1$ whenever $n \geq N$. Thus, for $n \geq N$, we have

$$|c_n x^n| = \left|\frac{c_n b^n x^n}{b^n}\right| = |c_n b^n| \left|\frac{x}{b}\right|^n < \left|\frac{x}{b}\right|^n$$

If $|x| < |b|$, then $|x/b| < 1$, so $\sum |x/b|^n$ is a convergent geometric series. Therefore, by the Comparison Test, the series $\sum_{n=N}^{\infty} |c_n x^n|$ is convergent. Thus, the series $\sum c_n x^n$ is absolutely convergent and therefore convergent.

Proof of 2 Suppose that $\sum c_n d^n$ diverges. If x is any number such that $|x| > |d|$, then $\sum c_n x^n$ cannot converge because, by part 1, the convergence of $\sum c_n x^n$ would imply the convergence of $\sum c_n d^n$. Therefore, $\sum c_n x^n$ diverges whenever $|x| > |d|$.

> **Theorem** For a power series $\sum c_n x^n$ there are only three possibilities:
> 1. The series converges only when $x = 0$.
> 2. The series converges for all x.
> 3. There is a positive number R such that the series converges if $|x| < R$ and diverges if $|x| > R$.

Proof Suppose that neither case 1 nor case 2 is true. Then there are nonzero numbers b and d such that $\sum c_n x^n$ converges for $x = b$ and diverges for $x = d$. Therefore, the set $S = \{x \mid \sum c_n x^n \text{ converges}\}$ is not empty. By the preceding theorem, the series diverges if $|x| > |d|$, so $|x| \leq |d|$ for all $x \in S$. This says that $|d|$ is an upper bound for the set S. Thus, by the Completeness Axiom (see Section 12.1), S has a least upper bound R. If $|x| > R$, then $x \notin S$, so $\sum c_n x^n$ diverges. If $|x| < R$, then $|x|$ is not an upper bound for S and so there exists $b \in S$ such that $b > |x|$. Since $b \in S$, $\sum c_n b^n$ converges, so by the preceding theorem $\sum c_n x^n$ converges.

> **3 Theorem** For a power series $\sum c_n (x - a)^n$ there are only three possibilities:
> 1. The series converges only when $x = a$.
> 2. The series converges for all x.
> 3. There is a positive number R such that the series converges if $|x - a| < R$ and diverges if $|x - a| > R$.

Proof If we make the change of variable $u = x - a$, then the power series becomes $\sum c_n u^n$ and we can apply the preceding theorem to this series. In case 3 we have convergence for $|u| < R$ and divergence for $|u| > R$. Thus, we have convergence for $|x - a| < R$ and divergence for $|x - a| > R$.

G Complex Numbers

A **complex number** can be represented by an expression of the form $a + bi$, where a and b are real numbers and i is a symbol with the property that $i^2 = -1$. The complex number $a + bi$ can also be represented by the ordered pair (a, b) and plotted as a point in a plane (called the Argand plane) as in Figure 1. Thus, the complex number $i = 0 + 1 \cdot i$ is identified with the point $(0, 1)$.

The **real part** of the complex number $a + bi$ is the real number a and the **imaginary part** is the real number b. Thus, the real part of $4 - 3i$ is 4 and the imaginary part is -3. Two complex numbers $a + bi$ and $c + di$ are **equal** if $a = c$ and $b = d$, that is, their real parts are equal and their imaginary parts are equal. In the Argand plane the horizontal axis is called the real axis and the vertical axis is called the imaginary axis.

The sum and difference of two complex numbers are defined by adding or subtracting their real parts and their imaginary parts:

$$(a + bi) + (c + di) = (a + c) + (b + d)i$$
$$(a + bi) - (c + di) = (a - c) + (b - d)i$$

For instance,

$$(1 - i) + (4 + 7i) = (1 + 4) + (-1 + 7)i = 5 + 6i$$

The product of complex numbers is defined so that the usual commutative and distributive laws hold:

$$(a + bi)(c + di) = a(c + di) + (bi)(c + di)$$
$$= ac + adi + bci + bdi^2$$

Since $i^2 = -1$, this becomes

$$(a + bi)(c + di) = (ac - bd) + (ad + bc)i$$

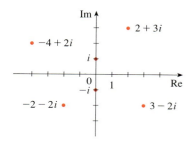

FIGURE 1
Complex numbers as points in the Argand plane

EXAMPLE 1

$$(-1 + 3i)(2 - 5i) = (-1)(2 - 5i) + 3i(2 - 5i)$$
$$= -2 + 5i + 6i - 15(-1) = 13 + 11i$$

Division of complex numbers is much like rationalizing the denominator of a rational expression. For the complex number $z = a + bi$, we define its **complex conjugate** to be $\bar{z} = a - bi$. To find the quotient of two complex numbers we multiply numerator and denominator by the complex conjugate of the denominator.

EXAMPLE 2 Express the number $\dfrac{-1 + 3i}{2 + 5i}$ in the form $a + bi$.

SOLUTION We multiply numerator and denominator by the complex conjugate of $2 + 5i$, namely $2 - 5i$, and we take advantage of the result of Example 1:

$$\frac{-1 + 3i}{2 + 5i} = \frac{-1 + 3i}{2 + 5i} \cdot \frac{2 - 5i}{2 - 5i} = \frac{13 + 11i}{2^2 + 5^2} = \frac{13}{29} + \frac{11}{29}i$$

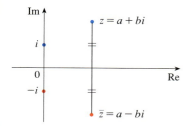

FIGURE 2

The geometric interpretation of the complex conjugate is shown in Figure 2: $\bar{z}$ is the reflection of z in the real axis. We list some of the properties of the complex conjugate in the following box. The proofs follow from the definition and are requested in Exercise 18.

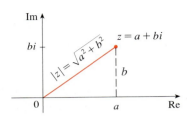

FIGURE 3

Properties of Conjugates

$$\overline{z+w} = \overline{z} + \overline{w} \qquad \overline{zw} = \overline{z}\,\overline{w} \qquad \overline{z^n} = \overline{z}^n$$

The **modulus**, or **absolute value**, $|z|$ of a complex number $z = a + bi$ is its distance from the origin. From Figure 3 we see that if $z = a + bi$, then

$$|z| = \sqrt{a^2 + b^2}$$

Notice that

$$z\bar{z} = (a + bi)(a - bi) = a^2 + abi - abi - b^2i^2 = a^2 + b^2$$

and so

$$z\bar{z} = |z|^2$$

This explains why the division procedure in Example 2 works in general:

$$\frac{z}{w} = \frac{z\overline{w}}{w\overline{w}} = \frac{z\overline{w}}{|w|^2}$$

Since $i^2 = -1$, we can think of i as a square root of -1. But notice that we also have $(-i)^2 = i^2 = -1$ and so $-i$ is also a square root of -1. We say that i is the **principal square root** of -1 and write $\sqrt{-1} = i$. In general, if c is any positive number, we write

$$\sqrt{-c} = \sqrt{c}\,i$$

With this convention, the usual derivation and formula for the roots of the quadratic equation $ax^2 + bx + c = 0$ are valid even when $b^2 - 4ac < 0$:

$$x = \frac{-b \pm \sqrt{b^2 - 4ac}}{2a}$$

EXAMPLE 3 Find the roots of the equation $x^2 + x + 1 = 0$.

SOLUTION Using the quadratic formula, we have

$$x = \frac{-1 \pm \sqrt{1^2 - 4 \cdot 1}}{2} = \frac{-1 \pm \sqrt{-3}}{2} = \frac{-1 \pm \sqrt{3}\,i}{2}$$

We observe that the solutions of the equation in Example 3 are complex conjugates of each other. In general, the solutions of any quadratic equation $ax^2 + bx + c = 0$ with real coefficients a, b, and c are always complex conjugates. (If z is real, $\bar{z} = z$, so z is its own conjugate.)

We have seen that if we allow complex numbers as solutions, then every quadratic equation has a solution. More generally, it is true that every polynomial equation

$$a_n x^n + a_{n-1} x^{n-1} + \cdots + a_1 x + a_0 = 0$$

of degree at least one has a solution among the complex numbers. This fact is known as the Fundamental Theorem of Algebra and was proved by Gauss.

Polar Form

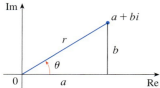

FIGURE 4

We know that any complex number $z = a + bi$ can be considered as a point (a, b) and that any such point can be represented by polar coordinates (r, θ) with $r \geq 0$. In fact,

$$a = r \cos \theta \qquad b = r \sin \theta$$

as in Figure 4. Therefore, we have

$$z = a + bi = (r \cos \theta) + (r \sin \theta)i$$

Thus, we can write any complex number z in the form

$$z = r(\cos \theta + i \sin \theta)$$

where $\qquad r = |z| = \sqrt{a^2 + b^2} \qquad$ and $\qquad \tan \theta = \dfrac{b}{a}$

The angle θ is called the **argument** of z and we write $\theta = \arg(z)$. Note that $\arg(z)$ is not unique; any two arguments of z differ by an integer multiple of 2π.

EXAMPLE 4 Write the following numbers in polar form.

(a) $z = 1 + i$
(b) $w = \sqrt{3} - i$

SOLUTION
(a) We have $r = |z| = \sqrt{1^2 + 1^2} = \sqrt{2}$ and $\tan \theta = 1$, so we can take $\theta = \pi/4$. Therefore, the polar form is

$$z = \sqrt{2}\left(\cos \frac{\pi}{4} + i \sin \frac{\pi}{4}\right)$$

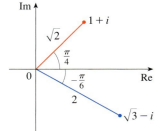

FIGURE 5

(b) Here we have $r = |w| = \sqrt{3 + 1} = 2$ and $\tan \theta = -1/\sqrt{3}$. Since w lies in the fourth quadrant, we take $\theta = -\pi/6$ and

$$w = 2\left[\cos\left(-\frac{\pi}{6}\right) + i \sin\left(-\frac{\pi}{6}\right)\right]$$

The numbers z and w are shown in Figure 5.

The polar form of complex numbers gives insight into multiplication and division. Let

$$z_1 = r_1(\cos \theta_1 + i \sin \theta_1) \qquad z_2 = r_2(\cos \theta_2 + i \sin \theta_2)$$

be two complex numbers written in polar form. Then

$$z_1 z_2 = r_1 r_2 (\cos \theta_1 + i \sin \theta_1)(\cos \theta_2 + i \sin \theta_2)$$
$$= r_1 r_2 [(\cos \theta_1 \cos \theta_2 - \sin \theta_1 \sin \theta_2) + i(\sin \theta_1 \cos \theta_2 + \cos \theta_1 \sin \theta_2)]$$

A44 |||| APPENDIX G COMPLEX NUMBERS

Therefore, using the addition formulas for cosine and sine, we have

$$\boxed{1} \qquad z_1 z_2 = r_1 r_2 [\cos(\theta_1 + \theta_2) + i \sin(\theta_1 + \theta_2)]$$

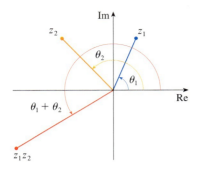

FIGURE 6

This formula says that *to multiply two complex numbers we multiply the moduli and add the arguments.* (See Figure 6.)

A similar argument using the subtraction formulas for sine and cosine shows that *to divide two complex numbers we divide the moduli and subtract the arguments.*

$$\frac{z_1}{z_2} = \frac{r_1}{r_2} [\cos(\theta_1 - \theta_2) + i \sin(\theta_1 - \theta_2)] \qquad z_2 \neq 0$$

In particular, taking $z_1 = 1$ and $z_2 = z$ (and therefore $\theta_1 = 0$ and $\theta_2 = \theta$), we have the following, which is illustrated in Figure 7.

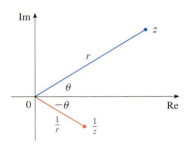

FIGURE 7

$$\text{If} \quad z = r(\cos \theta + i \sin \theta), \quad \text{then} \quad \frac{1}{z} = \frac{1}{r} (\cos \theta - i \sin \theta).$$

EXAMPLE 5 Find the product of the complex numbers $1 + i$ and $\sqrt{3} - i$ in polar form.

SOLUTION From Example 4 we have

$$1 + i = \sqrt{2} \left(\cos \frac{\pi}{4} + i \sin \frac{\pi}{4} \right)$$

and

$$\sqrt{3} - i = 2 \left[\cos\left(-\frac{\pi}{6}\right) + i \sin\left(-\frac{\pi}{6}\right) \right]$$

So, by Equation 1,

$$(1 + i)(\sqrt{3} - i) = 2\sqrt{2} \left[\cos\left(\frac{\pi}{4} - \frac{\pi}{6}\right) + i \sin\left(\frac{\pi}{4} - \frac{\pi}{6}\right) \right]$$

$$= 2\sqrt{2} \left(\cos \frac{\pi}{12} + i \sin \frac{\pi}{12} \right)$$

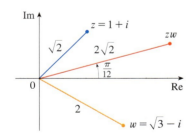

FIGURE 8

This is illustrated in Figure 8.

Repeated use of Formula 1 shows how to compute powers of a complex number. If

$$z = r(\cos \theta + i \sin \theta)$$

then

$$z^2 = r^2(\cos 2\theta + i \sin 2\theta)$$

and

$$z^3 = zz^2 = r^3(\cos 3\theta + i \sin 3\theta)$$

In general, we obtain the following result, which is named after the French mathematician Abraham De Moivre (1667–1754).

2 De Moivre's Theorem If $z = r(\cos\theta + i\sin\theta)$ and n is a positive integer, then
$$z^n = [r(\cos\theta + i\sin\theta)]^n = r^n(\cos n\theta + i\sin n\theta)$$

This says that *to take the nth power of a complex number we take the nth power of the modulus and multiply the argument by n.*

EXAMPLE 6 Find $\left(\frac{1}{2} + \frac{1}{2}i\right)^{10}$.

SOLUTION Since $\frac{1}{2} + \frac{1}{2}i = \frac{1}{2}(1 + i)$, it follows from Example 4(a) that $\frac{1}{2} + \frac{1}{2}i$ has the polar form
$$\frac{1}{2} + \frac{1}{2}i = \frac{\sqrt{2}}{2}\left(\cos\frac{\pi}{4} + i\sin\frac{\pi}{4}\right)$$

So by De Moivre's Theorem,
$$\left(\frac{1}{2} + \frac{1}{2}i\right)^{10} = \left(\frac{\sqrt{2}}{2}\right)^{10}\left(\cos\frac{10\pi}{4} + i\sin\frac{10\pi}{4}\right)$$
$$= \frac{2^5}{2^{10}}\left(\cos\frac{5\pi}{2} + i\sin\frac{5\pi}{2}\right) = \frac{1}{32}i$$

De Moivre's Theorem can also be used to find the nth roots of complex numbers. An nth root of the complex number z is a complex number w such that
$$w^n = z$$

Writing these two numbers in trigonometric form as
$$w = s(\cos\phi + i\sin\phi) \quad \text{and} \quad z = r(\cos\theta + i\sin\theta)$$

and using De Moivre's Theorem, we get
$$s^n(\cos n\phi + i\sin n\phi) = r(\cos\theta + i\sin\theta)$$

The equality of these two complex numbers shows that
$$s^n = r \quad \text{or} \quad s = r^{1/n}$$
and
$$\cos n\phi = \cos\theta \quad \text{and} \quad \sin n\phi = \sin\theta$$

From the fact that sine and cosine have period 2π it follows that
$$n\phi = \theta + 2k\pi \quad \text{or} \quad \phi = \frac{\theta + 2k\pi}{n}$$

Thus
$$w = r^{1/n}\left[\cos\left(\frac{\theta + 2k\pi}{n}\right) + i\sin\left(\frac{\theta + 2k\pi}{n}\right)\right]$$

Since this expression gives a different value of w for $k = 0, 1, 2, \ldots, n - 1$, we have the following.

3 Roots of a Complex Number Let $z = r(\cos\theta + i\sin\theta)$ and let n be a positive integer. Then z has the n distinct nth roots

$$w_k = r^{1/n}\left[\cos\left(\frac{\theta + 2k\pi}{n}\right) + i\sin\left(\frac{\theta + 2k\pi}{n}\right)\right]$$

where $k = 0, 1, 2, \ldots, n - 1$.

Notice that each of the nth roots of z has modulus $|w_k| = r^{1/n}$. Thus, all the nth roots of z lie on the circle of radius $r^{1/n}$ in the complex plane. Also, since the argument of each successive nth root exceeds the argument of the previous root by $2\pi/n$, we see that the nth roots of z are equally spaced on this circle.

EXAMPLE 7 Find the six sixth roots of $z = -8$ and graph these roots in the complex plane.

SOLUTION In trigonometric form, $z = 8(\cos\pi + i\sin\pi)$. Applying Equation 3 with $n = 6$, we get

$$w_k = 8^{1/6}\left(\cos\frac{\pi + 2k\pi}{6} + i\sin\frac{\pi + 2k\pi}{6}\right)$$

We get the six sixth roots of -8 by taking $k = 0, 1, 2, 3, 4, 5$ in this formula:

$$w_0 = 8^{1/6}\left(\cos\frac{\pi}{6} + i\sin\frac{\pi}{6}\right) = \sqrt{2}\left(\frac{\sqrt{3}}{2} + \frac{1}{2}i\right)$$

$$w_1 = 8^{1/6}\left(\cos\frac{\pi}{2} + i\sin\frac{\pi}{2}\right) = \sqrt{2}\,i$$

$$w_2 = 8^{1/6}\left(\cos\frac{5\pi}{6} + i\sin\frac{5\pi}{6}\right) = \sqrt{2}\left(-\frac{\sqrt{3}}{2} + \frac{1}{2}i\right)$$

$$w_3 = 8^{1/6}\left(\cos\frac{7\pi}{6} + i\sin\frac{7\pi}{6}\right) = \sqrt{2}\left(-\frac{\sqrt{3}}{2} - \frac{1}{2}i\right)$$

$$w_4 = 8^{1/6}\left(\cos\frac{3\pi}{2} + i\sin\frac{3\pi}{2}\right) = -\sqrt{2}\,i$$

$$w_5 = 8^{1/6}\left(\cos\frac{11\pi}{6} + i\sin\frac{11\pi}{6}\right) = \sqrt{2}\left(\frac{\sqrt{3}}{2} - \frac{1}{2}i\right)$$

All these points lie on the circle of radius $\sqrt{2}$ as shown in Figure 9.

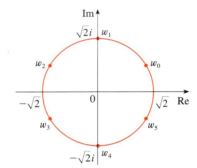

FIGURE 9
The six sixth roots of $z = -8$

Complex Exponentials

We also need to give a meaning to the expression e^z when $z = x + iy$ is a complex number. The theory of infinite series as developed in Chapter 12 can be extended to the case where the terms are complex numbers. Using the Taylor series for e^x (12.10.11) as our guide, we define

$$\boxed{4} \qquad e^z = \sum_{n=0}^{\infty} \frac{z^n}{n!} = 1 + z + \frac{z^2}{2!} + \frac{z^3}{3!} + \cdots$$

and it turns out that this complex exponential function has the same properties as the real exponential function. In particular, it is true that

$$\boxed{5} \qquad e^{z_1+z_2} = e^{z_1}e^{z_2}$$

If we put $z = iy$, where y is a real number, in Equation 4, and use the facts that

$$i^2 = -1, \quad i^3 = i^2 i = -i, \quad i^4 = 1, \quad i^5 = i, \quad \ldots$$

we get
$$e^{iy} = 1 + iy + \frac{(iy)^2}{2!} + \frac{(iy)^3}{3!} + \frac{(iy)^4}{4!} + \frac{(iy)^5}{5!} + \cdots$$

$$= 1 + iy - \frac{y^2}{2!} - i\frac{y^3}{3!} + \frac{y^4}{4!} + i\frac{y^5}{5!} + \cdots$$

$$= \left(1 - \frac{y^2}{2!} + \frac{y^4}{4!} - \frac{y^6}{6!} + \cdots\right) + i\left(y - \frac{y^3}{3!} + \frac{y^5}{5!} - \cdots\right)$$

$$= \cos y + i \sin y$$

Here we have used the Taylor series for $\cos y$ and $\sin y$ (Equations 12.10.16 and 12.10.15). The result is a famous formula called **Euler's formula**:

$$\boxed{6} \qquad \boxed{e^{iy} = \cos y + i \sin y}$$

Combining Euler's formula with Equation 5, we get

$$\boxed{7} \qquad e^{x+iy} = e^x e^{iy} = e^x(\cos y + i \sin y)$$

EXAMPLE 8 Evaluate: (a) $e^{i\pi}$ (b) $e^{-1+i\pi/2}$

SOLUTION

(a) From Euler's equation (6) we have $e^{i\pi} = \cos \pi + i \sin \pi = -1 + i(0) = -1$.

(b) Using Equation 7 we get

$$e^{-1+i\pi/2} = e^{-1}\left(\cos \frac{\pi}{2} + i \sin \frac{\pi}{2}\right) = \frac{1}{e}[0 + i(1)] = \frac{i}{e}$$

|||| We could write the result of Example 8(a) as

$$e^{i\pi} + 1 = 0$$

This equation relates the five most famous numbers in all of mathematics: 0, 1, e, i, and π.

Finally, we note that Euler's equation provides us with an easier method of proving De Moivre's Theorem:

$$[r(\cos \theta + i \sin \theta)]^n = (re^{i\theta})^n = r^n e^{in\theta} = r^n(\cos n\theta + i \sin n\theta)$$

G Exercises

1–14 |||| Evaluate the expression and write your answer in the form $a + bi$.

1. $(5 - 6i) + (3 + 2i)$
2. $(4 - \frac{1}{2}i) - (9 + \frac{5}{2}i)$
3. $(2 + 5i)(4 - i)$
4. $(1 - 2i)(8 - 3i)$
5. $\overline{12 + 7i}$
6. $\overline{2i(\frac{1}{2} - i)}$
7. $\dfrac{1 + 4i}{3 + 2i}$
8. $\dfrac{3 + 2i}{1 - 4i}$
9. $\dfrac{1}{1 + i}$
10. $\dfrac{3}{4 - 3i}$
11. i^3
12. i^{100}
13. $\sqrt{-25}$
14. $\sqrt{-3}\sqrt{-12}$

15–17 Find the complex conjugate and the modulus of the number.

15. $12 - 5i$ **16.** $-1 + 2\sqrt{2}\,i$ **17.** $-4i$

18. Prove the following properties of complex numbers.
(a) $\overline{z + w} = \overline{z} + \overline{w}$ (b) $\overline{zw} = \overline{z}\,\overline{w}$
(c) $\overline{z^n} = \overline{z}^n$, where n is a positive integer
[*Hint:* Write $z = a + bi$, $w = c + di$.]

19–24 Find all solutions of the equation.

19. $4x^2 + 9 = 0$ **20.** $x^4 = 1$
21. $x^2 + 2x + 5 = 0$ **22.** $2x^2 - 2x + 1 = 0$
23. $z^2 + z + 2 = 0$ **24.** $z^2 + \tfrac{1}{2}z + \tfrac{1}{4} = 0$

25–28 Write the number in polar form with argument between 0 and 2π.

25. $-3 + 3i$ **26.** $1 - \sqrt{3}\,i$
27. $3 + 4i$ **28.** $8i$

29–32 Find polar forms for zw, z/w, and $1/z$ by first putting z and w into polar form.

29. $z = \sqrt{3} + i$, $w = 1 + \sqrt{3}\,i$
30. $z = 4\sqrt{3} - 4i$, $w = 8i$
31. $z = 2\sqrt{3} - 2i$, $w = -1 + i$
32. $z = 4(\sqrt{3} + i)$, $w = -3 - 3i$

33–36 Find the indicated power using De Moivre's Theorem.

33. $(1 + i)^{20}$ **34.** $(1 - \sqrt{3}\,i)^5$
35. $(2\sqrt{3} + 2i)^5$ **36.** $(1 - i)^8$

37–40 Find the indicated roots. Sketch the roots in the complex plane.

37. The eighth roots of 1 **38.** The fifth roots of 32
39. The cube roots of i **40.** The cube roots of $1 + i$

41–46 Write the number in the form $a + bi$.

41. $e^{i\pi/2}$ **42.** $e^{2\pi i}$
43. $e^{i\pi/3}$ **44.** $e^{-i\pi}$
45. $e^{2+i\pi}$ **46.** $e^{\pi+i}$

47. If $u(x) = f(x) + ig(x)$ is a complex-valued function of a real variable x and the real and imaginary parts $f(x)$ and $g(x)$ are differentiable functions of x, then the derivative of u is defined to be $u'(x) = f'(x) + ig'(x)$. Use this together with Equation 7 to prove that if $F(x) = e^{rx}$, then $F'(x) = re^{rx}$ when $r = a + bi$ is a complex number.

48. (a) If u is a complex-valued function of a real variable, its indefinite integral $\int u(x)\, dx$ is an antiderivative of u. Evaluate
$$\int e^{(1+i)x}\, dx$$

(b) By considering the real and imaginary parts of the integral in part (a), evaluate the real integrals
$$\int e^x \cos x\, dx \quad \text{and} \quad \int e^x \sin x\, dx$$

(c) Compare with the method used in Example 4 in Section 8.1.

H Answers to Odd-Numbered Exercises

Chapter 5

Exercises 5.1 □ page 324

1. (a) 40, 52

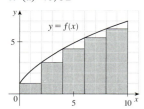 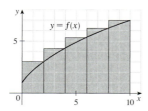

(b) 43.2, 49.2

3. (a) $\frac{77}{60}$, underestimate (b) $\frac{25}{12}$, overestimate

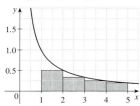 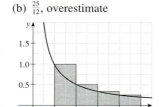

5. (a) 8, 6.875

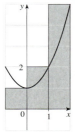 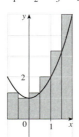

(b) 5, 5.375

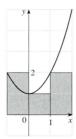 

(c) 5.75, 5.9375

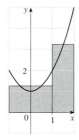 

(d) M_6

7. 1.9835, 1.9982, 1.9993; 2
9. (a) Left: 4.5148, 4.6165, 4.6366; right: 4.8148, 4.7165, 4.6966
11. 34.7 ft, 44.8 ft   **13.** 63.2 L, 70 L   **15.** 155 ft
17. $\lim\limits_{n \to \infty} \sum\limits_{i=1}^{n} \sqrt[4]{1 + 15i/n} \cdot (15/n)$
19. $\lim\limits_{n \to \infty} \sum\limits_{i=1}^{n} \left(\frac{i\pi}{2n} \cos \frac{i\pi}{2n} \right) \frac{\pi}{2n}$
21. The region under the graph of $y = \tan x$ from 0 to $\pi/4$
23. (a) $\lim\limits_{n \to \infty} \frac{64}{n^6} \sum\limits_{i=1}^{n} i^5$ (b) $\frac{n^2(n + 1)^2(2n^2 + 2n - 1)}{12}$ (c) $\frac{32}{3}$
25. $\sin b$, 1

Exercises 5.2 □ page 390

1. 0.25
The Riemann sum represents the sum of the areas of the two rectangles above the x-axis minus the sum of the areas of the two rectangles below the x-axis.

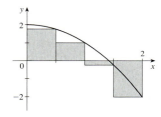

3. -0.856759
The Riemann sum represents the sum of the areas of the two rectangles above the x-axis minus the sum of the areas of the three rectangles below the x-axis.

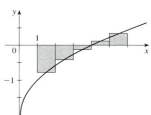

5. (a) 4 (b) 6 (c) 10   **7.** -475, -85 **9.** 124.1644
11. 0.3084   **13.** 0.30843908, 0.30981629, 0.31015563
15.

n	R_n
5	1.933766
10	1.983524
50	1.999342
100	1.999836

The values of R_n appear to be approaching 2.

17. $\int_0^\pi x \sin x \, dx$ **19.** $\int_1^8 \sqrt{2x + x^2} \, dx$ **21.** 42   **23.** $\frac{4}{3}$
25. 3.75   **29.** $\lim\limits_{n \to \infty} \sum\limits_{i=1}^{n} \frac{2 + 4i/n}{1 + (2 + 4i/n)^5} \cdot \frac{4}{n}$
31. $\lim\limits_{n \to \infty} \sum\limits_{i=1}^{n} \left(\sin \frac{5\pi i}{n} \right) \frac{\pi}{n} = \frac{2}{5}$
33. (a) 4 (b) 10 (c) -3 (d) 2   **35.** $-\frac{3}{4}$
37. $3 + \frac{9}{4}\pi$ **39.** 2.5   **41.** $-\frac{38}{3}$
43. 3   **45.** 22.5
47. $\int_{-1}^{5} f(x) \, dx$ **49.** 122   **55.** $\frac{1}{2} \leq \int_1^2 dx/x \leq 1$

57. $\frac{\pi}{12} \le \int_{\pi/4}^{\pi/3} \tan x \, dx \le \frac{\pi}{12}\sqrt{3}$

59. $2 \le \int_{-1}^{1} \sqrt{1 + x^4} \, dx \le 2\sqrt{2}$ **67.** $\int_0^1 x^4 \, dx$ **69.** $\frac{1}{2}$

Exercises 5.3 □ page 347

1. One process undoes what the other one does. See the Fundamental Theorem of Calculus, page 347.

3. (a) 0, 2, 5, 7, 3   (d)
(b) (0, 3)
(c) $x = 3$

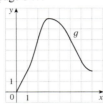

5. (a), (b) x^2

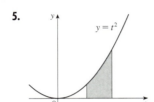

7. $g'(x) = \sqrt{1 + 2x}$ **9.** $g'(y) = y^2 \sin y$
11. $F'(x) = -\cos(x^2)$ **13.** $h'(x) = -\sin^4(1/x)/x^2$
15. $y' = \dfrac{\cos\sqrt{x}}{2x}$ **17.** $y' = \dfrac{3(1 - 3x)^3}{1 + (1 - 3x)^2}$ **19.** $\frac{364}{3}$
21. 138   **23.** $\frac{5}{9}$ **25.** $\frac{7}{8}$ **27.** Does not exist
29. $\frac{156}{7}$ **31.** 1   **33.** Does not exist   **35.** 10.7
37. $\frac{243}{4}$ **39.** 2   **41.** 3.75

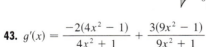

43. $g'(x) = \dfrac{-2(4x^2 - 1)}{4x^2 + 1} + \dfrac{3(9x^2 - 1)}{9x^2 + 1}$
45. $y' = 3x^{7/2}\sin(x^3) - (\sin\sqrt{x})/(2\sqrt[4]{x})$ **47.** $\sqrt{257}$
49. (a) $-2\sqrt{n}$, $\sqrt{4n - 2}$, n an integer > 0
(b) $(0, 1), (-\sqrt{4n - 1}, -\sqrt{4n - 3})$, and $(\sqrt{4n - 1}, \sqrt{4n + 1})$, n an integer > 0
(c) 0.74

51. (a) Loc. max. at 1 and 5; loc. min. at 3 and 7
(b) 9
(c) $(\frac{1}{2}, 2), (4, 6), (8, 9)$
(d) See graph at right.

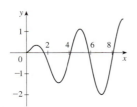

53. $\frac{1}{4}$ **59.** $f(x) = x^{3/2}$, $a = 9$

61. (b) Average expenditure over $[0, t]$; minimize average expenditure
63. ln 3   **65.** π **67.** $e^2 - 1$

Exercises 5.4 □ page 356

5. $4x^{1/4} + C$ **7.** $\frac{1}{4}x^4 + 3x^2 + x + C$
9. $2t - t^2 + \frac{1}{3}t^3 - \frac{1}{4}t^4 + C$
11. $4x - \frac{8}{3}x^{3/2} + \frac{1}{2}x^2 + C$ **13.** $\sec x + C$
15. $\frac{2}{5}x^{5/2} + C$

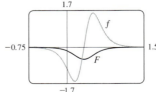

17. 18   **19.** 231   **21.** 52   **23.** $\frac{256}{15}$ **25.** $-\frac{63}{4}$
27. $\frac{55}{63}$ **29.** $2\sqrt{5}$ **31.** 8   **33.** $1 + \frac{1}{4}\pi$ **35.** $\frac{256}{5}$
37. 1   **39.** -3.5 **41.** 0, 1.32; 0.84   **43.** $\frac{4}{3}$
45. The increase in the child's weight (in pounds) between the ages of 5 and 10
47. Number of gallons of oil leaked in the first 2 hours
49. Increase in revenue when production is increased from 1000 to 5000 units
51. Newton-meters (or joules)   **53.** (a) $-\frac{3}{2}$ m   (b) $\frac{41}{6}$ m
55. (a) $v(t) = \frac{1}{2}t^2 + 4t + 5$ m/s   (b) $416\frac{2}{3}$ m
57. $46\frac{2}{3}$ kg   **59.** 1.4 mi   **61.** $58,000
63. (b) At most 40%; $\frac{5}{36}$ **65.** $\frac{1}{2}e^2 + e - \frac{1}{2}$
67. $\frac{1}{3}x^3 + x + \tan^{-1}x + C$

Exercises 5.5 □ page 365

1. $\frac{1}{3}\sin 3x + C$ **3.** $\frac{2}{9}(x^3 + 1)^{3/2} + C$
5. $-1/(1 + 2x)^2 + C$ **7.** $\frac{1}{5}(x^2 + 3)^5 + C$
9. $\frac{1}{63}(3x - 2)^{21} + C$ **11.** $2\sqrt{1 + x + 2x^2} + C$
13. $\dfrac{-3}{8(2y + 1)^4} + C$ **15.** $-\frac{2}{3}(4 - t)^{3/2} + C$
17. $-(1/\pi)\cos \pi t + C$ **19.** $2 \sin \sqrt{t} + C$
21. $\frac{1}{7}\sin^7\theta + C$ **23.** $\frac{1}{2}(1 + z^3)^{2/3} + C$
25. $-\frac{2}{3}(\cot x)^{3/2} + C$ **27.** $\frac{1}{3}\sec^3 x + C$
29. $\dfrac{2(b + cx^{a+1})^{3/2}}{3c(a + 1)} + C$
31. $\frac{4}{7}(x + 2)^{7/4} - \frac{8}{3}(x + 2)^{3/4} + C$
33. $\dfrac{-1}{6(3x^2 - 2x + 1)^3} + C$ **35.** $\frac{1}{4}\sin^4 x + C$

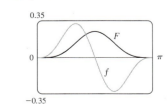

37. 0 **39.** $\frac{182}{9}$ **41.** 4 **43.** 0 **45.** 1 **47.** 3
49. $\frac{16}{15}$ **51.** Does not exist **53.** $\frac{1}{3}(2\sqrt{2}-1)a^3$
55. $\sqrt{3}-\frac{1}{3}$ **57.** 6π **59.** $\frac{5}{4\pi}\left(1-\cos\frac{2\pi t}{5}\right)$ L
61. 5 **67.** $-\frac{1}{3}\ln|5-3x|+C$ **69.** $\frac{1}{3}(\ln x)^3+C$
71. $\frac{2}{3}(1+e^x)^{3/2}+C$ **73.** $\ln|\ln x|+C$
75. $\ln|\sin x|+C$ **77.** $\tan^{-1}x+\frac{1}{2}\ln(1+x^2)+C$
79. $e-\sqrt{e}$ **81.** 2 **83.** $\pi^2/4$

Chapter 5 Review □ page 368

True-False Quiz

1. True **3.** True **5.** False **7.** True **9.** True
11. False **13.** False

Exercises

1. (a) 8 (b) 5.7

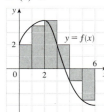
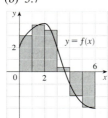

3. $\frac{1}{2}+\pi/4$ **5.** 3 **7.** $f=c, f'=b, \int_0^x f(t)\,dt=a$
9. 37 **11.** $\frac{9}{10}$ **13.** -76 **15.** $\frac{21}{4}$ **17.** Does not exist
19. $\frac{1}{3}\sin 1$ **21.** $\sqrt{x^2+4x}+C$ **23.** $[1/(2\pi)]\sin^2\pi t + C$
25. $\frac{1}{2}\sqrt{2}-\frac{1}{2}$ **27.** $\frac{23}{3}$ **29.** $2\sqrt{1+\sin x}+C$ **31.** $\frac{64}{5}$
33. $F'(x)=\sqrt{1+x^4}$ **35.** $g'(x)=3x^5/\sqrt{1+x^9}$
37. $y'=(2\cos x - \cos\sqrt{x})/(2x)$
39. $4 \le \int_1^3 \sqrt{x^2+3}\,dx \le 4\sqrt{3}$ **43.** 1.11
45. Number of barrels of oil consumed from Jan. 1, 2000, through Jan. 1, 2003
47. 72,400 **49.** $F(x)=\int_1^x t^2\sin(t^2)\,dt$
51. $(1+x^2)(x\cos x + \sin x)/x^2$

Problems Plus □ page 373

1. $\pi/2$ **5.** Does not exist **7.** -1 **9.** $[-1, 2]$
11. (a) $(n-1)n/2$
 (b) $\frac{1}{2}[\![b]\!](2b-[\![b]\!]-1) - \frac{1}{2}[\![a]\!](2a-[\![a]\!]-1)$
13. $f(x)=\frac{1}{2}x$ or $f(x)=0$ **17.** $2(\sqrt{2}-1)$

Chapter 6

Exercises 6.1 □ page 380

1. $\frac{32}{3}$ **3.** $\frac{4}{3}$ **5.** 19.5 **7.** $\frac{1}{6}$ **9.** $\frac{4}{3}$ **11.** $\frac{1}{3}$ **13.** 72
15. $\frac{59}{12}$ **17.** $\frac{9}{8}$ **19.** $\frac{8}{3}$ **21.** $\frac{1}{2}$ **23.** $2-\pi/2$ **25.** $\frac{3}{2}$

27. 6.5 **29.** $\frac{1}{2}$

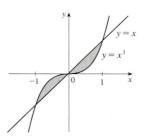

31. 0.6407 **33.** $\pm 1.02; 2.70$ **35.** $-0.72, 1.22; 1.38$
37. $12\sqrt{6}-9$ **39.** $117\frac{1}{3}$ ft
41. (a) Car A (b) The distance by which A is ahead of B after 1 minute (c) Car A (d) $t \approx 2.2$ min
43. $24\sqrt{3}/5$ **45.** $4^{2/3}$ **47.** ± 6
49. $\ln 2 - \frac{1}{2}$ **51.** $\pi - \frac{2}{3}$

Exercises 6.2 □ page 391

1. $\pi/5$

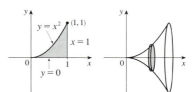

3. $\pi/2$

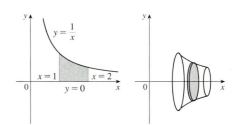

5. 8π

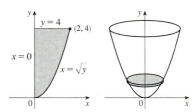

7. $3\pi/10$

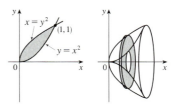

9. $64\pi/15$

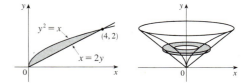

11. $\pi/6$

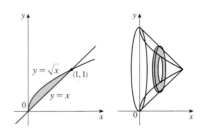

13. $208\pi/45$

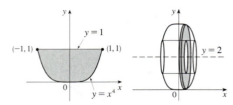

15. $16\pi/15$

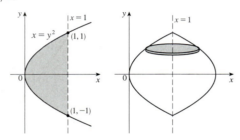

17. $29\pi/30$

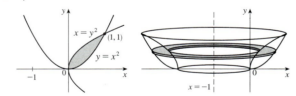

19. $\pi/7$ **21.** $\pi/10$ **23.** $\pi/2$ **25.** $7\pi/15$
27. $5\pi/14$ **29.** $13\pi/30$
31. $\pi \int_0^{\pi/4} (1 - \tan^3 x)^2 \, dx$ **33.** $\pi \int_0^{\pi} [1^2 - (1 - \sin x)^2] \, dx$
35. $\pi \int_{-\sqrt{8}}^{\sqrt{8}} \{[3 - (-2)]^2 - [\sqrt{y^2 + 1} - (-2)]^2\} \, dy$
37. $-0.72, 1.22; 5.80$ **39.** $\frac{11}{8}\pi^2$
41. Solid obtained by rotating the region $0 \le y \le \cos x$, $0 \le x \le \pi/2$ about the x-axis
43. Solid obtained by rotating the region $y^4 \le x \le y^2$, $0 \le y \le 1$ about the y-axis
45. 1110 cm^3 **47.** $\pi r^2 h/3$ **49.** $\pi h^2 [r - (h/3)]$
51. $2b^2 h/3$ **53.** 10 cm^3 **55.** 24 **57.** 2 **59.** 3
61. (a) $8\pi R \int_0^r \sqrt{r^2 - y^2} \, dy$ (b) $2\pi^2 r^2 R$
63. (b) $\pi r^2 h$ **65.** $\frac{5}{12}\pi r^3$
67. $8 \int_0^r \sqrt{R^2 - y^2} \sqrt{r^2 - y^2} \, dy$

Exercises 6.3 □ page 396

1. Circumference $= 2\pi x$, height $= x(x - 1)^2$; $\pi/15$

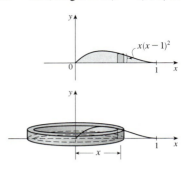

3. 2π

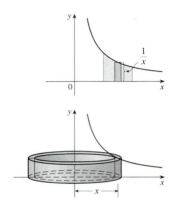

5. 8π **7.** 16π
9. $21\pi/2$

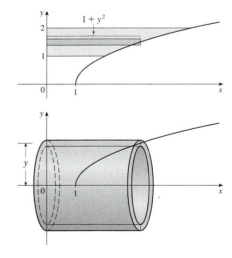

11. $768\pi/7$ **13.** $250\pi/3$
15. $17\pi/6$ **17.** $67\pi/6$ **19.** 24π **21.** $\int_{2\pi}^{3\pi} 2\pi x \sin x \, dx$
23. $\int_0^1 2\pi(x + 1)[\sin(\pi x/2) - x^4] \, dx$
25. $\int_0^{\pi} 2\pi(4 - y)\sqrt{\sin y} \, dy$ **27.** 1.142
29. Solid obtained by rotating the region $0 \le y \le x^4$, $0 \le x \le 3$ about the y-axis

31. Solid obtained by rotating the region bounded by
(i) $x = 1 - y^2$, $x = 0$, and $y = 0$, or (ii) $x = y^2$, $x = 1$, and $y = 0$
about the line $y = 3$
33. 0, 1.32; 4.05   **35.** $\frac{1}{32}\pi^3$ **37.** $81\pi/10$ **39.** $63\pi/2$
41. $4\pi/3$ **43.** $\frac{4}{3}\pi r^3$ **45.** $\frac{1}{3}\pi r^2 h$

Exercises 6.4 □ page 401

1. 7200 J   **3.** 9 ft-lb   **5.** 180 J
7. $\frac{15}{4}$ ft-lb   **9.** $\frac{25}{24} \approx 1.04$ J   **11.** 10.8 cm
13. (a) 625 ft-lb   (b) $\frac{1875}{4}$ ft-lb   **15.** 650,000 ft-lb
17. 3857 J   **19.** 2450 J   **21.** $\approx 1.06 \times 10^6$ J
23. $\approx 5.8 \times 10^3$ ft-lb   **25.** 2.0 m   **29.** $Gm_1 m_2[(1/a) - (1/b)]$

Exercises 6.5 □ page 405

1. $\frac{1}{3}$ **3.** $2/\pi$ **5.** $\frac{1}{15}(26^{3/2} - 1)$ **7.** $2/(5\pi)$
9. (a) 1   (b) 2, 4   (c)

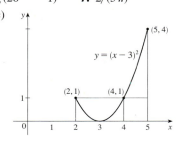

11. (a) $4/\pi$ (b) $\approx 1.24, 2.81$
(c)

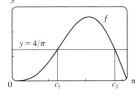

15. $38\frac{1}{3}$ **17.** $(50 + 28/\pi)°F \approx 59°F$
19. 6 kg/m   **21.** $5/(4\pi) \approx 0.4$ L

Chapter 6 Review □ page 406

Exercises

1. $\frac{125}{6}$ **3.** $\frac{1}{3}$ **5.** $\frac{4}{3} + 4/\pi$ **7.** $64\pi/15$ **9.** $1656\pi/5$
11. $\frac{4}{3}\pi(2ah + h^2)^{3/2}$ **13.** $\int_0^1 \pi[(1 - x^3)^2 - (1 - x^2)^2]\, dx$
15. (a) $2\pi/15$ (b) $\pi/6$ (c) $8\pi/15$
17. (a) 0.38   (b) 0.87
19. Solid obtained by rotating the region $0 \leq y \leq \cos x$, $0 \leq x \leq \pi/2$ about the y-axis
21. Solid obtained by rotating the region in the first quadrant bounded by $x = 4 - y^2$ and the axes about the x-axis
23. 36   **25.** $125\sqrt{3}/3$ m^3 **27.** 3.2 J
29. (a) $8000\pi/3 \approx 8378$ ft-lb   (b) 2.1 ft   **31.** $f(x)$

Problems Plus □ page 408

1. (a) $f(t) = 3t^2$ (b) $f(x) = \sqrt{2x/\pi}$ **3.** $\frac{32}{27}$
5. (b) 0.2261   (c) 0.6736 m
(d) (i) $1/(105\pi) \approx 0.003$ in/s   (ii) $370\pi/3$ s ≈ 6.5 min
9. $y = \frac{32}{9}x^2$
11. (a) $V = \int_0^h \pi[f(y)]^2\, dy$
(c) $f(y) = \sqrt{kA/(\pi C)}\, y^{1/4}$
Advantage: the markings on the container are equally spaced.
13. $B = 16A$

Chapter 7

Exercises 7.1 □ page 420

1. (a) See Definition 1.
(b) It must pass the Horizontal Line Test.
3. No   **5.** Yes   **7.** No   **9.** Yes   **11.** Yes   **13.** No
15. No   **17.** No   **19.** 2   **21.** 4
23. $F = \frac{9}{5}C + 32$; the Fahrenheit temperature as a function of the Celsius temperature; $[-273.15, \infty)$
25. $f^{-1}(x) = \frac{3}{2} - \frac{1}{2}x$ **27.** $f^{-1}(x) = -\frac{1}{3}x^2 + \frac{10}{3}$, $x \geq 0$
29. $y = \left(\dfrac{1 - x}{1 + x}\right)^2$, $-1 < x \leq 1$
31. $f^{-1}(x) = \sqrt{2/(1 - x)}$ **33.**

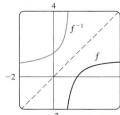

35. (b) $\frac{1}{12}$
(c) $g(x) = \sqrt[3]{x}$, domain $= \mathbb{R} =$ range
(e)

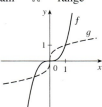

37. (b) $-\frac{1}{2}$
(c) $g(x) = \sqrt{9 - x}$, domain $= [0, 9]$, range $= [0, 3]$
(e)

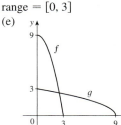

39. 1   **41.** $2/\pi$ **43.** $\frac{3}{2}$
45. $f^{-1}(x) =$
$-(\sqrt[3]{4}/6)(\sqrt[3]{D - 27x^2 + 20} - \sqrt[3]{D + 27x^2 - 20} + \sqrt[3]{2})$,
where $D = 3\sqrt{3}\sqrt{27x^4 - 40x^2 + 16}$; two of the expressions are complex.
47. (a) $g^{-1}(x) = f^{-1}(x) - c$ (b) $h^{-1}(x) = (1/c)f^{-1}(x)$

Exercises 7.2 □ page 431

1. (a) $f(x) = a^x, a > 0$ (b) $\mathbb{R}$ (c) $(0, \infty)$
(d) See Figures 4(c), 4(b), and 4(a), respectively.

3. 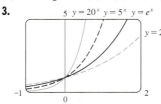 All approach 0 as $x \to -\infty$, all pass through $(0, 1)$, and all are increasing. The larger the base, the faster the rate of increase.

5. 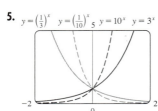 The functions with base greater than 1 are increasing and those with base less than 1 are decreasing. The latter are reflections of the former about the y-axis.

7. 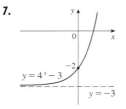 **9.**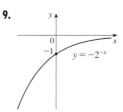

11.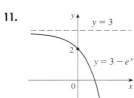

13. (a) $y = e^x - 2$ (b) $y = e^{x-2}$ (c) $y = -e^x$
(d) $y = e^{-x}$ (e) $y = -e^{-x}$
15. (a) $(-\infty, \infty)$ (b) $(-\infty, 0) \cup (0, \infty)$
17. $f(x) = 3 \cdot 2^x$ **21.** At $x \approx 35.8$ **23.** ∞ **25.** 1
27. 0   **29.** $f'(x) = x(x + 2)e^x$ **31.** $y' = 3ax^2 e^{ax^3}$
33. $f'(u) = (-1/u^2)e^{1/u}$ **35.** $F'(t) = e^{t \sin 2t}(2t \cos 2t + \sin 2t)$
37. $y' = 3e^{3x}/\sqrt{1 + 2e^{3x}}$ **39.** $y' = e^{e^x} e^x$
41. $y' = \dfrac{(ad - bc)e^x}{(ce^x + d)^2}$ **43.** $y = 2x + 1$
45. $y' = (1 - 2xye^{x^2y})/(x^2 e^{x^2y} - 1)$ **49.** $-4, -2$
51. $f^{(n)}(x) = 2^n e^{2x}$ **53.** (b) -0.567143
55. (a) 7.92 mg (b) -0.22 mg/yr (c) 56.6 yr
57. (a) 1 (b) $kae^{-kt}/(1 + ae^{-kt})^2$
(c) $t \approx 7.4$ h
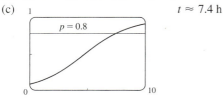

59. (a) $y \approx 100.012437 e^{-10.005531t}$ (b) $-670.63\ \mu A$
61. -1
63. (a) Inc. on $(-1, \infty)$; dec. on $(-\infty, -1)$
(b) CU on $(-2, \infty)$; CD on $(-\infty, -2)$ (c) $(-2, -2e^{-2})$
65. A. $\{x \mid x \neq -1\}$
B. y-int. $1/e$ C. None
D. HA $y = 1$; VA $x = -1$
E. Inc. on $(-\infty, -1), (-1, \infty)$
F. None
G. CU on $(-\infty, -1), (-1, -\tfrac{1}{2})$;
CD on $(-\tfrac{1}{2}, \infty)$; IP $(-\tfrac{1}{2}, 1/e^2)$
H. See graph at right
67. A. $\mathbb{R}$ B. y-int. 2 C. None
D. None
E. Inc. on $(\tfrac{1}{5} \ln \tfrac{2}{3}, \infty)$;
dec. on $(-\infty, \tfrac{1}{5} \ln \tfrac{2}{3})$
F. Loc. min.
$f(\tfrac{1}{5} \ln \tfrac{2}{3}) = (\tfrac{2}{3})^{3/5} + (\tfrac{2}{3})^{-2/5}$
G. CU on $(-\infty, \infty)$
H. See graph at right.
69. Loc. max. $f(-1/\sqrt{3}) = e^{2\sqrt{3}/9} \approx 1.5$;
loc. min. $f(1/\sqrt{3}) = e^{-2\sqrt{3}/9} \approx 0.7$;
IP $(-0.15, 1.15), (-1.09, 0.82)$
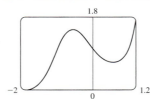

71. $\tfrac{1}{3}(1 - e^{-15})$ **73.** $\tfrac{2}{3}(1 + e^x)^{3/2} + C$ **75.** $x - e^{-x} + C$
77. $2e^{\sqrt{x}} + C$ **79.** 4.644 **81.** $\pi(e^2 - 1)/2$ **83.** $\tfrac{1}{2}$

Exercises 7.3 □ page 439

1. (a) It's defined as the inverse of the exponential function with base a, that is, $\log_a x = y \iff a^y = x$.
(b) $(0, \infty)$ (c) $\mathbb{R}$ (d) See Figure 1.
3. (a) 3 (b) $\tfrac{1}{2}$ **5.** (a) -2 (b) 15 **7.** (a) 2 (b) -1
9. $3 \log_2 x + \log_2 y - 2 \log_2 z$ **11.** $10 \ln u + 10 \ln v$
13. $\log_{10}(ac/b)$ **15.** $\ln 8$ **17.** $\ln \left[\sqrt{x}/(x^2 + 1)^5\right]$
19. (a) 0.402430 (b) 1.454240 (c) 1.651496
21.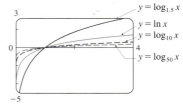

All graphs approach $-\infty$ as $x \to 0^+$, all pass through $(1, 0)$, and all are increasing. The larger the base, the slower the rate of increase.

23. 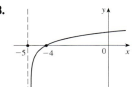   **25.**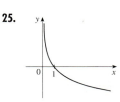

27.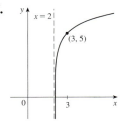

29. (a) $\sqrt{e}$ (b) $-\ln 5$ **31.** (a) $3 + 1/\log_{10} 5$ (b) 9999
33. e^e **35.** 2, 4   **37.** $(\ln C)/(a-b)$ **39.** 0.5210
41. 3.0949   **43.** (a) $x < \ln 10$ (b) $x > 1/e$
45. About 1,084,588 mi   **47.** 8.3
49. (a) $f^{-1}(n) = (3/\ln 2)\ln(n/100)$; the time elapsed when there are n bacteria   (b) After about 26.9 h
51. $-\infty$ **53.** 0   **55.** ∞ **57.** $(\frac{3}{5}, \infty)$, $\mathbb{R}$
59. (a) $(-\infty, \frac{1}{2}\ln 3]$ (b) $f^{-1}(x) = \frac{1}{2}\ln(3-x^2)$, $[0, \sqrt{3}]$
61. $y = e^x - 3$ **63.** $f^{-1}(x) = \sqrt[3]{\ln x}$
65. $y = \log_{10}[x/(1-x)]$ **67.** $(-\frac{1}{2}\ln 3, \infty)$
69. (b) $f^{-1}(x) = \frac{1}{2}(e^x - e^{-x})$ **71.** f is a constant function
75. $-1 \le x < 1 - \sqrt{3}$ or $1 + \sqrt{3} < x \le 3$

Exercises 7.4 □ page 449

1. The differentiation formula is simplest.
3. $f'(\theta) = -\tan\theta$ **5.** $f'(x) = 3/[(3x-1)\ln 2]$
7. $f'(x) = 1/[5x\sqrt[5]{(\ln x)^4}]$ **9.** $f'(x) = (2 + \ln x)/(2\sqrt{x})$
11. $F'(t) = 6/(2t+1) - 12/(3t-1)$
13. $g'(x) = -2a/(a^2 - x^2)$
15. $f'(u) = \dfrac{1 + \ln 2}{u[1 + \ln(2u)]^2}$ **17.** $h'(t) = 3t^2 - 3^t \ln 3$
19. $y' = (10x+1)/(5x^2 + x - 2)$ **21.** $y' = -x/(1+x)$
23. $y' = 5^{-1/x}(\ln 5)/x^2$ **25.** $y' = \ln x + 1$, $y'' = 1/x$
27. $y' = 1/(x\ln 10)$, $y'' = -1/(x^2 \ln 10)$
29. $f'(x) = \dfrac{2x - 1 - (x-1)\ln(x-1)}{(x-1)[1 - \ln(x-1)]^2}$; $(1, 1+e) \cup (1+e, \infty)$
31. $f'(x) = 2x\ln(1-x^2) - 2x^3/(1-x^2)$, $(-1, 1)$ **33.** 0
35. $x - ey = e$ **37.** $f'(x) = \cos x + 1/x$
39. $y' = (2x+1)^5(x^4 - 3)^6 \left(\dfrac{10}{2x+1} + \dfrac{24x^3}{x^4 - 3}\right)$
41. $y' = \dfrac{\sin^2 x \tan^4 x}{(x^2 + 1)^2}\left(2\cot x + \dfrac{4\sec^2 x}{\tan x} - \dfrac{4x}{x^2 + 1}\right)$
43. $y' = x^x(1 + \ln x)$ **45.** $y' = x^{\sin x}[(\sin x)/x + \ln x \cos x]$
47. $y' = (\ln x)^x(1/\ln x + \ln \ln x)$ **49.** $y' = e^x x^{e^x}(\ln x + 1/x)$
51. $y' = 2x/(x^2 + y^2 - 2y)$
53. $f^{(n)}(x) = (-1)^{n-1}(n-1)!/(x-1)^n$
55. 2.958516, 5.290718

57. CU on $(e^{8/3}, \infty)$, CD on $(0, e^{8/3})$, IP $(e^{8/3}, \frac{8}{3}e^{-4/3})$
59. A. All x in $(2n\pi, (2n+1)\pi)$ (n an integer)
 B. x-int. $\pi/2 + 2n\pi$ C. Period 2π D. VA $x = n\pi$
 E. Inc. on $(2n\pi, \pi/2 + 2n\pi)$; dec. on $(\pi/2 + 2n\pi, (2n+1)\pi)$
 F. Loc. max. $f(\pi/2 + 2n\pi) = 0$
 G. CD on $(2n\pi, (2n+1)\pi)$
 H.
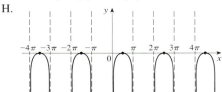

61. A. $\mathbb{R}$ B. y-int 0; x-int. 0
 C. About y-axis   D. None
 E. Inc. on $(0, \infty)$; dec. on $(-\infty, 0)$
 F. Loc. min. $f(0) = 0$
 G. CU on $(-1, 1)$; CD on $(-\infty, -1)$, $(1, \infty)$;
 IP $(\pm 1, \ln 2)$ H. See graph at right.
63. Inc. on $(0, 2.7)$, $(4.5, 8.2)$, $(10.9, 14.3)$;
 IP $(3.8, 1.7)$, $(5.7, 2.1)$, $(10.0, 2.7)$, $(12.0, 2.9)$
65. $3\ln 2$ **67.** $\frac{1}{3}\ln\frac{5}{2}$ **69.** $\frac{1}{2}e^2 + e - \frac{1}{2}$
71. $\frac{1}{3}\ln|6x - x^3| + C$ **73.** $\frac{1}{3}(\ln x)^3 + C$
75. $90/(\ln 10)$ **79.** $\pi \ln 2$ **81.** 45,974 J   **83.** $\frac{1}{3}$
85. $0 < m < 1$, $m - 1 - \ln m$

Exercises 7.2* □ page 458

1. $3\ln x + \ln y - 2\ln z$ **3.** $10\ln u + 10\ln v$
5. $\ln 8$ **7.** $\ln[\sqrt{x}/(x^2 + 1)^5]$
9. 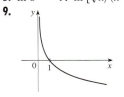   **11.**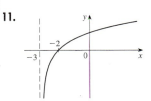

13. $f'(x) = (2 + \ln x)/(2\sqrt{x})$ **15.** $f'(\theta) = -\tan\theta$
17. $f'(x) = 1/[5x\sqrt[5]{(\ln x)^4}]$ **19.** $g'(x) = -2a/(a^2 - x^2)$
21. $f'(u) = \dfrac{1 + \ln 2}{u[1 + \ln(2u)]^2}$
23. $F'(t) = 6/(2t+1) - 12/(3t-1)$
25. $y' = (10x+1)/(5x^2 + x - 2)$ **27.** $y' = -6/[5(x^2 - 1)]$
29. $y' = \sec^2(\ln(ax+b))\dfrac{a}{ax+b}$
31. $y' = \dfrac{1}{x\ln x}$, $y'' = -\dfrac{1 + \ln x}{(x\ln x)^2}$
33. $f'(x) = \dfrac{2x - 1 - (x-1)\ln(x-1)}{(x-1)[1 - \ln(x-1)]^2}$; $(1, 1+e) \cup (1+e, \infty)$
35. $f'(x) = -\dfrac{1}{2x\sqrt{1 - \ln x}}$; $(0, e]$
37. 0   **39.** $f'(x) = \cos x + 1/x$ **41.** $y = 2x - 2$
43. $y' = 2x/(x^2 + y^2 - 2y)$

45. $f^{(n)}(x) = (-1)^{n-1}(n-1)!/(x-1)^n$
47. 2.958516, 5.290718
49. A. All x in $(2n\pi, (2n+1)\pi)$ (n an integer)
B. x-int. $\pi/2 + 2n\pi$ C. Period 2π D. VA $x = n\pi$
E. Inc. on $(2n\pi, \pi/2 + 2n\pi)$; dec. on $(\pi/2 + 2n\pi, (2n+1)\pi)$
F. Loc. max. $f(\pi/2 + 2n\pi) = 0$ G. CD on $(2n\pi, (2n+1)\pi)$
H.

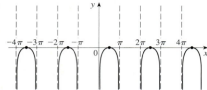

51. A. $\mathbb{R}$ B. y-int 0; x-int. 0
C. About y-axis   D. None
E. Inc. on $(0, \infty)$; dec. on $(-\infty, 0)$
F. Loc. min. $f(0) = 0$
G. CU on $(-1, 1)$; CD on $(-\infty, -1), (1, \infty)$;
IP $(\pm 1, \ln 2)$ H. See graph at right.
53. Inc. on $(0, 2.7), (4.5, 8.2), (10.9, 14.3)$;
IP $(3.8, 1.7), (5.7, 2.1), (10.0, 2.7), (12.0, 2.9)$
55. $y' = (2x+1)^5(x^4-3)^6\left(\dfrac{10}{2x+1} + \dfrac{24x^3}{x^4-3}\right)$
57. $y' = \dfrac{\sin^2 x \tan^4 x}{(x^2+1)^2}\left(2 \cot x + \dfrac{4 \sec^2 x}{\tan x} - \dfrac{4x}{x^2+1}\right)$
59. $3 \ln 2$ **61.** $\tfrac{1}{3} \ln \tfrac{5}{2}$ **63.** $\tfrac{1}{2}e^2 + e - \tfrac{1}{2}$
65. $\tfrac{1}{3} \ln|6x - x^3| + C$ **67.** $\tfrac{1}{3}(\ln x)^3 + C$ **71.** $\pi \ln 2$
73. 45,974 J   **75.** $\tfrac{1}{3}$ **77.** (b) 0.405
81. $0 < m < 1, m - 1 - \ln m$

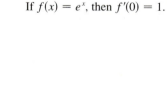

Exercises 7.3* □ page 465

1. (a) e is the number such that $\ln e = 1$.
(b) $e \approx 2.71828$
(c) If $f(x) = e^x$, then $f'(0) = 1$.

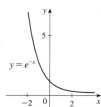

3. (a) $\sqrt{2}$ (b) 8   **5.** (a) $\sqrt{e}$ (b) $-\ln 5$ **7.** e^e
9. 2, 4   **11.** $(\ln C)/(a - b)$ **13.** 0.5210   **15.** 3.0949
17. (a) $x < \ln 10$ (b) $x > 1/e$
19.
21.

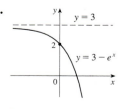

23. (a) $y = e^x - 2$ (b) $y = e^{x-2}$ (c) $y = -e^x$
(d) $y = e^{-x}$ (e) $y = -e^{-x}$
25. 0   **27.** 1   **29.** 0   **31.** $f'(x) = x(x+2)e^x$
33. $y' = 3ax^2 e^{ax^3}$ **35.** $f'(u) = (-1/u^2)e^{1/u}$
37. $F'(t) = e^{t \sin 2t}(2t \cos 2t + \sin 2t)$
39. $y' = 3e^{3x}/\sqrt{1 + 2e^{3x}}$ **41.** $y' = e^{e^x} e^x$
43. $y' = \dfrac{(ad - bc)e^x}{(ce^x + d)^2}$ **45.** $y = 2x + 1$
47. $y' = (1 - 2xye^{x^2y})/(x^2 e^{x^2y} - 1)$ **49.** $-4, -2$
51. $f^{(n)}(x) = 2^n e^{2x}$ **53.** (b) -0.567143
55. (a) 7.92 mg   (b) -0.22 mg/yr   (c) 56.6 yr
57. (a) 1   (b) $kae^{-kt}/(1 + ae^{-kt})^2$
(c) $t \approx 7.4$ h

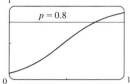

59. -1 **61.** $\left(-\tfrac{2}{3}, \infty\right)$
63. A. $\{x \mid x \neq -1\}$ B. y-int. $1/e$ C. None
D. HA $y = 1$; VA $x = -1$
E. Inc. on $(-\infty, -1), (-1, \infty)$
F. None
G. CU on $(-\infty, -1), \left(-1, -\tfrac{1}{2}\right)$;
CD on $\left(-\tfrac{1}{2}, \infty\right)$; IP $\left(-\tfrac{1}{2}, 1/e^2\right)$
H. See graph at right

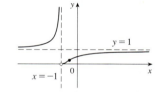

65. A. $\mathbb{R}$ B. y-int. 2   C. None
D. None
E. Inc. on $\left(\tfrac{1}{5} \ln \tfrac{2}{3}, \infty\right)$;
dec. on $\left(-\infty, \tfrac{1}{5} \ln \tfrac{2}{3}\right)$
F. Loc. min.
$f\left(\tfrac{1}{5} \ln \tfrac{2}{3}\right) = \left(\tfrac{2}{3}\right)^{3/5} + \left(\tfrac{2}{3}\right)^{-2/5}$
G. CU on $(-\infty, \infty)$
H. See graph at right.

67. Loc. max. $f(-1/\sqrt{3}) = e^{2\sqrt{3}/9} \approx 1.5$;
loc. min. $f(1/\sqrt{3}) = e^{-2\sqrt{3}/9} \approx 0.7$;
IP $(-0.15, 1.15), (-1.09, 0.82)$

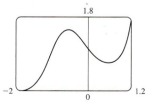

69. $\tfrac{1}{3}(1 - e^{-15})$ **71.** $\tfrac{2}{3}(1 + e^x)^{3/2} + C$ **73.** $x - e^{-x} + C$
75. $2e^{\sqrt{x}} + C$ **77.** 4.644   **79.** $\pi(e^2 - 1)/2$
81. $f^{-1}(x) = e^x - 3$ **83.** $\tfrac{1}{2}$

Exercises 7.4* □ page 475

1. (a) $a^x = e^{x \ln a}$ (b) $(-\infty, \infty)$ (c) $(0, \infty)$
   (d) See Figures 1, 2, and 3.   3. $e^{\sqrt{7} \ln 5}$ 5. $e^{x \ln \cos x}$
7. (a) 3 (b) -4
9. (a) 2 (b) $\sqrt{2}$
11. 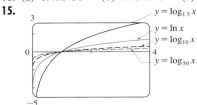 All approach 0 as $x \to -\infty$, all pass through $(0, 1)$, and all are increasing. The larger the base, the faster the rate of increase.

13. (a) 0.402430 (b) 1.454240 (c) 1.651496
15.

All graphs approach $-\infty$ as $x \to 0^+$, all pass through $(1, 0)$, and all are increasing. The larger the base, the slower the rate of increase.
17. $f(x) = 3 \cdot 2^x$ 19. (b) About 1,084,588 mi   21. 0
23. $h'(t) = 3t^2 - 3^t \ln 3$ 25. $y' = 5^{-1/x}(\ln 5)/x^2$
27. $f'(u) = 10 \ln 2(2^u + 2^{-u})^9(2^u - 2^{-u})$
29. $f'(x) = 2x/[(x^2 - 4) \ln 3]$ 31. $y' = x^x(\ln x + 1)$
33. $y' = x^{\sin x}[\cos x \ln x + (\sin x)/x]$
35. $y' = (\ln x)^x(\ln \ln x + 1/\ln x)$ 37. $y' = xe^x e^x(\ln x + 1/x)$
39. $y = (10 \ln 10)x + 10(1 - \ln 10)$ 41. $90/(\ln 10)$
43. $(\ln x)^2/(2 \ln 10) + C$ [or $\tfrac{1}{2}(\ln 10)(\log_{10} x)^2 + C$]
45. $3^{\sin \theta}/\ln 3 + C$ 47. $16/(5 \ln 5) - 1/(2 \ln 2)$
49. 0.600967   51. $y = \log_{10} x - \log_{10}(1 - x)$ 53. 8.3
55. $10^8/\ln 10$ dB/(watt/m^2)
57. (a) $y \approx 100.012437e^{-10.005531t}$ (b) -670.63 μA

Exercises 7.5 □ page 483

1. (a) $\pi/3$ (b) π 3. (a) $\pi/3$ (b) $-\pi/4$
5. (a) 0 (b) 5   7. $2/\sqrt{5}$ 9. $\tfrac{2}{3}\sqrt{2}$ 13. $x/\sqrt{1 + x^2}$
15.

The second graph is the reflection of the first graph about the line $y = x$.

23. $y' = 1/[2\sqrt{x}(1 + x)]$ 25. $y' = 1/\sqrt{-x^2 - x}$
27. $H'(x) = 1 + 2x \arctan x$ 29. $y' = -2e^{2x}/\sqrt{1 - e^{4x}}$
31. $y' = -(\sin \theta)/(1 + \cos^2 \theta)$ 33. $h'(t) = 0$

35. $y' = \sqrt{a^2 - b^2}/(a + b \cos x)$
37. $g'(x) = 2/\sqrt{1 - (3 - 2x)^2}$; [1, 2], (1, 2)   39. $\pi/6$
41. $f'(x) = \dfrac{e^{-x}}{1 + x^2} - e^{-x} \arctan x$ 43. $-\pi/2$ 45. $\pi/2$
47. At a distance $5 - 2\sqrt{5}$ from A 49. $\tfrac{1}{4}$ rad/s
51. A. $[-\tfrac{1}{2}, \infty)$
    B. y-int. 0; x-int. 0
    C. None
    D. HA $y = \pi/2$
    E. Inc. on $(-\tfrac{1}{2}, \infty)$
    F. None
    G. CD on $(-\tfrac{1}{2}, \infty)$
    H. See graph at right

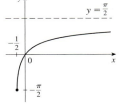

53. A. $\mathbb{R}$
    B. y-int. 0; x-int. 0
    C. About $(0, 0)$
    D. SA $y = x \pm \pi/2$
    E. Inc. on $\mathbb{R}$ F. None
    G. CU on $(0, \infty)$; CD on $(-\infty, 0)$;
    IP $(0, 0)$
    H. See graph at right.

55. Max. at $x = 0$, min. at $x \approx \pm 0.87$, IP at $x \approx \pm 0.52$
57. $x^2 + 5 \sin^{-1} x + C$ 59. π 61. $\pi/12$ 63. $\pi^2/72$
65. $\tfrac{1}{2} \ln(x^2 + 9) + 3 \tan^{-1}(x/3) + C$ 67. $\tfrac{1}{3} \sin^{-1} t^3 + C$
69. $2 \tan^{-1} \sqrt{x} + C$ 73. $\pi/2 - 1$

Exercises 7.6 □ page 491

1. (a) 0 (b) 1   3. (a) $\tfrac{3}{4}$ (b) $\tfrac{1}{2}(e^2 - e^{-2}) \approx 3.62686$
5. (a) 1 (b) 0
21. $\coth x = \tfrac{5}{4}$, $\mathrm{sech}\, x = \tfrac{3}{5}$, $\cosh x = \tfrac{5}{3}$, $\sinh x = \tfrac{4}{3}$, $\mathrm{csch}\, x = \tfrac{3}{4}$
23. (a) 1 (b) -1 (c) ∞ (d) $-\infty$ (e) 0 (f) 1
    (g) ∞ (h) $-\infty$ (i) 0
31. $f'(x) = x \sinh x + \cosh x$ 33. $h'(x) = 2x \cosh(x^2)$
35. $G'(x) = -(2 \sinh x)/(1 + \cosh x)^2$
37. $h'(t) = -(t \,\mathrm{csch}^2 \sqrt{1 + t^2})/\sqrt{1 + t^2}$
39. $H'(t) = e^t \,\mathrm{sech}^2(e^t)$
41. $y' = 3e^{\cosh 3x} \sinh 3x$ 43. $y' = 1/[2\sqrt{x}(1 - x)]$
45. $y' = \sinh^{-1}(x/3)$ 47. $y' = -1/(x\sqrt{x^2 + 1})$
49. (a) 0.3572 (b) 70.34°
51. (b) $y = 2 \sinh 3x - 4 \cosh 3x$ 53. $(\ln(1 + \sqrt{2}), \sqrt{2})$
55. $\tfrac{1}{3} \cosh^3 x + C$ 57. $2 \cosh \sqrt{x} + C$ 59. $-\mathrm{csch}\, x + C$
61. $\ln\left(\dfrac{6 + 3\sqrt{3}}{4 + \sqrt{7}}\right)$ 63. $\tanh^{-1} e^x + C$
65. (a) 0, 0.48 (b) 0.04

Exercises 7.7 □ page 501

1. (a) Indeterminate (b) 0 (c) 0
   (d) ∞, $-\infty$, or does not exist   (e) Indeterminate
3. (a) $-\infty$ (b) Indeterminate   (c) ∞ 5. -2 7. $\tfrac{9}{5}$

9. $-\infty$ **11.** ∞ **13.** p/q **15.** 0   **17.** $-\infty$ **19.** $\ln \frac{5}{3}$
21. $\frac{1}{2}$ **23.** ∞ **25.** 1   **27.** $\frac{1}{2}$ **29.** 0
31. 1   **33.** $-1/\pi^2$ **35.** $a(a-1)/2$ **37.** 0
39. 3   **41.** 0   **43.** $-2/\pi$ **45.** 0   **47.** $\frac{1}{2}$ **49.** ∞
51. 1   **53.** e^{-2} **55.** e^3 **57.** 1
59. $1/e$ **61.** $1/\sqrt{e}$ **63.** 5   **65.** $\frac{1}{4}$

67. A. $\mathbb{R}$ B. y-int. 0; x-int. 0
C. None   D. HA $y = 0$
E. Inc. on $(-\infty, 1)$, dec. on $(1, \infty)$
F. Loc. max. $f(1) = 1/e$
G. CU on $(2, \infty)$; CD on $(-\infty, 2)$
IP $(2, 2/e^2)$
H. See graph at right.

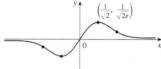

69. A. $\mathbb{R}$ B. y-int. 0; x-int. 0
C. About $(0, 0)$ D. HA $y = 0$
E. Inc. on $(-1/\sqrt{2}, 1/\sqrt{2})$; dec. on $(-\infty, -1/\sqrt{2}), (1/\sqrt{2}, \infty)$
F. Loc. min. $f(-1/\sqrt{2}) = -1/\sqrt{2e}$;
loc. max. $f(1/\sqrt{2}) = 1/\sqrt{2e}$
G. CU on $(-\sqrt{3/2}, 0), (\sqrt{3/2}, \infty)$; CD on $(-\infty, -\sqrt{3/2}), (0, \sqrt{3/2})$; IP $(\pm\sqrt{3/2}, \pm\sqrt{3/2}\,e^{-3/2}), (0, 0)$
H.

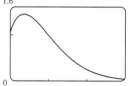

71. A. $(-1, \infty)$ B. y-int. 0; x-int. 0
C. None   D. VA $x = -1$
E. Inc. on $(0, \infty)$; dec. on $(-1, 0)$
F. Loc. min. $f(0) = 0$
G. CU on $(-1, \infty)$
H. See graph at right.

73. (a)

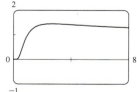

(b) $\lim_{x \to 0^+} x^{-x} = 1$
(c) Max. value $f(1/e) = e^{1/e} \approx 1.44$ (d) 1.0

75. (a)

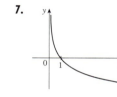

(b) $\lim_{x \to 0^+} x^{1/x} = 0$, $\lim_{x \to \infty} x^{1/x} = 1$
(c) Max. value $f(e) = e^{1/e} \approx 1.44$ (d) 0.58, 4.4

77.

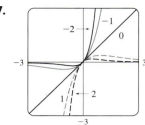

For $c > 0$, $\lim_{x \to \infty} f(x) = 0$ and $\lim_{x \to -\infty} f(x) = -\infty$. For $c < 0$, $\lim_{x \to \infty} f(x) = \infty$ and $\lim_{x \to -\infty} f(x) = 0$. As $|c|$ increases, the maximum and minimum points and the IPs get closer to the origin.
81. $\pi/6$ **83.** 56   **91.** $\frac{1}{3}$ **93.** (a) 0

Chapter 7 Review □ page 505

True-False Quiz
1. True   **3.** False   **5.** True   **7.** True   **9.** False
11. False   **13.** False   **15.** True   **17.** True

Exercises

1. No   **3.** (a) 7   (b) $\frac{1}{8}$
5.
7.

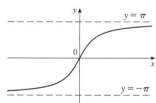

9.

11. (a) 9   (b) 2   **13.** $e^{1/3}$
15. $\ln \ln 17$ **17.** $\sqrt{1 + e}$
19. $\tan 1$ **21.** $f'(t) = t + 2t \ln t$
23. $h'(\theta) = 2 \sec^2(2\theta) e^{\tan 2\theta}$ **25.** $y' = 5 \sec 5x$
27. $y' = (1 + c^2) e^{cx} \sin x$ **29.** $y' = 2 \tan x$
31. $y' = e^{-1/x}(1 + 1/x)$ **33.** $y' = -(2 \ln 2) t 2^{-t^2}$
35. $H'(v) = (v/(1 + v^2)) + \tan^{-1} v$
37. $y' = 2x^2 \cosh(x^2) + \sinh(x^2)$
39. $y' = \cot x - \sin x \cos x$
41. $y' = -(1/x)[1 + 1/(\ln x)^2]$
43. $y' = 3 \tanh 3x$
45. $y' = (\cosh x)/\sqrt{\sinh^2 x - 1}$
47. $f'(x) = \dfrac{6x}{x^2 + 1} \sin^2(\ln(x^2 + 1)) \cdot \cos(\ln(x^2 + 1)) \cdot e^{\sin^3(\ln(x^2+1))}$

49. $f'(x) = g'(x)e^{g(x)}$ **51.** $f'(x) = g'(x)/g(x)$ **53.** $2^x(\ln 2)^n$
57. $y = -x + 2$ **59.** $(-3, 0)$
61. (a) $y = \frac{1}{4}x + \frac{1}{4}(\ln 4 + 1)$ (b) $y = ex$
63. 0   **65.** 0   **67.** $-\infty$ **69.** -1
71. π **73.** 8   **75.** 0   **77.** $\frac{1}{2}$

79. A. $\{x \mid x \neq 0\}$ B. None
C. About $(0, 0)$ D. HA $y = 0$
E. Dec. on $(-\infty, 0)$ and $(0, \infty)$
F. None
G. CU on $(0, \infty)$; CD on $(-\infty, 0)$
H. See graph at right.

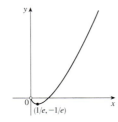

81. A. $(0, \infty)$ B. x-int. 1
C. None   D. None
E. Inc. on $(1/e, \infty)$; dec. on $(0, 1/e)$
F. Loc. min. $f(1/e) = -1/e$
G. CU on $(0, \infty)$
H. See graph at right.

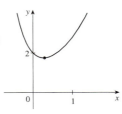

83. A. $\mathbb{R}$ B. y-int. 2
C. None   D. None
E. Inc. on $(\frac{1}{4}\ln 3, \infty)$, dec. on $(-\infty, \frac{1}{4}\ln 3)$
F. Loc. min. $f(\frac{1}{4}\ln 3) = 3^{1/4} + 3^{-3/4}$
G. CU on $(-\infty, \infty)$
H. See graph at right

85.

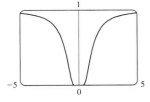

$(\pm 0.82, 0.22); (\pm\sqrt{2/3}, e^{-3/2})$

87. $v(t) = -Ae^{-ct}[\omega \sin(\omega t + \delta) + c \cos(\omega t + \delta)]$,
$a(t) = Ae^{-ct}[(c^2 - \omega^2) \cos(\omega t + \delta) + 2c\omega \sin(\omega t + \delta)]$
89. 4.32 days   **91.** $\frac{1}{4}(1 - e^{-2})$ **93.** $\ln 2 - \frac{7}{4}$
95. $2e^{\sqrt{x}} + C$ **97.** $\frac{1}{2}\ln|x^2 + 2x| + C$
99. $-\frac{1}{2}[\ln(\cos x)]^2 + C$
101. $2^{\tan\theta}/\ln 2 + C$
103. $\ln|1 + \sec\theta| + C$
107. $e^{\sqrt{x}}/(2x)$ **109.** $\frac{1}{3}\ln 4$
111. $\pi^2/4$ **113.** $\frac{2}{3}$
115. $2/e$ **119.** $f(x) = e^{2x}(2x + 1)/(1 - e^{-x})$

Problems Plus □ page 509
7. $1/\sqrt{2}$ **9.** $\frac{1}{2}$ **13.** $2\sqrt{e}$ **15.** $a \leq e^{1/e}$

Chapter 8

Exercises 8.1 □ page 516

1. $\frac{1}{2}x^2 \ln x - \frac{1}{4}x^2 + C$ **3.** $\frac{1}{5}x \sin 5x + \frac{1}{25}\cos 5x + C$
5. $2(r - 2)e^{r/2} + C$
7. $-\frac{1}{\pi}x^2 \cos \pi x + \frac{2}{\pi^2}x \sin \pi x + \frac{2}{\pi^3}\cos \pi x + C$
9. $\frac{1}{2}(2x + 1)\ln(2x + 1) - x + C$
11. $t \arctan 4t - \frac{1}{8}\ln(1 + 16t^2) + C$
13. $x(\ln x)^2 - 2x \ln x + 2x + C$
15. $\frac{1}{13}e^{2\theta}(2 \sin 3\theta - 3 \cos 3\theta) + C$
17. $y \cosh y - \sinh y + C$
19. $\pi/3$ **21.** $\frac{1}{2} - \frac{1}{2}\ln 2$ **23.** $\frac{1}{4} - \frac{3}{4}e^{-2}$
25. $(\pi + 6 - 3\sqrt{3})/6$ **27.** $\sin x (\ln \sin x - 1) + C$
29. $\frac{1}{2}x(\cos \ln x + \sin \ln x) + C$ **31.** $\frac{32}{5}(\ln 2)^2 - \frac{64}{25}\ln 2 + \frac{62}{125}$
33. $2(\sin \sqrt{x} - \sqrt{x} \cos \sqrt{x}) + C$ **35.** $-\frac{1}{2} - \pi/4$
37. $(x \sin \pi x)/\pi + (\cos \pi x)/\pi^2 + C$

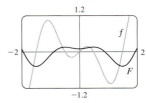

39. $(2x + 1)e^x + C$

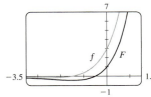

41. (b) $-\frac{1}{4}\cos x \sin^3 x + \frac{3}{8}x - \frac{3}{16}\sin 2x + C$
43. (b) $\frac{2}{3}, \frac{8}{15}$ **49.** $x[(\ln x)^3 - 3(\ln x)^2 + 6 \ln x - 6] + C$
51. $\frac{25}{4} - \frac{75}{4}e^{-2}$ **53.** 1.0475, 2.8731; 2.1828
55. $4 - 8/\pi$ **57.** $2\pi e$
59. $\frac{9}{2}\ln 3 - \frac{13}{9}$ **61.** $2 - e^{-t}(t^2 + 2t + 2)$ m   **63.** 2

Exercises 8.2 □ page 524

1. $\frac{1}{5}\cos^5 x - \frac{1}{3}\cos^3 x + C$ **3.** $-\frac{11}{384}$
5. $\frac{1}{5}\sin^5 x - \frac{2}{7}\sin^7 x + \frac{1}{9}\sin^9 x + C$ **7.** $\pi/4$ **9.** $3\pi/8$
11. $\frac{3}{2}\theta + 2 \sin \theta + \frac{1}{4}\sin 2\theta + C$ **13.** $(3\pi - 4)/192$
15. $(\frac{2}{7}\cos^3 x - \frac{2}{3}\cos x)\sqrt{\cos x} + C$
17. $\frac{1}{2}\cos^2 x - \ln|\cos x| + C$ **19.** $\ln(1 + \sin x) + C$
21. $\frac{1}{2}\tan^2 x + C$ **23.** $\tan x - x + C$
25. $\frac{1}{5}\tan^5 t + \frac{2}{3}\tan^3 t + \tan t + C$ **27.** $\frac{117}{8}$
29. $\frac{1}{3}\sec^3 x - \sec x + C$
31. $\frac{1}{4}\sec^4 x - \tan^2 x + \ln|\sec x| + C$
33. $\frac{1}{6}\tan^6\theta + \frac{1}{4}\tan^4\theta + C$ **35.** $\sqrt{3} - (\pi/3)$

37. $\frac{1}{3}\csc^3\alpha - \frac{1}{5}\csc^5\alpha + C$ **39.** $\ln|\csc x - \cot x| + C$
41. $\frac{1}{6}\sin 3x - \frac{1}{14}\sin 7x + C$ **43.** $\frac{1}{4}\sin 2\theta + \frac{1}{24}\sin 12\theta + C$
45. $\frac{1}{2}\sin 2x + C$ **47.** $\frac{1}{10}\tan^5(t^2) + C$
49. $-\frac{1}{5}\cos^5 x + \frac{2}{3}\cos^3 x - \cos x + C$

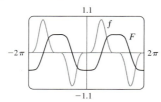

51. $\frac{1}{6}\sin 3x - \frac{1}{18}\sin 9x + C$

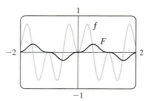

53. 0 **55.** $\frac{1}{3}$ **57.** 0 **59.** $\pi^2/4$ **61.** $2\pi + (\pi^2/4)$
63. $s = (1 - \cos^3\omega t)/(3\omega)$

Exercises 8.3 □ page 530

1. $\sqrt{x^2 - 9}/(9x) + C$ **3.** $\frac{1}{3}(x^2 - 18)\sqrt{x^2 + 9} + C$
5. $\pi/24 + \sqrt{3}/8 - \frac{1}{4}$ **7.** $-\sqrt{25 - x^2}/(25x) + C$
9. $\ln(\sqrt{x^2 + 16} + x) + C$ **11.** $\frac{1}{4}\sin^{-1}(2x) + \frac{1}{2}x\sqrt{1 - 4x^2} + C$
13. $\frac{1}{6}\sec^{-1}(x/3) - \sqrt{x^2 - 9}/(2x^2) + C$
15. $(x/\sqrt{a^2 - x^2}) - \sin^{-1}(x/a) + C$ **17.** $\sqrt{x^2 - 7} + C$
19. $\ln\left|(\sqrt{1 + x^2} - 1)/x\right| + \sqrt{1 + x^2} + C$ **21.** $\frac{64}{1215}$
23. $\frac{9}{2}\sin^{-1}((x - 2)/3) + \frac{1}{2}(x - 2)\sqrt{5 + 4x - x^2} + C$
25. $\frac{1}{3}\ln\left|3x + 1 + \sqrt{9x^2 + 6x - 8}\right| + C$
27. $\frac{1}{2}[\tan^{-1}(x + 1) + (x + 1)/(x^2 + 2x + 2)] + C$
29. $\frac{1}{4}\sin^{-1}x^2 + \frac{1}{4}x^2\sqrt{1 - x^4} + C$
33. $\frac{1}{6}(\sqrt{48} - \sec^{-1}7)$ **37.** 0.81, 2; 2.10
39. $r\sqrt{R^2 - r^2} + \pi r^2/2 - R^2 \arcsin(r/R)$ **41.** $2\pi^2 Rr^2$

Exercises 8.4 □ page 540

1. (a) $\dfrac{A}{x + 3} + \dfrac{B}{3x + 1}$ (b) $\dfrac{A}{x} + \dfrac{B}{x + 1} + \dfrac{C}{(x + 1)^2}$
3. (a) $\dfrac{A}{x + 4} + \dfrac{B}{x - 1}$ (b) $\dfrac{A}{x - 1} + \dfrac{Bx + C}{x^2 + x + 1}$
5. (a) $1 + \dfrac{A}{x - 1} + \dfrac{B}{x + 1} + \dfrac{Cx + D}{x^2 + 1}$
(b) $\dfrac{At + B}{t^2 + 1} + \dfrac{Ct + D}{t^2 + 4} + \dfrac{Et + F}{(t^2 + 4)^2}$
7. $x + 6\ln|x - 6| + C$
9. $2\ln|x + 5| - \ln|x - 2| + C$ **11.** $\frac{1}{2}\ln\frac{3}{2}$
13. $a\ln|x - b| + C$ **15.** $2\ln 2 + \frac{1}{2}$
17. $\frac{27}{5}\ln 2 - \frac{9}{5}\ln 3$ (or $\frac{9}{5}\ln\frac{8}{3}$)

19. $-\dfrac{1}{36}\ln|x + 5| + \dfrac{1}{6}\dfrac{1}{x + 5} + \dfrac{1}{36}\ln|x - 1| + C$
21. $2\ln|x| + 3\ln|x + 2| + (1/x) + C$
23. $\ln|x + 1| + 2/(x + 1) - 1/[2(x + 1)^2] + C$
25. $\ln|x - 1| - \frac{1}{2}\ln(x^2 + 9) - \frac{1}{3}\tan^{-1}(x/3) + C$
27. $\frac{1}{2}\ln(x^2 + 1) + (1/\sqrt{2})\tan^{-1}(x/\sqrt{2}) + C$
29. $\frac{1}{2}\ln(x^2 + 2x + 5) + \frac{3}{2}\tan^{-1}((x + 1)/2) + C$
31. $\frac{1}{3}\ln|x - 1| - \frac{1}{6}\ln(x^2 + x + 1) - \dfrac{1}{\sqrt{3}}\tan^{-1}\dfrac{2x + 1}{\sqrt{3}} + C$
33. $\frac{1}{3}\ln\frac{17}{2}$
35. $(1/x) + \frac{1}{2}\ln|(x - 1)/(x + 1)| + C$
37. $\dfrac{-1}{2(x^2 + 2x + 4)} - \dfrac{2\sqrt{3}}{9}\tan^{-1}\left(\dfrac{x + 1}{\sqrt{3}}\right) - \dfrac{2(x + 1)}{3(x^2 + 2x + 4)} + C$
39. $\ln\left|(\sqrt{x + 1} - 1)/(\sqrt{x + 1} + 1)\right| + C$
41. $2 + \ln\frac{25}{9}$ **43.** $\frac{3}{10}(x^2 + 1)^{5/3} - \frac{3}{4}(x^2 + 1)^{2/3} + C$
45. $2\sqrt{x} + 3\sqrt[3]{x} + 6\sqrt[6]{x} + 6\ln|\sqrt[6]{x} - 1| + C$
47. $\ln[(e^x + 2)^2/(e^x + 1)] + C$
49. $(x - \frac{1}{2})\ln(x^2 - x + 2) - 2x + \sqrt{7}\tan^{-1}((2x - 1)/\sqrt{7}) + C$
51. $-\frac{1}{2}\ln 3 \approx -0.55$
53. $\frac{1}{2}\ln|(x - 2)/x| + C$ **57.** $\frac{1}{5}\ln\left|\dfrac{2\tan(x/2) - 1}{\tan(x/2) + 2}\right| + C$
59. $\frac{1}{4}\ln|\tan(x/2)| + \frac{1}{8}\tan^2(x/2) + C$ **61.** $1 + 2\ln 2$
63. $t = -\ln P - \frac{1}{9}\ln(0.9P + 900) + C$, where $C \approx 10.23$
65. (a) $\dfrac{24{,}110}{4879}\dfrac{1}{5x + 2} - \dfrac{668}{323}\dfrac{1}{2x + 1} - \dfrac{9438}{80{,}155}\dfrac{1}{3x - 7} + \dfrac{1}{260{,}015}\dfrac{22{,}098x + 48{,}935}{x^2 + x + 5}$
(b) $\dfrac{4822}{4879}\ln|5x + 2| - \dfrac{334}{323}\ln|2x + 1| - \dfrac{3146}{80{,}155}\ln|3x - 7| + \dfrac{11{,}049}{260{,}015}\ln(x^2 + x + 5) + \dfrac{75{,}772}{260{,}015\sqrt{19}}\tan^{-1}\dfrac{2x + 1}{\sqrt{19}} + C$
The CAS omits the absolute value signs and the constant of integration.

Exercises 8.5 □ page 546

1. $\sin x + \ln|\csc x - \cot x| + C$
3. $4 - \ln 9$ **5.** $e^{\pi/4} - e^{-\pi/4}$
7. $\frac{243}{5}\ln 3 - \frac{242}{25}$ **9.** $\frac{1}{2}\ln(x^2 - 4x + 5) + \tan^{-1}(x - 2) + C$
11. $\frac{1}{8}\cos^8\theta - \frac{1}{6}\cos^6\theta + C$ (or $\frac{1}{4}\sin^4\theta - \frac{1}{3}\sin^6\theta + \frac{1}{8}\sin^8\theta + C$)
13. $x/\sqrt{1 - x^2} + C$ **15.** $1 - \frac{1}{2}\sqrt{3}$
17. $\frac{1}{4}x^2 - \frac{1}{2}x \sin x \cos x + \frac{1}{4}\sin^2 x + C$
(or $\frac{1}{4}x^2 - \frac{1}{4}x \sin 2x - \frac{1}{8}\cos 2x + C$)
19. $e^{e^x} + C$ **21.** $-\frac{1}{8}e^{-2t}(4t^3 + 6t^2 + 6t + 3) + C$
23. $\frac{4097}{45}$ **25.** $3x + \frac{23}{3}\ln|x - 4| - \frac{5}{3}\ln|x + 2| + C$
27. $\frac{1}{2}(\ln \sin x)^2 + C$ **29.** $15 + 7\ln\frac{2}{7}$
31. $\sin^{-1}x - \sqrt{1 - x^2} + C$
33. $2\sin^{-1}((x + 1)/2) + ((x + 1)/2)\sqrt{3 - 2x - x^2} + C$

35. 0 **37.** $\pi/8 - \frac{1}{4}$ **39.** $-\ln(1 + \sqrt{1-x^2}) + C$
41. $\theta \tan\theta - \frac{1}{2}\theta^2 - \ln|\sec\theta| + C$ **43.** $\frac{2}{3}(1+e^x)^{3/2} + C$
45. $-\frac{1}{3}(x^3+1)e^{-x^3} + C$
47. $\ln\sqrt{x^2+a^2} + \tan^{-1}(x/a) + C$
49. $\ln\left|\dfrac{\sqrt{4x+1}-1}{\sqrt{4x+1}+1}\right| + C$ **51.** $-\ln\left|\dfrac{\sqrt{4x^2+1}+1}{2x}\right| + C$
53. $(1/m)x^2\cosh(mx) - (2/m^2)x\sinh(mx) + (2/m^3)\cosh(mx) + C$
55. $3\ln(\sqrt{x+1}+3) - \ln(\sqrt{x+1}+1) + C$
57. $\frac{3}{7}(x+c)^{7/3} - \frac{3}{4}c(x+c)^{4/3} + C$
59. $e^{-x} + \frac{1}{2}\ln|(e^x-1)/(e^x+1)| + C$
61. $\frac{1}{20}\tan^{-1}(\frac{1}{4}x^5) + C$ **63.** $2(x - 2\sqrt{x}+2)e^{\sqrt{x}} + C$
65. $\frac{2}{3}[(x+1)^{3/2} - x^{3/2}] + C$
67. $\frac{2}{3}\sqrt{3}\pi - \frac{1}{2}\pi - \ln 2$ **69.** $e^x - \ln(1+e^x) + C$
71. $-\frac{1}{4}\ln(x^2+3) + \frac{1}{4}\ln(x^2+1) + C$
73. $\frac{1}{8}\ln|x-2| - \frac{1}{16}\ln(x^2+4) - \frac{1}{8}\tan^{-1}(x/2) + C$
75. $\frac{1}{24}\cos 6x - \frac{1}{16}\cos 4x - \frac{1}{8}\cos 2x + C$ **77.** $\frac{2}{3}\tan^{-1}(x^{3/2}) + C$
79. $\frac{1}{3}x\sin^3 x + \frac{1}{3}\cos x - \frac{1}{9}\cos^3 x + C$ **81.** $xe^{x^2} + C$

Exercises 8.6 □ page 551

1. $(-1/x)\sqrt{7-2x^2} - \sqrt{2}\sin^{-1}(\sqrt{2}x/\sqrt{7}) + C$
3. $(1/(2\pi))\sec(\pi x)\tan(\pi x) + (1/(2\pi))\ln|\sec(\pi x) + \tan(\pi x)| + C$
5. $\pi/4$ **7.** $\frac{1}{25}e^{-3x}(-3\cos 4x + 4\sin 4x) + C$
9. $-\sqrt{4x^2+9}/(9x) + C$ **11.** $e - 2$
13. $-\frac{1}{2}\tan^2(1/z) - \ln|\cos(1/z)| + C$ **15.** $\tan^{-1}(\sinh e^x) + C$
17. $((2y-1)/8)\sqrt{6+4y-4y^2} + \frac{7}{8}\sin^{-1}((2y-1)/\sqrt{7})$
$\quad - \frac{1}{12}(6+4y-4y^2)^{3/2} + C$
19. $\frac{1}{9}\sin^3 x\,[3\ln(\sin x) - 1] + C$ **21.** $\dfrac{1}{2\sqrt{3}}\ln\left|\dfrac{e^x+\sqrt{3}}{e^x-\sqrt{3}}\right| + C$
23. $\frac{1}{4}\tan x\sec^3 x + \frac{3}{8}\tan x\sec x + \frac{3}{8}\ln|\sec x+\tan x| + C$
25. $\frac{1}{2}(\ln x)\sqrt{4+(\ln x)^2} + 2\ln[\ln x + \sqrt{4+(\ln x)^2}] + C$
27. $\sqrt{e^{2x}-1} - \cos^{-1}(e^{-x}) + C$
29. $\frac{1}{5}\ln|x^5 + \sqrt{x^{10}-2}| + C$ **31.** $2\pi^2$
35. $-\frac{1}{4}x(5-x^2)^{3/2} + \frac{5}{8}x\sqrt{5-x^2} + \frac{25}{8}\sin^{-1}(x/\sqrt{5}) + C$
37. $-\frac{1}{5}\sin^2 x\cos^3 x - \frac{2}{15}\cos^3 x + C$
39. $\frac{1}{10}(1+2x)^{5/2} - \frac{1}{6}(1+2x)^{3/2} + C$
41. $-\ln|\cos x| - \frac{1}{2}\tan^2 x + \frac{1}{4}\tan^4 x + C$
43. $\dfrac{2^{x-1}\sqrt{2^{2x}-1}}{\ln 2} - \dfrac{\ln(\sqrt{2^{2x}-1}+2^x)}{2\ln 2} + C$
45. $F(x) = \frac{1}{2}\ln(x^2-x+1) - \frac{1}{2}\ln(x^2+x+1)$;
max. at -1, min. at 1; IP at -1.7, 0, and 1.7

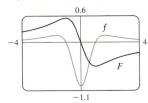

47. $F(x) = -\frac{1}{10}\sin^3 x\cos^7 x - \frac{3}{80}\sin x\cos^7 x + \frac{1}{160}\sin x\cos^5 x$
$\quad + \frac{1}{128}\sin x\cos^3 x + \frac{3}{256}\sin x\cos x + \frac{3}{256}x$;
max. at π, min. at 0; IP at 0.68, $\pi/2$, and 2.46

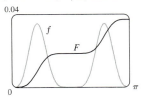

Exercises 8.7 □ page 563

1. (a) $L_2 = 6, R_2 = 12, M_2 \approx 9.6$
(b) L_2 is an underestimate, R_2 and M_2 are overestimates.
(c) $T_2 = 9 < I$ (d) $L_n < T_n < I < M_n < R_n$
3. (a) $T_4 \approx 0.895759$ (underestimate)
(b) $M_4 \approx 0.908907$ (overestimate)
$T_4 < I < M_4$
5. (a) 5.932957, $E_M \approx -0.063353$
(b) 5.869247, $E_S \approx 0.000357$
7. (a) 2.413790 (b) 2.411453 (c) 2.412232
9. (a) 0.146879 (b) 0.147391 (c) 0.147219
11. (a) 0.451948 (b) 0.451991 (c) 0.451976
13. (a) 2.031893 (b) 2.014207 (c) 2.020651
15. (a) -0.495333 (b) -0.543321 (c) -0.526123
17. (a) 1.064275 (b) 1.067416 (c) 1.074915
19. (a) $T_{10} \approx 0.881839$, $M_{10} \approx 0.882202$
(b) $|E_T| \leq 0.01\overline{3}$, $|E_M| \leq 0.00\overline{6}$
(c) $n = 366$ for T_n, $n = 259$ for M_n
21. (a) $T_{10} \approx 1.719713$, $E_T \approx -0.001432$;
$S_{10} \approx 1.718283$, $E_S \approx -0.000001$
(b) $|E_T| \leq 0.002266$, $|E_S| \leq 0.0000016$
(c) $n = 151$ for T_n, $n = 107$ for M_n, $n = 8$ for S_n
23. (a) 2.8 (b) 7.954926518 (c) 0.2894
(d) 7.954926521 (e) The actual error is much smaller.
(f) 10.9 (g) 7.953789422 (h) 0.0593
(i) The actual error is smaller. (j) $n \geq 50$

25.

n	L_n	R_n	T_n	M_n
4	0.140625	0.390625	0.265625	0.242188
8	0.191406	0.316406	0.253906	0.248047
16	0.219727	0.282227	0.250977	0.249512

n	E_L	E_R	E_T	E_M
4	0.109375	-0.140625	-0.015625	0.007813
8	0.058594	-0.066406	-0.003906	0.001953
16	0.030273	-0.032227	-0.000977	0.000488

Observations same as after Example 1.

27.

n	T_n	M_n	S_n
6	4.661488	4.669245	4.666563
12	4.665367	4.667316	4.666659

n	E_T	E_M	E_S
6	0.005178	−0.002578	0.000104
12	0.001300	−0.000649	0.000007

Observations same as after Example 1.
29. (a) 11.5 (b) 12 (c) $11.\overline{6}$
31. (a) 23.44 (b) $0.34\overline{13}$ **33.** 37.73 ft/s
35. 10,177 megawatt-hours   **37.** 828
39. 12.3251   **41.** 59.4
43.

Exercises 8.8 □ page 573

Abbreviations: C, convergent; D, divergent
1. (a) Infinite interval   (b) Infinite discontinuity
 (c) Infinite discontinuity   (d) Infinite interval
3. $\frac{1}{2} - 1/(2t^2)$; 0.495, 0.49995, 0.4999995; 0.5
5. $\frac{1}{12}$ **7.** D **9.** $2e^{-2}$ **11.** D **13.** 0 **15.** D
17. D **19.** $\frac{1}{25}$ **21.** D **23.** $\pi/9$
25. 1 **27.** $2\sqrt{3}$ **29.** D **31.** D **33.** $\frac{75}{4}$
35. D **37.** D **39.** $\frac{8}{3}\ln 2 - \frac{8}{9}$
41. e **43.** $\frac{2\pi}{3}$

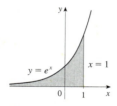 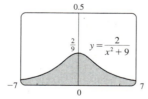

45. Infinite area

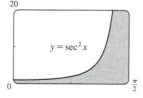

47. (a)

t	$\int_1^t [(\sin^2 x)/x^2]\,dx$
2	0.447453
5	0.577101
10	0.621306
100	0.668479
1,000	0.672957
10,000	0.673407

It appears that the integral is convergent.

(c)

49. C   **51.** C   **53.** D   **55.** π **57.** $p < 1, 1/(1-p)$
59. $p > -1, -1/(p+1)^2$ **65.** $\sqrt{2GM/R}$
67. (a)

(b) The rate at which the fraction $F(t)$ increases as t increases
(c) 1; all bulbs burn out eventually
69. 1000
71. (a) $F(s) = 1/s, s > 0$ (b) $F(s) = 1/(s-1), s > 1$
 (c) $F(s) = 1/s^2, s > 0$
77. $C = 1; \ln 2$

Chapter 8 Review □ page 576

True-False Quiz

1. False   **3.** False   **5.** False   **7.** False
9. (a) True   (b) False   **11.** False   **13.** False

Exercises

1. $5 + 10 \ln \frac{2}{3}$ **3.** $\ln 2$
5. $\frac{1}{9}\sec^9 x - \frac{3}{7}\sec^7 x + \frac{3}{5}\sec^5 x - \frac{1}{3}\sec^3 x + C$
7. $-\cos(\ln t) + C$ **9.** $\frac{64}{5}\ln 4 - \frac{124}{25}$
11. $\sqrt{3} - (\pi/3)$ **13.** $\ln|x| - \frac{1}{2}\ln(x^2 + 1) + C$
15. $\frac{1}{3}\sin^3\theta - \frac{2}{5}\sin^5\theta + \frac{1}{7}\sin^7\theta + C$
17. $x\sec x - \ln|\sec x + \tan x| + C$
19. $\frac{1}{18}\ln(9x^2 + 6x + 5) + \frac{1}{9}\tan^{-1}((3x+1)/2) + C$
21. $\ln|x - 2 + \sqrt{x^2 - 4x}| + C$
23. $-\frac{1}{12}(\cot^3 4x + 3\cot 4x) + C$
25. $\frac{3}{2}\ln(x^2+1) - 3\tan^{-1}x + \sqrt{2}\tan^{-1}(x/\sqrt{2}) + C$
27. $\frac{2}{5}$ **29.** 0   **31.** $6 - 3\pi/2$
33. $(x/\sqrt{4-x^2}) - \sin^{-1}(x/2) + C$
35. $4\sqrt{1+\sqrt{x}} + C$ **37.** $\frac{1}{2}\sin 2x - \frac{1}{8}\cos 4x + C$
39. $\frac{1}{8}e - \frac{1}{4}$ **41.** $\frac{1}{36}$ **43.** D
45. $4\ln 4 - 8$ **47.** D   **49.** $\pi/4$
51. $(x+1)\ln(x^2+2x+2) + 2\arctan(x+1) - 2x + C$
53. 0   **55.** $\frac{1}{2}[e^x\sqrt{1-e^{2x}} + \sin^{-1}(e^x)] + C$
57. $\frac{1}{4}(2x+1)\sqrt{x^2+x+1} + \frac{3}{8}\ln(x + \frac{1}{2} + \sqrt{x^2+x+1}) + C$
61. No
63. (a) 1.090608 (overestimate)   (b) 1.088840 (underestimate)
 (c) 1.089429 (unknown)
65. (a) $0.00\overline{6}, n \geq 259$ (b) $0.00\overline{3}, n \geq 183$

67. 8.6 mi   **69.** (a) 3.8   (b) 1.7867, 0.000646   (c) $n \geq 30$
71. C   **73.** 2   **75.** $3\pi^2/16$

Problems Plus □ page 580

1. About 1.85 inches from the center   **3.** 0
5. $f(\pi) = -\pi/2$ **9.** $(b^b a^{-a})^{1/(b-a)} e^{-1}$ **11.** $2 - \sin^{-1}(2/\sqrt{5})$

Chapter 9

Exercises 9.1 □ page 588

1. $3\sqrt{10}$ **3.** $\frac{46}{3}$ **5.** $\frac{2}{243}(82\sqrt{82} - 1)$ **7.** $\frac{1261}{240}$ **9.** $\frac{32}{3}$
11. $\ln(\sqrt{2} + 1)$ **13.** $\sinh 1$
15. $\sqrt{1 + e^2} - \sqrt{2} + \ln(\sqrt{1 + e^2} - 1) - 1 - \ln(\sqrt{2} - 1)$
17. $\int_0^{2\pi} \sqrt{1 + \sin^2 x}\, dx$ **19.** $\int_1^4 \sqrt{9y^4 + 6y^2 + 2}\, dy$
21. 5.115840   **23.** 1.569619
25. (a), (b) 3

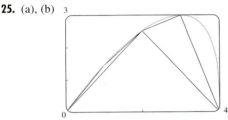

$L_1 = 4$,
$L_2 \approx 6.43$,
$L_4 \approx 7.50$

(c) $\int_0^4 \sqrt{1 + [4(3 - x)/(3(4 - x)^{2/3})]^2}\, dx$ (d) 7.7988
27. $\ln 3 - \frac{1}{2}$ **29.** 6

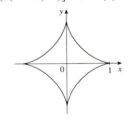

31. $s(x) = \frac{2}{27}[(1 + 9x)^{3/2} - 10\sqrt{10}]$
33. 209.1 m   **35.** 29.36 in.   **37.** 12.4

Exercises 9.2 □ page 595

1. $\int_1^3 2\pi \ln x \sqrt{1 + (1/x)^2}\, dx$
3. $\int_0^{\pi/4} 2\pi x \sqrt{1 + (\sec x \tan x)^2}\, dx$
5. $\pi(145\sqrt{145} - 1)/27$ **7.** $\pi(37\sqrt{37} - 17\sqrt{17})/6$
9. $\pi[1 + \frac{1}{4}(e^2 - e^{-2})]$ **11.** $21\pi/2$
13. $\pi(145\sqrt{145} - 10\sqrt{10})/27$ **15.** πa^2
17. 9.023754   **19.** 13.527296
21. $(\pi/4)[4 \ln(\sqrt{17} + 4) - 4 \ln(\sqrt{2} + 1) - \sqrt{17} + 4\sqrt{2}]$
23. $(\pi/6)[\ln(\sqrt{10} + 3) + 3\sqrt{10}]$
27. (a) $\pi a^2/3$ (b) $56\pi\sqrt{3}\, a^2/45$
29. $2\pi[b^2 + a^2 b \sin^{-1}(\sqrt{a^2 - b^2}/a)/\sqrt{a^2 - b^2}]$

31. $\int_a^b 2\pi[c - f(x)]\sqrt{1 + [f'(x)]^2}\, dx$
33. $4\pi^2 r^2$

Exercises 9.3 □ page 605

1. (a) 187.5 lb/ft^2 (b) 1875 lb   (c) 562.5 lb
3. 6000 lb   **5.** 6.5×10^6 N   **7.** 3.5×10^4 lb
9. $1000g\pi r^3$ N   **11.** 5.27×10^5 N
13. (a) 314 N   (b) 353 N
15. (a) $\approx 5.63 \times 10^3$ lb   (b) $\approx 5.06 \times 10^4$ lb
(c) $\approx 4.88 \times 10^4$ lb   (d) $\approx 3.03 \times 10^5$ lb
17. 2.5×10^5 N   **19.** $230; \frac{23}{7}$ **21.** $10; 1; (\frac{1}{21}, \frac{10}{21})$
23. $(0, 1.6)$ **25.** $(1/(e - 1), (e + 1)/4)$ **27.** $(\frac{2}{5}, \frac{1}{2})$
29. $((\pi\sqrt{2} - 4)/[4(\sqrt{2} - 1)], 1/[4(\sqrt{2} - 1)])$
31. $(2, 0)$ **33.** $\frac{4}{3}, 0, (0, \frac{2}{3})$ **35.** $(0.781, 1.330)$
39. $(0, \frac{1}{12})$ **41.** $\frac{1}{3}\pi r^2 h$

Exercises 9.4 □ page 610

1. \$38,000   **3.** \$43,866,933.33   **5.** \$407.25
7. \$12,000   **9.** 3727; \$37,753
11. $\frac{2}{3}(16\sqrt{2} - 8) \approx \$9.75$ million
13. 1.19×10^{-4} cm^3/s   **15.** $\frac{1}{9}$ L/s

Exercises 9.5 □ page 617

1. (a) The probability that a randomly chosen tire will have a lifetime between 30,000 and 40,000 miles
(b) The probability that a randomly chosen tire will have a lifetime of at least 25,000 miles
3. (a) $f(x) \geq 0$ for all x and $\int_{-\infty}^{\infty} f(x)\, dx = 1$
(b) $1 - \frac{3}{8}\sqrt{3} \approx 0.35$
5. (a) $f(x) \geq 0$ for all x and $\int_{-\infty}^{\infty} f(x)\, dx = 1$ (b) 5
7. $5 \ln 2 \approx 3.47$ min
9. (a) $e^{-4/2.5} \approx 0.20$ (b) $1 - e^{-2/2.5} \approx 0.55$
(c) If you aren't served within 10 minutes, you get a free hamburger.
11. $\approx 44\%$ **13.** ≈ 0.9545
15. (b) 0; a_0 (c) 1×10^{10}

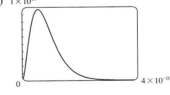

(d) $1 - 41e^{-8} \approx 0.986$ (e) $\frac{3}{2} a_0$

Chapter 9 Review □ page 619

Exercises

1. $\frac{15}{2}$ **3.** (a) $\frac{21}{16}$ (b) $41\pi/10$ **5.** 3.292287   **7.** $\frac{124}{5}$
9. ≈ 458 lb   **11.** $(-\frac{1}{2}, \frac{12}{5})$ **13.** $(2, \frac{2}{3})$
15. $2\pi^2$ **17.** \$7166.67
19. (a) $f(x) \geq 0$ for all x and $\int_{-\infty}^{\infty} f(x)\, dx = 1$
(b) ≈ 0.3455 (c) 5, yes

21. (a) $1 - e^{-3/8} \approx 0.31$ (b) $e^{-5/4} \approx 0.29$
(c) $8 \ln 2 \approx 5.55$ min

Problems Plus □ page 620

1. $2\pi/3 - \sqrt{3}/2$
3. (a) $2\pi r(r \pm d)$ (b) $\approx 3{,}360{,}000$ mi^2
(d) $\approx 78{,}400{,}000$ mi^2
5. (a) $P(z) = P_0 + g \int_0^z \rho(x)\, dx$
(b) $(P_0 - \rho_0 g H)(\pi r^2) + \rho_0 g H e^{L/H} \int_{-r}^{r} e^{x/H} \cdot 2\sqrt{r^2 - x^2}\, dx$
7. Height $\sqrt{2}\, b$, volume $\left(\frac{28}{27}\sqrt{6} - 2\right)\pi b^3$
9. 0.14 m   **11.** $2/\pi$, $1/\pi$

Chapter 10

Exercises 10.1 □ page 627

3. (a) ± 3 **5.** (b) and (c)
7. (a) It must be either 0 or decreasing
(c) $y = 0$ (d) $y = 1/(x + 2)$
9. (a) $0 < P < 4200$ (b) $P > 4200$
(c) $P = 0$, $P = 4200$
13. (a) At the beginning; stays positive, but decreases
(c)

Exercises 10.2 □ page 635

1. (a)
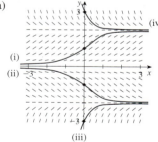
(b) $y = 0$, $y = 2$, $y = -2$

3. IV   **5.** III
7.
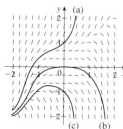
9.
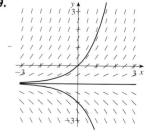

11.
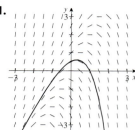
13.
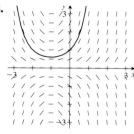

15.
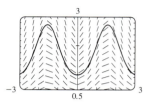

17. $-2 \le c \le 2$; $-2, 0, 2$
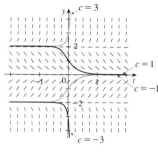

19. (a) (i) 1.4   (ii) 1.44   (iii) 1.4641
(b)
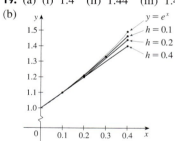
Underestimates

(c) (i) 0.0918   (ii) 0.0518   (iii) 0.0277
It appears that the error is also halved (approximately).
21. $-1, -3, -6.5, -12.25$ **23.** 1.7616
25. (a) (i) 3   (ii) 2.3928   (iii) 2.3701   (iv) 2.3681
(c) (i) -0.6321 (ii) -0.0249 (iii) -0.0022
(iv) -0.0002
It appears that the error is also divided by 10 (approximately).
27. (a), (d)   (b) 3
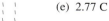
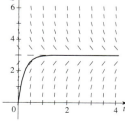
(c) Yes; $Q = 3$
(e) 2.77 C

Exercises 10.3 ▫ page 643

1. $y = Kx$ **3.** $y = K\sqrt{x^2 + 1}$
5. $y + \ln|\sec y| = \frac{1}{3}x^3 + x + C$
7. $y = \pm\sqrt{[3(te^t - e^t + C)]^{2/3} - 1}$
9. $u = Ae^{2t+t^2/2} - 1$ **11.** $y = \tan(x - 1)$
13. $\cos x + x \sin x = y^2 + \frac{1}{3}e^{3y} + \frac{2}{3}$
15. $u = -\sqrt{t^2 + \tan t + 25}$
17. $y = \dfrac{4a}{\sqrt{3}} \sin x - a$ **19.** $y = 7e^{x^4}$
21. (a) $\sin^{-1} y = x^2 + C$
(b) $y = \sin(x^2)$,
$-\sqrt{\pi/2} \leq x \leq \sqrt{\pi/2}$

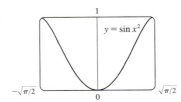

(c) No
23. $\cos y = \cos x - 1$

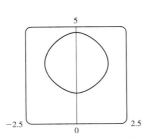

25. (a), (c)   (b) $y = \pm\sqrt{2(x + C)}$

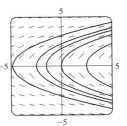

27. $x^2 + 2y^2 = C$ **29.** $y^3 = 3(x + C)$

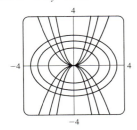 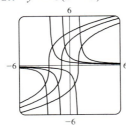

31. $Q(t) = 3 - 3e^{-4t}$; 3   **33.** $P(t) = M - Me^{-kt}$; M
35. (a) $x = a - 4/(kt + 2/\sqrt{a})^2$
(b) $t = \dfrac{2}{k\sqrt{a-b}}\left(\tan^{-1}\sqrt{\dfrac{b}{a-b}} - \tan^{-1}\sqrt{\dfrac{b-x}{a-b}}\right)$
37. (a) $C(t) = (C_0 - r/k)e^{-kt} + r/k$
(b) r/k; the concentration approaches r/k regardless of the value of C_0
21. (a) $\sin^{-1} y = x^2 + C$

39. (a) $15e^{-t/100}$ kg   (b) $15e^{-0.2} \approx 12.3$ kg   **41.** g/k
43. (a) $dA/dt = k\sqrt{A}\,(M - A)$
(b) $A(t) = M[(Ce^{\sqrt{M}kt} - 1)/(Ce^{\sqrt{M}kt} + 1)]^2$, where
$C = (\sqrt{M} + \sqrt{A_0})/(\sqrt{M} - \sqrt{A_0})$ and $A_0 = A(0)$

Exercises 10.4 ▫ page 656

1. About 235
3. (a) $500 \times 16^{t/3}$ (b) $\approx 20{,}159$
(c) 18,631 cells/h   (d) $(3 \ln 60)/\ln 16 \approx 4.4$ h
5. (a) 1508 million, 1871 million   (b) 2161 million
(c) 3972 million; wars in the first half of century, increased life expectancy in second half
7. (a) $Ce^{-0.0005t}$ (b) $-2000 \ln 0.9 \approx 211$ s
9. (a) $100 \times 2^{-t/30}$ mg   (b) ≈ 9.92 mg
(c) ≈ 199.3 years
11. ≈ 2500 years
13. (a) $\approx 137\,°F$ (b) ≈ 116 min
15. (a) $13.3\,°C$ (b) ≈ 67.74 min
17. (a) ≈ 64.5 kPa   (b) ≈ 39.9 kPa
19. (a) (i) \$3828.84   (ii) \$3840.25   (iii) \$3850.08
(iv) \$3851.61   (v) \$3852.01   (vi) \$3852.08
(b) $dA/dt = 0.05A$, $A(0) = 3000$
21. (a) $P(t) = (m/k) + (P_0 - m/k)e^{kt}$ (b) $m < kP_0$
(c) $m = kP_0, m > kP_0$ (d) Declining

Exercises 10.5 ▫ page 665

1. (a) 100; 0.05
(b) Where P is close to 0 or 100; on the line $P = 50$;
$0 < P_0 < 100$; $P_0 > 100$
(c)

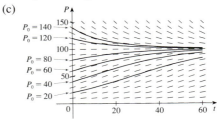

Solutions approach 100; some increase and some decrease, some have an inflection point but others don't; solutions with $P_0 = 20$ and $P_0 = 40$ have inflection points at $P = 50$
(d) $P = 0, P = 100$; other solutions move away from $P = 0$ and toward $P = 100$
3. (a) 3.23×10^7 kg   (b) ≈ 1.55 years
5. (a) $dP/dt = \frac{1}{265}P(1 - P/100)$, P in billions
(b) 5.49 billion   (c) In billions: 7.81, 27.72
(d) In billions: 5.48, 7.61, 22.41
7. (a) $dy/dt = ky(1 - y)$
(b) $y = y_0/[y_0 + (1 - y_0)e^{-kt}]$
(c) 3:36 P.M.
11. (a) Fish are caught at a rate of 15 per week.
(b) See part (d)   (c) $P = 250, P = 750$

(d)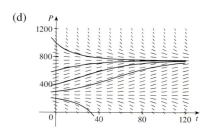

$0 < P_0 < 250: P \to 0;\ P_0 = 250: P \to 250;\ P_0 > 250: P \to 750$

(e) $P(t) = \dfrac{250 - 750ke^{t/25}}{1 - ke^{t/25}}$, where $k = \frac{1}{11}, -\frac{1}{9}$

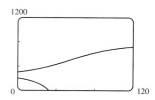

13. (b)

$0 < P_0 < 200: P \to 0;\ P_0 = 200: P \to 200;\ P_0 > 200: P \to 1000$

(c) $P(t) = \dfrac{m(K - P_0) + K(P_0 - m)e^{(K-m)(k/K)t}}{K - P_0 + (P_0 - m)e^{(K-m)(k/K)t}}$

15. (a) $P(t) = P_0 e^{(k/r)[\sin(rt - \phi) + \sin \phi]}$ (b) Does not exist

Exercises 10.6 □ page 672

1. No   **3.** Yes   **5.** $y = \frac{2}{3}e^x + Ce^{-2x}$
7. $y = x^2 \ln|x| + Cx^2$ **9.** $y = \frac{2}{3}\sqrt{x} + C/x$
11. $y = \frac{1}{2}x + Ce^{-x^2} - \frac{1}{2}e^{-x^2}\int e^{x^2}\,dx$
13. $u = (t^2 + 2t + 2C)/[2(t + 1)]$
15. $y = -x - 1 + 3e^x$ **17.** $v = t^3 e^{t^2} + 5e^{t^2}$
19. $y = -x \cos x - x$
21. $y = \sin x + (\cos x)/x + C/x$

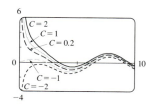

25. $y = \pm[Cx^4 + 2/(5x)]^{-1/2}$
27. (a) $I(t) = 4 - 4e^{-5t}$ (b) $4 - 4e^{-1/2} \approx 1.57$ A
29. $Q(t) = 3(1 - e^{-4t}),\ I(t) = 12e^{-4t}$

31. $P(t) = M + Ce^{-kt}$

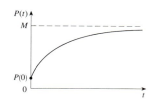

33. $y = \frac{2}{5}(100 + 2t) - 40{,}000(100 + 2t)^{-3/2};\ 0.2275$ kg/L
35. (b) mg/c (c) $(mg/c)[t + (m/c)e^{-ct/m}] - m^2 g/c^2$

Exercises 10.7 □ page 678

1. (a) $x =$ predators, $y =$ prey; growth is restricted only by predators, which feed only on prey.
(b) $x =$ prey, $y =$ predators; growth is restricted by carrying capacity and by predators, which feed only on prey.

3. (a) The rabbit population starts at about 300, increases to 2400, then decreases back to 300. The fox population starts at 100, decreases to about 20, increases to about 315, decreases to 100, and the cycle starts again.

(b)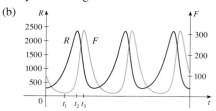

5.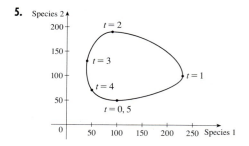

9. (a) Population stabilizes at 5000.
(b) (i) $W = 0, R = 0$: Zero populations
(ii) $W = 0, R = 5000$: In the absence of wolves, the rabbit population is always 5000.
(iii) $W = 64, R = 1000$: Both populations are stable.
(c) The populations stabilize at 1000 rabbits and 64 wolves.
(d)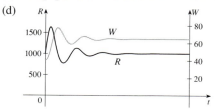

Chapter 10 Review □ page 684

True-False Quiz

1. True 3. False 5. True 7. True

Exercises

1. (a)

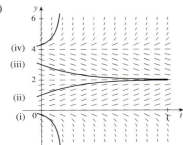

(b) $0 \le c \le 4$; $y = 0$, $y = 2$, $y = 4$

3. (a) $y(0.3) \approx 0.8$

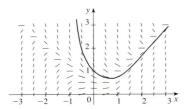

(b) 0.75676
(c) $y = x$ and $y = -x$; there is a local maximum or minimum
5. $y = (\frac{1}{2}x^2 + C)e^{-\sin x}$ 7. $y^3 + y^2 = \cos x + x \sin x + C$
9. $y = \sqrt{(\ln x)^2 + 4}$ 11. $y = e^{-x}(\frac{2}{3}x^{3/2} + 3)$
13. $y^2 - 2\ln|y| + x^2 = C$
15. (a) 1000×3^t (b) 27,000
(c) $27{,}000 \ln 3 \approx 29{,}663$ bacteria per hour
(d) $(\ln 2)/\ln 3 \approx 0.63$ h
17. (a) $C_0 e^{-kt}$ (b) ≈ 100 h
19. (a) $L(t) = L_\infty - [L_\infty - L(0)]e^{-kt}$
(b) $L(t) = 53 - 43e^{-0.2t}$
21. 15 days 23. $k \ln h + h = (-R/V)t + C$
25. (a) Stabilizes at 200,000
(b) (i) $x = 0$, $y = 0$: Zero populations
(ii) $x = 200{,}000$, $y = 0$: In the absence of birds, the insect population is always 200,000.
(iii) $x = 25{,}000$, $y = 175$: Both populations are stable.
(c) The populations stabilize at 25,000 insects and 175 birds.
(d)

27. (a) $y = (1/k)\cosh kx + a - 1/k$ or
$y = (1/k)\cosh kx - (1/k)\cosh kb + h$ (b) $(2/k)\sinh kb$

Problems Plus □ page 684

1. $f(x) = \pm 10e^x$ 5. $20°C$
7. (b) $f(x) = (x^2 - L^2)/(4L) - (L/2)\ln(x/L)$ (c) No
9. (a) 9.8 h
(b) $31{,}900\pi \approx 100{,}000$ ft²; 6283 ft²/h
(c) 5.1 h
11. $x^2 + (y - 6)^2 = 25$

Chapter 11

Exercises 11.1 □ page 692

1.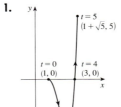

3.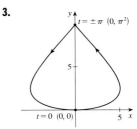

5. (a) 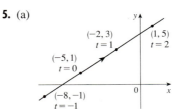 (b) $y = \frac{2}{3}x + \frac{13}{3}$

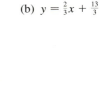

7. (a)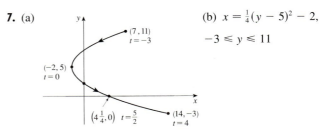

(b) $x = \frac{1}{4}(y - 5)^2 - 2$, $-3 \le y \le 11$

9. (a)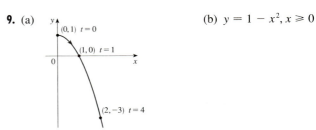

(b) $y = 1 - x^2$, $x \ge 0$

11. (a) $x^2 + y^2 = 1, x \geq 0$
(b)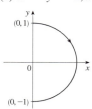

13. (a) $x + y = 1, 0 \leq x \leq 1$
(b)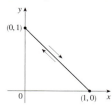

15. (a) $y = 1/x, x > 0$
(b)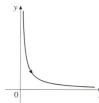

17. (a) $x^2 - y^2 = 1, x \geq 1$
(b)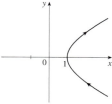

19. Moves counterclockwise along the circle $x^2 + y^2 = 1$ from $(-1, 0)$ to $(1, 0)$

21. Moves once clockwise around the ellipse $(x^2/4) + (y^2/9) = 1$, starting and ending at $(0, 3)$

23. It is contained in the rectangle described by $1 \leq x \leq 4$ and $2 \leq y \leq 3$.

25.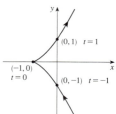

27.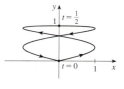

29.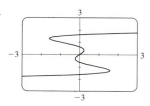

31. (b) $x = -2 + 5t, y = 7 - 8t, 0 \leq t \leq 1$
33. (a) $x = 2 \cos t, y = 1 - 2 \sin t, 0 \leq t \leq 2\pi$
(b) $x = 2 \cos t, y = 1 + 2 \sin t, 0 \leq t \leq 6\pi$
(c) $x = 2 \cos t, y = 1 + 2 \sin t, \pi/2 \leq t \leq 3\pi/2$
35. (a) $x = a \sin t, y = b \cos t, 0 \leq t \leq 2\pi$
(b)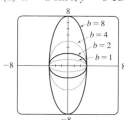
(c) As b increases, the ellipse stretches vertically.

39. $x = a \cos \theta, y = b \sin \theta$; $(x^2/a^2) + (y^2/b^2) = 1$, ellipse

41.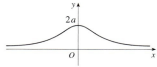

43. (a) Two points of intersection
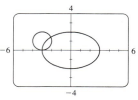
(b) One collision point at $(-3, 0)$ when $t = 3\pi/2$
(c) There are still two intersection points, but no collision point.
45. For $c = 0$, there is a cusp; for $c > 0$, there is a loop whose size increases as c increases.

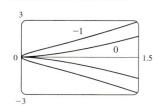 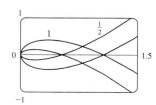

47. As n increases, the number of oscillations increases; a and b determine the width and height.

Exercises 11.2 □ page 702

1. $5/(3t^2 - 1)$ **3.** $y = -x$
5. $y = -(2/e)x + 3$ **7.** $y = -2x + 3$
9. $y = (\sqrt{3}/2)x - \frac{1}{2}$

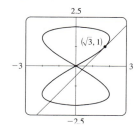

11. $1 + \frac{3}{2}t, 3/(4t), t > 0$ **13.** $-e^{-t}, e^{-t}/(1 - e^t), t < 0$
15. $-\frac{3}{2} \tan t, -\frac{3}{4} \sec^3 t, \frac{\pi}{2} < t < \frac{3\pi}{2}$
17. Horizontal at $(6, \pm 16)$, vertical at $(10, 0)$
19. Horizontal at $(\pm \sqrt{2}, \pm 1)$ (four points), vertical at $(\pm 2, 0)$
21. $(-0.25, 0.36), (-\frac{1}{4}, (1/\sqrt{2}) - \frac{1}{2} \ln 2)$
23.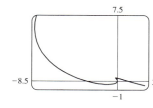
25. $y = x, y = -x$
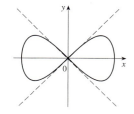

27. (a) $d \sin \theta/(r - d \cos \theta)$ **29.** $(-5, 6)$, $\left(-\frac{208}{27}, \frac{32}{3}\right)$
31. πab **33.** $(e^{\pi/2} - 1)/2$ **35.** $2\pi r^2 + \pi d^2$
37. $\int_1^2 \sqrt{1 + 4t^2}\, dt$ **39.** $\int_0^{2\pi} \sqrt{3 - 2\sin t - 2\cos t}\, dt$
41. $4\sqrt{2} - 2$
43. $-\sqrt{10}/3 + \ln(3 + \sqrt{10}) + \sqrt{2} - \ln(1 + \sqrt{2})$
45. $\sqrt{2}(e^\pi - 1)$

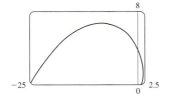

47. $e^3 + 11 - e^{-8}$

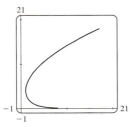

49. 612.3053   **51.** $6\sqrt{2}$, $\sqrt{2}$
55. (a) $t \in [0, 4\pi]$ (b) ≈ 294

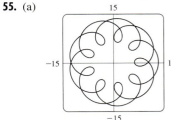

57. $\int_1^2 \frac{8}{3}\pi t^{3/2}\sqrt{1 + 4t^2}\, dt$ **59.** $2\pi(247\sqrt{13} + 64)/1215$
61. $6\pi a^2/5$ **63.** 59.101   **65.** $24\pi(949\sqrt{26} + 1)/5$
71. $\frac{1}{4}$

Exercises 11.3 □ page 713

1. (a)

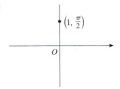

(b)

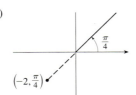

$(1, 5\pi/2), (-1, 3\pi/2)$ $(2, 5\pi/4), (-2, 9\pi/4)$

(c)

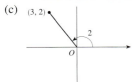

$(3, 2 + 2\pi), (-3, 2 + \pi)$

3. (a)

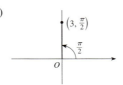

(b)

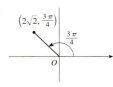

$(0, 3)$ $(-2, 2)$

(c)

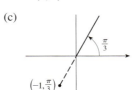

$\left(-\frac{1}{2}, -\sqrt{3}/2\right)$

5. (a) (i) $(\sqrt{2}, \pi/4)$ (ii) $(-\sqrt{2}, 5\pi/4)$
(b) (i) $(4, 11\pi/6)$ (ii) $(-4, 5\pi/6)$

7.

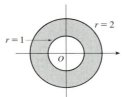

9.

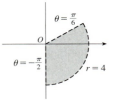

11.

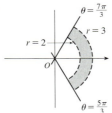

13. $\frac{1}{2}\sqrt{40 + 6\sqrt{6} - 6\sqrt{2}}$
15. Circle, center O, radius 2   **17.** Circle, center $\left(0, \frac{3}{2}\right)$, radius $\frac{3}{2}$
19. Horizontal line, 1 unit above the x-axis
21. $r = 3 \sec \theta$ **23.** $r = -\cot \theta \csc \theta$ **25.** $r = 2c \cos \theta$
27. (a) $\theta = \pi/6$ (b) $x = 3$

29.

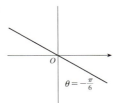

31.

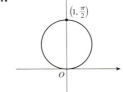

33.

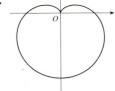

35.
37.
39.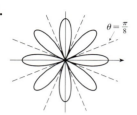
41.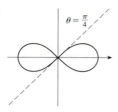
43.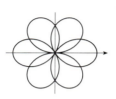
45.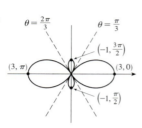
47.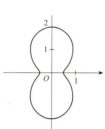
49.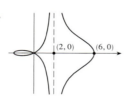
51.

53. (a) For $c < -1$, the loop begins at $\theta = \sin^{-1}(-1/c)$ and ends at $\theta = \pi - \sin^{-1}(-1/c)$; for $c > 1$, it begins at $\theta = \pi + \sin^{-1}(1/c)$ and ends at $\theta = 2\pi - \sin^{-1}(1/c)$.
55. $\sqrt{3}$ **57.** $-\pi$ **59.** -1
61. Horizontal at $(3/\sqrt{2}, \pi/4)$, $(-3/\sqrt{2}, 3\pi/4)$; vertical at $(3, 0)$, $(0, \pi/2)$
63. Horizontal at $(\frac{3}{2}, \pi/3)$, $(\frac{3}{2}, 5\pi/3)$, and the pole; vertical at $(2, 0)$, $(\frac{1}{2}, 2\pi/3)$, $(\frac{1}{2}, 4\pi/3)$
65. Horizontal at $(-1, 3\pi/2)$, $(-1, \pi/2)$, $(\frac{2}{3}, \alpha)$, $(\frac{2}{3}, \pi - \alpha)$, $(\frac{2}{3}, \pi + \alpha)$, $(\frac{2}{3}, 2\pi - \alpha)$, where $\alpha = \sin^{-1}(1/\sqrt{6})$; vertical at $(1, 0)$, $(1, \pi)$, $(-\frac{2}{3}, \beta)$, $(-\frac{2}{3}, \pi - \beta)$, $(-\frac{2}{3}, \pi + \beta)$, $(-\frac{2}{3}, 2\pi - \beta)$, where $\beta = \cos^{-1}(1/\sqrt{6})$
67. Center $(b/2, a/2)$, radius $\sqrt{a^2 + b^2}/2$

69.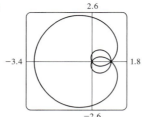
71.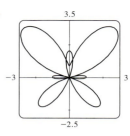
73.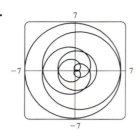

75. By counterclockwise rotation through angle $\pi/6$, $\pi/3$, or α about the origin
77. (a) A rose with n loops if n is odd and $2n$ loops if n is even
(b) Number of loops is always $2n$
79. For $0 < a < 1$, the curve is an oval, which develops a dimple as $a \to 1^-$. When $a > 1$, the curve splits into two parts, one of which has a loop.

Exercises 11.4 □ page 719

1. $\pi^2/64$ **3.** $\pi/12 + \sqrt{3}/8$ **5.** $\pi^3/6$ **7.** $41\pi/4$
9. $9\pi/4$ **11.** 4

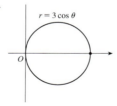

13. π **15.** 3π

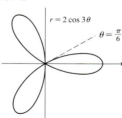

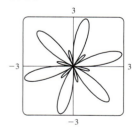

17. $\pi/8$ **19.** $9\pi/20$ **21.** $\pi - (3\sqrt{3}/2)$
23. $(4\pi/3) + 2\sqrt{3}$ **25.** $4\sqrt{3} - \frac{4}{3}\pi$ **27.** π
29. $(\pi - 2)/8$ **31.** $(\pi/2) - 1$
33. $(19\pi/3) - (11\sqrt{3}/2)$
35. $(\pi + 3\sqrt{3})/4$ **37.** $(1/\sqrt{2}, \pi/4)$ and the pole
39. $(\frac{1}{2}, \pi/3)$, $(\frac{1}{2}, 5\pi/3)$, and the pole
41. $(\sqrt{3}/2, \pi/3)$, $(\sqrt{3}/2, 2\pi/3)$, and the pole
43. Intersection at $\theta \approx 0.89, 2.25$; area ≈ 3.46

45. π **47.** $\frac{8}{3}[(\pi^2+1)^{3/2}-1]$ **49.** 29.0653
51. 9.6884
53. $\frac{16}{3}$

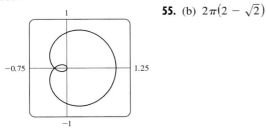

55. (b) $2\pi(2-\sqrt{2})$

Exercises 11.5 □ page 726
1. $(0,0)$, $(\frac{1}{8},0)$, $x=-\frac{1}{8}$ **3.** $(0,0)$, $(0,-\frac{1}{16})$, $y=\frac{1}{16}$

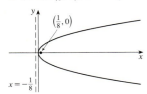 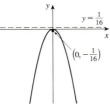

5. $(-2,3)$, $(-2,5)$, $y=1$ **7.** $(-2,-1)$, $(-5,-1)$, $x=1$

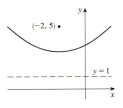 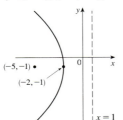

9. $x=-y^2$, focus $(-\frac{1}{4},0)$, directrix $x=\frac{1}{4}$
11. $(\pm 3, 0)$, $(\pm 2, 0)$ **13.** $(0, \pm 4)$, $(0, \pm 2\sqrt{3})$

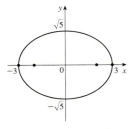 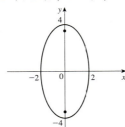

15. $(1, \pm 3)$, $(1, \pm\sqrt{5})$

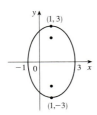

17. $\dfrac{x^2}{4}+\dfrac{y^2}{9}=1$, foci $(0, \pm\sqrt{5})$

19. $(\pm 12, 0)$, $(\pm 13, 0)$, $y = \pm\frac{5}{12}x$ **21.** $(0, \pm 2)$, $(0, \pm 2\sqrt{2})$, $y = \pm x$

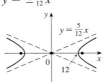

23. $(2\pm\sqrt{6}, 1)$, $(2\pm\sqrt{15}, 1)$, $y-1=\pm(\sqrt{6}/2)(x-2)$

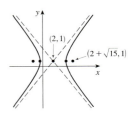

25. Parabola, $(0,-1)$, $(0,-\frac{3}{4})$ **27.** Ellipse, $(\pm\sqrt{2}, 1)$, $(\pm 1, 1)$
29. Hyperbola, $(0,1)$, $(0,3)$; $(0, -1\pm\sqrt{5})$
31. $x^2=-8y$ **33.** $y^2=-12(x+1)$
35. $y^2=16x$ **37.** $(x^2/25)+(y^2/21)=1$
39. $(x^2/12)+[(y-4)^2/16]=1$
41. $[(x-2)^2/9]+[(y-2)^2/5]=1$ **43.** $y^2-(x^2/8)=1$
45. $[(x-4)^2/4]-[(y-3)^2/5]=1$
47. $(x^2/9)-(y^2/36)=1$
49. $(x^2/3{,}763{,}600)+(y^2/3{,}753{,}196)=1$
51. (a) $(121x^2/1{,}500{,}625)-(121y^2/3{,}339{,}375)=1$
(b) ≈ 248 mi
55. (a) Ellipse (b) Hyperbola (c) No curve **57.** 9.69

Exercises 11.6 □ page 731
1. $r=42/(4+7\sin\theta)$ **3.** $r=15/(4-3\cos\theta)$
5. $r=8/(1-\sin\theta)$ **7.** $r=4/(2+\cos\theta)$
9. (a) 1 (b) Parabola (c) $y=1$
(d)

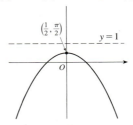

11. (a) $\frac{1}{4}$ (b) Ellipse (c) $y=-12$
(d)

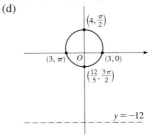

13. (a) $\frac{1}{3}$ (b) Ellipse (c) $x = \frac{9}{2}$
(d)
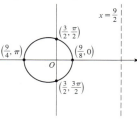

15. (a) 2 (b) Hyperbola (c) $x = -\frac{3}{8}$
(d)
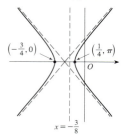

17. (a) $e = \frac{3}{4}$, directrix $x = -\frac{1}{3}$
(b) $r = 1/[4 - 3\cos(\theta - \pi/3)]$

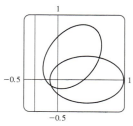

19. The ellipse is nearly circular when e is close to 0 and becomes more elongated as $e \to 1^-$. At $e = 1$, the curve becomes a parabola.

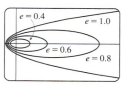

25. (b) $r = (1.49 \times 10^8)/(1 - 0.017 \cos \theta)$
27. 35.64 AU
29. 7.0×10^7 km   **31.** 3.6×10^8 km

Chapter 11 Review □ page 733

True-False Quiz

1. False   **3.** False   **5.** True   **7.** False   **9.** True

Exercises

1. $x = y^2 - 8y + 12$ **3.** $y = 1/x$

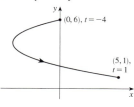   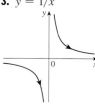

5. $x = t$, $y = \sqrt{t}$, $t \geq 0$; $x = t^4$, $y = t^2$;
$x = \tan^2 t$, $y = \tan t$, $0 \leq t < \frac{\pi}{2}$

7. 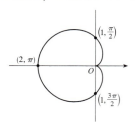   **9.**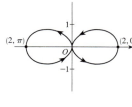

11. 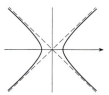   **13.**

15. $r = 2/(\cos \theta + \sin \theta)$

17.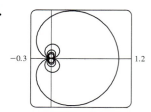

19. 2   **21.** -1
23. $(\sin t + t \cos t)/(\cos t - t \sin t)$, $(t^2 + 2)/(\cos t - t \sin t)^3$
25. $\left(\frac{11}{8}, \frac{3}{4}\right)$
27. Vertical tangent at $(3a/2, \pm\sqrt{3}\, a/2)$, $(-3a, 0)$; horizontal tangent at $(a, 0)$, $(-a/2, \pm 3\sqrt{3}\, a/2)$

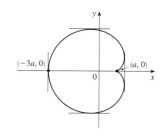

29. 18   **31.** $(2, \pm\pi/3)$ **33.** $(\pi - 1)/2$ **35.** $2(5\sqrt{5} - 1)$
37. $\dfrac{2\sqrt{\pi^2 + 1} - \sqrt{4\pi^2 + 1}}{2\pi} + \ln\left(\dfrac{2\pi + \sqrt{4\pi^2 + 1}}{\pi + \sqrt{\pi^2 + 1}}\right)$
39. $471{,}295\pi/1024$
41. All curves have the vertical asymptote $x = 1$. For $c < -1$, the curve bulges to the right. At $c = -1$, the curve is the line $x = 1$. For $-1 < c < 0$, it bulges to the left. At $c = 0$ there is a cusp at $(0, 0)$. For $c > 0$, there is a loop.
43. $(\pm 1, 0), (\pm 3, 0)$ **45.** $\left(-\frac{25}{24}, 3\right), (-1, 3)$

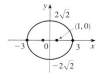

47. $x^2 = 8(y - 4)$
49. $5x^2 - 20y^2 = 36$
51. $(x^2/25) + ((8y - 399)^2/160{,}801) = 1$
53. $r = 4/(3 + \cos\theta)$
55. $x = a(\cot\theta + \sin\theta\cos\theta),\ y = a(1 + \sin^2\theta)$

Problems Plus □ page 735

1. $\ln(\pi/2)$ **3.** $\left[-\tfrac{3}{4}\sqrt{3}, \tfrac{3}{4}\sqrt{3}\right] \times [-1, 2]$
5. (a) At $(0, 0)$ and $\left(\tfrac{3}{2}, \tfrac{3}{2}\right)$
(b) Horizontal tangents at $(0, 0)$ and $\left(\sqrt[3]{2}, \sqrt[3]{4}\right)$;
vertical tangents at $(0, 0)$ and $\left(\sqrt[3]{4}, \sqrt[3]{2}\right)$
(d)   (g) $\tfrac{3}{2}$

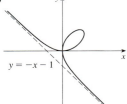

Chapter 12

Exercises 12.1 □ page 746

Abbreviations: C, convergent; D, divergent

1. (a) A sequence is an ordered list of numbers. It can also be defined as a function whose domain is the set of positive integers.
(b) The terms a_n approach 8 as n becomes large.
(c) The terms a_n become large as n becomes large.
3. 0.8, 0.96, 0.992, 0.9984, 0.99968   **5.** $-3, \tfrac{3}{2}, -\tfrac{1}{2}, \tfrac{1}{8}, -\tfrac{1}{40}$
7. 3, 5, 9, 17, 33
9. $a_n = 1/2^n$ **11.** $a_n = 5n - 3$ **13.** $a_n = \left(-\tfrac{2}{3}\right)^{n-1}$
15. D   **17.** 5   **19.** 0   **21.** 0
23. D   **25.** 0   **27.** 0   **29.** 0   **31.** 0
33. 1   **35.** 1   **37.** D   **39.** D   **41.** D
43. $\pi/4$ **45.** 0   **47.** 0
49. (a) 1060, 1123.60, 1191.02, 1262.48, 1338.23   (b) D
51. $-1 < r < 1$
53. Convergent by the Monotonic Sequence Theorem; $5 \leq L < 8$
55. Decreasing; yes   **57.** Not monotonic; yes
59. Decreasing; yes   **61.** 2   **63.** $(3 + \sqrt{5})/2$
65. (b) $(1 + \sqrt{5})/2$ **67.** (a) 0   (b) 9, 11

Exercises 12.2 □ page 756

1. (a) A sequence is an ordered list of numbers whereas a series is the *sum* of a list of numbers.
(b) A series is convergent if the sequence of partial sums is a convergent sequence. A series is divergent if it is not convergent.

3. $-2.40000, -1.92000$
$-2.01600, -1.99680,$
$-2.00064, -1.99987$
$-2.00003, -1.99999$
$-2.00000, -2.00000$
Convergent, sum $= -2$

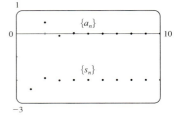

5. $1.55741, -0.62763,$
$-0.77018, 0.38764$
$-2.99287, -3.28388$
$-2.41243, -9.21214$
$-9.66446, -9.01610$
Divergent

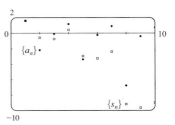

7. $0.64645, 0.80755,$
$0.87500, 0.91056,$
$0.93196, 0.94601,$
$0.95581, 0.96296,$
$0.96838, 0.97259$
Convergent, sum $= 1$

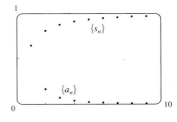

9. (a) C   (b) D
11. 9   **13.** D   **15.** 15   **17.** $\tfrac{1}{7}$ **19.** D   **21.** D
23. $\tfrac{3}{2}$ **25.** D   **27.** $\tfrac{3}{2}$ **29.** D   **31.** D   **33.** $\tfrac{7}{2}$
35. $\tfrac{2}{9}$ **37.** $\tfrac{1138}{333}$ **39.** $41{,}111/333{,}000$
41. $-3 < x < 3$; $x/(3 - x)$ **43.** $-\tfrac{1}{4} < x < \tfrac{1}{4}$; $1/(1 - 4x)$
45. all x; $\dfrac{2}{2 - \cos x}$ **47.** $\tfrac{1}{4}$
49. $a_1 = 0,\ a_n = 2/[n(n + 1)]$ for $n > 1$, sum $= 1$
51. (a) $S_n = D(1 - c^n)/(1 - c)$ (b) 5
53. $(\sqrt{3} - 1)/2$ **55.** $1/[n(n + 1)]$
57. The series is divergent.
63. $\{s_n\}$ is bounded and increasing.
65. (a) $0, \tfrac{1}{9}, \tfrac{2}{9}, \tfrac{1}{3}, \tfrac{2}{3}, \tfrac{7}{9}, \tfrac{8}{9}, 1$
67. (a) $\tfrac{1}{2}, \tfrac{5}{6}, \tfrac{23}{24}, \tfrac{119}{120}$; $[(n + 1)! - 1]/(n + 1)!$ (c) 1

Exercises 12.3 □ page 765

1. C

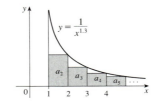

3. C   **5.** D   **7.** C   **9.** D   **11.** C   **13.** C   **15.** C

17. D   **19.** C   **21.** D   **23.** C   **25.** $p > 1$
27. $p < -1$ **29.** $(1, \infty)$
31. (a) 1.54977, error ≤ 0.1 (b) 1.64522, error ≤ 0.005
(c) $n > 1000$
33. 2.61   **39.** $b < 1/e$

Exercises 12.4 □ page 770

1. (a) Nothing   (b) C
3. C   **5.** C   **7.** D
9. C   **11.** D   **13.** C   **15.** C   **17.** D   **19.** C
21. D   **23.** C   **25.** D   **27.** C   **29.** C   **31.** D
33. 0.567975, error ≤ 0.0003 **35.** 0.76352, error < 0.001
45. Yes

Exercises 12.5 □ page 775

1. (a) A series whose terms are alternately positive and negative
(b) $0 < b_{n+1} \leq b_n$ and $\lim_{n \to \infty} b_n = 0$, where $b_n = |a_n|$
(c) $|R_n| \leq b_{n+1}$
3. C   **5.** C   **7.** D   **9.** C   **11.** C   **13.** D
15. C   **17.** C   **19.** D
21. 1.0000, 0.6464,
0.8389, 0.7139, 0.8033,
0.7353, 0.7893, 0.7451, 0.7821,
0.7505; error < 0.0275

23. 10   **25.** 7   **27.** 0.9721   **29.** 0.0676
31. An underestimate   **33.** p is not a negative integer
35. $\{b_n\}$ is not decreasing

Exercises 12.6 □ page 781

Abbreviations: AC, absolutely convergent;
CC, conditionally convergent

1. (a) D   (b) C   (c) May converge or diverge
3. AC   **5.** CC   **7.** D   **9.** AC   **11.** AC   **13.** AC
15. AC   **17.** CC   **19.** AC   **21.** D   **23.** AC
25. AC   **27.** D   **29.** D   **31.** (a) and (d)
35. (a) $\frac{661}{960} \approx 0.68854$, error < 0.00521 (b) $n \geq 11$, 0.693109

Exercises 12.7 □ page 784

1. D   **3.** C   **5.** C   **7.** D   **9.** C   **11.** C   **13.** C
15. C   **17.** D   **19.** C   **21.** C   **23.** D   **25.** C
27. C   **29.** C   **31.** D   **33.** C   **35.** C   **37.** C

Exercises 12.8 □ page 789

1. A series of the form $\sum_{n=0}^{\infty} c_n(x - a)^n$, where x is a variable and a and the c_n's are constants
3. $1, [-1, 1)$ **5.** $1, [-1, 1]$ **7.** $\infty, (-\infty, \infty)$ **9.** $\frac{1}{4}, \left(-\frac{1}{4}, \frac{1}{4}\right)$
11. $\frac{1}{2}, \left(-\frac{1}{2}, \frac{1}{2}\right]$ **13.** $4, (-4, 4]$ **15.** $1, (0, 2)$
17. $2, (-4, 0]$ **19.** $\infty, (-\infty, \infty)$ **21.** $b, (a - b, a + b)$

23. $0, \left\{\frac{1}{2}\right\}$ **25.** $\frac{1}{4}, \left[-\frac{1}{2}, 0\right]$ **27.** $\infty, (-\infty, \infty)$
29. (a) Yes   (b) No   **31.** k^k
33. (a) $(-\infty, \infty)$
(b), (c)

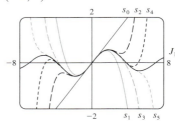

35. $(-1, 1), f(x) = (1 + 2x)/(1 - x^2)$ **39.** 2

Exercises 12.9 □ page 795

1. 10   **3.** $\sum_{n=0}^{\infty} (-1)^n x^n, (-1, 1)$ **5.** $\sum_{n=0}^{\infty} x^{3n}, (-1, 1)$

7. $-\sum_{n=0}^{\infty} \frac{1}{5^{n+1}} x^n, (-5, 5)$ **9.** $\sum_{n=0}^{\infty} (-1)^n \frac{1}{9^{n+1}} x^{2n+1}, (-3, 3)$

11. $\sum_{n=0}^{\infty} \left[\frac{(-1)^{n+1}}{2^{n+1}} - 1\right] x^n, (-1, 1)$

13. (a) $\sum_{n=0}^{\infty} (-1)^n (n + 1) x^n, R = 1$

(b) $\frac{1}{2} \sum_{n=0}^{\infty} (-1)^n (n + 2)(n + 1) x^n, R = 1$

(c) $\frac{1}{2} \sum_{n=2}^{\infty} (-1)^n n(n - 1) x^n, R = 1$

15. $\ln 5 - \sum_{n=1}^{\infty} \frac{x^n}{n5^n}, R = 5$

17. $\sum_{n=3}^{\infty} \frac{n - 2}{2^{n-1}} x^n, R = 2$

19. $\ln 3 + \sum_{n=1}^{\infty} \frac{(-1)^{n-1}}{n3^n} x^n, R = 3$

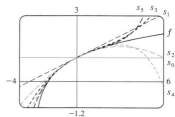

21. $\sum_{n=0}^{\infty} \frac{2x^{2n+1}}{2n + 1}, R = 1$

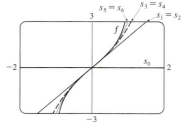

23. $C + \sum_{n=0}^{\infty} \dfrac{t^{8n+2}}{8n+2}$; 1

25. $C + \sum_{n=1}^{\infty} (-1)^{n+1} \dfrac{x^{2n-1}}{4n^2-1}$; 1

27. 0.199989   **29.** 0.000065   **31.** 0.09531
33. (b) 0.920   **37.** $[-1, 1], [-1, 1), (-1, 1)$

Exercises 12.10 □ page 806

1. $b_8 = f^{(8)}(5)/8!$ **3.** $\sum_{n=0}^{\infty} (-1)^n \dfrac{x^{2n}}{(2n)!}, R = \infty$

5. $\sum_{n=0}^{\infty} (-1)^n \dfrac{(n+1)(n+2)}{2} x^n, R = 1$

7. $\sum_{n=0}^{\infty} \dfrac{5^n}{n!} x^n, R = \infty$

9. $\sum_{n=0}^{\infty} \dfrac{x^{2n+1}}{(2n+1)!}, R = \infty$

11. $7 + 5(x-2) + (x-2)^2, R = \infty$

13. $\sum_{n=0}^{\infty} \dfrac{e^3}{n!} (x-3)^n, R = \infty$

15. $\sum_{n=0}^{\infty} (-1)^{n+1} \dfrac{1}{(2n)!} (x-\pi)^{2n}, R = \infty$

17. $\sum_{n=0}^{\infty} (-1)^n \dfrac{1 \cdot 3 \cdot 5 \cdot \cdots \cdot (2n-1)}{2^n \cdot 3^{2n+1} \cdot n!} (x-9)^n, R = 9$

23. $\sum_{n=0}^{\infty} (-1)^n \dfrac{\pi^{2n}}{(2n)!} x^{2n}, R = \infty$

25. $\sum_{n=0}^{\infty} (-1)^n \dfrac{1}{2n+1} x^{2n+2}, R = 1$

27. $\sum_{n=0}^{\infty} (-1)^n \dfrac{1}{n!} x^{n+2}, R = \infty$

29. $\sum_{n=1}^{\infty} \dfrac{(-1)^{n+1} 2^{2n-1} x^{2n}}{(2n)!}, R = \infty$

31. $\sum_{n=0}^{\infty} \dfrac{(-1)^n x^{2n}}{(2n+1)!}, R = \infty$

33. $1 + \dfrac{x}{2} + \sum_{n=2}^{\infty} (-1)^{n-1} \dfrac{1 \cdot 3 \cdot 5 \cdot \cdots \cdot (2n-3)}{2^n n!} x^n, R = 1$

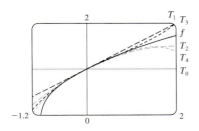

35. $\sum_{n=0}^{\infty} (-1)^n \dfrac{1}{(2n)!} x^{4n}, R = \infty$

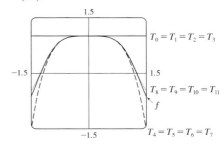

37. 0.81873   **39.** $C + \sum_{n=0}^{\infty} (-1)^n \dfrac{x^{6n+2}}{(6n+2)(2n)!}$

41. $C + x + \dfrac{x^4}{8} + \sum_{n=2}^{\infty} (-1)^{n-1} \dfrac{1 \cdot 3 \cdot 5 \cdot \cdots \cdot (2n-3)}{2^n n!(3n+1)} x^{3n+1}$

43. 0.440   **45.** 0.09998750   **47.** $\tfrac{1}{3}$ **49.** $\tfrac{1}{120}$

51. $1 - \tfrac{3}{2}x^2 + \tfrac{25}{24}x^4$ **53.** $1 + \tfrac{1}{6}x^2 + \tfrac{7}{360}x^4$

55. e^{-x^4} **57.** $1/\sqrt{2}$ **59.** $e^3 - 1$

Exercises 12.11 □ page 811

1. $1 + \dfrac{x}{2} + \sum_{n=2}^{\infty} (-1)^{n-1} \dfrac{1 \cdot 3 \cdot 5 \cdot \cdots \cdot (2n-3)}{2^n n!} x^n, R = 1$

3. $\sum_{n=0}^{\infty} (-1)^n \dfrac{(n+1)(n+2)}{2^{n+4}} x^n, R = 2$

5. $1 - 2x - \sum_{n=2}^{\infty} \dfrac{3 \cdot 7 \cdot \cdots \cdot (4n-5) \cdot 2^n}{n!} x^n, R = \tfrac{1}{8}$

7. $\tfrac{1}{2}x + \sum_{n=1}^{\infty} (-1)^n \dfrac{1 \cdot 3 \cdot 5 \cdot \cdots \cdot (2n-1)}{n! 2^{3n+1}} x^{2n+1}, R = 2$

9. $1 + \tfrac{3}{2}x$, $1 + \tfrac{3}{2}x - \tfrac{3}{8}x^2$, $1 + \tfrac{3}{2}x - \tfrac{3}{8}x^2 + \tfrac{5}{16}x^3$

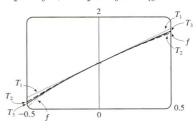

11. (a) $1 + \sum_{n=1}^{\infty} \dfrac{1 \cdot 3 \cdot 5 \cdot \cdots \cdot (2n-1)}{2^n n!} x^{2n}$

(b) $x + \sum_{n=1}^{\infty} \dfrac{1 \cdot 3 \cdot 5 \cdot \cdots \cdot (2n-1)}{(2n+1)2^n n!} x^{2n+1}$

13. (a) $1 + \dfrac{x}{3} + \sum_{n=2}^{\infty} (-1)^{n+1} \dfrac{2 \cdot 5 \cdot 8 \cdot \cdots \cdot (3n-4)}{3^n n!} x^n$

(b) 1.0033

15. (a) $\sum_{n=1}^{\infty} nx^n$ (b) 2

17. (a) $1 + \dfrac{x^2}{2} + \sum_{n=2}^{\infty} (-1)^{n-1} \dfrac{1 \cdot 3 \cdot 5 \cdot \cdots \cdot (2n-3)}{2^n n!} x^{2n}$

(b) 99,225

Exercises 12.12 □ page 819

1. (a) $T_0(x) = 1 = T_1(x)$, $T_2(x) = 1 - \frac{1}{2}x^2 = T_3(x)$,
$T_4(x) = 1 - \frac{1}{2}x^2 + \frac{1}{24}x^4 = T_5(x)$, $T_6(x) = 1 - \frac{1}{2}x^2 + \frac{1}{24}x^4 - \frac{1}{720}x^6$

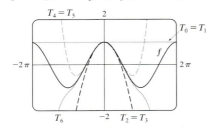

(b)

x	f	$T_0 = T_1$	$T_2 = T_3$	$T_4 = T_5$	T_6
$\frac{\pi}{4}$	0.7071	1	0.6916	0.7074	0.7071
$\frac{\pi}{2}$	0	1	-0.2337	0.0200	-0.0009
π	-1	1	-3.9348	0.1239	-1.2114

(c) As n increases, $T_n(x)$ is a good approximation to $f(x)$ on a larger and larger interval.

3. $(x - 1) - \frac{1}{2}(x - 1)^2 + \frac{1}{3}(x - 1)^3 - \frac{1}{4}(x - 1)^4$

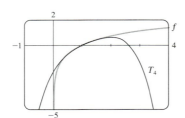

5. $\frac{1}{2} + \frac{\sqrt{3}}{2}\left(x - \frac{\pi}{6}\right) - \frac{1}{4}\left(x - \frac{\pi}{6}\right)^2 - \frac{\sqrt{3}}{12}\left(x - \frac{\pi}{6}\right)^3$

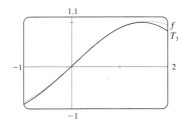

7. $x + \frac{1}{6}x^3$

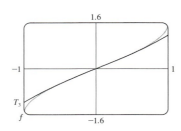

9. $x - 2x^2 + 2x^3$

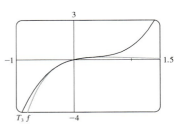

11. $T_8(x) = 1 + \frac{1}{2}x^2 + \frac{5}{24}x^4 + \frac{61}{720}x^6 + \frac{277}{8064}x^8$

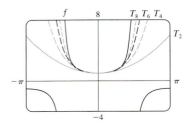

13. (a) $2 + \frac{1}{4}(x - 4) - \frac{1}{64}(x - 4)^2$ (b) 1.5625×10^{-5}
15. (a) $1 + \frac{2}{3}(x - 1) - \frac{1}{9}(x - 1)^2 + \frac{4}{81}(x - 1)^3$ (b) 0.000097
17. (a) $x + \frac{1}{3}x^3$ (b) 0.058
19. (a) $1 + x^2$ (b) 0.00006
21. (a) $x^2 - \frac{1}{6}x^4$ (b) 0.042
23. 0.57358 **25.** Four
27. $-1.037 < x < 1.037$ **29.** 21 m, no
35. (c) They differ by about 8×10^{-9} km.

Chapter 12 Review □ page 822

True-False Quiz

1. False **3.** True **5.** False **7.** False **9.** False
11. True **13.** True **15.** False **17.** True

Exercises

1. $\frac{1}{2}$ **3.** D **5.** 0 **7.** e^{12} **9.** 2 **11.** C **13.** C
15. D **17.** C **19.** C **21.** C **23.** CC **25.** AC
27. 8 **29.** $\pi/4$ **31.** e^{-e} **35.** 0.9721
37. 0.18976224, $|\text{error}| < 6.4 \times 10^{-7}$ **41.** $4, [-6, 2)$
43. $0.5, [2.5, 3.5)$
45. $\frac{1}{2}\sum_{n=0}^{\infty}(-1)^n\left[\frac{1}{(2n)!}\left(x - \frac{\pi}{6}\right)^{2n} + \frac{\sqrt{3}}{(2n+1)!}\left(x - \frac{\pi}{6}\right)^{2n+1}\right]$
47. $\sum_{n=0}^{\infty}(-1)^n x^{n+2}$, $R = 1$ **49.** $-\sum_{n=1}^{\infty}\frac{x^n}{n}$, $R = 1$
51. $\sum_{n=0}^{\infty}(-1)^n \frac{x^{8n+4}}{(2n+1)!}$, $R = \infty$
53. $\frac{1}{2} + \sum_{n=1}^{\infty}\frac{1 \cdot 5 \cdot 9 \cdot \cdots \cdot (4n - 3)}{n! 2^{6n+1}}x^n$, $R = 16$
55. $C + \ln|x| + \sum_{n=1}^{\infty}\frac{x^n}{n \cdot n!}$

57. (a) $1 + \frac{1}{2}(x-1) - \frac{1}{8}(x-1)^2 + \frac{1}{16}(x-1)^3$
(b) (c) 0.000006

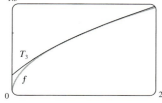

59. $-\frac{1}{6}$

Problems Plus □ page 825

1. $15!/5! = 10,897,286,400$
3. (b) 0 if $x = 0$, $(1/x) - \cot x$ if $x \neq k\pi$, k an integer
5. (a) $s_n = 3 \cdot 4^n$, $l_n = 1/3^n$, $p_n = 4^n/3^{n-1}$ (c) $2\sqrt{3}/5$
9. $(-1, 1)$, $(x^3 + 4x^2 + x)/(1 - x)^4$
11. $\ln \frac{1}{2}$

Appendixes

Exercises A □ page A9

1. 18   **3.** π **5.** $5 - \sqrt{5}$ **7.** $2 - x$
9. $|x + 1| = \begin{cases} x + 1 & \text{for } x \geq -1 \\ -x - 1 & \text{for } x < -1 \end{cases}$ **11.** $x^2 + 1$
13. $(-2, \infty)$ **15.** $[-1, \infty)$
17. $(3, \infty)$ **19.** $(2, 6)$
21. $(0, 1]$ **23.** $[-1, \frac{1}{2})$
25. $(-\infty, 1) \cup (2, \infty)$ **27.** $[-1, \frac{1}{2}]$
29. $(-\infty, \infty)$ **31.** $(-\sqrt{3}, \sqrt{3})$
33. $(-\infty, 1]$ **35.** $(-1, 0) \cup (1, \infty)$
37. $(-\infty, 0) \cup (\frac{1}{4}, \infty)$
39. $10 \leq C \leq 35$ **41.** (a) $T = 20 - 10h$, $0 \leq h \leq 12$
(b) $-30°C \leq T \leq 20°C$ **43.** $\pm \frac{3}{2}$ **45.** $2, -\frac{4}{3}$
47. $(-3, 3)$ **49.** $(3, 5)$ **51.** $(-\infty, -7] \cup [-3, \infty)$
53. $[1.3, 1.7]$ **55.** $[-4, -1] \cup [1, 4]$
57. $x \geq (a + b)c/(ab)$ **59.** $x > (c - b)/a$

Exercises B □ page A15

1. 5   **3.** $\sqrt{74}$ **5.** $2\sqrt{37}$ **7.** 2   **9.** $-\frac{9}{2}$
17.   **19.**

21. $y = 6x - 15$ **23.** $2x - 3y + 19 = 0$
25. $5x + y = 11$ **27.** $y = 3x - 2$ **29.** $y = 3x - 3$
31. $y = 5$ **33.** $x + 2y + 11 = 0$
35. $5x - 2y + 1 = 0$
37. $m = -\frac{1}{3}$, $b = 0$ **39.** $m = 0$, $b = -2$

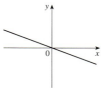

41. $m = \frac{3}{4}$, $b = -3$ **43.**

 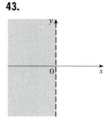

45.   **47.**

 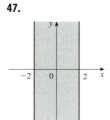

49.   **51.**

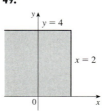 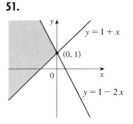

53. $(0, -4)$ **55.** (a) $(4, 9)$ (b) $(3.5, -3)$ **57.** $(1, -2)$
59. $y = x - 3$ **61.** (b) $4x - 3y - 24 = 0$

Exercises C □ page A23

1. $(x - 3)^2 + (y + 1)^2 = 25$ **3.** $x^2 + y^2 = 65$
5. $(2, -5)$, 4   **7.** $(-\frac{1}{2}, 0)$, $\frac{1}{2}$ **9.** $(\frac{1}{4}, -\frac{1}{4})$, $\sqrt{10}/4$

11. Parabola   **13.** Ellipse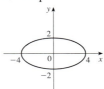

15. Hyperbola 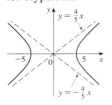   **17.** Ellipse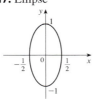

19. Parabola   **21.** Hyperbola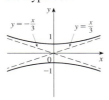

23. Hyperbola 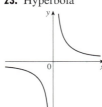   **25.** Ellipse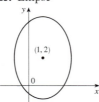

27. Parabola 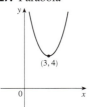   **29.** Parabola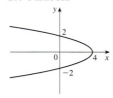

31. Ellipse 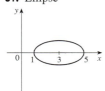   **33.**

35. $y = x^2 - 2x$

37.   **39.**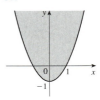

Exercises D □ page A32

1. $7\pi/6$ **3.** $\pi/20$ **5.** 5π **7.** $720°$ **9.** $75°$
11. $-67.5°$ **13.** 3π cm   **15.** $\frac{2}{3}$ rad $= (120/\pi)°$
17.   **19.**   **21.**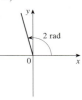

23. $\sin(3\pi/4) = 1/\sqrt{2}$, $\cos(3\pi/4) = -1/\sqrt{2}$, $\tan(3\pi/4) = -1$, $\csc(3\pi/4) = \sqrt{2}$, $\sec(3\pi/4) = -\sqrt{2}$, $\cot(3\pi/4) = -1$
25. $\sin(9\pi/2) = 1$, $\cos(9\pi/2) = 0$, $\csc(9\pi/2) = 1$, $\cot(9\pi/2) = 0$, $\tan(9\pi/2)$ and $\sec(9\pi/2)$ undefined
27. $\sin(5\pi/6) = \frac{1}{2}$, $\cos(5\pi/6) = -\sqrt{3}/2$, $\tan(5\pi/6) = -1/\sqrt{3}$, $\csc(5\pi/6) = 2$, $\sec(5\pi/6) = -2/\sqrt{3}$, $\cot(5\pi/6) = -\sqrt{3}$
29. $\cos\theta = \frac{4}{5}$, $\tan\theta = \frac{3}{4}$, $\csc\theta = \frac{5}{3}$, $\sec\theta = \frac{5}{4}$, $\cot\theta = \frac{4}{3}$
31. $\sin\phi = \sqrt{5}/3$, $\cos\phi = -\frac{2}{3}$, $\tan\phi = -\sqrt{5}/2$, $\csc\phi = 3/\sqrt{5}$, $\cot\phi = -2/\sqrt{5}$
33. $\sin\beta = -1/\sqrt{10}$, $\cos\beta = -3/\sqrt{10}$, $\tan\beta = \frac{1}{3}$, $\csc\beta = -\sqrt{10}$, $\sec\beta = -\sqrt{10}/3$
35. 5.73576 cm   **37.** 24.62147 cm   **59.** $(4 + 6\sqrt{2})/15$
61. $(3 + 8\sqrt{2})/15$ **63.** $\frac{24}{25}$ **65.** $\pi/3, 5\pi/3$
67. $\pi/4, 3\pi/4, 5\pi/4, 7\pi/4$ **69.** $\pi/6, \pi/2, 5\pi/6, 3\pi/2$
71. $0, \pi, 2\pi$ **73.** $0 \leq x \leq \pi/6$ and $5\pi/6 \leq x \leq 2\pi$
75. $0 \leq x < \pi/4, 3\pi/4 < x < 5\pi/4, 7\pi/4 < x \leq 2\pi$
77.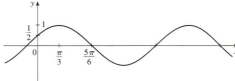

79.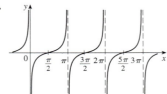

81.

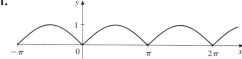

89. 14.34457 cm^2

Exercises E ◻ page A38

1. $\sqrt{1} + \sqrt{2} + \sqrt{3} + \sqrt{4} + \sqrt{5}$ **3.** $3^4 + 3^5 + 3^6$
5. $-1 + \frac{1}{3} + \frac{3}{5} + \frac{5}{7} + \frac{7}{9}$ **7.** $1^{10} + 2^{10} + 3^{10} + \cdots + n^{10}$
9. $1 - 1 + 1 - 1 + \cdots + (-1)^{n-1}$ **11.** $\sum_{i=1}^{10} i$
13. $\sum_{i=1}^{19} \frac{i}{i+1}$ **15.** $\sum_{i=1}^{n} 2i$ **17.** $\sum_{i=0}^{5} 2^i$ **19.** $\sum_{i=1}^{n} x^i$
21. 80   **23.** 3276   **25.** 0   **27.** 61   **29.** $n(n+1)$
31. $n(n^2 + 6n + 17)/3$
33. $n(n^2 + 6n + 11)/3$
35. $n(n^3 + 2n^2 - n - 10)/4$
41. (a) n^4 (b) $5^{100} - 1$ (c) $\frac{97}{300}$ (d) $a_n - a_0$
43. $\frac{1}{3}$ **45.** 14   **49.** $2^{n+1} + n^2 + n - 2$

Exercises G ◻ page A47

1. $8 - 4i$ **3.** $13 + 18i$ **5.** $12 - 7i$ **7.** $\frac{11}{13} + \frac{10}{13}i$
9. $\frac{1}{2} - \frac{1}{2}i$ **11.** $-i$ **13.** $5i$ **15.** $12 + 5i; 13$
17. $4i, 4$ **19.** $\pm\frac{3}{2}i$ **21.** $-1 \pm 2i$
23. $-\frac{1}{2} \pm (\sqrt{7}/2)i$ **25.** $3\sqrt{2}[\cos(3\pi/4) + i\sin(3\pi/4)]$
27. $5\{\cos[\tan^{-1}(\frac{4}{3})] + i\sin[\tan^{-1}(\frac{4}{3})]\}$
29. $4[\cos(\pi/2) + i\sin(\pi/2)], \cos(-\pi/6) + i\sin(-\pi/6),$
$\frac{1}{2}[\cos(-\pi/6) + i\sin(-\pi/6)]$
31. $4\sqrt{2}[\cos(7\pi/12) + i\sin(7\pi/12)],$
$(2\sqrt{2})[\cos(13\pi/12) + i\sin(13\pi/12)], \frac{1}{4}[\cos(\pi/6) + i\sin(\pi/6)]$
33. -1024 **35.** $-512\sqrt{3} + 512i$
37. $\pm 1, \pm i, (1/\sqrt{2})(\pm 1 \pm i)$ **39.** $\pm(\sqrt{3}/2) + \frac{1}{2}i, -i$

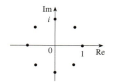 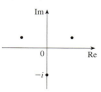

41. i **43.** $\frac{1}{2} + (\sqrt{3}/2)i$ **45.** $-e^2$

Credits

|||| This page constitutes an extension of the copyright page. We have made every effort to trace the ownership of all copyrighted material and to secure permission from copyright holders. In the event of any question arising as to the use of any material, we will be pleased to make the necessary corrections in future printings. Thanks are due to the following authors, publishers, and agents for permission to use the material indicated.

- **vii** □ Brian C. Betsill
- **viii** □ Brian C. Betsill
  © Ron Lindsey / Phototake
- **ix** □ © Gary Ladd 1984 / Photo Researchers, Inc.
  Brian C. Betsill
- **x** □ Brian C. Betsill
- **314** □ Brian C. Betsill
- **374** □ Brian C. Betsill
- **412** □ Brian C. Betsill
- **504** □ Title page courtesy of Thomas Fisher Rare Book Library.
- **510** □ © Ron Lindsey / Phototake
- **582** □ © Gary Ladd 1984 / Photo Researchers, Inc.
  Brian C. Betsill
- **622** □ (*photo*) © Jeff Lepore / Photo Researchers, Inc.
- **686** □ Brian C. Betsill
- **736** □ Membranes courtesy of National Film Board of Canada
  Mathematica models by Brian C. Betsill
- **821** □ © Luke Dodd / Photo Researchers, Inc.

Index

RP denotes Reference Page numbers.
Numbers followed by asterisks denote pages in Sections 7.2*, 7.3*, and 7.4*.

Absolutely convergent series, 776
Absolute value, A6, A42
Adaptive numerical integration, 562
Addition formulas for sine and cosine, A28, A29
Alternating harmonic series, 773, 776
Alternating series, 771
Alternating Series Estimation Theorem, 774
Alternating Series Test, 772
Analytic geometry, A10
Angle, A24
   negative, A25
   positive, A25
   standard position, A25
Aphelion, 732
Apolune, 726
Approximate integration, 554
Approximating cylinder, 385
Approximating surface, 591
Approximation:
   to e, 422
   by the Midpoint Rule, 554
   by Riemann sums, 327
   by Simpson's Rule, 558, 560
   by Taylor's Inequality, 800
   by the Trapezoidal Rule, 555
Archimedes' Principle, 409
Arc length, 583, 699, 700, 718
Arc length formula, 584, 585
Arc length function, 587
Area, 315
   of a circle, 527
   under a curve, 320, 326
   between curves, 375, 376, 378

of an ellipse, 527
under a parametric curve, 698
in polar coordinates, 715
of a sector of a circle, 715
surface, 701
of a surface of a revolution, 590, 596
Area function, 339
Area problem, 315
Argument of a complex number, A43
Arithmetic-geometric mean, 748
Asymptote of a hyperbola, 724, A20
Autonomous differential equation, 632
Average speed of molecules, 575
Average value of a function, 402, 403, 613
Axes, coordinate, A11
Axes of ellipse, A19

Bacterial growth, 647, 664
Barrow, Isaac, 340, 359
Baseball and calculus, 658
Base of cylinder, 382
Base of logarithm, 434, 472*
   change of, 436, 473*
Bernoulli, James, 637, 673
Bernoulli, John, 495, 504, 637
Bernoulli differential equation, 673
Bessel, Friedrich, 786
Bessel function, 736, 786
Bézier, Pierre, 705
Bézier curves, 690, 705
Binomial series, 808, 809, 812
Binomial Theorem, 808
   discovery by Newton, 812
Blackbody radiation, 821

Blood flow, 608
Bounded sequence, 744
Boyle's Law, 450
Brachistochrone problem, 691
Branches of hyperbola, 724, A20
Buffon's needle problem, 621

Cable (hanging), 481
Calculator, graphing, 690, 712
Calculus, 9
   invention of, 359
Cancellation equations, 416
   for inverse trigonometric functions, 478
   for logarithms, 434, 460*, 472*
Cantor, Georg, 758
Cantor set, 758
Capital formation, 611
Cardiac output, 609
Cardioid, 709
Carrying capacity, 659
Cartesian coordinate system, A11
Cartesian plane, A11
Cassini, Giovanni, 715
Catenary, 481
Cauchy, Augustin-Louis, 499
Cauchy's Mean Value Theorem, 500
Cavalieri's Principle, 392
Center of gravity, 600
Center of mass, 600
Centroid of a plane region, 601
Change of base, formula for, 436, 473*
Change of variables in integration, 361
Chebyshev polynomials, 581

Circle, A16
    area of, 527
Circular cylinder, 382
Cissoid, 694, 714
Closed interval, A3
Cochleoid, 733
Coefficient of inequality, 358
Coefficient of a power series, 785
Comets, orbits of, 732
Comparison properties of the
        integral, 335
Comparison Test, 572, 767
Comparison tests for series, 766
Comparison Theorem for integrals, 572
Completeness Axiom, 745
Complex conjugate, A41
Complex exponentials, A46
Complex numbers, A41
    argument of, A43
    division of, A41, A44
    equality of, A41
    imaginary part of, A41
    modulus of, A42
    multiplication of, A41, A44
    polar form, A43
    real part of, A41
    roots of, A46
Compound interest, 502, 654
Computer algebra system, 690
    for graphing sequences, 743
    for integration, 549
Computer, graphing with, 712
Conchoid, 692, 714
Conditionally convergent series, 777
Conic section, 720, 728
    directrix, 728
    eccentricity, 728
    focus, 728
    polar equation, 729
    shifted, 725, A21
Conjugates, properties of, A42
Consumer surplus, 607
Continued fraction expansion, 748
Continuous compounding of interest,
        502, 654
Continuous random variable, 611
Convergence:
    absolute, 776
    conditional, 777
    of an improper integral, 567, 570
    interval of, 787
    radius of, 787
    of a sequence, 739
    of a series, 750
Convergent improper integral, 567, 570
Convergent sequence, 739

Convergent series, 750
    properties of, 755
Coordinate(s), A2
    Cartesian, A11
    polar, 705
    rectangular, A11
Coordinate axes, A11
Cornu's spiral, 703
Cosine function, 736, A26
    graph, A31
    power series, 802
Cross-section, 383
Curvature, 704
Curve(s):
    Bézier, 690, 705
    length of, 583
    parametric, 687
    polar, 707
    smooth, 584
    swallowtail catastrophe, 695
Cusp, 692
Cycloid, 690
Cylinder, 382
Cylindrical shell, 393

Decay, law of natural, 648
Decay, radioactive, 652
Decreasing sequence, 744
Definite integral, 326
    properties of, 333, 334, 335
    Substitution Rule for, 363
Definite integration:
    by parts, 514
    by substitution, 363
Demand curve, 607
Demand function, 607
De Moivre, Abraham, A44
De Moivre's Theorem, A45
Density, liquid, 598
Derivative(s):
    of exponential functions, 426,
        445*, 468*
    of hyperbolic functions, 488
    of an integral, 342
    of an inverse function, 419
    of inverse trigonometric functions, 481
    of logarithmic functions, 441, 443, 445,
        452*, 473*
    of a power series, 791
Descartes, René, A11
Differential equation, 623, 629
    autonomous, 632
    Bernoulli, 673
    first-order, 637
    general solution of, 626
    linear, 668

    logistic, 624, 659
    logistic difference, 749
    order of, 625
    second-order, 625
    separable, 637
    solution of, 625
Differentiation:
    formulas for, RP5
    logarithmic, 446, 457*
    of a power series, 791
    term by term, 791
Direction field, 628, 629, 659
Directrix, 720, 728
Discontinuous integrand, 569
Disk method, 387
Displacement, 354
Distance between points in a plane, A11
Distance between real numbers, A7
Distance formula, A12
Distance problem, 322
Divergence:
    of an improper integral, 567, 570
    of an infinite series, 750
    of a sequence, 739
    test for, 754
Divergent improper integral, 567, 570
Divergent sequence, 739
Divergent series, 750
Division of power series, 805
Domain of a function, 12
Double-angle formulas, A29
Drumhead, vibration of, 736, 786
Dye dilution method, 609

e (the number), 427, 453*
    as a limit, 448, 474*
Eccentricity, 728
Electric circuit, 671
    to a flash bulb, 432, 476
Elementary function, 545
Elimination constant of a drug, 682
Ellipse, 722, 728, A19
    area, 527
    directrix, 728
    eccentricity, 728
    foci, 722, 728
    major axis, 722
    polar equation, 729
    reflection property, 723
    vertices, 722
Epicycloid, 695
Equation(s):
    of a circle, A17
    differential (see Differential equation)
    of an ellipse, A19
    of a graph, A10, A16

of a hyperbola, A20
of a line, A12, A13, A14, A16
linear, A14
logistic, 624, 659
logistic differential, 659, 749
Lotka-Volterra, 675
of a parabola, A18
parametric, 687
point-slope, A12
polar, 707
predator-prey, 675
second-degree, A16
slope-intercept, A13
two-intercept form, A16
Equilateral hyperbola, A21
Equilibrium point, 676
Equilibrium solution, 624, 675
Error:
    in approximate integration, 557
    in Taylor approximation, 813
Error bounds, 557, 561
Error estimate:
    for alternating series, 774
    for the Midpoint Rule, 557
    for Simpson's Rule, 561
    for the Trapezoidal Rule, 557
Escape velocity, 575
Estimate of the sum of a series, 762, 769, 774, 779
Eudoxus, 3
Euler's constant, 766
Euler's formula, A47
Euler's Method, 632, 661
Exponential decay, 647, 648
Exponential function(s), 421, 467*
    with base a, 421, 467*
    derivatives of, 426, 445, 468*
    graphs of, 423, 429, 469*
    integration of, 430, 470*
    limits of, 424, 429, 461*
    natural, 427, 429
    power series for, 798
    properties of, 424, 429, 468*
Exponential graphs, 429, 469*
Exponential growth, 428, 647, 648
Exponential integrals, 470*
Exponents, laws of, 424, 462*, 468*

Family of hypocycloids, 695
Family of solutions, 624
Fat circles, 589
Fermat, Pierre, 359
Fibonacci, 748
Fibonacci sequence, 738
First-order linear differential equation, 668
First-order optics, 818

Flux, 608
Focus:
    of a conic section, 728
    of an ellipse, 722, 728
    of a hyperbola, 724
    of a parabola, 720
Folium of Descartes, 735
Force, 398
    exerted by fluid, 598
Fourier sequence, 525
Four-leaved rose, 709
Fractions (partial), 532, 533
Fresnel, Augustin, 343
Fresnel function, 343
Frustum:
    of a cone, 392
    of a pyramid, 392
Function(s):
    arc length, 587
    average value of, 402, 403, 613
    Bessel, 736, 786
    demand, 607
    elementary, 545
    exponential, 416
    Fresnel, 343
    Gompertz, 667
    hyperbolic, 486
    inverse, 413, 415
    inverse hyperbolic, 488, 489
    inverse trigonometric, 477
    logarithmic, 434, 472*
    natural exponential, 427, 429, 460*
    natural logarithmic, 435, 451*
    one-to-one, 414
    probability density, 612
    rational, 532
    representation as a power series, 790
    sine integral, 349
    trigonometric, A26
Fundamental Theorem of Calculus, 340, 342, 344, 347

G (gravitational constant), 402
Gabriel's horn, 595
Galileo, 691, 699
Gause, G. F., 506, 664
Gauss, Karl Friedrich, A35
Gaussian optics, 818
Geometric series, 750
Gompertz function, 667
Gourdon, Xavier, 801
Graph(s):
    of an equation, A10, A16
    of exponential functions, 423, 429, 469*
    of logarithmic functions, 437, 472*

    of a parametric curve, 688
    polar, 707, 712
Graphing calculator, 690, 712
Gravitational acceleration, 398
Gravitation law, 402
Gregory, James, 794
Gregory's series, 794
Growth, law of natural, 648

Half-angle formulas, A29
Half-life, 426, 652
Hare-lynx system, 622, 678
Harmonic series, 753
Hecht, Eugene, 818
Height of a rocket, 510, 517
Hooke's Law, 399
Horizontal line, A13
Horizontal Line Test, 414
Huygens, 691
Hydrostatic pressure and force, 582, 597
Hyperbola, 723, 728, A20
    asymptotes, 724, A20
    branches, 724, A20
    directrix, 728
    eccentricity, 728
    equation, 724, A20
    equilateral, A21
    foci, 724, 728
    polar equation, 729
    reflection property, 727
    vertices, 724
Hyperbolic function(s), 486
    derivatives, 488
    inverse, 488, 489
Hyperbolic identities, 487
Hyperbolic substitution, 529
Hypocycloid, 695

i (imaginary number), A41
Improper integral, 566
Impulse of a force, 658
Increasing sequence, 744
Indefinite integral(s), 351
    table of, 351
Indeterminate difference, 498
Indeterminate forms of limits, 493
Indeterminate product, 497
Indeterminate power, 498
Index of summation, A34
Inequalities, rules for, A4
Infinite interval, 566, 567
Infinite sequence (see Sequence)
Infinite series (see Series)
Initial condition, 626
Initial point of a parametric curve, 688
Initial-value problem, 626

Insect population growth, 541
Integer, A2
Integral(s):
   approximations to, 332
   change of variables in, 360
   comparison properties of, 335
   definite, 326
   derivative of, 360
   evaluating, 329
   improper, 566
   indefinite, 351
   patterns in, 553
   properties of, 333
   of symmetric functions, 364
   table of, 542, RP6–10
   units for, 356
Integral Test, 760
Integrand, 327
   discontinuous, 569
Integration, 327
   approximate, 554
   by computer algebra system, 549
   of exponential functions, 430, 470*
   formulas, 542, RP 6–10
   indefinite, 351
   limits of, 327
   numerical, 554
   by partial fractions, 532
   by parts, 511, 512
   of a power series, 791
   strategy for, 541
   by substitution, 525
   tables, use of, 547
   term by term, 791
   by trigonometric substitution, 526
Integrating factor, 668, 669
Intercepts, A19
Interest, compounded continuously, 502, 654
Intersection of polar graphs, 717
Intersection of sets, A3
Interval, A3
Interval of convergence, 787
Inverse function(s), 413, 415
   cancellation equations for, 416
   derivative of, 419
   steps for finding, 416
Inverse hyperbolic functions, 488, 489
   derivatives of, 490
Inverse substitution, 525
Inverse trigonometric functions, 477
   cancellation equations for, 478
   derivatives of, 481
Involute, 704
Irrational number, A2

Joule, 398

Kepler's Laws, 728
Kinetic energy, 658
Kirchhoff's Laws, 660, 671

Lamina, 601
Laplace transform, 575
Law of Cosines, A33
Law of Gravitation, 402
Law of natural decay, 648
Law of natural growth, 648
Laws of exponents, 424, 462*, 468*
Laws of logarithms, 434, 452*
Learning curve, 628, 673
Least squares method, 28
Least upper bound, 745
Leibniz, Gottfried Wilhelm, 359, 637, 812
Length:
   of a curve, 611
   of a line segment, A7, A12
   of a parametric curve, 701
   of a polar curve, 718
l'Hospital, Marquis de, 495, 502, 504
L'Hospital's Rule, 494, 502
   origins of, 504
Limaçon, 713
Limit Comparison Test, 768
Limit Laws, A39
Limit Laws for sequences, 741
Limit(s), 70
   of exponential functions, 424, 429, 461*
   of a function, 922
   of integration, 327
   of logarithms, 435, 437, 453*
   of a sequence, 318, 739
Linear differential equation, 668
Linear equation, A14
Line(s) in the plane, A12
   equations of, A12, A13, A14
   horizontal, A13
   parallel, A14
   perpendicular, A14
   slope of, A12
Liquid force, 597, 598
Lissajous figure, 695
Lithotripsy, 723
Logarithm(s), 434, 451*, 467*
   laws of, 434, 452*
   natural, 435, 451*
   notation for, 435, 473*
Logarithmic differentiation, 446, 457*
Logarithmic function(s), 434, 472*
   with base a, 434, 472*

   cancellation equations for, 434, 460*, 472*
   derivatives of, 441, 443, 445, 452*, 473*
   graphs of, 434, 437, 453*, 472*
   limits of, 435, 437, 453*
   natural, 435, 451*
   properties of, 434, 452 *
Logistic difference equation, 749
Logistic differential equation, 624, 659
   analytic solution of, 662
Logistic model, 659
Logistic sequence, 749
LORAN system, 727
Lotka-Volterra equations, 675

Machine diagram of a function, 12
Maclaurin, Colin, 798
Maclaurin series, 796, 798
   table of, 803
Major axis of ellipse, 722
Marginal propensity to consume or save, 757
Mass, center of, 600
Mathematical induction, 746
Mean life of an atom, 575
Mean of a normal density function, 433, 466*
Mean of a probability density function, 614
Mean Value Theorem for integrals, 403
Mean waiting time, 614
Median of a probability density function, 615
Method of cylindrical shells, 393
Method of exhaustion, 85
Method of least squares, 28
Midpoint formula, A16
Midpoint Rule, 332, 554
   error in using, 556
Mixing problems, 641
Modeling:
   with differential equations, 623
   motion of a spring, 625
   population growth, 623, 648, 659, 665, 682
   vibration of membrane, 736, 786
Model(s), mathematical:
   comparison of natural growth vs. logistic, 664
   Gompertz function, 667
   predator-prey, 675
   seasonal growth, 667
   von Bertalanffy, 682
Modulus, A42

Moment:
    about an axis, 601
    of a lamina, 601
    of a mass, 600
    of a system of particles, 600
Momentum of an object, 658
Monotonic sequence, 744
Monotonic Sequence Theorem, 745
Movie theater seating, 485
Multiplication of power series, 805
Multiplier effect, 757

Napier's Inequality, 499
Natural exponential function, 427, 429, 460*
    cancellation equations for, 460*
    derivative of, 427, 462*, 463*
    graph of, 429
    integration of, 464*
    limits of, 461*
    properties of, 429, 461*, 462*
Natural growth law, 648
Natural logarithm function, 435, 451*
    derivative of, 443, 454*, 456*
    graph of, 437
    limits of, 437, 453*
    properties of, 452*, 453*
Negative angle, A25
Net area, 328
Net Change Theorem, 354
Net investment flow, 611
Newton, Sir Isaac, 340, 359, 812
Newton (unit of force), 398
Newton's Law of Cooling, 628, 653
Newton's Law of Gravitation, 402
Newton's Second Law, 398
Nicomedes, 692
Normal density function, 433, 466*
nth-degree Taylor polynomial, 799
Number:
    complex, A41
    integer, A2
    irrational, A2
    rational, A2
    real, A2
Numerical integration, 554

One-to-one function, 414
Open interval, A3
Optics:
    first-order, 818
    Gaussian, 818
    third-order, 818
Order of a differential equation, 625
Ordered pair, A10

Oresme, Nicole, 754
Origin, A2, A10
Orthogonal trajectories, 640
Ovals of Cassini, 715

Pappus, Theorem of, 604
Pappus of Alexandria, 604
Parabola, 720, 728, A18
    axis, 720
    directrix, 720
    equation, 721
    focus, 720, 728
    polar equation, 729
    vertex, 720
Parallelepiped, 382
Parallel lines, A14
Parameter, 687
Parametric curve, 687
Parametric equations, 687
Partial fractions, 532, 533
Partial integration, 511, 512
Partial sum of a series, 750
Parts, integration by, 511, 512
Patterns in integrals, 553
Perihelion, 732
Perilune, 726
Perpendicular lines, A14
Phase plane, 676
Phase portrait, 676
Phase trajectory, 676
Planck's Law, 821
Point-slope equation of a line, A12
Poiseuille's Laws, 609
Polar axis, 705
Polar coordinates, 705
Polar equation(s), 705
    of conics, 729
    graph of, 707
Polar form of a complex number, A43
Pole, 705
Polynomial, 29
Population growth:
    of bacteria, 446*, 647, 664
    model for, 623, 648
    world, 425, 470*, 650
Positive angle, A25
Potential, 578
Pound, 398
Power consumption, approximation of, 356
Power Rule, 447, 471*
Power series, 785
    coefficient of, 785
    differentiation of, 791
    division of, 805
    integration of, 791

interval of convergence, 787
multiplication of, 805
radius of convergence, 787
representations of functions as, 790
Predator, 674
Predator-prey model, 675
Pressure exerted by a fluid, 598
Prey, 674
Prime Number Theorem, 440
Principal square root of a complex number, A42
Principle of mathematical induction, A36
Probability density function, 612
    for customer waiting time, 617
    median, 615
Problem-solving principles, 58
Producer surplus, 610
Product formulas, A29
Projectile, 695
p-series, 761

Quadrant, A11

Radian measure, A24
Radiation from stars, 821
Radioactive decay, 652
Radiocarbon dating, 657
Radius of convergence, 787
Rational function, integration of, 532
Rationalizing substitution, 539
Rational number, A2
Ratio Test, 778
Rayleigh-Jeans Law, 821
Real line, A3
Real number, A2
Rearrangement of a series, 781
Rectangular coordinate system, A11
Rectilinear motion, 304
Reduction formula, 515
Reflection property:
    of an ellipse, 723
    of a hyperbola, 727
Region between two graphs, 375
Region under a graph, 315, 322
Relative growth rate, 649
Remainder estimates:
    for Alternating Series, 774
    for Comparison Test, 768
    for Integral Test, 763
    for Ratio Test, 778
Remainder of the Taylor series, 799
Revolution, solid of, 387
Revolution, surface of, 590
Richter scale, 439, 476*
Riemann, Georg Bernhard, 327

Riemann sum(s), 327
Right circular cylinder, 382
RMS voltage, 525
Roberval, Gilles de, 346, 699
Roots of a complex number, A46
Root Test, 780

Sample point, 320
Seasonal growth model, 667
Secant function, A26
    graph, A31
Second-order differential
    equation, 625
Sector of a circle, 715
Separable differential equation, 637
Sequence, 737
    bounded, 744
    convergent, 739
    decreasing, 744
    divergent, 739
    Fibonacci, 738
    graph of, 743
    increasing, 744
    limit of, 318, 739
    monotonic, 744
    of partial sums, 750
    term of, 737
Series, 749
    absolutely convergent, 776
    alternating, 771
    alternating harmonic, 773, 776
    binomial, 808, 809, 812
    coefficient of, 785
    conditionally convergent, 777
    convergent, 750
    divergent, 750
    geometric, 750
    Gregory's, 794
    harmonic, 753
    infinite, 749
    Maclaurin, 796, 798
    p-, 761
    partial sum of, 750
    power, 785
    rearrangement of, 781
    strategy for testing, 783
    sum of, 750
    Taylor, 796, 798
    term of, 749
    trigonometric, 785
Set, A3
Shell method, 393
Shifted conics, 725, A21
Shift of a function, 38

Sierpinski carpet, 758
Sieve of Erathosthenes, 440
Sigma notation, 320, A34
Simpson, Thomas, 559, 560
Simpson's Rule, 558, 560
    error bounds for, 561
Sine function, A26
    graph, A31
    power series, 802
Sine integral function, 349
Slope, A12
Slope-intercept equation of a line, A13
Slope field, 629
Smooth curve, 584
Smooth function, 584
Snowflake curve, 825
Solid, 382
    volume of, 384
Solid of revolution, 387
    rotated on a slant, 596
    volume of, 387, 394, 596
Solution of predator-prey
    equations, 675
Solution curve, 628
Spherical zones, 620
Spring constant, 399, 625
Squeeze Theorem for sequences, 741
Standard deviation, 433, 466*, 616
Stellar stereography, 575
Strategy:
    for integration, 541, 542
    for testing series, 783
    for trigonometric integrals, 520, 522
Strophoid, 719, 734
Substitution Rule, 360, 361, 363
Subtraction formulas for sine and
    cosine, A29
Sum:
    of a geometric series, 751
    of an infinite series, 750
    of partial fractions, 533
    Riemann, 327
    telescoping, 753
Sum Law of limits, 82
Summation notation, A34
Supply function, 610
Surface area, 592, 596
    of a parametric surface, 701
Surface of revolution, 590
    area of, 592, 596
Swallowtail catastrophe curve, 695
Symmetric functions, integrals of, 364
    in polar graphs, 710
Symmetry principle, 601

Table of differentiation formulas, RP5
Tables of integrals, 542, RP 6–10
    use of, 547
Tangent function, A26
    graph, A31
Tangent line(s):
    to a parametric curve, 696, 697
    to a polar curve, 710
Tautochrone problem, 691
Taylor, Brook, 798
Taylor polynomial, 799, 812
Taylor series, 796, 798
Taylor's Inequality, 799
Telescoping sum, 753
Term of a sequence, 737
Term of a series, 749
Terminal point of a curve, 688
Terminal velocity, 644
Term-by-term differentiation and
integration, 791
Tests for convergence and divergence of
    series:
    Alternating Series Test, 772
    Comparison Test, 767
    Integral Test, 760
    Limit Comparison Test, 768
    Ratio Test, 778
    Root Test, 780
    summary of tests, 783
    Test for Divergence, 754
Torricelli, Evangelista, 699
Torus, 392, 531
Trapezoidal Rule, 555
    error in, 556
Triangle Inequality, A8
Trigonometric functions, A26
    graphs of, A30
    integrals of, 518
    inverse, 477
Trigonometric identities, A28
Trigonometric integrals, 518
    strategy for evaluating, 520, 522
Trigonometric series, 785
Trigonometric substitutions, 525
    table of, 526
Trochoid, 694
Tschirnhausen cubic, 382

Ultraviolet catastrophe, 821
Union of sets, A3

Variables, change of, 361
Verhulst, Pierre-François, 624
Vertex of a parabola, 720

Vertices:
    of an ellipse, 722
    of a hyperbola, 724
Vibration of a rubber membrane, model of, 736
Viewing rectangle, 48
Volume, 382
    by cross-sections, 384
    by cylindrical shells, 393, 394
    by disks, 384, 387
    of a solid, 384, 388
    of a solid of revolution, 387, 596
    of a solid on a slant, 596
    by washers, 386, 387
Volterra, Vito, 675
Von Bertalanffy model, 682

Wallis, John, 4
Wallis product, 518
Washer method, 386
Weierstrass, Karl, 540
Weight, 398
Work, 398, 399
Wren, Sir Christopher, 701

x-axis, A10
x-coordinate, A10
x-intercept, A19

y-axis, A10
y-coordinate, A10
y-intercept, A19

Zone of a sphere, 596

REFERENCE PAGES

Special Functions

Power Functions $f(x) = x^a$

(i) $f(x) = x^n$, n a positive integer

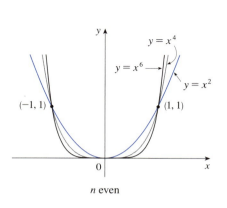

n even

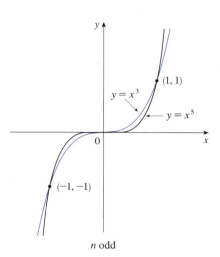

n odd

(ii) $f(x) = x^{1/n} = \sqrt[n]{x}$, n a positive integer

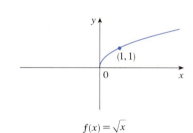

$f(x) = \sqrt{x}$

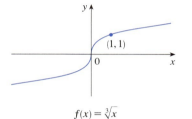

$f(x) = \sqrt[3]{x}$

(iii) $f(x) = x^{-1} = \dfrac{1}{x}$

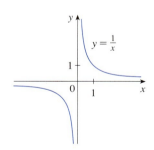

Inverse Trigonometric Functions

$\arcsin x = \sin^{-1} x = y \iff \sin y = x$ and $-\dfrac{\pi}{2} \leq y \leq \dfrac{\pi}{2}$

$\arccos x = \cos^{-1} x = y \iff \cos y = x$ and $0 \leq y \leq \pi$

$\arctan x = \tan^{-1} x = y \iff \tan y = x$ and $-\dfrac{\pi}{2} < y < \dfrac{\pi}{2}$

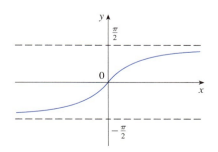

$y = \tan^{-1} x = \arctan x$

$\lim\limits_{x \to -\infty} \tan^{-1} x = -\dfrac{\pi}{2}$

$\lim\limits_{x \to \infty} \tan^{-1} x = \dfrac{\pi}{2}$

REFERENCE PAGES

Special Functions

Exponential and Logarithmic Functions

$\log_a x = y \iff a^y = x$

$\ln x = \log_e x$, where $\ln e = 1$

$\ln x = y \iff e^y = x$

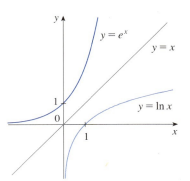

Cancellation Equations

$\log_a(a^x) = x \qquad a^{\log_a x} = x$

$\ln(e^x) = x \qquad e^{\ln x} = x$

Laws of Logarithms

1. $\log_a(xy) = \log_a x + \log_a y$
2. $\log_a\left(\dfrac{x}{y}\right) = \log_a x - \log_a y$
3. $\log_a(x^r) = r \log_a x$

$\lim\limits_{x \to -\infty} e^x = 0 \qquad \lim\limits_{x \to \infty} e^x = \infty$

$\lim\limits_{x \to 0^+} \ln x = -\infty \qquad \lim\limits_{x \to \infty} \ln x = \infty$

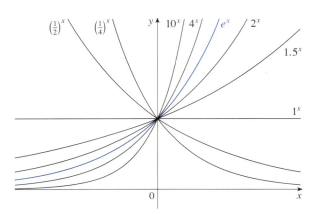

Exponential functions

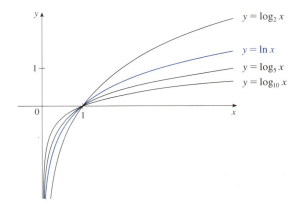

Logarithmic functions

Hyperbolic Functions

$\sinh x = \dfrac{e^x - e^{-x}}{2} \qquad \operatorname{csch} x = \dfrac{1}{\sinh x}$

$\cosh x = \dfrac{e^x + e^{-x}}{2} \qquad \operatorname{sech} x = \dfrac{1}{\cosh x}$

$\tanh x = \dfrac{\sinh x}{\cosh x} \qquad \coth x = \dfrac{\cosh x}{\sinh x}$

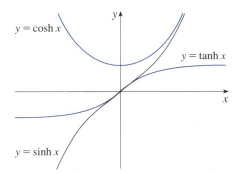

Inverse Hyperbolic Functions

$y = \sinh^{-1} x \iff \sinh y = x \qquad \sinh^{-1} x = \ln(x + \sqrt{x^2 + 1})$

$y = \cosh^{-1} x \iff \cosh y = x \text{ and } y \geq 0 \qquad \cosh^{-1} x = \ln(x + \sqrt{x^2 - 1})$

$y = \tanh^{-1} x \iff \tanh y = x \qquad \tanh^{-1} x = \tfrac{1}{2} \ln\left(\dfrac{1 + x}{1 - x}\right)$

REFERENCE PAGES

Differentiation Rules

General Formulas

1. $\dfrac{d}{dx}(c) = 0$

2. $\dfrac{d}{dx}[cf(x)] = cf'(x)$

3. $\dfrac{d}{dx}[f(x) + g(x)] = f'(x) + g'(x)$

4. $\dfrac{d}{dx}[f(x) - g(x)] = f'(x) - g'(x)$

5. $\dfrac{d}{dx}[f(x)g(x)] = f(x)g'(x) + g(x)f'(x)$ (Product Rule)

6. $\dfrac{d}{dx}\left[\dfrac{f(x)}{g(x)}\right] = \dfrac{g(x)f'(x) - f(x)g'(x)}{[g(x)]^2}$ (Quotient Rule)

7. $\dfrac{d}{dx}f(g(x)) = f'(g(x))g'(x)$ (Chain Rule)

8. $\dfrac{d}{dx}(x^n) = nx^{n-1}$ (Power Rule)

Exponential and Logarithmic Functions

9. $\dfrac{d}{dx}(e^x) = e^x$

10. $\dfrac{d}{dx}(a^x) = a^x \ln a$

11. $\dfrac{d}{dx}\ln|x| = \dfrac{1}{x}$

12. $\dfrac{d}{dx}(\log_a x) = \dfrac{1}{x \ln a}$

Trigonometric Functions

13. $\dfrac{d}{dx}(\sin x) = \cos x$

14. $\dfrac{d}{dx}(\cos x) = -\sin x$

15. $\dfrac{d}{dx}(\tan x) = \sec^2 x$

16. $\dfrac{d}{dx}(\csc x) = -\csc x \cot x$

17. $\dfrac{d}{dx}(\sec x) = \sec x \tan x$

18. $\dfrac{d}{dx}(\cot x) = -\csc^2 x$

Inverse Trigonometric Functions

19. $\dfrac{d}{dx}(\sin^{-1} x) = \dfrac{1}{\sqrt{1-x^2}}$

20. $\dfrac{d}{dx}(\cos^{-1} x) = -\dfrac{1}{\sqrt{1-x^2}}$

21. $\dfrac{d}{dx}(\tan^{-1} x) = \dfrac{1}{1+x^2}$

22. $\dfrac{d}{dx}(\csc^{-1} x) = -\dfrac{1}{x\sqrt{x^2-1}}$

23. $\dfrac{d}{dx}(\sec^{-1} x) = \dfrac{1}{x\sqrt{x^2-1}}$

24. $\dfrac{d}{dx}(\cot^{-1} x) = -\dfrac{1}{1+x^2}$

Hyperbolic Functions

25. $\dfrac{d}{dx}(\sinh x) = \cosh x$

26. $\dfrac{d}{dx}(\cosh x) = \sinh x$

27. $\dfrac{d}{dx}(\tanh x) = \text{sech}^2 x$

28. $\dfrac{d}{dx}(\text{csch}\, x) = -\text{csch}\, x \coth x$

29. $\dfrac{d}{dx}(\text{sech}\, x) = -\text{sech}\, x \tanh x$

30. $\dfrac{d}{dx}(\coth x) = -\text{csch}^2 x$

Inverse Hyperbolic Functions

31. $\dfrac{d}{dx}(\sinh^{-1} x) = \dfrac{1}{\sqrt{1+x^2}}$

32. $\dfrac{d}{dx}(\cosh^{-1} x) = \dfrac{1}{\sqrt{x^2-1}}$

33. $\dfrac{d}{dx}(\tanh^{-1} x) = \dfrac{1}{1-x^2}$

34. $\dfrac{d}{dx}(\text{csch}^{-1} x) = -\dfrac{1}{|x|\sqrt{x^2+1}}$

35. $\dfrac{d}{dx}(\text{sech}^{-1} x) = -\dfrac{1}{x\sqrt{1-x^2}}$

36. $\dfrac{d}{dx}(\coth^{-1} x) = \dfrac{1}{1-x^2}$

Table of Integrals

Basic Forms

1. $\int u\, dv = uv - \int v\, du$

2. $\int u^n\, du = \dfrac{u^{n+1}}{n+1} + C, \quad n \neq -1$

3. $\int \dfrac{du}{u} = \ln|u| + C$

4. $\int e^u\, du = e^u + C$

5. $\int a^u\, du = \dfrac{a^u}{\ln a} + C$

6. $\int \sin u\, du = -\cos u + C$

7. $\int \cos u\, du = \sin u + C$

8. $\int \sec^2 u\, du = \tan u + C$

9. $\int \csc^2 u\, du = -\cot u + C$

10. $\int \sec u\, \tan u\, du = \sec u + C$

11. $\int \csc u\, \cot u\, du = -\csc u + C$

12. $\int \tan u\, du = \ln|\sec u| + C$

13. $\int \cot u\, du = \ln|\sin u| + C$

14. $\int \sec u\, du = \ln|\sec u + \tan u| + C$

15. $\int \csc u\, du = \ln|\csc u - \cot u| + C$

16. $\int \dfrac{du}{\sqrt{a^2 - u^2}} = \sin^{-1}\dfrac{u}{a} + C$

17. $\int \dfrac{du}{a^2 + u^2} = \dfrac{1}{a}\tan^{-1}\dfrac{u}{a} + C$

18. $\int \dfrac{du}{u\sqrt{u^2 - a^2}} = \dfrac{1}{a}\sec^{-1}\dfrac{u}{a} + C$

19. $\int \dfrac{du}{a^2 - u^2} = \dfrac{1}{2a}\ln\left|\dfrac{u+a}{u-a}\right| + C$

20. $\int \dfrac{du}{u^2 - a^2} = \dfrac{1}{2a}\ln\left|\dfrac{u-a}{u+a}\right| + C$

Forms Involving $\sqrt{a^2 + u^2}, \ a > 0$

21. $\int \sqrt{a^2 + u^2}\, du = \dfrac{u}{2}\sqrt{a^2 + u^2} + \dfrac{a^2}{2}\ln(u + \sqrt{a^2 + u^2}) + C$

22. $\int u^2\sqrt{a^2 + u^2}\, du = \dfrac{u}{8}(a^2 + 2u^2)\sqrt{a^2 + u^2} - \dfrac{a^4}{8}\ln(u + \sqrt{a^2 + u^2}) + C$

23. $\int \dfrac{\sqrt{a^2 + u^2}}{u}\, du = \sqrt{a^2 + u^2} - a\ln\left|\dfrac{a + \sqrt{a^2 + u^2}}{u}\right| + C$

24. $\int \dfrac{\sqrt{a^2 + u^2}}{u^2}\, du = -\dfrac{\sqrt{a^2 + u^2}}{u} + \ln(u + \sqrt{a^2 + u^2}) + C$

25. $\int \dfrac{du}{\sqrt{a^2 + u^2}} = \ln(u + \sqrt{a^2 + u^2}) + C$

26. $\int \dfrac{u^2\, du}{\sqrt{a^2 + u^2}} = \dfrac{u}{2}\sqrt{a^2 + u^2} - \dfrac{a^2}{2}\ln(u + \sqrt{a^2 + u^2}) + C$

27. $\int \dfrac{du}{u\sqrt{a^2 + u^2}} = -\dfrac{1}{a}\ln\left|\dfrac{\sqrt{a^2 + u^2} + a}{u}\right| + C$

28. $\int \dfrac{du}{u^2\sqrt{a^2 + u^2}} = -\dfrac{\sqrt{a^2 + u^2}}{a^2 u} + C$

29. $\int \dfrac{du}{(a^2 + u^2)^{3/2}} = \dfrac{u}{a^2\sqrt{a^2 + u^2}} + C$

Table of Integrals

Forms Involving $\sqrt{a^2 - u^2}$, $a > 0$

30. $\displaystyle\int \sqrt{a^2 - u^2}\, du = \frac{u}{2}\sqrt{a^2 - u^2} + \frac{a^2}{2}\sin^{-1}\frac{u}{a} + C$

31. $\displaystyle\int u^2\sqrt{a^2 - u^2}\, du = \frac{u}{8}(2u^2 - a^2)\sqrt{a^2 - u^2} + \frac{a^4}{8}\sin^{-1}\frac{u}{a} + C$

32. $\displaystyle\int \frac{\sqrt{a^2 - u^2}}{u}\, du = \sqrt{a^2 - u^2} - a\ln\left|\frac{a + \sqrt{a^2 - u^2}}{u}\right| + C$

33. $\displaystyle\int \frac{\sqrt{a^2 - u^2}}{u^2}\, du = -\frac{1}{u}\sqrt{a^2 - u^2} - \sin^{-1}\frac{u}{a} + C$

34. $\displaystyle\int \frac{u^2\, du}{\sqrt{a^2 - u^2}} = -\frac{u}{2}\sqrt{a^2 - u^2} + \frac{a^2}{2}\sin^{-1}\frac{u}{a} + C$

35. $\displaystyle\int \frac{du}{u\sqrt{a^2 - u^2}} = -\frac{1}{a}\ln\left|\frac{a + \sqrt{a^2 - u^2}}{u}\right| + C$

36. $\displaystyle\int \frac{du}{u^2\sqrt{a^2 - u^2}} = -\frac{1}{a^2 u}\sqrt{a^2 - u^2} + C$

37. $\displaystyle\int (a^2 - u^2)^{3/2}\, du = -\frac{u}{8}(2u^2 - 5a^2)\sqrt{a^2 - u^2} + \frac{3a^4}{8}\sin^{-1}\frac{u}{a} + C$

38. $\displaystyle\int \frac{du}{(a^2 - u^2)^{3/2}} = \frac{u}{a^2\sqrt{a^2 - u^2}} + C$

Forms Involving $\sqrt{u^2 - a^2}$, $a > 0$

39. $\displaystyle\int \sqrt{u^2 - a^2}\, du = \frac{u}{2}\sqrt{u^2 - a^2} - \frac{a^2}{2}\ln\left|u + \sqrt{u^2 - a^2}\right| + C$

40. $\displaystyle\int u^2\sqrt{u^2 - a^2}\, du = \frac{u}{8}(2u^2 - a^2)\sqrt{u^2 - a^2} - \frac{a^4}{8}\ln\left|u + \sqrt{u^2 - a^2}\right| + C$

41. $\displaystyle\int \frac{\sqrt{u^2 - a^2}}{u}\, du = \sqrt{u^2 - a^2} - a\cos^{-1}\frac{a}{|u|} + C$

42. $\displaystyle\int \frac{\sqrt{u^2 - a^2}}{u^2}\, du = -\frac{\sqrt{u^2 - a^2}}{u} + \ln\left|u + \sqrt{u^2 - a^2}\right| + C$

43. $\displaystyle\int \frac{du}{\sqrt{u^2 - a^2}} = \ln\left|u + \sqrt{u^2 - a^2}\right| + C$

44. $\displaystyle\int \frac{u^2\, du}{\sqrt{u^2 - a^2}} = \frac{u}{2}\sqrt{u^2 - a^2} + \frac{a^2}{2}\ln\left|u + \sqrt{u^2 - a^2}\right| + C$

45. $\displaystyle\int \frac{du}{u^2\sqrt{u^2 - a^2}} = \frac{\sqrt{u^2 - a^2}}{a^2 u} + C$

46. $\displaystyle\int \frac{du}{(u^2 - a^2)^{3/2}} = -\frac{u}{a^2\sqrt{u^2 - a^2}} + C$

Table of Integrals

Forms Involving $a + bu$

47. $\displaystyle\int \frac{u\,du}{a+bu} = \frac{1}{b^2}(a + bu - a\ln|a+bu|) + C$

48. $\displaystyle\int \frac{u^2\,du}{a+bu} = \frac{1}{2b^3}\left[(a+bu)^2 - 4a(a+bu) + 2a^2\ln|a+bu|\right] + C$

49. $\displaystyle\int \frac{du}{u(a+bu)} = \frac{1}{a}\ln\left|\frac{u}{a+bu}\right| + C$

50. $\displaystyle\int \frac{du}{u^2(a+bu)} = -\frac{1}{au} + \frac{b}{a^2}\ln\left|\frac{a+bu}{u}\right| + C$

51. $\displaystyle\int \frac{u\,du}{(a+bu)^2} = \frac{a}{b^2(a+bu)} + \frac{1}{b^2}\ln|a+bu| + C$

52. $\displaystyle\int \frac{du}{u(a+bu)^2} = \frac{1}{a(a+bu)} - \frac{1}{a^2}\ln\left|\frac{a+bu}{u}\right| + C$

53. $\displaystyle\int \frac{u^2\,du}{(a+bu)^2} = \frac{1}{b^3}\left(a+bu - \frac{a^2}{a+bu} - 2a\ln|a+bu|\right) + C$

54. $\displaystyle\int u\sqrt{a+bu}\,du = \frac{2}{15b^2}(3bu - 2a)(a+bu)^{3/2} + C$

55. $\displaystyle\int \frac{u\,du}{\sqrt{a+bu}} = \frac{2}{3b^2}(bu - 2a)\sqrt{a+bu} + C$

56. $\displaystyle\int \frac{u^2\,du}{\sqrt{a+bu}} = \frac{2}{15b^3}(8a^2 + 3b^2u^2 - 4abu)\sqrt{a+bu} + C$

57. $\displaystyle\int \frac{du}{u\sqrt{a+bu}} = \frac{1}{\sqrt{a}}\ln\left|\frac{\sqrt{a+bu} - \sqrt{a}}{\sqrt{a+bu} + \sqrt{a}}\right| + C, \quad \text{if } a > 0$

$\displaystyle\phantom{\int \frac{du}{u\sqrt{a+bu}}} = \frac{2}{\sqrt{-a}}\tan^{-1}\sqrt{\frac{a+bu}{-a}} + C, \quad \text{if } a < 0$

58. $\displaystyle\int \frac{\sqrt{a+bu}}{u}\,du = 2\sqrt{a+bu} + a\int \frac{du}{u\sqrt{a+bu}}$

59. $\displaystyle\int \frac{\sqrt{a+bu}}{u^2}\,du = -\frac{\sqrt{a+bu}}{u} + \frac{b}{2}\int \frac{du}{u\sqrt{a+bu}}$

60. $\displaystyle\int u^n\sqrt{a+bu}\,du = \frac{2}{b(2n+3)}\left[u^n(a+bu)^{3/2} - na\int u^{n-1}\sqrt{a+bu}\,du\right]$

61. $\displaystyle\int \frac{u^n\,du}{\sqrt{a+bu}} = \frac{2u^n\sqrt{a+bu}}{b(2n+1)} - \frac{2na}{b(2n+1)}\int \frac{u^{n-1}\,du}{\sqrt{a+bu}}$

62. $\displaystyle\int \frac{du}{u^n\sqrt{a+bu}} = -\frac{\sqrt{a+bu}}{a(n-1)u^{n-1}} - \frac{b(2n-3)}{2a(n-1)}\int \frac{du}{u^{n-1}\sqrt{a+bu}}$

Table of Integrals

Trigonometric Forms

63. $\int \sin^2 u \, du = \frac{1}{2}u - \frac{1}{4}\sin 2u + C$

64. $\int \cos^2 u \, du = \frac{1}{2}u + \frac{1}{4}\sin 2u + C$

65. $\int \tan^2 u \, du = \tan u - u + C$

66. $\int \cot^2 u \, du = -\cot u - u + C$

67. $\int \sin^3 u \, du = -\frac{1}{3}(2 + \sin^2 u) \cos u + C$

68. $\int \cos^3 u \, du = \frac{1}{3}(2 + \cos^2 u) \sin u + C$

69. $\int \tan^3 u \, du = \frac{1}{2}\tan^2 u + \ln|\cos u| + C$

70. $\int \cot^3 u \, du = -\frac{1}{2}\cot^2 u - \ln|\sin u| + C$

71. $\int \sec^3 u \, du = \frac{1}{2}\sec u \tan u + \frac{1}{2}\ln|\sec u + \tan u| + C$

72. $\int \csc^3 u \, du = -\frac{1}{2}\csc u \cot u + \frac{1}{2}\ln|\csc u - \cot u| + C$

73. $\int \sin^n u \, du = -\frac{1}{n}\sin^{n-1} u \cos u + \frac{n-1}{n}\int \sin^{n-2} u \, du$

74. $\int \cos^n u \, du = \frac{1}{n}\cos^{n-1} u \sin u + \frac{n-1}{n}\int \cos^{n-2} u \, du$

75. $\int \tan^n u \, du = \frac{1}{n-1}\tan^{n-1} u - \int \tan^{n-2} u \, du$

76. $\int \cot^n u \, du = \frac{-1}{n-1}\cot^{n-1} u - \int \cot^{n-2} u \, du$

77. $\int \sec^n u \, du = \frac{1}{n-1}\tan u \sec^{n-2} u + \frac{n-2}{n-1}\int \sec^{n-2} u \, du$

78. $\int \csc^n u \, du = \frac{-1}{n-1}\cot u \csc^{n-2} u + \frac{n-2}{n-1}\int \csc^{n-2} u \, du$

79. $\int \sin au \sin bu \, du = \frac{\sin(a-b)u}{2(a-b)} - \frac{\sin(a+b)u}{2(a+b)} + C$

80. $\int \cos au \cos bu \, du = \frac{\sin(a-b)u}{2(a-b)} + \frac{\sin(a+b)u}{2(a+b)} + C$

81. $\int \sin au \cos bu \, du = -\frac{\cos(a-b)u}{2(a-b)} - \frac{\cos(a+b)u}{2(a+b)} + C$

82. $\int u \sin u \, du = \sin u - u \cos u + C$

83. $\int u \cos u \, du = \cos u + u \sin u + C$

84. $\int u^n \sin u \, du = -u^n \cos u + n \int u^{n-1} \cos u \, du$

85. $\int u^n \cos u \, du = u^n \sin u - n \int u^{n-1} \sin u \, du$

86. $\int \sin^n u \cos^m u \, du = -\frac{\sin^{n-1} u \cos^{m+1} u}{n+m} + \frac{n-1}{n+m}\int \sin^{n-2} u \cos^m u \, du$
$= \frac{\sin^{n+1} u \cos^{m-1} u}{n+m} + \frac{m-1}{n+m}\int \sin^n u \cos^{m-2} u \, du$

Inverse Trigonometric Forms

87. $\int \sin^{-1} u \, du = u \sin^{-1} u + \sqrt{1-u^2} + C$

88. $\int \cos^{-1} u \, du = u \cos^{-1} u - \sqrt{1-u^2} + C$

89. $\int \tan^{-1} u \, du = u \tan^{-1} u - \frac{1}{2}\ln(1+u^2) + C$

90. $\int u \sin^{-1} u \, du = \frac{2u^2-1}{4}\sin^{-1} u + \frac{u\sqrt{1-u^2}}{4} + C$

91. $\int u \cos^{-1} u \, du = \frac{2u^2-1}{4}\cos^{-1} u - \frac{u\sqrt{1-u^2}}{4} + C$

92. $\int u \tan^{-1} u \, du = \frac{u^2+1}{2}\tan^{-1} u - \frac{u}{2} + C$

93. $\int u^n \sin^{-1} u \, du = \frac{1}{n+1}\left[u^{n+1}\sin^{-1} u - \int \frac{u^{n+1} \, du}{\sqrt{1-u^2}}\right]$, $n \neq -1$

94. $\int u^n \cos^{-1} u \, du = \frac{1}{n+1}\left[u^{n+1}\cos^{-1} u + \int \frac{u^{n+1} \, du}{\sqrt{1-u^2}}\right]$, $n \neq -1$

95. $\int u^n \tan^{-1} u \, du = \frac{1}{n+1}\left[u^{n+1}\tan^{-1} u - \int \frac{u^{n+1} \, du}{1+u^2}\right]$, $n \neq -1$

Table of Integrals

Exponential and Logarithmic Forms

96. $\displaystyle\int ue^{au}\,du = \frac{1}{a^2}(au - 1)e^{au} + C$

97. $\displaystyle\int u^n e^{au}\,du = \frac{1}{a}u^n e^{au} - \frac{n}{a}\int u^{n-1}e^{au}\,du$

98. $\displaystyle\int e^{au}\sin bu\,du = \frac{e^{au}}{a^2 + b^2}(a\sin bu - b\cos bu) + C$

99. $\displaystyle\int e^{au}\cos bu\,du = \frac{e^{au}}{a^2 + b^2}(a\cos bu + b\sin bu) + C$

100. $\displaystyle\int \ln u\,du = u\ln u - u + C$

101. $\displaystyle\int u^n \ln u\,du = \frac{u^{n+1}}{(n+1)^2}[(n+1)\ln u - 1] + C$

102. $\displaystyle\int \frac{1}{u\ln u}\,du = \ln|\ln u| + C$

Hyperbolic Forms

103. $\displaystyle\int \sinh u\,du = \cosh u + C$

104. $\displaystyle\int \cosh u\,du = \sinh u + C$

105. $\displaystyle\int \tanh u\,du = \ln\cosh u + C$

106. $\displaystyle\int \coth u\,du = \ln|\sinh u| + C$

107. $\displaystyle\int \operatorname{sech} u\,du = \tan^{-1}|\sinh u| + C$

108. $\displaystyle\int \operatorname{csch} u\,du = \ln\left|\tanh\tfrac{1}{2}u\right| + C$

109. $\displaystyle\int \operatorname{sech}^2 u\,du = \tanh u + C$

110. $\displaystyle\int \operatorname{csch}^2 u\,du = -\coth u + C$

111. $\displaystyle\int \operatorname{sech} u\,\tanh u\,du = -\operatorname{sech} u + C$

112. $\displaystyle\int \operatorname{csch} u\,\coth u\,du = -\operatorname{csch} u + C$

Forms Involving $\sqrt{2au - u^2}$, $a > 0$

113. $\displaystyle\int \sqrt{2au - u^2}\,du = \frac{u-a}{2}\sqrt{2au - u^2} + \frac{a^2}{2}\cos^{-1}\left(\frac{a-u}{a}\right) + C$

114. $\displaystyle\int u\sqrt{2au - u^2}\,du = \frac{2u^2 - au - 3a^2}{6}\sqrt{2au - u^2} + \frac{a^3}{2}\cos^{-1}\left(\frac{a-u}{a}\right) + C$

115. $\displaystyle\int \frac{\sqrt{2au - u^2}}{u}\,du = \sqrt{2au - u^2} + a\cos^{-1}\left(\frac{a-u}{a}\right) + C$

116. $\displaystyle\int \frac{\sqrt{2au - u^2}}{u^2}\,du = -\frac{2\sqrt{2au - u^2}}{u} - \cos^{-1}\left(\frac{a-u}{a}\right) + C$

117. $\displaystyle\int \frac{du}{\sqrt{2au - u^2}} = \cos^{-1}\left(\frac{a-u}{a}\right) + C$

118. $\displaystyle\int \frac{u\,du}{\sqrt{2au - u^2}} = -\sqrt{2au - u^2} + a\cos^{-1}\left(\frac{a-u}{a}\right) + C$

119. $\displaystyle\int \frac{u^2\,du}{\sqrt{2au - u^2}} = -\frac{(u+3a)}{2}\sqrt{2au - u^2} + \frac{3a^2}{2}\cos^{-1}\left(\frac{a-u}{a}\right) + C$

120. $\displaystyle\int \frac{du}{u\sqrt{2au - u^2}} = -\frac{\sqrt{2au - u^2}}{au} + C$